Differential Equations:
A Modeling Approach

Differential Equations: A Modeling Approach

Robert L. Borrelli

Courtney S. Coleman

Harvey Mudd College
Claremont, California

Prentice-Hall, Inc., Englewood Cliffs, New Jersey 07632

Library of Congress Cataloging-in-Publication Data

BORRELLI, ROBERT L., (date)
  Differential equations.

  Includes index.
  1. Differential equations.   I. Coleman,
Courtney S., (date)   II. Title.
QA371.B74 1987   515.3'5   86–22601
ISBN 0–13–211533–6

Editorial/production supervision
    and interior design: *Kathleen M. Lafferty*
Cover design: *Bruce Kenselaar*
Manufacturing buyer: *John B. Hall*
Cover illustration: *Developed from the NASA photograph*
  "Earthrise from the Moon."

© 1987 by Prentice-Hall, Inc.
A Division of Simon & Schuster
Englewood Cliffs, New Jersey 07632

Printed in the United States of America

10 9 8 7 6 5 4 3 2 1

0-13-211533-6   01

Prentice-Hall International (UK) Limited, *London*
Prentice-Hall of Australia Pty. Limited, *Sydney*
Prentice-Hall Canada Inc., *Toronto*
Prentice-Hall Hispanoamericana, S.A., *Mexico*
Prentice-Hall of India Private Limited, *New Delhi*
Prentice-Hall of Japan, Inc., *Tokyo*
Prentice-Hall of Southeast Asia Pte. Ltd., *Singapore*
Editora Prentice-Hall do Brasil, Ltda., *Rio de Janeiro*

To Julie, David, Margaret, and Diane
and
To the memory of Jimmy and Johnny

Contents

CHAPTER 3

First-Order Differential Equations: Growth Processes   62

CHAPTER 4

Techniques for Solving First-Order Differential Equations: Newtonian Mechanics   99

CHAPTER 5

Second-Order Linear Differential Equations: Electrical Circuits   146

CHAPTER 6

Linear Algebra and Higher-Order Linear Differential Equations

206

CHAPTER 7

The Laplace Transform

254

CHAPTER 8

Linear Differential Equations with Nonconstant Coefficients

296

CHAPTER 9

Introduction to Systems   346

CHAPTER 10

Linear Systems of Differential Equations   403

CHAPTER 11

Nonlinear Systems and Stability   454

CHAPTER 12

Cycles and Bifurcations   478

CHAPTER 13

*Partial Differential Equations
and Fourier Series*   505

APPENDIX A

Basic Theory of Initial Value Problems   A1

Preface

Differential Equations: A Modeling Approach is intended as an introductory level textbook in differential equations for second- or third-year university students of science, mathematics, or engineering. As every practicing scientist or engineer knows, differential equations are an important tool in constructing mathematical models for phenomena which are dynamic in nature, that is, systems which evolve in time according to certain "laws." Indeed, differential equations can be used to portray the growth and decay of populations, the vibrations of a mechanical system, the response of an electrical circuit to an oscillating voltage, the motion of a plucked string, and the diffusion of heat, as well as a host of other time-varying phenomena. So widespread is their use and so well do differential equations perform their task that they must be considered as one of the most successful of modeling tools. Thus, it seems to us that a great deal can be gained by presenting differential equations within the context of a modeling environment.

One of the aims of a course in differential equations is to explain how formulas and approximations for solutions may be obtained. Much of the book is devoted to a presentation of the fundamental formulas and techniques for solution. Solution formulas and numerical approximations are often helpful in understanding the behavior of a solution, but sometimes a deeper understanding of an equation and its solutions can only be obtained by other methods. In many cases there are no known formulas for the solutions, nor is it likely there ever will be. Nevertheless, it is often possible to deduce how the solutions behave without having formulas. Many of these indirect techniques, including orbital geometry, computer graphics, intuition provided by models, and estimates of rates, play a valuable role and are presented at an elementary level in this book.

Differential equations is an area of theoretical and applied mathematics in which there are a large number of important but as yet unsolved problems. Chapter 11 and 12 give the flavor of some of the ideas of contemporary research in such areas as stability, bifurcation, and chaotic motion.

Chapter 13 introduces modeling and solution techniques for systems involving partial differential equations. Unlike the earlier part of the book in which ideas and techniques are introduced and developed gradually over several chapters, the basic modeling principles and methods of solution for partial differential equations are treated all in one chapter, and hence the pace is rather fast.

Design of Text

The main features of the book are outlined below:

- *Scope.* The traditional topics of an introductory course in differential equations are embedded in a context which supports geometric intuition, computing, modeling, and theory. Enough theory is given to provide a solid foundation for understanding the techniques and the applications. When proofs are given they are as complete and self-contained as possible. Some of the proofs are given in appendices, for example, the proof of the Existence and Uniqueness Theorem. The appendices also contain material supporting the text's presentation of numerical methods, complex solutions, and series solutions.

- *Prerequisites.* The prerequisites are calculus through multivariable calculus. A year of single-variable calculus is adequate except for those few sections depending upon partial differentiation, multiple integration, or the divergence and Green's theorems. A course in linear algebra is not assumed; linear concepts are developed as needed.

- *The Role of Computers.* A knowledge of computer programming is not required. A hand-held calculator, personal computer, or large computer system may be used to implement some of the techniques presented in this book. For those who want to introduce programming into an introductory course all the basic ideas needed for the construction of solution algorithms have been introduced. On the other hand, sophisticated interactive solver packages are now available with automatic features which do not require that the user be computer literate. For this reason we see no clear advantage to include either computer programs or flow charts in the text.

- *Geometry and Computer Graphics.* The solutions of an ordinary differential equation are functions whose graphs are curves in appropriate spaces. Pictures of these graphs may be computer-generated and provide compelling visual evidence of theoretical deductions and a clear understanding of complicated solution formulas. The text and the illustrations emphasize this visual connection with the theory.

- *Operator Concept.* Many problems in differential equations can be formulated in terms of operators, an approach which both organizes and simplifies theory and applications. The operator concept is introduced on an elementary level in Chapter 3 and developed as needed.

- *Pace.* Each section of Chapters 1 through 12 is designed for one or two lectures, depending on the desires of the instructor. Some of the sections of Chapter 13 may require two or three lectures each.

- *Style.* Each chapter begins with a brief abstract. The basic ideas of the chapter are presented in the first few sections, while later sections extend and apply these ideas. Fundamental principles are developed gradually and as appropriate over several sections and even chapters. The index is a guide to the development of each basic concept through the text.

- *Problem Sets.* Problems at the end of each section illustrate techniques developed in the section. Other problems extend the theory, relate to applications and models, or invoke computational work.

- *Modeling.* The models developed here have been chosen from three important areas: growth processes, motion of mechanical systems, and electrical circuits. Most models are developed as fully as possible in separate sections, and hence offer the reader some considerable flexibility in what material is actually covered in a first reading. In order that each modeling section may be read independently of others, we have had to tolerate a certain amount of repetition. Although we have made it convenient for the reader to acquire a knowledge of the relation of differential equations to their modeling needs, the text is written in such a way as to allow the reader a choice of how much the modeling environment intrudes on the theory and solution techniques of differential equations.

Possible Courses

The accompanying logical dependence chart may be used to design several courses (modeling sections are listed alongside the chart for convenience). We recommend that a comprehensive semester course cover the twenty-four sections in bold in the chart (approximately thirty lecture hours should suffice). Additional sections may be included to complete a basic course. An introductory quarter course could be designed by scaling back this material. On the other hand there is ample material for a full-year course. The second semester of a year course might have two components, an introduction to systems of ordinary differential equations (Chapters 9 to 12) and an introduction to series methods for ordinary differential equations (Chapter 8) and their application to separation of variable techniques for partial differential equations (Chapter 13).

In any course the instructor may want to include sections from Chapters 11, 12, or 13. Chapter 11 develops the important concept of stability, already introduced in earlier chapters. Chapter 12 is an introduction to cycles and bifurcations, areas of current research and applications. Chapter 13 is a basic introduction to the classical partial differential equations of applied physics and engineering.

Acknowledgments

As expected, with any work of this size and complexity, we owe a debt of gratitude to many people who in one way or another have contributed to the preparation of this text for publication. The project could not have been completed without their assistance, and to all of them we give our deepest thanks. Our students, S. Boettcher,

Dependency Chart

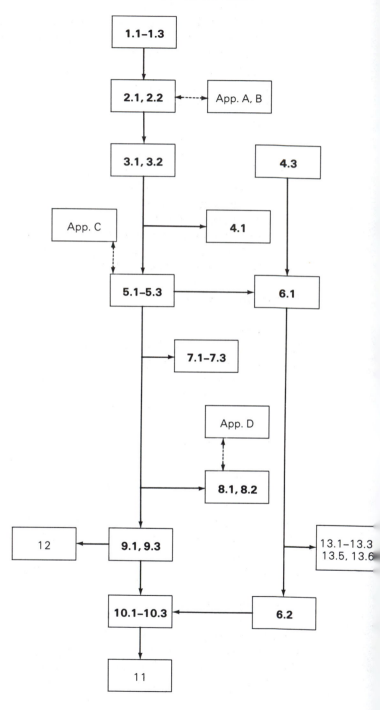

Modeling Sections

Modeling Principles,
  Process: 1.1 to 1.3
Radiocarbon Dating: 1.3
Numerical Solutions: 2.3 to 2.5
Growth Processes: 3.3
Reactors and Migration: 3.4
Destructive Competition: 4.1
Pursuit: 4.2
Newton's Laws of Motion: 4.3
Falling Bodies, Escape
  Velocity: 4.4
Electrical Circuits: 5.4, 5.6
Springs, Pendulum: 5.5, 5.6
Car Following: 7.4
Aging Springs: 8.1, 8.7
Potassium-Argon Dating: 9.2
The Uranium Series: 9.2
The Simple Pendulum: 9.3
Predator-Prey: 9.5
Low-Pass Filter: 10.5
Lead in the Human Body: 10.6
van der Pol Cycle: 12.1
Vibrating String: 13.2, 13.3
Shaking a String to Rest: 13.4
Heat Flow: 13.8
Optimal Depth of
  Wine Cellars: 13.8
Equilibrium Temperatures: 13.9

W. Consoli, G. Slaughter, A. Spellman, and K. Steinhoff, deserve special thanks for the vigorous and enlightened criticism they brought to the early versions of this book. We are also indebted to the MATHLIB™ crew at Harvey Mudd College, especially principal authors N. Freed and D. Newman, for designing the high-quality software package MATHLIB™, which was used to produce all the graphs of solution curves and orbits of the differential equations in this book. Thanks also go to our students R. Ngo, J. Pilliod, M. Ross, and to B. Staat of Harvey Mudd College for their expert use of MATHLIB™ to produce the excellent graphs for this text and its accompanying solutions manual.

We are grateful to many colleagues, but especially to P. Crooke (Vanderbilt University), R. Di Franco (University of the Pacific), L. Heitsch (University of Illinois at Chicago), and T. Helliwell (Harvey Mudd College) for reading portions of this book in various stages and for the thoughtful modifications suggested. We especially express our gratitude to Sue Cook for her careful preparation of the various versions of the text and for her patience with our foibles and crotchets. We also would like to express our gratitude to Kathleen Lafferty for inspired supervision of the editorial and production work, converting our manuscript into visually pleasing print. And last but not least, we acknowledge the patience and understanding of our departmental colleagues, our wives and children, and our friends, who have had to contend with our, at times, preposterous work and schedules.

We would appreciate having errors, misprints, and ambiguities called to our attention by writing to either of us at Harvey Mudd College, Claremont, CA 91711, USA. For information regarding the software package MATHLIB™, please send your query via electronic mail to the address QMATHLIB@YMIR.BITNET.

Robert L. Borrelli
Courtney S. Coleman

CHAPTER 1

Differential Equations and Models

Differential equations and the process of modeling have enjoyed a long history together—some 300 years now. When Isaac Newton invented the differential calculus in order to create a model to explain the motion of material bodies, a modeling tool of great power and scope came into being. In this brief introductory chapter we take a quick look at these partners. The chapters that follow go into the connection between modeling and differential equations more deeply.

1.1 INTRODUCTION TO DIFFERENTIAL EQUATIONS

Differential equations are equations involving unknown functions and their derivatives. Many processes in engineering and in the physical, biological, and social sciences are expressed by equations involving rates of change, velocities, slopes, gradients, curvatures, or accelerations, all of which are derivatives. When differential equations arise in this way they are said to "model" the phenomena with which they are associated. In this section we give examples of differential equations, define what is meant by solutions and solution curves, and give a simple technique for finding solutions. The modeling aspects are treated in the following sections.

We shall generally use the "prime" notation for derivatives, occasionally dropping specific reference to the independent variable if the name of that variable is clear from the context, for example,

$$y' \text{ or } y'(t) \quad \text{for} \quad \frac{dy(t)}{dt}, \qquad z'' \text{ or } z''(x) \quad \text{for} \quad \frac{d^2z(x)}{dx^2}$$

Partial derivative notation is used for functions of more than one variable [e.g., $\partial^2 u(x, t)/\partial x^2$]. The "variable" subscript notation for partial derivatives is often used, again dropping explicit reference to independent variables when no confusion results. For example, when $u = u(x, y, t)$, we may write

$$u_t \quad \text{or} \quad u_t(x, y, t) \quad \text{for} \quad \frac{\partial u\,(x, y, t)}{\partial t}$$

$$u_{xy} \quad \text{or} \quad u_{xy}(x, y, t) \quad \text{for} \quad \frac{\partial}{\partial y} \left[\frac{\partial u\,(x, y, t)}{\partial x} \right]$$

This text is mostly devoted to *ordinary* differential equations, that is, equations involving only derivatives of single variable functions. Equations involving partial derivatives, called *partial* differential equations, are treated only in the final chapter.

Listed below are some examples of differential equations.† Each is a mathematical model for the phenomenon indicated.

$$N'(t) = kN(t) \tag{1a}$$

(radioactive decay, population growth)

$$ms''(t) = -mg \tag{1b}$$

(vertical motion in a constant gravitational field)

$$Lq''(t) + Rq'(t) + \frac{1}{C}\,q(t) = A\,\cos t \tag{1c}$$

(charge on the plate of a capacitor in an electrical circuit)

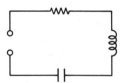

---

† We shall assume throughout this work that all constants, variables, and functions are real valued, not complex valued, unless specifically stated otherwise.

$$my''(t) + ke^{-at}y(t) = 0$$

(1d)

(vibrations of a mass attached to an aging spring)

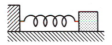

$$y'''(x) = y(x)y''(x) + k[(y'(x))^2 - 1]$$

(1e)

(steady-state velocity distribution in flow of fluids)

$$c^2 \frac{\partial^2 u(x, y, t)}{\partial x^2} + c^2 \frac{\partial^2 u(x, y, t)}{\partial y^2} = \frac{\partial^2 u(x, y, t)}{\partial t^2}$$

(1f)

(wave motion)

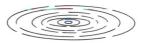

We could add hundreds of other important differential equations to the list. Alternatively, we can make up differential equations on our own by writing down derivatives more or less at random: for example,

$$y(x) = xy'(x) + [y'(x)]^2$$

(2a)

$$y^{10}(t)t^2 y^{(5)}(t) - \sin[1 + y^7(t)] = [y'(t)]^3$$

(2b)

The list above suggests the astonishing variety of differential equations. We cannot hope to find special techniques for solving all the "important" equations one by one. Instead, we shall group differential equations into natural categories and look for properties of solutions or solution techniques common to the equations in a given category. That is mostly what this book is about.

Classification and Notation

Differential equations are classified in a number of ways. The *order* of an equation is the order of the highest derivative of the unknown function that appears in the equation. For example, Eqs. (1a) and (2a) are first order in the respective unknown functions $N(t)$ and $y(x)$ [the square term in (2a) does not make that equation second order]. Equations (1b)–(1d) and (1f) are second order, (1e) is third order, and (2b) is fifth order (not tenth order). Equations of first or second order are particularly common in applications and we shall have much to say about them.

Another difference among Eqs. (1) is that all but (1f) are *ordinary differential equations*, each involving an unknown function of a single independent variable. Equation (1f) is a *partial differential equation*, involving as it does derivatives of an unknown function of more than one independent variable. Our main interest is with ordinary differential equations, and hereafter, the term "differential equation" will mean "ordinary differential equation" in this book, unless stated otherwise. Partial differential equations are taken up in the last chapter.

As a notational convenience, mathematicians define the general nth-order differential equation for the function $y(t)$ as

$$F(t, y(t), y'(t), y''(t), \ldots , y^{(n)}(t)) = 0 \tag{3}$$

where $F(\cdot, \cdot, \ldots, \cdot)$ is a well-behaved function of $n + 2$ variables. Because of the difficulty in constructing a theory for solutions of the most general differential equation (3), mathematicians usually start at a somewhat less general level. The assumption is usually made that (3) can be solved for $y^{(n)}(t)$ in terms of the other variables to obtain

$$y^{(n)}(t) = G(t, y(t), y'(t), \ldots , y^{(n-1)}(t)) \tag{4}$$

Equation (4) is said to be in *normal form* over a region described by the $n + 1$ variables $t, y, \ldots, y^{(n-1)}$ if G is continuous on that region. Equations in normal form are much easier to handle theoretically than are more general equations. For example, Eq. (1d) is equivalent to an equation in normal form,

$$y'' = -\frac{k}{m} e^{-at}y$$

It is more difficult to put (2a) in normal form; in fact, Eq. (2a) is not equivalent to a single equation in normal form—it takes two equations since we must use the quadratic formula to solve for y':

$$y' = \frac{-x - (x^2 + 4y)^{1/2}}{2} \quad \text{or} \quad y' = \frac{-x + (x^2 + 4y)^{1/2}}{2}$$

When Eq. (3) has the form

$$y^{(n)} + a_{n-1}(t)y^{(n-1)} + \cdots + a_1(t)y' + a_0(t)y = f(t)$$

it is called a *linear differential equation*. Examples are (1a)–(1d). Linear equations are easily seen to be equivalent to a single normal equation if $f(t)$ and all the coefficients $a_j(t)$ are continuous on a common interval.

Differential equations involving more than one unknown function occur quite often. Usually, the number of differential equations in such cases agrees with the number of unknown functions. Such a collection of differential equations that must be satisfied simultaneously by all the unknown functions is called a *system*. An example of a first-order system of two equations in two unknown functions I_1 and I_2 is

$$LI_1' + (R_1 + R_2)I_1 - R_2I_2 = 0$$

$$R_2I_2' - R_2I_1' + \frac{1}{C}I_2 = E'$$

(currents in coupled circuits)

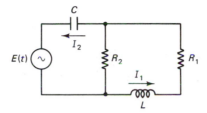

Higher-order systems involving higher-order derivatives also occur, but we shall have little occasion to consider them.

Solutions: Maximally Extended Solutions

Before beginning a serious study we need to have a precise definition of a solution of a differential equation.

> **Solution of a Differential Equation.**   A function $y(t)$ defined on an interval $a < t < b$ is a *solution* of a differential equation of order n if $y(t)$ and all its derivatives through order n exist and are continuous, and upon substituting $y(t)$ for the unknown function, the equation becomes an identity for all t in the interval $a < t < b$.

Example 1.1

Direct substitution shows that $N(t) = e^{2t}$ is a solution of the first-order linear differential equation

$$N'(t) = 2N(t) \tag{5}$$

on any interval $a < t < b$, where $-\infty \leq a < b \leq \infty$. In fact, for any constant C, $N = Ce^{2t}$, $a < t < b$, is a solution since $N' = 2Ce^{2t} = 2N$ on that interval. Since C, a, and b are arbitrary, there is considerable flexibility in what we can accept as a solution.

We can tighten up the definition of a solution considerably by eliminating a certain redundancy in the choice of the interval (a, b). For example, the fact that

e^{2t}, $-1 < t < 1$, is a solution of the differential equation in Example 1.1 is already contained in the fact that e^{2t}, $-\infty < t < \infty$, is a solution of the same equation. To avoid making trivial distinctions of this type, we extend the notion of a solution as follows:

> **Maximally Extended Solution.** A solution $f(t)$, $a < t < b$, of a differential equation is *maximally extended* if there is no other solution $g(t)$ that is identical to $f(t)$ on a $a < t < b$, and which is defined on an open interval containing, but larger than, (a, b).

Thus, in particular, we see that every solution of a differential equation is a "piece" of a maximally extended solution. From this point on, solutions are to be thought of as maximally extended unless specifically stated otherwise. See Figure 1.1 for a sketch of the graph of the solution $N = e^{2t}$, $-0.5 < t < 0.5$, of (5) and

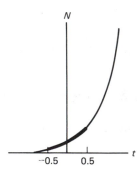

Figure 1.1 Solution and its maximal extension.

of its maximal extension, $N = e^{2t}$, $-\infty < t < \infty$. *Warning*: Sometimes the differential equation itself is restricted to an interval such as $a < t < b$. In this case a maximally extended solution is defined only on (a, b) or one of its subintervals.

A Simple Solution Technique

We cannot continue to solve differential equations by guessing a solution and then verifying that our guess satisfies the equation, as we did in Example 1.1. Surprisingly enough, once we find all the solutions of the simple differential equation

$$y'(t) = 0 \tag{6}$$

we will have a technique for finding all solutions of many other first-order equations. Examples below show why this is so. First, however, let us find solutions of (6). If C is any constant, then the constant function $y(t) = C$ is a solution. The following result shows that there are no other solutions.

> **Vanishing Derivative Theorem.**   Every solution of $dy/dt = 0$, $a < t < b$, has the form $y(t) = C$, $a < t < b$, where C is a constant. If the function f is continuous on an interval I, then all solutions of $y' = f(t)$ on I have the form $y(t) = F(t) + C$, where F is any (fixed) antiderivative of f and C is an arbitrary constant.

Proof.   This is just a formulation of the familiar "constant of integration" idea from integral calculus. The first assertion is proven via the Mean Value Theorem,† and for the second we need only observe that $F' = f$ and that the equation $y' = f$ can be written as $(y - F)' = 0$.

The Vanishing Derivative Theorem, as simple as it is, is the source of many solution techniques for differential equations and hence will be used frequently throughout this text. As a first application let us use the Vanishing Derivative Theorem to show that we have already found all solutions of the differential equation in Example 1.1.

Example 1.1 (*Revisited*)

Let $N(t)$, $a < t < b$, be any solution of $N' = 2N$. We want to show that $N(t) = Ce^{2t}$ for some constant C. Multiply each side of the differential equation by the nonvanishing function e^{-2t} to obtain (after rearrangement)

$$e^{-2t}N' - 2e^{-2t}N = 0, \qquad a < t < b \qquad (7)$$

Since $N(t)$ is continuously differentiable [$N(t)$ is continuous and $N'(t) = 2N(t)$], the Chain Rule may be used to rewrite (7) as

$$\frac{d}{dt}[e^{-2t}N(t)] = 0, \qquad a < t < b$$

By the Vanishing Derivative Theorem, we must have $e^{-2t}N(t) = C$ for some constant C. Thus $N(t) = Ce^{2t}$, $a < t < b$, and we are finished. Of course, since we are only considering maximally extended solutions, we might as well set $a = -\infty$, $b = \infty$.

Solution Curves

It will be very helpful to consider solutions of differential equations geometrically.

> **Solution Curves.**   A *solution curve* of a differential equation is the graph of a solution.

---

† Recall that the Mean Value Theorem states that if $f(t)$ is a continuous function on the closed interval $[c, d]$ and differentiable at each point of the open interval (c, d), then there exists a point t^* in (c, d) such that $f(d) - f(c) = f'(t^*)(d - c)$.

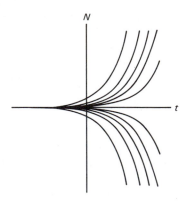

Figure 1.2   Solution curves of $N' = 2N$.

We shall consider only solution curves defined by maximally extended solutions. Our experience so far indicates that differential equations have many solution curves, and this is generally the case. In Figure 1.2 some of the solution curves of the differential equation $N' = 2N$ are sketched. Recall that the solutions of $N' = 2N$ are given by $N = Ce^{2t}$, $-\infty < t < \infty$, each value of C giving a different solution.

The reader may have gained the impression that maximally extended solutions are defined on the entire real line if the function G in the differential equation (4) is defined for all values of its variables. The next example shows that that is not the case.

Example 1.2 (*Finite Escape Time*)

Let us find all (maximally extended) solutions of

$$y'(t) = y^2(t) \tag{8}$$

There are no apparent restrictions on t in (8) itself, but as we shall see, the only solution defined for all t is the *trivial solution*, $y(t) \equiv 0$. If $y(t)$ is another solution of (8), it must be nonvanishing at some point t_0. Since $y(t)$ is continuous, there is an interval I containing t_0 on which $y(t)$ does not vanish and we may divide each side of (8) by $y^2(t)$ to obtain

$$y^{-2}(t)y'(t) = 1 \qquad \text{for all } t \text{ in } I \tag{9}$$

Since $[-y^{-1}]' = y^{-2}y'$ by the Chain Rule and since $t' = 1$, we can write (9) as $[-y^{-1}]' = [t]'$ or, rearranging,

$$[y^{-1} + t]' = 0 \qquad \text{for all } t \text{ in } I$$

The Vanishing Derivative Theorem implies that

$$y^{-1} + t = C \qquad \text{for all } t \text{ in } I \tag{10}$$

where C is some constant. Solving (10) for y, we have

$$y = \frac{1}{C - t} \qquad \text{for all } t \text{ in } I \tag{11}$$

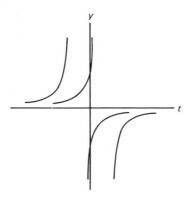

Figure 1.3 Solution curves of $y' = y^2$.

From (11) we see that $I = (-\infty, C)$ or $I = (C, \infty)$. In the former case $y \to +\infty$ as $t \to C^-$, and in the latter $y \to -\infty$ as $t \to C^+$. In either case the solution given by (11) is maximally extended, "escapes to infinity" as t tends to C, and hence cannot pierce the "barrier" $t = C$. It would have been difficult to predict from the differential equation (8) that no solution except the trivial solution is defined for all time t. See Figure 1.3 for sketches of some solution curves. The solutions $y = 1/(C - t)$, $t < C$, and $y = 1/(C - t)$, $t > C$, are distinct, maximally extended solutions.

Differential Equations with Discontinuous Data

Occasionally, we shall encounter differential equations involving discontinuous functions. When this happens we may have to allow "solutions" which are not everywhere differentiable, or which fail to satisfy the differential equation at a finite number of points, although we will continue to require solutions to be continuous. We make no general assertion about what to do in these pathological situations, but a simple example illustrates a possibility.

Example 1.3

Consider the first-order differential equation,

$$y'(t) = \begin{cases} -1, & t < 0 \\ +1, & t \geq 0 \end{cases} \tag{12}$$

where the function on the right-hand side has a discontinuity at $t = 0$. Let us solve the distinct equations,

$$y' = -1, \qquad t < 0 \tag{13}$$

$$y' = +1, \qquad t > 0 \tag{14}$$

and then try to "join" solutions at $t = 0$. The Vanishing Derivative Theorem (or direct integration) implies that the solutions of (13) and (14) are given, respectively, by

$$y = -t + C_1, \qquad t < 0$$
$$y = t + C_2, \qquad t > 0$$

where C_1 and C_2 are arbitrary constants. We can define an everywhere continuous function by setting $C_2 = C_1$, obtaining

$$y = \begin{cases} -t + C_1, & t < 0 \\ t + C_1, & t \geq 0 \end{cases} \tag{15}$$

For each C_1, the function defined by (15) satisfies the differential equation (12) for $t < 0$ and for $t > 0$, and it is continuous for all t, but it fails to be differentiable at $t = 0$. We see that (15) defines all continuous maximally extended solutions of (12). See Figure 1.4 for sketches of the corresponding "solution" curves. We shall say no more here about such pathologies. Whenever they do arise, it is usually clear from the context just how the definition of (maximally extended) solution needs to be broadened.

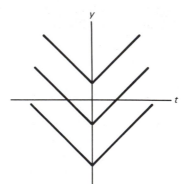

Figure 1.4   "Solution" curves of $y' = t/|t|$.

Comments

In this section we have defined what a differential equation is, what is meant by solutions and maximally extended solutions, and what solution curves are. There will be much more on these central ideas in later sections and chapters.

As we shall learn soon enough, we were fortunate in choosing for our examples differential equations that have closed-form solutions, that is, solutions that can be expressed in terms of elementary functions. Most differential equations do not have closed-form solutions. Fortunately, there is a great deal that we can say about solutions whether or not they exist in closed form. For many purposes exact solutions are not all that crucial; approximate solutions would do just as well, provided that the approximation technique is effective and guarantees an approximate solution with any prescribed degree of accuracy. All the points noted above will be elaborated upon throughout this work.

PROBLEMS

The following problem(s) may be more challenging: 3(c), 3(d), 4(e), 4(f), and 5(c).

1. Find the order of each differential equation listed below. Write out each equation in normal form (if possible). Describe the regions in the ty-plane or tyy'-space where the normal form is valid.
   (a) $(y')^2 - ty = 1$.
   (b) $(t^2 - 1)y'' + (\cos t)y' = y \ln(1 - t^2)$.
   (c) $(t + |t|)y'' - 2yy' + \sin t \, y^2 = e^t$.
   (d) $(y' \sin t)' + y \cot t = e^{-1/2}$.

2. Identify those among the following differential equations which can be written as linear differential equations. Give reasons.
   (a) $(t^2 + y^2)^{1/2} = y' + t$.

   (b) $\dfrac{dt}{dy} = \dfrac{1}{t^2 - ty}$.

   (c) $\dfrac{y'' - t^2y}{y'} = \sin t$.

   (d) $(e^t y')' - (1 + t^2)^{1/2}y = \ln|t|$.

3. In each case find all values of the constant that make the indicated function a solution of each of the accompanying differential equations.
   (a) $y = e^{rt}$ (a1) $y' + 3y = 0$.
       (r is a constant)   (a2) $y'' + 5y' + 6y = 0$.
                              (a3) $y^{(5)} - 3y^{(3)} + 2y' = 0$.

   (b) $y = ct^3$ (b1) $t^2y'' + 6ty' + 5y = 0$.
       (c is a constant)   (b2) $t^2y'' + 6ty' + 5y = 2t^3$,
                              (b3) $t^2y'' + 6ty' + 5y = t^3$.

   (c) $y = t^r$ (c1) $0 = t^4y^{(4)} + 7t^3y''' + 3t^2y'' - 6ty' + 6y$.
       (r is a constant)   (c2) $0 = t^2y'' + 4ty' + y$.

   (d) $x = a \sin \omega t$ $\begin{cases} x'' + 8x - 4y = 0. \\ y'' - 4y + 8x = 0. \end{cases}$
       $y = b \sin \omega t$
       ($a, b,$ and ω are constants)

4. In each case find a differential equation that has the family of functions as solutions. [*Hint*: Differentiate the function as often as necessary for eliminating the constants.] Verify your result.
   (a) $y = ct$, all c, $-\infty < t < \infty$. [For example: $y' = c$; so $y = ty'$.]
   (b) $y = c_1e^{-t} + c_2e^t$, all c_1, c_2, $-\infty < t < \infty$.
   (c) $y = 4ct^2$, all c, $-\infty < t < \infty$.
   (d) $y = (t^{3/2} + c)^2$, all $c \geq 0$, $t > 0$.
   (e) $y = (c_1 + c_2t)e^{2t}$, all c_1, c_2, $-\infty < t < \infty$.
   (f) $y = ce^{-t} \sin 2t$, all c, $-\infty < t < \infty$.
   (g) $y = ct + \sin c$, all c, $-\infty < t < \infty$.

5. Use the Vanishing Derivative Theorem, the methods of Example 1.1 (Revisited), and Example 1.2 to find all maximally extended solutions. Sketch solution curves.
   (a) $y' = 3y$.
   (b) $y' = 3y^2$.
   (c) $y' = 2|t|y$.

1.2 INTRODUCTION TO MODELING

Modeling loosely describes the process of recasting a problem or concept from its natural environment into a form that can be analyzed via techniques we understand and trust. Modeling is a device that aids the modeler in predicting or explaining the behavior of a phenomenon, experiment, or event. For example, say that we wish to launch a manned rocket and put it into orbit about the moon. Physical intuition alone cannot give us more than a rough idea of where to aim the rocket and what guidance strategy to employ once the rocket is launched. Since accuracy will be critical for the success of this mission, a mathematical model of the problem may be constructed utilizing applicable "laws" or "principles." The equations, constraints, and control elements in the model can be treated by mathematical techniques and then used to give a reasonably precise description of the orbital elements of the rocket in its course around the moon.

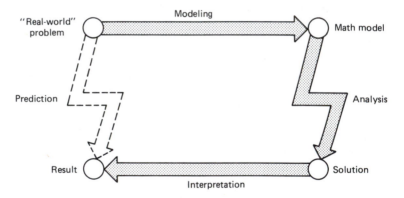

Figure 1.5   Problem solving via mathematical modeling.

The use of models in general to predict or explain outcomes of an observable situation is illustrated in Figure 1.5. It is important to stress that ad hoc models are almost never constructed for every specific problem. Instead, fairly broad "chunks" of the natural environment are "mathematized" by a general model in which all possible outcomes are described by a few basic principles. Some examples are the motion of material particles, the changing concentrations in a chemical reactor, the flow of fluids, the behavior of solid deformable media, the response of electrical circuits, radioactive decay, and population growth of competing species. Figure 1.5 illustrates that a specific problem in the environment is translated into a specific mathematical problem in the model. It is the specific mathematical problem that is solved—and definitely not the entire set of problems that the model will support.

Before discussing the modeling process in general, it would be helpful to have a few simple examples. The following examples should be familiar to the reader, and we can focus on the flow of ideas rather than the technical details.

Radioactive Decay

Certain elements, or their isotopes, are known to be unstable, decaying into isotopes of other elements by the emission of alpha particles (helium nuclei), beta particles (electrons), or photons. Such elements are said to be radioactive. For example, a radium atom might decay into a radon atom, giving up an alpha particle in the process, ^{226}Ra $\xrightarrow{\alpha}$ ^{222}Rn. The decay of a single radioactive nucleus is a random event, and the exact time of decay cannot be predicted with certainty. Nevertheless, something definite can be said about the decay process of a large number of radioactive nuclei.

We shall ask the following question:

How many nuclei are in a sample of a radioactive element at any given time?

Our system is the collection of radioactive nuclei in the sample, and the only measure of the system that concerns us is the number of radioactive nuclei present at any time. It is not obvious, however, what the laws governing the rate of decay are. There is a great deal of experimental evidence to suggest that the following *decay law* is true.

> **Radioactive Decay Law.**   In a sample containing a large number of radioactive nuclei, the decrease in the number of radioactive nuclei over a given interval of time is directly proportional to the length of the time interval and to the number of nuclei present at the start of the interval.

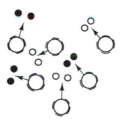

Denoting the number of radioactive nuclei in the sample at time t by $N(t)$, and a time interval by Δt, the law translates into

$$N(t + \Delta t) - N(t) = -kN(t)\,\Delta t \tag{1}$$

where k is the positive constant of proportionality. A reaction of this type is *first order* with *rate constant* k.

The mathematical model (1) of the decay law helps us to spot flaws in the law itself. $N(t)$ and $N(t + \Delta t)$ must be integers, but $k\,\Delta t$ need not be an integer, or even a fraction. If we want to keep the form of the law, we must transform the real phenomenon into an idealization in which a continuous rather than a discrete amount $N(t)$ undergoes decay. For example, measuring $N(t)$ in grams, say, will make it appear more "continuous" since the changes need not be multiples of unity.

Obviously, the number of nuclei can be calculated from the mass of the sample at any given time. Thus there is no need to be specific about the units for N or t at this point. Even if $N(t)$ is continuous, (1) could not hold for arbitrarily large Δt since $N(t + \Delta t) \to 0$ as $\Delta t \to \infty$. Nor does (1) make sense if Δt is so small that no nucleus decays in the time span Δt. The failure of the law for large Δt can be ignored since we are interested only in local behavior (in time). The difficulty with small Δt is more troublesome. We can only hope that the mathematical procedures we now introduce will lead to a mathematical model from which reasonably accurate predictions can be made concerning $N(t)$ at any given time.

If we divide both sides of Eq. (1) by Δt and let $\Delta t \to 0$ (ignoring the difficulty with small Δt mentioned above), we have the differential equation

$$N'(t) = \lim_{\Delta t \to 0} \frac{N(t + \Delta t) - N(t)}{\Delta t} = -kN(t) \tag{2}$$

The solution set of $N' = -kN$ is given by

$$N(t) = Ce^{-kt}, \qquad -\infty < t < \infty \tag{3}$$

where C is an arbitrary constant. Formula (3) can be derived by the procedure used in Example 1.1 (replace the 2 in the example by $-k$).

We cannot use (3) to make predictions about the value of $N(t)$ in the future until the constants C and k have been found. If $N(t)$ is known to have value N_0 at time t_0, then (3) implies that $N_0 = Ce^{-kt_0}$, $C = N_0 e^{kt_0}$, and

$$N(t) = N_0 e^{-k(t - t_0)} \tag{4}$$

The *half-life* $t_{1/2}$ of the radioactive nuclei can be used to determine k. The number $t_{1/2}$ is the time required for half of the collection of nuclei to decay. Curiously, $t_{1/2}$ is independent of the number of parent nuclei present, of the chemical or physical state of the nuclei, and of the time when the clock starts. The first and last of these properties follow from (4):

$$\frac{N(t + t_{1/2})}{N(t)} = \frac{1}{2} = \frac{N_0 e^{k(t_0 - t - t_{1/2})}}{N_0 e^{k(t_0 - t)}} = e^{-kt_{1/2}}$$

which is independent of N_0, t_0, and t. Taking logarithms, we have that

$$k = \frac{1}{t_{1/2}} \ln 2 \tag{5}$$

In the case of ^{226}Ra, time is measured in years and thus the rate constant k has units of (years)$^{-1}$. Formula (4), with k given by (5) in terms of $t_{1/2}$, can be used to make predictions about the value of $N(t)$ for values of $t \neq t_0$. These predictions can be checked against experimental determinations of $N(t)$ and the validity of the law and the model confirmed or questioned. Given the logical gaps mentioned above, it is surprising that (4) provides a remarkably accurate description of radioactive decay processes. But it is a fact of twentieth-century experimental physics that this

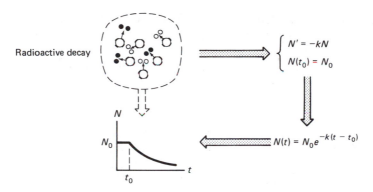

Figure 1.6   Modeling radioactive decay.

is so, at least for time spans, $t - t_0$, which are neither too long nor too short. The leap of faith made in ignoring the flaws of the law and the model is justified by the results. See Figure 1.6 for an interpretation of Figure 1.5 in the context of the phenomenon of radioactive decay.

Position of a Moving Ball

A ball is thrown with velocity v_0 at time t_0 in a vertical direction from a point s_0 units above the surface of the earth. Where will the ball be at any later time?†

The system is the moving ball and the measure of interest is the ball's location at any time after it is thrown. Numerous experiments and observations lead to the formulation of two laws governing the motion of the ball.

> **Laws of Vertical Motion.**   The ball moves along the vertical line through its initial position, and its acceleration is a constant, g, if air resistance is neglected. These laws are simple cases of Newton's First and Second Laws of Motion and of the Gravitational Law.

The statements of the problem and the laws suggest that suitable variables are distance s (measured in meters, say) above the ground and time t (measured in seconds, for example). Let $s(t)$, $t \geqq t_0$, denote the position of the ball at time t, and assume that $s(t)$ is twice continuously differentiable on an open interval, $t_0 < t < t_0 + T$. Assume, in addition, that the one-sided limits $s(t_0^+)$ and $s'(t_0^+)$ exist.‡

---

† We shall identify the ball's location with that of its center, which is reasonable enough if the ball is a small, homogeneous solid.

‡ The number $s(t_0^+)$ is the limiting value of $s(t)$ as t approaches t_0 through values greater than t_0, and similarly for $s'(t_0^+)$. If $s(t)$ is continuous at t_0, then $s(t_0^+) = s(t_0)$.

If we recall from calculus that acceleration is measured by the second time derivative of the position function, then a model which incorporates the statement of the problem and the laws is

$$s''(t) = -g, \qquad t_0 < t < T$$
$$s(t_0^+) = s_0 \tag{6}$$
$$s'(t_0^+) = v_0$$

The minus sign before the g arises because s increases in the upward direction, while the gravitational force acting on the ball is directed downward. In units of meters and seconds, the value of g near the ground is approximately 9.8 meters per second per second. The differential equation $s'' = -g$ is not in general valid for $t < t_0$ because we are not given any information about what the ball was doing before it got to its initial state at time t_0. It may have been carried to the top of a platform before being hurled vertically with velocity v_0. Or perhaps it was shot upward by a cannon in such a way that at exactly the instant t_0 the ball was at height s_0 with velocity v_0. The point is that the "history" of the ball before time t_0 is irrelevant to our simple model. Finally, the model is not likely to hold for times far in the future since we expect the ball eventually to return to the earth and stop. T denotes the time of impact, after which (6) ceases to have any validity.

To solve (6), we integrate each side of the differential equation with respect to t, obtaining

$$s'(t) = \int^t s''(r)\, dr = \int^t -g\, dr + C_1 = -gt + C_1$$

where C_1 is a constant.† Using the initial condition $s'(t_0^+) = v_0$ to evaluate C_1, we find that $C_1 = v_0 + gt_0$ and we have the first-order differential equation

$$s'(t) = -gt + v_0 + gt_0, \qquad t_0 < t < T \tag{7}$$

Integrating each side of (7) with respect to t, we obtain

$$s(t) = \int^t (-gr + v_0 + gt_0)\, dr + C_2 = -\tfrac{1}{2}gt^2 + (v_0 + gt_0)t + C_2 \tag{8}$$

Applying the initial condition $s(t_0^+) = s_0$ to (3), we find that

$$s_0 = -\tfrac{1}{2}gt_0^2 + (v_0 + gt_0)t_0 + C_2 \quad \text{and} \quad C_2 = s_0 + \tfrac{1}{2}gt_0^2 - (v_0 + gt_0)t_0$$

The unique solution of (6) is then

$$s(t) = -\tfrac{1}{2}gt^2 + (v_0 + gt_0)t + s_0 + \tfrac{1}{2}gt_0^2 - (v_0 + gt_0)t_0$$

or

$$s(t) = s_0 + v_0(t - t_0) - \tfrac{1}{2}g(t - t_0)^2, \qquad t_0 \le t \le T \tag{9}$$

---

† Equivalently, use the Vanishing Derivative Theorem of Section 1.1 after rewriting the differential equation as $(s' + gt)' = 0$. The theorem may also be used to solve (7) once it is rewritten in the form $[s + gt^2/2 - (v_0 + gt_0)t]' = 0$.

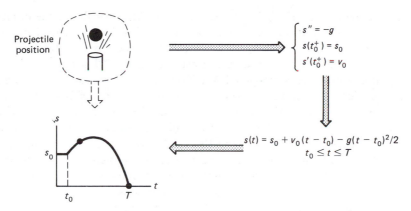

Figure 1.7   Modeling projectile motion.

which we obtain after rearranging the terms. From Eq. (9) we can calculate the time T of impact since $s(T)$ must be 0, but we leave the calculations to the problems. Knowing T, the velocity of impact can be calculated from Eq. (7). Of course, we can use (7) and (9) for any time t between t_0 and T to make predictions about the location and the velocity of the ball at that time. See Figure 1.7.

The validity of the laws of motion originally postulated for this problem can be checked against experimental observations by using (9). This has been done many times over the last three centuries. The model does have flaws. For example, it ignores air resistance and variations in the value of g with altitude and location on the earth. In Chapter 4 we enlarge the model to take these refinements into account. But Eqs. (6) and their solution (9) remain as the most widely known and used model for vertical motion near the ground in the earth's gravitational field. Figure 1.7 interprets the modeling process for this phenomenon of the moving ball.

Elements of a Model: The Modeling Process

The models described above bring to light some essential components of a model. First, the system to be modeled has associated with it a collection of dependent variables whose values depend on a single independent variable. Next, there is an underlying mathematical structure in which the natural laws or principles can be expressed after appropriate coordinates for the dependent variables are introduced. Sometimes these natural laws arise empirically (as was the case in our two examples), sometimes they seem to have intrinsic significance to the phenomenon being modeled (Newton's Laws of Motion, for example), and sometimes these natural laws are expressions of common-sense notions (e.g., "rate of accumulation = rate in minus rate out"). In any case, these laws or principles describe the way in which the system-dependent variables change as the independent variable changes.

To illustrate these concepts, let us return to the two examples treated earlier in this section. System-dependent variables for the moving ball are position, velocity, and acceleration of the ball, whereas for radioactive decay it is the number of radioactive nuclei and the rate of change of that number. The independent variable in both cases is time. The natural law for the falling body is the statement that "the ball has a constant acceleration directed toward the earth." The natural law for the radioactive decay example is embodied in (2), which states in effect that "the instantaneous rate of decay of the population of the radioactive nuclei is proportional to the size of the population at that instant." Both of these laws essentially arise empirically. Now comes the critical step in the modeling process. For the moving ball the s coordinate was introduced along the local vertical and the specific natural problem was translated into the specific mathematical problem (6), which was then solved and interpreted back in the natural world. No specific natural problem was posed for the radioactive decay process, so all solutions of the translated natural law (2) were found. The differential equations in (2) and in (6) are loosely spoken of as "modeling" the phenomena of radioactive decay and falling bodies near the earth, respectively. Added conditions such as $s(t_0^+) = s_0$ and $s'(t_0^+) = v_0$ which appear in (6) are called *initial conditions* and mathematical problems such as (6) are called *initial value problems*, for obvious reasons. We have more to say about these matters in the next section.

We have so far only hinted at the process in which models are constructed. Obviously, we need tools, some variables, and some natural laws to build a model, but one important ingredient in the modeling process has been overlooked. At the beginning of the modeling process the modeler needs to make some definitions and simplifying assumptions and to "discover" some "laws" or "principles" that "govern" or "explain" the behavior of the phenomenon at hand. Quotation marks are used on the terms above because the natural world neither knows nor cares about the modeler's efforts to understand its inner workings. The goal of the modeler is to generate a model that is general enough to explain the phenomenon at hand, but not too complicated to preclude analysis. Thus there are many trade-offs along the way. To have any confidence at all in the model, the modeler will solve a variety of special problems and check the results against experimental evidence. In this way the modeler builds confidence in the validity of the model and also learns something about its limits of applicability. This last step in the modeling process is usually called *validation* of the model. For example, if the differential equation in (6) were used to model the falling body when dropped from a height of 1000 miles above the earth, the inadequacy of the model would quickly be discovered.† Thus modelers sometimes speak of *ranges or regimes of validity*. Figure 1.8 is a schematic describing the entire modeling process.

As experienced practioners know well, models enjoy only a transitory existence. A model based on the empirical data of the day may become totally inadequate when a technological breakthrough produces better instruments, which, in turn, produce better data. In any field, moreover, models are constantly being examined for accuracy in predicting the phenomena modeled. This necessarily involves a careful

---

† If the law $s'' = -g$ were universally valid, it would be impossible for a rocket to leave the earth and never return.

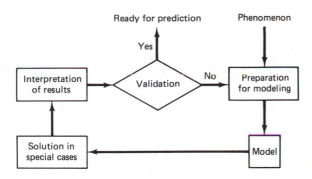

Figure 1.8   Schematic of the modeling process.

reconsideration of the basic assumptions that produced the model, as well as an analysis of the mathematical approximations used in the course of computation. When models are found to be deficient they are modified or supplanted by other models. The literature of science is a chronicle of this process.

PROBLEMS

The following problem(s) may be more challenging: 5(b), 7, and 11.

1. Radium decomposes at a rate proportional to the quantity of radium present. Suppose it is found that in 25 years 1.1% of a certain quantity of radium has decomposed. What is the half-life of radium?

2. If the half-life of a radioactive substance is 1000 years, what fraction of it is left after 100 years?

3. With time measured in years, the value of k in Eq. (5) for cobalt-60 is about 0.13. Estimate the half-life of cobalt-60.

4. A population grows exponentially with growth coefficient 0.03 per month. At some unknown time $T > 0$, circumstances cause the growth constant to change to 0.05 per month. The population doubles in 20 months. At what time T did the change in the growth constant occur?

5. A 600-g mass is thrown vertically upward from the ground with an initial velocity of 2000 cm/s. Find
   (a) The highest point reached and the time required to reach that point.
   (b) The distances from the starting point and the respective velocities attained 3 and 5 s after the motion began.
   (c) The time of impact with the ground.

6. Suppose that a ball is thrown vertically upward and at the time of release the ball is 2 m above the earth and traveling at 36 m/s. Find the time it takes for the ball to strike the earth. [*Hint*: See (9).]

7. A man drops a stone from the top of a building, waits 1.5 s, then hurls a baseball downward with an initial speed of 20 m/s.
   (a) If the ball and stone reach the ground together, how high is the building?
   (b) For a given initial speed of the ball, show that there is a maximum waiting time

after which the ball cannot reach the ground together with the stone, no matter how high the building.

8. Radioactive phosphorus, ^{32}P, is used as a tracer in biochemical studies. It has a half-life of 14.2 days. Upon completion of a tracer experiment with 8 curies (Ci) of ^{32}P, the researchers wanted to dispose of the contents of the experiment after the level of radioactivity had decreased to an acceptable level (1.0×10^{-5} Ci). Calculate the time for which the contents of the experiment should be stored so that the radioactivity has reached an acceptable level. [*Note*: One curie is the quantity of radioactive isotope undergoing 3.7×10^{10} disintegrations per second.]

9. Some financial institutions will cause the funds on deposit in a savings account to grow at a rate that is proportional to the instantaneous value of the account. If K is the constant of proportionality, this rate law is known as $100K\%$ *interest continuously compounded*.
   (a) What rate of interest payable annually is equivalent to 9% interest continuously compounded?
   (b) If the funds in a savings account earn interest continuously compounded and the funds double in 8 years, what is the interest rate?
   (c) How long will it take A dollars invested in a continuously compounded savings account to double if the interest rate is 5%; 9%; 12%?

10. Two common rules of thumb for business managers are the "Rule of 72" and the "Rule of 42." These rules say that the time it takes a sum of money invested at $r\%$ interest to double or to increase by 50%, respectively, is given by $72/r$ or $42/r$. Assume that the interest is compounded continuously [i.e., that $A'(t) = 0.01rA(t)$]. Show that these rules overestimate the time required.

11. During a steady snowfall, a man starts clearing a sidewalk at noon, shoveling at a constant rate and at a constant width. He shovels two blocks by 2 P.M., one block more by 4 P.M. When did the snow begin to fall? Explain your modeling process. [*Hint*: Additional assumptions are required to solve the problem. For example: Sidewalks have constant width, and the man does not turn back.]

1.3 STATE VARIABLES, DYNAMICAL SYSTEMS, AND INITIAL VALUE PROBLEMS

The first step in modeling a system in its natural environment is to identify an independent variable and a collection of variables dependent on that independent variable in the system. For example, in the system consisting of a vertically moving ball (Section 1.2) the independent variable that comes to mind is time. Some dependent variables would be the position, velocity, and acceleration of the ball, all as functions of time. Alternatively, we could have declared the position of the ball as the independent variable and have thought of the velocity and acceleration of the ball, and time, as dependent variables. Other arrangements are also possible. The reason time was chosen as the independent variable is that the dynamics of the ball's motion are most easily and naturally stated in those terms. Indeed, it is almost always the case that the laws or principles in a system with one degree of freedom describe the way in which the system evolves in time. Thus it does no great harm to think of the independent variable in most systems of interest to be time.

State Variables

> **State Variables.**   A set of system-dependent variables, called *state variables*, evaluated at some instant t is called the *state of the system* at time t if knowledge of the variables at that instant, together with the laws describing the evolution of the system, determine the values of the variables for all times in the past and the future where the laws apply.

As an example of these concepts, let us return first to the radioactive decay problem of Section 1.2.

Example 1.4

There are N_0 nuclei in a sample of a radioactive element at a given time t_0. How many radioactive nuclei are present at any later time? System-dependent variables are the number of radioactive nuclei present, denoted by $N(t)$, and the instantaneous rate of change of the radioactive nuclei, $N'(t)$. In Section 1.2 we saw that the law describing how the system evolves is given by $N' = -kN$, where the rate constant k must be determined experimentally. In addition to the law of evolution, $N(t)$ must also satisfy the condition $N(t_0) = N_0$. We see that the number of radioactive nuclei present in our particular sample must be a solution of the problem

$$N' = -kN$$
$$N(t_0) = N_0 \tag{1}$$

As we saw, problem (1) has precisely one solution for any given value for N_0, namely,

$$N(t) = N_0 e^{-k(t - t_0)} \tag{2}$$

If it is known how many radioactive nuclei are present at any given time t_0, the evolution law yields a unique value for the number of radioactive nuclei present at any other time where the evolution law holds. Accordingly, N is the state variable for the system. N' need not be included as a state variable in this case since knowledge of $N'(t_0)$ is not required to "solve" the system.

A single state variable is all that is needed for the radioactive decay problem, but two are required for the moving ball problem of the preceding section. Let us see why this is so.

Example 1.5

A small, homogeneous solid ball is thrown with velocity v_0 at time t_0 in a vertical direction when the ball is at a point s_0 above the earth's surface. What is the position of the ball at any later time? As in Section 1.2, we introduce the coordinate s along the local vertical through the point where the ball is released. Again, suppose that $s(t)$ is the coordinate of the ball at time t. System-dependent variables are position, velocity, and acceleration of the ball, denoted respectively by $s(t)$, $s'(t)$, and $s''(t)$. Then, as we saw earlier, the dynamics of the problem are modeled by the law of evolution, $s'' = -g$, where g is the gravitational constant. The conditions of the problem translate into $s(t_0^+) = s_0$, $s'(t_0^+) = v_0$. The position of the ball's center for $t_0 < t < T$ must be a solution of the problem

$$s'' = -g, \qquad t_0 < t < T$$
$$s(t_0^+) = s_0 \tag{3}$$
$$s'(t_0^+) = v_0$$

where T is the time of impact of the ball with the ground. According to Section 1.2, system (3) has a unique solution $s(t)$, with derivative $s'(t) \equiv v(t)$,

$$s(t) = s_0 + v_0(t - t_0) - \tfrac{1}{2}g(t - t_0)^2$$
$$v(t) = v_0 - g(t - t_0) \tag{4}$$

where $t_0 \leqq t \leqq T$. If s_0 and v_0 are known at time t_0, then $s(t)$ and $v(t)$ are uniquely defined by (4) for all time t from t_0 to the time of impact T. This implies that the position s and velocity v are the appropriate state variables for the problem of the vertically moving ball.

Dynamical Systems

> **Dynamical System.**   A physical system is a *dynamical system* if its law of evolution supports state variables.†

The phenomena of radioactive decay and of the vertically moving ball are examples of dynamical systems. The important thing to keep in mind about dynamical systems is the following. The values of the state variables of a dynamical system at a given instant of time uniquely determine the state variables at any time for which the system's evolution law is valid. The dynamical systems approach to modeling, involving as it does the concept of the state of the system through time, is a concept that is modern in its scope but has been around in one form or another for centuries. It is the basic approach we shall take in this book.

Some conventional terminology has evolved in speaking about various components of a dynamical system. Effects of the external environment on a system are sometimes called *input data* to the system. Values of the state variables at a given initial time t_0 are called *initial data*. Input data are often called *driving terms* or *source terms*, and such systems are said to be *driven*. For example, in a radioactive

---

† Dynamical systems are defined in different ways in different areas. For example, later we give a definition of a dynamical system in terms of the solution of the differential equation rather than in terms of the phenomenon modeled by the differential equation.

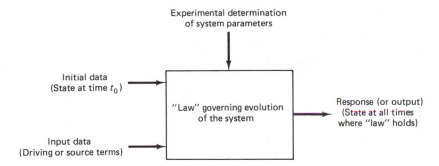

Figure 1.9 Schematic of a dynamical system.

decay problem, assume that the radioactive material is being replenished at a known rate, say $f(t)$. Then this problem is modeled by the differential equation

$$N' = -kN + f(t)$$

and $f(t)$ is the driving term in this case. The behavior of the state variables in time due to the input is called simply the *response* (or sometimes the *output*) of the system.

Figure 1.9 summarizes the basic components of a dynamic system. In examples of modeling in this book we will have an opportunity to interpret the schematic of Figure 1.9 in many specific cases.

Initial Value Problems

It will come as hardly any surprise to the reader that differential equations form the foundation of so many successful models for dynamical systems. Dynamical systems involve change over time, and instantaneous rates of change are interpreted conveniently by derivatives. As we have seen in the examples treated above, after selecting appropriate coordinates and state variables for a dynamical system, a mathematical model arises which consists of a differential equation and one or more conditions expressing values of the state variable at a given time t_0. As mentioned in Section 1.2, these mathematical models are called *initial value problems* and the conditions involving the initial data are called *initial conditions*. For example, (3) is an initial value problem consisting of the differential equation $s'' = -g$ and the initial conditions $s(t_0^+) = s_0$, $s'(t_0^+) = v_0$. Notice that the numbers of initial conditions in the initial value problems (1) and (3) agree with the orders of the differential equations in those problems.

There is an important distinction between initial value problems (1) and (3). The thrown ball problem (3) in its natural setting specifies initial data and requires the response of the system for times later than the initial time t_0 because the modeler has no confidence that the evolution law is valid for times prior to t_0. Problems such as (3) are called *forward initial value problems*. The initial conditions in such problems are expressed in terms of the limiting value of the state variables as $t \rightarrow t_0$ from above. The radioactive decay problem (1) in its natural setting, on the other hand, is a process that the modeler expects has persisted prior to an observation of the state variable N at time t_0. The mathematical model for problem (1) is a *"two-sided" initial value problem* in the sense that the initial time t_0 is in the interior of a time interval where the state variables are defined. To complete our characterization of initial value problems, we should also mention the *backward initial value problem*, the counterpart of the forward problem, where the initial conditions are imposed at a time t_0 and the state variables are sought for $t < t_0$.

There is an obvious relation among these three problems. Say that $y(t)$ is the solution to (3) as a forward problem and that $z(t)$ is the solution to (3) as a backward problem. Now certainly, $y(t_0^+) = z(t_0^-) = s_0$ and $y'(t_0^+) = z'(t_0^-) = v_0$ because of the initial conditions. If it turns out also that $y''(t_0^+) = z''(t_0^-)$, then the function

$$w(t) = \begin{cases} z(t), & t \leq t_0 \\ y(t), & t \geq t_0 \end{cases}$$

is the solution of the two-sided problem for (3). We will not make too much of these matters in a formal sense since the correct interpretation of initial value problems will be clear from the context.

An example will clarify these ideas.

Radiocarbon Dating

Living cells absorb carbon directly or indirectly from carbon dioxide (CO_2) in the air. The carbon atoms in some of this CO_2 are composed of a radioactive form of carbon, ^{14}C (rather than the common ^{12}C), which is produced by the collisions of cosmic rays (neutrons) with nitrogen in the atomsphere. The ^{14}C nuclei decay back to nitrogen atoms by emitting β-particles. Thus all living things, or things that were once alive, contain some radioactive carbon nuclei, ^{14}C. In the early 1960s, Willard Libby showed how a careful measurement of the ^{14}C decay rate in a fragment of dead tissue can be used to determine the number of years since its death. Let us consider a specific problem in order to focus our attention.

> *A Geiger counter is used to measure the current decay rate of ^{14}C in charcoal fragments found in the cave of Lascaux, France, where there are prehistoric wall paintings. The counter recorded about 1.69 disintegrations per minute per gram of carbon, while for living tissue the number of disintegrations was measured in 1950 to be 13.5 per minute per gram of carbon. How long ago was the charcoal formed (and, presumably, the paintings painted)?*

In any living organism the ratio of ^{14}C to ^{12}C in the cells is the same as that in the air. If the ratio in the air is constant in time and location, then so is the ratio in living tissue. After the organism is dead, ingestion of CO_2 ceases and only the radioactive decay continues. The half-life of ^{14}C is taken to be 5568 ± 30 years (since 1962 the internationally agreed-upon value). Let $q(t)$ be the amount of ^{14}C per gram of carbon at time t (measured in years) in the type of tissue represented by the charcoal sample. Let $t = 0$ be the current time and suppose that $T < 0$ is the time that the tissue represented by the charcoal fragment died. Then $q(t) \equiv q(T) \equiv q_T$ for all $t \leq T$, and for $t > T$ the ^{14}C nuclei decay via the first-order rate equation $q' = -kq$, where the rate constant k is computed via the half-life $t_{1/2}$ of ^{14}C by $kt_{1/2} = \ln 2$. Thus we see that $q(t)$ satisfies the backward initial value problem

$$q'(t) = \begin{cases} 0, & t \leq T \\ -kq(t), & T < t \leq 0 \end{cases} \tag{5}$$
$$q(0) = q_0$$

The right-hand side of the differential equation in (5) may have a discontinuity at $t = T$. We handle this in somewhat the same way as we did the problem in Example 1.3 by solving the separate problems

$$q'(t) = 0, \quad t \leq T \qquad q'(t) = -kq(t), \quad T < t \leq 0$$
$$q(T) = q_T \qquad\qquad q(0) = q_0$$

and then piecing together the solutions continuously at $t = T$ to obtain a continuous "solution." In this way, we obtain the solution

$$q(t) = \begin{cases} q_T, & t \leq T \\ q_T e^{-k(t-T)}, & T \leq t \leq 0 \end{cases} \tag{6}$$

which satisfies (5) for all $t \leq 0$ (except at $t = T$) and is continuous for all $t \leq 0$. Now using (6) and the initial condition in (5), we see that

$$q_0 = q(0) = q_T e^{kT} \tag{7}$$

Hence T is known as soon as q_0 and q_T are known. Since our model implies that $q'(t) = -kq(t)$ for $T < t < 0$, we see that

$$q'(0) = -kq(0) = -kq_0$$
$$q'(T+) = -kq(T) = -kq_T$$

Now we are given that the charcoal of the sample underwent 1.69 disintegrations per gram of carbon per minute. We are also told that in living tissue in 1950 when the paintings were dated there were 13.5 disintegrations per gram of carbon per minute. Since the number of disintegrations of ^{14}C per unit of time is proportional to the decay rate $q'(t)$, we have that $q'(0) = 1.69\alpha$ and $q'(T) = 13.5\alpha$, where α is a constant of proportionality. Thus using (7) and these observations, we have that

$$T = \frac{1}{k} \ln \frac{q_0}{q_T} = \frac{t_{1/2}}{\ln 2} \ln \frac{q'(0)}{q'(T+)}$$

$$= \frac{5568 \pm 30}{\ln 2} \ln \frac{1.69}{13.5} \approx -16{,}692 \pm 90 \text{ years}$$

See Figure 1.10 for the solution curve for this radioactive decay problem.

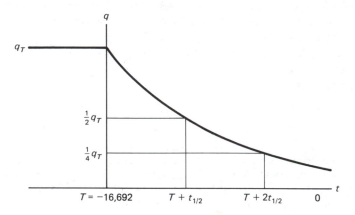

Figure 1.10   Solution curve for the radioactive decay problem.

The accuracy of the ^{14}C dating process depends on a knowledge of the exact ratio of radioactive ^{14}C to ^{12}C in the atmosphere. The ratio is now known to have changed over the years. First of all, there is a basic sinusoidal variation about the mean with a period of roughly 8000 years. On top of this variation, volcanic eruptions and, in the last two centuries, industrial smoke dump "dead" ^{14}C into the atmosphere and decrease the ratio. However, the most drastic change in recent times has occurred through the testing of nuclear weapons, resulting in an increase of 100% in the ratio, at least for some parts of the Northern Hemisphere. These variations are now factored into the dating process.

The procedure described above can be used in conjunction with any radioactive process that satisfies a first-order rate law. Since the experimental error in determining the rate constants for decay processes with long half-lives is large, we would not expect to use this process to date events in the very recent past. Recent events can be dated if a radioactive substance with a short half-life is involved. White lead, a pigment used by painters, contains a small amount of a radioactive isotope of lead with a half-life of 22 years. Using this fact, it has recently been shown that a number of paintings, purportedly from the seventeenth century, were actually painted in the 1940s. At the other extreme, radioactive substances such as uranium, with half-lives of billions of years, can be used to date the formation of the earth itself.

CONCLUSION

In this section we have introduced the general concepts of state variables, dynamical systems, and initial value problems. In the remainder of the book we elaborate on these ideas and give some indication of how deeply these notions are embedded in the contemporary study of phenomena of change.

PROBLEMS

The following problem(s) may be more challenging: 2 and 5.

1. Find a solution for each of the following initial value problems.
   (a) $y' = 0$, $y(2) = -5$.
   (b) $y' = t$, $y(2) = 9$.
   (c) $y' = -2y$, $y(0) = 3$.
   (d) $(t - 1)y' = 1$, $y(2) = 1$.
   (e) $y' = \cos t$, $y\left(\dfrac{\pi}{2}\right) = \frac{1}{2}$.
   (f) $y' = e^t + 1$, $y(0) = 1$.
   (g) $y' = 2ty$, $y(0) = 1$. [*Hint*: Write as $(\ln |y|)' = (t^2)'$.]
   (h) $y' = y^3$, $y(0) = 1$.
   (i) $e^y y' = t^2$, $y(0) = \frac{1}{2}$.

2. Use the substitution $y = z^{1/2}$ to solve the initial value problem

$$yy'' + (y')^2 = 1$$

$$y(0) = 1, \qquad y'(0) = 0$$

3. In 1977, the rate of ^{14}C radioactivity of a piece of charcoal found at Stonehenge in southern England was 8.2 disintegrations per minute per gram. Given that the rate of ^{14}C radioactivity (in 1950) of a living tree is 13.5 disintegrations per minute per gram and assuming that the tree which was burned to produce the charcoal was cut during the construction of Stonehenge, estimate the date of the construction.

4. An archeologist has found a seashell that contains 60% of the ^{14}C of a living shell. How old is the shell?

5. [*Newton's Law of Cooling*]. When an object is placed in a relatively large surrounding medium of constant temperature, according to *Newton's Law of Cooling* (or *Warming*) the temperature of the object changes at a rate directly proportional to the difference between the medium's temperature and the surrounding object's temperature. In taking the temperature of a sick horse, a veterinarian notes that the thermometer reads 82°F at the time of insertion. After 3 min the veterinarian observed a reading of 90°F, and 3 min later a reading of 94°F; however, a sudden convulsion destroyed the thermometer before a final reading could be obtained. What was the horse's temperature?

CHAPTER 2

Initial Value Problems and Their Approximate Solutions

The solution curves of a first-order differential equation are planar curves. The direction field of the differential equation may be used to sketch the curves directly without first solving the differential equation. Alternatively, a computer may be used to find approximate (but very accurate) solutions to initial value problems associated with the differential equation, and the values of the approximate solutions may be tabulated or graphed, again by the computer. The basic ideas underlying numerical methods for approximating solutions are sketched in this chapter, and the theoretical underpinnings of initial value problems are outlined.

2.1 SOLUTION CURVES AND DIRECTION FIELDS

The general first-order differential equation in normal form can be written as

$$y' = f(t, y) \tag{1}$$

where the function f is defined on some portion of the ty-plane. The early chapters in this text are devoted primarily to devising techniques for finding solutions $y(t)$ for (1) when $f(t, y)$ has a variety of special forms. Some techniques are explicit in the sense that all solutions $y(t)$ are characterized by a single formula involving t and an arbitrary constant. Other techniques are implicit and lead to equations involving y, t, and an arbitrary constant, equations that are often difficult to solve for y explicitly as a function of t and the constant.

We shall often sketch solution curves of Eq. (1). Well-drawn solution curves convey much information about the behavior of the solutions of differential equations, information that may be obscured by intricate solution formulas. If formulas for the solutions are available, sketching the solution curves may be carried out using techniques of analytic geometry. If solution formulas are too complicated to be helpful, or if there are no solution formulas at all, we need other ways to sketch solution curves.

One approach uses the computer to graph solution curves directly from solution formulas or to find approximate solutions of (1) and then to graph these approximations. The results appear on a screen or are sketched directly on paper if a hard-copy unit has been attached to the computer. Later in this chapter we have more to say on the use of computers to find approximate solutions and solution curves.

Geometry of Solution Curves

There is a way to view the solvability of differential equation (1) which appeals to geometric intuition and hence lends itself to a graphical approach to finding solution curves. Let $f(t, y)$ be defined over a region R† in the ty-plane (e.g., think of R as the interior of a rectangle with sides parallel to the axes, one or both dimensions

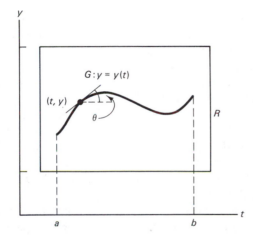

Figure 2.1 Solution curve for $y' = f(t, y)$. Note that $f(t, y) = \tan \theta$.

infinite). According to (1), at each point (t, y) in R, the value $f(t, y)$ gives the slope of the tangent line to any solution curve of (1) through that point. Now suppose that the graph, G, of the continuously differentiable function $y(t)$, $a < t < b$, lies in R. Then G is a solution curve for (1) if at each point (t, y) on G the slope of the tangent line to G has the value $f(t, y)$. Obviously, this is just the geometric equivalent of saying that $y'(t) = f(t, y(t))$, for each $a < t < b$, or that $y(t)$ is a solution for (1) (see Figure 2.1).

---

† A subset R of the ty-plane is a *region* if (1) R is an *open set* (i.e., for each point P in R there is a circle centered at P all of whose interior points lie in R), and (2) R is a *connected set* (i.e., any two points in R can be connected by a path that remains in R).

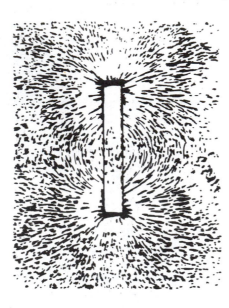

Figure 2.2 Magnetic field pattern for a bar magnet.

This slight change in point of view gives us a way to "construct" solutions for (1). By drawing short line segments (called *lineal elements*) with slope $f(t, y)$ at points (t, y) distributed throughout the region R, we obtain a diagram, called a *direction field*, which may give us some insight about the solution curves of (1). By sketching the curves that "follow" the line segments, the direction field may "suggest" curves in R with the property that at each point of each curve the tangent to the curve at that point is parallel to the line segment at the point. This process reveals solution curves of (1) in much the same way as iron filings sprinkled on paper held over the poles of a magnet reveal magnetic field lines (see Figure 2.2). If a particular direction field is not suggestive in this sense, then choosing a finer distribution of points in R and repeating the process may help. In Figures 2.3 and 2.4 we search for solution curves by using this method.

Although a computer was used to plot the direction fields in Figures 2.3 and 2.4, basically the only tools needed are a straightedge and graph paper with a fine rectangular grid. After coordinate axes and scales have been introduced, Eq. (1) is used to find the slopes of line segments at grid points, and these line segments are drawn. Notice that in each case, as the direction field is refined by inserting shorter and shorter line segments and more grid points, solution curves become more suggestive. Some shortcuts to plotting direction fields are described in the problems. This technique does not, of course, produce exact solutions (and is not without its pitfalls), but we can obtain a rough idea of how solutions of (1) behave.

As we shall see in Section 3.1, all solutions of the differential equation $y' = (1 - t)y - t$ associated with Figure 2.3 are given by the formula

$$y = e^{t - t^2/2}[C + \int_0^t (-s)e^{-s + s^2/2}\, ds]$$

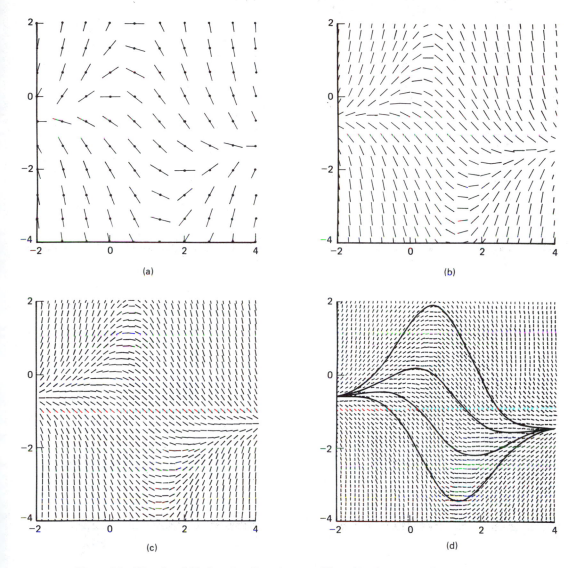

Figure 2.3 Direction fields for $y' = (1 - t) y - t$. The grid points are marked in part (a), but not in the successive refinements.

where C is an arbitrary constant. Note that knowing the solution formula is not as much help in understanding the behavior of the solution curves as is the direction-field approach of Figure 2.3. For the equation $y' = 3y \sin y + t$ of Figure 2.4 there is no known solution formula and the line segments of the direction field are all we have.

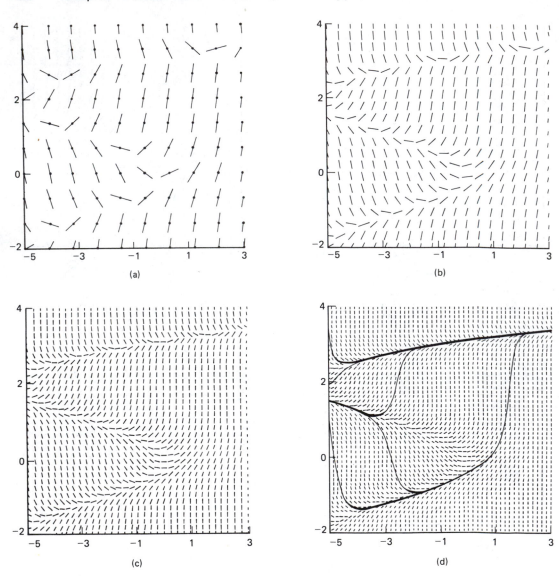

Figure 2.4 Direction fields for $y' = 3y \sin (y) + t$.

Comments

We have shown how the method of direction fields aids us in sketching solution curves for (1). Notice, in particular, that solution curves appear to "fill up" the region R with no intersections—this is guaranteed since we chose continuously differentiable functions $f(t, y)$ on region R and this implies that solution curves do not intersect (see Appendix A.1). Illustrating one of the pitfalls of the direction-field approach, observe in Figure 2.4 that many solution curves flow together so strongly that they appear to overlap—they do not, however. The problem is with the finite resolution and approximations inevitable with computed solutions and computer graphics.

The examples in Figures 2.3 and 2.4 are very instructive. All the direction fields illustrated were obtained using a computer, so the amount of labor involved was slight. The sample solution curves illustrated were obtained on a computer using a commercial differential equations solver. It is doubtful that one could "intuit" such accurate solution curves directly from the direction fields. The graphing aids mentioned in the problems are crucial if one needs to construct direction fields by hand.

PROBLEMS

The following problem(s) may be more challenging: 1(b), 1(d), 2(b), 2(c), and 4(f).

1. For each differential equation below, sketch some solution curves suggested by the direction field below the equation. Then solve the differential equation and compare the graphs of the exact solution curves with your sketch. [*Hint*: In each case the variables may be "separated" to find solution formulas. For example, the equation in part (a) may be written as $(1/y)y' + t = 0$ as long as $y \neq 0$. Since

$$\frac{d}{dt} (\ln |y|) = \frac{1}{y} \frac{dy}{dt}$$

by the Chain Rule and since $d(t^2/2)/dt = t$, the Vanishing Derivative Theorem implies that $\ln |y| + t^2/2 \doteq c$, where c is a constant. Thus $\ln |y| = c - t^2/2$, $|y| = e^c e^{-t^2/2}$, and hence $y = Ce^{-t^2/2}$, where $C = \pm e^c$. Note that $y \equiv 0$ is also a solution.]

(a) $y' = -ty$ (b) $y' = 2|t|y$

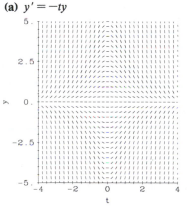
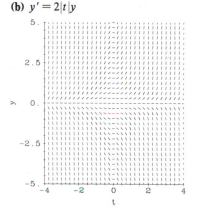

(c) $y' = 3y/t$

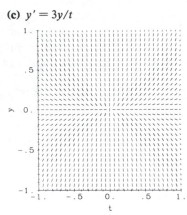

(d)† $y' = (\text{sgn } t)y$

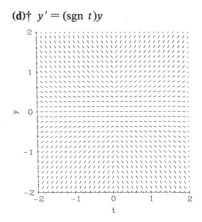

2. **(a)** Show that for any continuous function f on the real line, each member of the family of straight lines

$$y = ct + f(c), \qquad -\infty < c < \infty$$

is a solution of the *Clairaut equation*,

$$y = ty' + f(y')$$

(b) Find solutions to the differential equation

$$e^y + e^{ty'}y' = 0$$

(c) Write down an equation for the family of tangent lines to the parabola $y = t^2$. Find a first-order differential equation for which these lines are all solutions. Find another solution to this equation which is not a straight line. Can you infer anything in general about families of tangent lines to the graph of some function?

3. Recall that the first derivative can be used to show that a function is increasing, decreasing, or constant over an interval and that the second derivative can be used to show concavity or the existence of the inflection points of the graph of a function.

(a) For what region of the ty-plane are the solution curves of the equation $dy/dt = -y/t$? Rising ($y' > 0$)? Falling ($y' < 0$)? Concave upward ($y'' > 0$)? Concave downward ($y'' < 0$)?

(b) Using the information in part (a) one would obtain a rough sketch of the solution curves as seen shown in the figure on the next page. Check this sketch by solving the differential equation and graphing the actual solution curves.

---

† The function sgn t [short for *signum t*] is defined by sgn $t = \begin{cases} +1, & t > 0 \\ -1, & t < 0 \end{cases}$.

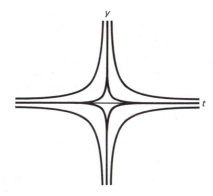

4. Use the method outlined in Problem 3 to sketch the graphs for the following equations. Your graphs should exhibit approximately the right slope and concavity.

(a) $\dfrac{dy}{dt} = y$.

(b) $\dfrac{dy}{dt} = \dfrac{2y}{t}$.

(c) $y' = (y+3)(y-2)$.

(d) $y' = 2t - y$.

(e) $\dfrac{dN}{dt} = rN\left(1 - \dfrac{N}{K}\right)$, $r > 0,\ K > 0$.

(f) $y' = (1-t)y$.

(g) $y' = y - t^2$.

5. When finding direction fields of first-order differential equations other than by computer, laborsaving devices are a necessity. One such device is known as the *method of isoclines*, where the slope of a lineal element is given and curves in the ty-plane are determined such that the slope of the lineal element at each point of the curve has the given value. Such curves are called *isoclines*. After a few parallel lineal elements are sketched along an isocline, a new slope is selected and the process repeated. The figure gives an isocline-generated field of lineal elements for the equation $y' = y$. Use the method of isoclines for the following equations.

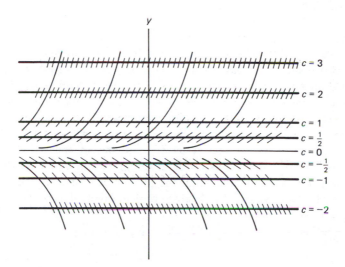

(a) $y' = y - t$.   (b) $y' = ty - 1$.   (c) $y' = y - t^2$.
(d) $y' = (1 + t)y$.   (e) $y' = (4t + 3y)/(3t + y)$.
(f) $y' = (t - y)/t$.   (g) $y' = t^2 + y^2$.

6. For each differential equation below, sketch the direction field over the interval specified. Sketch a solution curve of the equation which meets the initial condition.
   (a) $y' = 2t$, $y(2) = 2$, $-3 \le t \le 3$.
   (b) $y' = t + y$, $y(0) = 2$, $-2 \le t \le 2$.
   (c) $y' + ty = t^2$, $y(-2) = 1$, $-2 \le t \le 4$.
   (d) $ty' = 3y$, $y(0) = 0$, $-5 \le t \le 5$.

7. [*Diverging and Converging Flows*]. Consider the direction field for the equation $y' = f(t, y)$. The solution curves of the differential equation can be thought of as the *flow lines* of a fluid whose velocity at the point (t, y) is $f(t, y)$.

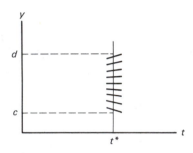

(a) If $\partial f/\partial y > 0$ at some $t = t^*$ and all $c < y < d$, verify that the flow directions across the line $t = t^*$ diverge in the manner indicated in the diagram. If $\partial f/\partial y > 0$ on some region in the ty-plane, what can be said about the vertical distance between solution curves in that region as t increases?
(b) Formulate an analogous property of flows in regions where $\partial f/\partial y < 0$.

2.2 INITIAL VALUE PROBLEMS

Reviewing the models in Chapter 1, we see that the natural problem is not to find all solutions of a differential equation but rather to find a solution of the equation which satisfies an initial condition. The differential equation/initial condition pair is called an *initial value problem* for the differential equation. The intimate connection between dynamical systems and initial value problems hinted at in Chapter 1 indicates the reason for the central position played by such problems. Modelers exploit this connection by using their mathematical knowledge of initial value problems to predict the behavior of dynamical systems, or to design special processes in the system's natural environment. Thus if our modeling efforts are to be successful, we ought to learn as much as possible about the properties of initial value problems.

The prototypical initial value problem involving first-order differential equations is given by

$$y' = f(t, y)$$

$$y(t_0) = y_0$$

(1)

where the function f is defined in some region R of the ty-plane and the point (t_0, y_0) is in R. By way of terminology we shall refer to the function $f(t, y)$ and the value y_0 collectively as *data* for the problem (1), although physicists and engineers use this term only for initial data and input data to a system. Our reason for this is that the function $f(t, y)$ in problem (1) may contain elements that are empirically determined or depend on approximations of circumstances in the system's external environment. Thus it is not inconsistent with our earlier characterization to classify $f(t, y)$ as "data" for the system and to ask how problem (1) behaves when the data $f(t, y)$ and y_0 are perturbed.

A function $y(t)$ is a solution to the initial value problem (1) if it is defined over an interval I containing t_0 and satisfies the conditions $y'(t) = f(t, y(t))$ for all t interior to I, and $y(t_0) = y_0$. Finding all maximally extended solutions of the differential equation in (1) for all initial points (t_0, y_0) in R is essentially the problem of finding the general solution of that differential equation in R. Sometimes it is possible to solve (1) by first finding an analytic expression for the general solution of $y' = f(t, y)$ and then using it to select solution(s) that meet the initial condition $y(t_0) = y_0$. This was the way we solved the initial value problems in Chapter 1. We shall use this technique whenever possible (and practical), but with the exception of a few special classes of functions $f(t, y)$, it is not possible to find an analytic expression for the general solution of the differential equation $y' = f(t, y)$. Thus other techniques for solving the initial value problem (1) must be devised.

Basic Questions for Initial Value Problems

In the analysis of any mathematical model there are several basic questions which always come up, questions phrased in terminology appropriate to the model. They are:

> *Existence*: What conditions on the data guarantee that (1) has at least one solution?
>
> *Uniqueness*: Under what conditions on the data will (1) have at most one solution?
>
> *Solvability*: Assuming the data are such that (1) has one solution, how can we find it?
>
> *Sensitivity*: Assuming that (1) has a unique solution for data within a class, how does the solution change as the data change?

Some simple conditions on the function f produce surprisingly satisfactory answers to all four questions above. The uniqueness and existence questions are treated in the theorem below, solvability is addressed by the solution techniques developed throughout this text, and the sensitivity question is addressed for the first-order linear equations in Chapter 3. A general approach to sensitivity is given in Appendix A.

> **Existence and Uniqueness Theorem.**   In (1) let f and $\partial f/\partial y$ be continuous
> on a region R of the ty-plane, and suppose that (t_0, y_0) is a point in R. Then
> problem (1) has a solution $y(t)$ on an interval I containing t_0 in its interior.
> Problem (1) cannot have more than one solution over any interval I containing
> t_0.

The following examples illustrate the theorem, but the proof of the theorem
is not given here (see Appendix A).

Example 2.1

Consider the initial value problem,

$$y' = y \cos t, \qquad y(0) = 1 \tag{2}$$

In this case $f(t, y) = y \cos t$, while $\partial f/\partial y = \cos t$. Both functions are continuous
on the entire ty-plane, which is itself a region. The Existence and Uniqueness
Theorem guarantees the existence of exactly one solution defined on some interval
containing $t_0 = 0$, but does not provide any assistance in finding a formula
for the solution. Problem (2), however, can be solved directly. If the differential
equation in (2) is multiplied by a nonzero factor, the set of solutions of the
equation, if there are any at all, remains unchanged. Multiplying by the factor
$e^{-\sin t}$ and rearranging terms, we have that

$$0 = e^{-\sin t}(y' - y \cos t) = (ye^{-\sin t})'$$

since $(ye^{-\sin t})' = y'e^{-\sin t} - ye^{-\sin t} \cos t$, as the Product and Chain Rules
for differentiation show. According to the Vanishing Derivative Theorem of
Chapter 1, there is a constant C such that

$$ye^{-\sin t} = C$$

Thus $y = Ce^{\sin t}$. Imposing the initial condition of (2), we see that $C = 1$.
Thus it appears that if $y(t)$ is a solution of (2), then

$$y(t) = e^{\sin t}, \qquad -\infty < t < \infty \tag{3}$$

That (3) actually defines a solution may be shown by substituting $y(t)$ from
(3) into the equation of (2) (see Figure 2.5).

The next example shows that there may be difficulties if the conditions of the
Existence and Uniqueness Theorem are not met.

Example 2.2

The function $f = 3y^{2/3}$ in the initial value problem

$$y' = 3y^{2/3}, \qquad y(0) = 0 \tag{4}$$

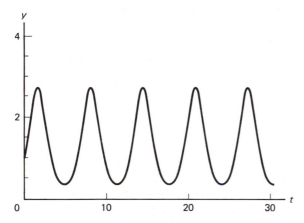

Figure 2.5   Solution of $y' = y \cos t$, $y(0) = 1$.

is continuous for all values of y (and t), but the partial derivative $\partial f/\partial y = 2y^{-1/3}$ is undefined when $y = 0$. Since the initial data are precisely $y_0 = 0$, we have advance warning that there may be a problem in solving (4).

The difficulty, curiously enough, is not in the existence of the solution, but in the uniqueness. Problem (4) has infinitely many solutions! First of all, it is easy to verify that $\bar{y}(t) \equiv 0$, $-\infty < t < \infty$, is a solution. To find other solutions we write the differential equation as $y^{-2/3}y'/3 - 1 = 0$ and observe that $(y^{1/3} - t)' = y^{-2/3}y'/3 - 1$. Using the Vanishing Derivative Theorem, we conclude that each nonvanishing solution $y(t)$ of the differential equation satisfies $y^{1/3}(t) - t = c$ for some constant c. Thus we have that

$$y = (t + c)^3, \qquad -\infty < t < \infty \tag{5}$$

is a solution of $y' = 3y^{2/3}$. Since both y and y' vanish at $t = -c$, the solution $\bar{y}(t) \equiv 0$ and that defined by (5) may be pieced together to form many other solutions of (4). For any t_0, and any numbers a and b with $a \leq t_0 \leq b$, we see that the functions

$$y_1(t) = \begin{cases} (t-a)^3, & t \leq a \\ 0, & a \leq t \leq b \\ (t-b)^3, & b \leq t \end{cases}$$

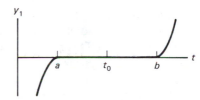

$$y_2(t) = \begin{cases} 0, & t \leq b \\ (t-b)^3, & b \leq t \end{cases}$$

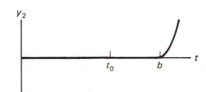

$$y_3(t) = \begin{cases} (t-a)^3, & t \le a \\ 0, & a \le t \end{cases}$$

are solutions of problem (4). Hence for any value of t_0, solutions of problem (4) are not unique. On the other hand, if the initial condition in (4) were replaced by $y(t_0) = y_0$, where $y_0 \ne 0$, that initial value problem would have a unique solution because $\partial f / \partial y$ is a continuous function on a region containing (t_0, y_0).

It is worth noting that although direction-field plots are useful in giving us some idea of what solution curves might look like, they are often not useful in deciding whether initial value problems have unique solutions. See, for example, Figure 2.4d.

Approximate Solutions

The direction-field approach can be used to generate techniques for finding approximate solutions for problem (1) to any desired degree of accuracy. We shall describe one such technique below—it is called *Euler's Method*—but defer to Appendix B the proof that it accomplishes its task.

Consider the first-order differential equation $y' = f(t, y)$, where f is defined over the entire ty-plane. Suppose that we know one point (t_1, y_1) on a solution curve $y(t)$. Then we could find an approximate value for the point $(t_2, y(t_2))$ on the solution curve at $t = t_2$ by following the (known) tangent line to the solution curve through (t_1, y_1) out to $t = t_2$. Since the slope of the tangent line to the solution curve at (t_1, y_1) is $f(t_1, y_1)$, we see that the value $y_2 = y_1 + (t_2 - t_1)f(t_1, y_1)$ is a reasonable approximation to $y(t_2)$ if $t_2 - t_1$ is small. [Of course, if (t_1, y_1) is not on the desired solution curve, the calculated value y_2 acquires an error from this source as well.] This simple calculation can be repeated as often as desired to produce a broken-line approximation to the solution curve through a given point of the ty-plane. If the sequence of points on the t-axis where this calculation is performed are equally spaced, this technique is called *Euler's Method*. Letting h be the step size, we define the equally spaced points $t_n = t_0 + nh$, $n = 0, 1, 2, \ldots, N$. Then from our description above we evidently have that

$$y_n = y_{n-1} + hf(t_{n-1}, y_{n-1}), \qquad n = 1, 2, \ldots, N, \quad \text{with } y_0 \text{ given} \qquad (6)$$

If the data $f(t, y)$ are smooth enough, it can be shown that the total error in calculating y_N from (6) can be made as small as desired by taking h small enough. An example will clarify this procedure.

Example 2.3

Let us consider the differential equation $y' = y \sin 3t$, and choose for (t_0, y_0) the point $(0, 1)$. We seek a solution curve that solves the initial value problem

$$y' = y \sin 3t$$
$$y(0) = 1 \qquad (7)$$

Choosing $h = 0.2$ and $N = 20$, then $t_n = 0 + (0.2)n$, $n = 0, 1, 2, \ldots, 20$, and (6) becomes

$$y_n = y_{n-1} + (0.2)(y_{n-1} \sin 3t_{n-1}), \qquad n = 1, 2, \ldots, 20, \quad \text{with } y_0 = 1 \qquad (8)$$

Problem (7) has the (unique) solution $y = \exp\left[\frac{1}{3}(1 - \cos 3t)\right]$, as we shall see in Section 3.1. The Euler approximation is generated by taking the points $(t_0, y_0), (t_1, y_1), \ldots, (t_{20}, y_{20})$ determined with the help of (8) and connecting them with linear segments. The results are graphed in Figure 2.6, where the solution of (7) is given by the dashed curve and the solid line is the Euler approximation.

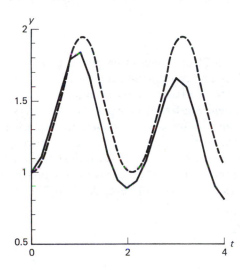

Figure 2.6 Euler's Method for $y' = y \sin (3t)$, $y(0) = 1$.

Extension of Solutions

If a solution $y(t)$ of $y' = f(t, y)$ is known on a closed interval $a \le t \le b$ and the solution curve $(t, y(t))$, $a \le t \le b$, lies in a region R of the ty-plane where f and $\partial f/\partial y$ are continuous, it is easy to see that $y(t)$ can be extended to a solution on a larger interval. Indeed, the solution of the initial value problem $z' = f(t, z)$, $z(b) = y(b)$, provides an extension of $y(t)$ to the right of $t = b$. This brings up the question of the relationship between the maximally extended solutions of $y' = f(t, y)$ and the region R where the data f are defined. The following result is useful in understanding the qualitative behavior of solutions of differential equations. For a proof of this principle, see Appendix A.

> **Extension Principle.** Let R be a region in the ty-plane on which f and $\partial f/\partial y$ are continuous. Let S be any closed bounded region† in R and consider the problem (1) with (t_0, y_0) an interior point of S. Then any solution of (1) can be extended both backward and forward in t until its solution curve exits through the boundary of S.

Example 2.4

The region R for the problem $y' = 3t^{1/2}y^{1/2}$, $y(1) = 1$, is the first quadrant, $t > 0$, $y > 0$. The maximally extended solution, $y = t^3$, $t > 0$, extends across any closed bounded coordinate rectangle in the first quadrant which contains the initial point $(1, 1)$, the solution entering across one edge and then leaving across another as t increases. Note that the domain of this solution cannot be extended to $t < 0$ since the rate function is no longer defined.

Comments

We have seen that direction-field plots can often give a good idea of what solution curves of $y' = f(t, y)$ look like, and can suggest procedures for generating approximate solutions. Also touched on was the basic theory of initial value problems (but proofs and a more complete treatment are found in Appendix A). We saw that in regions R in the ty-plane where f and $\partial f/\partial y$ are continuous, the equation $y' = f(t, y)$ has the following properties: There is a unique solution curve through each point (t_0, y_0) in R, and each solution curve can be extended forward and backward in t until it either exits via the boundary of R, or approaches the boundary of R "at infinity." In particular, solution curves of $y' = f(t, y)$ never "die" inside R (see, e.g., Example 1.2).

PROBLEMS

The following problem(s) may be more challenging: 3, 4, and 5(c).

1. Euler's method may be used to approximate the exact solution, $y(t) = e^t$, of the initial value problem $y' = y$, $y(0) = 1$, on the interval $0 \le t \le T$.
   (a) If the interval $[0, T]$ is split into N equal subintervals, show that $y_n = (1 + T/N)^n$, $n = 0, 1, \ldots, N$, where y_n is the Euler approximation to $y(t_n)$ and $t_n = nT/N$.
   (b) Show that $\lim_{N \to \infty} y_N = e^T$.

---

† A *closed region* contains all its boundary points. A *bounded region* in a plane is contained within some (sufficiently large) square.

2. (a) Why does the Existence Theorem not apply to the initial value problem $2tyy' = t^2 + y^2$, $y(0) = 1$?

   (b) Show that the problem $ty' = 2y$, $y(0) = 0$ has one solution $y_1 = t^2$, $-\infty < t < \infty$, and another solution

$$y_2 = \begin{cases} 0, & t < 0 \\ t^2, & t \geq 0 \end{cases}$$

   Why does this not contradict the Uniqueness Theorem?

3. Referring to Problem 1(d) in Section 2.1, find forward and backward solutions to the problem

$$y' = (\text{sgn } t)y, \qquad y(0) = 1$$

   where sgn $t = +1$ at $t = 0$ for this problem. Does the problem have a solution in a neighborhood of $t = 0$? If so, what is it? If not, what goes wrong?

4. Does the initial value problem

$$y' = (1 - y^2)^{1/2}, \qquad y(0) = 1$$

   have a unique solution? If there are multiple solutions, describe them.

5. Verify that there exists exactly one solution to each of the following initial value problems by checking the hypotheses of the Existence and Uniqueness Theorem (do not solve).

   (a) $y' = e^t y - y^3$, $y(0) = 0$.

   (b) $y' = |t|y^2 - \dfrac{1}{3y + t}$, $y(0) = 1$.

   (c) $y' = |t||y|$, $y(0) = 1$.

6. (a) Explain why the solution curve of the problem $y' = (4 - y^2)^{1/2}(4 - t^2)^{1/2}$, $y(0) = 0$, enters the square $-1 \leq t \leq 1$, $-1 \leq y \leq 1$ at some $t = t_1 < 0$ and exits at some $t_2 > 0$.

   (b) Explain why the solution curve in (a) never crosses the horizontal line $y = a$, for $|a| > 2$.

2.3 ONE-STEP NUMERICAL METHODS

The collection of differential equations whose solutions can be calculated analytically and displayed via elementary functions is remarkably small. Even when differential equations can be solved analytically, it often happens that the solution formula is not especially informative. Thus there is a clear need to have techniques for approximating solutions of differential equations to any given degree of accuracy. In this section we present a few useful numerical iterative processes for deriving approximate solutions of first-order differential equations. Advantages and pitfalls of using computers to implement these iterative processes are discussed briefly. Numerical solution of higher-order differential equations is touched on in Chapter 9, where such equations are viewed as first-order systems. For further details, see Appendix B.

Numerical Approximations to Solutions of Initial Value Problems

Let us assume throughout that the initial value problem to be solved is

$$y' = f(t, y_0), \qquad y(t_0) = y_0 \qquad (1)$$

where f and $\partial f / \partial y$ are continuous functions on the rectangle $R = \{a \leq t \leq b, c \leq y \leq d\}$ with (t_0, y_0) inside R. Then the Existence and Uniqueness Theorem in 2.2 implies that (1) has a unique solution on an interval I containing t_0.

In Section 2.2 we used the "direction-field" (also called the "flow-field") approach to obtain a rough idea of what the solution curves of $y' = f(t, y)$ look like in a given region of the ty-plane. This geometric notion of solution led to a method for producing approximate values for the solution of problem (1) on a set of equally spaced discrete points near the initial point $t = t_0$. Called *Euler's Method*, this technique produced approximations $y_0, y_1, y_2, \ldots$ to the true solution $y(t)$ to problem (1) on the discrete set of points $t_n = t_0 + nh$, $n = 0, 1, 2, \ldots$, for constant *step size* $h > 0$, in the following iterative way:

$$y_{n+1} = y_n + hf(t_n, y_n), \qquad n = 0, 1, 2, \ldots, \qquad y_0 \text{ given} \qquad (2)$$

We saw from Example 2.3 that Euler's Method may provide only a very rough approximation of the true solution of an initial value problem. In Appendix B we prove that, using exact arithmetic (i.e., no rounding off in performing calculations), Euler's Method provides arbitrarily close approximations to the solution of problem (1) when h is chosen small enough.

One-Step Methods

Euler's Method for producing approximate solutions of problem (1) is but one example of a class known as *one-step methods*. Such methods produce an approximate value for the solution of (1) at a selected point T in the interval I in the following way. Suppose that $T > t_0$ (every backward initial value problem can be converted to a forward problem, so that this assumption is not really a restriction). Select an increasing sequence $t_1, t_2, \ldots, t_N$ with $t_N = T$ and $t_0 < t_1$ and define the *step size* $h_n = t_n - t_{n-1}$ for $n = 1, 2, \ldots, N$ (in practice, the sequence t_n is usually chosen so that the h_n are all equal). Then a one-step method computes an approximation y_n to $y(t_n)$ for each $n = 1, 2, \ldots, N$ using the scheme

$$y_n = y_{n-1} + h_n A(t_{n-1}, y_{n-1}, h_n), \qquad n = 1, 2, \ldots, N \quad \text{with } y_0 \text{ as in (1)} \qquad (3)$$

where the (fixed) function $A(t, y, h)$ is called the *increment function* of the method. Notice that to compute y_n, only the value of y_{n-1} is required (and hence the reason for the name *one-step method*). Thus method (3) uses the given value y_0 to generate y_1, y_1 to generate y_2, and so on until the process terminates with the calculation of y_N, which is an approximation of $y(T)$. For Euler's Method the increment function $A(t_n, y_n, h_{n+1})$ is $f(t_n, y_n)$, and in practice the points t_n are equally spaced with step size $h = t_{n+1} - t_n$ constant for all n. Since from (3) we have $(y_n - y_{n-1})/h_n$

$= A(t_{n-1}, y_{n-1}, h_n)$, the function $A(t, y, h)$ is also called an *approximate slope function* for $y(t)$ at t_{n-1}.

Errors

It is useful to estimate how much the approximations $y_1, y_2, \ldots, y_N$ generated by a given one-step method deviate from the exact values $y(t_1), y(t_2), \ldots, y(t_N)$. If precise arithmetic (i.e., no rounding off or chopping of decimal strings) is used, the deviation $|y(t_n) - y_n| \equiv E_n$ is known as the *global* (or *accumulated*) *discretization error at the nth step*. There is a local version of the error due to discretization. When y_n is exact [i.e., $y(t_n) = y_n$] and y_{n+1} is computed from (3), $e_{n+1} \equiv |y(t_{n+1}) - y_{n+1}|$ is known as the *local discretization* (or *formula* or *truncation*) *error*. Thus the global discretization error at the nth step is due to two sources: the formula error at that step coupled with the inexact value of y_{n-1} for $y(t_{n-1})$. In practice, the formula error is easier to compute and is used in the analysis and control of errors. The *global discretization error* E_N is the most important of all the discretization errors since our goal is to use y_N to approximate $y(t_N) \equiv y(T)$.

The approximation to the solution $y(t)$ of problem (1) by the discrete one-step algorithm (3) has a simple interpretation. For each $j = 0, 1, \ldots, N - 1$ join the point (t_j, y_j) to (t_{j+1}, y_{j+1}) by a line segment. The polygonal path formed by these segments reaches from (t_0, y_0) to (t_N, y_N) and approximates the graph of the solution $y(t)$ (see Figure 2.7).

One-step methods may be classified according to the order of magnitude of the global discretization errors incurred when the method is applied.

> **Order of a One-Step Method.**   Given a slope function $f(t, y)$, a one-step method (3) for problem (1), a constant step size h, and a value $T = t_0 + Nh$, we say that the method is of *order p in the step size h* if there is a constant $M > 0$ such that $E_N \leq Mh^p$.

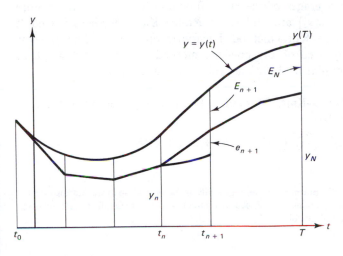

Figure 2.7   Polygonal approximation to the solution of problem (1).

It is usually an easy matter to adjust the step size when implementing a one-step method on a computer. Thus a reduction of the step size by one-half results in a 16-fold improvement in the upper bound on E_N for a fourth-order method, while a similar reduction for a first-order method gives only a twofold decrease in the upper bound. This suggests that the higher the order of a method, the more accurately it will approximate the solution of problem (1). However, in specific instances this may not hold true since the constant M may be larger for a higher-order method than for a lower-order algorithm. Also, higher-order methods usually involve more calculations and function evaluations, and the accompanying round-off errors may nullify the advantages of the higher order. Nevertheless, the order of a method is a significant indication of its accuracy. In Appendix B it is shown that Euler's Method is of first order.

Heun's Method

Euler's Method may be converted into a second-order method if we compute the increment function by averaging the slopes at t_n and at t_{n+1}. Euler's Method is used to find a first approximation to y_{n+1} so that the slope $f(t_{n+1}, y_{n+1})$ can be calculated. We then have a new one-step method called

> **Heun's Method.** The one-step method with constant step size h for problem (1) given by
>
> $$y_{n+1} = y_n + \tfrac{1}{2}[f(t_n, y_n) + f(t_{n+1}, y_n + hf(t_n, y_n))]h$$
>
> is called *Heun's Method*.

Heun's Method can be shown to be of second order.

Runge–Kutta Methods

One-step methods that use averages of the slope function $f(t, y)$ at two or more points over the interval $[t_n, t_{n+1}]$ are said to be *Runge–Kutta Methods*.† Heun's method is a second-order Runge–Kutta method. The fourth-order method given below is probably the most widely used of any one-step method. It involves a weighted average of slopes at the midpoint $t_n + h/2$ as well as at the endpoints.

> **Fourth-Order Runge–Kutta Method.** The one-step method in the constant step size h given by
>
> $$y_{j+1} = y_j + \tfrac{1}{6}(k_1 + 2k_2 + 2k_3 + k_4)h$$
>
> where

---

† C. D. T. Runge (1856–1927) did notable work not only in numerical analysis, but also on the Zeeman effect and in diophantine equations. M. W. Kutta (1867–1944) was an applied mathematician who contributed to the early theory of air foils.

$$k_1 = f(t_j, y_j) \qquad\qquad k_2 = f\left(t_j + \frac{h}{2}, y_j + k_1\frac{h}{2}\right)$$

$$k_3 = f\left(t_j + \frac{h}{2}, y_j + k_2\frac{h}{2}\right) \qquad k_4 = f(t_{j+1}, y_j + k_3 h)$$

is the *Fourth-Order Runge–Kutta Method.*

It may be shown that this method is a generalization of Simpson's Rule for approximating an integral (see the problem set).

Example 2.5

If the differential equation $y' = y$ is multiplied by the nonvanishing function e^{-t}, it assumes the form $(e^{-t}y)' = 0$. Thus all solutions of $y' = y$ have the form $y = Ce^t$ where C is an arbitrary real number. Hence the initial value problem

$$y' = y, \qquad y(0) = 1$$

has the unique solution $y = e^t$, $-\infty < t < \infty$. This solution is plotted in Figure 2.8 on the interval $0 \leq t \leq 1$, together with the approximating polygons of the one-step methods of Euler, Heun, and Runge–Kutta. As expected, the higher-order methods give more accuracy than those of lower order. Accuracy is observed to improve as the step size is reduced.

Round-off Errors

A major source of errors in any step-by-step process arises because precise arithmetic or precise evaluation of functions is not possible in general, not even on very large computers. Every computation made by machine (or human) chops long decimal strings (e.g., $\frac{1}{3}$ might be carried as 0.33333, π as 3.14159, and e^x might be evaluated as 2.71828 when $x = 1$). These are examples of *round-off errors*, individually small but often devastating in their cumulative effect.

The values $y_1, y_2, \ldots$ generated in a one-step method (3) cannot be computed with infinite precision, and hence the error $|y(t_n) - y_n|$ must be viewed as having a component due to round-off error as well as to discretization. Round-off has the nasty feature that the more operations that are to be performed, the more likely it is for round-off error to grow as it propagates through the scheme. Thus if we wish to compute an approximate value y^* for the solution of problem (1) at $t = t^* > t_0$, we are faced with the following trade-off. By taking the step size $h = (t^* - t_0)/N$ very small, we can make the discretization error at t^* as small as desired. But since we have many more operations to perform in order to approximate $y(t^*)$, this will frequently be at the expense of a buildup of round-off errors. Thus it is not easy to say just what step size would minimize the total of all the errors.

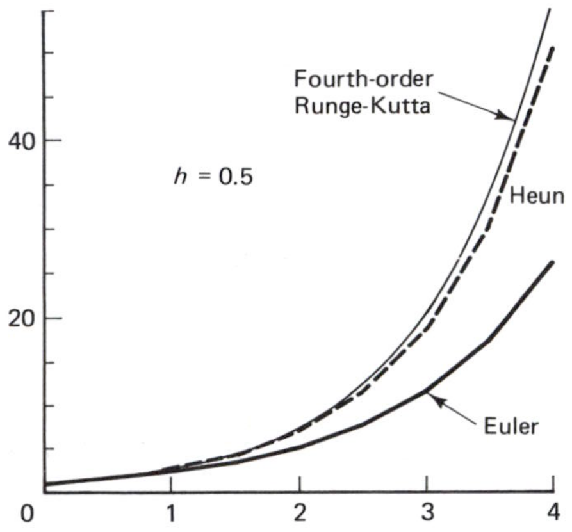

Time	Euler	Heun	Fourth-order Runge-Kutta	exp (t)
0.0	1.0000	1.0000	1.0000	1.0000
0.5	1.5000	1.6250	1.6454	1.6487
1.0	2.2500	2.6406	2.7174	2.7183
1.5	3.3750	4.2910	4.4796	4.4817
2.0	5.0625	6.9729	7.3840	7.3891
2.5	7.5933	11.3310	12.1720	12.1825
3.0	11.3906	18.4123	20.0643	20.0855
3.5	17.0359	29.9208	33.0756	33.1154
4.0	23.6239	48.5213	54.5230	54.5981

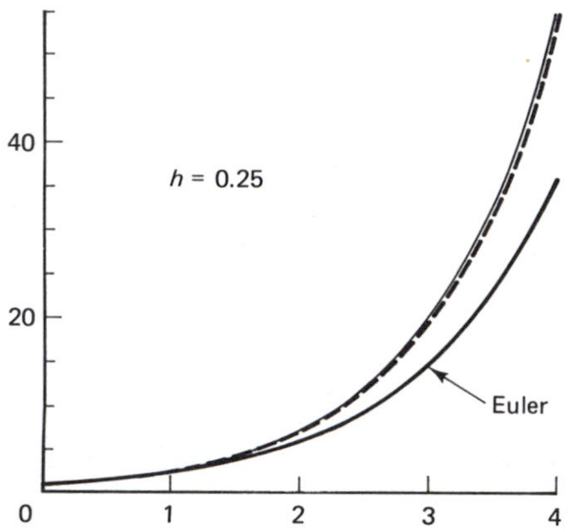

Time	Euler	Heun	Fourth-order Runge-Kutta	exp (t)
0.00	1.0000	1.0000	1.0000	1.0000
0.25	1.2500	1.2813	1.2840	1.2840
0.50	1.5625	1.6416	1.6487	1.6487
0.75	1.9531	2.1033	2.1170	2.1170
1.00	2.4414	2.6949	2.7182	2.7183
1.25	3.0518	3.4521	3.4902	3.4903
1.50	3.8147	4.4239	4.4815	4.4817
1.75	4.7684	5.5681	5.7543	5.7546
2.00	5.9605	7.2623	7.3887	7.3891
2.25	7.4506	9.3048	9.4872	9.4877
2.50	9.3132	11.9217	12.1817	12.1825
2.75	11.6415	15.2747	15.6415	15.6426
3.00	14.5519	19.5707	20.0839	20.0855
3.25	18.1899	25.0730	25.7881	25.7903
3.50	22.7374	32.1273	33.1124	33.1154
3.75	23.4217	41.1631	42.5169	42.5211
4.00	35.5271	52.7402	54.5924	54.5981

Figure 2.8 Approximating e^t (graph of e^t not shown).

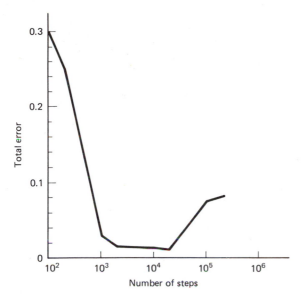

Figure 2.9   Decline and rise in total error with increase in number of steps.

Example 2.6

Suppose that Euler's Method is used to estimate $y(3)$, where $y(t)$ is the solution of the initial value problem $y' = y$, $y(0) = 1$. If the method is used for smaller and smaller step sizes (hence more and more steps), the total error decreases. However, as we see in Figure 2.9, the error begins to increase as the number of steps exceeds approximately 20,000, and hence round-off errors begin to overwhelm the drop in discretization error.

PROBLEMS

The following problem(s) may be more challenging: 3.

1. Estimate $y(1)$ if $y' = -y$, $y(0) = 1$, $h = 0.1$ by using **(a)** Euler's Method; **(b)** Heun's Method; **(c)** the Fourth-Order Runge–Kutta Method.

2. Repeat Problem 1(a) for step sizes **(a)** $h = 0.01$; **(b)** $h = 0.001$; **(c)** $h = 0.0001$. In each case calculate the error $|y(1) - e^{-1}|$. **(d)** If your calculated error does not decrease consistently with a decrease in step size, explain in the light of Example 2.6.

3. Estimate $y(1)$ if $y' = -y^3 + t^2$, $y(0) = 0$, with $h = 0.1$, $h = 0.01$, and finally $h = 0.001$, by using **(a)** Euler's Method; **(b)** Heun's Method; **(c)** the Fourth-Order Runge–Kutta Method.

4. Show that for each of the following initial value problems, $y_N(T) \to y(T)$ as $N \to \infty$, where $\{y_N(T)\}_{N=1}^{N=\infty}$ is the sequence of Euler approximations to $y(T)$ and T is fixed.
   (a) $y' = 2y$, $y(0) = 1$.   **(b)** $y' = -y$, $y(0) = 1$.

5. Simpson's Formula for approximating the integral $\int_a^b f(t)\,dt$ is

$$\frac{b-a}{6}\left[f(a) + 4f\left(\frac{a+b}{2}\right) + f(b)\right]$$

Show that the Fourth-Order Runge–Kutta Method for solving $y' = f(t)$, $y(t_0) = y_0$ reduces to Simpson's Formula applied at each step.

2.4 MULTISTEP NUMERICAL METHODS

There is another important class of methods for producing approximate solutions to the initial value problem

$$y' = f(t, y)$$
$$y(t_0) = y_0 \tag{1}$$

known collectively as *linear multistep methods*. One-step methods ignore the history of the approximation and use only numbers calculated in the immediately preceding step. In contrast, multistep methods recall numbers calculated in several earlier steps and use these numbers to predict the next approximation. Thus the approximation y_{n+k} (k fixed) to the solution of problem (1) at $t = t_{n+k}$ is given in terms of the k foregoing approximations $y_n, y_{n+1}, \ldots, y_{n-1+k}$ to the solution at $t_n, t_{n+1}, \ldots, t_{n+k-1}$. In general, a *linear k-step method* for the step size h has the general form (k a fixed positive integer)

$$y_{n+k} = \sum_{j=0}^{k-1} \alpha_j y_{n+j} + h \sum_{j=0}^{k} \beta_j f(t_{n+j}, y_{n+j}), \qquad n = 0, 1, 2, \ldots \tag{2}$$

where the α_j and β_j are given constants. Techniques for computing reasonable values for the constants α_j and β_j are given in Appendix B. Note that if $\beta_k \neq 0$, then (2) defines y_{n+k} only implicitly and hence such multistep methods are known as *implicit methods*. If $\beta_k = 0$, the method is called an *explicit method*. A disadvantage of multistep methods is that the data points $y_0, y_1, \ldots, y_{k-1}$ at $t_0, t_1, \ldots, t_{k-1}$ are needed to start the method, whereas problem (1) gives only the one data point y_0 at t_0. The other needed data points are usually generated by a one-step method (which, for this reason, is sometimes called a "starting" method).

We shall describe one of the most popular explicit multistep methods due to Adams and Bashforth and an improvement developed by Moulton.†

---

† J. C. Adams (1819–1892) was a noted English applied mathematician and astronomer who predicted in 1845 the existence and the orbit of the planet Neptune, his predictions being based on a mathematical analysis of perturbations in the orbit of Uranus. In later years Adams devoted much time to approximating the values of mathematical constants and functions. He and F. Bashforth made use of the algorithm now known by their names in a joint study of capillary action. During World War I, F. R. Moulton (1872–1952) adapted the algorithm to estimate the solutions of ballistics problems.

Three-Step and Four-Step Adams–Bashforth Methods

$$y_{n+1} = y_n + \frac{h}{12}(23f_n - 16f_{n-1} + 5f_{n-2}) \tag{3}$$

$$y_{n+1} = y_n + \frac{h}{24}(55f_n - 59f_{n-1} + 37f_{n-2} - 9f_{n-3}) \tag{4}$$

where $f_j \equiv f(t_j, y_j)$ and h is the common step size.

There is a k-step Adams–Bashforth method for each $k = 2, 3, \ldots$ (see Appendix B). The four-step method (4) is widely used, giving as it does reasonably good approximations without an excessive number of calculations. The *order of a multistep method* is defined analogously to that of one-step methods. As with all the Adams–Bashforth algorithms, the order of the method is equal to the number of steps if the slope function $f(t, y)$ is sufficiently differentiable. Thus the algorithm defined by (4) should be comparable in accuracy to the Fourth-Order Runge–Kutta Method. However, (4) has the added advantage of fewer calculations and function evaluations than the corresponding Runge–Kutta method since the numbers f_j, f_{j-1}, f_{j-2}, and f_{j-3} have already been calculated at previous steps. Thus multistep methods are generally more efficient than one-step methods of the same order. A drawback of every multistep Adams–Bashforth method is that it is not self-starting from the initial values t_0 and y_0. Adams–Bashforth methods are usually initialized by a Runge–Kutta algorithm (of the same order), which runs just long enough to produce the starting values needed.

Adams–Moulton Methods: Prediction and Correction

The Adams–Bashforth method for calculating y_{n+1} makes no use of the values of the slope function $f(t, y(t))$ for $t > t_n$, although one would expect improved accuracy if some of these values were used. Adams–Moulton methods do just that because they have the form (2) with the coefficient $\beta_k \neq 0$.

Three-Step Adams–Moulton Method

$$y_{n+1} = y_n + \frac{h}{24}(9f_{n+1} + 19f_n - 5f_{n-1} + f_{n-2}) \tag{5}$$

where $f_j = f(t_j, y_j)$ and h is the common step size.

See Appendix B for the derivation of this and other Adams–Moulton methods. The three-step method (5) determines y_{n+1} implicitly because it appears on both sides of the equation.

Equation (5) has the form $y_{n+1} = F(y_{n+1})$ and hence its solutions are called *fixed points* of the function F. The function F in this case might be quite a complicated function of y_{n+1} since it contains the term $f(x_{n+1}, y_{n+1})$, so the task of solving the relation $y_{n+1} = F(y_{n+1})$ for y_{n+1} explicitly is, in general, hopeless. A technique often used to calculate a fixed point x of the function F is to make an initial guess for x, call it $x^{(0)}$, and then calculate the sequence $x^{(m)}$ by the recursion relation $x^{(m+1)} = F(x^{(m)})$. Called the method of *simple iteration*, it can be shown that the *iterates* $x^{(m)}$ converge to the fixed point x if $|F'(x)| < 1$ and the initial guess $x^{(0)}$ is close enough to x. The derivative condition is nearly always satisfied for an Adams–Moulton method if the step size h is small enough, but how does one obtain a good initial guess for y_{n+1} from knowledge of the previous y_j? Moulton's idea was that an Adams–Bashforth method could be used for this purpose, so we have the basic apparatus for a *predictor–corrector method*.

The most widely used multistep predictor–corrector method uses the four-step Adams–Bashforth algorithm of Eq. (4) to calculate the "predicted" value $y_{n+1}^{(p)}$, which is then used as an initial guess for y_{n+1} in solving the Three-Step Adams–Moulton Method (5) by the method of successive iteration. This "correction" process is usually halted after one iteration (but may continue if error-control procedures so indicate), and this iterate is taken as the value of y_{n+1}. From here on the process is repeated. Thus we have the

> **Adams–Bashforth (Four-Step)–Moulton (Three-Step) Predictor–Corrector Method.**   Using a common step size h, first compute a "predicted" value for y_{n+1}, denoted by $y_{n+1}^{(p)}$, with
>
> $$y_{n+1}^{(p)} = y_n + \frac{h}{24}(55f_n - 59f_{n-1} + 37f_{n-2} - 9f_{n-3}) \tag{6}$$
>
> Then, setting $f_{n+1}^{(p)} = f(t_{n+1}, y_{n+1}^{(p)})$, the "corrected" value y_{n+1} is calculated with
>
> $$y_{n+1} = y_n + \frac{h}{24}(9f_{n+1}^{(p)} + 19f_n - 5f_{n-1} + f_{n-2}) \tag{7}$$
>
> The correction procedure can be repeated as often as required.

The Adams algorithms are used extensively in software packages for solving initial value problems. Both the step size h and the number of steps may be changed automatically to improve efficiency.

Example 2.7

The initial value problem

$$y' = y^2, \qquad y(0) = -1$$

has the exact solution $y = -(1 + t)^{-1}$, $t > -1$ (see Example 1.2). Table 2.1 compares the exact solution with approximate solutions computed by using the Adams–Bashforth–Moulton method above and by using the fourth-order Runge–Kutta algorithm (which was also used to start the Adams algorithm).

TABLE 2.1 COMPARISON OF METHODS
FOR SOLVING $y' = y^2$, $y(0) = -1$

t	Runge–Kutta	Adams–Bashforth–Moulton	Exact
0.0	−1.000000	−1.000000	−1.000000
0.5	−0.666677	−0.666667	−0.666667
1.0	−0.500029	−0.500000	−0.500000
1.5	−0.400025	−0.400000	−0.400000
2.0	−0.333353	−0.333333	−0.333333
2.5	−0.235730	−0.285714	−0.285714
3.0	−0.250012	−0.250000	−0.250000
3.5	−0.222222	−0.222222	−0.222222
4.0	−0.200003	−0.200000	−0.200000
4.5	−0.181825	−0.181818	−0.181818

PROBLEMS

The following problem(s) may be more challenging: 3.

1. Approximate $y(1)$ if $y' = -y$, $y(0) = 1$, where $h = 0.2$, by using **(a)** the Three-Step Adams–Bashforth Method; **(b)** the Four-Step Adams–Bashforth Method; **(c)** the Four-Step Adams–Bashforth–Moulton Method.

2. Repeat Problem 1 for the initial value problem $y' = -y^3 + t^2$, $y(0) = 1$, $h = 0.2$.

3. Find $y(1)$ for the initial value problem $y' = y^3 + t^2$, $y(0) = 0$ by using an Adams–Bashforth–Moulton method with $h = 0.001$.

2.5 IMPLEMENTING ALGORITHMS: USING THE COMPUTER

Highly reliable software packages are commercially available which contain very sophisticated differential equation solvers. These solvers have been coded by numerical analysts using the algorithms of earlier sections and have many automatic features of error control and error flags that are returned to the user. These solvers are far more efficient and accurate than any casual user could hope to write on the spur of the moment.

Software packages are not cheap, but they have reached such widespread acceptance that practically any computer center of any consequence subscribes to several of them. Thus our best advice is: Before becoming bogged down in coding an algorithm to meet an immediate need, check with your local computer center, which will almost surely have a package with solvers to meet that need. Thus every graph in this text could be reproduced in the reader's local environment if one of these packages were installed on the reader's computer.

Using Solvers

Keep in mind that many solvers have special features built in which may be costly in terms of computer time if used inappropriately. A routine that is designed for a certain type of problem may work perfectly well on another problem, but the special calculations it goes through are wasted. For example, some routines are designed to operate on problems when many evaluations of the function f are costly in terms of computer time—using such a package on simple, straightforward problems would mean having the program do unnecessary extra work. On the other hand, routines that use many function evaluations would not be practical when $f(t, y)$ is a complicated function.

In using a commercial solver on the initial value problem $y' = f(t, y)$, $y(t_0) = y_0$, we need to input the function $f(t, y)$, the initial data t_0 and y_0, and often the partial derivatives of f. Many routines do not require a step size to be specified. Instead, the user specifies a bound on the error and the routine calculates a step size at each step which will achieve this tolerance. Depending on the nature of the problem, either the relative or absolute error may be appropriate. If the solution varies by orders of magnitude, relative error control is probably more appropriate, but at the expense of slightly more computation.

There are usually some optional inputs which are given default values by the routine or ignored if left unspecified. For example, to prevent the use of too much computer time, the user may be able to specify a minimum allowable step size or a maximum number of function evaluations. If accuracy requirements are too high, these constraints may be exceeded and an error message returned. A useful feature when singularities are known to be present is the specification of a particular $\hat{t}$ which is to be avoided at all steps, including internal steps. Since some routines may overshoot the endpoint and then interpolate back to give their final value, this step may prevent any attempt to integrate through a singularity.

In addition to the actual solution, there may be output available from the routine which provides important information that could suggest changing some parameter. This information may include the actual step size used, the number of function evaluations, the number of interval steps used in one call to the routine, or an estimate of the local error in the solution accumulated during a call. Looking at these outputs may suggest that error tolerances be raised or lowered if the actual errors differ significantly from our worst-case estimates, or that a different technique be implemented which uses more or fewer function evaluations.

The relatively unsophisticated user is likely to find documentation for commercial software packages as confusing as it is helpful, simply because it was probably written by someone who understands perfectly all the intricacies of the program. Many of these routines are "prepared for anything" in some sense, and for normal, well-behaved equations, many of their options can be ignored. Sorting out just what is what may require more than just reading the documentation. If this is the case, it may be helpful to consult with someone connected with the computer center who is familiar with the routine.

Problems in Implementing Algorithms

Selection of one of the algorithms described in the preceding sections and coding it for a computer would seem to be an obvious way to find an approximate solution of an intitial value problem. But this approach will not always produce satisfactory results unless one is prepared to spend considerable time validating the output. We give some reasons for these words of caution.

Problems in Choice of Step Size.   As we saw in Example 2.6, too small a step size leads not only to inefficiency but possibly also to sizable round-off error. But as the next example shows, too large a step size may lead to sizable global errors obliterating features of the solution which are being sought. For an initial value problem whose solution is not well-behaved, the process of discretization may introduce some subtle errors that should make the modeler wary. When the solution "escapes to infinity" in finite time (see Example 1.2), the unadorned numerical approximation scheme considered so far may not be able to detect this fact.

Example 2.8

Consider the initial value problem

$$y' = -y^2, \qquad y(0) = -\tfrac{1}{2} \qquad\qquad (1)$$

which we solve numerically using Euler's Method with step size $h = 0.15$ (say). The result is displayed in Figure 2.10, which appears to be a fairly smooth curve. Since the flow field of the differential equation is well-behaved, we are inclined to accept this as an approximate solution of the problem (1). But, in fact, problem (1) can be solved easily (see Example 1.2) and we find the (unique) maximally extended solution $y = 1/(t - 2)$, $t < 2$. Observe that the solution $y(t)$ escapes to $-\infty$ as t tends toward 2 from below—our numerical solution

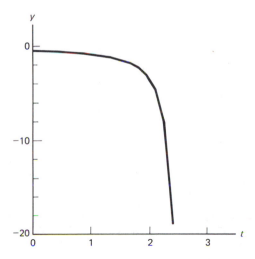

Figure 2.10   Numerical solution of (1) via Euler's Method, $h = 0.15$.

does not seem to be aware of that fact. What happened, of course, is that Euler's Method "stepped" across $t = 2$ and connected up with a neighboring solution curve. Inspection of the differential equation reveals nothing about "finite escape time," and if we did not know the true solution we would have been inclined to accept the approximate "solution" in Figure 2.10. The implications of this simple observation for equations that cannot be explicitly solved are obvious.

Behavior of Approximate Solutions: Chaos. As our next example shows, one must be cautious when using an approximate numerical solution for an initial value problem to describe qualitative features of the exact solution of that problem. Our example also shows that the qualitative behavior of approximate solutions may be significantly affected by the choice of step size.

Example 2.9

Consider the initial value problem for the *logistic equation*

$$y' = ry(1 - y), \qquad y(0) = y_0 \tag{2}$$

where r is a positive constant. As we shall see in Chapter 3, the logistic equation arises in growth models and can be solved explicitly. Typical solution curves for this equation are pictured in Figure 2.11. Now suppose that Euler's Method

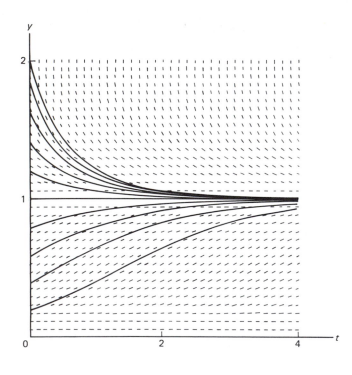

Figure 2.11 Solution curves for the logistic equation $y' = y(1 - y)$.

has been selected to generate approximate numerical solutions for problem (2). Referring to Section 2.3, we see that Euler's Method for problem (2) becomes

$$y_{n+1} = y_n + rhy_n(1 - y_n) \tag{3}$$

with step size h and initial value y_0. Since Euler's Method is of first order, we know that for any $T > 0$ and any tolerance $\epsilon > 0$ there is a step size h such that the values generated by (3) miss being the "true" values of the solution of problem (2) by at most ϵ over $0 \le t \le T$. Thus it would seem reasonable to infer that if h is chosen small enough (but not so small that round-off errors are significant), the Euler-generated approximation "looks like" the exact solution. That this is not quite the case is illustrated in Figures 2.12 to 2.14. In

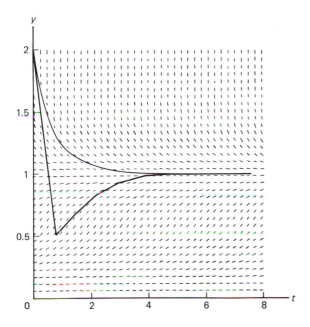

Figure 2.12 Euler versus "true" solution for (2) when $rh = 3/4$, $y_0 = 2$.

fact, we might infer from Figure 2.12 that the solution for (2) with $y_0 = 2$ falls initially and then rises monotonically toward the line $y = 1$ as $t \to \infty$. Figure 2.13 implies that the solution for problem (2) with $y_0 = 1.4$ is oscillatory about the line $y = 1$ with oscillations that die out as $t \to \infty$. Finally, Figure 2.14 would have us believe that the solution of problem (2) with $y_0 = 1.3$ oscillates in a chaotic fashion about the line $y = 1$ as $t \to \infty$ (i.e., with oscillations of seemingly arbitrary amplitude). Of course, none of these inferences about the exact solutions of the logistic equation is valid. The following assertion, however, gives us cause for concern.

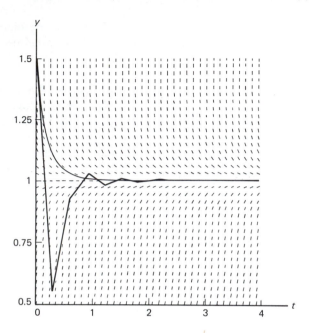

Figure 2.13 Euler versus "true" solution for (2) with $rh = 3/2$, $y_0 = 1.4$.

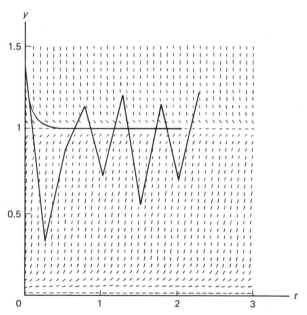

Figure 2.14 Euler versus "true" solution for (2) with $rh = 5/2$, $y_0 = 1.3$.

Given any step size $h > 0$, there exist values for the positive constant r and initial values y_0 with max $\{1, 1/rh\} < y_0 < 1 + 1/rh$ such that the Euler solution of problem (2) with that step size exhibits any of the behaviors given in Figures 2.12 to 2.14.† (4)

Assertion (4) in no way contradicts the fact that Euler's Method is of first order for a *fixed* problem (2), that is, for fixed values of r and y_0. Nevertheless, it does show that there is no "absolute" notion of a small step size, and that inferences about the qualitative behavior of the solution of an initial value problem from a numerical solution is dangerous without knowledge of the errors involved in the algorithm generating the numerical solution.

Numerical Instability.   Our next example shows that the wrong discretization process could lead to a numerically unstable algorithm. Some linear multistep methods are capable of rather peculiar behavior. Indeed, it can happen that the errors in the approximations $y_1, y_2, \ldots$ remain under control at first, but then suddenly begin to increase exponentially. When this happens the method is said to be *numerically unstable*.

Example 2.10 (*Numerical Instability*)

Consider the initial value problem

$$y' = -y, \qquad y(0) = 1 \tag{5}$$

whose true solution is easily seen to be $y(t) = e^{-t}$. In an attempt to discretize the differential equation in (5), let us evaluate both sides at $t = t_n$ and approximate $y'(t_n)$ by the central difference $(y(t_{n+1}) - y(t_{n-1}))/2h$, where we have assumed a uniform step size h. This central difference can, in turn, be approximated by $(y_{n+1} - y_{n-1})/2h$ and hence we arrive at the linear multistep method

$$y_{n+1} = y_{n-1} - 2hy_n, \qquad y_0 = 1 \tag{6}$$

Choosing $h = 0.06$ and using Euler's Method to calculate y_1 to start the scheme (6), we obtain the surprising result illustrated in Figure 2.15.

Comments

To the observations above we can add that

1. Computers can only recognize finitely many numbers, and hence underflow or overflow may result if computations are not arranged appropriately.
2. Inefficient code is unnecessarily expensive of computer time.

---

† The apparently random behavior shown in Figure 2.14 is known as *chaotic motion*. See R. L. Devany, *An Introduction to Chaotic Dynamical Systems* (Menlo Park, Calif.: Benjamin/Cummings, 1986).

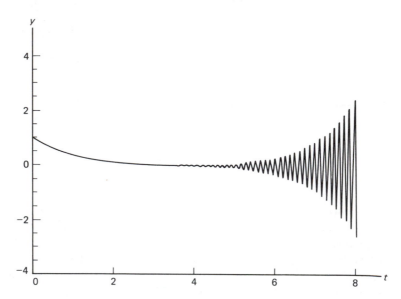

Figure 2.15   Numerical instability of the multistep method (6).

Thus we see that the theoretical fact that a method produces approximations to initial value problems which converge to the solution when exact arithmetic is used is a long way from a practical technique for finding such approximations.

The reader should not be too discouraged in spite of the pathological behavior exhibited by the elementary solvers described in the examples above. Indeed, the plots of solutions of differential equations in this text were produced by a very powerful and sophisticated package solver together with a graphics routine for displaying the results. So it *is* possible to achieve quite satisfactory results using the approximation algorithms described in the foregoing sections. The only point we wanted to make is that designing a good general-purpose solver involves more than the mere coding of a one-step (or multistep) method.

PROBLEMS

The following problem(s) may be more challenging: 3.

1. Illustrate the chaotic behavior which may occur when solving (3) with relatively large *rh* by graphing the solution in each of the following cases.
   (a) $y_0 = 2$, $rh = 0.75$, $h = 0.3$, $n = 0, \ldots, 13$.
   (b) $y_0 = 1.4$, $rh = 1.5$, $h = 0.3$, $n = 0, \ldots, 16$.
   (c) $y_0 = 1.3$, $rh = 2.5$, $h = 0.1$, $n = 0, \ldots, 20$.

2. Use a computer to verify that the solution curves of

$$(e^t \sin y - 2y \sin t) + (y^2 + e^t \cos y + 2 \cos t)y' = 0$$

are as described in Figure 4.18.

3. [*Numerical Instability*]. The following steps show that Figure 2.15 is indeed the polygonal graph of the approximate solution of $y' = -y$, $y(0) = 1$, obtained by using the *linear difference equation* (6), $y_{n+1} = y_{n-1} - 2hy_n$, with initial data $y_0 = 1$, $y_1 = 1 - h$, where y_1 is obtained by Euler's Method.

   (a) Explain why $y_2, y_3, \ldots, y_n \ldots$ are uniquely determined by (6) and the initial data.

   (b) Show that $y_n = C_1 r_1^n + C_2 r_2^n$, where C_1 and C_2 are arbitrary constants, $r_1 = -h + (h^2 + 1)^{1/2}$ and $r_2 = -h - (h^2 + 1)^{1/2}$, satisfies (6) for $n = 1, 2, \ldots$. [*Hint:* first set $y_n = r^n$ in (6) and then determine the unique pair of values r_1 and r_2 of r for which (6) holds. Then verify that $y_n = C_1 r_1^n + C_2 r_2^n$ is also a solution of (6) if r_1^n and r_2^n are solutions.]

   (c) Show that $y_n = C_1 r_1^n + C_2 r_2^n$, where r_1 and r_2 are given in (b), $C_1 = (1 - h - r_2)/(r_1 - r_2)$ and $C_2 = (1 - h - r_1)/(r_2 - r_1)$, satisfies (6) and the initial data.

   (d) According to the binomial theorem, $(h^2 + 1)^{1/2} \approx 1 + h^2/2$ for small h. Thus, $r_1 \approx 1 - h + h^2/2$, $r_2 \approx -1 - h - h^2/2$, or, to first order in h, $r_1 \approx 1 - h$ and $r_2 \approx -1 - h$. Let t be a fixed positive number, $t = nh$. Show that $r_1^n \to e^{-t}$ and $(-1)^n r_2^n \to e^t$ as $n \to \infty$. [*Hint:* $e^z = \lim\limits_{n\to\infty} \left(1 + \dfrac{z}{n}\right)^n$ for any z.]

   (e) On the basis of (d) we have that $y_n(t) \approx C_1 e^{-t} + C_2 e^t$ for large n and small h such that $nh = t$. Show that $C_1 \to 1$ and $C_2 \to 0$ as $n \to \infty$. Thus, show that the approximation to $y_n(t)$ tends to the exact solution $y = e^{-t}$ of $y' = -y$, $y(0) = 1$ as $n \to \infty$.

   (f) Explain why the approximation y_n may be considered to be the superposition of the exact solution of the initial value problem and an exponentially growing numerical error with alternating sign.

   (g) Reproduce Figure 2.15 using (6) with $h = 0.06$, $0 \le t \le 6$, $y_0 = 1$, $y_1 = 1 - h$.

CHAPTER 3

First-Order Differential Equations: Growth Processes

The main goal of this chapter and the next is to develop a collection of solution formulas and techniques for treating those first-order differential equations that arise most often in applications. In this chapter these solution techniques are applied to various models of growth processes. The basic properties of initial value problems for the linear first-order equation $y' + p(t)y = q(t)$ are considered here along with some first-order equations reducible to linear form by a change of variables.

3.1 LINEAR FIRST-ORDER DIFFERENTIAL EQUATIONS

In this section we find all solutions $y(t)$ of the *linear first-order differential equation*

$$y' + p(t)y = q(t) \tag{1}$$

where the given functions $p(t)$ and $q(t)$ are continuous on a t-interval I. In Section 3.2 we examine the properties of solutions of the *associated initial value problem*

$$
\begin{aligned}
y' + p(t)y &= q(t) \\
y(t_0) &= y_0
\end{aligned}
\tag{2}
$$

and illustrate these concepts with examples and models of physical phenomena.

Example 3.1 (*Salt Accumulation in a Vat*)

A large vat contains 100 gal of brine in which, initially, S_0 pounds of salt is dissolved. More brine is allowed to run into the vat through an inlet pipe at the rate of 10 gal/min. The concentration $C(t)$ (pounds of salt per gallon) in this incoming brine varies with time. Let us suppose that the solution in the

tank is thoroughly mixed and that brine flows out of the vat at the rate of 10 gal/min. How much salt is there in the solution in the vat at any time t?

The underlying law governing the accumulation of salt is given by the following basic principle:

| **Balance Law.**   Net rate of change of a substance $=$ rate in $-$ rate out   (3)

Let $S(t)$ denote the amount of salt in pounds in the vat at time t. Let $t_0 = 0$ and $S(t_0) = S_0$, the amount of salt in the brine in the vat at time 0. The rate at which salt is being added to the vat at time t is $10C(t)$ pounds per minute. The rate at which salt is leaving the vat is $10S(t)/100 = \frac{1}{10}S(t)$ since there is $S(t)/100$ lb of salt per gallon in the tank time t. Thus we have that the rate law (3) is modeled mathematically by

$$\frac{dS(t)}{dt} = \text{rate in} - \text{rate out} = 10C(t) - \tfrac{1}{10}S(t)$$

The corresponding initial value problem written in the usual form is

$$\frac{dS(t)}{dt} + \frac{1}{10}S(t) = 10C(t) \tag{4}$$

$$S(0) = S_0$$

Comparing (4) with (2), we see that in this case $p(t)$ is the constant function $\frac{1}{10}$ and $q(t) \equiv 10C(t)$. Problem (4) is solved explicitly in the next section, but the general approach for solving such problems is given below.

The Integrating Factor Approach

We have already solved a particular case of (1), the equation for radiocarbon decay,

$$y' = -ky \tag{5}$$

where k is a constant. A review of the way (5) is solved provides the clue for solving (1). Rearranging (5) in the form of (1) and multiplying by the "integrating factor," e^{kt}, we obtain the equivalent differential equation

$$e^{kt}[y' + ky] = \frac{d}{dt}[e^{kt}y] = 0$$

Thus the Vanishing Derivative Theorem (see Section 1.1) implies that $e^{kt}y = C$, where C is an arbitrary constant of integration, from which we obtain the solution formula, $y = Ce^{-kt}$, $-\infty < t < \infty$. This same process will work for (1) as well, but we will need to be more skillful in the choice of an integrating factor.

Let $P(t)$ be any antiderivative of the coefficient $p(t)$ of the y term in (1) on the interval I. [In other words, let $P(t)$ be such that $P'(t) \equiv p(t)$ on I.] Any choice of an antiderivative $P(t)$ for $p(t)$ will do, but once chosen, keep it fixed. [Note that any two antiderivatives of $p(t)$ differ only by a constant on I.] The exponential function exp $[P(t)]$ is called an *integrating factor* for equation (1). Now observe that for any continuously differentiable function $y(t)$ on I, we have the identity

$$\frac{d}{dt}[e^{P(t)}y] = e^{P(t)}y' + e^{P(t)}P'(t)y = e^{P(t)}[y' + p(t)y] \tag{6}$$

which is all we need to prove the following result.

General Solution Theorem.   Let the linear equation $y' + p(t)y = q(t)$ have continuous coefficients $p(t)$ and $q(t)$ on an interval I. Let $P(t)$ be any (fixed) antiderivative of $p(t)$ on I, and $Q(t)$ any (fixed) antiderivative of $e^{P(t)}q(t)$ on I. Then every solution $y(t)$ of the linear equation on the interval I has the form

$$y(t) = Ce^{-P(t)} + Q(t)e^{-P(t)} \tag{7}$$

for some constant C. Conversely, any function $y(t)$ of the form (7), where C is any constant, is a solution of the linear equation.

Proof. Multiplying (1) by the integrating factor exp $[P(t)]$, we have

$$e^{P(t)}[y' + p(t)y] = e^{P(t)}q(t)$$

which, because of the identity (6), can be written as

$$\frac{d}{dt}[e^{P(t)}y] = e^{P(t)}q(t) \tag{8}$$

Let $Q(t)$ be an antiderivative of $e^{P(t)}q(t)$ on I (any one will do); applying the Vanishing Derivative Theorem to (8), we obtain

$$e^{P(t)}y = Q(t) + C$$

where C is a constant. Thus we see that any solution $y(t)$ of (1) has the asserted form (7). The converse assertion that (8) is a solution of (1) for *any* constant C is shown directly (recall that $P'(t) = p(t)$ and $Q'(t) = q(t) \exp [P(t)]$), so we are done.

Observe that the formula (7) defines the family of *all* solutions of equation (1) over the interval I as the constant C takes on all possible values. For that reason (7) is called the *general solution* of the linear differential equation (1). The following uniqueness result is important.

> **Uniqueness Theorem.**   Distinct values of the constant C in (7) give rise to distinct solutions of Eq. (1). Moreover, if two solution curves of (1) share a point in common, then they coincide over the entire interval I.

Proof. Let $y_1(t)$ and $y_2(t)$ be solutions of (1) given by formula (7) corresponding to the distinct constants C_1 and C_2, respectively. If for some value t_0 in I we have $y_1(t_0) = y_2(t_0)$, then $(C_1 - C_2)e^{-P(t_0)} = 0$. But since exponential functions never vanish, it follows that $C_1 = C_2$, and the assertion is established.

The solutions of (1) given by the general solution (7) are classified as follows. If $q(t) \equiv 0$, then $y = e^{-P(t)}C$ is the general solution of the *homogeneous* (or *free*) *linear equation* $y' + p(t)y = 0$. On the other hand, if $C = 0$ in (7), then $y = Q(t)e^{-P(t)}$ is a *particular solution* of (1). Thus the general solution (7) of Eq. (1) is displayed as the sum of a particular solution of the nonhomogeneous equation (1) and the general solution of the homogenous equation for (1).

Example 3.2

The equation $y' + 2y = 3e^t$ is linear with $p = 2$ and $q = 3e^t$. Since p and q are continuous on the entire real line, each maximally extended solution is also defined on the entire real line. Since $P(t) = 2t$ is an antiderivative of $p(t) = 2$, an integrating factor for the equation is e^{2t}. Now $Q(t) = e^{3t}$ is an antiderivative for $q(t)e^{P(t)} \equiv 3e^{3t}$, and hence the general solution is

$$y = Ce^{-2t} + e^t, \qquad -\infty < t < \infty \qquad (9)$$

where C is an arbitrary constant. Each value of C gives a distinct solution. Imposing an initial condition will determine C. For example, if we want to find the solution for which $y = 2$ when $t = 0$, then from (9) we see that $2 = e^{-2\cdot 0}C + e^0$. Hence $C = 1$ and $y = e^{-2t} + e^t$, $-\infty < t < \infty$. The corresponding solution curve, together with those given by other values of C, is sketched in Figure 3.1.

The antiderivatives in (7) cannot always be evaluated in terms of elementary functions, as the next example shows. When this happens, (7) is not a very useful way to characterize solutions of Eq. (1). Note, however, that whether or not we

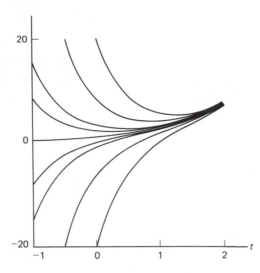

Figure 3.1 Solution curves of $y' + 2y = 3e^t$.

can compute them in elementary terms, these antiderivatives exist on any t-interval where p and q are continuous, and hence solutions of (1) are then in theory defined by (7) for all t in such an interval.†

Example 3.3

The linear equation $y' + 2ty = (1 + t^2)^{-1}$ has an integrating factor e^{t^2}. Multiplying by this factor, our equation becomes

$$e^{t^2}[y' + 2ty] = \frac{d}{dt}\,[e^{t^2}y] = e^{t^2}[1 + t^2]^{-1}$$

Antidifferentiation gives $e^{t^2}y = \int_{t_0}^{t} e^{s^2}(1 + s^2)^{-1}\,ds + C$, where t_0 is any (fixed) point on the real line. Thus the general solution of our equation is given by

$$y = e^{-t^2}C + e^{-t^2}\int_{t_0}^{t} e^{s^2}(1 + s^2)^{-1}\,ds \tag{10}$$

Since p and q are continuous for all t, the general solution is defined for all t; however, the integral in (1) is not expressible in terms of simple functions and we must leave the solutions in this somewhat unsatisfactory form.

The computational methods in Section 2.3 for approximating solutions of differential equations will produce tables of approximate values of y for various values of t given a specific value of C. This method will allow us to bypass

† Recall from the Fundamental Theorem of Calculus that for any continuous function $f(t)$ on an interval I, the definite integral $\int_{t_0}^{t} f(s)\,ds$ is an antiderivative for f on I for any choice of t_0 in I.

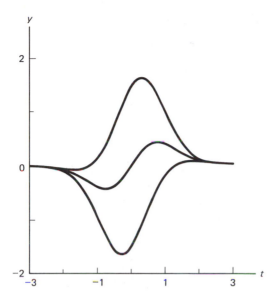

Figure 3.2 Solution curves for $y' + 2ty = (1 + t^2)^{-1}$.

(10), but at the price of obtaining approximations rather than exact solutions. Alternatively, a package solver with graphics output can be used to generate approximate graphs of the solution curves. That the solution curves in Figure 3.2 appear to touch is not a violation of the Uniqueness Theorem, only a limitation of computer graphics.

Equations Reducible to Linear Differential Equations

As the following examples show, some first-order differential equations are not linear in their original forms, but can be written in linear form by suitable rearrangement of the equation. The vast majority of first-order differential equations, however, are not linear, no matter how the equation is rewritten.

Example 3.4

The equation $du/dv = 1/(3e^u - 2v)$ is not linear in u and du/dv. However, recalling that $dv/du = [du/dv]^{-1}$, we can invert the equation to obtain $dv/du = 3e^u - 2v$, or $dv/du + 2v = 3e^u$, which is linear in v and dv/du. This equation has already been solved in Example 3.2.

Example 3.5†

Let $y'(t) = (a + by)(c(t) + d(t)y)$, where a and b are constants, $b \neq 0$, and $c(t)$ and $d(t)$ are continuous on some t-interval I. We shall introduce a new

---

† The differential equation is a special case of the Bernoulli and Riccati equations (see the problem set of Section 3.3).

dependent variable, z, by $z(t) = [a + by(t)]^{-1}$. Then we claim that $z(t)$ must satisfy the linear equation

$$\frac{dz(t)}{dt} = [ad(t) - bc(t)]z(t) - d(t) \tag{11}$$

Indeed, using the Chain Rule, we have that

$$z' = \frac{-by'}{(a + by)^2} = \frac{-b(c + dy)}{a + by} = -bz\left[c + d\left(\frac{1}{bz} - \frac{a}{b}\right)\right]$$

(since $z = [a + by]^{-1}$ and $y = 1/bz - a/b$), from which (11) follows. The integrating factor technique can be used to find a solution $z(t)$ of (11), and then $y(t) = 1/bz(t) - a/b$ is a solution of the original differential equation. We must, of course, remain within a t-interval on which $z(t)$ does not vanish.

Reduction via Varied Parameters

A technique often used to reduce the order of the *linear second-order equation*

$$y'' + a(t)y' + b(t)y = f(t) \tag{12}$$

if a solution $z(t)$ is known for the *associated homogeneous equation*

$$z'' + a(t)z' + b(t)z = 0 \tag{13}$$

is called the *Method of Reduction by Varied Parameters*. The technique is valid on any interval I where $z \neq 0$ and proceeds as follows. Let us look for a solution of (12) in the form $y = u(t)z(t)$, where $z(t)$ is a given solution of (13) and the "parameter" u varies with t. Direct substitution shows that (12) becomes

$$zu'' + (2z' + az)u' + (z'' + az' + bz)u = f$$

which simplifies to

$$zu'' + (2z' + az)u' = f \tag{14}$$

because $z(t)$ solves (13) by assumption. Now (14) is a first-order linear equation in u', and using the integrating factor approach, we see that (14) can be written as

$$(u'z^2 e^A)' = ze^A f \tag{15}$$

where A is an antiderivative of $a(t)$. Now if Q is any antiderivative of $ze^A f$, the Vanishing Derivative Theorem implies that $u' = e^{-A}z^{-2}Q$. Thus if we choose u to be any antiderivative of $e^{-A}z^{-2}Q$, then $y = u(t)z(t)$ is a particular solution of (12).

Example 3.6

It is easy to see that $z = t$ is a solution of the homogeneous equation

$$(1 - t^2)y'' - 2ty' + 2y = 0 \tag{16}$$

Comparing (16) with (12), we see that $a(t) = -2t/(1 - t^2)$ and $f \equiv 0$. Hence the substitution $y = u(t)z(t) \equiv u(t)t$ into (16) gives us [compare with (15)]

$$[u't^2(1 - t^2)]' = 0$$

since $A = \ln(1 - t^2)$ and $e^A = 1 - t^2$. Thus $u' = C/t^2(1 - t^2)$ for some (any) choice of the constant C. Since we seek just one nontrivial solution, we may as well take $C = 1$. Now using partial fractions, we find that

$$u = -\frac{1}{t} + \frac{1}{2}\ln\frac{1 + t}{1 - t}$$

so

$$y = tu = -1 + \frac{t}{2}\ln\frac{1 + t}{1 - t}$$

is another solution of (16) valid at least in the interval $I = (-1, 1)$. (Note: (16) is the Legendre Equation of order one. See Section 8.3.)

Nonnormal Equations

The method of integrating factors introduced in this section is an effective technique for solving the *normal linear equation*

$$y' + p(t)y = q(t)$$

where the word "normal" is used to indicate that the coefficient of y' is $+1$. The technique can also be applied to the equation $a(t)y' + b(t)y = c(t)$, but only after it is written in normal form

$$y' + \frac{b(t)}{a(t)}y = \frac{c(t)}{a(t)}$$

and if the quotients $b(t)/a(t)$ and $c(t)/a(t)$ are continuous on an interval I.

Example 3.7

The normal form of the equation $ty' + 3y = 2t$ is $y' + (3/t)y = 2$, but we must restrict attention to the interval $t > 0$ or to the interval $t < 0$ since the coefficient $p(t) = 3/t$ fails to be continuous (or even defined) at $t = 0$. Suppose, for example, that we take $I = (0, \infty)$. Then the integrating factor is $\exp[P(t)]$, where $P(t)$ is an antiderivative of $3/t$. For example, t^3 is an integrating factor, and the general solution of the normal equation on the interval $(0, \infty)$ is

$$y = \frac{C}{t^3} + \frac{t}{2}$$

PROBLEMS

The following problem(s) may be more challenging: 2(g), 2(h), 3(c), 3(f), 6, and 8.

1. Write the following equations in the normal linear form.

   (a) $\dfrac{dt}{dy} = \dfrac{1}{t^n - ty}$.

   (b) $t^2 y' + y \sin t = e^t$.

   (c) $\dfrac{dy}{dt} = \dfrac{1}{e^{-y} - t}$.

   (d) $e^{y'-3} = \dfrac{e^{(t+2)t}}{e^{6y}}$.

2. Find the general solution for each of the following equations by using an integrating factor.

   (a) $y' - 2ty = t$.

   (b) $y' - y = e^{2t} - 1$.

   (c) $y' = \sin t(1 - y)$.

   (d) $2y' + 3y = e^{-t}$.

   (e) $t(2y - 1) + 2y' = 0$.

   (f) $y' + y = te^{-t} + 1$.

   (g) $y' + y \cos t = 7$.

   (h) $7y' + y \sinh t = \cosh t$.

3. For each case below, find the general solution by using an integrating factor and give a maximal interval for which the solution is valid. Sketch solution curves.

   (a) $ty' + 2y = t^2$.

   (b) $(3t - y) + 2ty' = 0$.

   (c) $\dfrac{dy}{dt} = (\tan t)y + t \sin 2t$.

   (d) $\dfrac{dy}{dt} = \dfrac{y^2}{t + y}$.

   (e) $(y - 2) + (3t - y)\dfrac{dy}{dt} = 0$.

   (f) $y' = (2 + 4y)\left(\dfrac{2t^3 - 1}{4t} + t^2 y\right)$.

4. Find the general solution of the equation $dy/dt = (y/t) + t^n$, $t > 0$.

5. For each equation below, a solution $z(t)$ of the homogeneous equation is given. Find solutions of the nonhomogeneous equation, which involve two arbitrary constants.

   (a) $y'' - 2y' + y = t$, $z(t) = e^t$.

   (b) $t^2 y'' - 2ty' + 2y = 1$, $z(t) = t$.

6. Wilbert is saving to buy a new car which will cost him $12,400. Currently, he has $5800 saved. If he can earn 7% interest continuously compounded on his savings and has an income of $2600 per month, to what amount must Wilbert limit his monthly expenses if he is to have enough saved to buy the car in 1 year?

7. Industrial waste is pumped into a tank containing 1000 gal of water at the rate of 1 gal/min, and the well-stirred mixture leaves the tank at the same rate.

   (a) Find the concentration of waste in the tank at time t.

   (b) How long does it take for the concentration to reach 20%?

8. Find a substitution of the form $y = uz$ which converts (12) into an equation for u of the form $u'' + Q(t)u = f/z$ for some $Q(t)$.

9. Sketch the solution curves of the given differential equations for the given initial data and time span.

   (a) $y' = \sin t\, (1 - y)$, $y\,(-\pi/2) = -1, 0, 1$; $-\pi/2 \le t \le 2\pi$ [see also problem 2(c)].

   (b) $y' + y = te^{-t} + 1$; $y\,(0) = -1, 0, 1$; $-1 \le t \le 3$ [see also problem 2(f)].

   (c) $ty' + 2y = t^2$; $y\,(2) = 0, 1, 2$; $0 < t < 4$. Repeat with $y\,(-2) = 0,1,2$; $-4 < t < 0$ [see also problem 3(a)].

   (d) $y' - (\tan t)y = t \sin 2t$, $y\,(0) = -1, 0, 1$, where $-\pi/2 < t < \pi/2$ [see also problem 3(c)].

3.2 INITIAL VALUE PROBLEMS: THE OPERATOR CONCEPT

In Section 3.1 we used an integrating factor approach to show that the general solution of the first-order linear equation

$$y' + p(t)y = q(t) \tag{1}$$

with coefficients p and q continuous on an interval I is given by

$$y(t) = Ce^{-P(t)} + Q(t)e^{-P(t)} \tag{2}$$

where C is any constant, P is any (fixed) antiderivative of p, and Q is any (fixed) antiderivative of $q(t)e^{P(t)}$ on I.

There is an operational formulation of (1) which is not only useful in its own right but also gives rise to an effective solution technique for higher-order linear differential equations—a fact we exploit in later chapters.

Operator Terminology

An *operator* is a transformation (or mapping) which takes an object from a given set, called the *domain* of the operator, and assigns it to an object in another set, called the *codomain* of the operator. Using the symbol L to denote an operator, we express the fact that L "operates" on an element in its domain and "produces" an element in its codomain by writing the symbolic statement, L: Domain → Codomain. The *action* of the operator L acting on the element y in its domain is denoted symbolically by

$$L: y \rightarrow L[y] \qquad \text{for } y \text{ in the domain of } L\dagger$$

The subset of all objects in the codomain of an operator L having the form $L[y]$, for some y in the domain of L, is called the *range* of L. Note that the domain and range of an operator are very precisely defined, but the codomain can be any set that contains the range.

For our present purposes the function classes defined below will serve as domains or codomains of our operators.

> **The Continuity Classes $C^0(I)$ and $C^1(I)$.**   Let $C^0(I)$ denote the class of all continuous functions on the interval I, and let $C^1(I)$ denote the class of all continuous functions on I whose first derivatives are also continuous on I.

Example 3.8

Let I be the open interval $(0, 2)$; then the functions $y_1(t)$ and $y_2(t)$ defined in Figure 3.3 *both* belong to the continuity class $C^1((0, 2))$. Observe, however, that when I is the *closed* interval $[0, 2]$, $y_1(t)$ does *not* belong to the class $C^1([0, 2])$, but $y_2(t)$ does.

---

† Notice that square brackets are used to enclose the operand. Following convention, we may write Ly instead of $L[y]$ unless the brackets are needed for clarity.

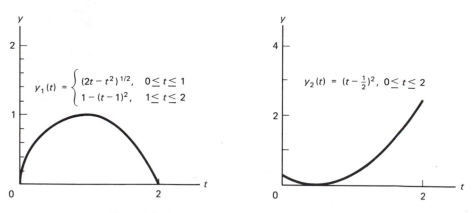

Figure 3.3 Examples of functions in continuity classes.

The reason L is called an "operator" instead of the perfectly correct name "function" is because the domains and codomains of the operators we consider here will be function classes such as $C^0(I)$ and $C^1(I)$, so confusion is minimized if L is called an operator.

Example 3.9

Notice that if y is any function in the class $C^1(I)$, then y' is an element of $C^0(I)$. Thus the *differentiation operator* $D: C^1(I) \rightarrow C^0(I)$ and has the action $D: y \rightarrow Dy \equiv y'$.

Now let $p(t)$ be a continuous function on an interval I, and let L be the operator whose action is

$$L: y \rightarrow y' + p(t)y \qquad \text{for } y \text{ in } C^1(I) \tag{3}$$

Thus the domain of L is $C^1(I)$, and a codomain of L is evidently $C^0(I)$. It is often useful to write the operator L in (3) as $L = D + p(t)$, where we understand that

$$L[y] \equiv (D + p(t))[y] \equiv Dy + p(t)y$$

for any y in $C^1(I)$. Notice now that the differential equation (1) has the operator formulation

$$(D + p(t))[y] = q(t) \tag{4}$$

where q is a given function in $C^0(I)$.

We show now how to "extract" the solution formula (2) from the operator form (4) of Eq. (1). Let $P(t)$ be any (fixed) antiderivative of $p(t)$, and observe the operator identity

$$D[ye^{P(t)}] = e^{P(t)}(D + p(t))y \qquad \text{for any } y \text{ in } C^1(I) \tag{5}$$

Now let $y(t)$ be any function in $C^1(I)$ that solves (4). Then multiplying both sides of the operator equation (4) by $\exp[P(t)]$ and using the identity (5), we have

$$D[ye^{P(t)}] = q(t)e^{P(t)}$$

Thus if $Q(t)$ is any (fixed) antiderivative of $q(t) \exp[P(t)]$, the solution formula (2) results from the Vanishing Derivative Theorem. Observe that for any constant C we have

$$(D + p(t))[Ce^{-P(t)} + Q(t)e^{-P(t)}] = q(t) \text{ on } I$$

Hence the range of $D + p(t)$ is precisely $C^0(I)$, but there are infinitely many distinct elements in $C^1(I)$ which map into a given q in $C^0(I)$ under the operator $D + p(t)$, one for each value of the constant C.

Initial Value Problems

Using the general solution (2), we can easily solve the initial value problem associated with the first-order differential equation (1).

Solution of Initial Value Problem.   Let $p(t)$ and $q(t)$ be continuous on an interval I containing t_0, and let y_0 be any constant. Then the initial value problem

$$y' + p(t)y = q(t), \qquad y(t_0) = y_0$$

has exactly one solution, which is given by the formula

$$y(t) = e^{-P_0(t)}y_0 + e^{-P_0(t)} \int_{t_0}^{t} e^{P_0(s)}q(s)\,ds, \qquad t \text{ in } I \qquad (6)$$

where $P_0(t) = \int_{t_0}^{t} p(r)\,dr$.

Proof. Note first that $y(t)$ in (6) is a solution of the initial value problem since it has the form of (2) and $y(t_0) = y_0$ [note that $P_0(t_0) = \int_{t_0}^{t_0} p(r)\,dr = 0$]. The Uniqueness Theorem shows that no other solution of (1) can satisfy the initial condition $y(t_0) = y_0$, and hence the assertion follows.

Rather than memorizing (6), the reader should use the operator approach outlined above to solve the initial value problem, computing the required antiderivatives by integrations from t_0 to t. As the following example again verifies, the effect of imposing the initial condition on the differential equation in problem (1) is to select one value for the constant C in the general solution (2).

Example 3.10

To solve the initial value problem $y' + 2ty = t$, $y(1) = 2$, we first find an antiderivative of $2t$, say t^2 (any one will do). Then exp $[t^2]$ is an integrating factor for the differential equation, and hence multiplying both sides of the equation by this factor, the equation becomes

$$e^{t^2}[y' + 2ty] = \frac{d}{dt}\,[e^{t^2}y] = te^{t^2}$$

Integrating this equation from 1 to t and using the initial conditions on $y(t)$, we have that

$$\int_1^t \frac{d}{ds}\,[e^{s^2}y(s)]\,ds = e^{t^2}y(t) - 2e = \int_1^t se^{s^2}\,ds = \frac{1}{2}(e^{t^2} - e)$$

Solving this equation for $y(t)$, we arrive at the solution of the given initial value problem

$$y(t) = \tfrac{1}{2} + \tfrac{3}{2}e^{1-t^2}, \qquad -\infty < t < \infty$$

The corresponding solution curve is sketched in Figure 3.4.

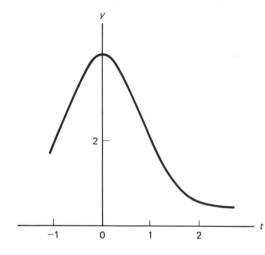

Figure 3.4 Solution curve for $y' + 2ty = t$, $y(1) = 2$.

Basic Questions

We have already answered three of the four basic questions posed in Section 2.2 for the initial value problem

$$y' + p(t)y = q(t)$$
$$y(t_0) = y_0$$

$$(7)$$

We have seen that for any continuous $p(t)$ and $q(t)$ on an interval I and any initial datum y_0, the initial value problem (7) has a unique solution given by formula (6). Thus problem (7) models a dynamical system, the initial datum y_0 represents the state of the system at time t_0, and the value of the solution $y(t)$ of (7) represents the state of the system at time t.

What can we say about the fourth question concerning the dependence of the solution of the initial value problem (7) on the data? To answer this question, we first need to digress a bit. We shall consider the data to include the *initial datum* y_0 and the *driving term* (or *input*) $q(t)$. The solution $y(t)$ of problem (7) is called the *response* (or *output*). Observe that the solution formula (6) for problem (7) can be viewed as characterizing the state of the system at time t in terms of the state of the system at time t_0 [for a given driving term $q(t)$]. It is useful to analyze the properties of the two terms in the solution formula (6). Writing

$$y(t) \equiv z(t) + Z(t) \equiv e^{-P_0(t)}y_0 + e^{-P_0(t)} \int_{t_0}^{t} e^{P_0(s)}q(s)\, ds$$

we see that $z(t)$ is the unique solution of the initial value problem

$$z' + p(t)z = 0, \qquad z(t_0) = y_0$$

while $Z(t)$ is the unique solution of the initial value problem

$$Z' + p(t)Z = q(t), \qquad Z(t_0) = 0$$

The function $z(t)$ is the *zero input response* and measures the response of the system to a zero input and to the initial state y_0. $Z(t)$ is the *zero state response* and denotes the system's response to zero initial state $y(t_0) = 0$ and to the input $q(t)$. If the interval I is $t_0 \leq t < \infty$ and $\lim_{t \to \infty} z(t) = 0$, then $z(t)$ is said to be a *transient*. [It can be shown that $z(t)$ is a transient if and only if $\lim_{t \to \infty} \int_{t_0}^{t} p(r)\, dr = \infty$.] Thus (6) can be rephrased as follows:

total response of system	=	response due to initial datum with zero input	+	response due to driving term with zero initial datum

Example 3.11

We now solve the initial value problem mentioned in the salt accumulation model of Section 3.1. The integration factor is $\exp\left[\int_0^t \frac{1}{10}\, ds\right] = e^{t/10}$, and hence the equation becomes

$$\frac{d}{ds}[S(t)e^{t/10}] = 10C(t)e^{t/10}$$

Integrating from 0 to 10 and using the initial condition, we have that

$$S(t) = e^{-t/10}S_0 + e^{-t/10}\int_0^t e^{s/10}[10C(s)]\,ds \tag{8}$$

that is,

$$S(t) = \text{response to initial amount} + \text{response to input}$$

The first term on the right-hand side is the response to the initial amount S_0 of salt in the tank and corresponds to $C(t) \equiv 0$ (pure water running into the tank at the rate of 10 gal/min). As expected, this response decays exponentially to zero as time tends to infinity. The second term on the right-hand side of (8) measures the amount of salt in the tank at time t due solely to the incoming salt brine. In the long run the input response term will dominate, and the effect of the initial amount S_0 on the response will be insignificant. We can illustrate all this by graphing the terms of (8). To be specific, let us take $S_0 = 15$ lb and assume an input concentration $C(t)$ which varies periodically about a mean of 1 lb of salt per gallon. For example, suppose that $C(t) = 1 + \frac{1}{10}\sin t$ (pounds of salt per gallon). Then using a table of integrals, (8) becomes

$$S(t) = 15e^{-t/10} + \left[100 + \frac{1}{1.01}\left(\frac{1}{10}\sin t - \cos t\right) - \frac{1}{0.0101}e^{-t/10}\right]$$

$$\begin{matrix}\text{total} \\ \text{response}\end{matrix} = \begin{matrix}\text{response to} \\ \text{initial datum}\end{matrix} + \qquad\qquad \text{response to input}$$

The graphs of the responses are sketched in Figure 3.5. Observe that the response to the input eventually dominates.

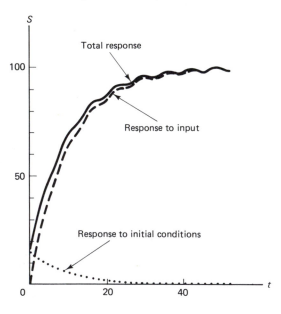

Figure 3.5 Solution of the problem $S' + \frac{1}{10}S = 10 + \sin t$, $S(0) = 15$.

From formula (8) we see precisely how $S(t)$ depends on the initial data and the input. From that formula it would be possible to calculate the effect on $S(t)$ of changes in S_0 and $C(t)$. We have the following result along these lines.

Continuity with Respect to the Data.   Let $T > t_0$ in I be given, and suppose that $y(t)$ and $\bar{y}(t)$ are the responses of the system (7) to the data y_0, $q(t)$ and $\bar{y}_0$, $\bar{q}(t)$, respectively. Then there are positive constants K_1 and K_2, independent of the data, such that

$$|y(t) - \bar{y}(t)| \le K_1|y_0 - \bar{y}_0| + K_2 \max_{t_0 \le t \le T} |q(t) - \bar{q}(t)| \tag{9}$$

for all $t_0 \le t \le T$.

Proof. First write $y(t)$ and $\bar{y}(t)$ in terms of their zero-input and zero-state responses. Thus $y = z + Z$ and $\bar{y} = \bar{z} + \bar{Z}$. Now put

$$m = \min_{t_0 \le t \le T} P_0(t), \qquad M = \max_{t_0 \le t \le T} P_0(t)$$

where $P_0(t) = \int_{t_0}^{t} p(s)\, ds$, $t_0 \le t \le T$. Thus for any $t_0 \le t \le T$, we may use the Triangle Inequality and the standard estimate for the magnitude of an integral† to derive the following inequality:

$$|y(t) - \bar{y}(t)| \le |z(t) - \bar{z}(t)| + |Z(t) - \bar{Z}(t)| \le e^{-m}|y_0 - \bar{y}_0| + e^{M-m}\int_{t_0}^{t} |q(s) - \bar{q}(s)|\, ds$$

$$\le e^{-m}|y_0 - \bar{y}_0| + e^{M-m}|T - t_0| \max_{t_0 \le t \le T} |q(t) - \bar{q}(t)|$$

The estimate (9) follows with $K_1 = e^{-m}$ and $K_2 = e^{M-m}|T - t_0|$.

Now since the constants K_1 and K_2 in (9) are independent of the data, we see that the more the data $\bar{y}_0$, $\bar{q}(t)$ look like y_0, $q(t)$, the more the response $\bar{y}(t)$ looks like $y(t)$. That is, "small" changes in y_0 and in $q(t)$ produce no more than small changes in $y(t)$. Such behavior is usually associated with the term "continuity," which explains the title of the theorem just proved. Thus a satisfactory answer to the fourth question posed in Section 2.2 has been found by means of the estimate (9) on $|y(t) - \bar{y}(t)|$.

The estimate (9) provides a rather crude upper bound on the changes in the total response due to changes in the data. In a specific problem such as the salt accumulation model it may be easier to work directly with a solution formula.

† Triangle Inequality:

$$|a + b| \le |a| + |b|, \qquad \text{for all numbers } a, b.$$

Integral Estimate:

$$\left| \int_a^b f(t)\, dt \right| \le \int_a^b |f(t)|\, dt \le (b - a)M, \ M \ge |f(t)| \qquad \text{for all } a \le t \le b$$

Discontinuous Driving Term and Coefficients

Discontinuous data is present in many applications and must be taken into account. The following definition covers the most common type of discontinuity.

> **Piecewise Continuity.** A function $q(t)$ is said to be piecewise continuous on a finite interval I if the one-sided limits $q(t^+)$ and $q(t^-)$ exist for all t in I (one-sided limits are used at endpoints of I contained in I), and if $q(t^+) = q(t^-)$ except (possibly) at a finite number of points of I. If I is an infinite interval then $q(t)$ is said to be piecewise continuous on I if it has that property on every finite subinterval.

Step functions and square waves are notable examples of piecewise continuous functions which are not continuous (continuous functions on closed intervals are automatically piecewise continuous).

If $p(t)$ is in $C^0(I)$, but $q(t)$ is only piecewise continuous on I, it is a remarkable fact that problem (7) is still uniquely solvable if the domain of $D + p(t)$ is enlarged to be all those continuous functions on I whose first derivaties are piecewise continuous on I. Indeed, the solution formula (6) is still valid in this case and has the stated properties, but we omit the proof.

The next two examples show some of the peculiarities that may occur when the coefficient $p(t)$ in a first-order linear equation (1) is discontinuous. Observe that any nonzero constant multiple $Ae^{P(t)}$, $A \neq 0$, of an integrating factor $e^{P(t)}$ of (1) is again an integrating factor. In Examples 3.12 and 3.13 we will see that it is convenient to choose $A = -1$ on some t-intervals, $A = +1$ on others.

Example 3.12

The equation $ty' + y = 2t$ can be written in normal form

$$y' + \frac{1}{t}y = 2 \tag{10}$$

on any t-interval not containing 0. For $t > 0$ or for $t < 0$, $p = 1/t$ and $q = 2$ are continuous, and solutions are defined on either interval. An integrating factor is given by $e^{\ln|t|} = |t|$, $t \neq 0$. For $t > 0$ we shall use the integrating factor $|t| = t$, while for $t < 0$ we use the integrating factor $-|t| = t$.

For $t > 0$, or for $t < 0$, multiply equation (10) by the integrating factor t to obtain the equation

$$t\left[y' + \frac{1}{t}y\right] = ty' + y = \frac{d}{dt}[ty] = 2t$$

Antidifferentiation gives $ty = t^2 + C$, or

$$y = t + \frac{C}{t}, \qquad t > 0 \tag{11}$$

The reader may show that (11) defines the general solution of (10) for $t < 0$ as well. Each value of C gives two distinct solutions of (10); one is defined on the t-interval $(0, \infty)$, while the other is defined on the interval $(-\infty, 0)$. If $C \neq 0$, none of these solutions can be extended to $t = 0$ since $y \rightarrow \pm\infty$ as $t \rightarrow 0$. On the other hand, if $C = 0$, then $y = t$ is easily seen to be a solution of the original differential equation for all t. This is the only solution of that equation defined for all t. Some of the solution curves are sketched in Figure 3.6. Note that the initial value problem $ty' + y = 2t$, $y(0) = 1$, has no solution.

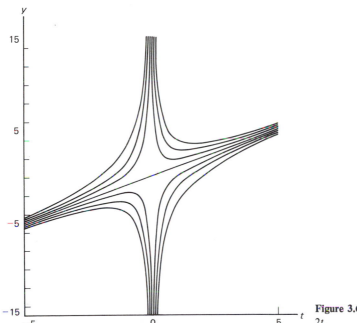

Figure 3.6   Solution curves for $ty' + y = 2t$.

Example 3.13

The equation $ty' - y = t^2 \cos t$ has normal form $y' - (1/t)y = t \cos t$ on the intervals $t > 0$ and $t < 0$. On each interval, $p(t) = -1/t$ and $q(t) = t \cos t$ are continuous. An integrating factor is given by $e^{-\ln|t|} = 1/|t|$. On the interval $t > 0$, we take $1/|t| = 1/t$ as the integrating factor, while on the interval $t < 0$, we use $-1/|t| = 1/t$. On either interval we have that

$$\frac{1}{t}\left[y' - \frac{1}{t}y\right] = \frac{d}{dt}\left[\frac{1}{t}y\right] = \frac{1}{t}[t \cos t] = \cos t$$

Antidifferentiation gives $(1/t)y = \sin t + C$, or

$$y = t \sin t + Ct, \qquad t > 0 \quad \text{or} \quad t < 0 \tag{12}$$

Formula (12) defines the general solution of the differential equation for all t, not just for $t > 0$ or for $t < 0$, since $(t \sin t + Ct)' = t \cos t + \sin t + C$, and hence

$$ty' - y = t(t \cos t + \sin t + C) - (t \sin t + Ct) = t^2 \cos t$$

Thus it turns out that the "singularity" at $t = 0$, where the differential equation is not normal, does not cause trouble after all. All solutions are defined for all t and all satisfy $y = 0$ when $t = 0$. This means that all solution curves cut the y axis at the origin, but nowhere else (see Figure 3.7). Note that the initial value problem $ty' - y = t^2 \cos t$, $y(0) = 0$, has infinitely many solutions.

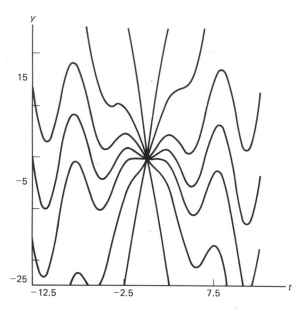

Figure 3.7   Solution curves for $ty' - y = t^2 \cos t$.

Examples 3.12 and 3.13 suggest that the behavior of the solutions of a nonnormal equation $ay' + by = c$ near values of t where $a(t)$ vanishes cannot easily be predicted. One may first write the equation in normal form, solve by the integrating factor technique on intervals where $p \equiv b/a$ and $q \equiv c/a$ are continuous, and then analyze the resulting solutions to see what happens near a zero of $a(t)$ [or near a discontinuity of $p(t)$ or $q(t)$].

Comments

In summary, in this and the previous section we have shown how to construct and use an integrating factor to find a formula for all solutions of a linear first-order differential equation. An operator technique was introduced to perform this task. We have also seen something of the variety of shapes that solution curves may assume. We have addressed the four critical questions posed in Section 2.2 for the linear first-order initial value problem (7).

PROBLEMS

The following problem(s) may be more challenging: 3, 5, 6, 7, 8, and 11.

1. For each case, find the solution of the given initial value problem and state the maximal interval in which the solution is valid. Sketch the solution curve.

   (a) $ty' + 2y = \sin t$,　　　　　　　$y(\pi) = 1/\pi$.
   (b) $(\sin t)y' + (\cos t)y = 0$,　　　$y(3\pi/4) = 2$.
   (c) $y' + y(\cot t) = 2 \cos t$,　　　　$y(\pi/2) = 3$.
   (d) $y' + (2/t)y = (\cos t)/t^2$,　　　$y(\pi) = 0$.
   (e) $(1 + t^2)y' + 2ty = e^{-t}$,　　　$y(-2) = 0$.
   (f) $y' = kt + y$,　　　　　　　　　$y(0) = 1 - k$.
   (g) $y' + aty = bt$,　　　　　　　　$y(0) = 2b/a$.
   (h) $ty' + 2y = t^2 + t - 2$,　　　　$y(1) = \frac{1}{2}$.

2. For Problem 1 (a), (d), and (h), describe the behavior of the solution as $t \to 0$, for various values of the constant of integration. Sketch several members of the family of solution curves.

3. (a) Solve the initial value problem and sketch the solution:

$$y' + 2y = q(t), \qquad y(0) = 0$$

   where

$$q(t) = \begin{cases} 1, & 0 \le t \le 1 \\ 0, & t > 1 \end{cases}$$

   (b) Solve the initial value problem

$$y' + p(t)y = 0, \qquad y(0) = 1$$

   where

$$p(t) = \begin{cases} 2, & 0 \le t \le 1 \\ 1, & t > 1 \end{cases}$$

4. Show that the initial value problem

$$(\cos t)y' + (\cot t)y = 1$$
$$y(0) = 1$$

   does not have a solution on the interval $(-\pi/2, 3\pi/2)$. Why is this not a contradiction to the solution principle?

5. Suppose that we want the output $y(t)$ of a system, modeled by $y'(t) + p(t)y(t) = q(t)$, $y(0) = y_0$, to be maintained at the constant level y_0 for $t \ge 0$. $p(t)$ is given and $q(t)$ is to be determined. How do we choose the driving term $q(t)$ to accomplish this task?

6. A tank initially contains 50 gal of water. Starting at time $t = 0$, a salt solution containing 2 lb of salt per gallon flows into the tank at a rate of 3 gal/min. The mixture is well-stirred. At time $t = 3$ min the mixture begins to flow out of the tank at a rate of 3 gal/min.

  (a) How much salt is in the tank at $t = 2$ min?

  (b) How much salt is in the tank at $t = 25$ min?

  (c) How much salt remains in the tank as $t \to \infty$?

7. [*Flow through a Drain*]. A vertical cylindrical tank with a base radius of 1 m and a height of 2 m is filled with water. A circular drain, which has a diameter of 4 cm, is located at the bottom of the tank. Given that the volume flow rate in m³/s through the drain is $dV/dt = -kA\sqrt{2gh}$, where k is a positive constant, A the cross-sectional area of the drain, g the acceleration due to gravity, and h the height of the water in the tank, how long will it take for the tank to empty?

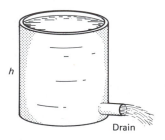

Drain

8. [*Uranium Series*]. Suppose that element A undergoes a series of radioactive decays where A → B → C. The rates of decay of elements A and B are proportional to the amounts present. Further, the rate constants k_A and k_B correspond to the decays A → B and B → C, respectively, and $k_A \neq k_B$.

  (a) If the amounts A_0, B_0, and C_0 are initially present at $t = 0$, find the amount of element C for any later time $(t > 0)$.

  (b) Now let A, B, and C represent uranium-234, thorium-230, and radium-226, respectively. Uranium-234 has a half-life of approximately 2×10^5 years and thorium-230 has a half-life of approximately 8×10^4 years. How much radium-226 will be present after 1×10^6 years if initially 238 g of uranium is present but no thorium and radium?

9. What can be said about the solvability of initial value problems for the differential equations in Examples 3.12 and 3.13? Explain why the initial value problem $y(0) = y_0 \neq 0$ in Example 3.13 has no solution. How many solutions does the initial value problem $y(0) = 0$ in Example 3.13 have? Why does the Solution of Initial Value Problem Theorem not apply in these cases?

10. Let $z(t)$ and $\bar{z}(t)$ be the zero-input responses of the dynamical system (7) due to the initial states y_0 and $\bar{y}_0$, respectively. Show that for any constants a and b, $az(t) + b\bar{z}(t)$ is the zero-input response to the initial state $ay_0 + b\bar{y}_0$. Show that the zero-state response $Z(t)$ of the dynamical system (7) exhibits a similar property if the input is $aq(t) + b\bar{q}(t)$.

11. Let L_0 be the operator with the same action as $D + p(t)$, where p is in $C^0(I)$, but whose domain consists of those functions in $C^1(I)$ which vanish at some fixed t_0 in I.

  (a) Show that the range of L_0 is $C^0(I)$ and that the equation $L_0 y = q$ is uniquely solvable for each q in $C^0(I)$.

(b) Construct an operator $G : C^0(I) \to C^1(I)$ with the property that

$$G[L_0 y] = y \qquad \text{for all } y \text{ in the domain of } L_0$$
$$L_0[Gq] = q \qquad \text{for all } q \text{ in } C^0(I)$$

[*Hint*: G is an integral operator.]

3.3 GROWTH PROCESSES

The U.S. Bureau of the Census annually predicts population trends. To do this, demographers generally use known population data to formulate "laws" of population change. Each law is converted into a mathematical formula that can then be used to make the predictions. Since there is a good deal of uncertainty about which "law" best describes actual population shifts, several are derived, each proceeding from different assumptions about the conditions influencing the birth and death rates (and immigration and emigration). Similarly, ecologists make predictions about changes in the numbers of fish and animal populations, and biologists formulate "laws" for the changing densities of bacteria growing in cultures. In this section we introduce some general laws of population change, solve the corresponding mathematical models, and interpret the results in terms of the growth, stabilization, or decline of a population. The effect on population levels of the harvesting of a species is outlined in the problems.

Let $P(t)$ denote the population at time t of a species in a "community." The values of $P(t)$ are integers and change by integer amounts as time goes on. However, for a large population an increase by one or two over a short time span is "infinitesimal" relative to the total, and we may think of the population as changing continuously instead of by discrete jumps. Once we assume that $P(t)$ is continuous, we might as well go all the way, smooth off any corners on the graph of $P(t)$, and assume that the function is differentiable. Figure 3.8 shows this smoothing process. If we had let $P(t)$ denote the population *density* (i.e., the number per unit area or volume of habitat), the continuity and differentiability of $P(t)$ would have seemed more natural. However, we shall continue to interpret $P(t)$ as the size of the population, rather than density.

The underlying principle of the simple Balance Law of Section 3.1,

$$\text{net rate of change} = \text{rate in} - \text{rate out} \qquad (1)$$

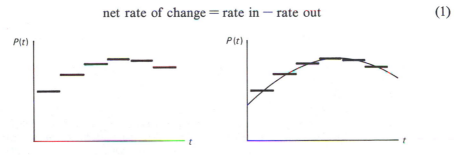

Figure 3.8   Smoothing a population curve.

applies to the changing population $P(t)$ just as it does to the salt solution problem of Sections 3.1, 3.2. For a population, the "rate in" term is the sum of the birth and the immigration rates, while the "rate out" term is the sum of the death and the emigration rates. Let us regroup the rates into an "internal" birth minus death rate and "external" migration rate M (immigration minus emigration). Averaged over all classes of age, sex, and fertility, a "typical" individual makes a net contribution R to the internal rate of change. The internal rate of change at time t is, then, $RP(t)$, where

$$RP = (\text{individual's contribution}) \times (\text{number of individuals})$$

The *intrinsic rate coefficient R* will differ from species to species, but it always denotes the average individual contribution to the rate. Thus (1) becomes

$$P'(t) = RP(t) + M \tag{2}$$

and the study of population changes becomes the problem of solving (2) given the rate coefficient R and the migration rate M.

 R and M may depend on P and t, but in the simplest cases that dependence will either disappear or be linear. If simple models suffice for observed growth processes, there is little need for more complex assumptions. The guiding principle of modeling in this, as in most situations, is that of Ockham's Razor: *What can be accounted for by fewer assumptions is explained in vain by more*.†

Malthusian Growth

> It may safely be pronounced, therefore, that population, when unchecked, goes on doubling itself every twenty-five years, or increases in a geometric ratio.
>
> Malthus‡

 The Malthusian principle of explosive growth of human populations has become one of the classic "laws" of population change. The principle follows directly from (2) if we set $M = 0$ and $R =$ a positive constant, say r. In this case, (2) is linear and has the exponential solution

$$P(t) = P_0 e^{rt} \tag{3}$$

where P_0 is the population at the time $t = 0$. We can see from (3) that the *doubling time* of a species is given by $T = (\ln 2)/r$ since if $P(t + T) = 2P(t)$, then

---

† William of Ockham (1285–1349) was an English theologian and philosopher who applied the Razor to arguments of every kind. The principle is called the Razor because Ockham used it so often and so sharply.

 ‡ Thomas Robert Malthus (1766–1834) was a professor of history and political economy in England. The quotation is from "An Essay on the Principle of Population As It Affects the Future Improvement of Society." Malthus's views have had a profound effect on nineteenth- and twentieth-century Western thought. Both Darwin and Wallace have said that it was reading Malthus which led them to the theory of evolution.

$P_0 \exp[r(t + T)] = 2P_0 \exp[rt]$ implies that $\exp[rT] = 2$. The reader may observe the close connection with the half-life of a radioactive element (see Chapter 1). Malthus claimed a doubling time of 25 years for the human population, which implies that the rate coefficient $r = (1/T) \ln 2 = \frac{1}{25} \ln 2 \cong 0.0277$, corresponding to a 2.8% annual increase in population. Malthus's figure for r is too high for our late-twentieth-century world. However, individual countries, such as Mexico or Sri Lanka, have intrinsic rate coefficients which exceed 0.0277 and may even be as high as 0.033. The corresponding doubling time shrinks from 25 years to 21 years when $r = 0.033$.

Logistic Growth

> The positive checks to population are extremely various and include . . . all unwholesome occupations, severe labor and exposure to the seasons, extreme poverty, bad nursing of children, great towns, excesses of all kinds, the whole train of common diseases and epidemics, wars, plague, and famine.[†]

The unbridled growth of a population as predicted by the simple Malthusian law of exponential increase cannot continue forever. Malthus claimed that resources grow at most arithmetically (i.e., the net increase in resources each year does not exceed a fixed constant). A geometric increase in the size of a population must soon outstrip the resources available to support that population. The resulting hardships would increase the death rate and put a damper on growth.

The simplest way to model restricted growth within the context of the rate equation (2) with no net migration is to assume that the rate coefficient R has the form $r_0 - r_1 P$, where r_0 and r_1 are positive constants. In effect, we are now taking the first two terms of a Taylor series expansion of R as a function of population, rather than just the first term as in the earlier model. A negative coefficient r_1 represents a restraint on the growth rate, whereas r_1 positive would lead to accelerated growth. It is customary to write the rate coefficient as $r(1 - P/K)$, where r and K are positive constants, rather than as $r_0 - r_1 P$. We have the *logistic equation*,[‡] with initial condition

$$P' = r\left(1 - \frac{P}{K}\right)P$$

$$P(0) = P_0$$

(4)

Observe that $P(t) \equiv 0$ and $P(t) \equiv K$ are solutions of the logistic equation, the *stationary* or *equilibrium* solutions. As we shall soon see, the first is related to extinction and the second is the *carrying capacity* of the species, or the *saturation population*.

The logistic equation is not linear, but the change of variable $z = 1/P$ transforms the equation into a linear equation (see Example 3.5). Of course, we would have to

---

[†] Malthus, op. cit.

[‡] The logistic equation and its solutions were introduced in the 1840s by the Belgian statistician and astronomer, Quetelet.

keep away from $P = 0$, but given the meaning of $P(t)$, it is no restriction to assume that $P > 0$. We have that

$$z' = -\frac{1}{P^2} P' = \frac{-1}{P^2} r \left(1 - \frac{P}{K} \right) P = -z^2 r \left(1 - \frac{1}{Kz} \right) \frac{1}{z}$$

or

$$z' = -rz + \frac{r}{K} \tag{5}$$

which is linear in z and z'. Solving (5) by the integrating factor technique and setting $z(0) = 1/P_0$, we have that

$$z(t) = e^{-rt} \left(\frac{1}{P_0} - \frac{1}{K} \right) + \frac{1}{K}$$

Thus we have that

$$P(t) = \frac{1}{z(t)} = \frac{P_0 k}{P_0 + (K - P_0)e^{-rt}} \tag{6a}$$

or

$$P(t) = \frac{K}{1 + Ce^{-rt}}, \qquad \text{where } C = \frac{K}{P_0} - 1 \tag{6b}$$

Since $r > 0$, $P(t) \rightarrow K$ as $t \rightarrow \infty$. The reader may verify that $P(t)$ is strictly increasing if $0 < P_0 < K$ and strictly decreasing if $P_0 > K$. A population curve defined by (6) has an inflection point if $P = K/2$, since

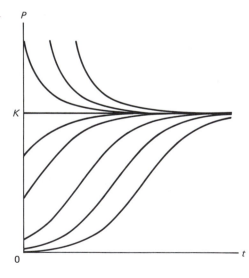

Figure 3.9   Logistic population curves.

$$P'' = \left[r \left(1 - \frac{P}{K} \right) P \right]'$$

$$= r \left(1 - \frac{2P}{K} \right) P'$$

$$= r \left(1 - \frac{2P}{K} \right) r \left(1 - \frac{P}{K} \right) P$$

which changes sign at $P = 0$, $K/2$, and K. The elongated S-shaped (or sigmoid) population curves of the logistic equation are called *logistic curves*. Figure 3.9 illustrates the appropriateness of the term "carrying capacity" or "saturation population" for $P = K$. The resources of the community can support a population of size K, and this is precisely the asymptotic limit of all the nonvanishing population curves.

The population of the United States from 1790 through 1910 followed a logistic

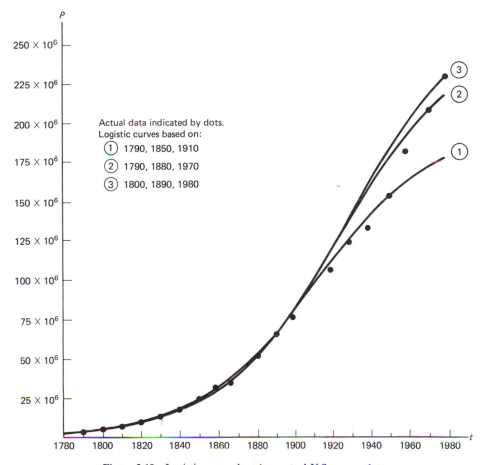

Figure 3.10   Logistic curves based on actual U.S. census data.

curve quite closely. The census figures of 1790, 1850, and 1910 can be used to determine the parameters y_0, r, and K. The resulting logistic curve is sketched in Figure 3.10 together with the actual census figures for the decades from 1790 to 1980. It is remarkable that the curve fits the data so well during the nineteenth century, given that the model completely ignores the massive immigration from Europe from 1850 onward. There is a noticeable discrepancy in 1960, 1970, and 1980. The other two population curves in Figure 3.10 have been determined by using the census figures of 1790, 1880, and 1970 and 1800, 1890, and 1980, but these logistic curves also reveal large deviations between the actual and the predicted. It appears that a single logistic model is no longer appropriate for the population of the United States. Alternatively, one might use one logistic model of 1790–1910 and another for 1910–1980 (i.e., allow for time-varying coefficients r and K).

Comments

Logistic laws may be better suited to laboratory or other isolated populations for which there is greater reason to believe that K and r are constants. Beginning with the experiments of G. F. Gause in the early 1930s, there have been numerous experiments with colonies of protozoa growing under controlled laboratory conditions. The results of these experiments generally confirm the logistic model.†

PROBLEMS

The following problem(s) may be more challenging: 4, 6, and 8(f).

1. If the population of the earth in 1960 was 3 billion (3×10^9) and the population in 1970 was 4 billion, and if the earth can only support a maximum population of 10 billion, in what year will the limit be reached? In what year will the limit be reached if the maximum population that the earth can support is 20 billion? (Assume that the rate of growth is proportional to the population present.)

2. If the population of a third-world country doubled in the past 25 years and the present population is 150,000, when will the country have a population of 560,000? (Assume a Malthusian growth rate.)

3. A colony of bacteria grows according to the logistic law, with a limiting population of 5×10^8 individuals and with natural growth coefficient $r = 0.01/h$.
   (a) What will the population be after 2 h if initially it is 1×10^8 individuals?
   (b) What will the population be after 3 h if initially it is 1×10^3 individuals?

4. Suppose that a population has a birth rate a and a death rate b so that the growth of the population is $dN/dt = (a - b)N$. After 15 years of steady growth, the population members simply stopped reproducing; that is, a became zero. If initially $a = 0.06$ birth/

---

† See J. H. Vandermeer, "The competitive structure of communities: an experimental approach with protozoa," *Ecology 50* (1969), 362–371.

year and $b = 0.04$ death/year, how long after reproduction stops will it take for the population

(a) to return to its original level?

(b) to reach 50% of its original level?

(c) to decrease to 30% of its population when a became zero?

5. Nutrients flow into a cell at a constant rate of R molecules per unit time, and leave it at a rate proportional to the concentration, with constant of proportionality K. Let N be the concentration at time t. Assume the cell volume V is fixed.

(a) Write the mathematical expression for the rate of change of the concentration of the nutrients in the process.

(b) Will the concentration of nutrients reach an equilibrium?

(c) Sketch graphs of $N(t)$ for $t \geq 0$ for various values of $N(0) > 0$ including $N(0) = R/K$.

6. [*Harvesting a Population*]. The equation $dP/dt = r(1 - P/K)P - H$ models a population with natural rate coefficient r, carrying capacity K, and a constant harvest rate H. Assume that r, K, and H are positive constants, and that the population P_0 at time 0 is positive.

(a) Show that $dP/dt < 0$ for all $P > 0$ if $H > rK/4$, the *critical harvest rate*. The population becomes extinct if H exceeds this rate.

(b) Plot three portraits of solution curves in the quadrant $P > 0$, $t > 0$ if there is no harvesting ($H = 0$); subcritical harvesting ($H = rK/8$); supercritical harvesting ($H = rK/2$). Let $r = 0.01$, $K = 1000$, and choose various values of P_0.

(c) The rate equation can be solved by separating the variables. Factor the rate function in part (a) symbolically as $-(r/K)(P - P_1)(P - P_2)$. Show that for $0 < H < rK/4$, P_1 and P_2 are positive numbers between 0 and K. What happens at $H = rK/4$? What happens if $H > rK/4$?

7. [*Bernoulli's Equation*]. Bernoulli's equation is $dy/dt + p(t)y = q(t)y^a$, where a is a real number. The equation is linear if $a = 0$ or 1, nonlinear otherwise. A change of variable reduces the equation to a first-order linear equation, which may then be solved by using an integrating factor. The process is outlined in the following steps.

(a) Show that the logistic equation is a Bernoulli equation with $a = 2$.

(b) Let a be a fixed real number, $a \neq 0, 1$. Divide Bernoulli's equation by y^a and show that the resulting equation is

$$\frac{d}{dt}(y^{1-a}) + (1-a)p(t)y^{1-a} = (1-a)q(t)$$

(c) Introduce the new variable z by $z = y^{1-a}$. Show that z satisfies a first-order linear equation. If $z(t)$ is known, how can $y(t)$ be recovered?

(d) Find all solutions of $dy/dt + t^{-1}y = y^{-4}$, $t > 0$. Sketch some solutions. [*Hint*: Set $z = y^{1-(-4)} = y^5$ and solve $dz/dt + 5t^{-1}z = 5$.]

(e) Find all solutions of $dy/dt - t^{-1}y = -y^{-1}/2$, $t > 0$. Sketch some solutions.

(f) Find all solutions of $t(dy/dt) + y = y^2 \ln t$, $t > 0$. Sketch some solutions.

8. [*Riccati's Equation*]. Riccati's equation is $dy/dt = a(t)y + b(t)y^2 + F(t)$. If $F(t) \equiv 0$, Riccati's equation reduces to a special case of Bernoulli's equation (Problem 7). If $F(t) \not\equiv 0$, Riccati's equation may be reduced to a first-order linear equation if one solution of the equation is known and a new variable is introduced. The following steps outline the process.

(a) Let $y = g(t)$ be a known solution of Riccati's equation [i.e., $dg/dt = ag + bg^2 + $

F]. Let a new variable z be defined by $z = [y - g]^{-1}$. Show that $dz/dt + (a + 2bg)z = -b$, which is a first-order linear equation in z. Suppose that $z(t)$ has been found. How can $y(t)$ be determined?

(b) The Riccati equation $dy/dt = (1 - 2t)y + ty^2 + t - 1$ has a particular solution $y(t) \equiv 1$. (Show.) Let $z = (y - 1)^{-1}$ and show that $dz/dt = -z - t$. Solve for all functions $z(t)$ and then find the general solution $y(t)$ of the Riccati equation.

(c) Find all solutions of $dy/dt = e^{-t}y^2 + y - e^t$. [*Hint*: First show that $y = e^t$ is a solution.]

(d) Show that $y = t$ is one solution of $dy/dt = t^3(y - t)^2 + yt^{-1}$, $t > 0$, and then find all solutions of the equation.

(e) Show that the differential equation of Problem 6 is a Riccati equation. Solve the equation if $K = 1600$, $r = 0.01$, $H = 3$. [*Hint*: Show that $P(t) \equiv 1200$ is one solution of the equation.]

(f) Suppose that $g_1(t)$ and $g_2(t)$ are two different solutions of the Riccati equation $y' = a(t)y + b(t)y^2 + F(t)$. Show that every other solution y, $y \neq g_1$, $y \neq g_2$ satisfies the equation

$$\frac{y - g_1}{y - g_2} = c \exp\left[\int^t b(g_1 - g_2)\, ds\right]$$

(g) Use the technique of part (f) to solve $dy/dt = (t - 1)^{-1}y^2 - t(t - 1)^{-1}y + 1$, $t > 1$. [*Hint*: Show that $y = 1$ and $y = t$ are solutions.]

3.4 DEPENDENCE ON DATA: REACTORS AND MIGRATION

In Section 2.2 four basic questions were posed for an initial value problem.

1. Are there any solutions?
2. How many?
3. How can they be found?
4. How do solutions change if the data change?

For a linear first-order problem we have seen in Sections 3.1 and 3.2 that all four questions have simple, and direct, answers. Now we look more deeply into the fourth question and derive the basic Bounded Input–Bounded Output Principle of engineering practice in this first-order setting. We also consider the data dependence of some special nonlinear problems such as logistic growth with migration. The effects of changing data are of considerable importance since any well-designed system such as a chemical or nuclear reactor, an electrical circuit, or a space probe should be relatively impervious to the inevitable small disturbances and random shocks which will act upon it.

Bounded Input–Bounded Output

As we saw in Section 3.2, if $p(t)$ and $q(t)$ are continuous on an interval I containing t_0, the linear initial value problem

$$y' + p(t)y = q(t)$$
$$y(t_0) = y_0$$

(1)

has the unique solution

$$y(t) = e^{-P(t)}y_0 + e^{-P(t)} \int_{t_0}^{t} e^{P(s)}q(s)\, ds \tag{2}$$

where $P(t) = \int_{t_0}^{t} p(r)\, dr$. Suppose that the scalar function $y(t)$ satisfying (1) is a measure of the deviation from the "ideal" operating state of some system; that is, $y(t) \equiv 0$ indicates ideal operation, while $y(t) \neq 0$ indicates that a disturbance is acting, or has acted. If a disturbance is bounded in magnitude and acts only over a brief span of time, we might expect its effect to attenuate [i.e., $y(t) \rightarrow 0$ as t increases]. A persistent disturbance, even though small in magnitude, may have a cumulative effect which in time will tend to destroy the system in the sense that $|y(t)|$ could approach unacceptably high values. Can this possible state of affairs be predicted and then avoided?

Suppose that the function $p(t)$ in (1) is a design element under our control, while y_0 denotes the initial deviation from the ideal state of $y = 0$ and $q(t)$ represents the disturbances, which may persist over time. Since $q(t)$ need never vanish, we cannot always hope to choose $p(t)$ so that $|y(t)| \rightarrow 0$ as $t \rightarrow \infty$. Our goal is the more modest one of finding conditions on $p(t)$ which will ensure that $|y(t)|$ remains bounded and, more practically, to estimate these bounds. For simplicity, we set $t_0 = 0$. We have the following result:

> **Bounded Input–Output Principle.**   Suppose that for $t \geq 0$ $p(t)$ and $q(t)$ are continuous functions satisfying (a) $p(t) \geq p_0 > 0$ and (b) $|q(t)| \leq M$. Then for all $t \geq 0$ the solution $y(t)$ of (1) satisfies
>
> $$|y(t)| \leq e^{-p_0 t}|y_0| + \frac{M}{p_0}(1 - e^{-p_0 t}) \leq \max\left\{|y_0|, \frac{M}{p_0}\right\}$$

The hypotheses require that the positive function $p(t)$ be "bounded away" from 0 by a positive constant p_0 and that the input $q(t)$ be bounded in magnitude by a constant M. The conclusion states that the output $y(t)$ is bounded in magnitude and, in addition, gives estimates for the bound.

Proof of the Principle. First note that the hypothesis on $p(t)$ implies that $-p(t) \leq -p_0 < 0$, and hence that

$$-P(t) = \int_0^t -p(r)\, dr \leq \int_0^t -p_0\, dr = -p_0 t \tag{3}$$

$$-P(t) + P(s) = \int_0^t -p(r)\, dr + \int_0^s p(r)\, dr = \int_s^t -p(r)\, dr$$

$$\leq -p_0(t - s) \qquad \text{if } 0 \leq s \leq t \tag{4}$$

We shall use these two inequalities below. Since the solution $y(t)$ of (1) is given by (2), the following inequalities may be derived as in the proof of the Continuity with Respect to the Data Theorem in Section 3.2:

$$|y(t)| \leqq |e^{-P(t)}y_0| + |e^{-P(t)} \int_0^t e^{P(s)}q(s)\,ds|$$

$$\leqq e^{-P(t)}|y_0| + \int_0^t e^{-P(t)+P(s)}|q(s)|\,ds$$

$$\leqq e^{-p_0 t}|y_0| + \int_0^t e^{-p_0(t-s)}M\,ds$$

$$= e^{-p_0 t}|y_0| + \frac{M}{p_0}e^{-p_0 t}[e^{p_0 s}]_{s=0}^{s=t}$$

$$= e^{-p_0 t}|y_0| + \frac{M}{p_0}(1 - e^{-p_0 t})$$

This is the first inequality of the Bounded Input–Bounded Output Principle.

The second inequality is obtained by observing that the function

$$B(t) \equiv e^{-p_0 t}|y_0| + \frac{M}{p_0}(1 - e^{-p_0 t}), \qquad t \geqq 0$$

is monotone since $e^{-p_0 t}$ is a strictly decreasing function. If $|y_0| \geqq M/p_0$, $B(t)$ is nonincreasing and has maximal value $B(0) = |y_0|$, while if $|y_0| < M/p_0$, then $B(t)$ is increasing and tends to the upper bound M/p_0 as $t \to \infty$. This shows the validity of the second inequality of the principle.

The hypotheses and the conclusions of the Bounded Input–Bounded Output Principle are the best possible in the following senses:

1. If $p(t) \equiv p_0 > 0$ and $q(t) \equiv M > 0$, the inequalities in the conclusion become equalities.
2. If we have that $p(t) > 0$ but not $p(t) \geqq p_0 > 0$, (1) may have unbounded solutions.
3. If we have an unbounded input $q(t)$, the output $y(t)$ may also be unbounded.

See the problem set for examples that illustrate these points.

From a design point of view we see that to obtain a tighter bound on the magnitude of $|y(t)|$, we should make $p(t)$ (hence p_0) as large as possible. Alternatively, if p_0 is large, the system can tolerate shocks of large magnitude, that is, the upper bound M on $|q(t)|$ may be large, without running the risk of producing excessively large values of $|y(t)|$.

The Bounded Input–Bounded Output Principle has been interpreted in terms of the design of a dynamical system. As with any mathematical theorem, however, the principle can be applied to all phenomena with the same mathematical model covered by the theorem. We have seen before just how effective this universality of mathematical theorems and models can be, and the following example illustrates the point once again.

Example 3.14 (*Chemical Reactor*)

A substance A is gradually and irreversibly converted into a different substance B in a chemical reactor, which is a tank of V liters filled with a solution of A, B, and perhaps, other chemicals and catalysts. How much of A remains in the reactor t units of time after the reaction begins? Let [A] denote the *concentration* of A in, say, moles per liter of solution and suppose that a *first-order rate equation* models the reaction A $\rightarrow$ B,

$$\frac{d[A]}{dt} = -k[A] \tag{5}$$

where k is the positive rate constant.† With the passage of time, [A(t)] will decay exponentially from its initial value: $[A(t)] = [A(0)]e^{-kt}$.

In real life, however, we have to contend with leaky valves, open stopcocks, and so on, which permit small amounts of A to dribble into the vat even while the reaction is taking place. Let $a(t)$ denote the rate at which A drips into the reactor. Then the basic Balance Law:

$$\text{rate of accumulation} = \text{rate in} - \text{rate out}$$

implies that

$$\frac{d[A]}{dt} = \frac{a(t)}{V} - k[A]$$

If we estimate that $a(t) \leqq a_0$, the Bounded Input–Bounded Output Principle implies that

$$[A(t)] \leqq e^{-kt}[A(0)] + \frac{a_0}{kV}(1 - e^{-kt}) \leqq \max\left\{[A(0)], \frac{a_0}{kV}\right\} \tag{6}$$

Suppose that we require that [A(t)] never exceed some constant value K. Such a restriction might be necessary to ensure safe operation of the reactor, for example. By (6) the restrictions $[A(0)] \leqq K$ and $a_0 \leq kVK$ will guarantee that $[A(t)] \leqq K$. The first restriction is obvious, the second less so. In any event, the inequalities of (6) give us practical limits on the concentration [A(t)] over the course of the reaction.

---

† To a chemist, (5) is first order, while rate equations such as $d[A]/dt = -k[A]^2$ or $d[A]/dt = -k[A][B]$ are second order, involving as they do, products of two concentrations. Mathematically, we say that all of these are first order since only the first derivatives are involved.

The Bounded Input–Bounded Output Principle works for a linear equation in large part because of the exponentially decaying term $e^{-p_0 t}$. What happens if the equation is nonlinear?

Nonlinear Equations with Bounded Inputs

Every linear first-order differential equation has the same form, but nonlinear equations have nothing in common but their nonlinearity. It is not surprising that there is no general extension of the Bounded Input–Bounded Output Principle to nonlinear equations. Restricted versions of the principle do apply, however, to certain nonlinear equations, as we shall see below. For other nonlinear equations, everything goes wrong and bounded inputs result in unbounded solutions. We shall illustrate both points with nonlinear equations of the form

$$y' = f(y) + q(t)$$

where we assume that $f(y)$ is not linear [i.e., $f(y)$ is not of the form $ay + b$]. We shall call $q(t)$ the *input* or *driving term* in accord with the terminology for the linear case. Example 3.15 shows that the Bounded Input-Output Principle applies at least to the logistic growth model.

Example 3.15 (*Logistic Growth with Migration*)

What happens to the population curves when a migration term is added to the model of logistic growth of Section 3.3? The corresponding initial value problem is

$$P'(t) = r\left[1 - \frac{P(t)}{K}\right]P(t) + q(t)$$

$$P(0) = P_0$$

$$(7)$$

where r, K, and P_0 are positive constants, and $q(t)$ denotes the net rate of migration, that is, the rate of immigration minus the rate of emigration. Recall from Section 3.3 that $P(t)$ denotes the size of the population at time t and that the model is assumed to be valid only for $P \geq 0$ and $t \geq 0$. Let us assume that the migration rate $q(t)$ is bounded, $|q(t)| \leq Q$. Although $P' = r(1 - P/K)P + q$ cannot generally be transformed into a first-order linear equation by a change of variables, we can estimate $P(t)$ by techniques quite different from any used up to this point.

We shall show that for each initial value $P_0 > 0$, there are positive constants P^* and T^* such that

$$0 < P(t) \leq P^*$$

$$(8)$$

for all $0 \leq t < T^*$, where either

$$T^* = +\infty \quad \text{or} \quad P(T^*) = 0$$

$$(9)$$

In other words, bounded $|q(t)|$, $t \geq 0$, implies that the solution of (7), the population $P(t)$, either remains forever bounded or else becomes extinct in finite time. This is a special nonlinear version of the Bounded Input–Bounded Output Principle.

To prove that (8) and (9) hold, we proceed as follows. Since r, K, and Q are positive constants, the roots P_1 and P_2 of the quadratic polynomial

$$R(P) = r\left(1 - \frac{P}{K}\right)P + Q = -\frac{r}{K}P^2 + rP + Q$$

are real and of opposite sign:

$$P_1 = \frac{K}{2} - \frac{1}{2}\left(K^2 + \frac{4KQ}{r}\right)^{1/2} < 0; \qquad P_2 = \frac{K}{2} + \frac{1}{2}\left(K^2 + \frac{4KQ}{r}\right)^{1/2} > 0$$

Since the coefficient of P^2 in $R(P)$ is negative, $R(P)$ is negative for all values of P larger than P_2. Suppose that for some positive τ, $P(\tau) > P_2$. Then $P'(\tau)$ is negative, for

$$P'(\tau) = r\left[1 - \frac{P(\tau)}{K}\right]P(\tau) + q(\tau)$$

$$\leq r\left[1 - \frac{P(\tau)}{K}\right]P(\tau) + Q$$

$$= R(P(\tau)) < 0$$

since $P(\tau) > P_2$. Thus any solution curve of the differential equation of (7) which intersects a horizontal line $P = P_3$, where $P_3 > P_2$, must move downward across that line (see Figure 3.11).

Now suppose that the initial population P_0 lies somewhere in an open interval $0 < P_0 < P_3$, where P_3 is larger than P_2. The solution curve defined

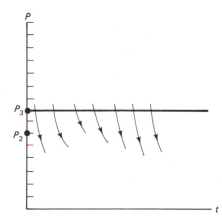

Figure 3.11   Falling solution curves.

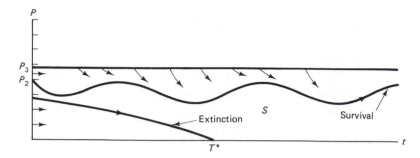

Figure 3.12 Population curves.

by $P(t)$ enters the region S of Figure 3.12 from the left. As time goes on, the solution curve cannot "escape" from S by crossing the upper boundary, $P = P_3$, since $P'(t)$ is always negative on that boundary. By the Extension Principle of Section 3.3, we see that either $P(t)$ is defined for all $t \geqq 0$ and the solution curve remains in S forever, or else $P(t^*) = 0$ for some finite T^* and the solution curve hits the lower edge, the extinction line $P = 0$ (see Figure 3.12).

Suppose that $P_0 > P_2$. By the same argument as that used above to show that every solution curve intersecting the line $P = P_3$ must be falling at the point of intersection, we can show that the same must be true for every solution curve crossing the line $P = P_0$. Hence the solution curve defined by $P(t)$ enters the region S of Figure 3.13a or b from the upper left-hand corner and either terminates at some finite time T^* on the extinction line (Figure 3.13a) or remains in the region for all positive time (Figure 3.13b). The Extension Principle once again implies that this must be so.

Thus we have verified (8) and (9) with P^* defined to be max $\{P_0, P_2\}$. In the language of a growth process, we have shown that limited migration cannot produce unlimited growth if the basic dynamics is logistic.

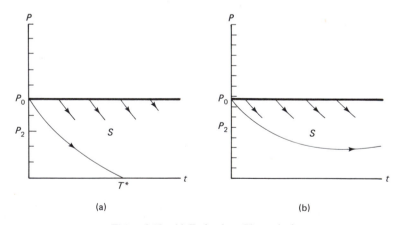

(a)   (b)

Figure 3.13   (a) Extinction; (b) survival.

In spite of the positive results of Example 3.15, we cannot hope to have a general result along these lines. See Problem 6 for an example of just how badly things may go wrong in the presence of a different type of nonlinearity.

Comments

We have shown in this section that the Bounded Input–Bounded Output Principle provides an effective estimate of the dependence of the solution of a linear initial value problem on the input. On the other hand, nonlinear equations pose problems. There are special versions of the principle valid for special equations, but there is no general result.

PROBLEMS

The following problem(s) may be more challenging: 2 and 6.

1. A decaying population is replenished from time to time with new stock. Suppose that the rate law is $P'(t) = -r(t)P(t) + R(t)$, where all we know about r and R is that $0 < r_0 < r(t)$ and that $|R(t)| \leq R_0$, $t \geq 0$. If $P(0) = P_0$, find a reasonable upper bound for $|P(t)|$, $t \geq 0$.

2. A large vat contains 100 gal of brine in which initially 5 lb of salt is dissolved. More brine runs into the vat at a rate of r gallons per minute with a concentration of $C(t)$ pounds of salt per gallon. The solution is thoroughly mixed and runs out at a rate of r gallons per minute. For safety reasons, the concentration in the tank must never exceed 0.1 lb of salt per gallon. If all that is known about r and $c(t)$ is that $0 < r_0 \leq r \leq r_1$ and $0 \leq C(t) \leq C_0$, what conditions must r_0 and C_0 satisfy?

3. Show that the inequality

$$y(t) \leq e^{-p_0 t}|y_0| + \frac{M}{p_0}(1 - e^{-p_0 t})$$

of the Bounded Input–Bounded Output Principle cannot be improved. [*Hint*: Solve the initial value problem $y' + p_0 y = M$, $y(0) = y_0$, where p_0, M, and y_0 are positive constants.]

4. Show that the hypothesis $p(t) \geq p_0 > 0$ in the Bounded Input–Bounded Output Principle cannot be extended to the condition $p(t) > 0$. [*Hint*: Show that the problem $y' + t^{-2}y = e^{1/t}$, $t \geq 1$, has unbounded solutions even though $p(t) > 0$ and $|q(t)| \leq 3$ for $t \geq 1$.]

5. Show that the equation $y' + y = t$ has unbounded solutions on the interval $0 \leq t$, thus showing that the hypothesis $|q(t)| \leq M$ cannot be dropped from the Bounded Input–Bounded Output Principle.

6. The following shows how badly things may go wrong in the presence of a nonlinearity:
   (a) Show that every solution of the undriven equation $y' = -y/(1 + y^2)$ tends to zero as $t \to \infty$. [*Hint*: Solve the equation by separating the variables.]
   (b) Using calculus, show that every solution $y(t)$ of $y' = -y/(1 + y^2) + 1$ satisfies the inequality $y'(t) \geq \frac{1}{2}$, for all $t \geq 0$.

(c) Upon integrating the inequality, we have that $y(t) \geq \frac{1}{2}t + C$, where C is a constant of integration. Show that the Bounded Input–Bounded Output Theorem fails to hold in this nonlinear situation. [*Remark*: The underlying cause of this behavior is that although all solutions of the undriven equation $y' = y/(1 + y^2)$ decay, they do so very slowly and the addition of the bounded input $q(t) = 1$ is enough to destabilize the equation.]

(d) Sketch some representative solution curves for the differential equations in parts (a) and (b).

Techniques for Solving First-Order Differential Equations: Newtonian Mechanics

In this chapter we consider techniques for solving the first-order equation $y' = f(t, y)$ under various assumptions about the form of f. We shall not assume that f is linear in y, nor that the equation is reducible to a linear differential equation as we did in the preceding Chapter 3. The principal models in this chapter have to do with motion in force fields (i.e., Newtonian dynamics), although rate equations also appear. We also show how first-order techniques can be applied to some second-order equations that arise in Newtonian models. The reader should note that we alternate between using t and x as the independent variable and carefully identify which is being used in a given section or example.

4.1 SEPARABLE EQUATIONS: DESTRUCTIVE COMPETITION

After linear equations, the next most commonly encountered first-order differential equation is the *separable equation*

$$N(y)y'(x) + M(x) = 0 \tag{1}$$

so named because the variables x and y are separated as indicated. We shall show how to find solutions of (1) and then give a curious and controversial application to models of destructive competition.

Suppose that M is continuous on the x-interval I, and that N is continuous on the y-interval J. Then M and N have antiderivatives defined on I and J, denoted by

$$F(x) = \int^{x} M(s)\,ds, \qquad G(y) = \int^{y} N(s)\,ds$$

which is just another way of saying that $dF/dx = M(x)$, all x in I, and $dG/dy = N(y)$, all y in J. Then if $y(x)$ is any solution of (1) whose solution curve remains in the xy-rectangle defined by I and J, it follows from the Chain Rule that (1) can be rewritten as

$$[G(y(x)) + F(x)]' = 0 \qquad (2)$$

Thus the Vanishing Derivative Theorem (see Section 1.1) implies via (2) that there is a constant C such that

$$G(y(x)) + F(x) = C \qquad (3)$$

for all x on the interval where $y(x)$ is defined. Conversely, suppose that $y(x)$ is a continuously differentiable function whose graph lies in the xy-rectangle determined by I and J, and that $y(x)$ satisfies Eq. (3) for some constant C. Then differentiation of (3) via the Chain Rule easily shows that $y(x)$ is a solution of the differential equation (1). Hence (3), for C an arbitrary constant, can be regarded as defining the "general solution" of Eq. (1) in the xy-rectangle determined by I and J.

Example 4.1

The equation $y' = 2xe^{y}(1 + x^2)^{-1}$ can be written in separable form as $-e^{-y}y' + 2x(1 + x^2)^{-1} = 0$. Therefore, solutions $y = y(x)$ of the differential equation satisfy the equation $e^{-y} + \ln(x^2 + 1) = C$, C a constant, which can be solved for y in terms of x to obtain $y = -\ln |C - \ln(x^2 + 1)|$. Clearly, $y(x)$ is defined on an x-interval for which $C - \ln(x^2 + 1) > 0$. Observe that we must have $C > 0$. See Figure 4.1 for some of the solution curves.

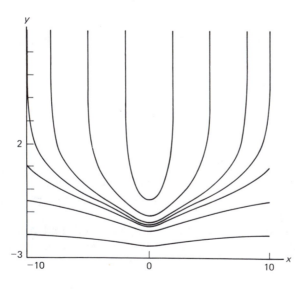

Figure 4.1 Solution curves for $y' = 2xe^{y}(1 + x^2)^{-1}$.

It is not always possible to solve (3) for y in terms of x, as the next example shows. In such a case, the level sets defined by (3) are *integral curves* of (1) and a solution curve is then a subarc of an integral curve.

Example 4.2

Using the methods described above, the separable equation $(\cos y + e^y)y' = 7x^6 + 1$ is seen to have solutions $y(x)$ satisfying the equation $\sin y(x) + e^{y(x)} - x^7 - x = C$, where C is a constant. However, this transcendental equation cannot be solved for y explicitly in terms of elementary functions of x. In a case like this, we must be satisfied with an equation that defines y implicitly in terms of x. Some of the integral curves defined by the transcendental equation above are sketched in Figure 4.2.

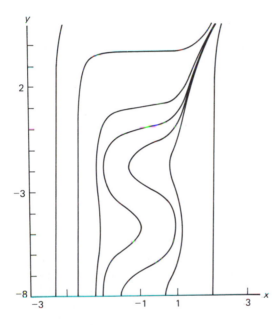

Figure 4.2 Integral curves for $(\cos y + e^y)y' = 7x^6 + 1$.

One can use the foregoing procedure to solve initial value problems for separable equations. Suppose that one wants to find a solution to the initial value problem

$$N(y)y' + M(x) = 0, \qquad y(x_0) = y_0 \qquad (4)$$

This is easily accomplished by choosing the constant $C = 0$ and the special antiderivatives

$$G(y) = \int_{y_0}^{y} N(s)\, ds, \qquad F(x) = \int_{x_0}^{x} M(s)\, ds$$

in the solution formula (3). Indeed, it is easy to see directly from the Chain Rule that a solution $y(x)$ of the problem (4) satisfies the equation

$$\int_{y_0}^{y(x)} N(u)\,du + \int_{x_0}^{x} M(s)\,ds = 0 \tag{5}$$

It is also instructive to derive the solution formula (5) directly from (4) in the following way. Integrating both sides of the differential equation in (4) from x_0 to x, we have

$$\int_{x_0}^{x} N(y(s))y'(s)\,ds + \int_{x_0}^{x} M(s)\,ds = 0$$

Making the substitution $u = y(s)$ in the first integral above yields (5) directly.

Example 4.3

Consider the initial value problem

$$yy' - x = 0, \qquad y(2) = -1$$

Any solution $y(x)$ must satisfy the equation

$$\int_{-1}^{y} u\,du - \int_{2}^{x} s\,ds = 0$$

or $x^2 - y^2 = 3$, which defines a hyperbola through the initial point $(2, -1)$. The particular branch $y(x)$ of the hyperbola passing through $(2, -1)$ is defined by

$$y = -(x^2 - 3)^{1/2}, \qquad x > 3^{1/2}$$

The hyperbolic integral curve contains four (maximally extended) solution curves, which are defined by the equations $y = \pm(x^2 - 3)^{1/2}$, $x > 3^{1/2}$, or $x < -3^{1/2}$ (see Figure 4.3).

Sometimes, as in Example 4.1, a first-order differential equation is not in separable form but can be written in separable form by appropriate algebraic manipulation. One must be careful when doing this so as not to "lose" solutions in the process. The next example shows what might happen if one is careless.

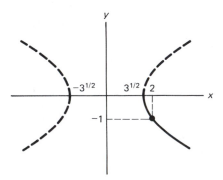

Figure 4.3   Solution curve for $yy' = x$, $y(2) = -1$ (bold face line).

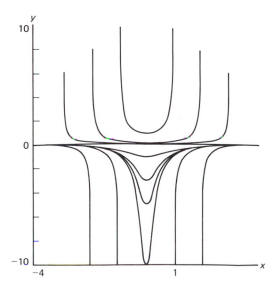

Figure 4.4   Solution curves of $y' = 2xy^2$.

Example 4.4

The equation $y' = 2xy^2$ can be written in separable form as $y'y^{-2} = 2x$, but at the price of "losing" the solution $y(x) \equiv 0$ of the original equation. For $y \neq 0$, we can solve the separated equation, obtaining $-y^{-1} = x^2 + C$ or $y = -(x^2 + C)^{-1}$, where C is an arbitrary constant. Thus the collection of all solution curves of $y' = 2xy^2$ is given via

$$y = \frac{-1}{x^2 + C}, \qquad C \text{ any constant, and } y \equiv 0 \qquad (6)$$

In the first case, if $C \leqq 0$, x must be restricted to an interval on which $x^2 + C \neq 0$. For example, if $C = -1$, the first equation of (6) defines three solutions: one for $x < -1$, another for $|x| < 1$, and a third for $x > 1$. See Figure 4.4 for a sketch of some of the solution curves.

With the examples above in mind, the reader should use the method of solving separable equations with some caution, being careful not to lose solutions and to choose the right "branch" when solving an initial value problem.

Destructive Competition: Models of Combat

Satietie of sleepe and love, satietie of ease,
Of musicke, can find place, yet harsh warre still must please
Past all these pleasures, even past these.†

† From *The Iliad of Homer*, trans. G. Chapman, 2 vols. (London; J. R. Smith, 1857). In contemporary English rather than Chapman's poetic Elizabethan, the lines might be read as:

  Men grow tired of sleep, love, singing, and dancing
  Sooner than of war.

In 1916, F. W. Lanchester described some mathematical models for air warfare. These have been extended to a general combat situation, not just air warfare. We consider a particular case of the models here, others in the problem set.

An "x-force" and a "y-force" are engaged in combat. Let $x(t)$ and $y(t)$ denote the respective strengths of the forces at time t, where t is measured in days from the start of the combat. It is not easy to quantify "strength," including, as it does, the numbers of combatants, their battle readiness, the nature and number of the weapons, the quality of the leadership, and a host of psychological and other intangible factors difficult even to describe, much less to turn into numbers. We shall take the easy way out and identify the strengths $x(t)$ and $y(t)$ with the numbers of combatants.

We shall assume that $x(t)$ and $y(t)$ vary continuously and even differentiably as functions of time. This is, of course, an idealization of the true state of affairs since the strength must be an integer and change only by integer amounts as time goes on. However, as we did earlier with models of radioactive decay and growth, one might argue that when the numbers are large, an increase by one or two is infinitesimal compared to the total, and we might as well allow $x(t)$ and $y(t)$ to change continuously over small time spans. Once we have made the idealization to continuous functions, we "smooth" off any "corners" on the graphs of $x(t)$ and $y(t)$ versus t; thus we can reasonably take $x(t)$ and $y(t)$ to be continuous and differentiable.

Although we may not have a specific formula for $x(t)$, say, as a function of t, we may have a good deal of information about the *noncombat loss rate* of the x-force (i.e., the loss rate due to the inevitable diseases, desertions, and other noncombat mishaps), the *combat loss rate* due to encounters with the y-force, and the *reinforcement* (or *supply*) *rate*. We shall assume that the net rate of change in $x(t)$ is given by the basic law

$$x'(t) = \text{reinforcement rate} - (\text{noncombat loss rate} + \text{combat loss rate})$$

A similar equation applies to the y-force. The problem is to find appropriate formulas for these rates for each force and then to analyze the solutions $x(t)$ and $y(t)$ of the respective differential equations to determine who "wins" the combat.

According to Lanchester, a model for a pair of conventional combat forces operating in the open (relatively speaking) with negligible noncombat losses is

$$\begin{aligned} x'(t) &= -by(t) + R_1(t) \\ y'(t) &= -ax(t) + R_2(t) \end{aligned} \tag{7}$$

where a and b are positive constants and R_1 and R_2 are the rates of reinforcement (e.g., in numbers per day). In this model the reinforcement rates are assumed to depend only on time and not on the strength of either force (a dubious assumption). The combat loss rates, $-by(t)$ and $-ax(t)$, introduce actual combat into the model since without these terms neither force has any effect whatsoever on the other. Lanchester argues for the form of the combat loss rate terms as follows. Every member of the conventional force is assumed to be within range of the enemy. It is also assumed that as soon as a conventional force suffers a loss, fire is concentrated on the remaining combatants. Thus the combat loss rate of the x-force is proportional to the number of the enemy and is given by $-by(t)$. The coefficient b is a measure of the average effectiveness in combat of each member of the y-force. A similar argument applies to the term $-ax(t)$.

We cannot say much about solving the two equations of (7) at this point since the equations are coupled. However, consider the situation of a pair of isolated conventional forces with no reinforcements. In this setting (7) reduces to

$$x' = -by, \qquad y' = -ax \tag{8}$$

a system in which time no longer explicitly appears. Let $x = x(t)$ and $y = y(t)$ be solutions of (8) that lie inside the first quadrant of the xy-plane (i.e., $x > 0$, $y > 0$). Since $x'(t) = -by$ never vanishes in this quadrant, we can apply the Inverse Function Theorem of calculus and theoretically solve $x = x(t)$ for t in terms of x, obtaining $t = t(x)$. Moreover, $dt(x)/dx = [dx(t)/dt]^{-1}$. Thus, using the Chain Rule to find $d[y(t(x))]/dx$, we have

$$\frac{dy}{dx} = \frac{d}{dx}[y(t(x))] = \frac{dy}{dt}\frac{dt}{dx} = \frac{dy/dt}{dx/dt} = \frac{-ax}{-by}$$

or

$$by\frac{dy}{dx} = ax \tag{9}$$

which is separable. In effect, we have eliminated time as the independent variable and replaced it by x. The only assumptions made above are that $x > 0$ and $y > 0$ and, of course, that is no restriction at all for the phenomenon being modeled.

Let $x_0 > 0$ and $y_0 > 0$ be the strengths of the two forces at the start of combat. Integrating (9), we have that $\int_{y_0}^{y} bu\,du = \int_{x_0}^{x} as\,ds$, or

$$by^2(t) - ax^2(t) = by_0^2 - ax_0^2 \equiv k \tag{10}$$

where the constant k is determined by the initial data. Equation (10) is known as the *square law of conventional combat*. Although we cannot tell from (10) just how $x(t)$ and $y(t)$ change in time, (10) implies that the point $(x(t), y(t))$ moves along an arc of a hyperbola as time runs on. The hyperbolas defined by (10) are sketched in Figure 4.5 for various values of k. The arrowheads on the curves show the direction of changing strengths as time passes. Since $dx/dt < 0$ and $dy/dt < 0$ whenever $x(t) > 0$ and $y(t) > 0$, the directions of the arrowheads are as indicated.

Who "wins" in such a combat? Let us say that one force wins if the other force vanishes first. For example, y wins if $k > 0$, since according to (10), y never vanishes in this case, while the x-force has been annihilated by the time $y(t)$ has

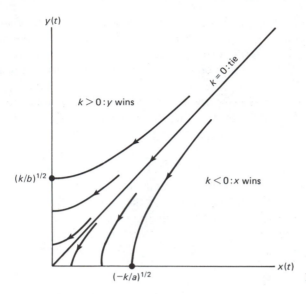

Figure 4.5 Hyperbolas of the square law.

decreased to $(k/b)^{1/2}$. Thus the y-force seeks to establish a combat setting in which $k > 0$; that is, the y-force wants the following inequality to hold:

$$\left(\frac{y_0}{x_0}\right)^2 > \frac{a}{b} \tag{11}$$

Because of the quadratic magnification in (11), a small increase in the ratio of the initial forces could convert a predicted loss for the y-force $[(y_0/x_0)^2 < a/b]$ to a win $[(y_0 + \epsilon)^2/x_0^2 > a/b]$.

The simplified model solved above is unrealistic. If noncombat loss rates and reinforcement rates are included, however, a certain element of realism enters and one might actually compare the model with historical battles. Studies along these lines for the Battle of the Ardennes and the Battle of Iwo Jima, both in World War II, have been carried out.† These studies give results reasonably close to the actual combat statistics once the coefficients a and b are determined. Whether the effectiveness coefficients could ever be estimated with any accuracy before an engagement is an open question. Terms such as "combat readiness" are used but would be hard to quantify.

Mathematical Models of Combat: The Dark Side of Modeling

Models like these raise ethical questions about the uses made of mathematics. G. H. Hardy, one of the best known English "pure" mathematicians of the first half

---

† See, e.g., C. S. Coleman, "Combat models," Chap. 8 in *Modules of Applied Mathematics*, vol. 1 (ed. W. Lucas, M. Braun, C. Coleman, and D. Drew (New York: Springer-Verlag, 1983)).

of the twentieth century, has spoken on this matter: "So a real mathematician has his conscience clear; there is nothing to be set against any value his work may have; mathematics is . . . a "harmless and innocent," occupation."† One wonders if it is really all that simple. Mathematics is not a world apart from the cultures and societies of the mathematicians who create or discover it. Since war and the preparation for war continue to be a principal preoccupation of humankind, it is not surprising that mathematics is applied to its study and analysis.

PROBLEMS

The following problem(s) may be more challenging: 1(a), 2(c), 4, and 6(a).

1. Find the general solution to the following separable equations. Sketch solution curves for (a)–(f).

   (a) $y' = \dfrac{y+1}{x+1}$.

   (b) $y' = \dfrac{-y}{x^2-4}$.

   (c) $y' = -e^{-(x+y)}x$.

   (d) $\dfrac{\log y}{\log x}\,dy - \dfrac{x^4}{y^2}\,dx = 0$.

   (e) $\tan^2 y\,dy = \sin^3 x\,dx$.

   (f) $(1-x)y' = y^2$.

   (g) $dr = b(\cos\theta\,dr + r\sin\theta\,d\theta)$.

   (h) $a^2\,dx = x\sqrt{x^2-a^2}\,dy$.

2. Solve the following separable equations subject to the indicated constraint.

   (a) $y' = \dfrac{y^2}{x}$, $y(1) = 1$.

   (b) $y' = \dfrac{(x^2+2)(y+1)}{xy}$, $y(1) = \sqrt{e}-1$.

   (c) $y' = ay - by^2$, $y(0) = \dfrac{a}{2b}$, $a \neq 0, b \neq 0$.

   (d) $y' = ye^{-x}$, $y(0) = e$.

   (e) $y' = \dfrac{3t^2}{1+t^3}$, $y(0) = 2$.

   (f) $\dfrac{dy}{dx} = \dfrac{-x}{y}$, $y = 2$ where $x = 1$.

   (g) $y' = xe^{y-x^2}$, $y = 0$ where $x = 0$.

3. For the following equations, obtain the particular solution satisfying the initial condition indicated. In each case draw a graph of the solution.

   (a) $\dfrac{dr}{dt} = -4rt$; when $t = 0, r = r_0$

   (b) $2xyy' = 1 + y^2$; when $x = 2, y = 3$

   (c) $2y\,dx = 3x\,dy$; when $x = 2, y = 1$

4. The solutions of the separable differential equation

$$y' = \frac{(A-x)y}{B+Cx}$$

† See G. H. Hardy, *A Mathematician's Apology*, 2nd ed. (Cambridge: Cambridge University Press, 1967), pp. 140–141.

give many important distributions of statistics for appropriate choices of the constants A, B, and C. Solve the differential equation in each case below; then sketch some solution curves. Find $y(x)$ such that its integral from 0 to ∞ [$-\infty$ as in (a)] equals 1.

(a) $C = 0$, $B > 0$, and A is arbitrary (normal distribution where A is the mean and B is the standard deviation)

(b) $A = B = 0$, and $C > 0$ (exponential distribution)

(c) $B = 0$, $C > 0$, and A/C is a nonnegative integer n (gamma distribution)

5. Consider the conventional combat model

$$x'(t) = -by(t) + R_1$$
$$y'(t) = -ax(t) + R_2$$

where R_1 and R_2 are positive constants.

(a) Since the system is coupled, eliminate time as the independent variable. Further, solve the newly obtained equation to show that $x(t)$ and $y(t)$ satisfy the equation

$$b\left(y - \frac{R_1}{b}\right)^2 - a\left(x - \frac{R_2}{a}\right)^2 = C$$

where C is a constant determined by the initial data.

(b) Take $a = b = 1$, $R_1 = 2$, and $R_2 = 3$ and sketch the curves in part (a) when (i) $x(t_0) = 3$, $y(t_0) = 1$; (ii) $x(t_0) = 2$, $y(t_0) = 2$; (iii) $x(t_0) = 4$, $y(t_0) = 2$.

(c) By considering the original differential system, assign arrowheads to the curves indicating the direction of increasing time.

(d) Who wins in each of the cases in part (b)?

(e) What happens in part (a) when $x(t_0) = R_2/a$ and $y(t_0) = R_1/b$? Explain in terms of the development of the combat.

6. [*Conventional versus Guerrilla Combat*]. Suppose that we modify the combat situation studied in this section so that one of the forces is a guerrilla force. Let $x(t)$ denote the strength of the conventional force and $y(t)$ the strength of the guerrilla force. Assume no reinforcements or noncombat losses. As given in the text, the loss rate inflicted on a conventional force by its opponent force is given by $x'(t) = -ay(t)$. Now the situation is very different for a guerrilla force. A guerrilla force is usually invisible to its opponent force and occupies a region R of fixed area.

(a) Argue why the combat loss rate for the guerrilla force y may be modeled by cxy.

(b) Given the combat equations

$$x'(t) = -ay(t)$$
$$y'(t) = -cx(t)y(t)$$

derive the "parabolic law"

$$cx^2 = 2ay + Q, \qquad \text{where } Q = cx_0^2 - 2ay_0$$

(c) Assume that $a > 0$, $c > 0$. Sketch qualitatively some typical curves for various values of c. Indicate with arrows the direction of increasing time on each curve.

(d) Who wins when $Q > 0$? $Q = 0$? $Q < 0$?

(e) Show that in order for the conventional force to win, $cx_0^2 > 2ay_0$.

7. Find all functions $f(t)$ such that the area of the region bounded by the lines $x = 0$,

$x = t$, $y = 0$, and $y = f(t)$ is equal to $1/f^2(t)$. [*Hint*: $\int_0^t f(x)\,dx = 1/f^2(t)$). Differentiate with respect to t to find a differential equation for $f(t)$.]

4.2 CHANGE OF VARIABLES: PURSUIT MODELS

Some first-order differential equations can be transformed by a change of variable into separable equations for which the methods of the preceding section apply. We shall show just how this is done for one such class of differential equations which arise in pursuit problems and then apply the technique to find the path of a bird in a crosswind.

Change of Variables

What happens to a differential equation when the dependent variable is changed? The immediate result is to obtain a new differential equation in a new variable, which is hardly surprising. For example, suppose that we begin with a first-order differential equation written in normal form

$$y'(x) = f(x, y(x)) \tag{1}$$

where y is the dependent variable. Introduce a new dependent variable $z = z(x)$ by the equation

$$y(x) = G(x, z(x)) \tag{2}$$

where, say, G is a given continuously differentiable function of x and z in some region A of the xz-plane. To replace (1) by a differential equation in z and x, instead of y and x, we need to express $y'(x)$ in terms of z. To do this, differentiate each side of (2) with respect to x, using the Chain Rule for differentiating a function of several variables:

$$y' = \frac{\partial G}{\partial x} + \frac{\partial G}{\partial z} z'$$

Then Eq. (1) becomes

$$\frac{\partial G(x, z)}{\partial x} + \frac{\partial G(x, z)}{\partial z} z' = f(x, G(x, z))$$

Rearranging, we have that

$$z' = \frac{f(x, G(x, z)) - \partial G(x, z)/\partial x}{\partial G(x, z)/\partial z} \tag{3}$$

This is our transformed differential equation in z instead of y. To avoid trouble with division by zero, we assume that $\partial G/\partial z$ never vanishes in A. If $z(x)$ is a solution of (3), reversing the steps and using (2), we see that $y(x) \equiv G(x, z(x))$ is a solution of (1).

The new differential equation (3) does not look any simpler than the old (1). In fact, it looks, and often is, more complicated. But in a handful of critically important cases, (3) does, in fact, turn out to be simpler than (1), so much simpler that solutions of (3) can be found directly by known techniques. The reader should not memorize (3), but make any change of variable directly.

Example 4.5 (*Bernoulli and Riccati Equations*)

As we showed in the problem set of Section 3.3, the Bernoulli equation, $y' + py = qy^a$, becomes to the linear equation $z' + (1 - a)pz = (1 - a)q$ by the change of variable $z = y^{1-a}$ (i.e., $y = z^{1/(1-a)}$). Similarly, the Riccati equation, $y' = ay + by^2 + F$, is transformed to the linear equation $z' + (a + 2bg)z = -b$ by the change of variable $y = z^{-1} + g(t)$, where $g(t)$ is one solution of the Riccati equation.

Other changes of variable simplify other equations. Consider, for example, Eq. (1) where $f(x, y)$ has the following property:

> **Homogeneous Function of Order One.**   A continuous function $f(x, y)$ defined on $\mathbb{R}^2$ is said to be *homogeneous of order 0* if for any $(x, y) \neq (0, 0)$ and any $k \neq 0$ we have $f(kx, ky) = f(x, y)$.

For example, $f(x, y) = (x^2 + y^2)^{1/2}/(x + y)$ is homogeneous of order 0, but $f(x, y) = (x + 1)/(x + y)$ is not.

If $f(x, y)$ is homogeneous of order 0, the change of dependent variable $y = xz$ converts the homogeneous equation (1) into a separable equation. Indeed we have that

$$xz' + z = y' = f(x, xz) = f(x \cdot 1, x \cdot z) = f(1, z)$$

since f is homogeneous of order 0 (x is playing the role of k in this setting). Rearranging this differential equation, we have the separable equation

$$\frac{1}{f(1, z) - z} z' = \frac{1}{x} \tag{4}$$

Equation (4) can be solved by the method outlined in the preceding section:

$$\int^z \frac{1}{f(1, s) - s} \, ds = \int^x \frac{1}{t} \, dt + C = \ln |x| + C$$

where C is a constant of integration. Let $F(z)$ be an antiderivative of $[f(1, z) - z]^{-1}$. We have that $F(z) = \ln |x| + C$, and since $z = y/x$,

$$F\left(\frac{y}{x}\right) = \ln |x| + C \tag{5}$$

Thus a solution $y(x)$ of the first-order equation (1) with $f(x, y)$ homogeneous of order 0 satisfies (5) on some x-interval for some choice of C.

This leaves open two questions: How can the antiderivative F be found, and how can (5) be solved for y in terms of x? These are the same questions that were posed in the preceding section. For some equations, F can be expressed in terms of elementary functions, for others it cannot. When F is known, sometimes (5) can easily be solved for y in terms of x, but often this is not possible. The following examples illuminate some of the advantages and limitations of the method of change of variables for equations of this kind.

Example 4.6

The function $f(x, y) = (x^2 + y^2)/2xy$ is homogeneous of order 0. Hence to solve

$$y' = \frac{x^2 + y^2}{2xy}, \qquad x \neq 0, \quad y \neq 0 \tag{6}$$

we introduce the new variable z by setting $y = zx$ to obtain the equation

$$xz' + z = \frac{x^2 + x^2z^2}{2zx^2} = \frac{1 + z^2}{2z}$$

Separating the variables, we have

$$\frac{2z}{1 - z^2} z' = \frac{1}{x}, \qquad z \neq \pm 1$$

Antidifferentiating, we have

$$-\ln|1 - z^2| = \ln|x| + C, \qquad z \neq \pm 1$$

where C is the constant of integration. Rearranging and returning to the original variables y and x by setting $z = y/x$, we have

$$\ln|x| + \ln\left|1 - \frac{y^2}{x^2}\right| = \ln\left|x\left(1 - \frac{y^2}{x^2}\right)\right| = -C, \qquad y \neq \pm x$$

Exponentiating, we have

$$\left|x\left(1 - \frac{y^2}{x^2}\right)\right| = e^{-C}, \qquad y \neq \pm x \tag{7}$$

By direct substitution into (6) we see that $y = \pm x$ are each solutions of (6) on any interval not containing $x = 0$. Taking this fact into account, we can remove the absolute value signs from (7), replace e^{-C} by a constant K, which can have any real value, and obtain, after multiplying by x, the following equation defining solutions of (6):

$$x^2 - y^2 = Kx, \qquad x \neq 0, \qquad y \neq 0 \tag{8}$$

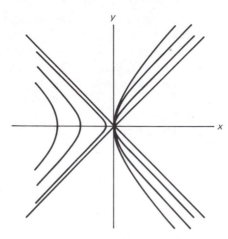

Figure 4.6 Graphs of $x^2 - y^2 = Kx$, $K \leq 0$.

Equation (8) defines a hyperbola for each K. See Figure 4.6 for some integral curves. Observe that along $y = 0$ (but $x \neq 0$), we might as well assume that the solution curves have infinite slope.

Flight Paths

A bird attempts to fly back to its roost, but a constant crosswind is blowing. A ferryboat is to sail directly across a river, but there is a strong crosscurrent. Different phenomena, but the same problem: What course is followed?

We shall formulate everything in terms of a flying bird, but the analysis applies equally well to the ferryboat. Suppose that the wind is out of the south at constant speed w, while the bird's speed in still air is the constant b (see Figure 4.7). An eagle keeps a constant heading somewhat to the southwest, so that the net result of its own efforts and the force of the wind would take it home along a straight path from A to the roost. A goose, on the other hand, keeps heading toward the roost, but, of course, the wind blows it off course. Can the goose ever get home? If it does, what is its path?

First, after a little reflection, we realize that neither bird has a chance of making it home at all unless the bird's speed b exceeds the speed w of the wind. Hence we assume that $b > w$. Suppose that the path (as yet unknown) of the goose is given by the parametric equations $x = x(t)$, $y = y(t)$, where t is time, $(x(t), y(t))$ is the location of the goose at time t, and $(x(0), y(0)) = A(a, 0)$ and $(x(T), y(T)) = (0, 0)$ (see Figure 4.8). Time T is the time of arrival at the roost, but, of course, it is not known. Let θ denote the angle of the heading of the goose as indicated in Figure 4.8; θ will change in time as the bird is blown off course.

The rate of change of $x(t)$ is the component of the goose's velocity in the x direction,

$$\frac{dx(t)}{dt} = -b \cos \theta = \frac{-bx}{(x^2 + y^2)^{1/2}}$$

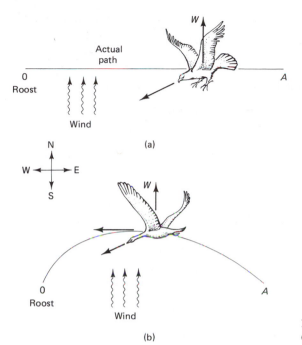

(a)

(b)

Figure 4.7   Bird, wind, and flight paths:
(a) path of the eagle; (b) path of a goose.

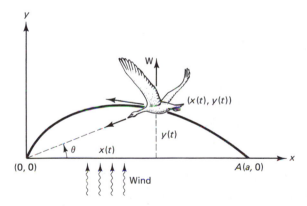

Figure 4.8   Path of the goose.

The rate of change of $y(t)$ is obtained similarly, except that the wind's effect must
be included:

$$\frac{dy(t)}{dt} = -b \sin \theta + w = \frac{-by}{(x^2 + y^2)^{1/2}} + w$$

Now if we divide these two equations, we have a first-order equation in x and y,
where the time t does not appear at all:

$$\frac{dy}{dx} = \frac{by - w(x^2 + y^2)^{1/2}}{bx} \equiv f(x, y)$$

This equation is homogeneous, even though it may not appear to be, for

$$f(kx, ky) = \frac{bkx - w(k^2x^2 + k^2y^2)^{1/2}}{bkx} = \frac{bx - w(x^2 + y^2)^{1/2}}{bx} = f(x, y)$$

Now we set $y = zx$ and obtain (with $v = w/b$ for simplicity)

$$\frac{dy}{dx} = \frac{dz}{dx}x + z = \frac{zx - v(x^2 + z^2x^2)^{1/2}}{x} = z - v(1 + z^2)^{1/2}$$

Thus $(dz/dx)x = -v(1 + z^2)^{1/2}$ and we have the separable equation

$$\frac{1}{(1 + z^2)^{1/2}}\frac{dz}{dx} = \frac{-v}{x}$$

with solutions defined by $\ln [z + (1 + z^2)^{1/2}] = -v \ln x + C$, where C is a constant of integration. We do not need absolute value signs inside the logarithms since $x > 0$ and $z > 0$. $C = v \ln a$ since $z = y = 0$ when $x = a$. Hence $\ln [z + (1 + z^2)^{1/2}] = -v \ln x + v \ln a = \ln (x/a)^{-v}$, and exponentiating,

$$z + (1 + z^2)^{1/2} = \left(\frac{x}{a}\right)^{-v}$$

Solving for z involves writing the equation above as $(1 + z^2)^{1/2} = (x/a)^{-v} - z$, squaring to obtain

$$1 + z^2 = \left(\frac{x}{a}\right)^{-2v} - 2\left(\frac{x}{a}\right)^{-v}z + z^2$$

and then solving for z:

$$z = \frac{1}{2}\left[\left(\frac{x}{a}\right)^{-v} - \left(\frac{x}{a}\right)^{v}\right]$$

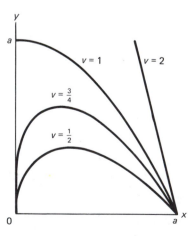

Figure 4.9 Flight paths of goose for various values of $v = w/b$.

Since $z = y/x$, we have that the equation of the path followed by the goose is

$$y = \frac{a}{2}\left[\left(\frac{x}{a}\right)^{1-v} - \left(\frac{x}{a}\right)^{1+v}\right]$$

In Figure 4.9 this path is plotted for each of several different values of $v = w/b$. When $v > 1$, the goose is gone with the wind, but if $v < 1$, the bird will eventually reach its roost.

Comments

If $f(x, y)$ is homogeneous of order 0 we can write $f(x, y)$ as $f(x \cdot 1, x \cdot y/x) = f(1, y/x)$ and hence see that the dominant variable in the equation $y' = f(x, y)$ is neither x nor y, but their ratio. This is why homogeneous equations often arise in models of pursuit, where it is the heading (i.e., y/x) that is the critical variable. Whether it is the path of a bird, a boat, a plane, a dog, or a bug (see the problem set), a homogeneous equation often describes the nature of that path. Other change-of-variables techniques appear in the problems.

PROBLEMS

The following problem(s) may be more challenging: 1(e), 1(f), 5, 6, 7, and 8.

1. Solve the following differential equations and sketch some solution or integral curves.
   (a) $\dfrac{y+x}{x} = y'$.
   (b) $(x - y)\,dx + (x - 4y)\,dy = 0$.
   (c) $(x^2 - xy + y^2)\,dx - xy\,dy = 0$.
   (d) $(x^2 - 2y^2)\,dx + (xy)\,dy = 0$.
   (e) $x^2 y' = 4x^2 + 7xy + 2y^2$.
   (f) $x \tan(y/x) + yy' = 0$.

2. Show that the differential equation

$$\frac{dy}{dx} = \frac{4y^2 - x^4}{4xy}$$

is nonseparable but becomes separable upon making the substitution $y = vx$. Use this to find the solution of the original equation.

3. (a) Let $f(x, y)$ be homogeneous of order 0. Introduce polar coordinates $x = r\cos\theta$ and $y = r\sin\theta$, and show that the differential equation $y' = f(x, y)$ in polar form is separable. [*Hint*: $dx = \cos\theta\,dr - r\sin\theta\,d\theta$, $dy = \sin\theta\,dr + r\cos\theta\,d\theta$.]
   (b) Use this technique to solve $y' = (-x^2 + y^2)/xy$.

4. As demonstrated by Problem 2, a nonseparable differential equation can become separable by changing the dependent variable. For each of the following cases, demonstrate this process and solve the resultant equation. Then find the solution of the original equation.
   (a) $\dfrac{dy}{dx} = \cos(x + y)$. [*Hint*: Let $z = x + y$.]

(b) $(2x + y + 1)\,dx + (4x + 2y + 3)\,dy = 0$. [*Hint*: Let $z = 2x + y$.]

(c) $(x + 2y - 1)\,dx + 3(x + 2y)\,dy = 0$.

(d) $e^{-y}(y' + 1) = xe^x$.

5. A dog at the center of a circular pond sees a duck at the edge of the pond and decides to catch it. The duck begins to swim at speed V around the circumference of the pond. The dog moves so as to keep its nose pointed at the duck at all times. The dog swims at speed $3V/4$. Find and sketch the path of the dog. Show that the dog does not catch the duck, but eventually circles the pond three-fourths of the way from the center to the edge. How can a smart dog catch the duck? [*Hint*: Use polar coordinates].

6. Four bugs are at the corners of a square table whose sides are each of length a. They all begin to move at the same instant, each walking at the same rate directly toward the bug on its right. Find and sketch the path of each bug. Do the bugs ever meet?

7. Set up and solve the flight path problem if for the goose the wind is from the southeast and of speed $w = b/2^{1/2}$. Sketch the path. [*Hint*: $x' = -b \cos \theta - b/2$, $y' = -b \sin \theta + b/2$. Use tables of integrals.]

8. A Coast Guard boat is hunting a smuggler's launch in a dense fog. The fog lifts momentarily and the guardsmen spot the smugglers 4 mi away. The Coast Guard boat has a top speed three times that of the smuggler's launch. What path does the Coast Guard boat follow? Explain your model and sketch the path. [*Hint*: Use polar coordinates after the first 3 mi (see the sketch). Assume both boats operate at top speed, the smugglers move away on a straight path of unknown direction after the sighting, and the Coast Guard initially heads directly toward the sighting spot.]

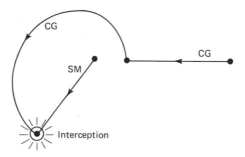

4.3 VECTORS: NEWTON'S LAWS OF MOTION

For a process of growth or decay, we have seen that the basic principle leading to a differential equation is the Balance Law: net rate of accumulation = rate in − rate out. We have made much use of this principle in constructing models for a variety of phenomena—radioactive decay, population growth, accumulation of salt. However, the principle will not help us to model the motion of a falling body, a vibrating spring, or the earth in its orbit around the sun. In these dynamical systems there is no change in amount of mass, no growth or decay, but there are changes in position and in velocity. A new principle is needed, a principle that relates these changes in the motion of a body to the action of its surroundings. Building on the work of Galileo, Isaac Newton† saw to the heart of the matter and formulated three fundamental laws of motion which relate the acceleration of a material body to the "mass" of the body and the "resultant force" acting on that body. To do this, Newton in effect introduced the *vector concept* as a modeling device in order to express his Laws of Dynamics with an elegant simplicity that transcends any particular frame of reference or coordinates in these frames.

The Vector Concept in Mechanics

In introductory physics courses, students are introduced to the geometric vector approach in mechanics where basic physical concepts such as "velocity" and "acceleration" are thought of as directed line segments, or arrows, in the familiar 3-space of Euclid. It is common to reinforce the idea that a symbol stands for a geometric vector by placing a small arrow over it, but for simplicity we shall drop the arrow and only use boldface letters. Two vectors **v** and **w** are considered identical if and only if they can be made to coincide by translations (which preserve length and direction of vectors). Thus two parallel vectors of equal length, but directed oppositely, are not identical. (Geometric vectors are sometimes called "free" because we do not need to have a new name for any vector that can be identified with another vector which already has a name.) The length of a vector **v** is denoted by $\|\mathbf{v}\|$. To be complete we define the *zero vector*, denoted by **0**, to be the vector of zero length (note that direction of **0** is irrelevant).

The velocity and acceleration vectors of a material body each have been observed to have the following property: If a body has been measured to have the velocities (or accelerations) **v** and **w** in two directions at the same time, then it has the resultant velocity (or acceleration), denoted by **v** + **w**, defined by the *parallelogram law* as follows. Find a vector equivalent to **v** whose "tail" coincides with **w**'s "head," then **v** + **w** is the vector whose "tail" is **w**'s and whose "head" is **v**'s (a diagonal of the parallelogram formed by **v** and **w**) (see Figure 4.10). This inspires the following definition:

> **Basic Operations for Geometric Vectors.**   Let **v** and **w** be any two geometric vectors and r any real number. The *sum* **v** + **w** is the vector produced by the parallelogram law. The *product* $r\mathbf{v}$ is the vector of length $|r|\,\|\mathbf{v}\|$ which points in the direction of **v** if $r > 0$, and in the opposite direction to **v** if $r < 0$.

Note that $r\mathbf{v} = \mathbf{0}$ if either $r = 0$, or **v** = **0**. Also note that $(+1)\mathbf{v} = \mathbf{v}$, and that $(-1)\mathbf{v}$ has the property that $\mathbf{v} + (-1)\mathbf{v} = \mathbf{0}$, for any **v**. The vector $(-1)\mathbf{v}$ is commonly denoted by $-\mathbf{v}$. Note that $-\mathbf{v}$ has the same length as **v** but points in the opposite direction. When a number of vectors are to be added together, the parallelogram

---

† Isaac Newton was born on Christmas Day, 1642 (the year Galileo died), and died in 1727. In addition to setting forth the basic laws of motion, Newton discovered the law of universal gravitation and invented the calculus (at the same time as, but independently of, Leibniz).

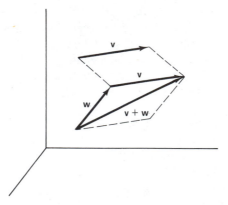

Figure 4.10 Parallelogram Law for vector addition.

law is used in succession—the order in which this is done can be shown to be immaterial (see Figure 4.11).

When geometric notions have served their purpose in constructing a model, the idea of coordinates for vectors is introduced to facilitate computation. A (*coordinate*) *frame* is a triple of vectors, denoted commonly by $\{i, j, k\}$, which are mutually orthogonal and all of unit length. Now a little thought (and some trigonometry) reveals that every vector can be uniquely written as the sum of vectors parallel to **i**, **j**, and **k**. Thus for each vector **v** there is a unique set of real numbers v_1, v_2, and v_3 such that $\mathbf{v} = v_1\mathbf{i} + v_2\mathbf{j} + v_3\mathbf{k}$. The elements of the ordered triple (v_1, v_2, v_3) are called the *coordinates* (or *components*) of **v** in the frame $\{i, j, k\}$ (see Figure 4.12). To put together the vector **v** from its coordinates in a given frame, one must have a way of associating each coordinate with the appropriate frame vector. This is normally done by writing coordinates in the same order of occurrence as vectors in the frame $\{i, j, k\}$.

Other computations with vectors are defined geometrically as follows:

1. Two nonzero, nonparallel vectors **u** and **v** form a plane. In this plane two vectors equivalent to **u** and **v** can be found which have coincident "tails." The angle less than or equal to 180° between these vectors is said to be the *angle*

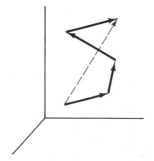

Figure 4.11 Addition of several vectors.

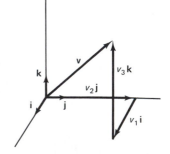

Figure 4.12 Resolution of a vector as a sum.

between **u** and **v**. The angle between nonzero parallel vectors is either 0° or 180°.

2. The *dot product* of two vectors **u** and **v**, denoted by **u** · **v**, is the real number $\|\mathbf{u}\|\ \|\mathbf{v}\|\ \cos\theta$ if neither **u** nor **v** is the zero vector and θ is the angle between them. When either **u** or **v** is the zero vector, their dot product is defined to be zero. Two vectors are orthogonal if and only if their dot product is zero.†

3. If the vector $\mathbf{u} = \mathbf{u}(t)$ depends on t in an interval of the real axis, then the *derivative $d\mathbf{u}/dt$* (or $\mathbf{u}'(t)$) is defined in the usual way as the limit of a difference quotient: namely,

$$\mathbf{u}'(t) = \frac{d\mathbf{u}}{dt} = \lim_{h \to 0} \frac{\mathbf{u}(t+h) - \mathbf{u}(t)}{h} \tag{1}$$

where the limit exists if the difference between the vectors $d\mathbf{u}/dt$ and $[\mathbf{u}(t+h) - \mathbf{u}(t)]/h$ has length which tends to zero as $h \to 0$. In particular, if $\mathbf{u}(t)$ and $\mathbf{v}(t)$ are two differentiable vector functions and $r(t)$ is a real-valued differentiable function, then we have the identities

$$(\mathbf{u} \cdot \mathbf{v})' = \mathbf{u}' \cdot \mathbf{v} + \mathbf{u} \cdot \mathbf{v}', \qquad (r(t)\mathbf{u}(t))' = r'\mathbf{u} + r\mathbf{u}'$$

very reminiscent of the usual product differentiation rule. If **u** is a constant vector, note that $\mathbf{u}' = 0$.‡

There are many frames of reference in the 3-space of our experience. One can imagine frames that move in space or frames that are fixed. Suppose it is known that {**i**, **j**, **k**} is a fixed frame and suppose that a particle moves in a manner described by the *position vector*

$$\mathbf{R} = \mathbf{R}(t) = x(t)\mathbf{i} + y(t)\mathbf{j} + z(t)\mathbf{k}$$

where the *xyz*-coordinates are defined relative to the fixed frame (see Figure 4.13). If **R** is differentiable, it follows from (1) that

$$\mathbf{R}'(t) = x'(t)\mathbf{i} + y'(t)\mathbf{j} + z'(t)\mathbf{k} \tag{2}$$

where the primes indicate time derivatives. It is shown in elementary physics that $\mathbf{R}'(t) = \mathbf{v}(t)$ is the velocity vector of the particle at time t, and that $\mathbf{v}(t)$ is tangential to the path of the particle's motion at the point $\mathbf{R}(t)$. Furthermore, if $\mathbf{R}'(t)$ is differentiable,

$$\mathbf{R}''(t) = \mathbf{v}'(t) = \mathbf{a}(t) = x''(t)\mathbf{i} + y''(t)\mathbf{j} + z''(t)\mathbf{k} \tag{3}$$

is the acceleration vector for the body's center of mass (see Figure 4.13). (Computation of the velocity and acceleration vectors in given coordinate frames when the path of the motion is known is called *kinematics*.)

---

† See the problem set for properties satisfied by the dot products.

‡ In the physics and engineering literature it is more conventional to use the "dot" notation, **u̇**, instead of the "prime" notation, **u'**, to denote the derivative of the vector function of time, **u**(t). For simplicity, however, we will always use the prime notation for derivatives of a single variable function.

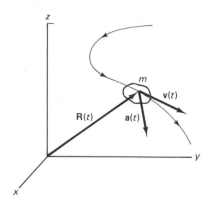

Figure 4.13   Kinematics of motion.

Forces, Newton's Laws

Newton carried further Galileo's central idea that the environment creates "forces" which act on bodies causing them to accelerate. Galileo's principle (stated also by Newton as his *First Law*) states that

> **Newton's First Law.**   A body remains in a state of rest or of uniform motion in a straight line if there are no external forces acting on it.

Now forces can be measured (or computed, as we shall see) without taking into account frames of reference. Scientists of Newton's time were able to show experimentally that forces behave like geometric vectors (i.e., that they satisfy the parallelogram law).

The effect of Newton's First Law is to identify frames of reference which are either fixed in space or undergoing a translation at constant velocity with respect to a fixed frame. Such frames are called *inertial* and they are extremely important in the modeling of moving bodies. Practically speaking, how do we know when we are dealing with an inertial frame? According to Newton's First Law, a frame is inertial if and only if a body is unaccelerated with respect to that frame whenever the resultant (sum) force acting on the body vanishes.

Next we have the basic and central principle in dynamics,

> **Newton's Second Law.**   For a body in any inertial frame we have that
>
> $$m\mathbf{a} = \mathbf{F} \tag{4}$$
>
> where m denotes the mass (assumed to be constant), $\mathbf{a}$ is the acceleration of the body, and $\mathbf{F}$ is the net resultant of all external forces acting on the body, where these quantities are measured in a consistent system of units† (see Figure 4.14).

† This form (4) of Newton's Second Law is for so-called "point" masses. For some "distributed" masses it is permissible to apply (4) as if all the mass of the body were concentrated at its center of mass.

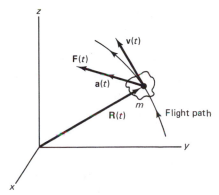

Figure 4.14 Geometry of Newton's Second Law for constant mass.

The terms in Newton's Second Law must be quantified in some system of units. In the *mks system* (meter, kilogram, second) the unit of mass is 1 kilogram; (4) is then used to define a unit of force as that which will accelerate a 1-kg mass at 1 m/s². This unit of force is called a *newton*. In the *cgs system* (centimeter, gram, second) the unit of force is a *dyne* (1 dyne = 10⁻⁵ newton) and is that which accelerates a 1-g mass at 1 c/s². In these metric units, mass, length, and time are considered to be fundamental quantities, while force is secondary. In this text we avoid the use of any particular system of units as much as possible. When units must be used, they will be explicitly labeled.

The Third Law of Newton gives us some insight as to how we may treat multiple body systems.

Newton's Third Law.   If body A exerts a force $\mathbf{F}$ on body B, then body B exerts a force $-\mathbf{F}$ on body A.

Since acceleration is defined to be the rate of change of velocity, Newton's Second Law can be written in the form

$$m\,\mathbf{a} = m\,\frac{d\mathbf{v}}{dt} = \frac{d(m\mathbf{v})}{dt} = \mathbf{F} \tag{5}$$

where v is the velocity of the body and, once again, the mass is assumed to be constant. The product mv is called the *momentum* of the body. Newton himself expressed the Second Law in terms of the rate of change of the "quantity of motion," his term for momentum. Thus Newton's Second Law states that "the rate of change of the momentum of a body equals the resultant external force acting on the body." To give this simply stated principle some operational content, we would need to specify ways of calculating the "momentum" of a body and the "external force" acting on a body in an inertial frame of reference. These techniques form the central core of the field known as dynamics, and we cannot do more here than give the barest of introductions to this, the oldest branch of physics. As we shall see, a major consideration in the application of Newton's Second Law is the choice of a frame of reference and a coordinate system within that frame.

Nothing can be done to solve the first-order differential equation (5) and find the velocity and, eventually, the position of a body as functions of time until some functional form is given to **F**. This, of course, depends on the nature of the forces acting on the body—gravitational or electromagnetic forces, for example. Many other forces, such as spring forces, tensions in cables, and the normal force exerted by surfaces in contact with a body, are actually electromagnetic forces at a fundamental level. Nuclear forces, on the other hand, are so short-ranged that they cannot be used in Newton's Equations (quantum mechanics must be used for these forces).

The examples considered below will give the reader a general idea of how Newton's Second Law is applied.

Gravitation

Using observations of falling bodies and the extensive astronomical work and empirical laws of Tycho Brahe and Johannes Kepler concerning the orbits of the moon and of the planets, Newton focused attention on just one force, gravity. His law of universal gravitation has to do with the gravitational effect of one body upon another:

> **Newton's Law of Universal Gravitation.**   The force **F** between any two particles having masses m_1 and m_2 and separated by a distance r is attractive, acts along the line joining the particles (i.e., tends to pull the particles together), and has magnitude
>
> $$|F| = G \frac{m_1 m_2}{r^2} \tag{6}$$
>
> where G is a universal constant independent of the nature and masses of the particles. In the mks system, $G = 6.67 \times 10^{-11}$ n-m²/kg².

Newton also showed that massive bodies affect one another as if the mass of each body were concentrated at its "center of mass," if the masses of those bodies are distributed in a spherically symmetric way (it is not true in general). In this case, r is the distance between the centers of mass.

A small object (e.g., a ball of lead) is dropped and falls toward the surface of the earth. How do its velocity and its location change with time? We shall suppose that the only significant forces acting on the body are the gravitational attraction of the earth and the resistive force, if any, of the medium through which the body falls. We shall fix a z-coordinate line on the surface of the earth directly below the body and pointing upward. A positive value of z indicates a location above the surface of the earth. A negative value of $v = dz/dt$ means that the body is moving downward in the direction of the negative z-axis. A negative value of a force acting on the body means that the force acts in the direction of the negative z-axis, and so on. We shall suppose that at time $t = 0$ the body is at height $z(0) = h$ and has velocity v_0. If the body is falling, then $v_0 < 0$. In calculating the gravitational attraction of the earth on this body, we imagine that the mass of the earth is concentrated at its center. This means that the earth attracts the body with a force whose magnitude is

$$|F| = \frac{GMm}{(R+z)^2}$$

where M is the mass of the earth, m the mass of the falling body, R the radius of the earth, and z the distance from the earth's surface to the center of mass of the falling body. During the course of the fall the factor $GM/(R+z)^2$ changes from $GM/(R+h)^2$ to GM/R^2. For motion near the surface of the earth with, say, h no more than $(0.001)R$, the relative change in this gravitational coefficient is

$$\frac{GM/R^2 - GM/(R+h)^2}{GM/(R+h)^2} = \frac{(R+h)^2}{R^2} - 1 = \left(1 + \frac{h}{R}\right)^2 - 1$$

$$= \frac{2h}{R} + \left(\frac{h}{R}\right)^2 \leq 0.002 + 0.000001 = 0.002001$$

We shall ignore this small variation and set the gravitational coefficient equal to the constant $g = GM/R^2$. The gravitational force acting on the body is then $-mg$, where $g \approx 9.8 \ m/s^2$ in the mks system.

Comment

In the next section and Chapter 5 we use Newton's Laws to find the motion of falling bodies, pendula, and mass–spring systems.

PROBLEMS

The following problem(s) may be more challenging: 2(f).

1. [*Coordinate Version of Dot Product*]. Let $\{i, j, k\}$ be a given (fixed) frame in Euclidean 3-space. Then every vector u can be written uniquely as a sum $u_1 i + u_2 j + u_3 k$, where the scalars are known as the *coordinates* of u in the given frame. Care should be taken to use the subscript 1 always to denote i-coordinates, the subscript 2 for j-coordinates, and so on. Thus, when coordinates are listed as an ordered triple (u_1, u_2, u_3), the vector u can be recovered from its coordinates.
   (a) Show that $\|u\|^2 = u_1^2 + u_2^2 + u_3^2$, for any vector u.
   (b) Show that $u \cdot v = u_1 v_1 + u_2 v_2 + u_3 v_3$, for any vectors u, and v. [*Hint*: Imagine u and v to have their tails at the origin (as in the diagram), then use the Cosine Law, the definition of $u \cdot v$, and part (a).]

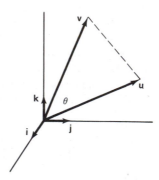

[*Remark*: It is worth nothing that parts (a) and (b) can be used to calculate $\|\mathbf{u}\|$ or $\mathbf{u} \cdot \mathbf{v}$ in *any* reference frame.]

2. [*Properties of Dot Product*]. Verify that the dot product $\mathbf{u} \cdot \mathbf{v}$ satisfies the following properties. [*Hint*: Use the geometric definition of $\mathbf{u} \cdot \mathbf{v}$ or the coordinate version in Problem 1].

   (a) [*Symmetry*] $\mathbf{u} \cdot \mathbf{v} = \mathbf{v} \cdot \mathbf{u}$ for all $\mathbf{u}, \mathbf{v}$.

   (b) [*Bilinearity*] $(\alpha\mathbf{u} + \beta\mathbf{w}) \cdot \mathbf{v} = \alpha\mathbf{u} \cdot \mathbf{v} + \beta\mathbf{w} \cdot \mathbf{v}$ for all scalars α, β and $\mathbf{u}, \mathbf{v}$, and $\mathbf{w}$.

   (c) [*Positive Definiteness*] $\mathbf{u} \cdot \mathbf{u} \geqq 0$ for all $\mathbf{u}$, and $\mathbf{u} \cdot \mathbf{u} = 0$ if and only if $\mathbf{u} = 0$.

   (d) $\mathbf{u} \cdot \mathbf{u} = \|\mathbf{u}\|^2$ for all $\mathbf{u}$.

   (e) [*Parallelogram Property of Norms*] $\|\mathbf{u} + \mathbf{v}\|^2 + \|\mathbf{u} - \mathbf{v}\|^2 = 2\|\mathbf{u}\|^2 + 2\|\mathbf{v}\|^2$ for all $\mathbf{u}, \mathbf{v}$. Why is this called the Parallelogram Property?

   (f) [*Cauchy–Schwartz Inequality*] $|\mathbf{u} \cdot \mathbf{v}| \leq \|\mathbf{u}\| \, \|\mathbf{v}\|$ for all vectors $\mathbf{u}$ and $\mathbf{v}$.

   (g) $\mathbf{u} \cdot \mathbf{v} = \frac{1}{4}\{\|\mathbf{u} + \mathbf{v}\|^2 - \|\mathbf{u} - \mathbf{v}\|^2\}$ for all $\mathbf{u}, \mathbf{v}$.

   (h) Show that if $\mathbf{u} + \mathbf{v} + \mathbf{w} = 0$, but each vector is orthogonal to the other two, then $\mathbf{u} = \mathbf{v} = \mathbf{w} = 0$.

3. Let $\{\mathbf{i}, \mathbf{j}, \mathbf{k}\}$ be a reference frame in Euclidean 3-space. If $\mathbf{u} = u_1\mathbf{i} + u_2\mathbf{j} + u_3\mathbf{k}$, and $\mathbf{v} = v_1\mathbf{i} + v_2\mathbf{j} + v_3\mathbf{k}$, are given non-zero vectors, find all vectors $\mathbf{w}$ orthogonal to both $\mathbf{u}$ and $\mathbf{v}$.

4. Let $\{\mathbf{i}, \mathbf{j}, \mathbf{k}\}$ be a reference frame, and consider the two vectors

$$\mathbf{u} = 3\mathbf{i} - \mathbf{j} + 2\mathbf{k}, \qquad \mathbf{v} = 2\mathbf{i} + \mathbf{j} - \mathbf{k}$$

   (a) Determine the conditions on the coordinates of $\mathbf{w} = w_1\mathbf{i} + w_2\mathbf{j} + w_3\mathbf{k}$ which guarantee that $\mathbf{w}$ is orthogonal to $\mathbf{u}$, or that $\mathbf{w}$ lies in the plane determined by $\mathbf{u}$ and $\mathbf{v}$.

   (b) Find a scalar α and a vector $\mathbf{w}$ such that $\mathbf{u} = \alpha\mathbf{v} + \mathbf{w}$, where $\mathbf{w}$ is orthogonal to $\mathbf{v}$. Are α and $\mathbf{w}$ unique?

5. (a) Verify that the following set of vectors $\{\mathbf{i}', \mathbf{j}', \mathbf{k}'\}$ is a reference frame.

$$\mathbf{i}' = \frac{\sqrt{2}}{4}\mathbf{i} + \frac{\sqrt{6}}{4}\mathbf{j} + \frac{\sqrt{2}}{2}\mathbf{k}$$

$$\mathbf{j}' = \frac{\sqrt{2}}{4}\mathbf{i} + \frac{\sqrt{6}}{4}\mathbf{j} - \frac{\sqrt{2}}{2}\mathbf{k}$$

$$\mathbf{k}' = -\frac{\sqrt{3}}{2}\mathbf{i} + \frac{1}{2}\mathbf{j}$$

(b) Find the coordinates of the vector $\mathbf{u} = 3\mathbf{i} - \mathbf{j} + \mathbf{k}$ in the frame $\{\mathbf{i}', \mathbf{j}', \mathbf{k}'\}$ of part (a).

6. A small airplane, during a part of its flight that begins at a point P in space, flies 20 mi due south, then turns left 90° and immediately goes into a climb 8 mi long at angle of 10° with the horizontal, and finally turns left and flies horizontally 42 mi due north. What is the final position of the airplane relative to P?

7. At a certain moment an aircraft is observed by radar to be 80 mi away at 60°10′ east of north and 83°40′ from the local vertical. Find the northward, eastward, and vertical components of its positions.

8. According to the Ideal Gas Law, the pressure P, volume V, temperature T, and the number of moles (1 mol = 6.02×10^{23} molecules) n of a gas in an enclosed container are related according to the equation $PV = nRT$, where R is a universal constant. Now suppose that we have a cylinder containing an ideal gas with a piston of mass m on top (see the diagram). Assuming that the only forces acting on the piston are gravity and gas pressure: Find the differential equation for the position of the piston measured from the bottom of the cylinder. Assume that $T = $ constant and

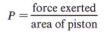

$$P = \frac{\text{force exerted}}{\text{area of piston}}$$

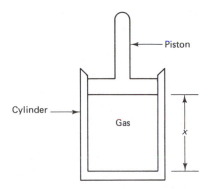

9. Two particles of masses $m_1 = 10$ gm and $m_2 = 40$ gm are located in a rectangular coordinate system. The location of particle m_1 is (4 cm, 0, 0) and the location of particle m_2 is (−4 cm, 0, 0).

   (a) Find the rectangular components of the total gravitational force exerted by the two particles on a 1.0 gm particle placed at the point (−5 cm, 0, 0).

   (b) At the point (0, 0, 6 cm).

   (c) At what point on the x-axis is the force on a 1.0 gm particle zero?

4.4 REDUCTION METHODS: FALLING BODIES, ESCAPE VELOCITY

Observation convinces us that objects in free fall near the earth's surface move along the local vertical. We saw in Section 4.3 that the gravitational force on a body of mass m near the earth's surface has the magnitude mg, where g is a constant, directed downward along the local vertical. If this were the only force acting on the body, Newton's Second Law implies that the height $z(t)$ of the body above the earth's surface at time t satisfies the differential equation $mz'' = -mg$. Thus if the earth had no atmosphere, then, as we saw in Section 1.2, all objects would move in the same way regardless of their size, shape, or mass. But the resistive effect our atmosphere has on moving objects makes a vast difference in the way objects fall, a fact obvious even to the casual observer. We construct below a model that describes this phenomenon to an acceptable degree of accuracy. The differential equations that arise in the model involve derivatives up to second order, but we show how they can be solved by a change of variable that reduces them to first-order equations.

Feathers and Snowflakes

When a body falls toward the surface of the earth through a vacuum, the velocity steadily increases without limit until the time of impact (see Section 1.2). The velocity of a body falling through a resistive medium, however, tends to a finite limit. Let us show this when that resistive medium is our atmosphere. Experiments show that for a body of low density and extended rough surface (e.g., a feather or a snowflake) the resistance of the air exerts a force on the body proportional to the magnitude of the velocity but, of course, acting opposite to the direction of motion. Let y be distance measured along the local vertical, with down as the positive direction, and let $y' \equiv v$ be the velocity of the object.† Thus if the body is falling, the force of resistance has magnitude kv, where k is a positive constant of proportionality, and the resistive force will act upward, opposing the fall. Similarly, if the body were rising through the air, the resistance would have magnitude $-kv$ and would act in a downward direction (see Figure 4.15).

Suppose that the object has mass m and is released from rest at a height h above the earth's surface. Then by Newton's Second Law the motion of the object solves the initial value problem

$$my'' = mg - kv, \qquad y(0) = -h, \quad v(0) = 0 \tag{1}$$

where g is the gravitational constant. Replacing y'' by v', the differential equation in (1) becomes first order and linear in v and v',

$$v' + \left(\frac{k}{m}\right)v = g, \qquad v(0) = 0$$

which can be solved via the integrating factor technique (see Section 3.1) to obtain

$$v(t) = \frac{mg}{k}\left(1 - e^{-kt/m}\right) \tag{2}$$

---

† Note that y points *down* toward the center of the earth, whereas z in Section 4.3 points *up* away from the center.

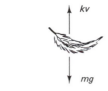

$$y = 0 \rule{3cm}{0.4pt} \text{Surface of earth}$$

$$y$$

Figure 4.15　Feather falling through air.

Naturally, this model ceases to have any validity beyond the time of impact, T. From (2) we see that as t tends to infinity, $v(t)$ approaches the limiting velocity of mg/k. Although the falling body never quite reaches the limiting velocity before it hits the surface, its velocity quickly approaches and remains very near this limiting value. This is evident as one observes the slow fall of a feather or a snowflake at an apparently constant speed.

　　To determine $y(t)$, we replace $v(t)$ by $y'(t)$ in (1) to obtain the first-order equation

$$y' = \frac{mg}{k}(1 - e^{-kt/m}), \qquad y(0) = -h$$

whose solution is found by antidifferentiation to be

$$y(t) = -h + \frac{mg}{k}t - \frac{m^2g}{k^2}(1 - e^{-kt/m}), \qquad 0 \le t \le T$$

Observe that the exact calculation of the moment of impact, T, is difficult since it would require solving the transcendental equation

$$0 = -h + \frac{mg}{k}T - \frac{m^2g}{k^2}(1 - e^{-kT/m})$$

for T in terms of the other parameters.

Raindrops and Baseballs

When a dense body falls through air (e.g., a raindrop, baseball, or bullet), the resistive force of the air is no longer proportional to the velocity but to its square. This is known as *Newtonian damping* (see Figure 4.16). Using down as the positive direction again, we thus have the initial value problem

$$my'' = mg \pm kv^2, \qquad y(0) = -h, \quad v(0) = 0 \tag{3}$$

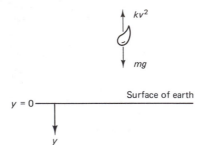

Figure 4.16 Falling raindrop.

where the negative sign is used if the body is falling, while the positive sign applies when the body is rising. We shall restrict our analysis to the case of a falling body and use (3) only with the minus sign.

Replacing y'' by v' in (3), we obtain the first-order initial value problem

$$v' = g - \left(\frac{k}{m}\right) v^2, \qquad v(0) = 0$$

This differential equation is separable and can be solved using the techniques of Section 4.1. Indeed, if the variables are separated, we would have

$$\int_0^v \frac{1}{g - (k/m)s^2} \, ds = \int_0^t dr = t$$

Using partial fractions or a table of integrals to carry out the left-hand integration, we have that

$$\frac{1}{2} \left(\frac{m}{gk}\right)^{1/2} \ln \left| \frac{(mg/k)^{1/2} + v}{(mg/k)^{1/2} - v} \right| = t \tag{4}$$

Solving (4) for v in terms of t by taking exponentials, and then doing some algebraic manipulation, we have that

$$v(t) = \left(\frac{mg}{k}\right)^{1/2} \frac{1 - e^{-At}}{1 + e^{-At}} \tag{5}$$

where $A = 2(gk/m)^{1/2}$. Observe that as $t \to \infty$, $v(t) \to (mg/k)^{1/2}$, but this limiting velocity is not attained because the body hits the ground.

To find y as a function of t, we replace $v(t)$ by $y'(t)$ in (5) and solve the first-order initial value problem

$$y' = \left(\frac{mg}{k}\right)^{1/2} \frac{1 - e^{-At}}{1 + e^{-At}}, \qquad y(0) = -h \tag{6}$$

Observe that

$$\frac{1 - e^{-At}}{1 + e^{-At}} = \frac{e^{At/2} - e^{-At/2}}{e^{At/2} + e^{-At/2}} = \frac{\sinh At/2}{\cosh At/2} = \tanh \frac{At}{2}$$

and thus (6) has the unique solution

$$y(t) + h = \left(\frac{mg}{k}\right)^{1/2} \int_0^t \tanh\left(\tfrac{1}{2} As\right) ds$$

or

$$y(t) = -h + \frac{m}{k} \ln \cosh\left[\left(\frac{gk}{m}\right)^{1/2} t\right] \tag{7}$$

where we have used a table of integrals and the fact that $\cosh 0 = 1$. We must, of course, restrict t to the interval $0 \leq t \leq T$, where T (the time of impact) can be calculated from (7) by setting $y = 0$. Although the height of the falling raindrop above the surface of the earth at any time t before impact is given by (7), the formula is not particularly simple and we shall carry our analysis no further.

Validation of Models for Falling Bodies

Why does a feather fall to earth more slowly than a large raindrop? For the feather we had shown above that the limiting velocity is mg/k_f, while for the raindrop it is $(mg/k_r)^{1/2}$, where k_r and k_f are the k-values in the modes corresponding to these phenomena. If the raindrop and the feather have the same mass m, it must be that the differences in the limiting velocities are due solely to the somewhat amorphous constants k_f and k_r. In fact, a little algebra will show that

$$|v_f| < |v_r| \quad \text{if and only if} \quad k_r mg < k_f^2$$

One can argue about the validity of the last inequality in terms of the larger cross-sectional area of a feather and hence the larger k_f. However, the k-factor depends on too many other aspects of the falling body for us to be able to push the discussion much further.

The phenomenon of a free-falling human being (e.g., a parachutist before the chute opens) has been studied in great detail. For this case of a relatively dense falling body, Newtonian damping provides a good model. If m is approximately 120 kg (e.g., a parachutist with equipment), then k turns out to be approximately 0.1838 kg/m and the limiting velocity has magnitude approximately 80 m/s. This value has been compared with the actual limiting velocity of free-falling parachutists, and remarkably good agreement has been observed.

Reduction of Order, I

As we saw above, the second-order differential equations (1) and (3) can be solved by reducing them to first-order equations by suitable choice of a new variable. More generally, suppose that we have a second-order equation of the form

$$y'' = F(t, y') \tag{8}$$

in which the dependent variable appears only in derivative form. Let $v = y'$. Then (8) becomes a first-order equation in v,

$$v' = F(t, v) \tag{9}$$

to which we may be able to apply one of the solution techniques already introduced. Once $v(t)$ has been found by solving (9), $y(t)$ itself can be found as an antiderivative,

$$y(t) = \int^t v(s)\, ds + C_2$$

where C_2 is a constant of integration. We use C_2 since, presumably, the expression for $v(t)$ will also include an arbitrary constant C_1 obtained when (9) is solved. (Generally speaking, solving an nth-order differential equation will involve n constants, which we can think of as "constants of integration.") If the initial conditions $y(t_0) = y_0$ and $v(t_0) = v_0$ are imposed, the corresponding initial value problem for (8) reduces to the pair of first-order initial value problems

$$v' = F(t, v), \qquad v(t_0) = v_0$$

$$y' = \frac{dy}{dt} = v(t), \qquad y(t_0) = y_0$$

C_1 and C_2 will be determined by the initial data. The special cases of free-falling bodies worked out above illustrate these reduction-of-order techniques.

Reduction of Order, II

The method of reduction of order discussed above applies to Newtonian equations in which the size of the forces acting on the moving body do not depend on the location of the body, although they may depend on time or the body's velocity. There is another method that applies when the forces acting on the body depend on its position and velocity, but not on time, and it is this second method that we now present.

The dependent and independent variables in the second-order initial value problem

$$y'' = F(y, y'), \qquad y(t_0) = y_0, \quad y'(t_0) = v_0 \tag{10}$$

are, ostensibly, y and t. Since t does not appear explicitly in the differential equation, another independent variable might be more suitable. In fact, we shall introduce y itself as the new independent variable and $v = y'$ as a new dependent variable. Using the Chain Rule to rewrite y'' in the new variables, we have

$$y'' = \frac{d^2 y}{dt^2} = \frac{dv}{dt} = \frac{dv}{dy}\frac{dy}{dt} = \frac{dv}{dy}\, v$$

and (10) is transformed to the pair of first-order initial value problems

$$v \frac{dv}{dy} = F(y, v), \qquad v = v_0 \quad \text{when} \quad y = y_0 \tag{11a}$$

$$\frac{dy}{dt} = v(y), \qquad y = y_0 \quad \text{when} \quad t = t_0 \tag{11b}$$

Observe that we have used the new variables for (11a) and the original variables for (11b). Once the solution $v(y)$ of (11a) is known, (11b) can be solved by separating variables and integrating.

Example 4.7

The initial value problem

$$y'' = \frac{1}{y} (y')^2 - \frac{y'}{y}, \qquad y = 1 \quad \text{and} \quad y' = 2 \quad \text{when} \quad t = 0$$

can be solved by the method given above. We shall assume that $y > 0$. Introducing y and $v = y'$ as variables, we first solve the first-order problem

$$v \frac{dv}{dy} = \frac{1}{y} v^2 - \frac{v}{y}, \qquad v = 2 \quad \text{when} \quad y = 1$$

The solution $v = 0$ of the differential equation is not of interest and we cancel the factor v to obtain the linear problem in v and dv/dy,

$$\frac{dv}{dy} = \frac{1}{y} v - \frac{1}{y}, \qquad v = 2 \quad \text{when} \quad y = 1$$

The integrating factor technique of Section 3.1 leads to $v = 1 + y$ as the solution to this problem (y must be positive). The second first-order problem to be solved is

$$\frac{dy}{dt} = v(y) = 1 + y, \qquad y = 1 \quad \text{when} \quad t = 0$$

This linear problem in y and dy/dt has the solution

$$y(t) = 2e^t - 1$$

where we must restrict t to be greater than $-\ln 2$ so that y remains positive, and the original problem is solved.

Example 4.7 illustrates the technique, but it is deceptive. The two first-order equations cannot usually be solved this simply in terms of elementary functions.

Escape Velocities

Hydrogen is one of the more abundant elements on the surface of the earth, yet there are today only minute traces of free hydrogen in the atmosphere. In itself that is not surprising since hydrogen gas is lighter than air and would tend to rise. What is surprising is that there is almost no hydrogen even in the topmost layers of the atmosphere. There must have been an abundance of atmospheric hydrogen in the early millenia of the earth's evolution. What has happened?

The disappearance of the gas can be modeled in the following way. Let us suppose that a particle of mass m is in motion above the surface of the earth. Since we shall only be interested in the vertical movements of the particle and the forces acting in a vertical direction, we shall employ a z-axis pointing upward from the earth's surface. According to Newton's Second Law and his Universal Law of Gravita-

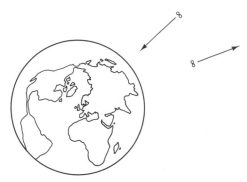

tion and ignoring all forces (frictional, magnetic, or other) except for gravity, the vertical motion of the particle is modeled by the second-order initial value problem

$$mz'' = \frac{-mMG}{(z+R)^2}, \qquad z(0) = 0, \quad z'(0) = v_0 \tag{12}$$

where M is the mass of the earth, G the universal gravitational constant, R the radius of the earth, and we shall assume that v_0 is positive.

Since the dynamical equation of (12) does not involve the variable t explicitly, we may use the second method of reduction of order to find its solutions. If we set $z'' = v \, dv/dz$, we have that [see (11a)]

$$v \frac{dv}{dz} = -\frac{MG}{(z+R)^2}, \qquad v(0) = v_0 \tag{13}$$

Separating variables and solving, we have that

$$v^2 = \left(v_0^2 - \frac{2MG}{R}\right) + \frac{2MG}{z+R} \tag{14}$$

From (14) we see that v^2 remains positive for all $z \geqq 0$ as long as $v_0^2 > 2MG/R$. The constant $(2MG/R)^{1/2}$ is called the *escape velocity* of the particle and is independent of the particle's mass m. The escape velocity from the earth's surface turns out to be approximately 11.179 km/s.

If the particle has a positive initial velocity in excess of the escape velocity, its velocity never vanishes and the particle's upward motion never ceases. In addition, the particle moves infinitely far away from the earth as time increases. To see this, suppose that v_1^2 denotes the positive constant $v_0^2 - 2MG/R$. Then from (14) we have that

$$z'(t) = v(t) = \left[v_1^2 + \frac{2MG}{z(t) + R} \right]^{1/2} \geq v_1 \tag{15}$$

Hence, upon integrating the inequality, $z(t) \geq v_1$ from 0 to t, we have that $z(t) \geq v_1 t$ and $z(t) \to \infty$ as t increases. The particle escapes completely from the earth and moves into space.

A hydrogen molecule H_2 or an atom H in the upper layers of the atmosphere at some point in the earth's history would have been subjected to thermal and other excitations. If an excitation transmits enough energy to the particle to raise its component of velocity in the upward direction above the escape velocity, the particle escapes. That is exactly what has happened over the millenia to free atmospheric hydrogen.

A lightweight molecule such as H_2 is particularly susceptible to this process, in contrast, for example, to the nitrogen molecule N_2, the major component of air and 14 times as heavy as H_2. A given amount E_0 of energy when entirely converted into the kinetic energy $\frac{1}{2} m v_0^2$ of motion of an H_2 particle of mass m produces the velocity $v_0 = (2E_0/m)^{1/2}$. The same amount of energy produces a velocity of $0.27 v_0$ in the nitrogen molecule $[(14)^{-1/2} \approx 0.27]$. That is why the heavier components such as nitrogen and oxygen in our atmosphere tend to stay, whereas the lighter elements escape.

The escape velocity $(2MG/R)^{1/2}$ from the surface of a body depends only on its mass and radius. Escape velocities for all the larger satellites in the solar system have been calculated. Some of these are listed in Table 4.1 together with the corresponding radii and masses, all expressed in "earth" units.†

TABLE 4.1 ESCAPE VELOCITIES FROM BODIES IN THE SOLAR SYSTEM

Body	Radius	Mass	Escape velocity
Earth	1	1	1
Jupiter	10.49	317.929	5.375
Saturn	9.137	95.066	3.223
Neptune	3.565	17.177	2.1875
Uranus	3.6837	14.521	1.9821
Venus	0.9500	0.8617	0.9286
Mars	0.5306	0.10734	0.4489
Pluto	0.8946	0.1806	0.4482
Mercury	0.3819	0.0532	0.3723
Titan	0.3734	0.0199	0.2309
(moon of Saturn)			
Moon (of Earth)	0.27283	0.01230	0.2116

† See R. C. Weast (ed.), *Handbook of Chemistry and Physics* (CRC Press, 1981), for values of the escape velocities in kilometers per second.

Our moon has no atmosphere; the lunar escape velocity is low and whatever gases once were there have long since escaped. Titan does have a measurable atmosphere (probably methane gas, CH_4, eight times heavier than H_2) which is visible in the "computer photos" taken by the *Voyager* space craft. Mercury has no atmosphere at all, even though it has a higher escape velocity than Titan. Mercury is so close to the sun that the solar wind (streams of protons, i.e., hydrogen nuclei, from the sun) long ago imparted high energies to the particles in Mercury's primordial atmosphere and all escaped. Pluto presumably has an atmosphere, but little is known about that dim and distant planet. Mars has a tenuous atmosphere of carbon dioxide. All the more massive planets have atmospheres (mostly of methane). The asteroids and the other moons of the solar system have escape velocities much too low to support atmospheres.

We have interpreted escape velocities in terms of small particles, but we could also apply the results to projectiles or rockets. However, for the latter we should take into account the declining mass of the rocket as the fuel burns, and for both projectiles and rockets, friction will play a significant role. What this means is that Eq. (12) should be modified to take these factors into account before escape velocities are calculated.

Comments

In this section we have given yet another technique for solving, or at least simplifying, a differential equation. The technique and the type of differential equation may seem to be highly specialized, but both are of central importance in Newtonian mechanics. We have more to say about these matters in later chapters.

PROBLEMS

The following problem(s) may be more challenging: 4, 5(b), and 7(b)–(d).

1. Solve the following differential equations by reduction of order. Sketch some solution curves.
   (a) $ty'' - y' = 3t^2$.
   (b) $y'' - y = 0$.
   (c) $yy'' + (y')^2 = 1$.
   (d) $y'' + 2ty' = 2t$.

2. Solve the given initial value problems.
   (a) $2yy'' + (y')^2 = 0$, $y(0) = 1$, $y'(0) = -1$.
   (b) $y'' = y'(1 + 4/y^2)$, $y(0) = 4$, $y'(0) = 3$.
   (c) $y'' = -g - y'$, $y(0) = h$, $y'(0) = 0$, g, h are positive constants.

3. Solve the initial value problem of the "falling" raindrop if the raindrop is rising upward, subject to the initial conditions $y(0) = 0$, $y'(0) = v_0$.

4. (a) A grain of sand, a feather, and a raindrop, all of the same mass, are released from rest at the same height. The grain of sand falls without resistance through a vacuum

chamber. The feather and raindrop fall through the air with resistance negatively proportional to the velocity and to the square of the velocity, respectively [that is, their respective force laws are given by (1) and by (3) with constants m, g, k_f and k_r]. Show that the Maclaurin expansions in powers of t of the three velocities v_s, v_f, and v_r have the form $v = gt +$ [higher powers of t with coefficients depending on the type of resistive force].

(b) Why would one expect the three velocities to be roughly equal and satisfy $v_s > v_r > v_f$ for small $t > 0$?

5. A straight tunnel is bored through the earth as pictured. An object of mass m is dropped into the tunnel.

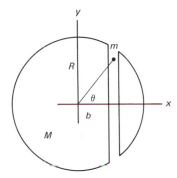

(a) Neglecting friction, write the equation of motion for the object if the force of gravity in the direction of motion is $F_y = GmM \sin \theta/(b^2 + y^2)$.

(b) Solve the equation for the general solution. [*Note*: Answer may be in implicit form.]

6. An astronaut making measurements of the force of gravity on a planet finds that the force of gravity is

$$F_g = \frac{\tilde{G}mM}{(y + R)^3}$$

where m is mass of the object, M the mass of the planet, $\tilde{G}$ a new universal constant, and R the radius of the planet.

(a) What is the escape velocity for this planet?

(b) What is the ratio of this planet's escape velocity $\tilde{v}_0$ to that of the planet's Newtonian escape velocity?

7. [*Einstein's Field Equations of General Relativity*]. These equations are a complicated set of nonlinear coupled partial differential equations. Nevertheless, in 1922, the Soviet mathematician Alexander Alexandrovich Friedmann succeeded in obtaining cosmological solutions governing the behavior of the universe as a whole. He assumed that on a broad scale the matter in the universe is homogeneous and isotropic. The equations reduce to the single nonlinear initial value problem

$$2RR'' + R'^2 + kc^2 = 0, \qquad R(0) = R_0 > 0, \quad R'(0) = v_0 > 0 \qquad (*)$$

where R is the "radius of the universe," $k = +1$ (spherical space), 0 (Euclidean space), or -1 (pseudospherical space), c is the speed of light, and the derivatives are with respect to time.

(a) Using ingenuity and guesswork, reduce (*) to a first-order differential equation. [*Hint*: Multiply (*) by R' and write as $(\cdots)' + kc^2R' = 0$. Then antidifferentiate to obtain $R(R')^2 + kc^2R = R_0v_0^2 + kc^2R_0$.].

(b) Separate the variables in part (a) for the case $k = 0$ and solve. Express R as a function of t, R_0, v_0, k, and c. Your answer should have a $\pm$ sign in it. Show that the minus sign leads to the eventual collapse of the universe (the "Big Crunch"), whereas the plus sign corresponds to perpetual expansion. Explain why in both cases as $R_0 \downarrow 0$ a "Big Bang" cosmology results. Sketch solution curves.

(c) Solve (*) when $k = +1$ and sketch the solution curves. [*Hint*: Assume that $v_0 < c$. Let $R = R_0c^{-2}(v_0^2 + kc^2)\sin^2 u$ and rewrite the differential equation with u as the dependent variable; solve to find t in terms of u. The equations for R and t in terms of u define a cycloid with u as the parameter (show). This corresponds to periodic collapse and rebirth.]

Crunch/bang   Crunch/bang   Crunch/bang

(d) Solve (*) when $k = -1$, proceeding as in part (c), but with $v_0 > c$ and use $\sinh^2 u$ rather than $\sin^2 u$. Sketch the solutions. Interpret your answers.

4.5 IMPLICIT METHODS: EXACT EQUATIONS

The techniques introduced so far for solving first-order differential equations have a common property: Each method transforms a differential equation of some special form into a differential equation which can be solved by integration. For example, a change of variable in the pursuit equation of Section 4.2 reduces it to a separable equation, whose solutions can be found (implicitly) by integrating. In this section we give a simple test that reveals whether a first-order differential equation can be "solved" by integration. Moreover, if an equation "passes" the test, we show just how (in theory) the solutions can be defined by an implicit equation.

Exactness

We shall consider first-order differential equations of the form

$$M(x, y) + N(x, y)y' = 0 \tag{1}$$

where M and N are continuously differentiable functions in a region R of the xy-plane. Although the form of (1) seems special, it is not. Any normal first-order equation, for example, can be written this way: $y' = f(x, y)$ becomes $-f(x, y) + y' = 0$. In fact, an equation can usually be put into the form of (1) in more than one way. For instance, the equation

$$y' = \frac{-2xy}{1+x^2} \tag{2}$$

can be written as

$$2xy + (1+x^2)y' = 0 \tag{3}$$

or as

$$\frac{2xy}{1+x^2} + y' = 0 \tag{4}$$

each of which has the form of (1), but with differing coefficients M and N.

> **Exact Equations.**   Equation (1) is said to be *exact* in the region R if there is a continuously differentiable function $F(x, y)$ such that
>
> $$\frac{\partial F}{\partial x} = M(x, y) \quad \text{and} \quad \frac{\partial F}{\partial y} = N(x, y) \qquad \text{for all } (x, y) \text{ in } R \tag{5}$$
>
> F is called an *exactness function* for equation (1).

For example, Eq. (3) is exact since the function $F = y + x^2 y$ has the property that $\partial F/\partial x = 2xy$, and $\partial F/\partial y = 1 + x^2$. But even though (4) is an equation with the same solutions as (3), we show later that (4) is *not* exact (see Example 4.9). Exactness depends on the form of the equation.

As our next result shows, solutions of an exact equation can be characterized implicitly in a single step once an exactness function for the equation is known.

> **Solutions of an Exact Equation.**   Suppose that $M + Ny' = 0$ is an exact equation in a region R, and that $F(x, y)$ is any exactness function for that equation in R. Then $y(x)$, $a < x < b$, is a solution of $M + Ny' = 0$ if and only if $(x, y(x))$ lies in R for all $a < x < b$ and for some constant C,
>
> $$F(x, y(x)) = C, \qquad \text{all } a < x < b \tag{6}$$

Proof.   Suppose that $y(x)$, $a < x < b$, is a solution of (1) in R. Then

$$0 = M(x, y(x)) + N(x, y(x))y'(x) = \frac{\partial F(x, y(x))}{\partial x} + \frac{\partial F(x, y(x))}{\partial y} y'(x)$$

$$= \frac{d}{dx} [F(x, y(x))] \qquad \text{for all } a < x < b$$

where the last equality follows from the Chain Rule for differentiation of functions of two variables. Thus from the Vanishing Derivative Theorem there is a constant C such that (6) holds.

Conversely, suppose that $y(x)$, $a < x < b$, is a continuously differentiable function, $(x, y(x))$ in R, and $F(x, y(x)) = C$, $a < x < b$, for some constant C. Then the foregoing steps can be reversed to show that $M(x, y(x)) + N(x, y(x))y'(x) = 0$, $a < x < b$, and hence $y(x)$ is a solution of (1). This proves the result.

Now the graph of the relation $F(x, y) = C$ in the xy-plane for some constant C is called a *contour* or a *level curve* of the function F. Thus the characterization theorem above states that if (1) is exact, every solution curve for (1) lies on some level curve for an exactness function for (1), and conversely, every portion of such a level curve that does not "double back on itself" is a solution curve for (1).

The difficulty in using (6) to "solve" an exact equation (1) is that it may be difficult or impossible to find y as an explicit function of x. Often we must either be content with (6) itself as a "solution" of (1) or resort to some numerical procedure to find approximate solution curves. Some examples will clarify these ideas.

Example 4.8

Equation (3), $2xy + (1 + x^2)y' = 0$, is exact with an exactness function $F(x, y) = y + x^2y$, as we saw above. Hence solutions $y(x)$ of this equation satisfy the implicit relation $y(x) + x^2y(x) = C$ for any constant C. Solving for y, we have $y = C/(1 + x^2)$, where C is a constant. In this case the region R in which $M = 2xy$ and $N = 1 + x^2$ are continuously differentiable is the entire xy-plane, and as we see, $y(x)$ is defined for all x. Some of the solution curves are sketched in Figure 4.17.

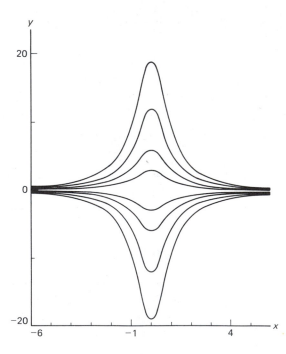

Figure 4.17 Solution curves of $2xy + (1 + x^2)y' = 0$.

A Test for Exactness

What we need now is a simple test for the exactness of the differential equation $M + Ny' = 0$ and a method for constructing an exactness function F.

Test for Exactness.   Let $M(x, y)$ and $N(x, y)$ be continuously differentiable in a rectangle $R: a < x < b, c < y < d$. Then $M + Ny' = 0$ is exact if and only if

$$\frac{\partial M}{\partial y} = \frac{\partial N}{\partial x} \qquad \text{throughout } R \qquad (7)$$

Proof.   The verification of the exactness test has two parts. First, we shall assume that $M + Ny' = 0$ is exact and show that $\partial M / \partial y = \partial N / \partial x$. Exactness implies that there is a function $F(x, y)$ for which $\partial F / \partial x = M$ and $\partial F / \partial y = N$. By hypothesis, all the first-order partial derivatives of M and N exist and are continuous. Thus the second-order partial derivatives of F,

$$\frac{\partial^2 F}{\partial x^2} \equiv \frac{\partial M}{\partial x}, \qquad \frac{\partial^2 F}{\partial y\, \partial x} \equiv \frac{\partial M}{\partial y}, \qquad \frac{\partial^2 F}{\partial x\, \partial y} \equiv \frac{\partial N}{\partial x}, \qquad \frac{\partial^2 F}{\partial y^2} \equiv \frac{\partial N}{\partial y}$$

exist and are continuous. A theorem of multivariable calculus about the equality of "mixed" partial derivatives then implies that

$$\frac{\partial^2 F}{\partial y\, \partial x} = \frac{\partial^2 F}{\partial x\, \partial y}$$

which is the same as saying that (7) holds.

The important aspect of the test is the other way around, of course, and we now prove the converse.

Finding an Exactness Function

Suppose that $\partial M / \partial y = \partial N / \partial x$ throughout the rectangle R. We shall show that $M + Ny' = 0$ is exact by actually constructing an exactness function F. Thus at one and the same time we show exactness and construct the solutions, a procedure greatly to be desired when attempting to solve a differential equation. We must find a function $F(x, y)$ for which $\partial F / \partial x = M$ and $\partial F / \partial y = N$. A function F will satisfy the first of these equations if and only if

$$F(x, y) = \int^x M(s, y)\, ds + g(y) \qquad (8)$$

where $g(y)$ is an arbitrary function of y and the antidifferentiation is intended for fixed y. We now determine the function $g(y)$ so that $\partial F / \partial y \equiv N$. We have

$$\frac{\partial F}{\partial y} = \frac{\partial}{\partial y}\left[\int^x M(s,y)\,ds\right] + g'(y) = N(x,y)$$

Therefore,

$$g'(y) = N(x,y) - \frac{\partial}{\partial y}\int^x M(s,y)\,ds \tag{9}$$

The function $g(y)$ can be determined from this equation by an antidifferentiation, provided that the right-hand side of (9) is independent of x. This will be the case if the derivative of the right-hand side of (9) with respect to x is identically zero. Computing this derivative, we find

$$\frac{\partial}{\partial x}\left[N - \frac{\partial}{\partial y}\int^x M(s,y)\,ds\right] = \frac{\partial N}{\partial x} - \frac{\partial}{\partial x}\left[\frac{\partial}{\partial y}\int^x M(s,y)\,ds\right]$$

From the equality of the mixed partial derivatives we have that†

$$\frac{\partial}{\partial x}\left[\frac{\partial}{\partial y}\int^x M(s,y)\,ds\right] = \frac{\partial}{\partial y}\left[\frac{\partial}{\partial x}\int^x M(s,y)\,ds\right] = \frac{\partial M(x,y)}{\partial y}$$

Therefore, remembering that $\partial M/\partial y = \partial N/\partial x$, we have

$$\frac{\partial}{\partial x}\left[N - \frac{\partial}{\partial y}\int^x M(s,y)\,ds\right] = \frac{\partial N}{\partial x} - \frac{\partial M}{\partial y} = 0$$

and by the Vanishing Derivative Theorem, the right-hand side of (9) really is independent of x. Hence $g(y)$ is determined up to a constant by integrating each side of (9):

$$g(y) = \int^y N(x,t)\,dt - \int^y\left[\frac{\partial}{\partial t}\int^x M(s,t)\,ds\right]dt \tag{10}$$

where dummy variables of integration, s and t, have been introduced.

From (8) and (10) we construct the exactness function F:

$$F(x,y) = \int^x M(s,y)\,ds + \int^y N(x,t)\,dt - \int^y\left[\frac{\partial}{\partial t}\int^x M(s,t)\,ds\right]dt \tag{11}$$

Solutions $y(x)$ of the exact equation $M + Ny' = 0$ then satisfy (6) for some constant C, with F given by (11).

We have thus shown that the exactness test is valid and in the process, con-

---

† The Fundamental Theorem of Calculus asserts that

$$\frac{d}{dx}\left[\int^x f(s)\,ds\right] = f(x)$$

or, in this case,

$$\frac{\partial}{\partial x}\left[\int^x M(s,y)\,ds\right] = M(x,y)$$

structed the exactness function F by means of (11). As is usual in this field, it is not worthwhile remembering formula (11). Rather, one should remember the exactness test and the general procedure for constructing F once an equation has been shown to be exact.

Example 4.9

The equation

$$(e^x \sin y - 2y \sin x) + (y^2 + e^x \cos y + 2 \cos x)y' = 0 \qquad (12)$$

is exact throughout the xy-plane since

$$\frac{\partial}{\partial y}(e^x \sin y - 2y \sin x) = e^x \cos y - 2 \sin x$$

and

$$\frac{\partial}{\partial x}(y^2 + e^x \cos y + 2 \cos x) = e^x \cos y - 2 \sin x$$

We can find an exactness function $F(x, y)$ as follows. Since $\partial F / \partial x = M = e^x \sin y - 2y \sin x$,

$$F(x, y) = \int^x M = \int^x (e^s \sin y - 2y \sin s)\, ds + g(y)$$

$$= e^x \sin y + 2y \cos x + g(y)$$

Since we also must have $\partial F / \partial y = N = y^2 + e^x \cos y + 2 \cos x$, we see that

$$\frac{\partial}{\partial y}[e^x \sin y + 2y \cos x + g(y)] = y^2 + e^x \cos y + 2 \cos x$$

Hence

$$e^x \cos y + 2 \cos x + g'(y) = y^2 + e^x \cos y + 2 \cos x$$

or $g'(y) = y^2$. Hence $g(y) = y^3/3$, where we have not included a constant of integration in calculating g since that will come in anyway when we set $F(x, y) = C$. Thus

$$F(x, y) = e^x \sin y + 2y \cos x + \frac{y^3}{3}$$

and any solution $y(x)$ satisfies, for some constant C, the equation

$$e^x \sin (y(x)) + 2y(x) \cos x + \frac{y^3(x)}{3} = C \qquad (13)$$

We cannot express $y(x)$ explicitly in terms of elementary functions of x. See Figure 4.18 for a sketch of some of the integral curves defined by (13). These curves

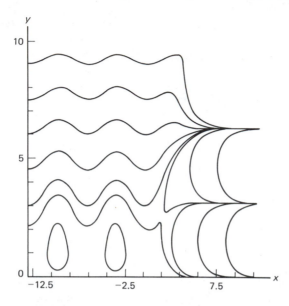

Figure 4.18   Integral curves of (12).

were constructed by a computer, (13) being much too complicated to attempt to draw the curves manually.

If we were interested in finding not all solutions of an exact equation but, say, only that solution (or solutions) passing through a fixed point (x_0, y_0), then the constant C in (6) must be determined by $F(x_0, y_0) = C$. For example, if a solution $y(x)$ of (12) passes through the point $(0, \pi)$, we must have from (13) that $e^0 \sin \pi + 2\pi \cos 0 + \pi^3/3 = C$ and $y(x)$ satisfies the implicit equation

$$e^x \sin (y(x)) + 2y(x) \cos x + \frac{y^3(x)}{3} = 2\pi + \frac{\pi^3}{3}$$

Comments

We have seen how the concept of exactness leads to a solution technique that characterizes the solutions of (1) implicitly. Observe that all separable equations written in the form $M(x) + N(y)y' = 0$ are exact since $\partial M(x)/\partial y \equiv 0 \equiv \partial N(y)/\partial x$. But written in some other form, separable equations may not be exact. Thus exactness is highly dependent on the form in which the equation is presented. There are other equations that are not themselves exact but which can be made exact by multiplying a suitable factor. For example, the inexact equation given in (4) is made exact by multiplying by $1 + x^2$, which is then called an *integrating factor* for the equation. The existence and the construction of integrating factors for inexact equations is taken up in the problem set.

PROBLEMS

The following problem(s) may be more challenging: 5(b), 8(c), (d), and (e).

1. Determine if each of the following equations is exact. If so, give the exactness function and solve. Sketch the integral curves.
   (a) $2xy + 3y^2 + (x^2 + 6xy + 3y^2)y' = 0$.
   (b) $y + xy' = 0$.
   (c) $ye^x + e^x y' = 0$.
   (d) $y^2 e^y + 2x + xe^y (y^2 + 2y)y' = 0$.
   (e) $\cos x \cos y - y' \sin x \sin y = 0$.
   (f) $e^{-kx}[e^x(k-1) + k] + y + xy' = 0$, where k is a real constant.

2. Test each of the following equations for exactness and solve the equation. (The equations that are not exact may be solved by methods discussed in the preceding sections.)
   (a) $(x + 2y)\,dx + (2x + y)\,dy = 0$.
   (b) $(2xy - 3x^2)\,dx + (x^2 + 2y)\,dy = 0$.
   (c) $(\cos 2y - 3x^2 y^2)\,dx + (\cos 2y - 2x \sin 2y - 2x^3 y)\,dy = 0$.
   (d) $(y^2 + x^2)\,dx - 2xy\,dy = 0$.
   (e) $(2x - 3y)\,dx + (2y - 3x)\,dy = 0$.
   (f) $y' = \dfrac{2xy^2 + 2y + 1}{-2(x^2 y + x)}$.

3. Solve each of the following initial value problems. Sketch the integral curves.
   (a) $(2y - x)y' = y - 2x$, $y(1) = 2$.
   (b) $2xy\,dx + (x^2 + 1)\,dy = 0$, $y(1) = -3$.
   (c) $\dfrac{dx}{dy} = \dfrac{x \sec^2 y}{\sin 2x - \tan y}$, $y(\pi) = \dfrac{\pi}{4}$.
   (d) $2xy^3 + 3x^2 y^2 y' = 0$, $y(1) = 1$.
   (e) $3x^2 + 4xy + (2y + 2x^2)y' = 0$, $y(0) = 1$.
   (f) $(ye^{xy} - 2y^3)\,dx + (xe^{xy} - 6xy^2 - 2y)\,dy = 0$, $y(0) = 2$.

4. Find the value of α for which each of the following equations is exact, and solve it using the value of α.
   (a) $(xy^2 + \alpha x^2 y)\,dx + (x + y)x^2\,dy = 0$.
   (b) $x + ye^{2xy} + \alpha xe^{2xy}y' = 0$.

5. [*Exactness and Simple Connectedness*]. Consider the differential equation

$$\frac{x}{x^2 + y^2}\,dy - \frac{y}{x^2 + y^2}\,dx = 0$$

and let A be the region between two concentric circles with center at the origin and radii $\frac{1}{2}$ and 1.
   (a) Let R be any coordinate rectangle contained in A. Show that the equation above is exact in R, and find a function f with continuous first and second derivatives such that

$$f_x = \frac{-y}{x^2 + y^2}, \qquad f_y = \frac{x}{x^2 + y^2} \qquad \text{in all of } R$$

   (b) Observe for this differential equation M and N are continuously differentiable in A and $M_y \equiv N_x$ in A. Show that in spite of this, the differential equation is *not* exact in A. [*Remark*: This problem shows that when using the exactness theorem, we should avoid annular regions such as A, and strongly suggests that trouble will develop for any regions with "holes" in them. This turns out to be true; such regions are said to be *not simply connected.*]

Integrating Factors.   A function $\mu(x, y)$ is an *integrating factor* for the equation $M + Ny' = 0$ in a region R of the xy-plane if (1) μ is continuously differentiable in R, (2) $\mu \neq 0$ in R, and (3) the equation $\mu M + \mu Ny' = 0$ is exact in R. Since $\mu \neq 0$ in R, the equations $M + Ny' = 0$ and $\mu M + \mu Ny' = 0$ have exactly the same solutions. The following problems have to do with finding integrating factors for inexact equations.

---

6. Show that by multiplying through by the indicated function μ, the nonexact equation becomes exact. Solve.
   (a) $\cos x\, dy - (2y \sin x - 3)\, dx = 0,\quad \mu(x) = \cos x,\quad -\pi/2 < x < \pi/2$.
   (b) $y\, dx + (4x - y^2)\, dy = 0,\quad \mu(y) = y^3, y > 0$.

7. Show that $\mu = \exp\left[\int^x p(s)\, ds\right]$ is an integrating factor for the general first-order linear equation $p(x)y - q(x) + y' = 0$.

8. Show that if M and N have the special properties given in the left-hand column of the following table, the equation $M + Ny' = 0$ has the integrating factor given in the right-hand column.

	If:	Then there is an integrating factor of the form:
(a)	$\dfrac{1}{N}\left(\dfrac{\partial M}{\partial y} - \dfrac{\partial N}{\partial x}\right) = a(x)$	$e^{\int a(x)\,dx}$
(b)	$\dfrac{1}{M}\left(\dfrac{\partial M}{\partial y} - \dfrac{\partial N}{\partial x}\right) = b(y)$	$e^{-\int b(y)\,dy}$
(c)	$\left.\begin{array}{l} M(tx, ty) = t^n M(x, y) \\ N(tx, ty) = t^n N(x, y) \end{array}\right\}$ all t	$\dfrac{1}{xM + yN}$
(d)	$\begin{array}{l} M = yf(xy),\ N = xg(xy) \\ f(xy) \neq g(xy) \end{array}$	$\dfrac{1}{xy[f(xy) - g(xy)]}$
(e)	$\dfrac{1}{xM - yN}\left(\dfrac{\partial N}{\partial x} - \dfrac{\partial M}{\partial y}\right) = F(xy)$	$e^{\int F(u)\,du}$, where $u = xy$

9. Use the table of Problem 8 if necessary to solve each of the following equations.
   (a) $(2xy^4 e^y + 2xy^3 + y)\, dx + (x^2 y^4 e^y - x^2 y^2 - 3x)\, dy = 0$.
   (b) $(x^4 + y^4)\, dx - xy^3\, dy = 0$.
   (c) $(y^2 \cos x - y)\, dx + (x + y^2)\, dy = 0$.
   (d) $(\sec x + y \tan x)\, dx + dy = 0$.
   (e) $x^2 y^3 + 4y + (4x - 4x^3 y^2)y' = 0$.
   (f) $(y^3 - 2x^2 y)\, dx + (2xy^2 - x^3)\, dy = 0$.

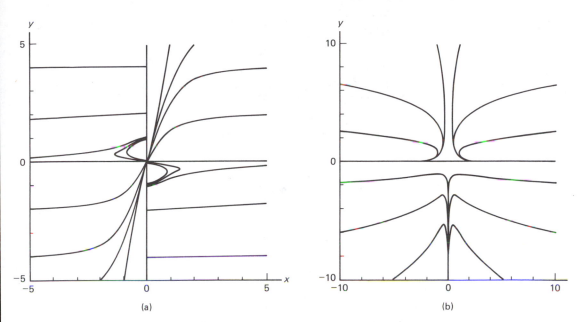

Figure 4.19   (a) Integral curves for Problem 10a. (b) Integral curves for Problem 10b.

(g) $(y^2 + xy + 1) \, dx + (x^2 + xy + 1) \, dy = 0$.

(h) $(2x^3y^2 + 4x^2y + 2xy^2 + xy^4 + 2y) \, dx + 2(y^3 + x^2y + x) \, dy = 0$.

10. (a) For $x > 0$ (or $x < 0$), solve $ye^{y/x} - (xe^{y/x} + 2x^2y)y' = 0$ (see Figure 4.19a).

(b) If M and N are polynomials, it sometimes happens that $M + Ny' = 0$ has an integrating factor $\mu = x^a y^b$, where a and b are constants. Find an integrating factor and solve $y + x^2y^2 + (3x^3y - 2x)y' = 0$ (see Figure 4.19b).

11. [*Euler's Theorem for Homogeneous Functions*]. A continuously differentiable function $f(x, y)$ is *homogeneous of order* n if $f(tx, ty) = t^n f(x, y)$, for all t, x, y. For example, $x^2 - 3xy + 7y^2$ is homogeneous of order 2, but $xy^3 - x + 1$ is not homogeneous. Show that if f is homogeneous of order n, then $x(\partial f/\partial x) + y(\partial f/\partial y) = nf(x, y)$. [*Hint*: Use the Chain Rule for functions of several variables to calculate $\partial f(tx, ty)/\partial t \equiv (\partial/\partial t)[t^n f(x, y)]$ and then set $t = 1$.] The functions M and N in 8(c) are homogeneous of order n.

CHAPTER 5

Second-Order Linear Differential Equations: Electrical Circuits

Because they model so many important physical phenomena, we devote this chapter to a treatment of linear second-order differential equations and leave higher-order linear equations to Chapter 6. We describe as completely as possible the solutions of both the driven and undriven equations. Because of the simplicity and utility they bring to the task at hand we also briefly discuss the differentiation properties of complex-valued functions of a single real variable. Kirchhoff's Laws for electrical circuits are introduced, and together with Newton's Second Law are used in a variety of applications.

5.1 OPERATORS, INITIAL VALUE PROBLEMS

The *normal linear second-order differential equation* is given by

$$y'' + a(t)y' + b(t)y = f(t) \tag{1}$$

where the coefficient functions $a(t)$ and $b(t)$ and the function $f(t)$ are all continuous on a given interval I. Equation (1) is called *nonhomogeneous* (or *forced*) if the *forcing term* $f(t)$ does not vanish identically, and *homogeneous* (or *free* or *reduced*) if $f(t)$ does vanish identically on I. When the coefficients a and b are constants, Eq. (1) is called a *linear second-order differential equation with constant coefficients*. To be precise, we define the notion of a solution for Eq. (1).

> **Solution.** If the given functions $f(t)$, $a(t)$, and $b(t)$ are continuous on an interval I, then $y(t)$ is a *solution for Eq.* (1) on I if $y(t)$ has continuous derivatives up to, and including, order 2 on I and satisfies Eq. (1) for all t in I.

Observe that the second-order linear differential equation (1) is a straightforward generalization of the first-order linear differential equations investigated quite thoroughly in Section 3.1. The purpose of this chapter is to give a treatment of second-order linear differential equations. Higher-order linear equations are left for Chapter 6.

Role of Linear Equations

Observe that the differential equation of a falling body with Newtonian damping (derived in Section 4.4) is nonlinear. Indeed, our experience with model building in earlier chapters indicates that no matter what physical system is considered, a model can be found whose mathematical formulation will be nonlinear. Practitioners sometimes describe this state of affairs by saying that "nature is inherently not linear." Interestingly enough, however, we will see that linear differential equations arise in the analysis of many physical systems and the models which they make possible agree rather well with physical reality.† Thus linear differential equations occupy a very special place in the construction of mathematical models. But linear differential equations are also important theoretically because a knowledge of the behavior of their solutions is useful in the investigation of the behavior of solutions of more general differential equations (as we will see in Chapter 11).

Operational Notation

We now extend the operational notation introduced in Section 3.2 as an aid in finding all solutions of higher-order linear differential equations. Recall that we used the symbol D to denote the operator that takes a differentiable function $y(t)$ into its derivative, $y'(t)$. Also recall that for any continuous function $r(t)$ on an interval I, we defined the operator $L = D + r(t)$ which takes any function $y(t)$ in $C^1(I)$ into the function $Ly \equiv Dy + r(t)y$ which lies in $C^0(I)$. Thus $L: C^1(I) \rightarrow C^0(I)$, and we saw in Section 3.2 how to find all solutions y in $C^1(I)$ of the equation $Lu = f$ for a given f in $C^0(I)$.

We need to generalize this operational notion in two ways. First note that if $y(t)$ is a complex-valued function with real and imaginary parts $f(t)$ and $g(t)$, then $y(t) = f(t) + ig(t)$, and

$$Dy \equiv y' = D[f + ig] = Df + iDg$$

(See Appendix B for a review of complex numbers and the calculus of complex-valued functions of a real variable.)

---

† In this connection the following paper contains many interesting and thought-provoking ideas: E. P. Wigner, "The unreasonable effectiveness of mathematics in the natural sciences," *Comm. Pure Appl. Math.* 13 (1960) 1–14.

Example 5.1

Recall that for any real number β the complex-valued function $e^{i\beta t}$ is defined as $e^{i\beta t} \equiv \cos \beta t + i \sin \beta t$. Thus

$$D[e^{i\beta t}] = D[\cos \beta t] + iD[\sin \beta t] = -\beta \sin \beta t + i\beta \cos \beta t = i\beta e^{i\beta t} \qquad (2)$$

More generally, let $r = \alpha + i\beta$ be any complex number and recall that $e^{rt} = e^{(\alpha+i\beta)t} = e^{\alpha t}e^{i\beta t}$. Then using (2) and the product rule for differentiation, we have that

$$D[e^{rt}] = D[e^{\alpha t}]e^{i\beta t} + e^{\alpha t}D[e^{i\beta t}] = re^{rt} \qquad (3)$$

The differentiation formula (3) has much to do with the reason the exponentiation symbol was used to denote that particular complex-valued function.

Next, observe that if $y(t)$ is such that Dy is differentiable, then D can be applied to Dy. The operator of applying D twice in succession is denoted by the symbol D^2. Thus $D[Dy] \equiv D^2y$. The set of functions $C^2(I)$ defined below is a natural domain for the operator D^2.

> **$C^2(I)$.** $C^2(I)$ denotes the collection of all real-valued (or more generally, complex-valued) functions on the interval I which have at least two continuous derivatives on I. [This is consistent with how the "continuity sets" $C^0(I)$ and $C^1(I)$ were defined; see Section 3.2.]

Operators that will occupy a great deal of our attention have a special property called linearity.

> **Linear Operators on Continuity Sets.**   An operator L whose domain and codomain are continuity sets (i.e., $C^0(I)$, $C^1(I)$, or $C^2(I)$, for some interval I in **R**) is called a *linear operator* if for any pair of elements u, v in the domain and any scalars α, β (i.e., real or complex numbers according as the continuity sets consist of real- or complex-valued functions) we have that
>
> $$L[\alpha u + \beta v] = \alpha L[u] + \beta L[v]$$

For any given real- (or complex-) valued functions a_0, a_1, and a_2 in $C^0(I)$, we shall use the notation

$$P(D) = a_2 D^2 + a_1 D + a_0 \qquad (4)$$

For the *polynomial operator* $P(D)$: $C^2(I) \to C^0(I)$, which acts on a function y in $C^2(I)$ to produce $P(D)[y] \equiv a_2 D^2 y + a_1 Dy + a_0 y$. Whether the domain $C^2(I)$ of $P(D)$ consists of complex-valued or real-valued functions will be clear from context if not stated explicitly. A basic property of polynomial operators is as follows:

Linearity of $P(D)$.   For any polynomial operator $P(D)$, any constants c_1 and c_2, and any functions y_1 and y_2 in $C^2(I)$, we have that

$$P(D)[c_1 y_1 + c_2 y_2] = c_1 P(D)[y_1] + c_2 P(D)[y_2]$$

Proof. Observe that the linearity property holds for the special cases $P(D) \equiv D$ and $P(D) \equiv D^2$, so the general case follows by direct calculation.

For a given function f in $C^0(I)$, the second-order differential equation $P(D)[y] = f$, with $P(D)$ as in (4), is precisely the equation with which this chapter is concerned. (We consider later the effect of choosing a driving term that is only piecewise continuous on I.) The equation is said to be *linear* because of the Linearity Property satisfied by $P(D)$. An important concept is the

Null Space of $P(D)$.   For $P(D)$ as in (4), the set of all functions in $C^2(I)$ carried into the zero function by the action of $P(D)$ is said to be the *null space of $P(D)$* and is denoted by $N(P(D))$. Thus $y(t)$ is in $N(P(D))$ if and only if $P(D)[y] = 0$.

The linear second-order differential equation $P(D)[y] = 0$ is homogeneous and is the main topic of Section 5.2. Note that $N(P(D))$, the null space of $P(D)$, is the solution set of the equation $P(D)[y] = 0$. An immediate consequence of the Linearity Property of $P(D)$ is the

Null Space Closure Property for $P(D)$.   For any $P(D)$ as in (4), any functions $w(t)$ and $z(t)$ in $N(P(D))$, and any constants c_1 and c_2, we have that $c_1 w + c_2 z$ is also in $N(P(D))$.

To gain some insight into the structure of solutions of the equation $P(D)[y] = f$, we use the Linearity and Null Space Closure Properties of $P(D)$ as follows. Suppose that by some means or another we have managed to find a solution $u(t)$ of $P(D)[y] = f$. The linearity of $P(D)$ implies that if $v(t)$ is any other solution of $P(D)[y] = f$, then $v - u$ is a solution of the homogeneous equation $P(D)[y] = 0$. In other words, $v - u$ is in $N(P(D))$, and hence v has the form $u(t) +$ an element in $N(P(D))$. On the other hand, linearity of $P(D)$ implies that $u(t) +$ any element in $N(P(D))$ is a solution of $P(D)[y] = f$. Thus all solutions of $P(D)[y] = f$ are known as soon as just one (any one) solution and $N(P(D))$ are known. Accomplishing this task will occupy our attention in Section 5.2.

Example 5.2

One easily verifies that $y = t$ is a solution of $y'' + 4y = 4t$, and that $\sin 2t$ and $\cos 2t$ are solutions of the homogeneous equation $y'' + 4y = 0$. Thus the Null Space Closure Property implies that $c_1 \sin 2t + c_2 \cos 2t$ is an element of $N(D^2 + 4)$ for any values of the constants c_1 and c_2. The linearity of $D^2 + 4$ implies that $y = t + c_1 \sin 2t + c_2 \cos 2t$ is a real-valued solution of $y'' + 4y = 4t$ for any real constants c_1 and c_2. (We will see that there are no other real-valued solutions.)

Initial Value Problems

From our experience with Newton's Laws of Mechanics in Section 4.3, we see that initial conditions on the state variable y and its derivative y' are required so that the second-order differential equation $P(D)[y] = f$ has precisely one solution. Our next result, stated without proof, verifies this hunch.

Dynamical System Property for $P(D)[y] = f$.   Let $P(D)[y] \equiv y'' + a(t)y' + b(t)y$ for coefficients $a(t)$ and $b(t)$ continuous functions on an interval I. Then for any driving term $f(t)$ continuous on I, and any t_0 in I and any constants y_0, v_0, the initial value problem

$$y'' + a(t)y' + b(t)y = f$$
$$y(t_0) = y_0, \qquad y'(t_0) = v_0 \tag{5}$$

has a unique solution $y(t)$ in $C^2(I)$. When the functions $a(t)$, $b(t)$, $f(t)$ and the constants y_0 and v_0 are real valued, the solution of problem (5) is also real valued.

There are several features of this result that require special mention. The assertion is that the initial value problem (5) has a unique solution on the entire interval I, not just on some neighborhood of t_0. Also note that $P(D)$ has the *normalized form*

$$P(D) = D^2 + a(t)D + b(t) \tag{6}$$

That is, the coefficient $a_2(t)$ in (4) has been set equal to unity. This is easy to achieve in practice since the equation $a_2 y'' + a_1 y' + a_0 y = g$ can be divided through by the leading coefficient $a_2(t)$ to produce the equation in (5). The "hidden" assumption is therefore that the functions $a(t) \equiv a_1(t)/a_2(t)$, $b(t) \equiv a_0(t)/a_2(t)$, and $f(t) \equiv g(t)/a_2(t)$ are continuous on I. A problem only arises, of course, when $a_2(t)$ vanishes at points in I (i.e., when the differential equation is nonnormal). We saw from examples in Sections 3.1 and 3.2 that initial value problems for nonnormal first-order linear equations are difficult to describe (see Examples 3.7, 3.12, and 3.13).

Finally, the Dynamical System Property obviously implies that the range of any polynomial operator $P(D)$ as in (6) is all of $C^0(I)$.

Example 5.3

As an illustration of the Dynamical System Property, let us examine the initial value problem $y'' + 4y = 4t$, $y(0) = 1$, $y'(0) = 0$. Referring to Example 5.2 we see that the differential equation has as solutions the functions $y = t +$

$c_1 \sin 2t + c_2 \cos 2t$, where c_1 and c_2 are arbitrary constants. Thus we might search through this family of solutions for one that satisfies the initial conditions. The condition $y(0) = 0$ requires that $c_2 = 1$, and the condition $y'(0) = 0$ requires that $0 = 1 + 2c_1$, or $c_1 = -\frac{1}{2}$. Thus $y = t - \frac{1}{2} \sin 2t + \cos 2t$ is a solution of the initial value problem (and we know from the Dynamical System Property that it is the only solution).

Example 5.4

It is easy to check that the differential equation $t^2 y'' - 2ty' + 2y = 0$ has the solution $y = t^2$, and hence the initial value problem

$$t^2 y'' - 2ty' + 2y = 0$$
$$y(0) = 0, \qquad y'(0) = 0 \tag{7}$$

has infinitely many solutions: $y = Ct^2$, where C is an arbitrary constant. This might appear to contradict the unicity portion of the Dynamical System Property, but closer examination shows that is not the case. Putting the given equation into normal form, we have $y'' - 2t^{-1}y' + 2t^{-2}y = 0$, and since the coefficients $a(t) = -2t^{-1}$ and $b(t) = 2t^{-2}$ are not continuous for any interval I containing the origin, the Dynamical System Property does not apply and hence there is no contradiction.

Vanishing Data

The Dynamical System Property has an immediate consequence which is often useful in simplifying calculations.

> **Vanishing Data Theorem.**   The initial value problem (5) with $f \equiv 0$ and $y_0 = v_0 = 0$ has only the trivial solution $y \equiv 0$ for any t_0.

Proof. Since problem (5) with vanishing data evidently has the trivial solution $y \equiv 0$, we infer from the Dynamical System Property that this is the only solution to the problem.

Example 5.5

There is no nontrivial solution curve of problem (5) with $f \equiv 0$ which is tangential to the t-axis. For if this were the case, $y_0 = v_0 = 0$, and hence the solution must be the trivial one.

Real-Valued versus Complex-Valued Solutions

Until now we have not made much of the fact that the coefficients $a(t)$ and $b(t)$, the driving term $f(t)$, or the data y_0 and v_0 in problem (5) may be complex valued (and hence that the solution may be complex valued). As we shall see presently, the concept of complex-valued solutions of differential equations can be very useful as a computational device in much the same way that complex numbers make it easier to discuss the roots and factorizability of polynomials.

If the differential equation $y'' + a(t)y' + b(t)y = f(t)$ arises in a mathematical model, then, of course, the coefficients $a(t)$ and $b(t)$ and the driving term $f(t)$ are all real-valued functions, and the solutions we ultimately seek are real-valued. If the solution set of complex-valued solutions for that equation is known, all real-valued solutions must appear among them. Another result that is useful in identifying real-valued solutions is as follows.

Real- and Complex-Valued Solutions. Let the polynomial operator $P(D)$ in (6) have real-valued coefficients, and suppose that $f(t)$ and $g(t)$ are given real-valued continuous functions. Then $y(t) = \alpha(t) + i\beta(t)$ for real-valued functions $\alpha(t)$ and $\beta(t)$ is a solution of $P(D)[y] = f(t) + ig(t)$ if and only if $P(D)[\alpha] = f$ and $P(D)[\beta] = g$. Thus any solution of the homogeneous equation $P(D)[y] = 0$ has the property that its real and imaginary parts are real-valued solutions of that equation.

Proof. From the Linearity Property of $P(D)$ we have that $P(D)[\alpha + i\beta] = P(D)[\alpha] + iP(D)[\beta]$. Since the coefficients in $P(D)$ are real valued, it follows that $P(D)[\alpha]$ and $P(D)[\beta]$ are real-valued functions, and hence identification of the real and imaginary parts of $P(D)[\alpha + i\beta]$ and $f + ig$ yields the result.

Example 5.6

Consider $P(D) = D^2 + 4$. The differentiation formula (3) shows that $P(D)[e^{2it}] = 0$ and $P(D)[e^{-2it}] = 0$. Thus $c_1 e^{2it} + c_2 e^{-2it}$ is a complex-valued solution of the homogeneous equation $P(D)[y] = 0$ for any complex constants c_1 and c_2. Since $e^{2it} = \cos 2t + i \sin 2t$, we see from the result above that $\cos 2t$ and $\sin 2t$ are real-valued solutions of $P(D)[y] = 0$ (see Example 5.2). These real solutions also arise by choosing special values for c_1 and c_2 in the family of complex-valued solutions above. Indeed, since $e^{-2it} = \cos 2t - i \sin 2t$, we see that $\cos 2t = \frac{1}{2}e^{2it} + \frac{1}{2}e^{-2it}$ and $\sin 2t = -(i/2)e^{2it} + (i/2)e^{-2it}$.

PROBLEMS

The following problem(s) may be more challenging: 7 and 9.

1. For the polynomial operator $P(D) \equiv D^2$, find:
   (a) The null space of $P(D)$.
   (b) The general solution of the equation $P(D)[y] = \sin t$.
2. For the polynomial operator $P(D) \equiv tD^2 + D$:
   (a) Find the null space of $P(D)$ over the interval $I = (0, \infty)$. [*Hint:* Write $P(D)[y]$ as $(ty')'$.]

(b) Show that the initial value problem $P(D)[y] = 2t$, $y(0) = a$, $y'(0) = b$, has a unique solution on any interval I containing the origin provided that b and a are chosen suitably.

(c) Why does the result in part (b) not contradict the Dynamical System Property?

3. Find the null space of the operator $P(D) \equiv (1 - t^2)D^2 - 2tD$ on the interval $[-\frac{1}{2}, \frac{1}{2}]$.

4. Solve the following initial value problems given the indicated information. [*Hint:* Follow Example 5.3.]

(a) $y'' - y' - 2y = 2e^t$, $y(0) = 0$, $y'(0) = 1$, given that $y_1 = e^{2t}$ and $y_2 = e^{-t}$ are solutions of the homogeneous equation and that $y_p = -e^t$ is a solution of the nonhomogeneous equation.

(b) $y'' + 2y' + 2y = 2t$, $y(1) = 1$, $y'(1) = 0$, given that $y_1 = e^{-t} \cos t$ and $y_2 = e^{-t} \sin t$ are solutions of the homogeneous equation and that $y_p = t - 1$ is a solution of the nonhomogeneous equation.

5. Verify that the given functions y_1 and y_2 are solutions of the differential equation and find all solutions that satisfy the initial conditions. Sketch the solution curves.

(a) $t^2 y'' + ty' - y = 0$; $y_1 = t^{-1}$, $y_2 = t$; $y(1) = 0$, $y'(1) = -1$.

(b) $t^2 y'' - ty' + y = 0$; $y_1 = t$, $y_2 = t \ln t$; $y(1) = 1$, $y'(1) = 0$.

(c) $t^2 y'' + ty' - 4y = 0$; $y_1 = t^2$, $y_2 = t^{-2}$; $y(0) = 0$, $y'(0) = 0$.

6. Consider the polynomial operator $P(D) = D^2 + aD + b$, where a and b are real (or complex) constants. Show that for any real (or complex) constant r we have $P(D)[e^{rt}] = (r^2 + ar + b)e^{rt} = P(r)e^{rt}$.

7. (a) Show for any $n = 1, 2, \ldots$, any constant r_0, and any n-times differentiable function z that $(D - r_0)^n[ze^{r_0t}] = e^{r_0t}D^n[z]$. [*Hint:* Do for $n = 1$, and then iterate.]

(b) Show for any polynomial $p(t)$ with degree $n - 1$ that the function $y = p(t)e^{r_0t}$ is a solution of the nth-order differential equation $(D - r_0)^n[y] = 0$ for any constant r_0.

8. Find as many elements as possible in the null spaces of the following operators. [*Hint:* See Problem 6.]

(a) $D^2 - 2D - 3$.   **(c)** $D^2 - 4i$.

(b) $D^2 + 4D + 13$.   **(d)** $D^2 + (1 + i)D + i$.

9. Find real-valued solutions of the following equations. Sketch several solution curves. [*Hint:* Use Euler's Formula and Problem 6.] Sketch several solutions.

(a) $y'' + 2y' + 2y = \sin t$.   **(b)** $y'' + 4y = e^{-t} \cos 2t$.

(c) $y'' - 2y' + 2y = 0$.

5.2 LINEAR DEPENDENCE, WRONSKIANS, AND HOMOGENEOUS EQUATIONS

As we shall see, the concept of linear dependence is useful in constructing $N(P(D))$ for any polynomial operator.

Linear Dependence.   The pair of functions $\{f_1, f_2\}$ in $C^0(I)$ is said to be (*linearly*) *dependent* if and only if there exist constants c_1 and c_2, not both zero, such that

$$c_1 f_1(t) + c_2 f_2(t) = 0 \qquad \text{for all } t \text{ in } I \tag{1}$$

(In other words, $\{f_1, f_2\}$ is dependent if and only if one function is a constant multiple of the other on I.) A function pair is *(linearly) independent* if and only if it is not dependent.

Example 5.7

The pair $\{t^3, |t|^3\}$ is independent on any interval I that contains the origin in its interior. Indeed, say that there are constants c_1, c_2 such that $c_1 t^3 + c_2 |t|^3 = 0$, for all t in I. Let $a > 0$ be any point in I such that $-a$ is also in I. Then, by substitution in the dependence relation, we see that $c_1 a^3 + c_2 a^3 = 0$ and $-c_1 a^3 + c_2 a^3 = 0$, and hence $c_1 = c_2 = 0$. Thus the pair cannot be dependent.

Example 5.8

The zero function $z(t)$, together with any other function $f(t)$, is always a dependent pair because of the dependency relation $(1)z(t) + (0)f(t) = 0$ for all t where f is defined.

Example 5.9

The pair $\{e^{r_1 t}, e^{r_2 t}\}$ is independent over any interval for any choice of the constants r_1 and r_2 as long as $r_1 \neq r_2$. The constants r_1 and r_2 may even be complex. Assume that c_1 and c_2 are constants (real or complex) such that $c_1 e^{r_1 t} + c_2 e^{r_2 t} = 0$ for all t in I. Differentiating this identity produces the new identity, $r_1 c_1 e^{r_1 t} + r_2 c_2 e^{r_2 t} = 0$ for all t in I. Subtracting r_1 times the first identity from the second, we obtain that $(r_2 - r_1)c_2 e^{r_2 t} = 0$ for all t in I. Since $r_2 \neq r_1$ and since $e^{r_2 t}$ never vanishes, we conclude that $c_2 = 0$. Similarly, we show that $c_1 = 0$. Thus no dependency relation (1) can exist and the pair is independent.

The Wronskian function is useful in testing a pair of functions for dependence.

Wronskian.† For any pair of functions $\{f, g\}$ in $C^1(I)$ the determinant

$$\begin{vmatrix} f(t) & g(t) \\ f'(t) & g'(t) \end{vmatrix} \equiv f(t)g'(t) - g(t)f'(t)$$

is called the *Wronskian* of the pair and is denoted by $W[f, g]$, or $W[f, g](t)$ if its dependence on the variable t is under consideration.

Example 5.10

We calculate some Wronskians. Observe that

(a) $W[e^{r_1 t}, e^{r_2 t}] = \begin{vmatrix} e^{r_1 t} & e^{r_2 t} \\ r_1 e^{r_1 t} & r_2 e^{r_2 t} \end{vmatrix} = (r_2 - r_1)e^{(r_1 + r_2)t}$

for any constants r_1 and r_2.

† Höene Wronski (1778–1853) started out as a soldier and then went on to become, successively, a mathematician, a philosopher, and insane.

(b) $W[e^{2t}, t^2] = \begin{vmatrix} e^{2t} & t^2 \\ 2e^{2t} & 2t \end{vmatrix} = 2te^{2t}(1 - t).$

(c) $W[t^3, |t|^3] = \begin{vmatrix} t^3 & |t|^3 \\ 3t^2 & 3t|t| \end{vmatrix} \equiv 0$ for all t.

Thus the Wronskian of a function pair may (a) never vanish on an interval, (b) vanish at some points and not at others, or (c) vanish identically on an interval.

However, when the two functions whose dependence or independence is in question are both elements of the null space of the same normal operator $P(D) = D^2 + aD + b$, we can say much more about their Wronskian. The following result shows the connection.

Wronskian and Independence.   Let w and z belong to the null space of $D^2 + a(t)D + b(t)$ on the interval I, where a and b are continuous. Then $\{w, z\}$ is an independent set if and only if $W[w, z](t)$ does *not* vanish for any t in I.

Proof. We shall prove the contrapositive form of this result; that is, we shall show that $\{w, z\}$ is dependent if and only if $W[w, z]$ vanishes for at least one value of t in I. Suppose first that $\{w, z\}$ is dependent on I. Then there are constants c_1 and c_2, not both zero, such that

$$c_1 w(t) + c_2 z(t) = 0, \qquad \text{all } t \text{ in } I \qquad (2a)$$

We may as well suppose that $c_1 \neq 0$. Differentiating (2a), we have that

$$c_1 w'(t) + c_2 z'(t) = 0, \qquad \text{all } t \text{ in } I \qquad (2b)$$

Multiplying (2a) by z', (2b) by z, and subtracting, we have that

$$0 = c_1(wz' - w'z) = c_1 W[w, z](t)$$

for all t in I. Since $c_1 \neq 0$, the Wronskian must vanish for all t.

Now suppose that $W[w, z](t_0) = 0$ for some t_0 in I. We must show that $\{w, z\}$ is dependent on I. Since $W[w, z](t_0) = 0$, the system

$$\begin{aligned} Aw(t_0) + Bz(t_0) &= 0 \\ Aw'(t_0) + Bz'(t_0) &= 0 \end{aligned} \qquad (3)$$

has a solution $(A, B) \neq (0, 0)$ (see Problem 5). Put $y(t) = Aw(t) + Bz(t)$ for this pair (A, B) and observe that $y(t)$ solves the initial value problem $P(D)[y] = 0$, $y(t_0) = 0$, $y'(t_0) = 0$. Thus by the Vanishing Data Theorem of Section 5.1, $y \equiv 0$ on I and hence the pair $\{w, z\}$ is dependent.

The theorem above has some interesting consequences. It implies that $\{w, z\}$ in $C^1(I)$ is independent if $W[w, z](t)$ is nonzero at some point t_0 in I. Note that in the proof that the Wronskian vanishes if $\{w, z\}$ is dependent we never used the fact that w and z belong to the null space of $P(D)$. Thus this half of the result holds whether or not we are dealing with $P(D)$. We see from the result above and Example 5.10(c) that the function pair $\{t^3, |t|^3\}$ cannot be solutions of the same differential equation $P(D)[y] = 0$ with $P(D)$ in the normalized form $D^2 + a(t)D + b(t)$. As a further example, the equation of Example 5.4 has the two solutions $w = t^2$ and $z = t$. Since $W[t^2, t] = -t^2$ vanishes at $t = 0$, we might at first glance think that this contradicts the Wronskian and Independence Theorem. But notice that Eq. (7) of Example 5.4 cannot be normalized to the form $(D^2 + a(t)D + b(t))[y] = 0$ with $a(t)$, $b(t)$ continuous on an interval I containing the origin, and hence the theorem does not apply in this case.

Null Space of P(D)

We shall now use the Wronskian and Independence Theorem to characterize the null space of the normalized polynomial operator $P(D) = D^2 + a(t)D + b(t)$. In other words, we will characterize the solution set of the homogeneous equation $y'' + a(t)y' + b(t)y = 0$, where $a(t)$ and $b(t)$ are in $C^0(I)$ for some interval I. But first we show that $P(D)[y] = 0$ always has a pair of independent solutions on I for any choice of $a(t)$, $b(t)$. Indeed, let t_0 be in I, and let y_1 be the solution of the initial value problem

$$P(D)[y] = 0, \qquad y(t_0) = 1, \quad y'(t_0) = 0$$

and y_2 the solution of the initial value problem

$$P(D)[y] = 0, \qquad y(t_0) = 0, \quad y'(t_0) = 1$$

guaranteed by the Dynamical System Property. Then because $W[y_1, y_2](t_0) = y_1(t_0)y_2'(t_0) - y_1'(t_0)y_2(t_0) = 1$, we see that the pair of solutions $\{y_1, y_2\}$ of $P(D)[y] = 0$ is an independent set.

Null Space of $P(D)$.   Let $P(D) = D^2 + a(t)D + b(t)$ with $a(t)$, $b(t)$ in $C^0(I)$. If the domain of $P(D)$ is $C^2(I)$ regarded as *complex*-valued functions, then $N(P(D))$ is

$$y = Ay_1 + By_2, \qquad A, B \text{ arbitrary } complex \text{ numbers} \qquad (4a)$$

where y_1, y_2 is any (fixed) independent pair in $N(P(D))$. If $a(t)$ and $b(t)$ are real valued and the domain of $P(D)$ is $C^2(I)$ regarded as a space of *real*-valued functions, then $N(P(D))$ is given by

$$y = Ay_1 + By_2, \qquad A, B \text{ arbitrary } real \text{ numbers} \qquad (4b)$$

where the *real*-valued pair $\{y_1, y_2\}$ is any (fixed) independent pair in $N(P(D))$.

Proof. We shall prove only (4b), the proof of (4a) being essentially the same. Let $\{y_1, y_2\}$ be an independent pair of real-valued solutions of $P(D)[y] = 0$, which we know to exist from the discussion preceding the theorem. Now if $z(t)$ is any real-valued solution of $P(D)[y] = 0$ we shall show that it can be expressed in the form (4b). Choose any point t_0 in I, and observe that $W[y_1, y_2](t_0) \neq 0$. Now the system of linear algebraic equations in the unknowns A and B,

$$y_1(t_0)A + y_2(t_0)B = z(t_0)$$
$$y_1'(t_0)A + y_2'(t_0)B = z'(t_0)$$
(5)

has a unique (real) solution (A_0, B_0) since $y_1(t_0)y_2'(t_0) - y_1'(t_0)y_2(t_0) \equiv W[y_1, y_2]$ $(t_0) \neq 0$. Putting $y(t) \equiv A_0y_1(t) + B_0y_2(t)$, we see from (5) that both $y(t)$ and $z(t)$ are solutions of the same initial value problem for $P(D)[y] = 0$. Thus the Dynamical System Property implies that $y(t) \equiv z(t)$ on I and hence $z = A_0y_1 + B_0y_2$, as asserted in (4b).

Example 5.11

In Example 5.2 we saw that $\{\sin 2t, \cos 2t\}$ is a pair of solutions of $(D^2 + 4)[y] = 0$. Now since $W[\sin 2t, \cos 2t] = -2$ we see that *all* real-valued solutions of the equation $y'' + 4y = 0$ are given by $y = A \sin 2t + B \cos 2t$, for arbitrary real numbers A, B. It is instructive to compare this result with the following one. In Example 5.6 we saw that $\{e^{2it}, e^{-2it}\}$ is a pair of complex-valued solutions of the equation $y'' + 4y = 0$. Now since $W[e^{2it}, e^{-2it}] = -4i \neq 0$, we see that the pair is independent. Thus according to the Null Space Theorem, we see that *all* complex-valued solutions of the equation $y'' + 4y = 0$ are given by $y = Ae^{2it} + Be^{-2it}$, where A, B are arbitrary complex numbers. Observe that the real-valued solutions are all contained among the complex-valued ones [because the set of complex-valued functions $C^2(I)$ contains the set of real-valued functions $C^2(I)$].

Because the representation (4) gives all solutions of $P(D)[y] = 0$ it is sometimes called the *general solution* of that equation. In using (4) one must be careful to distinguish the *general (complex-valued) solution* (4a) from the *general (real-valued) solution* (4b) when appropriate (see Example 5.11). More examples and illustrations of this point will be given as we go along.

Equations with Constant Coefficients

We turn now to homogeneous second-order linear differential equations with constant coefficients. Thus the polynomial operators we consider have the form $P(D) = D^2 + aD + b$, but with a and b constants. If in $P(D)$ the symbol D is formally replaced by the algebraic variable r, the resulting polynomial

$$P(r) \equiv r^2 + ar + b$$
(6)

is called the *characteristic polynomial* associated with the polynomial operator $P(D)$. Observe that if $P(D)$ is applied to e^{rt}, where r is any real (or complex) constant, then by direct calculation,

$$P(D)[e^{rt}] \equiv (r^2 + ar + b)e^{rt} \equiv P(r)e^{rt} \qquad (7)$$

where $P(r)$ is precisely the characteristic polynomial (6) associated with the operator $P(D)$. Using (7), it can be shown that there is a one-to-one correspondence between polynomials in r and polynomial operators.

If r_1 is any constant and $y(t)$ is any function in $C^2(I)$, note that $(D + r_1)[y]$ is an element of $C^1(I)$. Thus if r_2 is constant, the operator $D + r_2$ may be applied to $(D + r_1)[y]$, and this particular succession of operators applied to a function y is denoted by $(D + r_2)(D + r_1)$. Now suppose that the constant-coefficient polynomial $P(r) = r^2 + ar + b$ is written in the factored form $(r - r_2)(r - r_1)$. Then we claim that the polynomial operator $P(D) \equiv D^2 + aD + b$ can be written in the form $(D - r_2)(D - r_1)$. Indeed, if $Q(D)$ is the "factorized" polynomial operator $(D - r_2)(D - r_1)$, then using successively the fact that

$$(D - r)e^{st} = (s - r)e^{st} \qquad \text{for any constant } r, s \qquad (8)$$

we see that

$$Q(D)[e^{rt}] = (D - r_2)[(r - r_1)e^{rt}] = (r - r_2)(r - r_1)e^{rt}$$

Thus $Q(D) \equiv P(D)$ since they both have the same characteristic polynomial. In the same way we can show that $P(D) = (D - r_1)(D - r_2)$, and hence

$$(D - r_1)(D - r_2) = (D - r_2)(D - r_1) \qquad \text{for all constants } r_1 \text{ and } r_2$$

Now we are ready for the

General (Complex-Valued) Solution Theorem.   Let a and b be any real (or complex) constants and suppose that r_1 and r_2 are the roots of the polynomial $r^2 + ar + b$. Then the general solution of the homogeneous linear second-order equation

$$y'' + ay' + b = 0 \qquad (9)$$

is given by

$$K_1 e^{r_1 t} + K_2 e^{r_2 t} \qquad \text{if } r_1 \neq r_2 \qquad (10a)$$

$$K_1 e^{r_1 t} + K_2 t e^{r_1 t} \qquad \text{if } r_1 = r_2 \qquad (10b)$$

where K_1 and K_2 are arbitrary complex constants.

Proof. Writing (9) in the form $(D - r_1)(D - r_2)[y] = 0$ and using (8), we see that $e^{r_1 t}$ and $e^{r_2 t}$ are solutions. Since $W[e^{r_1 t}, e^{r_2 t}] = (r_2 - r_1)e^{(r_1 + r_2)t} \neq 0$ if $r_1 \neq r_2$, we see that $\{e^{r_1 t}, e^{r_2 t}\}$ is independent in that case and (10a) follows by (4a) in the Null Space Property. Now when $r_1 = r_2$ the procedure above fails to produce a second solution of (9) independent of $e^{r_1 t}$. In this case we look for a second solution in the form $z(t)e^{r_1 t}$ for some $z(t)$. Now using the identity $(D - r_1)[h(t)e^{r_1 t}] = e^{r_1 t}h'$ twice in succession, we see that $(D - r_1)^2[ze^{r_1 t}] = e^{r_1 t}z''$, and hence we must require that z satisfy the equation $z'' = 0$. We conclude $(A + Bt)e^{r_1 t}$ must be a solution of (9) in this case for any choice of constants A, B, and hence in particular $te^{r_1 t}$ is a solution of (8). Since $W[e^{r_1 t}, te^{r_1 t}] \neq 0$, we conclude from the Null Space Property that this independent set of solutions generates all solutions in the manner shown in (10b).

Example 5.12

The characteristic polynomial of the linear equation

$$y'' + 2y' + 5y = 0 \tag{11}$$

has the roots $r_1 = -1 + 2i$ and $r_2 = -1 - 2i$, and hence (10) gives the general complex-valued solution

$$y = K_1 e^{(-1+2i)t} + K_2 e^{(-1-2i)t} \tag{12}$$

for arbitrary complex numbers K_1 and K_2.

Example 5.13

The equation $y'' + 4y' + 4y = 0$ has the characteristic polynomial $r^2 + 4r + 4$ with roots $r_1 = r_2 = 2$. Thus (10) gives the general (complex-valued) solution as $y = (K_1 + K_2 t)e^{2t}$, where K_1 and K_2 are arbitrary complex numbers. But since the solutions e^{2t} and te^{2t} are real valued as well as independent, the Null Space Property (4b) implies that the general (real-valued) solution is $y = (C_1 + C_2 t)e^{2t}$, where C_1 and C_2 are arbitrary *real* numbers.

Real-Valued Solutions

If the differential equation (9) has real coefficients, we would want to find the *real-valued general solution* of that equation, that is, all solutions which are *real-valued* functions. Of course, the complex-valued general solution (10) must contain all real-valued solutions of (9), and hence it is only a matter of recognizing them within this much larger class of solutions. An example will help explain how this goes.

Example 5.14

We shall use the general complex-valued solution (12) for Eq. (11) to find the general real-valued solution of that equation. Recall that $e^{(\alpha+i\beta)t} = e^{\alpha t}\cos \beta t + ie^{\alpha t}\sin \beta t$, for any real numbers α and β, and hence the exponentials in (12) are complex conjugates of one another. Thus taking the constants $K_1 = K_2 = \frac{1}{2}$ in (12), we have the real-valued solution $e^{-t}\cos 2t$, whereas if we put $K_1 = -i$ and $K_2 = i/2$ in (12), the real-valued solution $e^{-t}\sin 2t$ results. Now since $W[e^{-t}\cos 2t, e^{-t}\sin 2t] \neq 0$, we have an independent pair of *real*-valued solutions of (11) and hence by (4b) we have the general real-valued solution for (11):

$$y = Ae^{-t} \cos 2t + Be^{-t} \sin 2t, \qquad A, B \text{ arbitrary reals}$$

General (Real-Valued) Solution Theorem.   Suppose the coefficients of $P(D) = D^2 + aD + b$ are real numbers and that the characteristic polynomial $P(r)$ has the roots r_1 and r_2. Then the general (real-valued) solution of the homogeneous equation $P(D)[y] = 0$ is given by

$$
\begin{array}{ll}
k_1 e^{r_2 t} + k_2 e^{r_2 t} & \text{if } r_1, r_2 \text{ real, } \quad r_1 \neq r_2 \\
k_1 e^{\alpha t} \cos \beta t + k_2 e^{\alpha t} \sin \beta t, & \text{if } r_1 = \bar{r}_2 = \alpha + i\beta, \quad \beta \neq 0 \qquad (13) \\
k_1 e^{r_1 t} + k_2 t e^{r_1 t} & \text{if } r_1 = r_2
\end{array}
$$

where k_1 and k_2 are arbitrary real numbers.

Proof. Only the case $r_1 = \bar{r}_2 = \alpha + i\beta$, $\beta \neq 0$ needs any argument. Since $e^{r_1 t}$ is a (complex-valued) solution of $P(D)[y] = 0$, and since

$$e^{(\alpha + i\beta)t} = e^{\alpha t} \cos \beta t + i e^{\alpha t} \sin \beta t$$

we see from an earlier result that the functions $e^{\alpha t} \cos \beta t$ and $e^{\alpha t} \sin \beta t$ are (real-valued) solutions of that same equation. Since $W[e^{\alpha t} \cos \beta t, e^{\alpha t} \sin \beta t] \neq 0$, the pair is independent and the assertion follows from (4b) in the Null Space Property.

PROBLEMS

The following problem(s) may be more challenging: 1(a), 1(f), 8, 11, 12, 15, 17(c), 18(a), 19(a) and 20(a).

1. Determine whether the following pairs of functions are dependent or independent on the real line.
   (a) $\{1 + t, 1 - |t|\}$.
   (b) $\{e^{it}, e^{-it}\}$.
   (c) $\{\sin^2 t, 1 - \cos 2t\}$.
   (d) $\{\cos t - \sin t, \cos t + \sin t\}$.
   (e) $\{\sin 2t - i \cos 2t, 3e^{2it}\}$.
   (f) $\{t^2, t|t|\}$.

2. Find the null spaces of each of the following operators.
   (a) $D^2 - D - 2$ acting on *real*-valued functions.
   (b) $D^2 - 4D + 5$ acting on *real*-valued functions.
   (c) $D^2 + 2D + 1$ acting on *complex*-valued functions.
   (d) $D^2 + 2D + 2$ acting on *complex*-valued functions.
   (e) $2D^2 - 2D + 4$ acting on *complex*-valued functions.
   (f) $t^2 D^2 - 2tD + 2$ acting on *real*-valued functions on $t > 0$. [*Hint:* Try $y = t^r$.]

3. For each case below write a differential equation with real coefficients for which the given mathematical expression is a solution.
   (a) $e^{-5t} - e^{-2t}$. [*Hint:* Consider the operator $(D + 5)(D + 2)$.]

(b) $3e^{4t} + 2te^{4t}$. [*Hint*: Consider $(D - 4)^2$.]

(c) $8e^{-4t} \sin t$.

(d) $e^{(3-5i)t}$.

(e) $(1 + it)e^{(3-5i)t}$.

(f) $\cos t + e^{(2-i)t}$.

4. Give all the real-valued solutions (if any) of the following differential equations.

(a) $y'' + (1 + i)y' + iy = 0$.

(b) $y'' + (-1 + 2i)y' - (1 + i)y = 0$.

5. Show that the system (3) always has a solution $(A, B) \neq (0, 0)$ if $W[w, z](t_0) = 0$. [*Hint*: If either $z(t_0) \neq 0$ or $w(t_0) \neq 0$, show that $A = z(t_0)$, $B = -w(t_0)$ solves (3). If $w(t_0) = z(t_0) = 0$, there is only one equation to solve.]

6. Find the solution of each of the following initial value problems. Sketch the solution curves.

(a) $y'' + 2y' + 2y = 0$, $y(0) = 1$, $y'(0) = 0$.

(b) $y'' + 4y' + 4y = 0$, $y(1) = 2$, $y'(1) = 0$.

(c) $y'' + y' - 6y = 0$, $y(-1) = 1$, $y'(-1) = -1$.

(d) $y'' + 2y' + 2y = e^t$, $y(0) = 0$, $y'(0) = 0$. [*Hint*: Use (7) to find a particular solution.]

(e) $ty'' - y' = 0$, $y(1) = 1$, $y'(1) = -1$.

7. Find all solutions of the following *boundary value problems*.

(a) $y'' - 3y' + 2y = 0$, $y(0) - y'(0) = 1$, $y'(1) = -2$.

(b) $y'' = 2$, $2y(0) + y'(0) = 0$, $2y(1) - y'(1) = 0$.

(c) $y'' = 2$, $y(0) = 0$, $y(1) + y'(1) = 0$.

8. Show that $y_1 = t^3$, $y_2 = |t|^3$ is a pair of solutions of the equation $t^2 y'' - 4ty' + 6y = 0$ on the interval $-\infty < t < \infty$. In Example 5.10 we saw that $W[t^3, |t|^3] \equiv 0$ for all t, and in Example 5.7 that $\{t^3, |t|^3\}$ is independent. Why do these facts not provide a counterexample to the Wronskian and Independence Theorem?

9. There are many pairs of solutions of the equation $y'' - y' - 2y = 0$ that generate $N(D^2 - D - 2)$ as in (4b) of the Null Space Property. One such pair $\{e^{-t}, e^{2t}\}$ is as given in the General Solution Theorem. Show that the pair $\{ae^{-t} + be^{2t}, ce^{-t} + de^{2t}\}$ for any choice of constants a, b, c, d such that $ad - bc \neq 0$ also generates $N(D^2 - D - 2)$.

10. **(a)** Show that the solutions of $y'' + ay' + by = 0$ for complex constants a and b are given by (10). [*Hint*: Repeat the steps used in the real coefficient case, making appropriate changes.]

(b) Find the general solution of the equation $y'' + iy' + 2y = 0$.

(c) Does the equation in part (b) have nontrivial real-valued solutions?

(d) Find the general solution of $y'' + iy = 0$. [*Hint*: Use De Moivre's Formula to find the roots of the characteristic polynomial.]

11. **(a)** Show that if the real numbers a, b are such that $a > 0$, $b > 0$, then every solution of $y'' + ay' + by = 0$ tends to zero as $t \to \infty$.

(b) Show that the converse of part (a) holds: If every solution tends to zero as $t \to \infty$, then $a > 0$ and $b > 0$.

12. **(a)** A solution $y(t)$ of $y'' + ay' + by = 0$, a and b real numbers, is said to be *positively bounded* if there is a positive constant M such that $|y(t)| \leq M$ for all $t \geq 0$. Show that all solutions are positively bounded if $a \geq 0$, $b \geq 0$, and $a^2 + b^2 \neq 0$.

(b) Show the converse: If all solutions are positively bounded, then $a \geq 0$, $b \geq 0$, and $a^2 + b^2 \neq 0$.

13. [*Reduction of Order Technique*]. Let $u(t)$ be a solution of $y'' + a(t)y' + b(t)y = 0$, where $a(t)$ and $b(t)$ are in $C^0(I)$.

    (a) Show that $y = u(t)z(t)$ is also a solution of that equation if $z(t)$ satisfies the equation $uz'' + (2u' + a(t)u)z' = 0$.

    (b) Show that

$$z(t) = \int_{t_0}^{t} \left\{ u^{-2}(s) \exp\left[-\int^{s} a(r)\, dr \right] \right\} ds$$

    is a solution of the z-equation in part (a) if $u \neq 0$ on I.

    (c) Show that the pair of solutions $\{u, uz\}$ constructed in parts (a) and (b) is independent.

    (d) Find all solutions of $ty'' - (t + 2)y' + 2y = 0$ for $t > 0$ given that e^t is one solution.

14. [*Wronskian Reduction of Order*]. Let $P(D)$ be the polynomial operator in (6), and let $W(t) = W[y_1, y_2](t)$ be the Wronskian of any pair of solutions $\{y_1, y_2\}$ of $P(D)[y] = 0$.

    (a) Show that $W(t)$ satisfies the first-order linear equation $W' = -a(t)W$, and hence W is given by *Abel's Formula*

$$W(t) = W(t_0) \exp\left[-\int_{t_0}^{t} a(s)\, ds \right] \text{ for } t_0, t \text{ in } I$$

    [*Hint*: Differentiate W directly.]

    (b) Use Abel's Formula to show that if $u(t) \neq 0$ is a solution of $P(D)[y] = 0$, then a second solution $v(t)$ independent of $u(t)$ can be found by solving for v in the first-order linear equation

$$u(t)v' - u'(t)v = \exp\left[-\int^{t} a(s)\, ds \right]$$

    (c) Given that e^t is a solution of $ty'' - (t + 2)y' + 2y = 0$, find a second independent solution. [*Hint*: Remember to normalize the equation first.]

    (d) Show that this method and the method of problem 13 are essentially the same.

15. Suppose that L is the second-order linear operator $D^2 + p(t)D + q(t)$, where $p(t)$ and $q(t)$ are real-valued and continuous for all t. Suppose further that $\{y_1(t), y_2(t)\}$ is an independent subset of the null space of L. Which of the following sketches of the graphs of $y_1(t)$ and $y_2(t)$ are *not* logically possible? Explain.

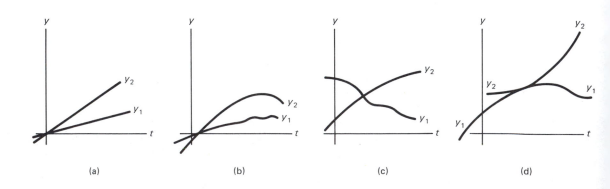

    (a)   (b)   (c)   (d)

The following problems give properties of

$$y'' + a(t)y' + b(t)y = 0 \qquad (*)$$

where $a(t)$ and $b(t)$ are continuous for t in a common open interval I. A number α in I is a *zero* of a solution $y(t)$ if $y(\alpha) = 0$. A nontrivial solution is *oscillatory* on I if it has infinitely many zeros in I. It can be shown that any nontrivial solution of (*) has at most finitely many zeros on any closed, bounded subinterval of I. Two distinct zeros of a solution are *consecutive* if there are no other zeros in between. The zeros of two solutions *interlace* if between consecutive zeros of either there is a zero of the other. The following problems illustrate these ideas:

16. (a) Show that nontrivial solutions of $y'' + 9y = 0$ are oscillatory on the real line and that any two independent solutions interlace. [*Hint*: Use the identity $A \cos \theta + B \sin \theta = C \sin (\theta + \delta)$, $C = (A^2 + B^2)^{1/2}$, $\cos \delta = B/C$, $\sin \delta = A/C$.]
    (b) Repeat part (a) for the equation $y'' + 2y' + 5y = 0$.
    (c) Show that no solution of $y'' + 3y' + 2y = 0$ is oscillatory on any interval.

17. (a) Prove that if α is a zero of a nontrivial solution of $y(t)$, then $y'(\alpha) \neq 0$. [*Hint*: Recall that the trivial solution of (*) vanishes together with its derivative at α.]
    (b) Show that no two independent solutions of (*) have a common zero. [*Hint*: What must the Wronskian of the two solutions be at a common zero?]
    (c) Show that no sequence of distinct zeros of a nontrivial solution of (*) can converge to a point of I.

18. (a) [*Interlace Theorem*]. Show that the zeros of any two independent solutions of (*) interlace.
    (b) [*Interlace and Oscillation*]. Show that if (*) has one oscillatory solution on I, all nontrivial solutions are oscillatory on I, whereas if one solution has exactly n zeros in I, every solution has $n - 1$, n, or $n + 1$ zeros in I.

19. (a) [*Comparison Theorem*]. Let u and v be respective nontrivial solutions of $u'' + q(t)u = 0$, $v'' + r(t)v = 0$ (v is the *comparison equation*), where q and r are continuous and $q(t) > r(t)$, t in I. Show that between consecutive zeros of $v(t)$ there is a zero of $u(t)$. [*Hint*: If not so, then $u(t)$ has no zeros between some pair α, β, $\alpha < \beta$ of consecutive zeros of $v(t)$. Suppose that $u(t)$ and $v(t) > 0$, $\alpha < t < \beta$, and let $W = W[u, v](t)$. Show that $W' > 0$, for all t with $\alpha < t < \beta$, and that $W(\beta) \leq 0 \leq W(\alpha)$: a contradiction.]
    (b) Use part (a) to show that all nontrivial solutions of $u'' + (4 + 3e^{-t})u = 0$ are oscillatory on I. [*Hint*: Compare with $v'' + 4v = 0$.]
    (c) [*Oscillation and Comparison*]. Let u be a nontrivial solution of $u'' + q(t)u = 0$, where $q(t) \geq k^2 > 0$, t in an infinite interval I. Show that u is oscillatory on I with consecutive zeros less than π/k apart. [*Hint*: Use part (a) and the comparison equation $v'' + k^2v = 0$.]

20. (a) [*Normal Form*]. Show that the change of variable $y = \rho u$, where $\rho(t) = \exp \left[- \int^t \frac{1}{2}a(s) \, ds \right]$, changes $y'' + a(t)y' + b(t)y = 0$ to its *normal form*, $u'' + q(t)u = 0$, where $q = b - a^2/4 - a'/2$, and a is assumed to be differentiable. Show that the zeros of $y(t)$ and of $u(t)$ coincide. Note that this definition of normal form differs from that used previously.

(b) [*Bessel's Equation of Order 0*]. Show that every nontrivial solution of $y'' + t^{-1}y' + y = 0$ is oscillatory on the interval $t > 0$ and that consecutive zeros are less than π apart. [*Hint*: Change to normal form and compare to $v'' + v = 0$.]

5.3 NONHOMOGENEOUS EQUATIONS

In Section 5.1 we learned a great deal about the polynomial operator

$$P(D) = D^2 + a(t)D + b(t) \tag{1}$$

where the coefficients are in $C^0(I)$ for some interval I on the t-axis. We saw that $P(D)$ satisfies a linearity property and that its null space is closed with respect to taking sums and multiplying by a constant. We stated the fundamental Dynamical System Property that for each f in $C^0(I)$ and each t_0 in I the differential equation $P(D)[y] = f(t)$ has precisely one solution on I that satisfies the initial conditions $y(t_0) = y_0$, $y'(t_0) = v_0$, where y_0 and v_0 are any given constants. Using this property we were able to give a useful characterization of $N(P(D))$—or in other words, to characterize the general solution of the homogeneous equation $P(D)[y] = 0$. In this section we examine the nonhomogeneous equation $P(D)[y] = f$ more closely and give several techniques for producing the general solution of the driven equation if $N(P(D))$ is known. A key result for this purpose is the

---

Decomposition of Solutions Property.   Let $P(D)$ be any polynomial operator, and let y_p be any given (fixed) solution of the equation $P(D)[y] = f$. Then the general solution of $P(D)[y] = f$ is given by $y = y_p + z$, with z any function in $N(P(D))$.

Proof. Observe that y is a solution of $P(D)[y] = f$ if and only if the difference $z = y - y_p$ satisfies the equation $P(D)[z] = P(D)[y - y_p] = P(D)[y] - P[y_p] = f - f = 0$, that is, if and only if z solves the equation $P(D)[z] = 0$.

Example 5.15

Suppose that we have somehow found the solution $y_p = (-\frac{3}{4} - \frac{1}{2}t)e^t$ to the equation

$$(D + 1)(D - 2)[y] = (1 + t)e^t \tag{2}$$

(see Example 5.15 (revisited) for one way to do this). Then from the Decomposition Property the general solution of equation (2) is $y = (-\frac{3}{4} - \frac{1}{2}t)e^t + z$, where z is the general solution of the associated homogeneous equation

$$(D + 1)(D - 2)[z] = 0 \tag{3}$$

But from Section 5.2 we know how to find the general solution of (3), and the general solution of (2) is

$$y = (-\tfrac{3}{4} - \tfrac{1}{2}t)e^t + K_1 e^{-t} + K_2 e^{2t}, \qquad K_1, K_2 \text{ arbitrary constants}$$

Thus the Decomposition Property for solutions of $P(D)[y] \equiv f$ implies that the search for the general solution essentially comes down to finding just one particular solution of the nonhomogeneous equation $P(D)[y] = f$ (any one will do) if $N(P(D))$ is known [recall that finding the general solution of the homogeneous equation $P(D)[z] = 0$ is a fairly routine matter in case P has constant coefficients]. There are several methods that are widely used for doing this. If $P(D)$ has constant coefficients, some of these methods are (a) the *Method of Undetermined Coefficients* (discussed below), (b) a direct operator method (treated in Section 6.4), and (c) *Laplace Transforms* (treated in Chapter 7). The *Method of Variation of Parameters* is employed when the general solution of the homogeneous equation $P(D)[z] = 0$ is known (treated below).

Exponential Driving Terms

Observe from (7) of Section 5.2 that $P(D)[Ce^{\omega t}] = CP(\omega)e^{\omega t}$ for any constants C and ω (real or complex). If $P(\omega) \neq 0$, we easily find a particular solution to the equation $P(D)[y] = Ae^{\omega t}$ for a given constant A, by taking $y_p = [A/P(\omega)]e^{\omega t}$.

Example 5.16

Consider $P(D) = D^2 - D + 1$ and the equation $P(D)[y] = -2e^{-it}$. Now since $P(-i) = (-i)^2 - (-i) + 1 = i$, we see that

$$y_p = \frac{-2}{i}\, e^{-it} = 2ie^{-it}$$

is a particular solution for that equation.

To find a (real-valued) particular solution for the equation $P(D)[y] = Ae^{\alpha t}\cos \beta t$ when A, α, and β are real constants and $P(D)$ has *real* coefficients, we proceed as follows. First note that $Ae^{\alpha t}\cos \beta t$ is the real part of $Ae^{\omega t}$, where $\omega = \alpha + i\beta$. Then from the relationship between real and complex solutions we see that if y_p is any solution of the equation $P(D)[y] = Ae^{\omega t}$, then $z_p = \mathrm{Re}\,[y_p]$ is a real-valued solution to the equation $P(D)[y] = Ae^{\alpha t}\cos \beta t$.

Example 5.17

Consider the equation

$$y'' + 4y' + 3y = 3e^{-t}\cos 3t \tag{4}$$

Note that $3e^{-t}\cos 3t = \mathrm{Re}\,[\beta e^{(-1+3i)t}]$, and hence we look at the equation

$$(D^2 + 4D + 3)[y] = 3e^{(-1+3i)t} \tag{5}$$

Putting $P(D) \equiv D^2 + 4D + 3$, we see that $P(-1 + 3i) \neq 0$, and hence (5) has the particular solution

$$y_p = \frac{3}{P(-1+3i)} e^{(-1+3i)t} = -\left(\frac{3}{13} + \frac{2}{13} i\right) e^{-t}(\cos 3t + i \sin 3t)$$

Thus $z_p = \text{Re}\,[y_p] = (-\frac{3}{13}\cos 3t + \frac{2}{13}\sin 3t)e^{-t}$ is a particular solution of (4).

Method of Undetermined Coefficients

The *Method of Undetermined Coefficients* deals with a "matching-up-the-coefficients" technique for finding a particular solution of the equation $P(D)[y] = f$, where the polynomial operator $P(D)$ has constant coefficients, and the forcing term has the *polynomial-exponential* form $f(t) = \sum_{j=1}^{n} p_j(t)e^{\omega_j t}$ with each p_j a polynomial (real or complex coefficients) and each ω_j a real or complex number. The Linearity Property of $P(D)$ implies that if y_j is such that $P(D)[y_j] = p_j(t)e^{\omega_j t}$ for each $j = 1, 2, \ldots,$ n, then $y = \sum_{j=1}^{n} y_j$ is a solution of $P(D)[y] = \sum_{j=1}^{n} p_j(t)e^{\omega_j t}$. Thus we lose nothing by considering only the special case

$$P(D)[y] = p(t)e^{\lambda t} \tag{6}$$

for a given polynomial $p(t)$ and constant λ. Since derivatives of polynomial-exponential functions are obviously again polynomial-exponential functions (with the same exponentials), it is not unreasonable to look for a solution of (6) in the form $y = q(t)e^{\lambda t}$ for some polynomial q. Notice that

$$P(D)[q(t)e^{\lambda t}] \equiv D^2[qe^{\lambda t}] + aD[qe^{\lambda t}] + bqe^{\lambda t} \equiv e^{\lambda t}\{q'' + (a + 2\lambda)q' + P(\lambda)q\}$$

and hence $qe^{\lambda t}$ is a solution of (6) if and only if

$$q'' + (a + 2\lambda)q' + P(\lambda)q = p(t) \tag{7}$$

Let the characteristic polynomial $P(r)$ have the factorization $(r - r_1)(r - r_2)$. Now if $\lambda \neq r_1, r_2$, then $P(\lambda) \neq 0$ and hence if $p = \sum_{k=0}^{n} b_k t^k$, it is not difficult to see that (7) has a unique solution of the form $q = \sum_{k=0}^{n} a_k t^k$.

Example 5.15 (*Revisited*)

How did we find the particular solution of the equation $(D + 1)(D - 2)[y] = (1 + t)e^t$ of Example 5.15? Using (7) with $\lambda = 1$ we have that if $q(t)$ is a polynomial such that $q'' + q' - 2q = 1 + t$, then $y = qe^t$ is a particular solution of the given equation. Supposing that q has the form $q(t) = a + bt$, then $(D^2 + D - 2)[q] = b - 2a - 2bt$, which equals $1 + t$ if $b - 2a = 1$ and $-2b = 1$. Thus $b = -\frac{1}{2}$, $a = -\frac{3}{4}$, yielding the particular solution $y_p = (-\frac{3}{4} - \frac{1}{2}t)e^t$.

Notice, however, that if $\lambda = r_1$ but $r_1 \neq r_2$, then (7) becomes $q'' + (r_1 - r_2)q' = p$ because $P(r_1) = 0$ and $a = -r_1 - r_2$. Now for any polynomial q we see that $\deg(q'' + (r_1 - r_2)q') = \deg(q) - 1$, and hence to solve (7) we must take for q a polynomial of degree $= 1 + \deg(p)$.† That is, if $p = \sum_{k=0}^{n} b_k t^k$, then take $q = \sum_{k=0}^{n+1} a_k t^k$ in (7). Since a_0 contributes nothing to $q'' + (r_1 - r_2)q'$, a_0 may be considered to be arbitrary and we may as well put $a_0 = 0$. An example will clarify this.

Example 5.18

Let us find a particular solution of the equation $(D + 1)(D - 1)[y] = 3t^2 e^t$. Now in this case $\lambda = 1$, $r_1 = 1$, and $r_2 = -1$ and hence (7) becomes $q'' + 2q' = 3t^2$. From the discussion above we take $q = c_1 t + c_2 t^2 + c_3 t^3$. Inserting this in (7) and comparing coefficients, we have that $6c_3 = 3$, $4c_2 + 6c_3 = 0$, and $2c_1 + 2c_2 = 0$. Thus $c_3 = \frac{1}{2}$, $c_2 = -\frac{3}{4}$, and $c_1 = \frac{3}{4}$, so the function $y_p = (t/4)(3 - 3t + 2t^2)e^t$ is a particular solution of the given equation.

Finally, in the case where $\lambda = r_1 = r_2$, we see that (7) becomes $q'' = p$. Now since q arises by antidifferentiating p twice, we see that $\deg(q) = 2 + \deg(p)$, and that the constant and first-order terms in q play no role and hence may as well be dropped. The three cases discussed above are summarized in the following theorem.

Undetermined Coefficients Theorem.   Let the constant-coefficient operator $P(D) = D^2 + aD + b$ have the factorization $(D - r_1)(D - r_2)$. Then the equation $P(D)[y] = p(t)e^{\lambda t}$, where p is a polynomial and λ a constant, has a particular solution $y_p = q(t)e^{\lambda t}$, where $q(t)$ is a polynomial whose form is indicated below.

If $\deg(p) = n$, and:	Then take
$\lambda \neq r_1$, $\lambda \neq r_2$	$q(t) = \sum_{k=0}^{n} a_k t^k$, where $q'' + (a + 2\lambda)q' + P(\lambda)q = p$
$\lambda = r_1 \neq r_2$	$q(t) = t \sum_{k=0}^{n} a_k t^k$, where $q'' + (r_1 - r_2)q' = p$
$\lambda = r_1 = r_2$	$q(t) = t^2 \sum_{k=0}^{n} a_k t^k$, where $q'' = p$

---

† Deg $(q(x))$, where $q(x)$ is a polynomial, denotes the degree of $q(x)$. For example, we have that $\deg(3x^2 - 7x + 1) = 2$.

Example 5.19

We find a real-valued particular solution of the equation

$$(D^2 - 2D + 2)[y] = 4te^t \sin t + 3 \cos 2t \tag{8}$$

by considering the two separate equations,

$$(D^2 - 2D + 2)[y] = 4te^t \sin t \tag{9a}$$

$$(D^2 - 2D + 2)[y] = 3 \cos 2t \tag{9b}$$

Observe that if y_p^1 is a solution of (9a) and y_p^2 is a solution of (9b), then $y_p = y_p^1 + y_p^2$ is a solution of (9) because of the linearity property of $P(D) = D^2 - 2D + 2$. Treating (9a) first, note that the coefficients in $P(D)$ are real and that $4te^t \sin t = \text{Im } [4te^{(1+i)t}]$. Thus if we find a (complex-valued) solution z_p of

$$(D^2 - 2D + 2)[z] = 4te^{(1+i)t} \tag{9a'}$$

then Im $[z_p]$ is a solution of (9a) and may be taken as y_p^1. Now (9a') is in a form treated by the Undetermined Coefficients Theorem. Note that $\lambda = r_1 = 1 + i$, $r_2 = 1 - i$, and hence $P(\lambda) = 0$. Thus to find a polynomial solution to $q'' + 2iq' = 4t$ we must take $q = c_1 t + c_2 t^2$. Substituting and matching coefficients, we see that $2ic_1 + 2c_2 = 0$, $4ic_2 = 4$. Thus $c_2 = -i$, $c_1 = 1$, and hence $z_p = (t - it^2)e^{(1+i)t}$ is a solution of (9a'). It follows that $y_p^1 = \text{Im } [z_p] = (t \sin t - t^2 \cos t)e^t$ is a solution of (9a). Similarly, since $3 \cos 2t = \text{Re } [3e^{2it}]$, we find a solution z_p of

$$(D^2 - 2D + 2)[z] = 3e^{2it} \tag{9b'}$$

and observe that Re $[z_p]$ is a solution of (9b) and hence may be taken for y_p^2. Note that $\lambda = 2i$ and that $P(2i) = -2 - 4i$. Therefore, $z_p = [3/(-2 - 4i)]e^{2it}$ is a solution of (9b'). Since $3/(-2 - 4i) = -\frac{3}{10} + \frac{3}{5}i$ and $e^{2it} = \cos 2t + i \sin 2t$, it follows that $y_p^2 = \text{Re } [z_p] = -\frac{3}{10} \cos 2t - \frac{3}{5} \sin 2t$ is a solution of (9b'). Thus a real-valued particular solution y_p for (9) is

$$y_p = y_p^1 + y_p^2 = (t \sin t - t^2 \cos t)e^t - \tfrac{3}{8} \cos 2t - \tfrac{3}{4} \sin 2t$$

Variation of Parameters

Now we turn to the task of finding a particular solution of the linear differential equation $P(D)[y] = f$ for $P(D)$ as in (1) and f any function in the class $C^0(I)$. The Method of Undetermined Coefficients is successful only when $P(D)$ has constant coefficients and $f(t)$ is a polynomial-exponential function. Thus the method we are about to describe is more general than that of undetermined coefficients. Our approach assumes that we are given an independent pair in the null space $N(P(D))$, and then uses this pair to construct a particular solution of $P(D)[y] = f$. The construction of the general solution then follows, as before, from the Decomposition of Solutions Property.

For any independent pair $\{y_1, y_2\}$ in $N(P(D))$ we know that $c_1 y_1 + c_2 y_2$ is a solution of the homogeneous equation $P(D)[y] = 0$. We shall show that it is possible to "vary the parameters" c_1 and c_2 by making them functions of t in such a way that $c_1(t) y_1 + c_2(t) y_2$ becomes a solution of the nonhomogeneous equation $P(D)[y] = f$. This is the essence of the *Method of Variation of Parameters*.

Variation of Parameters.   Let $P(D)$ be the polynomial operator in (1), and suppose that $\{y_1, y_2\}$ is a given independent pair in $N(P(D))$ with Wronskian $W[y_1, y_2](t) \equiv W(t)$. Let t_0 be any point in I, and f any function in $C^0(I)$. Then the function

$$y_p = c_1(t) y_1(t) + c_2(t) y_2(t) \tag{10}$$

where

$$c_1(t) = \int_{t_0}^{t} -\frac{y_2(s) f(s)}{W(s)}\, ds, \qquad c_2(t) = \int_{t_0}^{t} \frac{y_1(s) f(s)}{W(s)}\, ds \tag{11}$$

is the unique solution of the initial value problem

$$P(D)[y] = f, \qquad y(t_0) = 0, \quad y'(t_0) = 0 \tag{12}$$

Proof. Our approach is simple and direct. We substitute y_p in (10) into the equation $P(D)[y] = f$ and determine the "constants" c_1 and c_2 so that the equation is satisfied. As we shall see, it is helpful to assume that $c_1(t)$ and $c_2(t)$ satisfy the further condition that $c_1' y_1 + c_2' y_2 \equiv 0$ on I (the condition seems arbitrary, but "the end justifies the means"). Substituting y_p in (10) into the equation $P(D)[y] = f$, we obtain (after some calculation) that $c_1' y_1' + c_2' y_2' = f$. Thus the coefficients c_1 and c_2 must satisfy the conditions

$$c_1' y_1 + c_2' y_2 = 0, \qquad c_1' y_1' + c_2' y_2' = f, \qquad \text{for all } t \text{ in } I \tag{13}$$

Since the coefficient matrix of this system of equations in the unknown functions c_1' and c_2' is $W(t) \equiv W[y_1, y_2](t)$, which we know never vanishes on I, we solve (13) uniquely for c_1' and c_2' to obtain

$$c_1' = \frac{-y_2 f}{W}, \qquad c_2' = \frac{y_1 f}{W}$$

Thus any antiderivatives of $-y_2 f / W$ and $y_1 f / W$ will yield candidates for c_1 and c_2 which, when substituted into (10), will provide a particular solution for $P(D)[y] = f$. The expressions in (11) are the special antiderivatives that vanish at $t = t_0$. Using (11) to construct y_p, a direct calculation shows that the initial conditions $y_p(t_0) = 0$, $y_p'(t_0) = 0$ are satisfied, and hence y_p is the unique solution of the initial value problem (12).

Notice that if the expressions for c_1 and c_2 in (11) are substituted into (10), we have the alternative expression for y_p,

$$y_p(t) = \int_{t_0}^{t} K(t, s)f(s)\, ds \qquad (14a)$$

where the *kernel function* $K(t, s)$ is the ratio of two determinants†

$$K(t, s) = \frac{\det \begin{bmatrix} y_1(s) & y_2(s) \\ y_1(t) & y_2(t) \end{bmatrix}}{\det \begin{bmatrix} y_1(s) & y_2(s) \\ y_1'(s) & y_2'(s) \end{bmatrix}} \qquad (14b)$$

The denominator in $K(t, s)$ is the Wronskian $W[y_1, y_2]$. Observe that the kernel $K(t, s)$ in (14) does *not* depend on the driving term $f(t)$. Thus if $g(t)$ is another driving term, a particular solution for $P(D)[y] = g$ can be calculated via (14) *without* recalculating $K(t, s)$, a significant advantage for some applications. Observe also that $K(t, s)$ is the same whatever independent set $\{y_1, y_2\}$ in $N(P(D))$ is used in the definition. This fact is far from obvious (see Problem 10), but allows for considerable flexibility in the calculation of K.

Example 5.20

Let us find a particular solution of the equation

$$t^2 u'' - tu' + u = F(t) \qquad (15)$$

on $t > 0$ for F in $C^0(0, \infty)$. The function $y_p(t)$ given by (14) would be such a solution if $t_0 = 1$ (for example) and $\{y_1, y_2\}$ is an independent pair in the solution space of $t^2 u'' - tu' + u = 0$. To apply (14a) we must first normalize (15) to obtain

$$u'' + t^{-1}u' + t^{-2}u = \frac{F(t)}{t^2}$$

A direct calculation shows that we can take $\{y_1, y_2\}$ to be the pair $\{t, t \ln t\}$. Thus (14) provides the particular solution to (15),

$$u_p(t) = \int_{1}^{t} \frac{\det \begin{bmatrix} s & s \ln s \\ t & t \ln t \end{bmatrix}}{\det \begin{bmatrix} s & s \ln s \\ 1 & 1 + \ln s \end{bmatrix}} \frac{F(s)}{s^2}\, ds = \int_{1}^{t} t \ln \left(\frac{t}{s}\right) \frac{F(s)}{s^2}\, ds \qquad (16)$$

Initial Value Problems

Now that we have some practical methods for finding particular solutions of nonhomogeneous linear differential equations, we have a convenient way of finding the solution of the initial value problem

---

† The 2×2 determinant $\det \begin{bmatrix} a & c \\ b & d \end{bmatrix}$ is defined to be $ad - bc$. (See Chapter 6 for further discussion.)

$$P(D)[y] = f, \qquad y(t_0) = y_0, \quad y'(t_0) = v_0 \tag{17}$$

where f is in $C^0(I)$, t_0 is in I, and y_0 and v_0 are arbitrary constants. Our approach is as follows. First find an independent pair $\{y_1, y_2\}$ in $N(P(D))$, and then find a particular solution y_p of $P(D)[y] = f$ (either through the Method of Undetermined Coefficients or by means of Variation of Parameters). Then the Decomposition of Solutions Property implies that the general solution of $P(D)[y] = f$ is given by $y = k_1 y_1 + k_2 y_2 + y_p$, where k_1 and k_2 are arbitrary constants (real or complex, as appropriate). The constants k_1 and k_2 are then determined so that the initial conditions are satisfied. If y_p is determined via Variation of Parameters [i.e., via (14)], then since $y_p(t_0) = y_p'(t_0) = 0$, calculation of k_1 and k_2 from the initial data becomes that much easier.

Example 5.21

Using the particular solution (16) we see that the general solution of (15) is

$$u = k_1 t + k_2 t \ln t + \int_1^t t \ln\left(\frac{t}{s}\right) \frac{F(s)}{s^2} \, ds$$

where k_1, k_2 are arbitrary constants. Thus if we impose the initial conditions $u(1) = -1$, $u'(1) = 1$, then we must choose k_1, k_2 such that $-1 = k_1$, and $1 = k_1 + k_2$. Hence the solution of the initial value problem

$$t^2 u'' - tu' + u = F(t)$$
$$u(1) = -1, \qquad u'(1) = 1$$

is given by

$$u = -t + 2t \ln t + \int_1^t t \ln\left(\frac{t}{s}\right) \frac{F(s)}{s^2} \, ds \tag{18}$$

$$= \textit{response to} \; + \textit{response to input}$$
$$\textit{initial data}$$

This decomposition of the solution into the sum of an element in the null space that satisfies the initial data and a particular solution that satisfies zero initial conditions is one of the advantages of Variation of Parameters. It allows us to identify just how the system responds to the initial data whatever the input may be and how the system responds to the input independently of the values of the initial data.

Although Variation of Parameters has many advantages, other methods of finding a particular solution are often easier to use for particular input functions.

Example 5.22

To solve the initial value problem

$$y'' + y = t^3, \qquad y(0) = 1, \quad y'(0) = -1$$

we use the Undetermined Coefficients to find the particular solution $y_p = t^3 - 6t$ of the equation $y'' + y = t^3$. We know that the solution is real valued, so we may consider the general real-valued solutions to $(D^2 + 1)[y] = t^3$. Now $\{\cos t, \sin t\}$ is an independent pair of real-valued solutions of $(D^2 + 1)[y] = 0$. Thus $y = k_1 \cos t + k_2 \sin t + (t^3 - 6t)$ is the real-valued general solution of $y'' + y = t^3$. Imposing the initial conditions we find that $k_1 = 1$, $k_2 = 5$, and hence the solution is $y = \cos t + 5 \sin t + (t^3 - 6t)$. Observe that the particular solution constructed this way does not satisfy trivial initial conditions $y_p(0) = y_p'(0) = 0$. Thus, unlike the case when a particular solution is constructed by Variation of Parameters, the solution of the given initial value problem does not immediately separate into two parts as in (18).

Example 5.23

Consider the initial value problem

$$I'' + 2I' + 10I = 3 \cos 3t$$
$$I(0) = I_0, \qquad I'(0) = I_0' \tag{19}$$

The roots of the characteristic polynomial $r^2 + 2r + 10$ are $r_1 = 1 + 3i$ and $r_2 = 1 - 3i$. Thus the real-valued solutions of $I'' + 2I' + 10I = 0$ are described by the independent pair $\{e^{-t} \cos 3t, e^{-t} \sin 3t\}$. Hence using (14) to find a particular solution of the driven equation, we see that the general solution of the driven equation is

$$I(t) = k_1 e^{-t} \cos 3t + k_2 e^{-t} \sin 3t + \int_0^t e^{-(t-s)} \sin 3(t - s) \cos 3s \, ds \tag{20}$$

Imposing the initial conditions, we now find that $k_1 = I_0$, $k_2 = (I_0' + I_0)/3$, and if these values are inserted into (20), we have the solution to problem (19). It is interesting to carry out the integration to calculate explicitly the particular solution $y_p \equiv \int_0^t e^{-(t-s)} \sin 3(t - s) \cos 3s \, ds$ in (20). Using the identity $\sin \alpha \cos \beta = [\sin (\alpha + \beta) + \sin (\alpha - \beta)]/2$, we have

$$y_p = \tfrac{1}{2} e^{-t} \int_0^t e^s \{\sin 3t + \sin (3t - 6s)\} \, ds$$

$$= \tfrac{1}{2} e^{-t}(e^t - 1) \sin 3t + \tfrac{1}{2} e^{-t} \int_0^t e^s \sin (3t - 6s) \, ds$$

Using integral tables to evaluate the last integral, we obtain

$$y_p = \tfrac{1}{37} (18 \sin 3t + 3 \cos 3t - 19e^{-t} \sin 3t - 3e^{-t} \cos 3t) \tag{21}$$

Notice that the last two terms of (21) are actually solutions of the corresponding homogeneous equation and that the sum of the first two terms is another particular solution of the driven equation (and is the particular solution the Method of Undetermined Coefficients would have produced).

The form of (21) is typical if Variation of Parameters is used to construct a particular solution. The particular solution found this way will be the sum of another particular solution and a particular element of the null space needed so that the sum will both satisfy the differential equation and have trivial initial data.

Piecewise-Continuous Driving Terms

For $P(D)$ as in (1), we sometimes encounter in practice the equation $P(D)[y] = f$, where the driving term is only piecewise continuous on the interval I and not continuous. In this case $y(t)$ is said to be a solution of the equation if y is in $C^1(I)$, y'' is piecewise continuous on I, and the equation $P(D)[y] = f$ is satisfied wherever f is continuous. (Observe that the value of f at a discontinuity is immaterial.)

An advantage of the particular solution y_p to problem (12) provided by (14) is that it still provides a solution to (12) even when $f(t)$ is only piecewise continuous, albeit in the generalized sense described above (see Example 1.3). The proof is a straightforward but long calculation and we omit it.

Example 5.24

The problem

$$u'' = \begin{cases} 2, & t < 1 \\ -2, & t > 1 \end{cases}$$

$$u(0) = u'(0) = 0$$

has the solution

$$u(t) = \begin{cases} t^2, & t < 1 \\ -t^2 + 4t - 2, & t \geq 1 \end{cases}$$

Note that $u(t)$ and $u'(t)$ are continuous for all t, while $u''(t)$ is discontinuous at $t = 1$ but continuous otherwise [note that $u''(1^-) = 2$, $u''(1^+) = -2$].

Properties of the Kernel Function

The kernel function $K(t, s)$ can be interpreted as the response at t of the system determined by the operator $P(D)$ to a "unit impulse" input located at s, where s is between t_0 and t, $t_0 \leq s \leq t$. We can make this more precise as follows. Fix T, $t_0 < T < t$, and let r be such that $t_0 < T - r < T < T + r < t$. Let $g(r, T, s)$ be the "unit function"

$$g = \begin{cases} \dfrac{1}{2r} & \text{if } T - r \leq s \leq T + r \\ 0 & \text{all other values of } s \end{cases} \tag{22}$$

(See Figure 5.1.) The response to this input is

$$y_p(t) = \int_{t_0}^{t} K(t, s) g(r, T, s)\, ds = \int_{T-r}^{T+r} K(t, s) \frac{1}{2r}\, ds$$

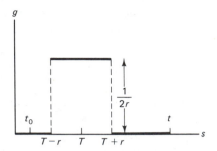

Figure 5.1   Unit function.

where (22) has been used. Now the *Mean Value Theorem†* for Integrals implies that for some value $s*$ between $T - r$ and $T + r$, we have that

$$y_p(t) = \int_{T-r}^{T+r} K(t, s) \frac{1}{2r} ds = K(t, s*) \frac{1}{2r}[(T + r) - (T - r)] = K(t, s*)$$

Now let $r \to 0$, which forces $s*$ to approach T. Thus since $K(t, s)$ is continuous,

$$\lim_{r \to 0} y_p(t) = \lim_{r \to 0} K(t, s*) = K(t, T)$$

Note that as $r \to 0$, the graph of the unit function $g(r, T, s)$ narrows and elongates so that the area under the graph remains 1. In the limit as $r \to 0$, it has "infinite height," "infinitesimal width," but retains the area 1 property. Obviously no longer a function in the conventional sense, this limit is called a *unit impulse* located at T. Thus $K(t, T)$ is the response at t to a unit impulse located at T.

This result has practical implications. Often the underlying action operator $P(D)$ of a physical system is not known. However, the Green's kernel $K(t, s)$ can be numerically approximated by driving the system with unit impulses at selected values of s and measuring the responses at selected values of $t > s$.

Note that the solution of $P(D)[y] = f$, $y(t_0) = 0$, $y'(t_0) = 0$ is obtained by multiplying $K(t, s)$ by $f(s)$ and then "summing" (i.e., integrating) as s ranges from t_0 to t. This can be interpreted as asserting that the response of the system at t to the input $f(s)$, $t_0 \leqq s \leqq t$, is the "superposition" (i.e., integral) of the responses at t to a continuum of impulsive inputs at s scaled by $f(s)$ as s ranges from t_0 to t.

PROBLEMS

The following problem(s) may be more challenging: 3(j), 4(e), 8, and 10.

1. Express the following functions as polynomial-exponential functions. [*Hint*: Recall the identity $e^{i\theta} = \cos \theta + i \sin \theta$.]
   (a) $\cos 2t - \sin t$.   (b) $t \sin^2 t$.

---

† The *MVT for Integrals* asserts that $\int_a^b f(s) \, ds = f(s*)(b - a)$ for some $s*$ between a and b if f is continuous.

(c) $t^2 \sin 2t - (1+t) \cos^2 t$.   **(d)** $\sin^3 t$.

(e) $(1-t)e^{it} \cos 3t$.   **(f)** $(i + t - t^2)e^{(3+i)t} \sin^2 3t$.

2. Express each of the functions in Problem 1(a)–(d) as the real part of some polynomial-exponential function.

3. Use the Method of Undetermined Coefficients to find all solutions of the following differential equations. Sketch the solutions for all but (d) and (f) if $y(0) = 0$, $y'(0) = -1$, 0, 1.

(a) $y'' - y' - 2y = 2 \sin 2t$.   **(b)** $y'' - y' - 2y = t^2 + 4t$.

(c) $y'' - 2y' + y = -te^t$.   **(d)** $y'' + 4y = e^{2it}$.

(e) $y'' - 2y' + y = 2e^t$.   **(f)** $y'' + y' + 2y = 2e^{it}$.

(g) $y'' + 2y' + y = e^t \cos t$.   **(h)** $y'' + y' + y = \sin^2 t$.

(i) $y'' + 4y' + 5y = e^{-t} + 15t$.   **(j)** $y'' - y' - 2y = (1 + t^2)e^{2t} - te^{-t}$.

4. Find the general solution of each of the following equations.

(a) $y'' + y = \dfrac{1}{\cos t}$.   **(b)** $4y'' + 4y' + y = te^{-t/2} \sin t$.

(c) $y'' + y' - 12y = \dfrac{(e^{2t} + 1)^2}{e^{2t}}$ **(d)** $y'' + 2y' + y = \sin 3t + te^{-t}$.

(e) $y'' + 4y = t^2 + 3t \cos 2t$.

5. For which of the equations in Problem 4 does the Method of Undetermined Coefficients fail? Give reasons.

6. Solve each of the following initial value problems.

(a) $y'' - 4y = 2 - 8t$, $y(0) = 0$, $\quad y'(0) = 5$.

(b) $y'' + 9y = 81t^2 + 14 \cos 4t$, $y(0) = 0$, $\quad y'(0) = 3$.

(c) $y'' + y = 10e^{2t}$, $y(0) = 0$, $\quad y'(0) = 0$.

(d) $y'' - y = e^{-t}(2 \sin t + 4 \cos t)$, $y(0) = 1$, $\quad y'(0) = 1$.

(e) $y'' - 3y' + 2y = 8t^2 + 12e^{-t}$, $y(0) = 0$, $\quad y'(0) = 2$.

7. Solve the initial value problem

$$\begin{cases} y'' + y = f(t) \\ y(0) = y'(0) = 0 \end{cases}$$

8. Find all solutions of the boundary value problem

$$y'' - 3y' + 2y = \sin t, \qquad y(0) - y'(0) = 1, \quad y'(1) = -2$$

9. Show that if a, b are real numbers and f is in $C^0(I)$, the real-valued general solution of $(D^2 + aD + b)[y] = f$ is

$$y(t) = k_1 y_1(t) + k_2 y_2(t) + \int_{t_0}^{t} K(t, s)f(s)\, ds, \qquad t \text{ in } I$$

where t_0 is any point in I, and the functions $y_1(t)$, $y_2(t)$ and the *kernel* $K(t, s)$ are determined as follows. Let r_1 and r_2 be the roots of the characteristic polynomial $r^2 + ar + b$.

Case 1: $r_1 \neq r_2$, r_1 and r_2 real numbers

$$y_1(t) = e^{r_1 t}, \quad y_2(t) = e^{r_2 t}, \qquad K(t, s) = \frac{e^{r_2(t-s)} - e^{r_1(t-s)}}{r_2 - r_1}$$

Case 2: $r_1 = r_2$ (and hence real)

$$y_1(t) = e^{r_1 t}, \quad y_2(t) = te^{r_1 t}, \qquad K(t, s) = (t - s)e^{r_1(t-s)}$$

Case 3: $r_1 = \alpha + i\beta$, $r_2 = \alpha - i\beta$, $\beta \neq 0$

$$y_1(t) = e^{\alpha t} \cos \beta t, \quad y_2(t) = e^{\alpha t} \sin \beta t, \qquad K(t, s) = \frac{e^{\alpha(t-s)} \sin \beta(t - s)}{\beta}$$

10. Let the function $K(t, s)$ be the kernel defined in (14) for the polynomial operator $P(D)$ in (1).
    (a) What is the domain of $K(t, s)$ in the ts-plane?
    (b) Prove that $K(t, t) \equiv 0$ and $K_t(t, t) \equiv 1$, for all t in I. [K_t denotes the partial derivative of $K(t, s)$ with respect to t.]
    (c) Show that for each fixed s in I the function $y(t) = K(t, s)$ is a solution on I of the initial value problem $P(D)[y] = 0$, $y(s) = 0$, $y'(s) = 1$.
    (d) Use part (c) and the uniqueness part of the Dynamical System Property of the preceding section to show that $K(t, s)$ is the same for every $\{y_1(t), y_2(t)\}$ used to construct the null space of $P(D)$.

5.4 ELECTRICAL CIRCUITS: KIRCHHOFF'S LAWS

Among the many physical processes that can be modeled by second-order linear ordinary differential equations, one of the most important is the flow of electrical energy in a circuit.

Electrical Units

A *circuit* consists of basic circuit elements called *resistors*, *capacitors*, and *inductors* (described below) connected by conducting wires. See Figure 5.2 for a schematic diagram of a simple circuit.

The *current I* in a circuit is proportional to the number of free electrons (each with a constant negative charge) moving through (any) given point in the conductor per second. According to common practice, however, current flow is described in

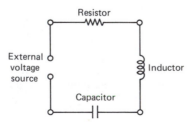

Figure 5.2 Simple *RLC* circuit.

terms of positive charge carriers whose movement is opposite to that of electrons; that is, if I is positive when the circuit orientation of Figure 5.2 is clockwise, the actual movement of the electrons is counterclockwise. Current is measured in *amperes*,† the basic electrical unit in the mks (*m*eter-*k*ilogram-*s*econd) system, in terms of which all other units are defined. [Actually, one ampere corresponds to 6.2420×10^{18} positive carriers moving past a given point in one second.]

The unit of charge is the *coulomb*, defined to be the amount of charge flowing through a cross section of wire in 1 second when a current of 1 ampere is flowing. Thus one ampere is one coulomb per second. If the current I is not constant in time, an instantaneous current $I(t)$ flowing past a point in a circuit is defined in the same way as other instantaneous physical quantities (e.g., velocity). Thus if $I(t)$ is a continuous function, the amount of charge flowing past a point in time interval $[a, b]$ is given by $\int_a^b I(t)\, dt$.

As current moves through a circuit (at constant velocity, incidentally) the charge carriers either impart energy to a circuit element or receive energy from a circuit element. The way in which physicists usually account for this phenomenon is by defining a potential function V throughout a circuit. The energy per coulomb of charge that has been imparted (or received) by the charge carriers between circuit points a and b (denoted by V_{ab}) is computed as $V_{ab} = V_a - V_b$, where V_a and V_b are the values of the potential function V at points a and b of the circuit. The difference V_{ab} is called the *voltage drop* or *potential difference* and is measured in joules per coulomb or *volts*.

Some circuit elements are devices which have the peculiar property that they can maintain a prescribed potential difference (or voltage), denoted by E, between two terminals. Such devices behave like an external force when connected to a circuit, and in fact are known as sources of *electromotive force* (EMF). The electromotive force E is also measured in volts. Examples are batteries and electric generators. For batteries, the higher potential terminal is labeled with a plus sign and the lower level terminal with a minus sign. The internal chemical energy supplied by the battery imparts a constant amount of energy per coulomb as positive charge carriers move through it thus raising the potential function V by the voltage rating of the battery.

Circuit Elements

Each circuit is composed of three basic circuit elements: the resistor, capacitor, and inductor, which we describe briefly below.

---

† André Marie Ampère (1775–1836), a French mathematician and natural philosopher, is known for his contributions to electrodynamics. The names of many early researchers in electricity have been used for various electrical units. Thus we have ohms [Georg Simon Ohm (1787–1854), German physicist], henries [Joseph Henry (1797–1878), American physicist], farads [Michael Faraday (1791–1867), English scientist], volts [Alessandro Volta (1745–1827), Italian physicist], and coulombs [Charles Auguste de Coulomb (1736–1806), French physicist].

Resistor. As a current flows through a conducting substance the charge carriers lose energy (which then appears as other energy forms such as heat and light), and hence the potential where the current emerges is lower than the potential where the current enters the conductor. Although every conductor has this property, the more efficient among them are called *resistors* and are symbolized by the schematic shown here. Common examples of resistors are heating coils, toaster coils, and light-

$$a \circ\!\!-\!\!-\!\!-\!\!\!\!\underset{R}{\text{WW}}\!\!-\!\!-\!\!\circ b$$

bulb filaments. The instantaneous voltage drop between the two terminals *a* and *b* of a resistor is found experimentally to be proportional to the current flowing through the resistor at that instant. If the current is measured in amperes, then the constant of proportionality, denoted by R, is said to be measured in *ohms* (the ohm is denoted by Ω). Thus we have

Ohm's law

$$V_{ab} = RI$$

and R is called the *resistance* of the resistor. Thus if the current is directed from a to b then the voltage at b is lower than the voltage at a and hence $V_{ab} = RI > 0$. If the current is directed from b to a, then $V_{ab} = -RI < 0$.

Inductor. Variable electrical current $I(t)$ creates a changing magnetic field near the conductor which induces a voltage drop between the ends of the conductor which opposes the change in current. Conductors which are efficient in this respect are called *inductors* and are denoted by the symbol shown here. Coils of wire are

$$a \circ\!\!-\!\!-\!\!-\!\!\underset{L}{\text{000}}\!\!-\!\!-\!\!\circ b$$

an example of inductors. It has been found experimentally that the voltage drop across an inductor, V_{ab}, is proportional to the instantaneous rate of change of the current through the inductor. If V_{ab} is measured in volts and dI/dt in amperes per second, then this constant of proportionality, denoted by L, is said to be measured in *henries*. Thus we have

Law of Induction

$$V_{ab} = L \frac{dI}{dt}$$

and L is called the *inductance* of the inductor. Note here also that the voltage through an inductor drops in the direction of the current flow.

Capacitor. A *capacitor* consists of two conductors separated by an insulator (such as air). Capacitors are often visualized as a pair of conducting plates separated by a gap, represented schematically as shown. If terminals a and b of a capacitor

$$a \circ\!\!-\!\!-\!\!-\!\!-\!\!|\!|\!-\!\!-\!\!-\!\!-\!\!\circ b \quad {}^{C}$$

are connected to a voltage source, a negative charge will build up on the plate connected to the negative terminal and a positive charge on the plate connected to the positive terminal. We speak of the total *charge* $q(t)$ on the capacitor and observe that if $q(t_0)$ is the initial charge, then [since $dq/dt \equiv I(t)$]

$$q(t) = q(t_0) + \int_{t_0}^{t} I(s)\,ds \qquad \text{for } t \geq t_0 \tag{1}$$

It has been observed experimentally that the instantaneous voltage drop across a capacitor is proportional to the charge on the capacitor. When the charge is measured in coulombs and V_{ab} in volts, the constant of proportionality is $1/C$, where the *capacitance* C is measured in *farads*. Thus we have

Coulomb's Law

$$V_{ab}(t) = \frac{1}{C} q(t) = \frac{1}{C} \{ q(t_0) + \int_{t_0}^{t} I(s)\,ds \}$$

Note again that the voltage drops in the direction of the current flow.

The capacitance C depends on the area of the plates, their separation, and the nature of the insulator between. As might be expected, it takes a high voltage to build up charge on the capacitor and thus for a given charge the capacitance C is usually very small (on the order of 10^{-5} or 10^{-6} farads).

Strictly speaking, the coefficients of resistance, inductance, and capacitance are not constant but depend on the magnitudes of the current and the voltage drop, and hence their values can vary over time. We shall ignore this fact, however, since over the range of standard engineering applications these coefficients are essentially constant.

Kirchhoff's Voltage Law; Single Loop Circuits

One of the simplest circuits involving resistors, capacitors, and inductors is the series circuit in Figure 5.2, called the *simple RLC series circuit*, which is driven by a given external voltage source. More complicated circuits are easy to imagine (see the problem set), but we will not treat them very extensively here. The modeler should keep in mind that resistance, inductance, and capacitance are circuit characteristics which are not always accounted for by devices called resistors, inductors, and capacitors. Sometimes these characteristics arise through the physical properties of the circuit as a whole (e.g., all conducting wire offers some resistance to current flow). Therefore, elementary circuit elements sometimes appear in a circuit description in an effort to "lump" together the effects of these distributed characteristics. *Lumping* in this sense is an important and frequently used modeling tool.

What is needed now for finding the current flowing in a circuit is a relation between the voltage drops across the various components of a circuit. The following "conservation" law has been observed to hold:

> **Kirchhoff's Voltage Law.**   Let the points $a_1, a_2, \ldots, a_n$ be identified in a circuit such that the oriented sequence $a_1 a_2 \cdots a_n a_1$ forms a closed loop in the circuit. Then
>
> $$V_{a_1 a_2} + V_{a_2 a_3} + \cdots + V_{a_n a_1} = 0$$
>
> where $V_{a_i a_{i+1}} \equiv V_{a_i} - V_{a_{i+1}}$ is the voltage drop between points a_i and a_{i+1}.

Because of the way voltage is defined, Kirchhoff's Voltage Law is essentially a conservation-of-energy law. For example, in Figure 5.3, the sequence of points $abcda$ forms an oriented closed loop for the simple *RLC* series circuit.

We now apply Kirchhoff's Voltage Law to the simple *RLC* series circuit to derive an equation that governs how the current in that simple loop evolves with time. For this purpose, it will be helpful to introduce some notation into the simple loop of Figure 5.2. In what follows refer to Figure 5.3. First we label the polarity of the external source with "plus" and "minus" signs in order to be precise about the way the external source is connected to the circuit [the labeling shown indicates how the voltage source $E(t)$ graphed in Figure 5.3 is connected to the circuit]. Next, observe that the current flows in the same direction and has the same value through every circuit element, so we may as well *assume* the positively directed current $I(t)$ moves clockwise around the circuit (as indicated in Figure 5.3). Should it turn out that the current actually moves counterclockwise at a certain t_0 then the value of $I(t_0)$ will turn out to be a negative number. Now as the current flows through the external voltage source the voltage will increase, but on each of the other circuit elements the voltage drops in the (assumed) direction of the current. Finally, we introduce reference points a, b, c, and d as shown. Now observe that (refer to Figure 5.3) Ohm's Law, the Law of Induction, and Coulomb's Law yield the voltage drops

$$V_{ab} = RI(t), \qquad V_{bc} = L\frac{dI}{dt}, \qquad V_{cd} = \frac{1}{C}\left[q(t_0) + \int_{t_0}^{t} I(s)\,ds\right]$$

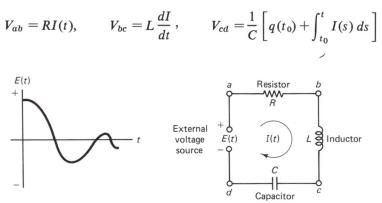

Figure 5.3   Simple RLC series loop prepared for modeling.

and hence application of Kirchhoff's Voltage Law yields the integrodifferential equation

$$RI(t) + L\frac{dI}{dt} + \frac{1}{C}\left[q(t_0) + \int_{t_0}^{t} I(s)\,ds\right] = E(t) \qquad (2)$$

for the unknown current $I(t)$. We can convert (2) to a differential equation in two ways. In terms of the charge $q(t)$ on the capacitor, we have that

$$Lq'' + Rq' + \frac{1}{C}q = E(t) \qquad (3)$$

since $q = q(t_0) + \int_{t_0}^{t} I(s)\,ds$. It is often more convenient to use $I(t)$ rather than $q(t)$ as the unknown; if $E(t)$ is differentiable, we can differentiate (2) with respect to t and obtain

$$LI'' + RI' + \frac{1}{C}I = E' \qquad (4)$$

The relevant initial value problem for the circuit equation in the form (4) is derived as follows: If q_0, the initial charge, and I_0, the initial current, are known, then from (2) we have that $RI_0 + LI'(t_0) + (1/C)q_0 = E(t_0)$ which immediately determines $I'(t_0)$. Denoting $I'(t_0)$ by I_0' we are thus led to solve the initial value problem

$$LI'' + RI' + \frac{1}{C}I = E'(t)$$

$$I(t_0) = I_0, \qquad I'(t_0) = I_0' \equiv [E(t_0) - RI_0 - q_0/C]/L \qquad (5)$$

A Special Case: E ≡ Constant

If $E(t)$ is constant in time, then we can use the techniques of Section 5.2 to find all solutions of (4). Indeed, let us assume that $E \equiv E_0$, a constant. Then (4) becomes the homogeneous equation

$$LI'' + RI' + \frac{1}{C}I = 0$$

where we assume that L, R, and C, are positive constants. We can think of this as modeling a circuit in which a battery is allowed to charge up a capacitor and then at some time (say $t_0 = 0$) the battery is removed from the circuit (see Figure 5.4). Thus we may assume that q_0 is known and that $I(t_0) = 0$. Thus we see from (5) that the initial value problem to be solved in this case is given by

$$LI'' + RI' + \frac{1}{C}I = 0$$

$$I(t_0) = 0, \qquad I'(t_0) = (E_0 - q_0/C)/L \qquad (6)$$

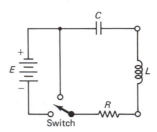

Figure 5.4 Series circuit.

From the quadratic formula the roots r_1 and r_2 of the characteristic polynomial $P(r) = Lr^2 + Rr + 1/C$ are seen to be $[-R \pm (R^2 - 4L/C)^{1/2}]/2L$. Thus r_1 and r_2 always have negative real parts for any positive constants R, L, and C. Hence all solutions of Eq. (5) are *transients* (i.e., die out with increasing time), but their exact nature depends on the relative sizes of these constants. We use the General Solution Theorem (see Section 5.2) to distinguish three qualitatively distinct cases. If $R^2 > 4L/C$, the circuit is *overdamped* and in that case every solution is a linear combination of the decaying exponentials $\exp [r_1 t]$ and $\exp [r_2 t]$. If $R^2 = 4L/C$, the circuit is *critically damped* and every solution is the product of the decaying exponential $\exp [-Rt/2L]$ and a linear polynomial $k_1 + k_2 t$. If $R^2 < 4L/C$, the circuit is *underdamped*, and since r_1 and r_2 have the form $\alpha \pm i\beta$ where $\alpha = -R/2L$, $\beta = (4LC - R^2C^2)^{1/2}/2LC$, every solution of Eq. (5) is the product of a sinusoid $\cos (\beta t + \delta)$ with frequency β cycles per second and a decaying exponential $A \exp [\alpha t]$ for constants δ and A. Contrasting these three cases, we observe that in the overdamped and critically damped cases, solutions essentially decay steadily to zero, whereas in the underdamped case, solutions oscillate back and forth about zero. Thus no matter what the values of I_0 and I_0', the solution of initial value problem (6) is a transient that behaves in one of the three ways described above.

Finally, observe that if the resistance in our constant-EMF-driven *RLC* circuit is negligible, we might expect that the current could be modeled by the equation $LI'' + (1/C)I = 0$. Since the characteristic roots of this equation are $\pm i\beta$, with $\beta = 1/(LC)^{1/2}$, every solution has the sinusoidal form $A \cos (\beta t + \delta)$, where A and δ are arbitrary constants. Note that all these solutions are periodic and hence do not decay as $t \rightarrow \infty$. Thus such a circuit would be able to sustain undiminished oscillations forever, and for this reason the circuit is known as an *LC oscillator*. In general, $x'' + k^2 x = 0$ is known as the *harmonic oscillator equation* and its sinusoidal solutions are generally referred to as *(simple) harmonic motion*. By contrast, the damped oscillatory solutions of Eq. (5) with $R > 0$ are sometimes referred to as *damped harmonic motion*.

In practice, however, the resistance in a circuit may be very small, but it can never be zero. Hence, from what we have just seen, it would be a serious mistake to infer too much about an *RLC* circuit with very small R by setting $R = 0$ in Eq. (6). Thus an *LC* oscillator is a mathematical idealization of a physically unrealizable phenomenon.

Driven (or Forced) Circuits

We discussed above the *free solutions* of the simple RLC series circuit. Now we turn to the *forced solutions*, that is, solutions of (3) or (4) when the external voltage source $E(t)$ is nonconstant. We know from the Decomposition of Solutions Property (see Section 5.3) that all forced solutions of (4) are known if just one forced solution and all free solutions are known. Let us turn to an example.

Example 5.25

We shall determine all solutions of the simple RLC series circuit Eq. (3) when $L = 20$ H, $R = 80$ Ω, $C = 10^{-2}$ F, and the external voltage is $E(t) = 100 \cos 2t$. Thus Eq. (3) becomes

$$q'' + 4q' + 5q = 5 \cos 2t \qquad (7)$$

with characteristic polynomial $P(r) = r^2 + 4r + 5$, which has roots $r_1 = -2 + i$ and $r_2 = -2 - i$. Thus, by the General Solution Theorem (see Section 5.2), the free solutions of Eq. (7) have the form $k_1 e^{-2t} \cos t + k_2 e^{-2t} \sin t$, where k_1 and k_2 are arbitrary real constants. Now we need to find a forced solution for Eq. (7). We use the Method of Undetermined Coefficients. Observe that $5 \cos 2t = \mathrm{Re}\,[5e^{2it}]$, so if q_p^* is any (complex-valued) solution of the equation

$$P(D)q \equiv q'' + 4q' + 5q = 5e^{i2t}$$

then $\mathrm{Re}\,[q_p^*] \equiv q_p$ is a (real-valued) solution of Eq. (7). We try a solution of the form $q_p^* = ce^{2it}$ for some complex number c. Now $P(D)(ce^{2it}) = cP(D)e^{2it} = cP(2i)e^{2it}$ by property (7) of Section 5.2 for $P(D)$. Thus we must choose $c = 5/P(2i) = 5/(1 + 8i)$. Since $(1 + 8i)(1 - 8i) = 65$, we see that

$$c = \frac{5}{1 + 8i}\frac{1 - 8i}{1 - 8i} = \frac{1}{13}(1 - 8i)$$

and hence

$$q_p = \mathrm{Re}\,[ce^{2it}] = \tfrac{1}{13}(\cos 2t + 8 \sin 2t) \qquad (8)$$

is a particular solution of Eq. (7).

We shall employ an often-used trick to rewrite the solution in (8) as a single sinusoid. Picking out the coefficients of $\cos 2t$ and $\sin 2t$ in the parentheses, we form the point $(8, 1)$ in the Cartesian plane. Multiplying the coordinates by the scalar $1/(1^2 + 8^2)^{1/2} = 1/\sqrt{65}$, we obtain $(8/\sqrt{65}, 1/\sqrt{65})$, a point on the unit circle. Hence from trigonometry we know that there is an angle δ with $|\delta| \le \pi$ such that

$$\cos \delta = \frac{8}{\sqrt{65}}, \qquad \sin \delta = \frac{1}{\sqrt{65}} \qquad (9)$$

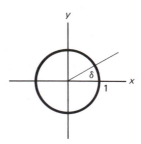

Tables indicate that $\delta \cong 7.1°$. Inserting (9) into (8), we have

$$q_p = \frac{5}{\sqrt{65}} (\sin \delta \cos 2t + \cos \delta \sin 2t) = \frac{5}{\sqrt{65}} \sin (2t + \delta) \qquad (10)$$

Thus q_p is revealed to be sinusoidal with the same frequency as the driving term but out of phase slightly. From (10) we see that

$$I_p = \frac{dq_p}{dt} = \frac{10}{\sqrt{65}} \cos (2t + \delta)$$

is a particular solution of Eq. (4). Now since every solution of Eq. (3) is a free solution (which decays to zero) plus q_p, we infer that after a very short time the only charge we "see" in the circuit is q_p, and for that reason q_p is called the *steady-state solution* of Eq. (3); the free solutions are called *transients* because they all decay to zero as $t \to \infty$.

Using this example as a guide, we find all solutions of Eq. (3) where $R \neq 0$ and the driving term $E(t) = E_0 \cos \omega t$ for given real numbers E_0 and ω. The free solutions of Eq. (3) have been treated earlier, so we need only a single particular solution to complete our analysis. Toward this end we use the device introduced in Example 5.25 to first find a particular (complex-valued) solution q_p^* of

$$Lq'' + Rq' + \frac{1}{C} q = E_0 e^{i\omega t} \qquad (11)$$

and then observe that $q_p = \text{Re}\,[q_p^*]$ is a particular solution of Eq. (3) with $E(t) = E_0 \cos \omega t$. From the quadratic formula, we see that the roots of the characteristic polynomial $P(r) = Lr^2 + Rr + 1/C$ must always have real parts strictly negative (since $R \neq 0$). Thus $P(i\omega) \neq 0$, so $q_p^* = E_0 e^{i\omega t}/P(i\omega)$ is a particular solution of Eq. (11). Since $P(i\omega) = -L\omega^2 + 1/C + iR\omega$, we have that $1/P(i\omega) = [(-L\omega^2 + 1/C) - iR\omega]/[(-L\omega^2 + 1/C)^2 + R^2\omega^2]$, and hence

$$q_p = \text{Re}\,[q_p^*] = E_0 \frac{(1/C - L\omega^2) \cos \omega t + \omega R \sin \omega t}{(1/C - L\omega^2)^2 + \omega^2 R^2} \qquad (12)$$

Using another trick from Example 5.25, let δ be an angle such that

$$\sin \delta = \frac{1/C - L\omega^2}{[(1/C - L\omega^2)^2 + \omega^2 R^2]^{1/2}}, \qquad \cos \delta = \frac{\omega R}{[(1/C - L\omega^2)^2 + \omega^2 R^2]^{1/2}}$$

Then (12) can be written as the sinusoid

$$q_p = \frac{E_0 \sin(\omega t + \delta)}{[(1/C - L\omega^2)^2 + \omega^2 R^2]^{1/2}} \tag{13}$$

Differentiating Eq. (13), we find that

$$I_p = \frac{E_0 \cos(\omega t + \delta)}{[(\omega L - 1/(\omega C))^2 + R^2]^{1/2}} \tag{14}$$

is the steady-state current for the simple RLC series circuit with a sinusoidal driving term. Observe that the steady-state response has the same frequency as the driving term but is shifted in phase. The denominator in Eq. (14) is called the *impedance* of the circuit; it plays a role similar to the one played by the resistance, R, in Ohm's Law. Note that Eq. (13) could have been written with cosine instead of sine since $\sin(\omega t + \delta) = \cos[\omega t + \delta - (\pi/2)]$. Both are called sinusoids.

Multiloop Circuits; Kirchhoff's Current Law

Much more complicated circuits than the simple RLC series circuit can be modeled via differential equations for the currents flowing in loops that compose the circuit. To do this we will need, in addition to the Voltage Law, a law governing currents in circuit elements.

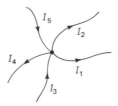

> **Kirchhoff's Current Law.**   At each point of a circuit the sum of currents entering is equal to the sum of the currents leaving

Now to apply Kirchhoff's laws to multiloop circuits it is important to keep the following observation in mind: It can not be assumed in general that the current through every circuit element of a given loop will be the same. Thus the first thing we must do is *assume* a direction for the current through every circuit element; the current through that element will be positive if it flows in the direction specified, and negative otherwise. If our assumption of (positive) current flow is incorrect for a given current element then our circuit equations will ultimately inform us that the current has a negative value for that element, indicating current flow in the opposite direction. Next introduce the reference points a, b, c, and d, and finally we decide on the polarity of the given external source by indicating the positive and negative connectors by "+" and "−" signs. We can best illustrate the use of Kirchhoff's laws in a multiloop circuit by means of an example.

Example 5.26

Consider the two loop circuit pictured in Figure 5.5 where I_1, I_2, I_3, and I_4 are the positively directed currents through the elements indicated. From Kirchhoff's Current Law at node point b, the current through the coil with inductance

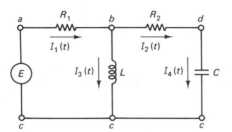

Figure 5.5   Two-loop circuit.

L must be $I_1 - I_2$. Applying Kirchhoff's Voltage Law to the two inner loops, we have that (assuming $q_0 = 0$ on the capacitor)

$$R_1 I_1 + L(I_1' - I_2') = E(t)$$

$$R_2 I_2 + \frac{1}{C} \int_0^t I_2(s)\, ds + L(I_2' - I_1') = 0$$

Differentiating the second equation and rearranging, we have that

$$L(I_1' - I_2') + R_1 I_1 = E(t)$$

$$-L(I_1'' - I_2'') + R_2 I_2' + \frac{1}{C} I_2 = 0$$

We leave the treatment of such coupled linear systems to later chapters.

Comments

If $E(t)$ is not a continuous function then (4) is not an alternative way to solve (3). To see this, imagine, for example, that E is piecewise constant. We shall pursue this situation further in the problems. Although second-order linear constant coefficient ordinary differential equations model circuits quite well, there are some difficulties. For example, what if the circuit elements are not passive but change in time? What if the elements are distributed through the circuit and cannot be lumped as we have assumed? The latter case is exemplified by the current and voltage drop in a long transmission line and has a partial differential equation model. These complications are beyond the scope of this text.

PROBLEMS

The following problem(s) may be more challenging: 6(b) and 7.

1. (a) Solve the initial value problem (5) if $R = 20\ \Omega$, $L = 10$ H, $C = 0.05$ F, $E_0 = 12$ V, $q_0 = 0.6$ Coulomb, and $I_0 = 1$ A.
   (b) Show that a damped oscillation occurs. Sketch the graph of $I = I(t)$.
   (c) If the switch in Figure 5.4 is kept closed, what is the limiting value of $q(t)$? Sketch $q = q(t)$.
   (d) To what value must the resistance be increased in order to reach the overdamped case?

2. Suppose that a switch in an RLC series circuit is open and there is an initial charge $q_0 = 10^{-3}$ Coulomb on the capacitor. Find the charge on the capacitor and the current flowing in the circuit after the switch is closed for each of the following cases.
   (a) $L = 0.3$ H, $R = 15\ \Omega$, $C = 3 \times 10^{-2}$ F.
   (b) $L = 1$ H, $R = 1000\ \Omega$, $C = 4 \times 10^{-4}$ F.
   (c) $L = 2.5$ H, $R = 500\ \Omega$, $C = 10^{-6}$ F.

3. Two capacitors ($C_1 = 10^{-6}$ F and $C_2 = 2 \times 10^{-6}$ F) and a resistor ($R = 3 \times 10^6\ \Omega$) are arranged in a circuit as shown. The capacitor C_1 is initially charged to a voltage E_0 volts with polarity as shown. The switch S is closed at time $t = 0$. Determine the current that flows through the capacitor C_1 as a function of time.

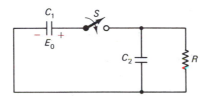

4. Find the charge $q(t)$ on the capacitor of the RLC circuit in Figure 5.3 assuming zero initial charge and current if $R = 20\ \Omega$, $L = 10$ H, $C = 0.01$ F, and $E(t) = 30 \cos 2t$ volts.

5. A simple RLC circuit has a capacitor of 0.25 F, a resistor of $7 \times 10^4\ \Omega$, and an inductor of 2 H. The initial charge on the capacitor is zero. If an impressed voltage of 60 V is connected to the circuit and the circuit is closed at $t = 0$, determine the charge on the capacitor for $t > 0$. Estimate the charge when $t = 0.1$ s.

6. (a) Show that if $R = 0$ in the simple RLC circuit and the impressed voltage is of the form $E_0 \cos \omega t$, the charge on the capacitor will become unbounded as $t \to \infty$ if $\omega = 1/\sqrt{LC}$. This is the phenomenon of resonance.
   (b) Show that the charge will always be bounded, no matter what the choice of ω, provided that there is some resistance in the circuit.

7. Set up, but do not solve, three loop equations for the circuit shown on the next page.

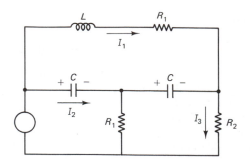

5.5 SPRINGS AND PENDULUMS

In this section we consider two important mechanical systems which have only *one degree of freedom*, that is, systems whose states can be completely described by one position coordinate and its time derivative. Vectors and Newton's Second Law are used to derive the equations of the motion and then those equations are expressed in terms of a coordinate system which makes use of the one-degree-of-freedom nature of the motion.

Spring–Mass System

Consider the motion of a mass m suspended at the end of a spring and subject to an *external force* acting in the vertical direction. The mass will also be acted upon by a *gravitational force* of magnitude mg directed toward the center of the earth, a *spring force* due to the displacement of the spring from a rest position, and a *frictional* (or *damping*) *force* opposing the motion (*viscous damping*). See Figure 5.6, where the damping is indicated by the *dashpot* (or *shock absorber*), which is sometimes imagined to be a cylinder of fluid through which a piston moves. The problem now is to describe the displacement of the mass m from static equilibrium through time under the action of all these forces. Clearly, the motion of the mass will take place in the vertical direction, so it would simplify matters considerably if Newton's Second Law were interpreted in a Cartesian coordinate frame with one axis along the local vertical. Let us call this axis the x-axis and take its positive direction to be downward (see Figure 5.6). Before we can apply Newton's Second Law, however, we must make some assumptions about the spring and damping forces. Common assumptions are

> **Hooke's Law.**   The force exerted by a spring has magnitude proportional to the displacement of the spring from its equilibrium position.

and the

> **Viscous Damping Assumption.**   The damping force experienced by the mass m has magnitude proportional to the velocity of the mass and always acts in a direction opposite to the motion.

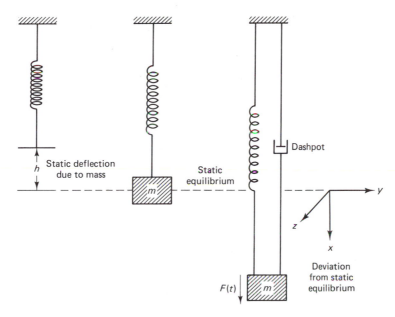

Figure 5.6   Forced damped spring.

Using these assumptions (and the fact that the mass m is constant in time), Newton's Second Law implies that

$$mx''(t) = -k(x(t) + h) - cx'(t) + mg + F(t) \qquad (1)$$

where $k > 0$ is the *stiffness* or *spring constant*, $c \geq 0$ is the *damping constant* ($c = 0$ if there is no damping), g is the gravitational constant, and $h > 0$ is the static deflection due to hanging the mass on the spring (see Figure 5.6). It is easy to compute a value for h since for the mass to hang at rest at the end of the spring, we must have that the force on the mass exerted by the spring must be equal and opposite to the gravitational force on the mass [i.e., $-(-kh) = mg$]. Using this fact in Eq. (1), we see that the differential equation modeling the motion of the mass is given by

$$mx'' + cx' + kx = F(t) \qquad (2)$$

It is interesting to observe that the gravitational constant has disappeared from our model. Thus we are led to conclude that our spring–mass system would basically behave the same way on the moon or on Mars as it would on the earth.

Comparison of Spring–Mass and Circuit Models

Observe that Eq. (2) has exactly the same form as the differential equations modeling the charge on the capacitor and the current in a simple *RLC* circuit responding to an external voltage (see Section 5.4). In fact, Table 5.1 tabulates correspondences

TABLE 5.1 CORRESPONDENCE BETWEEN SPRING
AND *RLC*-CIRCUIT MODELS

Spring:	Circuit:
$mx'' + cx' + kx = F(t)$	$Lq'' + Rq' + \dfrac{1}{C}q = E(t)$
	$LI'' + RI' + \dfrac{1}{C}I = E'(t)$

Mass m	Inductance L
Displacement x	Charge q or current I
Damping constant c	Resistance R
Spring constant k	Elastance $\dfrac{1}{C}$
	(reciprocal of capacitance C)
External force $F(t)$	External voltage $E(t)$ or
	voltage rate $E'(t)$

between the parameters of the circuit and the parameters of the spring. Thus the properties of solutions presented for the simple *RLC* circuit equation in Section 5.4 carry over to the spring–mass case, Eq. (2). The correspondence in Table 5.1 explains why a mechanical system can be "modeled" by an electrical circuit, and conversely. From this point on the reader may interpret at will any equation of the form $\alpha x'' + \beta x' + \gamma x = f(t)$, where α, β, and γ are positive constants ($\beta = 0$, possibly), in terms of a vibrating spring, or a changing charge on a capacitor or a variable current in a circuit.

Free Vibrations of a Spring

Recall from Section 5.3 that all possible motions (vibrations) of the mass at the end of a spring are known when all solutions of the homogeneous equation $mx'' + cx' + kx = 0$ (the free solutions) and a single particular solution (a forced solution) of Eq. (2) are known. Free solutions behave differently when damping is present than when it is not. In the undamped case free solutions satisfy the *harmonic oscillator* equation

$$mx'' + kx = 0 \tag{3}$$

Using Section 5.2, all solutions of Eq. (3) are found to be the sinusoids

$$x(t) = A \cos(\omega t + \delta)$$

where A and δ are arbitrary real numbers, and $\omega = \sqrt{k/m}$ cycles per second, called the *natural frequency* of the harmonic oscillators. Thus motions described by Eq. (3) are periodic with natural frequency ω, and persist forever. Such motion is called *simple harmonic motion*.

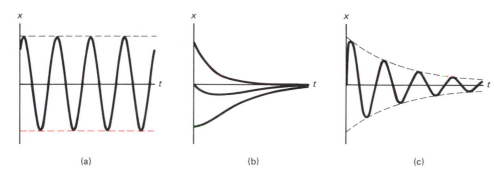

Figure 5.7 Damped and undamped free solutions: (a) simple harmonic motion; (b) over and critically damped motion; (c) underdamped motion.

Of course, no physical spring can sustain simple harmonic motion since some damping is always present because of the imperfect stretching of the spring. Indeed, free solutions in the damped case can be found by first computing the roots r_1 and r_2 of the characteristic polynomial $P(r) \equiv mr^2 + cr + k$. Using the quadratic formula we have that r_1 and r_2 are given by $(-c + \sqrt{c^2 - 4mk})/2m$. Since m, c, and k are all positive numbers, we see that the real parts of r_1 and r_2 are always negative whatever the actual values of m, c, and k. Thus we see via the General Solution Theorem (Section 5.2) that all damped free solutions are *transients* (i.e., die out with increasing time), but their exact nature depends on the relative sizes of the constants m, c, and k. There are three qualitatively different cases that can arise. If $c^2 > 4mk$, the system is *overdamped* and every free solution is a linear combination of the decaying exponentials $\exp[r_1 t]$ and $\exp[r_2 t]$. If $c^2 = 4mk$, the system is *critically damped* and every free solution is the product of a decaying exponential $\exp[-ct/2m]$ and a linear polynomial $k_1 + k_2 t$. If $c^2 < 4mk$, the system is *under-damped*, and since r_1 and r_2 have the form $\alpha + i\beta$ where $\alpha = -c/2m$, $\beta = (4mk - c^2)^{1/2}/2m$, every underdamped free solution is the product of a sinusoid $\cos(\beta t + \delta)$ with frequency β cycles per second, and a decaying exponential $A \exp[\alpha t]$ for constants α and A. See Figure 5.7 for a graphical depiction of typical free solutions in each of these cases. Observe that the critically damped case is border-line between oscillatory (underdamped) motion of the mass about the equilibrium position and motion which decays steadily to the equilibrium position essentially without oscillation (the overdamped case). Underdamped free solutions are sometimes referred to as *damped harmonic motion*. The above discussion is, of course, virtually the same as that in Section 5.4 for circuits.

The Pendulum

We shall use Newton's Second Law to model the motion of a suspended bob of mass m in the earth's gravitational field. For simplicity let us assume that the bob is suspended at the end of a weightless rod of fixed length L pivoted at the other

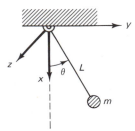

Figure 5.8 Geometry for a pendulum with one degree of freedom.

end. See Problem 6 for the pendulum of variable length. Although it is not immediately evident, if the pivot is positioned at the origin of a Cartesian frame and the center of mass of the bob is in the xy-plane (see Figure 5.8), the bob will oscillate back and forth in the xy-plane if it is released with an initial velocity in that plane (see Problem 5). Thus the polar angle θ is sufficient to track the motion of the bob in that case, so the system has one degree of freedom.

Let us assume that the moving pendulum experiences a viscous damping force **f** proportional to its linear velocity and opposing the motion, and that the other forces acting on the bob are the tension **T** in the rod and gravity (see Figure 5.9).

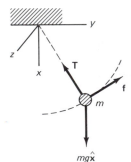

Figure 5.9 Force diagram for bob.

Now to Apply Newton's Second Law we shall use kinematics to derive an expression for the acceleration of the bob relative to the (inertial) xyz-frame,† compute the sum of the external forces acting on the bob, and then substitute these quantities into Newton's Second Law. This procedure is not as easy to carry out as it sounds if one proceeds in a mindless way. First, it may not be so easy to compute the acceleration and external force vectors in terms of Cartesian coordinates in the xyz-frame of reference. Second, Newton's Law is a statement about vectors in the inertial xyz-frame, and before it can be solved it must be converted to a statement about the components of these vectors in some frame of reference. Any frame will do for this purpose (the virtue of vectors!); hence the original xyz-frame need not be used. The real skill of the modeler comes through in the choice of a frame of reference

---

† For the Cartesian xyz-frame it is common in physics to denote the unit vectors in the positive x, y, and z directions by $\hat{\mathbf{x}}$, $\hat{\mathbf{y}}$, and $\hat{\mathbf{z}}$, respectively.

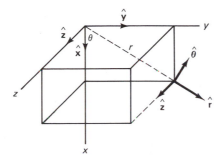

Figure 5.10 Cylindrical coordinates.

in which to express Newton's Second Law—a poor choice could present some formidable technical difficulties in obtaining a solution.

Using cylindrical coordinates, r, θ, and z in the xyz-frame (see Figure 5.10) we see that the position vector $\mathbf{R}(t)$ of the center of the bob can be expressed as $\mathbf{R} = L(\cos\theta\hat{\mathbf{x}} + \sin\theta\hat{\mathbf{y}})$. Unit vectors $\hat{\mathbf{r}}$, $\hat{\boldsymbol{\theta}}$, and $\hat{\mathbf{z}}$ in cylindrical coordinates are related to the Cartesian unit vectors $\hat{\mathbf{x}}$, $\hat{\mathbf{y}}$, $\hat{\mathbf{z}}$ as follows (see Figure 5.10):

$$\hat{\mathbf{r}} = \cos\theta\hat{\mathbf{x}} + \sin\theta\hat{\mathbf{y}}$$
$$\hat{\boldsymbol{\theta}} = -\sin\theta\hat{\mathbf{x}} + \cos\theta\hat{\mathbf{y}} \tag{4}$$
$$\hat{\mathbf{z}} = \hat{\mathbf{z}}$$

(Note that $\hat{\mathbf{r}}$, $\hat{\boldsymbol{\theta}}$, $\hat{\mathbf{z}}$ form an orthogonal system and that $\hat{\boldsymbol{\theta}}$ "points" in the direction of increasing θ.) Observe that $\mathbf{R} = L\hat{\mathbf{r}}$. Using the differentiation rule for vectors written in an inertial frame (see Section 4.3), we see that

$$\frac{d(\hat{\mathbf{r}})}{dt} = (-\sin\theta\hat{\mathbf{x}} + \cos\theta\hat{\mathbf{y}})\theta' = \theta'\hat{\boldsymbol{\theta}}$$

$$\frac{d(\hat{\boldsymbol{\theta}})}{dt} = (-\cos\theta\hat{\mathbf{x}} - \sin\theta\hat{\mathbf{y}})\theta' = -\theta'\hat{\mathbf{r}}$$

Thus we have that

$$\mathbf{v} \equiv \frac{d\mathbf{R}}{dt} = \frac{d(L\hat{\mathbf{r}})}{dt} = L\theta'\hat{\boldsymbol{\theta}}$$

$$\tag{5}$$

$$\mathbf{a} \equiv \frac{d^2\mathbf{R}}{dt^2} = \frac{d(L\theta'\hat{\boldsymbol{\theta}})}{dt} = L\theta'\frac{d\hat{\boldsymbol{\theta}}}{dt} + L\theta''\hat{\boldsymbol{\theta}} = -L(\theta')^2\hat{\mathbf{r}} + L\theta''\hat{\boldsymbol{\theta}}$$

Now for the external forces, note that

$$\mathbf{T} = -T\hat{\mathbf{r}} \qquad\qquad \text{for some scalar } T > 0$$
$$\mathbf{f} = -c\mathbf{v} = -cL\theta'\hat{\boldsymbol{\theta}} \qquad \text{for some constant } c > 0 \tag{6}$$
$$\text{force of gravity} = mg\hat{\mathbf{x}} = mg(-\sin\theta\hat{\boldsymbol{\theta}} + \cos\theta\hat{\mathbf{r}})$$

where we have used (4) and (5). See Figure 5.11 for a geometric visualization of these force decompositions. Now substituting (5) and (6) into Newton's Second Law

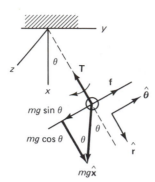

Figure 5.11   Decomposition of forces.

and equating coefficients of $\hat{\mathbf{r}}$ and $\hat{\boldsymbol{\theta}}$ ($\hat{\mathbf{z}}$ does not appear) on both sides of the equation, we obtain the two equations

$$-mL\,(\theta')^2 = -T + mg\,\cos\,\theta \qquad \text{(from } \hat{\mathbf{r}}\text{)}$$
$$mL\,\theta'' = -cL\,\theta' - mg\,\sin\,\theta \qquad \text{(from } \hat{\boldsymbol{\theta}}\text{)} \qquad (7)$$

If the pendulum were being driven by a given external force F of the form $F(t)\hat{\boldsymbol{\theta}}$, the second equation in (7) would be

$$mL\,\theta'' + cL\,\theta' + mg\,\sin\,\theta = F(t) \qquad (8)$$

called the *equation of the damped simple pendulum* with driving force $F(t)$. Equation (8) is a nonlinear second-order differential equation, but for "small" motions it is equivalent to a linear one. Indeed, Taylor's Theorem implies that $\sin\,\theta$ is approximately θ for small θ; hence for motions where θ is small, the term $\sin\,\theta$ in (8) may be replaced by θ without appreciably affecting its solutions over limited time periods. This is implied by a Continuity in the Data Property for second-order differential equations similar to the one mentioned in Appendix A, but we omit the proof. When this approximation is made we have the equation of the driven *linearized pendulum with damping*,

$$mL\,\theta'' + cL\,\theta' + mg\,\theta = F(t) \qquad (9)$$

Note the similarity of Eq. (9) with Eq. (2). Thus we see again the power of the modeling approach: Even though the linearized pendulum is a different physical system from the spring–mass system considered earlier (not to mention the electrical circuits of Section 5.4), the methods of Section 5.3 suffice to give a satisfactory explanation of the motions of both systems.

PROBLEMS

The following problem(s) may be more challenging: 3, 5, 7(b), 8, and 9.

1. Consider a simple undamped linearized undriven pendulum.
   (a) Find the period T of the pendulum in terms of its length L and g (the acceleration due to gravity at ground level).

(b) Find the length of a pendulum whose period is exactly 1 s.

2. If a simple undamped linearized undriven pendulum is 8 ft long and swings with an amplitude of 1 rad, compute
   (a) The angular velocity of the pendulum at its lowest point.
   (b) Its acceleration at the ends of its path.

3. Consider the "backlash" model sketched here, where x measures the displacement of the mass m from a point midway between the plates of the springs in equilibrium position.

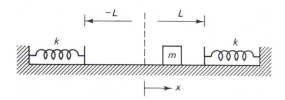

Suppose that m moves on a smooth strip with springs of spring constant k on either end. Suppose that at $t = 0$, $x(0) = L + a$ and $x'(0) = 0$. Show that the period of the resulting motion is

$$T = 2\pi \left(\frac{m}{k}\right)^{1/2} \left[1 + \frac{2L}{\pi a}\right]$$

4. A magnet is suspended by a spring above a large fixed iron plate, as shown in the diagram.

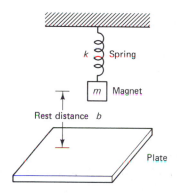

Let x be the vertical displacement of the magnet from its rest position (a distance b from the plate). Assume that the force of magnetic attraction is inversely proportional to the distance between the magnet and the plate, and that the spring force is directly proportional to the displacement x. Find the differential equation describing the motion of the magnet.

5. Referring to Figure 5.8 let $\mathbf{R}(t)$ denote the radius vector from the pendulum support to the bob and suppose that at $t = 0$,

$$\hat{\mathbf{z}} \cdot \mathbf{R}(0) = 0, \qquad \hat{\mathbf{z}} \cdot \mathbf{R}'(0) = 0$$

Show that for $t \geq 0$ the bob will oscillate back and forth in the xy-plane [i.e., $\hat{\mathbf{z}} \cdot \mathbf{R}(t) \equiv 0$]. [*Hint*: Show that $w(t) \equiv \hat{\mathbf{z}} \cdot \mathbf{R}(t)$ satisfies a homogeneous linear differential equation and has vanishing initial data.]

6. [*Variable Length Pendulum*]. Show that if the length of the pendulum is a function of time $L(t)$, the equation of angular motion of the pendulum is, not (8), but

$$mL\theta'' + (2mL' + cL)\theta' + mg \sin\theta = F$$

7. [*Archimedes' Buoyancy Principle*]. According to Archimedes' Principle, the buoyant force acting on a body wholly or partly immersed in a fluid equals the weight of the fluid displaced.

   (a) Ignoring friction, show that a flat wooden block of face L^2 square feet and depth h feet floating half-submerged in water will act as a harmonic oscillator of period $2\pi\sqrt{h/2g}$ if it is depressed slightly. [*Hint*: Show that if x is a vertical coordinate from the midpoint of the block, $L^2 h(\rho/2)x'' + L^2\rho g x = 0$, where ρ is the density of water.]

   (b) Assuming no friction, replace the block in part (a) by a buoyant sphere of radius R floating half-submerged, and show that if the initial displacement is slight, the sphere will act as a harmonic oscillator of period $2\pi\sqrt{2R/3g}$. Show that if the displacement is large, the nonlinearities in the differential equation modeling the motion cannot be ignored.

8. In a spring–mass system, let the spring constant be 1.01 N/m and the weight of the mass be 1 kg. Find the response of the system if at $t = 0$ it is at rest in its equilibrium state, the mass is acted on by viscous damping with coefficient equal to 0.2 and driven by the periodic force (measured in newtons) described in the figure.

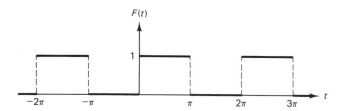

9. A mass m sliding on a rough surface is connected to rigid walls by springs with total effective constant k (see the sketch). There is a frictional force of constant magnitude F which opposes the motion (*Coulomb friction*).

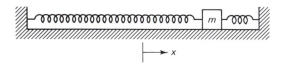

   (a) Show that the system can be modeled by the equation

$$mx'' = \begin{cases} -kx + F & \text{if } x' < 0 \\ -kx - F & \text{if } x' > 0 \\ -kx & \text{if } x' = 0 \quad \text{and} \quad k|x| > F \\ 0 & \text{if } x' = 0 \quad \text{and} \quad k|x| \le F \end{cases} \qquad (*)$$

[Note that (*) is not a linear equation—but it is linear for motion where the velocity is bounded away from zero.]

(b) A function $x(t)$ is a solution of (*) on an interval I if x is continuously differentiable on I and $x(t)$ satisfies (*) on I except possibly where $x' = 0$ (at such points x'' may be discontinuous). Show that if $x(t)$ is a solution of (*), the point $P(t) = (x(t), x'(t))$ traces out a curve in the xx' *phase plane* like one of those sketched here, where

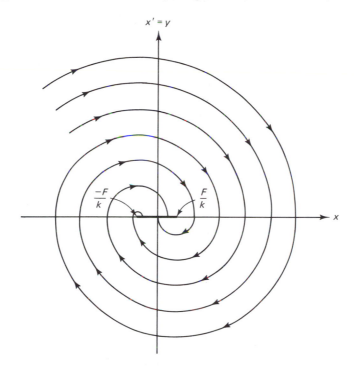

we assume for simplicity that $m = k$. We denote x' by y throughout. Each time-oriented arc in the upper half-plane is an arc of a circle (of radius A, say) centered at $(-F/k, 0)$, the subsequent arc in the lower half-plane is the arc of a circle of radius $A - 2F/k$ centered at $(F/k, 0)$, then followed by an arc in the upper half-plane of a circle of radius $A - 4F/k$ centered at $(-F/k, 0)$, and so on. No arc can be extended through the segment $[-F/k, F/k]$ of the x-axis. [*Hint*: Start at time 0 at a point in the upper half-plane; $x' > 0$ and the second equation of (*) is used. Solve this equation for $x(t)$, calculate $x'(t)$, and observe that the parametric equations, $x = x(t)$, $y = x'(t)$, describe a circle centered at $(-F/k, 0)$. When $y = 0$ and the "orbit" moves into the plane $y < 0$, the orbit is at the point $P = (A - F/k, 0)$. Now solve the first equation of (*) using as "initial" point P. Continue this process.]

(c) The x-interval $[-F/k, F/k]$ is called the *dead zone* and corresponds to the mass at rest at some displacement, the friction force F being larger than the Hooke's Law restoring force at that displacement: $F > k|x|$ if x is in the dead zone. Show that for any initial (x_0, y_0), $(x(t), y(t))$ reaches the dead zone in finite time (the *decay time*). Estimate this decay time as a function of (x_0, y_0).

(d) Show that the arcs of circles in part (b) are replaced by arcs of ellipses if $k \neq m$.

5.6 RESONANCE

In Section 5.3 we developed techniques for finding the response of a linear second-order equation to a driving force. When the free solutions of the linear equation are oscillatory in character, periodic driving forces produce peculiar responses which scientists say occur by the phenomenon of *resonance*. The excitation of a tuning fork at rest, in response to another tuning fork that has been struck is an example of resonance. Tuning circuits in radio receivers are also a familiar example. The remarkable phenomenon of resonance is central to the operation of a great many devices and instruments used not only for scientific research but also by society in general. It is noteworthy that resonance can be explained mathematically in terms of the analysis of Section 5.3. For simplicity, we take up undamped systems first and then the more realistic damped systems.

Beats and Resonance in Undamped Systems

Let us consider the differential equation

$$x'' + k^2 x = A \cos \omega t \tag{1}$$

where k, A, and ω are positive constants. As we have seen in Section 5.5, Eq. (1) arises in modeling an undamped spring–mass system driven by a sinusoidal force of *amplitude A* and *circular frequency ω*.† Of course, Eq. (1) can arise in many other contexts (e.g., a simple RLC circuit with no resistance and an impressed sinusoidal voltage source; see Section 5.4), but our analysis will use only Eq. (1) and no other feature of the physical system that underlies the model. Thus our conclusions will apply to any system modeled by Eq. (1)—a very important attribute of the modeling approach to problem solving.

For convenience we distinguish special cases in treating Eq. (1). Observe that every free solution of Eq. (1) is periodic with frequency k. The driving term has frequency ω, and the character of the solutions of Eq. (1) depends on whether or not $\omega = k$.

Case I: $\omega = k$.   Using the Method of Undetermined Coefficients (see Section 5.3) we find the (real-valued) general solution of Eq. (1) by first solving for the (complex-valued) general solution of the equation

$$x'' + \omega^2 x = A e^{i\omega t} \tag{2}$$

Observe that the characteristic polynomial $P(r) \equiv r^2 + \omega^2$ is such that $P(i\omega) = 0$, but $P'(i\omega) \neq 0$ and hence we look for a particular solution of Eq. (2) in the form $x_p(t) = Cte^{i\omega t}$. To calculate the constant C, we substitute $x_p(t)$ into Eq. (2) and find that $C = -Ai/2\omega$. Thus a real-valued particular solution of Eq. (1) when $\omega = k$ is $\mathrm{Re}\,[(-Ai/2\omega)te^{i\omega t}] = (A/2\omega)t \sin \omega t$. Hence the general solution of Eq. (1) when $\omega = k$ is

---

† The frequency of the sinusoid $\cos \omega t$ in *cycles per second* is $\omega/2\pi$. *Note:* The unit designation *hertz* denotes 1 cycle per second.

$$x(t) = C_1 \cos \omega t + C_2 \sin \omega t + \frac{A}{2\omega} t \sin \omega t$$

where C_1 and C_2 are arbitrary (real) constants. Hence every solution of Eq. (1) when $\omega = k$ is the superposition of a *free oscillation* $C_1 \cos \omega t + C_2 \sin \omega t$ with natural frequency ω, and a *forced solution* $(A/2\omega)t \sin t$ which has the form of a sinusoid with amplitude that grows with time. Observe that the forced solution is the response of the system to a periodic external force whose frequency matches the natural frequency of the system. When an undamped system is driven with a periodic external force having the system's natural fequency, the unbounded oscillation that results is said to be due to *pure resonance*. Observe that in the case of pure resonance, the system responds to a bounded input with an unbounded output.

Case II: $\omega \neq k$. This time the Method of Undetermined Coefficients gives as the general solution of Eq. (1),

$$x(t) = C_1 \cos kt + C_2 \sin kt + \frac{A}{k^2 - \omega^2} \cos \omega t \qquad (3)$$

where C_1 and C_2 are arbitrary constants. Thus, $x(t)$ in Eq. (3) is the superposition of a *free oscillation* $C_1 \cos kt + C_2 \sin kt$ with *natural fequency* k, and a *forced oscillation* $[A/(k^2 - \omega^2)] \cos \omega t$ representing the system's response to the externally applied force, $A \cos \omega t$. Observe that the frequency of the forced oscillation exactly matches that of the external force. We shall leave to the problems the proof that the response $x(t)$ is periodic if and only if ω/k is a rational number.

Let us find that unique solution of Eq. (1) which starts out from rest, that is, a solution $x(t)$ such that $x(0) = x'(0) = 0$. Using the general solution given by Eq. (3), we see that these initial conditions imply that $C_1 = A(\omega^2 - k^2)^{-1}$, $C_2 = 0$ and we obtain the solution

$$x(t) = \frac{A}{\omega^2 - k^2}(\cos kt - \cos \omega t)$$

Using the identity

$$\cos \alpha - \cos \beta = -2 \sin \frac{\alpha + \beta}{2} \sin \frac{\alpha - \beta}{2}$$

to rewrite this solution, we have

$$x(t) = \left(\frac{2A}{k^2 - \omega^2} \sin \frac{k - \omega}{2} t \right) \sin \frac{k + \omega}{2} t \qquad (4)$$

which clearly has the form of a sinusoid of frequency $(k + \omega)/2$ with an amplitude that varies periodically in time with frequency $|k - \omega|/2$. This observation leads to a very interesting interpretation when $|k - \omega|$ is very small with respect to $k + \omega$, the phenomenon of *beats*. For in this case, the system's response to being driven by a sinusoid of frequency ω is essentially a sinusoid of frequency $(k + \omega)/2$ but

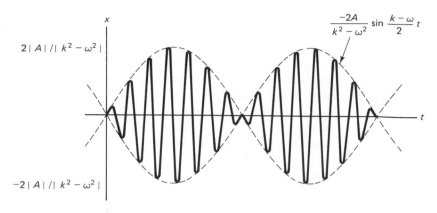

Figure 5.12   Beats. Plot of equation (4).

with a relatively slowly varying amplitude, called a *beat*, which is a sinusoid of *beat frequency* $|k - \omega|/2$ (see Figure 5.12). Beats can be experienced, for example, as the alternate fading and swelling in amplitude heard when two tuning forks vibrate at not quite the same frequency; this simple device permits the human ear to detect frequency differences as low as 0.06%. The beat phenomenon is used in heterodyne detection, where an incoming radio signal is mixed with the signal of an oscillator in the receiver to give a difference frequency down in the audio range. This signal is then applied to a detector whose output will drive a loudspeaker. Finally, observe that the maximum amplitude of the sinusoid in Eq. (4) tends to ∞ as $\omega \to k$; this is another manifestation of the phenomenon of resonance.

Resonance in Damped Systems

Now let us turn our attention to the more realistic case of the driven damped harmonic oscillator which is modeled by the differential equation

$$x'' + cx' + k^2x = A \cos \omega t \tag{5}$$

where A, c, k, and ω are positive constants. In order that the free solutions of Eq. (5) be oscillatory in character, we shall also assume throughout that $c^2 < 4k^2$.† Then using the methods of Section 5.2 it is easy to see that the free (i.e., undriven) response of Eq. (5) is $e^{-ct/2}(C_1 \cos \beta t + C_2 \sin \beta t)$, where $\beta = (4k^2 - c^2)^{1/2}/2$, and C_1 and C_2 are arbitrary (real) constants. Thus every free response is a transient and dies out with increasing time (see Figure 5.13). Using the Method of Undetermined

---

† It should be noted here that strictly speaking, the term *resonance* need not be applied only to equations such as Eq. (5) where $c^2 < 4k^2$. In fact, the term can be applied when $c \geq 0$, $k > 0$ are such that the maximal amplitude of the steady-state response obtained by undetermined coefficients exceeds the amplitide A of the input. Observe that this definition includes the pure resonance case mentioned earlier and the case of beats, the latter only when $|\omega^2 - k^2| < 2$ (see Figure 5.12).

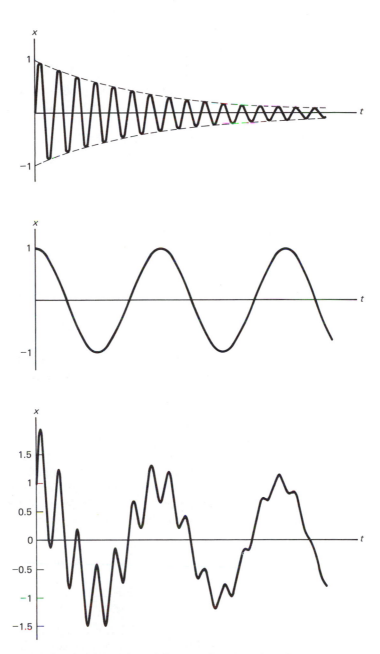

Figure 5.13 Superposition of free and forced motions for $x'' + cx' + k^2x = A \cos \omega t$ with $0 < c < 2k$.

Coefficients, we see that the steady-state forced response of the system modeled by Eq. (5) is the particular solution

$$x_p(t) = A\left[\frac{c\omega}{(k^2 - \omega^2)^2 + \omega^2 c^2}\sin \omega t + \frac{k^2 - \omega^2}{(k^2 - \omega^2)^2 + \omega^2 c^2}\cos \omega t\right] \qquad (6)$$

Now the general solution of Eq. (5) consists of the superposition of the steady-state forced response $x_p(t)$ in Eq. (6) and the transients mentioned earlier (see Figure 5.13).

The *phase angle* δ of the forced response is the unique angle $|\delta| < \pi/2$ such that

$$\sin \delta = \frac{c\omega}{[(k^2 - \omega^2)^2 + \omega^2 c^2]^{1/2}}, \qquad \cos \delta = \frac{k^2 - \omega^2}{[(k^2 - \omega^2)^2 + \omega^2 c^2]^{1/2}}$$

The phase-angle concept is useful in that it allows us to use the cosine addition formula to rewrite $x_p(t)$ in Eq. (6) as the sinusoid

$$x_p(t) \equiv \frac{A}{[(k^2 - \omega^2)^2 + \omega^2 c^2]^{1/2}}\cos(\omega t - \delta) \qquad (7)$$

This shows that the forced response of a sinusoidally driven damped system is always a sinusoid with the same frequency as the driving term (regardless of the values of the other system parameters) but with a phase lag δ (which does depend on system parameters).

To obtain some feeling for how the forced response $x_p(t)$ in Eq. (7) behaves as the system parameters are changed, let us first rewrite it as

$$x_p(t) = B(\omega)\cos(\omega t - \delta)$$

where $B(\omega)$ is the amplitude of the response, $A[(k^2 - \omega^2)^2 + \omega^2 c^2]^{-1/2}$. For convenience let us imagine that the "restoring force coefficient" k is given and measure ω in units defined by the frequency of undamped free oscillations, $\omega_0 \equiv k$, and the damping coefficient c in units defined by the critical damping coefficient $c_{cr} \equiv 2k$ (note that our assumption that $c^2 < 4k^2$ implies that $c/c_{cr} < 1$). Thus we shall rewrite the amplitude $B(\omega)$ and the phase angle δ in terms of the ratios ω/ω_0 (the *frequency ratio*) and c/c_{cr} (the *damping ratio*) to obtain

$$B(\omega) = \frac{A/k^2}{[(1 - (\omega/\omega_0)^2)^2 + 4(\omega/\omega_0)^2(c/c_{cr})^2]^{1/2}} \qquad (8)$$

$$\tan \delta = \frac{2(c/c_{cr})(\omega/\omega_0)}{1 - (\omega/\omega_0)^2} \qquad (9)$$

Observe that the numerator, A/k^2, in Eq. (8) can be interpreted in a spring–mass system as the static deflection of the mass due to a constant force of value A. The quantity

$$M = \frac{1}{[(1 - (\omega/\omega_0)^2)^2 + 4(\omega/\omega_0)^2(c/c_{cr})^2]^{1/2}} \qquad (10)$$

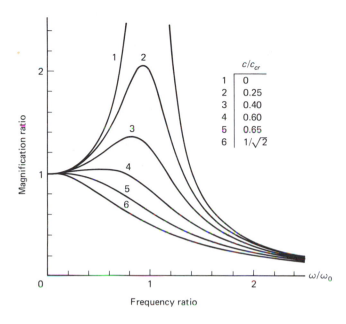

Figure 5.14 Plots of magnification ratio versus frequency ratio.

which appears in Eq. (8) is called the *magnification ratio*, for obvious reasons. Note that $M \to 1$ as $\omega/\omega_0 \to 0$ and $M \to 0$ as $\omega/\omega_0 \to +\infty$ for any given value of c/c_{cr}. Elementary calculus shows that M has a unique maximum value when $\omega/\omega_0 = [1 - 2(c^2/c_{cr}^2)]^{1/2} < 1$, provided that $0 < c/c_{cr} < 1/\sqrt{2}$. Thus the value $\omega^* = \omega_0[1 - 2(c^2/c_{cr}^2)]^{1/2} \equiv (k^2 - c^2/2)^{1/2}$ is known as the *resonant frequency* of the system modeled by Eq. (5). Plots of M against the frequency ratio ω/ω_0 for various values of the damping ratio c/c_{cr} are given in Figure 5.14. Observe that when there is no damping (i.e., $c/c_{cr} = 0$) it follows that $M \to +\infty$ as $\omega/\omega_0 \to 1$. Whenever damping is present, however, M is a bounded continuous function for $\omega/\omega_0 \geq 0$.

From Eq. (9) we see that the phase angle δ of the forced response can be obtained as a principal value of the inverse tangent function

$$\delta = \tan^{-1} \frac{(c/c_{cr})(\omega/\omega_0)}{1 - (\omega/\omega_0)^2} \tag{11}$$

Plots of the phase angle δ against the frequency ratio ω/ω_0 are given in Figure 5.15 for various values of the damping ratio c/c_{cr}. Note that when there is no damping present, $\delta = 0$ no matter what the frequency of the driving sinusoid is.

Engineers use a different way of displaying the information contained in Figures 5.14 and 5.15. The *Bode plot* of the steady-state response consists of the graphs of the phase angle and the logarithm of the magnification factor versus the logarithm of the frequency ratio. The *Nyquist plot* combines the phase angle and the magnification factor in one diagram.

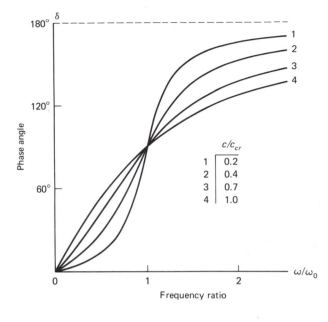

Figure 5.15 Plots of phase angle versus frequency ratio.

Applications of Resonance in Damped Systems

The applications of pure resonance mentioned earlier carry over in an obvious way to the damped case. Here is another application. Imagine the *RLC* oscillator in a radio receiver as a circuit driven by the signals $\sum_{n=1}^{N} A_n \cos (\omega_n t - \phi_n)$ from N different broadcasting stations using the distinct frequencies $\omega_1, \ldots, \omega_N$. The steady-state output is again a combination signal of much the same form. By *tuning* the radio, we can vary the capacitance (or the inductance) to match the resonance frequency of the circuit with the frequency of the input signal from one of the stations, thereby allowing us to amplify a single input frequency (see Problem 3).

It has been observed that certain structures such as bridges or buildings have resonant frequencies. An external oscillating force could then conceivably induce an oscillation of destructive amplitude. One theory of the collapse of the Tacoma Narrows bridge in 1940 while a strong breeze was blowing is that the support work along the roadway created sinusoidal eddies whose frequency matched the natural frequency of the bridge. Tachometers, seismometers, and vibrometers all depend on the resonance phenomenon for their operation. For example, the *Frahm tachometer* is a box containing rows of cantilever-mounted steel reeds of varying natural frequencies. When the box is placed on a vibrating machine, only the reeds with natural frequencies close to the machine's vibrational frequency will be seen to move.

PROBLEMS

The following problem(s) may be more challenging: 3(d) and 6(a).

1. Let $k\omega > 0$. Show that $\cos kt + \cos \omega t$ is periodic if and only if k/ω is rational.

2. Show that the phenomenon of beats can be observed for the equation $x'' + k^2 x = A \sin \omega t$, where k, ω, and A are positive constants with $k \neq \omega$, for every set of initial conditions. [*Hint*: Use the identity

$$A \sin \alpha + B \sin \beta = (A^2 + B^2 + 2AB \cos (\alpha - \beta))^{1/2} \sin (\alpha + \beta - \phi)$$

where ϕ is a certain phase angle.]

3. (a) [*Impedance*]. Show that the steady-state solution of $x' + ax' + bx = Ae^{\lambda t}$, where a and b are positive constants and λ is real, is

$$x = \frac{A}{\lambda^2 + a\lambda + b} e^{\lambda t}$$

[$P(\lambda) = \lambda^2 + a\lambda + b$ is called the *impedance*. Thus the ratio of input to output in this case is the reciprocal of the impedance. We assume that $P(\lambda) \neq 0$.]

(b) Repeat part (a) but now with the input $Ae^{i\omega t}$ and show that the steady-state output is

$$x = \frac{A}{(i\omega)^2 + a(i\omega) + b} e^{i\omega t} = \frac{A}{P(i\omega)} e^{i\omega t}$$

[The quantity $P(i\omega)$ is called the *complex impedance*.]

(c) Relate the conclusion in part (b) to Eq. (6).

(d) Carry out the details of tuning an *RLC* circuit to show that the broadcast signal of frequency ω_j will receive maximal amplification if the capacitor is tuned appropriately.

4. Show that the free vibration of a damped oscillator $x'' + 2nx' + p^2 x = F \sin \omega t$, subject to the initial conditions $x(0) = x_0$, $x'(0) = v_0$, is given by

$$x = e^{-nt} \left(x_0 \cos Qt + \frac{v_0 + nx_0}{Q} \sin Qt \right)$$

where $Q = \sqrt{p^2 - n^2}$.

5. (a) Show that the amplitude of the frequency response curve $B(\omega)$ has a maximum provided that $c/c_{cr} < 1/\sqrt{2}$.

(b) Verify the results given in the text for the location and value of this maximum.

6. In a damped spring system the static deflection is 0.5 in., while damped vibrations decay in amplitudes from 0.4 to 0.1 in. in 20 cycles. Assume a mass of 1 k.

(a) What is the damping constant?

(b) Find the resonant frequency ω^* and calculate the maximal amplitude of the steady-state response of the system to a driving force $A \cos \omega^* t$.

CHAPTER 6

Linear Algebra and Higher-Order Linear Differential Equations

In Section 4.3 we saw that the geometric vector concept was very useful as a modeling tool in mechanics. In this chapter we extend this concept of "vector" to a more general setting and apply it to characterizing solutions of linear differential equations of order n, where n is any positive integer. Geometric vectors will continue to be denoted by boldface characters, but vectors in this more general setting will not. The real power of the vector approach as a modeling tool becomes most impressive when the states of the system involve a large number of dependent variables or degrees of freedom. In Chapter 5 we concentrated on second-order linear differential equations and we show here that those solution techniques are easily extended to cover the nth-order case quite well.

6.1 LINEAR SPACES

It is a straightforward matter to verify that the set of all geometric vectors defined in Section 4.3 satisfies all the properties listed below provided that we remember to identify geometric vectors which are translations of one another. Note that for geometric vectors *scalars* are understood to be in **R**, the real numbers.

Vector Space Properties. In the nonempty collection of objects V (called *vectors*) there are two operations defined which satisfy the properties listed below. Let us denote by "+" the operation of *addition* which assigns a vector $u + v$ to every pair of vectors u and v; the operation of *multiplication by scalars* assigns to every *scalar* α and every vector u the vector αu. V with these operations is such that

(A1) *Addition is order independent.* A finite collection of vectors can be added together in any order without affecting the sum. [This follows once it is known that for any vectors u, v, and w we have $u + v = v + u$ and $u + (v + w) = (u + v) + w$.]

(A2) *Existence of zero element.* † There is an additive *zero* element in V, denoted by 0, with the property that $u + 0 = u$ for all vectors u.

(A3) *Existence of additive inverses.* Every vector u has an *additive inverse*, denoted by $-u$, with the property that $u + (-u) = 0$.

(M1) *Multiplication by scalars is distributive.* For any vectors u and v and any scalars α and β, we have that $\alpha(u + v) = \alpha u + \alpha v$ and $(\alpha + \beta)u = \alpha u + \beta u$.

(M2) *Multiplication by scalars is associative.* For any vector u and any scalars α and β, we have that $\alpha(\beta u) = (\alpha\beta)u$.

(M3) For the scalar 1 we have that $1u = u$, for all vectors u.

Turning the tables around, we may use the Vector Space Properties above to characterize the abstract notion of a *linear* (or *vector*) *space*. This is precisely the thrust of the following definition.

> **Linear Space.**   Let V be a set of objects and **F** a set of scalars (**F** is either **R**, the real numbers, or **C**, the complex numbers) with two operations, "addition" and "multiplication by a scalar," defined in V such that the Vector Space Properties hold. Then V is a *real linear* (or *vector*) *space* if **F** = **R**, and a *complex linear* (or *vector*) *space* if **F** = **C**. The elements of V are sometimes called *vectors*.

Thus since geometric vectors evidently satisfy the Vector Space Properties with **F** = **R**, we conclude that they form a real linear space. A feature of this abstract approach to linear spaces is that we can deduce from the defining properties (A1)–(A3) and (M1)–(M3) some additional properties which must then hold in *any* linear space (and hence for geometric vectors in particular). Some of them are as follows:

Further Properties of Linear Spaces.   Let V be any linear space.
1. There is only one zero vector in V.
2. $\alpha u = 0$ if either the scalar $\alpha = 0$ or the vector $u = 0$. Conversely, if $\alpha u = 0$, then either $\alpha = 0$ or $u = 0$.
3. Every vector v in V has a *unique* additive inverse, $-v$; moreover, $-v = (-1)v$ for any vector v.

---

† It is traditional to use the same symbol for the zero scalar and the zero vector—context distinguishes one from the other.

Proof.

1. Let 0 and $0'$ be two zero vectors for V, then by (A1) and (A2) we see that $0 = 0 + 0' = 0'$, so there is only one zero in V.

2. For any u note that $u = (1 + 0)u = u + 0u$ using (M1) and (M3). Adding $-u$ to both sides of this equation, we obtain by (A3) that $0 = 0u$, as asserted. Next, we assume that $\alpha \neq 0$. Then for any vector u, $\alpha 0 + u = \alpha 0 + (\alpha \cdot \alpha^{-1})u = \alpha(0 + \alpha^{-1}u) = \alpha(\alpha^{-1}u) = 1u = u$, where we have used (M1)–(M3). Thus $\alpha 0$ acts like a zero vector, and since there is only one such vector, we must have $\alpha 0 = 0$. Finally, suppose that $\alpha u = 0$ for some scalar α and some vector u. If $\alpha \neq 0$, then $u = \alpha^{-1}(\alpha u) = \alpha^{-1}0 = 0$, and we are done.

3. Let v and w be two additive inverses for the vector u. Then $v = v + 0 = v + (u + w) = (v + u) + w = 0 + w = w$ where we have used (A1) and (A2). Hence u has a unique inverse. Now since $0 = 0u = (1 - 1)u = 1u + (-1)u = u + (-1)u$, we see that $-u = (-1)u$, finishing the proof.

It is worth noting that there is a linear space V with only one element. In view of the result above, we see that that single element must be the zero vector, and hence V is written as $\{0\}$ and is referred to as the *trivial* linear space.

Example 6.1

Let $\mathbf{R}^n$ for any $n = 1, 2, \ldots$ be the collection of all ordered n-tuples of real numbers, $x = (x_1, x_2, \ldots, x_n)$. Note that x here denotes the n-tuple and the x_i are real numbers. Two elements x and y in $\mathbf{R}^n$ are identified if and only if $x_i = y_i$, $i = 1, 2, \ldots, n$. Using the real numbers $\mathbf{R}$ as scalars and defining the operations of addition and multiplication by scalars component-wise by

$$x + y = (x_1 + y_1, \ldots, x_n + y_n), \qquad \alpha x = (\alpha x_1, \ldots, \alpha x_n) \qquad (1)$$

for all x, y in $\mathbf{R}^n$ and all scalars α, it is a straightforward exercise to verify that $\mathbf{R}^n$ is a real linear space. Note that $\mathbf{R}^1$ can be identified with $\mathbf{R}$, and hence $\mathbf{R}$ itself can be considered a real linear space.

Example 6.2

Let $\mathbf{C}^n$ for any $n = 1, 2, \ldots$ be the collection of all ordered n-tuples of complex numbers $z = (z_1, z_2, \ldots, z_n)$. Using the complex numbers $\mathbf{C}$ as scalars and defining operations of addition and multiplication by scalars as in (1), it is straightforward to show that $\mathbf{C}^n$ is a complex linear space. Note that $\mathbf{C}^1$ can be identified with $\mathbf{C}$.

Linear Spaces of Functions

Perhaps more surprising examples are linear spaces of functions. We shall see in later sections that such linear spaces are useful in characterizing the general solution of a linear differential equation. For example, let $F(I)$ be the set of all real-valued functions defined on an interval I. Using **R** as the scalars and defining addition and multiplication by scalars by

$$(f+g)(t)=f(t)+g(t) \qquad (\alpha f)(t)=\alpha f(t), \quad \text{all } t \text{ in } I \tag{2}$$

for any f, g in $F(I)$ and scalars α, we can show that $F(I)$ is a real linear space. But $F(I)$ is too big for our purposes, so we define the continuity classes.

> **Continuity Classes $C^k(I)$.**   Let I be an interval and $k \geqq 0$ an integer. Then the *continuity class* $C^k(I)$ is the class of all real-valued functions (or the class of all complex-valued functions) that have at least k continuous derivatives on I.† By convention $C^0(I)$ is the class of all continuous functions on I, and $C^\infty(I)$ is the class of all functions common to all $C^k(I)$, $k \geqq 0$. [This is consistent with how $C^0(I)$, $C^1(I)$, and $C^2(I)$ were defined earlier.]

Example 6.3

It is not difficult to show that for any integer $k \geqq 0$ the class $C^k(I)$ of real-valued functions is a real linear space if real scalars and the operations (2) are used. If $C^k(I)$ consists of complex-valued functions, then $C^k(I)$ is a complex linear space if complex scalars and the operations (2) are used.

Subspaces

Let A be a subset of a linear space V. Then for any finite collection of vectors a_1, $a_2, \ldots, a_m$ in A and any scalars $\alpha_1, \alpha_2, \ldots, \alpha_m$, the sum $\alpha_1 a_1 + \cdots + \alpha_m a_m$ is said to be a *finite linear combination* over A. The *span* of A, denoted by Span (A), is the collection of all finite linear combinations over A. Span (A) is contained in V, and if we use the operations of addition and multiplication by scalars from V, we see that Span (A) is itself a linear space. Span (A) is called the *subspace of V spanned by A*. The set A is said to span V if Span (A) = V. In general, the subset A of a linear space V is a *subspace* of V if Span (A) = A. For example, the subsets consisting of the zero vector alone, or all of V, are both subspaces—called the *trivial subspaces*.

Example 6.4

There is a simple device for "visualizing" vectors in $\mathbf{R}^3$. Associate with each vector (x_1, x_2, x_3) in $\mathbf{R}^3$ the geometric vector $x_1\mathbf{i} + x_2\mathbf{j} + x_3\mathbf{k}$, where $\{\mathbf{i}, \mathbf{j}, \mathbf{k}\}$ is an orthogonal frame in Euclidean 3-space. Then in the Cartesian coordinates generated by that frame, (x_1, x_2, x_3) are the coordinates of the head of the geometric vector and the origin is the tail. The sum of vectors in $\mathbf{R}^3$ is associated with the sum of the associated geometric vectors via the parallelogram law. Thus we can use our geometric intuition in Euclidean 3-space to "see" the

---

† Note that the same symbol $C^k(I)$ is used to denote both the class of real-valued functions *and* the class of complex-valued functions—context reveals which one is intended.

following: (a) The span of a single nonzero vector x is the line through the origin determined by x. (b) The span of two nonparallel vectors x and y is the plane through the origin determined by those two vectors. (c) Let three vectors have the property that one vector does not lie in the plane determined by the other two. Then the span of those three vectors must be all of $\mathbf{R}^3$. Thus it follows that every nontrivial subspace of $\mathbf{R}^3$ is either a line or a plane through the origin. (Lines or planes *not* containing the origin are *not* subspaces of $\mathbf{R}^3$; they are translations of subspaces.)

Subspace Criterion.   The subset A of a linear space V is a subspace of V if and only if any linear combination of any two elements of A is a vector in A. (In other words, the criterion states that A is *closed with respect to finite linear combinations*).

Proof. To show that the criterion implies that $\text{Span}(A) = A$, notice first that $\text{Span}(A)$ always includes A. The criterion says that $\text{Span}(A)$ is contained in A. Thus $\text{Span}(A)$ must be precisely A, and we are done.

Example 6.5

Observe that $C^{n+1}(I)$ is a subspace of $C^n(I)$ for any $n = 0, 1, 2, \ldots$ and that $C^\infty(I)$ is a subspace of all the $C^n(I)$, which, in turn, are subspaces of $F(I)$, the linear space of all functions on the interval I (see Figure 6.1).

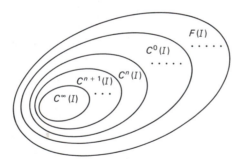

Figure 6.1   Relation between continuity spaces.

Example 6.6

Consider the homogeneous linear differential equation with constant coefficients

$$y^{(n)} + a_{n-1}y^{(n-1)} + \cdots + a_1y' + a_0y = 0 \tag{3}$$

As we show in the next section, for any positive integer n all solutions of Eq. (3) belong to the class $C^\infty(\mathbf{R})$ (we have seen this for $n = 1$ in Chapter 3 and for $n = 2$ in Chapter 5). For any pair of solutions y_1 and y_2 of Eq. (3) and any scalars α_1 and α_2, we see by direct substitution that $\alpha_1 y_1 + \alpha_2 y_2$ is also a solution of Eq. (3). Thus by the Subspace Criterion we see that the set of all solutions of Eq. (3) is a subspace of $C^\infty(\mathbf{R})$. Furthermore, the set of all solutions of Eq. (3) that satisfy a condition such as $y(0) = 3y(1)$ is a subspace of the linear space of solutions of Eq. (3) and hence of $C^\infty(\mathbf{R})$.

Example 6.7

Observe that the vectors in $\mathbf{R}^3$ which solve the system

$$3x_1 - x_2 + 2x_3 = 0$$
$$x_1 + 4x_2 - x_3 = 0$$

(4)

satisfy the Subspace Criterion, and hence form a subspace of $\mathbf{R}^3$. Thus the solution space of (4) must either be a trivial subspace or a line or plane through the origin—there are no other possibilities (see Example 6.4).

Basis

For a set A in a linear space V let us put Span $(A) = W$ and observe, as before, that W is a subspace of V. In words, we say that A *spans* the subspace W, or that A is a *spanning set* for W. With the single exception of the trivial subspace $\{0\}$, every subspace W of V has infinitely many spanning sets. For example, from Example 6.4 we see that the subspace spanned by the set of unit vectors $\{\mathbf{j}, \mathbf{k}\}$ is also spanned by *any* pair of nonparallel vectors in that plane. As the next example shows, vectors in Span (A) can sometimes be written in more than one way as a finite linear combination over A.

Example 6.8

In $\mathbf{R}^2$ consider the set $A = \{(1, 2), (-1, 1), (5, 1)\}$. We claim that Span$(A) = \mathbf{R}^2$. Indeed, let (α, β) be any element of $\mathbf{R}^2$. Then since

$$\frac{\alpha + \beta}{3}(1, 2) + \frac{\beta - 2\alpha}{3}(-1, 1) = (\alpha, \beta)$$

(5)

we see that the claim is verified [and we did not have to use the third vector $(5, 1)$ at all]. On the other hand, notice that (α, β) can also be written as

$$\left(\frac{\alpha + \beta}{3} - 2\gamma\right)(1, 2) + \left(\frac{\beta - 2\alpha}{3} + 3\gamma\right)(-1, 1) + \gamma(5, 1) = (\alpha, \beta)$$

where γ is any scalar. Thus vectors in $\mathbf{R}^2$ can be written in more than one way as a finite linear combination over A.

If every vector in Span (A) can be written in only one way as a finite linear combination over A, then we will say that the spanning set A has the *unique representation property*. Such spanning sets play a fundamental role in the applications of linear spaces.

Basis.   If the set A spans a linear space V and has the unique representation property, A is called a *basis* for V.

Since it is a bit awkward to check directly that a set A in a linear space V satisfies the unique representation property, we will give an equivalent condition which in some ways is easier to deal with. If there are two different finite linear combinations over A for the same vector in Span (A), there must exist a finite linear combination over A which adds up to the zero vector and whose scalar coefficients are not all zero. Conversely, if the zero vector can be represented by a finite linear combination over A whose scalar coefficients are not all zero, A cannot satisfy the unique representation property. We shall formalize this concept. For convenience we shall refer to a finite linear combination of vectors with at least one nonzero scalar coefficient as a *nontrivial* finite linear combination. The case where all scalar coefficients vanish is said to be *trivial*.

Linear Independence Property.   The set A in a linear space is said to be *linearly independent* if and only if there is no nontrivial finite linear combination over A whose sum is the zero vector. (The similarity of this definition and the "independence" property of Section 5.2 is no accident.)

Thus we see that a set A in a linear space has the unique representation property if and only if A is linearly independent. This proves the following result.

Basis Property.   A spanning set A for a linear space W is a basis if and only if A is linearly independent.

A linearly independent set in a linear space is a basis for the linear space which it spans. Also note that if A is a linearly independent set and if $x_1, \ldots, x_n$ are in A and $\sum_1^n \alpha_i x_i = \sum_1^n \beta_i x_i$, then $\alpha_i = \beta_i$, $i = 1, \ldots, n$. For otherwise $\sum_1^n (\alpha_i - \beta_i) x_i = 0$, but not all the coefficients $\alpha_i - \beta_i$ vanish. Observe that any nonempty subset of an independent set is also independent.

Note that bases are not unique. In fact, if v belongs to a basis A of a linear space V, one can construct a different basis by replacing v by αv, $\alpha \neq 0, 1$. More generally, one could replace v and w in A by $\alpha v + \beta w$, $\gamma v + \delta w$ as long as $\alpha\delta - \beta\gamma \neq 0$ and still have a basis. Finally, let the linear space V have a finite basis $A = \{x_1, \ldots, x_n\}$. If v is an element of V, then, as we know, there is a unique set of scalars $\{\alpha_1, \ldots, \alpha_n\}$ such that $v = \sum_{i=1}^n \alpha_i x_i$; the α_i's are the *components* (or *coordinates*) of v with respect to the basis A. The value α_i is said to be the x_i *component of* v.

Example 6.9

The set $A_1 = \{(1, 0), (0, 1)\}$ spans $\mathbf{C}^2$, and since A_1 is linearly independent, it is a basis for $\mathbf{C}^2$. The same is true for the set $A_2 = \{(1, 1), (2, i)\}$, although it is not as evident. Given (z_1, z_2) in $\mathbf{C}^2$, we ask whether there are (complex) scalars c_1, c_2 such that

$$c_1(1, 1) + c_2(2, i) = (z_1, z_2) \tag{6}$$

But (6) is equivalent to the system

$$\begin{aligned} c_1 + 2c_2 &= z_1 \\ c_1 + ic_2 &= z_2 \end{aligned} \tag{7}$$

Subtracting the second equation from the first, we see that necessarily $c_2 = (z_1 - z_2)(2 - i)^{-1}$, and back substitution yields the value $c_1 = z_1 - 2(z_1 - z_2)(2 - i)^{-1}$. Our steps in solving the system (7) for c_1, c_2 are all reversible, and hence these values uniquely solve (6) for *any* (z_1, z_2) in $\mathbf{C}^2$. Thus A_2 is a basis for $\mathbf{C}^2$. Note that A_2 is linearly independent since the only representation via (6) for the zero vector $(z_1, z_2) = (0, 0)$ is $c_1 = 0$, $c_2 = 0$, the trivial solution. On the other hand, the set $A_3 = \{(1 - i, 1 + i), (2, 2i)\}$ does not span $\mathbf{C}^2$ and is not linearly independent. Note that

$$(1 + i)(1 - i, 1 + i) - (2, 2i) = (0, 0)$$

and hence the zero vector has a nontrivial linear representation over A_3.

Example 6.10

The set of n-tuples $A = \{(1, 0, \ldots, 0), \ldots, (0, \ldots, 0, 1)\}$ is a basis for both $\mathbf{R}^n$ and $\mathbf{C}^n$. A is called the *standard basis* of $\mathbf{R}^n$ or $\mathbf{C}^n$; its elements are commonly denoted by $e_1, \ldots, e_n$.

Example 6.11

One way of showing that a collection of functions in $C^0(I)$ is linearly independent is to assume that some linear combination of them produces the zero vector (i.e., the zero function on I), and then show that this linear combination must have been the trivial one. The following are two "tricks" that are useful for this purpose. Consider the set $\{\sin t, \cos t, e^t\}$ in $C^0(\mathbf{R})$. Suppose that c_1, c_2, and c_3 are constants such that $c_1 \sin t + c_2 \cos t + c_3 e^t = 0$ for all t in $\mathbf{R}$. Then inserting the particular points $t = 0$, $t = \pi/2$, $t = \pi$ into this relation, we obtain the conditions

$$c_2 + c_3 = 0, \qquad c_1 + c_3 e^{\pi/2} = 0, \qquad -c_2 + c_3 e^{\pi} = 0$$

The only constants that can satisfy all three of these conditions are $c_1 = c_2 = c_3 = 0$, and hence the set is linearly independent. Another "trick" is to differentiate the dependency relation twice to produce the three identities,

$$c_1 \sin t + c_2 \cos t + c_3 e^t = 0, \qquad \text{all } t \text{ in } \mathbf{R}$$
$$c_1 \cos t - c_2 \sin t + c_3 e^t = 0, \qquad \text{all } t \text{ in } \mathbf{R}$$
$$-c_1 \sin t - c_2 \cos t + c_3 e^t = 0, \qquad \text{all } t \text{ in } \mathbf{R}$$

Inserting $t = 0$ into these identities, we obtain the conditions

$$c_2 + c_3 = 0, \qquad c_1 + c_3 = 0, \qquad -c_2 + c_3 = 0$$

Since these relations can only be satisfied by $c_1 = c_2 = c_3 = 0$, we conclude (again) that the set of given functions are linearly independent in $C^0(\mathbf{R})$.

Example 6.12

In the real linear space $C^\infty(0, 1)$ the infinite set of polynomial functions $A = \{1, x, x^2, \ldots, x^n, \ldots\}$ is linearly independent but does not span $C^\infty(0, 1)$ and hence is not a basis for $C^\infty(0, 1)$. To see that A is linearly independent, suppose to the contrary that some nontrivial finite linear combination over A sums to the zero element in $C^\infty(0, 1)$, the zero function. Then we would have a nontrivial polynomial with infinitely many roots—an impossibility. Hence the constants in the finite linear combination must all be zero, so the set A is linearly independent. Since Span(A) is precisely the set of all polynomials, it cannot exhaust all elements of $C^\infty(0, 1)$, and A cannot be a basis for this space.

Dimension

A linear space is said to be *finite-dimensional* if it has a spanning set which is finite. Every finite spanning set A of a finite-dimensional linear space V contains a basis in the sense that a subset B of A can be found which spans V and is also linearly independent. Thus B is a basis for V. To see this, we reason as follows. If A is not linearly independent, one of its vectors can be written as a linear combination of the other vectors in A. Deleting this one vector from A still leaves a subset A_1 which spans V. If A_1 is linearly independent, we are done, for then A_1 is a basis for V. If A_1 is linearly dependent, we repeat the process to produce a subset A_2 of A_1 which still spans V, and so on. Since A is a finite set, this process must terminate. Thus we have shown that every finite-dimensional linear space has a basis. More remarkable, however, is the next result, whose proof is omitted,

---

Dimension Theorem. The number of elements in one basis for a given finite-dimensional linear space is the same as the number of elements in any other basis.

This leads us to define the important concept of the dimension of a linear space.

Dimension.   Let V be any nontrivial finite-dimensional linear space. The *dimension* of V is the number of elements in a basis and is denoted by dim V. The dimension of the trivial linear space $\{0\}$ is defined to be zero.

Although it is not immediately evident, it is a fact that in a linear space of dimension n no linearly independent set can have more than n members. This, in turn, implies that every linearly independent set in a finite-dimensional linear space can be completed to a basis. Indeed, if A is a linearly independent set in a finite-dimensional linear space V, then if A spans V, it follows that A is a basis and the claim is established. If Span(A) $\neq V$, then select any element in V not in Span(A) and adjoin it to A to form the set A_1. It is not difficult to see that A_1 is a linearly independent set. Now repeat the argument above with A_1 replacing A. This process must eventually terminate in a basis for V because, as we have stated, no linearly independent set in V can have more than dim V members.

Example 6.13

$\mathbf{R}^n$ and $\mathbf{C}^n$, $n = 1, 2, \ldots$ have dimension n because the set

$$B = \{(1, 0, \ldots, 0), (0, 1, 0, \ldots, 0), \ldots, (0, \ldots, 0, 1)\}$$

is a basis for each of them. The linear space of infinitely differentiable functions on the interval $(0, 1)$, denoted by $C^\infty(0, 1)$, is not finite-dimensional. We have shown that the set $A = \{1, x, x^2, \ldots\}$ is a linearly independent set in $C^\infty(0, 1)$. But A has infinitely many members, and $C^0(0, 1)$ cannot be finite-dimensional.

We conclude this section with a very useful result in constructing bases. A consequence of the Dimension Theorem is the following result:

Basis Construction Principle.   Let B be a subset of a linear space V with dim $V = n \geq 1$. If B is a linearly independent set containing n vectors, then B spans V (i.e., B is a basis for V).

Thus since dim $\mathbf{R}^n = n$, we see that any set of n linearly independent vectors in $\mathbf{R}^n$ must be a basis for that space. As a further illustration of this principle, we need only recall how in Chapter 5 we constructed the general solution of a homogeneous second-order linear differential equation:

Example 6.14

In Section 5.2 we showed that all (say, real-valued) solutions of $y'' - y' - 2y = 0$ can be generated by taking all (real) linear combinations of any given "independent" (real-valued) pair in $N(D^2 - D - 2)$. Observe that "independence" of a pair $\{y_1, y_2\}$ in $C^2(I)$ as defined in Section 5.2 corresponds precisely to the notion of linear independence as defined in this section for a pair in the (real) linear space $C^2(I)$. Thus, in the language of this section, $N(D^2 -$

$D - 2$) is a two-dimensional subspace of $C^2(\mathbf{R})$ whose elements can be *uniquely* represented by taking linear combinations of any (given) linearly independent pair in that null space. Note that (as expected) $N(D^2 - D - 2)$ does not have a unique basis. It may be shown that the pair $y_1 = ae^{-t} + be^{2t}$, $y_2 = ce^{-t} + de^{2t}$ is a basis for $N(D^2 - D - 2)$ for any (real) constants a, b, c, and d such that $ad - bc \neq 0$.

PROBLEMS

The following problem(s) may be more challenging: 7 and 8.

1. Find all solutions of each of the following systems and characterize them in terms of geometric vectors.
   (a) $x_1 - 2x_2 = 0$, $4x_2 - 2x_1 = 0$.
   (b) The system (4).
   (c) $x_1 - 2x_2 - x_3 = 0$, $x_2 + x_3 = 0$, $x_1 + x_2 + 2x_3 = 0$.
   (d) $3x_1 + x_2 + 2x_3 = 1$, $x_1 + 4x_2 - x_3 = -1$.

2. Which of the following are subspaces of $C^0[-1, 1]$? If not, why not?
   (a) All polynomials of degree 2.
   (b) All polynomials of degree greater than 3.
   (c) All odd functions $[f(-x) = -f(x)]$.
   (d) All nonnegative functions.

3. Determine if each of the sets of functions below is linearly dependent or independent in $C^0(\mathbf{R})$.
   (a) $\{e^{-t}, -3e^t, \cosh t\}$.   (b) $\{e^t, te^t, -t^2e^t\}$.
   (c) $\{e^{-t}\cos t, e^{-t} \sin t\}$.   (d) $\{1, t - 1, 3t^2 + t + 1, 1 - t^2\}$.

4. Show that the set $\{1, \sin t, \cos t, \sin 2t, \cos 2t, \ldots, \sin nt, \cos nt\}$ is linearly independent in the linear space of real-valued functions, $C^0[-\pi, \pi]$. [*Hint*: Let the reals $a_0, \ldots,$
   $a_n, b_1, \ldots, b_n$ be such that $a_0 + \sum\limits_{1}^{n} (a_k \cos kt + b_k \sin kt) \equiv 0$. Multiply by $\sin jt$ or $\cos jt$ and integrate from $-\pi$ to π.]

5. Let U and W be subspaces of a linear space V.
   (a) Show that the set of all vectors common to U and W is also a subspace of V.
   (b) If $V = \mathbf{R}^3$, U the span of vectors $(2, 1, 0)$, $(1, 0, -1)$, and W the span of $(-1, 1, 1)$ and $(0, -1, 1)$, describe the subspace of all vectors common to U and W. What is its dimension?

6. Determine explicitly each of the subsets of $C^2(\mathbf{R})$ described below. If it is a subspace, state its dimension.
   (a) $\{y : y'' \equiv 0,\ 2y(0) + y'(0) = 0,\ 2y(1) - y'(1) = 0\}$.
   (b) $\{y : y'' \equiv 2,\ y(0) = 0,\ y(1) = 0\}$.
   (c) $\{y : y'' \equiv 2,\ 2y(0) + y'(0) = 0,\ 2y(1) - y'(1) = 0\}$.

7. In $\mathbf{R}^2$ let us define new operations of addition and multiplication by scalars, denoted by $\oplus$ and $\odot$, which are different from the standard ones. Using $\mathbf{R}$ for scalars, let us choose a fixed nonzero element $r = (r_1, r_2)$ in $\mathbf{R}^2$ and put

$$x \oplus y = (x_1 + y_1 - r_1, x_2 + y_2 - r_2), \qquad \text{all } x, y \text{ in } \mathbf{R}^2$$
$$a \odot x = (ax_1 + (1-a)r_1, ax_2 + (1-a)r_2), \qquad \text{all } x \text{ in } \mathbf{R}^2, \quad \text{all scalars } a$$

(a) Show that $\mathbf{R}^2$ with the operations $\oplus$ and $\odot$ is a real linear space.

(b) Find the dimension of the linear space in part (a).

(c) If two vectors x and y are linearly independent in $\mathbf{R}^2$ with the standard operations, are they also linearly independent in $\mathbf{R}^2$ with $\oplus$ and $\odot$? Explain your answer.

8. [*Abstract Scalar Product*]. Let V be a linear space, and suppose that $\langle \cdot, \cdot \rangle$ is a function which associates with every ordered pair of vectors u, v the scalar $\langle u, v \rangle$. Refer to Problem 2 of Section 4.3 for relevant definitions [if $\langle \cdot, \cdot \rangle$ satisfies properties $(a) - (c)$ it is said to be an *Abstract Scalar Product* on V].

(a) Suppose that the scalars are real numbers and that properties (a)–(c) are satisfied with $\langle u, v \rangle$ replacing $\mathbf{u} \cdot \mathbf{v}$. Defining $\|\cdot\|$ in terms of $\langle \cdot, \cdot \rangle$ via (d), prove that properties (e)–(g) hold. [*Hint*: For (f) use fact that $\|au - \beta v\|^2 \geq 0$ for *all* scalars a and β.]

(b) Suppose that the scalars are complex numbers, and properties (b), (c), and

(a′) $\qquad\qquad\qquad \langle u, v \rangle = \overline{\langle v, u \rangle} \qquad$ for all u, v

hold with $\langle u, v \rangle$ replacing $\mathbf{u} \cdot \mathbf{v}$. Defining $\|\cdot\|$ in terms of $\langle \cdot, \cdot \rangle$ via (d), show that (e)–(g) hold.

(c) Let I be the interval $a \leq t \leq b$, and put $\langle f, g \rangle = \int_a^b fg \, dt$ for all (real-valued) f, g in $C^0(I)$. Show that (a)–(g) hold with $\langle u, v \rangle$ replacing $\mathbf{u} \cdot \mathbf{v}$.

6.2 MATRICES: SYSTEMS OF LINEAR ALGEBRAIC EQUATIONS

To solve linear differential equations it is useful to have a technique for finding all solutions of a collection of linear algebraic equations. In this section we present several techniques for finding all solutions of the linear algebraic system of m equations in n unknowns

$$a_{11}x_1 + \cdots + a_{1n}x_n = b_1$$
$$a_{21}x_1 + \cdots + a_{2n}x_n = b_2 \tag{1}$$
$$\vdots \qquad\qquad \vdots$$
$$a_{m1}x_1 + \cdots + a_{mn}x_n = b_m$$

where the coefficients† a_{ij}, $1 \leq i \leq m$, $1 \leq j \leq n$, and the right-hand side b_i, $1 \leq i \leq m$, are all given real (or complex) numbers. By combining the equations of (1) in a systematic manner it is not difficult to see that a set of linear equations with the same solution set as (1) can be found whose solutions can practically be read off by inspection. This procedure is most easily treated by using the so-called "matrix" formulation of system (1) described below.

† By convention each coefficient carries a double index ij, where i refers to the row and j the column in which the coefficient occurs.

An $m \times n$ array (i.e., m rows and n columns) of real or complex numbers a_{ij} is called a *matrix*. It is customary to use the symbol A to denote the matrix

$$\begin{bmatrix} a_{11} & \cdots & a_{1n} \\ & & \\ \vdots & & \vdots \\ & & \\ a_{m1} & \cdots & a_{mn} \end{bmatrix}$$

Sometimes the abbreviation $[a_{ij}]$ is used. Two matrices A and B, are equal if and only if they have the same size and the corresponding components are equal (i.e., $a_{ij} = b_{ij}$, for all i, j). If we agree to write elements of $\mathbf{R}^n$ (or $\mathbf{C}^n$) as $n \times 1$ *column matrices* and define the "product" of an $m \times n$ matrix by an $n \times 1$ matrix by

$$\begin{bmatrix} a_{11} & \cdots & a_{1n} \\ \vdots & & \vdots \\ & & \\ a_{m1} & \cdots & a_{mn} \end{bmatrix} \begin{bmatrix} x_1 \\ \vdots \\ x_n \end{bmatrix} = \begin{bmatrix} \sum_{j=1}^{n} a_{1j}x_j \\ \vdots \\ \sum_{j=1}^{n} a_{mj}x_j \end{bmatrix} \tag{2}$$

we see that the system (1) takes the matrix form

$$\begin{bmatrix} a_{11} & \cdots & a_{1n} \\ a_{21} & \cdots & a_{2n} \\ \vdots & & \\ a_{m1} & \cdots & a_{mn} \end{bmatrix} \begin{bmatrix} x_1 \\ x_2 \\ \vdots \\ x_n \end{bmatrix} = \begin{bmatrix} b_1 \\ b_2 \\ \vdots \\ b_m \end{bmatrix}$$

When the matrix A is defined from a system of equations such as (1), it is called the *coefficient matrix* of the system. Denoting the column matrices by the symbols x and b, (1) assumes the even simpler form $Ax = b$.

The literature on matrices and their properties is very large indeed. The immense value of matrices in both computational techniques and theoretical investigations in all areas of applied mathematics is an unquestioned fact. We only have space to mention briefly some aspects of matrices that are useful for treating the material covered in this text.

Solution of Linear Algebraic Systems

To solve (1) we shall use three *elementary operations* that can be performed on the equations of the system:

$E(i \leftrightarrow k)$ Interchange the ith and kth equations.
$E(k + \alpha : i)$ To the kth equation add the scalar α times the ith equation.
$E(\alpha : i)$ Multiply the ith equation by the scalar α.

When any of these elementary operations is performed on system (1), the resulting new system has the same solution set as the original system. A significant notational simplification results if we suppress the explicit appearance of the unknowns x_1, . . . , x_n, and apply the elementary operations to the rows of the augmented matrix

$$\begin{bmatrix} a_{11} & \cdots & a_{1n} & b_1 \\ \cdot & & \cdot & \cdot \\ \cdot & & \cdot & \cdot \\ \cdot & & \cdot & \cdot \\ a_{m1} & \cdots & a_{mn} & b_m \end{bmatrix} \tag{3}$$

for there is a close relation between the modified system and the modified matrix that arises from (1) and (3) under the same sequence of elementary operations. The strategy in selecting elementary row operations to apply to (3) is to produce a *row echelon matrix*, a matrix such that

1. The first nonzero entry in each nonzero row is 1 (called a *leading* 1).
2. The leading 1 in each nonzero row is to the right of the leading 1's in previous rows.
3. Zero rows occur after all nonzero rows.
4. All other entries in the column containing a leading 1 are zeros.

Notice that at any stage in the application of these row operations we can recover (by inspection) an equivalent system whose solution set is the same as the original one. A system of linear equations whose augmented matrix is in echelon form can be solved by inspection. The process of solving a linear system by systematic row reduction of its augmented matrix by elementary row operations to echelon form is called *Gauss–Jordan elimination*. An example will clarify this procedure.

Example 6.15

Starting with the system whose augmented matrix has the form on the top left and applying elementary row operations to this matrix, we have

$$
\begin{bmatrix}
1 & 3 & 2 & 1 & | & -1 \\
-1 & 1 & 0 & 2 & | & 1 \\
3 & 1 & 2 & -3 & | & -3
\end{bmatrix}
\xrightarrow[E(3-3\,:\,1)]{E(2+1\,:\,1)}
\begin{bmatrix}
1 & 3 & 2 & 1 & | & -1 \\
0 & 4 & 2 & 3 & | & 0 \\
0 & -8 & -4 & -6 & | & 0
\end{bmatrix}.
$$

$$
\xrightarrow[E(3+2\,:\,2)]{E(1-\frac{3}{4}\,:\,2)}
\begin{bmatrix}
1 & 0 & \frac{1}{2} & -\frac{5}{4} & | & -1 \\
0 & 4 & 2 & 3 & | & 0 \\
0 & 0 & 0 & 0 & | & 0
\end{bmatrix}
\xrightarrow{E(\frac{1}{4}\,:\,2)}
\begin{bmatrix}
1 & 0 & \frac{1}{2} & -\frac{5}{4} & | & -1 \\
0 & 1 & \frac{1}{2} & \frac{3}{4} & | & 0 \\
0 & 0 & 0 & 0 & | & 0
\end{bmatrix}
$$

which is in row echelon form. Inserting the unknowns x_1, x_2, x_3, and x_4, we now have the equivalent system

$$
\begin{aligned}
x_1 + \tfrac{1}{2}x_3 - \tfrac{5}{4}x_4 &= -1 \\
x_2 + \tfrac{1}{2}x_3 + \tfrac{3}{4}x_4 &= 0
\end{aligned}
$$

which can immediately be solved for x_1 and x_2 in terms of x_3 and x_4. Letting $x_3 = a$ and $x_4 = b$ take on any arbitrary values, we have the solutions $x_1 = -1 - \tfrac{1}{2}a + \tfrac{5}{4}b$, $x_2 = -\tfrac{1}{2}a - \tfrac{3}{4}b$, or in column vector form

$$
\begin{bmatrix} x_1 \\ x_2 \\ x_3 \\ x_4 \end{bmatrix}
=
\begin{bmatrix} -1 \\ 0 \\ 0 \\ 0 \end{bmatrix}
+ a
\begin{bmatrix} -\frac{1}{2} \\ -\frac{1}{2} \\ 1 \\ 0 \end{bmatrix}
+ b
\begin{bmatrix} \frac{5}{4} \\ -\frac{3}{4} \\ 0 \\ 1 \end{bmatrix}
\tag{4}
$$

In this form we see that the solution set in $\mathbf{R}^4$ of the given system is a two-dimensional subspace [with basis $\{(-\tfrac{1}{2}, -\tfrac{1}{2}, 1, 0), (\tfrac{5}{4}, -\tfrac{3}{4}, 0, 1)\}$] translated by the constant vector $(-1, 0, 0, 0)$. In general, the solution set of system (2) has the form of a subspace of $\mathbf{R}^4$ translated by a constant vector.

Matrix Operations

Several common operations defined for matrices are listed below.

> **Addition and Multiplication by a Scalar.**   Let $A = [a_{ij}]$, $B = [b_{ij}]$ be two $m \times n$ matrices, and α a scalar. Then the operations of "addition" and "multiplication by scalars" are given by
>
> $$A + B \equiv [a_{ij} + b_{ij}], \qquad \alpha A \equiv [\alpha a_{ij}]$$
>
> **Matrix Products.**   If $A = [a_{ij}]$ is an $m \times n$ matrix and $B = [b_{ij}]$ is an $n \times k$ matrix , the product AB is the $m \times k$ matrix
>
> $$AB = [c_{ij}], \qquad \text{where } c_{ij} = \sum_{s=1}^{n} a_{is} b_{sj}$$

Observe that c_{ij} is obtained by taking the ith row of A and the jth column of B, considering them to be vectors in $\mathbf{R}^n$, and finding their dot product (see Problem 1 of Section 4.3).

If A is an $m \times n$ matrix and B an $n \times 1$ matrix, AB defined above agrees precisely with what we defined above as a matrix "acting" on a column matrix. If B is a general $n \times k$ matrix, it can be partitioned into k column vectors c_1, c_2, $\ldots$, c_k, each with n entries. Then the k columns of the $m \times k$ product AB are given (in order) by Ac_1, Ac_2, $\ldots$, Ac_k. In symbols,

$$A[c_1|c_2|\cdots|c_k] = [Ac_1|Ac_2|\cdots|Ac_k] \tag{5}$$

where the partitioning symbol "|" is used to separate columns. Usually, the meaning is clear without the symbol.

Example 6.16

Let

$$A = \begin{bmatrix} 2 & 1 & -1 \\ 0 & 3 & 1 \end{bmatrix}, \qquad B = \begin{bmatrix} -3 & 5 \\ 4 & 0 \\ 2 & -1 \end{bmatrix}$$

Then

$$AB = \left[A \begin{bmatrix} -3 \\ 4 \\ 2 \end{bmatrix} \;\middle|\; A \begin{bmatrix} 5 \\ 0 \\ -1 \end{bmatrix} \right] = \begin{bmatrix} -4 & 11 \\ 14 & -1 \end{bmatrix}$$

Observe that BA is a 3×3 matrix and does not coincide with AB.

If the matrices A, B, and C have compatible dimensions, some careful work with indices shows that

$$\begin{aligned} A(BC) &= (AB)C & &(\textit{associativity}) \\ A(B+C) &= AB + AC & &(\textit{distributivity}) \end{aligned} \tag{6}$$

but, in general, notice that $AB \neq BA$.

Extending the notion of linearity for operators given in Section 5.1, we say that an operator $L: \mathbf{R}^n \to \mathbf{R}^m$ is *linear* if $L[\alpha x + \beta y] = \alpha Lx + \beta Ly$, for all x, y in $\mathbf{R}^n$, and all reals α, β. A similar definition holds for an operator $L: \mathbf{C}^n \to \mathbf{C}^m$. If A is an $m \times n$ matrix of real numbers then a standard linear operator from $\mathbf{R}^n$ to $\mathbf{R}^m$ (also denoted by A) is associated with A as follows: Writing elements of $\mathbf{R}^n$ and $\mathbf{R}^m$ as column matrices the operator $A: \mathbf{R}^n \to \mathbf{R}^m$ has the action $A: x \to Ax$, where Ax is the product of matrices A and x as defined by (2). Using the standard operations in $\mathbf{R}^n$ and $\mathbf{R}^m$ and the properties of matrix products it is straightforward to verify that A is a linear operator. A similar result holds if A has complex coefficients, and elements of $\mathbf{C}^n$ and $\mathbf{C}^m$ are written as column matrices.

The special $n \times n$ matrix with $a_{ii} = 1$ for all i and $a_{ij} = 0$ for $i \neq j$ is called the $n \times n$ *unit matrix* and is denoted by E. Evidently, E has the property that

$$E = \begin{bmatrix} 1 & 0 & . & . & . & 0 \\ 0 & 1 & . & & & . \\ . & . & . & . & & . \\ . & & . & . & . & . \\ . & & & . & . & 0 \\ 0 & . & . & . & 0 & 1 \end{bmatrix}$$

$$EB = B \qquad \text{for any } n \times k \text{ matrix } B$$
$$BE = B \qquad \text{for any } k \times n \text{ matrix } B \tag{7}$$

The special $m \times n$ matrix all of whose entries are zero is denoted by $Z(m, n)$ and is called the $m \times n$ *zero matrix*. Observe that

$$Z(m, n) + B = B + Z(m, n) = B \qquad \text{for any } m \times n \text{ matrix } B$$

Finally, given any $m \times n$ matrix A, the matrix obtained from A by interchanging the rows and columns is called the *transpose* of A and is denoted by A^T. Thus A^T has the form

$$A^T = [a_{ji}] = \begin{bmatrix} a_{11} & . & . & . & a_{m1} \\ . & & & & . \\ . & & & & . \\ . & & & & . \\ a_{1n} & . & . & . & a_{mn} \end{bmatrix}$$

Transposition has the useful properties listed below.

Properties of Matrix Transposition.   For any matrices A, B for which $A + B$ and AB are defined, and any scalar c, we have that

$$(A^T)^T = A \quad \text{and} \quad (cA)^T = cA^T \tag{8}$$

$$(A + B)^T = A^T + B^T \tag{9}$$

$$(AB)^T = B^T A^T \tag{10}$$

Proof. We prove only (10). Put $C = [c_{ij}] = AB$, and let $A = [a_{ij}]$, $B = [b_{ij}]$. Defining $a_{ij}^T = a_{ji}$ and $b_{ij}^T = b_{ji}$ for all indices i, j, we see that $A^T = [a_{ij}^T]$

and $B^T = [b_{ij}^T]$. Now $c_{ij}^T = c_{ji} = \sum_k a_{jk} b_{ki} = \sum_k b_{ik}^T a_{kj}^T$, which is the (ij)th element in $B^T A^T$. This proves (10).

Nonsingular Matrices

The class of square matrices contains some rather special members:

> **Nonsingular Matrices.** An $n \times n$ matrix A is said to be *nonsingular* (or *invertible*) if and only if there exists an $n \times n$ matrix B such that $AB = E$. Square matrices which are not nonsingular are said to be *singular*.

Example 6.17

Consider the two 2×2 matrices

$$A = \begin{bmatrix} 3 & 1 \\ 5 & 2 \end{bmatrix}, \qquad B = \begin{bmatrix} 2 & -1 \\ -5 & 3 \end{bmatrix}$$

Since $AB = E$, it follows from the definition above that A is nonsingular. Observe that also $BA = E$, and hence B is nonsingular, too.

Invertible matrices have many important properties. First, observe that if an $n \times n$ matrix A is invertible, the associated matrix operator $A : \mathbf{R}^n \to \mathbf{R}^n$ expressed by (2) has $\mathbf{R}^n$ for its precise range. Indeed, since there is a matrix B such that $AB = E$, we see that for any b in $\mathbf{R}^n$ the matrix equation $Ax = b$ has the solution $x = Bb$. It is a remarkable fact that if the operator associated with an $n \times n$ matrix A has $\mathbf{R}^n$ for its range, A is nonsingular. To see this, let e_i be the column n-vector with 1 in the ith position and zeros elsewhere. Now for each $i = 1, 2, \ldots, n$ the equation $Ax = e_i$ is solvable since the range of A is $\mathbf{R}^n$. Denote this solution by b_i. Now the $n \times n$ matrix B whose ith column is b_i has the property that $AB = E$, and hence A is nonsingular.

Thus invertible $n \times n$ matrices are the only $n \times n$ matrices whose associated operators have $\mathbf{R}^n$ for their range. This observation brings to mind several techniques for testing a matrix for invertibility. One way would be by applying elementary row operations to the matrix A. The range condition above easily implies that A is invertible if and only if it can be reduced to the echelon matrix E by elementary row operations. Another technique makes use of the following fact. Denoting the columns of an $n \times n$ matrix A by $a_1, a_2, \ldots, a_n$ so that $A = [a_1 \mid a_2 \mid \cdots \mid a_n]$, then for any column n-vector x we have $Ax = x_1 a_1 + x_2 a_2 + \cdots + x_n a_n$. Thus the range of A is precisely the span of its n columns in $\mathbf{R}^n$. Let us now define the *column rank* of an $m \times n$ matrix A as the dimension of the subspace of $\mathbf{R}^m$ spanned by the n columns of A. Thus an $n \times n$ matrix is invertible if and only if its column rank is n.

There are other useful equivalent conditions for the invertibility of an $n \times n$ matrix, but to prove them we must digress to present the following result (without proof):

Rank-Nullity Theorem for Matrices.   Let A be an $m \times n$ matrix of real numbers. Denote by $A(\mathbf{R}^n)$ the set of all y in $\mathbf{R}^m$ such that the equation $Ax = y$ is solvable, and by $N(A)$ the set of all x in $\mathbf{R}^n$ such that $Ax = 0$. Then $A(\mathbf{R}^n)$ is a subspace of $\mathbf{R}^m$, and $N(A)$ is a subspace of $\mathbf{R}^n$, and

$$\dim A(\mathbf{R}^n) + \dim N(A) = n$$

[Note that $\dim A(\mathbf{R}^n)$ is the column rank of A. $N(A)$ is called the *null space* of A, and $\dim N(A)$ the *nullity* of A.]

Now we are ready for the conditions of invertibility for matrices.

Invertibility Theorem.   The $n \times n$ matrix A is invertible if and only if any of the following conditions is satisfied:
(a) The matrix equation $Ax = y$ is solvable for all y in $\mathbf{R}^n$.
(b) A has column rank n.
(c) The matrix equation $Ax = 0$ has only the trivial solution $x = 0$.
(d) There exists an $n \times n$ matrix B such that $BA = E$.

Proof. That conditions (a) and (b) are equivalent to the invertibility of A was proven above, and that condition (c) is equivalent to condition (a) is a consequence of the Rank-Nullity Theorem. To show condition (d), assume first that A is invertible and hence there exists an $n \times n$ matrix B such that $AB = E$. Thus it follows that $Bx = 0$ has only the trivial solution $x = 0$, and hence by condition (c), B is invertible. Hence there is an $n \times n$ matrix C such that $BC = E$. But since $A = AE = A(BC) = (AB)C = EC = C$, it follows that $BA = E$. Conversely, if condition (d) holds, the same argument with B and A reversed shows that $AB = E$, and hence that A is invertible, finishing the proof.

In proving condition (d) in the Invertibility Theorem, we proved more. We actually showed that if $AB = E$ for two $n \times n$ matrices A and B, then also $BA = E$. This shows that if A is a nonsingular $n \times n$ matrix, there is one, and only one, matrix B such that $AB = E$ (*and* such that $BA = E$). Indeed, if also $AB' = E$, then since $BA = E$, we see that $B = BE = B(AB') = (BA)B' = EB' = B'$, establishing the claim. Thus the following definition makes sense.

Inverse of a Nonsingular Matrix.   The *inverse* of a nonsingular $n \times n$ matrix A is the unique matrix B such that $AB = E$ (or $BA = E$) and is denoted by the symbol A^{-1}.

Example 6.18

The matrix A in Example 6.17 is invertible and $A^{-1} = B$. Note that $BA = E$.

There is an effective technique based on elementary row operations which will both reveal whether or not a square matrix is invertible and construct the inverse matrix if it exists (see Problem 5).

Matrix inversion has the useful properties listed below; the proofs are omitted.

Properties of Matrix Inversion.   Let A and B be nonsingular $n \times n$ matrices. Then (a) $A^{-1}A = AA^{-1} = E$; (b) A^{-1} is nonsingular and $(A^{-1})^{-1} = A$; (c) AB is nonsingular and $(AB)^{-1} = B^{-1}A^{-1}$.

Using the properties of nonsingular matrices in the Invertibility Theorem, we can infer the following about the solvability of the matrix equation $Ax = y$.

Solution of an $n \times n$ Linear System.   Let A be an $n \times n$ matrix, and y an $n \times 1$ matrix. (a) When A is nonsingular, $Ax = y$ has the unique solution $x = A^{-1}y$. In particular, the equation $Ax = 0$ has only the trivial solution $x = 0$. (b) When A is singular, $Ax = y$ cannot be solvable for all y, and if a solution exists, it is not unique. In particular, the equation $Ax = 0$ has nontrivial solutions (i.e., solutions other than $x = 0$).

Other Types of Matrices

We shall list some other types of matrices which will be useful later. An $n \times n$ matrix $A = [a_{ij}]$ with $a_{ij} = 0$, $i \neq j$, is called a *diagonal* matrix. The elements $a_{11}, a_{22}, \ldots, a_{nn}$ constitute the *principal diagonal* of A. The $n \times n$ matrix with block square matrices B_i appearing along the principal diagonal,

$$
A = \begin{bmatrix} B_1 & & & \\ & \cdot & \cdot & 0 \\ 0 & & \cdot & \\ & & & B_k \end{bmatrix}
$$

is called a *block diagonal matrix*. The $n \times n$ matrix

$$
A = \begin{bmatrix} a_{11} & \cdot & \cdot & & a_{1n} \\ 0 & \cdot & & & \cdot \\ \cdot & \cdot & \cdot & & \cdot \\ \cdot & & \cdot & \cdot & \cdot \\ \cdot & & & \cdot & \cdot \\ 0 & \cdot & & \cdot & 0 & a_{nn} \end{bmatrix}, \qquad a_{ij} = 0 \quad \text{if } i > j
$$

is called an *upper triangular matrix*. Lower triangular matrices are defined similarly.

Determinants

The determinant of a square matrix is a useful concept in working with systems of linear algebraic equations. We outline the basic definitions and elementary properties of determinants.

Every square matrix A with real (or complex) entries has a real (or complex) value associated with it called the *determinant* of A, or det A, and defined by the recursive scheme described below. First, for $n = 2$ we have

$$\det \begin{bmatrix} a_{11} & a_{12} \\ a_{21} & a_{22} \end{bmatrix} = a_{11}a_{22} - a_{12}a_{21}$$

Assume now that det A has been defined for all square matrices of size $(n - 1) \times (n - 1)$ or smaller. Let A be an $n \times n$ matrix. The ijth *minor matrix*, denoted by A_{ij}, is the $(n - 1) \times (n - 1)$ matrix obtained by striking out the ith row and the jth column of A (see Fig. 6.2). The number $(-1)^{i+j} \det A_{ij}$ is called the ijth *cofactor* and will be denoted by cof A_{ij}.

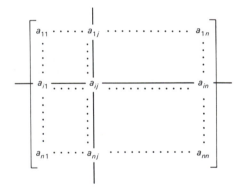

Figure 6.2 Minor matrix A_{ij}.

> **Cofactor Expansion of Det A.** Let A be an $n \times n$ matrix; then the *cofactor expansions of det A on the ith row or jth column, respectively, are given by*
>
> $$\det A = \sum_{k=1}^{n} a_{ik} \operatorname{cof} A_{ik}, \qquad \det A = \sum_{k=1}^{n} a_{kj} \operatorname{cof} A_{kj}$$

All cofactor expansions yield the same number denoted by det A, but we shall not prove that here. The advantage of this cofactor approach to determinants is that it allows us to calculate the determinant of an $n \times n$ matrix A by looking instead at determinants of $(n - 1) \times (n - 1)$ submatrices of A, these in turn being reduced to determinants of $(n - 2) \times (n - 2)$ submatrices, and so on. We state without proof some other useful properties of determinants.

Properties of Determinants.   Let A be an $n \times n$ matrix. (a) If any row or column of a matrix A is multiplied by a scalar λ, the determinant of that matrix is $\lambda \det A$. (b) If B is the matrix that arises from A by adding a multiple of a row (or column) to a different row (column), $\det A = \det B$. (c) If B is the matrix which arises from A by interchanging two rows (or two columns), then $\det A = -\det B$. (d) $\det A = \det A^T$.

Product Theorem.   For any $n \times n$ matrices A and B,

$$\det AB = (\det A)(\det B) = \det BA \qquad (11)$$

Determinant of Special Matrices.   The determinant of an (upper or lower) triangular matrix is the product of the elements on the principal diagonal. The determinant of a block matrix is the product of the determinants of the blocks.

Singularity Theorem.   A square matrix A is singular if and only if either of the following conditions is satisfied: (a) $\det A = 0$, or (b) the set of rows or columns of A form a linearly dependent set.

Example 6.19

Many zeros in a matrix (i.e., a *sparse matrix*) make its determinant easier to calculate. We can often introduce zeros by adding a multiple of one column or row to another; do this as often as time and energy warrant and then carry out a cofactor expansion using the row or column containing the most zeros. For example,

$$\det \begin{bmatrix} 2 & -1 & -6 \\ -3 & -2 & -15 \\ 2 & 0 & -5 \end{bmatrix} = \det \begin{bmatrix} 2 & -1 & -6 \\ -7 & 0 & -3 \\ 2 & 0 & -5 \end{bmatrix} \qquad \text{(first row multiplied by } -2 \\ \text{and added to second row)}$$

$$= -(-1) \det \begin{bmatrix} -7 & -3 \\ 2 & -5 \end{bmatrix} = 41 \qquad \text{(cofactor expansion on the second column)}$$

Cramer's Rule

The determinant idea leads to an elegant formula for A^{-1} if A has an inverse. First we define the *classical adjoint matrix* of the $n \times n$ matrix A to be the matrix $\text{adj } A \equiv [\text{cof } A_{ij}]^T$. We then have the following result.

Calculation of A^{-1}.   Let A be an $n \times n$ matrix. Then A is invertible if and only if $\det A \neq 0$, and in this case

$$A^{-1} = \frac{1}{\det A} \text{ adj } A \qquad (12)$$

Proof. The method of cofactors shows that $(\det A)E = A(\text{adj } A) = (\text{adj } A)A$, whether or not $\det A = 0$. Now (12) follows immediately from the definition of A^{-1}.

We can now deduce a useful rule for solving n linear equations in n unknowns when the coefficient matrix is nonsingular.

Cramer's Rule.   Let A be an $n \times n$ matrix with $\det A \neq 0$, and denote the columns of A by $a_1, a_2, \ldots, a_n$. Then the unique solution of $Ax = b$ is

$$x_i = \frac{\det [a_1| \cdots |b| \cdots |a_n]}{\det A}, \qquad i = 1, 2, \ldots, n$$

where the column vector b takes the place of a_i.

Proof. Now $x = A^{-1}b$, and let us denote the entries of A^{-1} by c_{ij} so that $A^{-1} = [c_{ij}]$. Then (12) implies that $c_{ij} = (\text{cof } A_{ji})/\det A$. Hence, using the method of cofactors, we have for each $i = 1, 2, \ldots, n$,

$$x_i = \sum_{j=1}^{n} \frac{\text{cof } A_{ji}}{\det A} b_j = \frac{\det [a_1| \cdots |b| \cdots |a_n]}{\det A}$$

and we are done.

Although Cramer's Rule is not bad for 2×2 or 3×3 systems, it is unwieldy for larger n, and of course it works only when A is nonsingular. The method of Gaussian elimination is computationally efficient and can be used whether or not A is singular.

PROBLEMS

The following problem(s) may be more challenging: 6(b).

1. Find all solutions (if there are any) of the following systems.
   (a) $3x_1 - x_2 + x_3 + 2x_4 - 2x_5 = 1$
   $\quad\ \ x_1 + 2x_2 \qquad\ - x_4 + x_5 = -1$
   $\quad\qquad\quad x_2 + 2x_3 - 5x_4 + 2x_5 = 2$

   (b) $\quad x_1 - 2x_2 + 4x_3 - x_4 = -1$
   $\quad\ 2x_1 - x_2 - x_3 + x_4 = 1$
   $\quad\ 5x_1 \qquad\quad + 2x_3 + x_4 = -1$

2. In the solution formula (4), verify that the first column vector is a particular solution of the given linear system, while the other two column vectors lie in the null space of the matrix operator associated with the system.

3. Show that a 2×2 upper triangular matrix is invertible if and only if the elements on its principal diagonal are all nonzero, and that the inverse (if it exists) must also be upper triangular.

4. Show that if a square matrix P is invertible, so is P^T and $(P^{-1})^T = (P^T)^{-1}$.

5. Let A be an $n \times n$ matrix and let e_j be the column vector with 1 in the jth place and zero elsewhere. Show that if x_j is a column vector such that $Ax_j = e_j$, for *every* $j = 1$,

2, . . . , n, the matrix $B = [x_1|x_2| \cdots |x_n]$ is the inverse of A. [*Hint:* Use (5).] How can elementary row operations be used to construct A^{-1} if it exists?

6. Evaluate the following determinants.

$$\text{(a) det} \begin{bmatrix} 3 & 5 & 7 & 2 \\ 2 & 4 & 1 & 1 \\ -2 & 0 & 0 & 0 \\ 1 & 1 & 3 & 4 \end{bmatrix} \qquad \text{(b) det} \begin{bmatrix} 1 & -2 & 3 & -2 & -2 \\ 2 & -1 & 1 & 3 & 2 \\ 1 & 1 & 2 & 1 & 1 \\ 1 & -4 & -3 & -2 & -5 \\ 3 & -2 & 2 & 2 & -2 \end{bmatrix}$$

7. Calculate adj A if $A = \begin{bmatrix} 2 & 1 & 0 \\ 0 & 3 & 1 \\ 1 & 0 & 4 \end{bmatrix}$, and solve the system $Ax = [0 \quad -1 \quad -2]^T$.

6.3 GENERAL LINEAR DIFFERENTIAL EQUATIONS

In Chapter 5 we characterized all solutions of the linear second-order equation $y'' + a_1(t)y' + a_0(t)y = f$, where the coefficients a_0, a_1 and the driving term f are continuous functions on some interval. We now embark on a path that will give us a similar characterization for the *general linear differential equation*

$$y^{(n)} + a_{n-1}(t)y^{(n-1)} + \cdots + a_1(t)y' + a_0(t)y = f(t) \tag{1}$$

where all the coefficients and the driving term f belong to the continuity class $C^0(I)$ for some interval I on the t-axis. Using the nth *order polynomial operator* denoted by

$$P(D) = D^n + a_{n-1}(t)D^{n-1} + \cdots + a_1(t)D + a_0(t) \tag{2}$$

which has (by convention) the action

$$P(D): y \longrightarrow y^{(n)} + a_{n-1}(t)y^{(n-1)} + \cdots + a_1(t)y' + a_0(t)y$$

for any function y in $C^n(I)$, we see that $P(D): C^n(I) \to C^0(I)$ and that the differential equation (1) can be written in the convenient operator form

$$P(D)[y] = f \tag{3}$$

The nth order polynomial operator $P(D)$ obviously satisfies a linearity property analogous to the one given for second order polynomial operators in Section 5.1, a fact from which Eq. (1) derives its name.

It is a fact that the null space of the linear operator $P(D)$ is an n-dimensional subspace of $C^n(I)$, and that $P(D)[y] = f$ has solutions for any f in $C^0(I)$; these properties are consistent with those of the second-order case treated in Chapter 5. These facts are implied (as in Chapter 5) by imposing *initial conditions* on a solution of $P(D)[y] = f$ at some point t_0 in I, and deriving the following important property.

Dynamical System Property for $P(D)[y] = f$. Let the functions $f(t)$ and $a_j(t)$, $j = 0, 1, \ldots, n - 1$, all be continuous on a common interval I. Then for any values of the constants $b_0, b_1, \ldots, b_{n-1}$, and any t_0 in I, the initial value problem

$$y^{(n)} + a_{n-1}(t)y^{(n-1)} + \cdots + a_1(t)y' + a_0(t)y = f$$
$$y^{(k)}(t_0) = b_k, \qquad k = 0, 1, \ldots, n - 1 \tag{4}$$

has a unique solution $y(t)$ in $C^n(I)$. If the coefficients $a_j(t)$, the driving term $f(t)$, and the data b_k are all real valued, the solution of (4) is real valued.

A proof of the existence of a solution $y(t)$ in $C^n(I)$ for the initial value problem (4) can be patterned after the existence proof for the first-order initial value problems in Appendix A, but we omit it.

The Null Space of P(D)

With $P(D)$ as in (2), we define the function $y_j(t)$ in $C^n(I)$, $j = 0, 1, \ldots, n - 1$, as the unique solution (which we know exists by the Dynamical System Property) of the initial value problem

$$P(D)[y] = 0$$
$$y^{(k)}(t_0) = \delta_{jk}, \qquad k = 0, 1, \ldots, n - 1 \tag{5}$$

where $\delta_{jk} = 1$ if $j = k$ and 0 otherwise.†

Basis for $N(P(D))$. The set of functions $B = \{y_0, y_1, \ldots, y_{n-1}\}$ defined by (5) is a basis for the null space of $P(D)$, and hence dim $N(P(D)) = n$.

Proof. First we show that Span $(B) = N(P(D))$. Any linear combination of the functions $y_0, y_1, \ldots, y_{n-1}$ must satisfy the equation $P(D)[y] = 0$ since $P(D)$ is a linear operator. On the other hand, let $y(t)$ be any soluton of $P(D)[y] = 0$, and put $b_k = y^{(k)}(t_0)$, $k = 0, 1, \ldots, n - 1$. Then the function $z = b_0 y_0 + b_1 y_1 + \cdots + b_{n-1} y_{n-1}$ is a solution of the initial value problem

$$P(D)[w] = 0$$
$$w^{(k)}(t_0) = b_k, \qquad k = 0, 1, \ldots, n - 1 \tag{6}$$

But $y(t)$ is also a solution of (6). By the Dynamical System Property, the problem (6) can only have one solution, and hence $y(t) = z(t)$. It follows that B spans $N(P(D))$. Next we show that B is a linearly independent set in $C^n(I)$. Say that $c_0, c_1, \ldots, c_{n-1}$ are any scalars with

$$c_0 y_0(t) + c_1 y_1(t) + \cdots + c_{n-1} y_{n-1}(t) = 0 \qquad \text{for all } t \text{ in } I \tag{7}$$

---

† δ_{jk} is called the *Kronecker delta function*.

Inserting $t = t_0$ in (7), we obtain that $c_0 = 0$. In general, if (7) is differentiated k times, $k = 1, 2, \ldots, n - 1$, and evaluated at $t = t_0$ we obtain that $c_k = 0$. Thus B is independent and we are done.

Example 6.20

The equation $P(D)[y] \equiv t^3 y''' + 4t^2 y'' = 0$, $t > 0$, has solutions $y_1 = 1$, $y_2 = t$, $y_3 = t^{-2}$. Now since the triple $\{1, t, t^{-2}\}$ is a linearly independent set in $N(P(D))$, we conclude that every solution of $P(D)[y] = 0$, $t > 0$, can be written as a (unique) linear combination of this triple. In particular, let us find the solution of $P(D)[y] = 0$ for which $y(1) = 0$, $y'(1) = 1$, $y''(1) = 0$. Setting $y = c_1 + c_2 t + c_3 t^{-2}$ and imposing the initial conditions we obtain that $c_1 + c_2 + c_3 = 0$, $c_2 - 2c_3 = 1$, $6c_3 = 0$, which has the unique solution $c_1 = -1$, $c_2 = 1$, and $c_3 = 0$. Thus $y = -1 + t$ is the solution of our initial value problem.

Wronskians

As for the case $n = 2$, we will find it convenient to use Wronskian determinants to identify bases in $N(P(D))$. The arguments and results are analogous to those presented in Section 5.2, so at times our presentation here will be sketchy.

If $y_1(t), \ldots, y_n(t)$ are any n functions in $C^{n-1}(I)$ for some interval I, then their *Wronskian Determinant* (or, simply, *Wronskian*) is defined by

$$W(t) \equiv W[y_1, \ldots, y_n](t) \equiv \det \begin{bmatrix} y_1 & \cdots & y_n \\ \cdot & & \cdot \\ \cdot & & \cdot \\ \cdot & & \cdot \\ y_1^{(n-1)} & \cdots & y_n^{(n-1)} \end{bmatrix}$$

It is not difficult to see that if $\{y_1(t), \ldots, y_n(t)\}$ is a dependent set on I, then $W[y_1, \ldots, y_n](t) = 0$ for all t in I (see Problem 8). Thus if the Wronskian of the set of functions $\{y_1, y_2, \ldots, y_n\}$ in $C^{n-1}(I)$ fails to vanish at any point t_0 in I, the set is linearly independent. It does *not* follow in general that the vanishing of the Wronskian of a set $S = \{y_1, \ldots, y_n\}$ identically in an interval I implies that the set S is linearly dependent. The situation becomes much simpler in case the functions $\{y_1, \ldots, y_n\}$ are in $N(P(D))$ for an nth-order polynomial operator $P(D)$.

Let $y_1(t), y_2(t), \ldots, y_n(t)$ be n solutions of the homogeneous nth-order linear differential equation $P(D)[y] = 0$, with $P(D)$ as in (2). We know that $W[y_1, y_2, \ldots, y_n] \equiv 0$ in I if the set $\{y_1, \ldots, y_n\}$ is linearly dependent in $C^{n-1}(I)$. If $\{y_1, \ldots, y_n\}$ is a basis for $N(P(D))$, then $W[y_1, \ldots, y_n]$ is nonvanishing on I. We have the following basis test, whose proof is omitted since it is a direct extension of that given in Chapter 5 for the second-order case.

> **Wronskian Basis Test.** Let $P(D)$ be as in (2). Then a set $S = \{y_1, \ldots, y_n\}$ of solutions of $P(D)[y] = 0$ is a basis for $N(P(D))$ if and only if $W[y_1, \ldots, y_n](t)$ never vanishes in I.

Equations with Constant Coefficients

Here we consider polynomial operators $P(D)$ as in (2) but with constant coefficients. If $P(D)$ is applied to the function e^{rt}, where r is any real (or complex) constant, then

$$P(D)[e^{rt}] = (r^n + a_{n-1}r^{n-1} + \cdots + a_1 r + a_0)e^{rt} = P(r)e^{rt} \tag{8}$$

The algebraic polynomial $P(r)$ in (8) is related in an obvious way to the polynomial operator $P(D)$ in (2). This polynomial $P(r)$ is called the *characteristic polynomial* of the operator $P(D)$ and can be used to generate $P(D)$ in a routine way. Conversely, the characteristic polynomial of a polynomial operator $P(D)$ can be generated by applying $P(D)$ to e^{rt} and using (8). There is a one-to-one relation between polynomials and polynomial operators.

A fundamental result of algebra is that any polynomial $P(r)$ can be factorized completely into linear factors if complex numbers are used, and in only one way if $P(r)$ is a monic polynomial (i.e., the coefficient of the top degree term in P is unity). For example,

$$r^3 - 6r^2 + 11r - 6 = (r-1)(r-2)(r-3)$$
$$r^5 + 8r^3 + 16r = r(r^2 + 4)^2 = r(r - 2i)^2(r + 2i)^2$$

(Note that the order of the factors is immaterial). Thus we shall assume $P(r)$ is in factored form, although this is not always easy to do in practice.† Since everything we have to say depends on it, we may as well assume that (monic) $P(r)$ is written in the factored form

$$P(r) = (r - r_1)^{n_1}(r - r_2)^{n_2} \cdots (r - r_m)^{n_m} \tag{9}$$

where the distinct roots r_j may be real or complex numbers, and the n_j are nonnegative integers with $\Sigma\, n_j = n$. By convention, n_j is called the *multiplicity of the root* r_j. Note that the order of the factors in (9) is immaterial.

If $P(r)$ has the factored form (9), it follows that the corresponding polynomial operator $P(D)$ has the similar factorization

$$P(D) = (D - r_1)^{n_1}(D - r_2)^{n_2} \cdots (D - r_m)^{n_m} \tag{10}$$

---

† To factorize polynomials one must have a way of precisely computing roots of any polynomial. If $n = 1$ or 2, there is no difficulty in finding these roots. If $n = 3$ or 4, it can still be done explicitly, but not easily, in terms of radicals—the formulas were first discovered in the sixteenth century and are given in most mathematical handbooks. However, if $n \geq 5$, it has been proven that no general formulas for expressing the roots in terms of radicals can exist.

where again the order of the factors is immaterial. Indeed, if $Q(D)$ is the "factorized" polynomial operator in (10), then using the identity (8) repeatedly, we obtain that

$$Q(D)[e^{rt}] = [(r - r_1)^{n_1}(r - r_2)^{n_2} \cdots (r - r_m)^{n_m}]e^{rt} \equiv P(r)e^{rt}$$

Thus $Q(D) \equiv P(D)$ because they both have the same characteristic polynomial.

Let us note that for any positive integer n, any h in $C^n(\mathbf{R})$ and any real or complex number r_0, we have that

$$(D - r_0)^n[he^{r_0 t}] = e^{r_0 t}D^n[h] \tag{11}$$

The proof (by induction) is omitted. (See Problem 7 of Section 5.1).

Example 6.21

Notice that (11) implies that $(D - r)^3[t^2 e^{rt}] = e^{rt}D^3[t^2] \equiv 0$. Thus $t^2 e^{rt}$ is a solution of the linear differential equation $(D - r)^3[y] = 0$. Similarly (11) implies that

$$(D - 1)^3 D^2[p(t)e^t] = D^2(D - 1)^3[p(t)e^t] = D^2[e^t D^3[p(t)]] \equiv 0$$

if and only if $p(t)$ is a polynomial of degree ≤ 2.

Polynomial-Exponential Solutions of $P(D)[y] = 0$.   For $P(D)$ as in (10), we have that for each $j = 1, 2, \ldots, m$,

$$P(D)[p_j(t)e^{r_j t}] = 0 \qquad \text{for any polynomial } p_j \text{ of degree } \leq n_j - 1$$

Proof. Using the "integrating factor property" (11) for any factor $(D - r_j)^{n_j}$ of $P(D)$ and any polynomial $p_j(t)$ of degree $\leq n_j - 1$, we have, as asserted, that for any $j = 1, 2, \ldots, m$,

$$(D - r_j)^{n_j}[p_j(t)e^{r_j t}] = e^{r_j t}D^{n_j}[p_j] \equiv 0$$

Polynomial-exponential functions have the following useful property whose proof we omit (see the problem set).

Independence of Polynomial-Exponentials.   If for some polynomials $p_1, \ldots, p_m$ and some distinct constants $\lambda_1, \ldots, \lambda_m$,

$$\sum_{j=1}^{m} p_j(t)e^{\lambda_j t} \equiv 0 \qquad \text{for all } t$$

then $p_j \equiv 0$ for all $j = 1, 2, \ldots, m$.

Now we have seen that every constant coefficient linear homogeneous differential equation has polynomial-exponential solutions. Our next result shows that there are no other maximally extended solutions within the continuity class C^n.

General Solution of $P(D)[y] = 0$.   For $P(D)$ as in (10) the general solution of $P(D)[y] = 0$ is given by

$$y = p_1(t)e^{r_1 t} + p_2(t)e^{r_2 t} + \cdots + p_m(t)e^{r_m t} \tag{12}$$

for all polynomials p_j of degree $\leq n_j - 1$, $j = 1, 2, \ldots, m$.

Proof. For $P(D)$ as in (10) we see from the Polynomial-Exponential Solutions Theorem that the n elements in the set

$$B = \{e^{r_1 t}, te^{r_1 t}, \ldots, t^{(n_1-1)}e^{r_1 t}, \ldots, e^{r_m t}, te^{r_m t}, \ldots, t^{(n_m-1)}e^{r_m t}\} \tag{13}$$

belong to the null space $N(P(D))$. From the Independence of Polynomial Exponentials Theorem we see that B is a linearly independent set of $N(P(D))$. Now since $\dim N(P(D)) = n$ it follows that B spans $N(P(D))$, and we are done.

Example 6.22

We shall determine all solutions of the homogeneous equation $y^{(5)} + 8y^{(3)} + 16y = 0$. The characteristic polynomial of this equation is $P(r) = r^5 + 8r^3 + 16r$, which has the factorization $P(r) = r(r - 2i)^2(r + 2i)^2$. Thus the differential equation can be written in factorized operator form $D(D - 2i)^2(D + 2i)^2 y = 0$. From the results above, the solution space of this equation is spanned by the basis $B = \{1, e^{-2it}, te^{-2it}, e^{2it}, te^{2it}\}$, that is, $y = C_1 + C_2 e^{-2it} + C_3 te^{-2it} + C_4 e^{2it} + C_5 te^{2it}$, where C_j, $j = 1, \ldots, 5$ are arbitrary complex numbers. The solution space is a five-dimensional subspace of $C^\infty(\mathbf{R})$.

Real-Valued Solutions of $P(D)[y] = 0$

The nonreal roots of a polynomial with real coefficients always occur in conjugate pairs and with the same multiplicities. Let P have the nonreal conjugate pairs r_j, $\bar{r}_j$, each with multiplicity n_j, $j = 1, 2, \ldots, k$, and the real roots s_j with multiplicity m_j, $j = 1, 2, \ldots, l$. Then the differential equation $P(D)[y] = 0$ takes the form

$$(D - r_1)^{n_1}(D - \bar{r}_1)^{n_1} \cdots (D - r_k)^{n_k}(D - \bar{r}_k)^{n_k}$$

$$(D - s_1)^{m_1} \cdots (D - s_l)^{m_l}[y] = 0 \tag{14}$$

where $2(n_1 + \cdots + n_k) + (m_1 + \cdots + m_l) = n$. Because $P(D)$ is a linear operator and has real coefficients, it follows that the real-valued solutions of $P(D)[y] = 0$ form an n-dimensional subspace of the real linear space $C^\infty(\mathbf{R})$. As the next result shows, the real and imaginary parts of complex-valued solutions provide enough real-valued solutions to span $N(P(D))$.

General (Real-Valued) Solution of $P(D)[y] = 0$.   Let $P(D)$ have the factorization (14) and put $r_j = \alpha_j + i\beta_j$, $j = 1, 2, \ldots, k$. Then the real-valued general solution of $P(D)[y] = 0$ is given by

$$y = h_1(t)e^{s_1 t} + \cdots h_l(t)e^{s_l t}$$
$$+ p_1(t)e^{\alpha_1 t}\cos\beta_1 t + q_1(t)e^{\alpha_1 t}\sin\beta_1 t + \cdots \qquad (15)$$
$$+ p_k(t)e^{\alpha_k t}\cos\beta_k t + q_k(t)e^{\alpha_k t}\sin\beta_k t$$

for all polynomials h_j, p_j, q_j with real coefficients such that deg $h_j \leq m_j - 1$, for $j = 1, 2, \ldots, l$, and deg $p_j \leq n_j - 1$, deg $q_j \leq n_j - 1$, for $j = 1$, $2, \ldots, k$. There are precisely n arbitrary constants in the representation (15) and the real-valued solutions of $P(D)[y] = 0$ form an n-dimensional subspace of the real linear space $C^\infty(\mathbf{R})$.

Proof. That the n real-valued functions

$$e^{s_1 t}, \ldots, t^{m_l - 1}e^{s_l t}; e^{\alpha_1 t}\cos\beta_1 t, e^{\alpha_1 t}\sin\beta_1 t, \ldots;$$

$$t^{n_k - 1}e^{\alpha_k t}\cos\beta_k t, t^{n_k - 1}e^{\alpha_k t}\sin\beta_k t \qquad (16)$$

are all real-valued solutions of $P(D)[y] = 0$ is immediate. It is not difficult to show that the set (16) is linearly independent (see the problem set), and since we know that dim $N(P(D)) = n$, (15) follows.

Example 6.23

$P(D)$ in Example 6.22 has real coefficients, and using the notation of the theorem above: $s_1 = 0$, with multiplicity $m_1 = 1$, and $r_1 = 2i$ with multiplicity $n_1 = 2$. Thus the real-valued general solution of $P(D)[y] = 0$ is

$$y = c_1 + c_2\cos 2t + c_3\sin 2t + c_4 t\cos 2t + c_5 t\sin 2t$$

where c_1, c_2, c_3, c_4, and c_5 are arbitrary real numbers.

Example 6.24

Turning the problem around, we find a polynomial $P(r)$ with real coefficients such that the function $t^3 e^{-t}\cos 2t$ is a solution of $P(D)[y] = 0$. Observe that $t^3 e^{-t}\cos 2t = \text{Re}\ [t^3 e^{(-1+2i)t}]$. Thus $r_1 = -1 + 2i$ needs to be a root of $P(r)$ with multiplicity at least 4, and to ensure that $P(r)$ has real coefficients $\bar{r}_1 = -1 - 2i$ must also be a root of $P(r)$ with the same multiplicity. Thus

$$P(r) = [r - (-1 + 2i)]^4 [r - (-1 - 2i)]^4 = (r^2 + 2r + 5)^4$$

will suffice for the stated purpose.

Example 6.25

To solve the initial value problem

$$P(D)[y] \equiv y^{(5)} + 8y^{(3)} + 16y = 0$$

$$y(0) = 1 = y'(0), \qquad y^{(2)}(0) = y^{(3)}(0) = y^{(4)}(0) = 0 \tag{17}$$

first observe that all solutions of $P(D)[y] = 0$ are given by the linear span of the set $\{1, \cos 2t, \sin 2t, t\cos 2t, t\sin 2t\}$ (see Examples 6.22 and 6.23). Thus we need to find coefficients $c_1, \ldots, c_5$ such that $y = c_1 + c_2 \cos 2t + c_3 \sin 2t + c_4 t \cos 2t + c_5 t \sin 2t$ satisfies the initial conditions. That is, we must solve the linear system $c_1 + c_2 = 1$, $2c_3 + c_4 = 1$, $-4c_2 + 4c_5 = 0$, $-8c_3 - 12c_4 = 0$, $-16c_2 - 32c_5 = 0$. Solving, we have $c_1 = 1$, $c_2 = c_5 = 0$, $c_3 = \frac{3}{4}$, $c_4 = -\frac{1}{2}$, and hence the initial value problem (17) has the unique solution $y = 1 + \frac{3}{4} \sin 2t - \frac{1}{2} t \cos 2t$.

Undetermined Coefficients (Method of Annihilators)

The method of undetermined coefficients was introduced in Section 5.3 for finding a particular solution of the second-order constant coefficient equation $P(D)[y] = f$ for forcing terms f which are sums of polynomial-exponential functions. We could generalize this technique to equations of arbitrary order, but now that we know more about polynomial operators we can use another approach, called the *Method of Annihilators*. Suppose that

$$f(t) = p_1(t)e^{r_1 t} + \cdots + p_m(t)e^{r_m t}$$

where the p_j are polynomials and the r_j are distinct complex numbers. Because $P(D)$ is a linear operator, we see that if $P(D)[y_j] = p_j e^{r_j t}$ for each $j = 1, 2, \ldots, m$, then $y = y_1 + \cdots + y_m$ is a solution of $P(D)[y] = f$. Thus it suffices to describe the method when f has the special form $p(t)e^{rt}$. If $Q(D)$ is any polynomial operator such that $Q(D)[p(t)e^{rt}] = 0$, then for any particular solution y_p of $P(D)[y] = p(t)e^{rt}$, it follows that $Q(D)P(D)[y_p] = 0$. Thus y_p is a function in the null space of the operator $Q(D)P(D)$. Since we know the form of all solutions of the equation $Q(D)P(D)[y_p] = 0$, we can use an undetermined coefficients method to identify particular solutions of $P(D)[y] = p(t)e^{rt}$. An example will clarify the procedure.

Example 6.26

Find a particular solution of the equation

$$y''' - 3y' + 2y = 6(3t - 1)e^t \tag{18}$$

Putting $P(D) = D^3 - 3D + 2$, the equation can be written as $P(D)[y] = 6(3t - 1)e^t$. Note that $P(D) = (D - 1)^2(D + 2)$, and in the notation above,

$p(t) = 6(3t - 1)$ and $r = 1$. Observe that $Q(D) \equiv (D - 1)^2$ is such that $Q(D)[6(3t - 1)e^t] = 0$. Thus any particular solution of (18) satisfies the equation $Q(D)P(D)[y] = 0$. Hence every particular solution y_p of (18) has the form $y_p = h(t)e^t + Ae^{-2t}$ for some constant A and a polynomial h of degree ≤ 3. Using (11), we see that

$$P(D)[y_p] = (D - 1)^2(D + 2)[he^t] + (D - 1)^2(D + 2)[Ae^{-2t}] = e^t D^2(D + 3)[h]$$

Thus we see that $D^2(D + 3)[h] = 6(3t - 1)$, and if we put $h = k_0 + k_1 t + k_2 t^2 + k_3 t^3$ and compare coefficients, we have that $12k_3 = 18$ and $4k_2 + 6k_3 = -6$. Hence $k_3 = \frac{3}{2}$ and $k_2 = -\frac{15}{4}$, and $y_p = \frac{3}{4}(-5t^2 + 2t^3)e^t$.

Comments on Constant-Coefficients Method

A tacit assumption in our approach throughout has been that the roots of the characteristic polynomial $P(r)$ are known precisely together with their multiplicities. As observed earlier, such precise information is, in general, not possible if deg $P \geqq 5$. So what is the value of our solution formulas and procedures if we cannot explicitly find the precise information about the characteristic polynomial required by our approach? Under appropriate circumstances, if we have suitably close bounds on the characteristic roots, the solution formulas are useful in obtaining approximate solutions or in obtaining information on the qualitative properties of solutions of $P(D)[y] = f$. Many commercial software packages contain efficient solvers for finding close approximations of all roots of a polynomial, and hence we will consider the case where $P(D)$ has constant coefficients as essentially solved.

Nonhomogeneous Equations: Variation of Parameters

We now address the problem of solving the nonhomogeneous equation $P(D)[y] = f$, where $P(D)$ is as in (2), and the driving term f is any function in $C^0(I)$. This problem was solved in Section 5.3 for the case $n = 2$, and our approach here will be essentially the same. We assume that a basis $\{y_1, \ldots, y_n\}$ for $N(P(D))$ is known and then "vary the parameters" in the general solution $y = c_1 y_1 + \cdots + c_n y_n$ of the homogeneous equation $P(D)[y] = 0$ so as to produce a particular solution y_p of the nonhomogeneous equation. Because $P(D)$ satisfies the Linearity Property, we then construct the general solution of $P(D)[y] = f$ by taking $y = y_p + c_1 y_1 + \cdots + c_n y_n$, for arbitrary constants $c_1, \ldots, c_n$. We shall only sketch the proof of the next result because it so closely parallels the analogous result in Section 5.3 (except for technical details of dealing with general linear algebraic systems).

Variation of Parameters. Let the operator $P(D)$ be as in (2), $\{y_1(t), \ldots, y_n(t)\}$ a basis of $N(P(D))$, and let $W(t)$ denote the Wronskian of this basis. If $\Delta(s, t)$ is the determinant

$$\Delta(s, t) = \det \begin{bmatrix} y_1(s) & \cdots & y_n(s) \\ y_1'(s) & \cdots & y_n'(s) \\ \cdot & & \cdot \\ \cdot & & \cdot \\ \cdot & & \cdot \\ y_1^{(n-2)}(s) & \cdots & y_n^{(n-2)}(s) \\ y_1(t) & \cdots & y_n(t) \end{bmatrix} \tag{19}$$

and t_0 any point in I, then for any element f of $C^0(I)$, the function

$$u_p(t) = \int_{t_0}^{t} \frac{\Delta(s, t)}{W(s)} f(s)\, ds \tag{20}$$

is the unique solution of the initial value problem

$$\begin{aligned} P(D)[y] &= f \\ y^{(j)}(t_0) &= 0, \qquad j = 0, 1, \ldots, n-1 \end{aligned} \tag{21}$$

Proof (Sketch). Searching for u_p in the form $u_p = c_1(t)y_1(t) + \cdots + c_n(t)y_n(t)$, we impose the following $n-1$ conditions on the derivatives $c_1'(t), \ldots, c_n'(t)$:

$$c_1'(t)y_1^{(k)}(t) + \cdots + c_n'(t)y_n^{(k)}(t) = 0 \qquad \text{for } k = 0, \ldots, n-2$$

Substituting u_p into $P(D)[u] = f$ and using these conditions, we derive the additional condition

$$c_1'(t)y_1^{(n-1)}(t) + \cdots + c_n'(t)y_n^{(n-1)}(t) = f(t)$$

Using Cramer's Rule to solve the conditions above for $c_1'(t), \ldots, c_n'(t)$, integrating from t_0 to t to find $c_1, \ldots, c_n$, and then inserting them into the assumed form for u_p (and using the properties of determinants), we obtain (20).

Example 6.27

Let us use Variation of Parameters to find a solution $u_p(t)$ of the equation

$$P(D)[u] \equiv (D^3 + 4D)[u] = f(t), \qquad \text{where } f \text{ is in } C^0(\mathbf{R})$$

Since $\{1, \cos 2t, \sin 2t\}$ is a basis of $N(P(D))$, we have from (20) that

$$u_p(t) = \int_{t_0}^{t} \frac{\det \begin{bmatrix} 1 & \cos 2s & \sin 2s \\ 0 & -2\sin 2s & 2\cos 2s \\ 1 & \cos 2t & \sin 2t \end{bmatrix}}{\det \begin{bmatrix} 1 & \cos 2s & \sin 2s \\ 0 & -2\sin 2s & 2\cos 2s \\ 0 & -4\cos 2s & -4\sin 2s \end{bmatrix}} f(s)\, ds$$

$$= \int_{t_0}^{t} \tfrac{1}{4}[1 - \cos 2(s - t)] f(s)\, ds$$

where t_0 is any real number and we have used the identity $\cos \alpha \cos \beta + \sin \alpha \sin \beta = \cos(\alpha - \beta)$.

Using the linearity property of the operator $P(D)$, we have the following result.

General Solution of $P(D)[u] = f$.   Let $\{y_1, \ldots, y_n\}$ be a basis of the solution space of $P(D)[u] = 0$. Let $u^*(t)$ be any particular solution of $P(D)[u] = f$ [e.g., the solution $u_p(t)$ obtained by Variation of Parameters]. Then the general solution of $P(D)[u] = f$ is $u = c_1 y_1 + \cdots + c_n y_n + u^*$, where $c_1, \ldots, c_n$ are arbitrary constants.

It is a routine matter to solve the general initial value problem (4) if we know a basis $\{y_1, \ldots, y_n\}$ for the associated homogeneous equation $P(D)[y] = 0$. Let u_p be the particular solution of $P(D)[y] = f$ given by (20). Then the constants in the general solution $y = c_1 y_1 + \cdots + c_n y_n + u_p$ can be uniquely determined so that the resulting solution $y(t)$ satisfies the initial conditions. Note that since $u_p^{(j)}(t_0) = 0$, for $j = 0, 1, 2, \ldots, n - 1$, the imposed initial conditions take the form

$$c_1 y_1^{(j)}(t_0) + \cdots + c_n y_n^{(j)}(t_0) = b_j \qquad \text{for } j = 0, 1, \ldots, n - 1 \qquad (22)$$

The determinant of the linear system (22) is $W[y_1, \ldots, y_n](t_0)$, which is nonzero since $\{y_1, \ldots, y_n\}$ is a basis for $N(P(t, D))$. An example will illustrate the procedure.

Example 6.28

Let us solve the initial value problem

$$y''' + 4y' = \tan 2t, \qquad y(0) = 0, \quad y'(0) = 1, \quad y''(0) = -1 \qquad (23)$$

on the interval I: $-\pi/4 < t < \pi/4$. Using the results of Example 6.27 to find a particular solution, we put $t_0 = 0$, $f(t) = \tan 2t$ and after a straightforward integration obtain

$$y_p(t) = \tfrac{1}{8} \ln|\sec 2t| - \tfrac{1}{8} \sin 2t \, \ln|\sec 2t + \tan 2t| + \tfrac{1}{8} - \tfrac{1}{8} \cos 2t$$

Now since $y_1 = 1$, $y_2 = \cos 2t$, $y_3 = \sin 2t$ is a basis for $N(D^3 + 4D)$, we see that (22) becomes

$$c_1 + c_2 = 0, \qquad 2c_3 = 1, \qquad -4c_2 = -1 \tag{24}$$

Notice that since the determinant of the system (24) is nonzero, it has a unique solution. The system is solvable by inspection (so we need not use the full power of the methods in Section 6.2): $c_1 = -\frac{1}{4}$, $c_2 = \frac{1}{4}$, $c_3 = \frac{1}{2}$, and hence the solution of problem (23) is

$$y = \frac{-1 + \cos 2t + 2 \sin 2t}{4} + y_p(t)$$

Piecewise-Continuous Inputs. The Variation of Parameters formula (20) can still be used if the input f is only piecewise continuous on I, since piecewise-continuous functions are integrable. However, at a point of discontinuity t_1 of f it will happen that t_1 is a point of discontinuity of $P(D)[u_p]$. If we are willing to tolerate a solution $u_p(t)$ whose nth derivative is piecewise continuous only at the points of discontinuity of f (at which points the equation $P(D)[u_p] = f$ may not hold), the results of this section are still valid. With these reservations in mind, we can allow piecewise-continuous inputs f. This is equivalent to enlarging the domain of the operator $P(D)$ to include functions u in $C^{n-1}(I)$, with $u^{(n)}$ possibly only a piecewise-continuous function on I.

PROBLEMS

The following problem(s) may be more challenging: 7, 10, 11, 12, and 13.

1. Find the real-valued general solution of each of the following differential equations.
   (a) $y'''' + 81y = 0$.
   (b) $(D^3 + 1)[y] = 0$. Sketch solutions if $y(0) = y'(0) = 1$, $y''(0) = 0, \pm 1$.
   (c) $y''' + 3y'' - y' - 3y = 0$.
   (d) $y'''' + 18y'' + 81y = 0$.
   (e) $(D - 1)^3(D + 1)^3[y] = t$.

2. Find the complex-valued general solution of each of the following differential equations.
   (a) $(D - i)^3[y] = t^2$.
   (b) $(D^2 - 4D + 5)^3(D^2 + 1)^2[y] = 0$.
   (c) $(D^2 - 4D + 3)^2(D^2 + 2D + 2)^3[y] = 0$.

3. Find a constant-coefficient homogeneous differential equation with real coefficients which includes $t^3 e^{-3t}$ and $te^t \sin 2t$ among its solutions.

4. Find the solution of each of the following initial value problems.
   (a) $(D^3 + 1)y = 0$, $y(0) = 0$, $y'(0) = 0$, $y''(0) = 1$.
   (b) $y''' - y'' + 3y' + 5y = 0$, $y(0) = 1$, $y'(0) = y''(0) = 0$.
   (c) $(D^2 + 1)(D + 2)^2 y = 0$, $y(0) = y'(0) = 0$, $y''(0) = y'''(0) = 1$.

5. Find and sketch the unique solution of each initial value problem.

   (a) $y''' - y'' - y' + y = t + 1$, $y(0) = 5$, $y'(0) = -1$, $y''(0) = 5$.
   (b) $y''' + y'' - 2y' = 3 + \sin t$, $y(0) = y'(0) = 0$, $y''(0) = 1$.

6. Find all solutions of each of the following problems [f in $C^0(\mathbf{R})$].
   (a) $u''' + 2u'' - u' - 2u = f(t)$.
   (b) $u''' + 2u'' - u' - 2u = f(t)$, $u(0) = 1$, $u'(0) = 0$.
   (c) $t^3 u''' + 3t^2 u'' = f(t)$, $u(1) = 0$, $t > 0$. [*Hint:* $\{1, t^{-1}, t\}$ is a set of three solutions of $t^3 u''' + 3t^2 u'' = 0$.]

7. Solve the problem $u'' + 4u = f(t) = \begin{cases} 1, & t < 1 \\ 3, & t \geq 1 \end{cases}$, $u(0) = 0$, $u'(0) = 0$. Comment on the behavior of the "solution" at $t = 1$. Sketch the graphs of $u(t)$ and of $u'(t)$.

8. Let $S = \{y_1(t), \ldots, y_n(t)\}$ be a set of functions belonging to $C^{n-1}(I)$. Suppose that S is dependent on I; that is, suppose there exist constants $c_1, \ldots, c_n$, not all zero, such that $c_1 y_1(t) + \cdots + c_n y_n(t) = 0$ for all t in I. Show that $W[y_1, \ldots, y_n](t) = 0$ for all t in I.

9. For any integer $n \geq 1$ and any given f in $C^0(I)$ show that the general solution of the equation $D^n y = f$ is

$$y(t) = p(t) + \int_{t_0}^{t} \frac{(t - s)^{n-1}}{(n - 1)!} f(s) \, ds \qquad \text{for } t \text{ in } I$$

where $p(t)$ is an arbitrary polynomial with degree $\leq n - 1$ and t_0 is any (fixed) point in I.

10. Prove the Independence of Polynomial-Exponentials Theorem.

11. [*Abel's Formula*]. For the operator $P(D)$ in (2), let $y_1, \ldots, y_n$ be any n solutions of $P(D)[y] = 0$. Then show that

$$W[y_1, \ldots, y_n](t) = W[y_1, \ldots, y_n](t_0) \exp\left[-\int_{t_0}^{t} a_{n-1}(s) \, ds\right]$$

for all t_0 and t in I. [*Hint:* Use the fact that the derivative of an $n \times n$ determinant Δ of differentiable functions is the sum, $\Delta_1 + \cdots + \Delta_n$, where Δ_k coincides with Δ except that all entries in the kth row have been differentiated once.]

12. Let $A = \{y_1(t), \ldots, y_n(t)\}$ and $B = \{z_1(t), \ldots, z_n(t)\}$ be any two bases of the solution space of $y^{(n)} + a_{n-1}(t)y^{(n-1)} + \cdots + a_1(t)y' + a_0(t)y = 0$, where $a_j(t)$ in $C^0(\mathbf{R})$, $j = 0, 1, \ldots, n - 1$. Show that the Wronskian of A is a constant multiple of the Wronskian of B.

13. Let $P(D)$ in (2) have constant real coefficients and the factorization (14). Show that the set of n functions in (16) is a basis for $N(P(D))$. [*Hint:* Note that $e^{\alpha t} \cos \beta t = (e^{(\alpha+i\beta)t} + e^{(\alpha-i\beta)t})/2$ and $e^{\alpha t} \sin \beta t = i(e^{(\alpha-i\beta)t} - e^{(\alpha+i\beta)t})/2$ for any reals α and β. Use this fact and the Independence of Polynomial-Exponentials Theorem.]

6.4 LINEAR OPERATORS: GREEN'S FUNCTIONS

In developing techniques for solving linear differential equations, we found it useful to employ the formalism of operators. We viewed a linear differential equation as asking for solutions of the operator equation $P(D)[y] = f$, where $P(D)$ was a polynomial operator. In this section we continue our development of the operator approach as a problem-solving tool in applied mathematics. The enormous literature on operator theory indicates the great value this concept has in applications.

Basic Problem of Applied Mathematics

It is truly remarkable how many problems in applied mathematics come down to an operator equation

$$Lu = f \tag{1}$$

where the operator L is given and f is a given element in a codomain of L. The problem is to find the set of all values for u in the domain of L which satisfy the operator equation (1), called the solution set of the equation. But the questions asked of problem (1) are always the same, regardless of how the operator equation arises:

> *Existence*: For a given f in the codomain of L, does $Lu = f$ have any solutions at all?
>
> *Uniqueness*: How many solutions does $Lu = f$ have for a given f?
>
> *Solvability*: How can all the solutions be found?
>
> *Sensitivity to the data*: Assuming unique solvability of $Lu = f$ for all f in the range of L, how sensitive is the solution of $Lu = f$ to "small changes" in the data, f?

We have seen these questions before in Section 2.2. They address the fundamental properties of the *well-posedness* of problem (1), and we deal with these questions throughout this book in a wide variety of contexts. The real power of the operator approach stems from two sources:

1. Properties of operators characterized abstractly must hold without further proof for any realization of the operator, regardless of the context.
2. The operator formulation of systems sometimes permits a type of "calculus" which simplifies and extends the mathematical analysis of such systems.

The material of this chapter and the next provide some simple examples of these two points.

Products of Operators, Inverse of an Operator

Operators may be combined by performing the operations in succession, assuming that this is permitted. Suppose that $L: y \rightarrow Ly$, and that for all y in the domain of L, Ly lies in the domain of an operator $M: z \rightarrow Mz$. Then we may combine L and M to form a new operator called the *product* of M and L, denoted by ML, where $ML: y \rightarrow (ML)y = M(Ly)$, for all y in the domain of L. Even if the products LM and ML are both defined in this sense, it should not be assumed that LM and ML are the same operators.

Example 6.29

Observe that $(D - 1)(D - 2) = D^2 - 3D + 2 = (D - 2)(D - 1)$. However, note that $(D - t)(D - 2t) = D^2 - 2Dt - tD + 2t^2 = D^2 - 2tD - 2 - tD + 2t^2 = D^2 - 3tD + (2t^2 - 2)$, but that $(D - 2t)(D - t) = D^2 - Dt - 2tD + 2t^2 = D^2 - tD - 1 - 2tD + 2t^2 = D^2 - 3tD + (2t^2 - 1) \neq (D - t)(D - 2t)$.

Invertibility, Inverse Operators.   The operator L is said to be *invertible* if there is an operator M on the range of L taking values in the domain of L such that

$$MLv = v \qquad \text{for all } v \text{ in the domain of } L \tag{2}$$

When L is invertible there is an unique M which satisfies condition (2); this operator M is called the *inverse* of L and is denoted by L^{-1}.

The virtue of dealing with an invertible operator L is that if an operator M is known which satisfies condition (2), the operator equation $Lv = w$ for a given w is easy to solve for v. Applying M to both sides of the equation $Lv = w$, we obtain $v = Mw$.

Invertibility Theorem. L is an invertible operator if and only if it is one-to-one. For an invertible operator L there is only one operator M satisfying condition (2).

Proof. When condition (2) holds, then L is one-to-one; that is, L sends distinct points into distinct points. If $v^1 \neq v^2$, we must have $Lv^1 \neq Lv^2$, because if $Lv^1 = Lv^2$, then (2) implies that $v^1 = MLv^1 = MLv^2 = v^2$. Conversely, if L is one-to-one, there is an operator M which satisfies (2). M simply sends each point in the range of L into the unique point that it "came from." There cannot be two operators M and M' which both satisfy (2). Let w be in the range of L. Then there is a v in the domain of L such that $w = Lv$. Thus from (2), $Mw = MLv = v$ and $M'w = M'Lv = v$, so $Mw = M'w$ for all w in the range of L. Hence $M = M'$.

Example 6.30

Let $L = P(D) \equiv D + a(t)$, where $a(t)$ is continuous on an interval I. Thus $L: C^1(I) \rightarrow C^0(I)$ is the operator whose action is given by

$$L: y \rightarrow y' + a(t)y \qquad \text{for } y \text{ in } C^1(I) \tag{3}$$

Hence the operator equation $Ly = h(t)$ for some given h in $C^0(I)$ is just another way of writing the linear first-order equation $y' + a(t)y = h(t)$, whose solutions, we know from Section 3.1, are all given by the formula

$$y(t) = k \exp\left[-A_0(t)\right] + \exp\left[-A_0(t)\right] \int_{t_0}^{t} \exp\left[A_0(s)\right] h(s) \, ds \qquad (4)$$

where t_0 is any point in I, k is an arbitrary constant, and

$$A_0(t) = \int_{t_0}^{t} a(r) \, dr$$

Thus for any constant k, L applied to the function y in (4) yields $h(t)$, and hence the range is precisely $C^0(I)$ since $h(t)$ was an arbitrarily chosen function. L is a many-to-one mapping since it takes infinitely many distinct functions into any given $h(t)$. Now let us consider an operator L_0 which has the same action as L, but the smaller domain of functions in $C^1(I)$ which vanish at some given t_0 in I. Thus the operator equation $L_0 y = h$ now translates into the initial value problem $y' + a(t)y = h(t)$, $y(t_0) = 0$, which we know (via Section 3.2) has the unique solution

$$y(t) = \exp\left[-A_0(t)\right] \int_{t_0}^{t} \exp\left[A_0(s)\right] h(s) \, ds \qquad (5)$$

Hence L_0 is a one-to-one map of its domain onto $C^0(I)$. Now let us consider the operator $G: C^0(I) \to C^1(I)$ whose action is

$$G: h \longrightarrow \exp\left[-A_0(t)\right] \int_{t_0}^{t} \exp\left[A_0(s)\right] h(s) \, ds \qquad \text{for } h \text{ in } C^0(I) \qquad (6)$$

Now from (5) we know that Gh is in the domain of L_0, no matter which $h(t)$ is chosen in $C^0(I)$. Moreover, G "undoes" the action of L_0 in the sense that if $h = L_0 y$ for some y, then Gh gives us back this function $y(t)$ again. In operator terminology this is written as

$$G[L_0 y] = y \qquad \text{for all } y \text{ in the domain of } L_0$$

In a similar way we have that

$$L_0[Gh] = h \qquad \text{for all } h \text{ in } C^0(I)$$

Thus we have given an explicit construction of an operator G which inverts the differential operator L_0. It comes as no surprise that G is an integral operator.

Linear Operators

Operators that will occupy our attention for a great deal of the remainder of this book have a special property called linearity.

Linear Operator. An operator L whose domain and codomain are linear spaces is called a *linear operator* if for any pair of elements u and v in the domain of L and any scalars a and b, we have that

$$L[au + bv] = aLu + bLv \qquad (7)$$

Example 6.31

The operators L and G defined via (3) and (6) are both linear operators. So is the operator L_0 in Example 6.30, which has the same action as L but on a smaller domain. Note that $Ly = h$ is a first-order linear differential equation, $L_0y = h$ is an initial value problem, and $Gh = f$ is an integral equation.

Example 6.32

If we agree to write elements of $\mathbf{C}^n$ and $\mathbf{C}^m$ as $n \times 1$ and $m \times 1$ column matrices and A is an $m \times n$ matrix, we can define an operator $A: \mathbf{C}^n \to \mathbf{C}^m$ whose action is defined by

$$A: x \to Ax = \begin{bmatrix} a_{11} & \cdots & a_{1n} \\ \vdots & & \vdots \\ a_{m1} & \cdots & a_{mn} \end{bmatrix} \begin{bmatrix} x_1 \\ \vdots \\ x_n \end{bmatrix} = \begin{bmatrix} \sum_{j=1}^{n} a_{1j}x_j \\ \vdots \\ \sum_{j=1}^{n} a_{mj}x_j \end{bmatrix} \qquad (8)$$

Using the standard operations in $\mathbf{C}^n$ and $\mathbf{C}^m$, it is a straightforward calculation to show that A is a linear operator. It is interesting to note that for a given b in $\mathbf{C}^m$, the operator equation $Ax = b$ becomes a system of m equations in n unknowns:

$$a_{11}x_1 + \cdots + a_{1n}x_n = b_1$$
$$\vdots \qquad \qquad \vdots \qquad \vdots$$
$$a_{m1}x_1 + \cdots + a_{mn}x_n = b_m$$

Such linear algebraic systems were treated in Section 6.2.

When the operator L is linear, the task of finding solutions of the operator equation $Lu = f$ is facilitated by a simple observation. Recall that the zero vector belongs to each linear space and is denoted simply by the symbol 0. The set of all solutions to the operator equation $Lu = 0$ is called the *null space* of L. Notice that if L is a linear operator, the null space of L is a linear subspace of the domain of L. A specific, single solution u of the operator equation $Lu = f$, for a given f, goes by the unimaginative name of *a particular solution*. Now we state the

---

Characterization of Solutions for Linear Equations. Let u_p be any fixed particular solution of the equation $Lu = f$, where L is a linear operator. Then the solution set of the equation $Lu = f$ has the form $u = v + u_p$, where v is any element in the null space of L.

Proof. If v is any element in the null space of L and u_p is any particular solution of $Lu = f$, then because L is a linear operator, (7) implies that $L(v + u_p) = Lv + Lu_p = 0 + f = f$. Conversely, suppose that u_p and w_p are any two particular solutions of $Lu = f$. Then (7) implies that $L(w_p - u_p) = Lw_p - Lu_p = f - f = 0$, so $w_p - u_p = v$ must be in the null space of L, and the claim is established.

Example 6.33

Let the matrix $A = \begin{bmatrix} 1 & 2 & -1 \\ 3 & 7 & -2 \end{bmatrix}$ determine the linear operator $A: \mathbf{R}^3 \to \mathbf{R}^2$ with action given by (8). Using techniques of Section 6.2, we see that the solution set of the operator equation $Ax = \begin{bmatrix} -1 \\ -2 \end{bmatrix}$ is $x = (-3, 1, 0) + a(3, -1, 1)$, where a is an arbitrary real number. Notice that $x_p = (-3, 1, 0)$ is a particular solution of the equation and that the vector $(3, -1, 1)$ is a basis for the null space of A.

Example 6.34

For any $a(t)$ in $C^0(I)$ the linear operator $L = D + a(t)$ has a one-dimensional null space spanned by the function $\exp[-A(t)]$, where $A(t)$ is any antiderivative of $a(t)$ on I. The function

$$y_p(t) = \exp[-A(t)] \int^t \exp[A(s)]h(s)\, ds$$

is a particular solution of the operator equation $(D + a(t))y = h$. Thus by the Characterization Theorem the general solution of the operator equation $(D + a(t))y = h$ is given by $y = k \exp[-A(t)] + y_p(t)$, where k is an arbitrary constant (see Example 6.30).

The range and null space of an operator $L: V \to W$ are sometimes denoted by the symbols $L(V)$ and $N(L)$, respectively. Note that if L is a linear operator (and thus by definition V and W are linear spaces), then $N(L)$ is a subspace of V and $L(V)$ is a subspace of W. The following useful property of linear operators is stated without proof (see a special case of this result in Section 6.2).

Rank-Nullity Theorem. Let $L: V \to W$ be a linear operator from a finite-dimensional linear space V to another (finite- or infinite-dimensional) linear space W. Then

$$\dim[N(L)] + \dim[L(V)] = \dim V$$

The Rank-Nullity Theorem shows, in particular, that the dimension of $L(V)$ cannot be larger than $\dim V$, and hence linear operators do not "expand" dimension.

Furthermore, if L "collapses" dimension [i.e., dim $L(V) \neq$ dim V], this can only be at the expense of increasing the dimension of the null space by the shortfall.

The Invertibility Theorem does not assume that the operator L is linear. The theorem below, given without proof, cites some convenient conditions for the invertibility of linear operators defined on finite-dimensional spaces.

Invertibility Theorem for Linear Operators.   A linear operator $L: V \to W$ on a finite-dimensional space V is invertible if any of the conditions below are satisfied. Conversely, if L is invertible, all the conditions below are satisfied.
(a) The null space of L is trivial (i.e., it does not contain a nonzero element).
(b) dim $L(V) =$ dim V.
(c) If $\{e_1, \ldots, e_n\}$ is a basis of V, then $\{Le_1, \ldots, Le_n\}$ is a basis of $L(V)$.

Example 6.35

The matrix $A = \begin{bmatrix} 1 & 1 \\ 1 & -1 \end{bmatrix}$ determines a linear operator $A: \mathbf{R}^2 \to \mathbf{R}^2$ in the manner described in Example 6.32. Thus a column vector $\begin{bmatrix} x_1 \\ x_2 \end{bmatrix}$ is in the null space of A if and only if $A \begin{bmatrix} x_1 \\ x_2 \end{bmatrix} = \begin{bmatrix} 0 \\ 0 \end{bmatrix}$, or equivalently, $x_1 + x_2 = 0$ and $x_1 - x_2 = 0$. This system has the unique solution $x_1 = x_2 = 0$, and hence A is invertible by criterion (a) of the Invertibility Theorem for Linear Operators.

Linear Differential Operators

As we have seen earlier, the *differentiation operator* is represented by the symbol D and has the action $D: y \to Dy \equiv y'$. Repeated application of the operator k times produces the derivative $y^{(k)}$; the operator that takes y into $y^{(k)}$ is denoted by D^k. Now the polynomial in the operator D

$$P(D) = D^n + a_{n-1}(t)D^{n-1} + \cdots + a_1(t)D + a_0(t)$$

where the coefficients $a_k(t)$ are all continuous on some common interval I on the t-axis, was defined (in Section 6.2) to be the *polynomial operator* with domain $C^n(I)$ and action

$$P(D): y \to D^n y + a_{n-1}(t)D^{n-1}y + \cdots + a_1(t)Dy + a_0(t)y \tag{9}$$

The continuity space $C^0(I)$ is a codomain for $P(D)$. Also, observe that $L \equiv P(D)$ satisfies the identity (7), and hence $P(D)$ is a linear operator.

We can form the "product" of two polynomial operators $P(D)$ and $Q(D)$ to obtain the operator $L \equiv P(D)Q(D)$ by first applying $Q(D)$ to y, and then $P(D)$ to the function $Q(D)y$ (i.e., $L[y] = P(D)[Q(D)[y]]$). When the polynomials P and Q have constant coefficients, we showed that $L = P(D)Q(D)$ is the polynomial operator

$R(D)$, where R is the usual algebraic product of the polynomials P and Q. Thus when P and Q have constant coefficients, we have that $P(D)Q(D) = Q(D)P(D)$.

Properties of $P(D)$.   For any polynomial operator $P(D)$ and any real (or complex) number r, we have the operator identities

$$P(D)[e^{rt}y] \equiv e^{rt}P(D + r)[y] \tag{10}$$

$$e^{rt}P(D)[y] \equiv P(D - r)[e^{rt}y] \tag{11}$$

for all y in the domain of $P(D)$.

Proof.   First show that (10) and (11) hold for $P(D) = D^k$, and then the general result follows from the linearity of $P(D)$.

Example 6.36

Let us find the general solution of

$$D^2(D - 1)^3[y] = 1 \tag{12}$$

Let $y(t)$ be a solution of (12). Rewriting (12) as $(D - 1)^3 D^2[y] = 1$ and multiplying both sides of the equation by e^{-t}, we apply (11) with $P(D) \equiv (D - 1)^3$ to obtain

$$D^3[e^{-t}D^2y] = e^{-t} \tag{13}$$

Now $-e^{-t}$ is a particular solution of $D^3z = e^{-t}$, and the null space of D^3 is the set of all polynomials of degree ≤ 2. Thus from (13) we have that $e^{-t}D^2y = p(t) - e^{-t}$, or

$$D^2y = p(t)e^t - 1 \tag{14}$$

where p is a polynomial of degree ≤ 2. Hence a particular solution of (14) is $q(t)e^t - \frac{1}{2}t^2$, where q is a polynomial with deg $p = $ deg q. Thus we have shown that every solution of (12) must have the form

$$y = h(t) + q(t)e^t - \frac{1}{2}t^2 \tag{15}$$

where h and q are polynomials with deg $h \leq 1$, deg $q \leq 2$. A direct computation shows that (15) is a solution of (12) for *any* such polynomials h, q:

$$D^2(D - 1)^3[h + qe^t - \frac{1}{2}t^2] = (D - 1)^3 D^2[h] + D^2(D - 1)^3[qe^t]$$
$$- \frac{1}{2}(D - 1)^3 D^2[t^2]$$
$$= 0 + e^t(D + 1)^2 D^3[q] - \frac{1}{2}(D - 1)^3[2]$$
$$= 0 + 0 - (D - 1)^3[1] = -(0 - 1)^3 = 1$$

where we have used the linearity of $P(D)$ and the operational formula (10) repeatedly.

P(D) as an Operator: Green's Functions

The operator $P(D)$: $C^n(I) \to C^0(I)$ as defined in (9) is a linear operator, and from the General Solution Theorem we see that the range of $P(D)$ is precisely $C^0(I)$. However, since dim $N(P(D)) = n > 0$, we see that $P(D)$ is not an invertible operator. We construct below an invertible operator M with the same action and range as $P(D)$ but whose domain is a subspace of $C^n(I)$.

Let t_0 be any point in I and put

$$\mathcal{M} = \{u \text{ in } C^n(I): u^{(j)}(t_0) = 0, \qquad j = 0, 1, \ldots, n-1\}$$

It is easy to see that $\mathcal{M}$ is a subspace of $C^n(I)$ and that the operator M: $\mathcal{M} \to C^0(I)$ whose action coincides with $P(D)$ is a linear operator. Thus M is the restriction of $P(D)$ to the linear space $\mathcal{M}$. Observe that the problem (21) of Section 6.3 is equivalent to the operator equation $M[u] = f$, which according to the Variation of Parameters Theorem has the unique solution

$$u_p(t) = \int_{t_0}^{t} \frac{\Delta(s, t)}{W(s)} f(s) \, ds$$

for any f in $C^0(I)$, where $\Delta(s, t)$ is defined by (19) in Section 6.3 using a basis $\{y_1, \ldots, y_n\}$ for $N(P(D))$. Thus the range of M is $C^0(I)$ and M is invertible. One marked advantage of the formula for u_p is that the *kernel*

$$G(t, s) \equiv \frac{\Delta(s, t)}{W(s)} \tag{16}$$

can be computed as soon as a basis for $N(P(D))$ is known. Although it is not obvious, the kernel $G(t, s)$ depends in no way on the particular basis chosen for $N(P(D))$. However, it is clear from (16) that $G(t, s)$ does not depend on the input function, f. Observe that the linear (integral) operator G with domain $C^0(I)$ and action

$$G: f \to \int_{t_0}^{t} G(t, s) f(s) \, ds$$

where kernel G defined via (16) has the property that

$$G(Mu) = u \qquad \text{for all } u \text{ in the domain of } M$$
$$M(Gf) = f \qquad \text{for all } f \text{ in the domain of } G$$

Hence G inverts the action of M.

> **Green's Function.**   The kernel function $G(t, s)$ in (16) is called the *Green's function*† for the operator M. Observe that G is continuous on the square $I \times I = \{(t, s) : t \text{ in } I, s \text{ in } I\}$. Finally, observe in the second-order case that the Green's function is precisely the kernel function $K(t, s)$ of Section 5.3.

† George Green (1793–1841) was an English miller and self-taught mathematician who also made contributions to the theory of electricity and magnetism, introducing the term "potential."

Example 6.37

Let $P(D) = D^2 + \omega^2$, where ω is a positive constant. Since $\{\cos \omega t, \sin \omega t\}$ is a basis for $N(P(D))$, we have that the corresponding Green's function is given by

$$G(t, s) = \frac{\det \begin{bmatrix} \cos \omega s & \sin \omega s \\ \cos \omega t & \sin \omega t \end{bmatrix}}{\det \begin{bmatrix} \cos \omega s & \sin \omega s \\ -\omega \sin \omega s & \omega \cos \omega s \end{bmatrix}} = \frac{1}{\omega} \sin \omega (t - s)$$

Example 6.38

The Green's function corresponding to the operator $P(D) = D^2 - (1/t)D + 1/t^2$, $t > 0$, is

$$G(t, s) = \frac{\det \begin{bmatrix} s & s \ln s \\ t & t \ln t \end{bmatrix}}{\det \begin{bmatrix} s & s \ln s \\ 1 & 1 + \ln s \end{bmatrix}} = t \ln (t/s)$$

since $\{t, t \ln t\}$ is a basis for $N(P(D))$.

Characterizations of Green's Functions

Green's functions can be characterized in other ways than as the quotient of two determinants. The result below (given without proof) cites two alternative characterizations.

Characterizations of Green's Functions.   Let $P(D)$ be the operator defined by (9). Then (a) $G(t, s) = \Delta(s, t)/W(s)$ is the unique element of $C^0(I \times I)$ such that for every f in $C^0(I)$ the solution of problem (21) of Section 6.3 is given by

$$u(t) = \int_{t_0}^{t} G(t, s)f(s) \, ds$$

(b) For each fixed s in I the function $G(t, s) = \Delta(s, t)/W(s)$ is a solution to the initial value problem†

$$P(D)G(t, s) \equiv \frac{d^n}{dt^n} G(t, s) + \cdots + a_0(t)G(t, s) = 0, \qquad \text{all } t \neq s$$

$$G(s^+, s) = \frac{dG}{dt}\bigg|_{t=s^+} = \cdots = \frac{d^{n-2}G}{dt^{n-2}}\bigg|_{t=s^+} = 0, \qquad \frac{d^{n-1}G}{dt^{n-1}}\bigg|_{t=s^+} = 1 \qquad (17)$$

---

† It is important to note that the characterization of $G(t, s)$ via (b) is a *forward* initial value problem with initial point $t_0 = s$.

If $P(D)$ is a constant-coefficient operator, a simplification results.

Green's Function, Constant Coefficients.   The Green's function of the constant-coefficient operator $P(D) = D^n + a_{n-1}D^{n-1} + \cdots + a_0$, is $g(t - s)$, where $g(r)$ is the solution of

$$P(D)[g(r)] = 0, \quad g(0) = 0, \quad \ldots, \quad g^{(n-2)}(0) = 0, \quad g^{(n-1)}(0) = 1 \quad (18)$$

Proof. Let $g(r)$ be the solution of (18) and let s be fixed, $t \geq s$. Then

$$\frac{dg(t-s)}{dt} \equiv \frac{dg(r)}{dr}, \ldots, \frac{d^{(n)}g(t-s)}{dt^n} \equiv \frac{d^{(n)}g(r)}{dr^n}, \qquad \text{where } r = t - s$$

Hence $G^*(t, s) \equiv g(t - s)$ satisfies the first condition of (17), $P(D)G^*(t, s) = 0$. The boundary conditions of (17) are also satisfied since g satisfies the boundary conditions of (18).

Example 6.39

The Green's function $g(t - s)$ of the constant-coefficient operator $P(D) = D^3 + 4D$ is found by solving

$$g'''(r) + 4g'(r) = 0, \qquad g(0) = g'(0) = 0, \quad g''(0) = 1$$

Using the methods of Section 6.3, we find that $g(r) = \frac{1}{4} - \frac{1}{4} \cos 2r$. Hence the Green's function is $g(t - s) = \frac{1}{4} - \frac{1}{4} \cos 2(t - s)$.

PROBLEMS

The following problem(s) may be more challenging: 11 and 13.

1. Show that the range of any linear operator is a linear subspace of its codomain.
2. The matrix

$$A = \begin{bmatrix} 1 & 3 & 2 & 1 \\ -1 & 1 & 0 & 2 \\ 3 & 1 & 2 & -3 \end{bmatrix}$$

   determines a linear operator $A: \mathbf{R}^4 \to \mathbf{R}^3$ whose action is given by (8).
   (a) Find the range of A.
   (b) Find the null space of A.
   (c) Verify the Rank-Nullity Theorem for A.
3. Show that the zero polynomial is the only polynomial solution of the operator equation $(D - r)^n[p] = 0$ for any constant $r \neq 0$ and any positive integer n.
4. Let L be the linear operator whose action is given by $P(D) = D + 2$ and whose domain is those functions in $C^1(\mathbf{R})$ which vanish at $t_0 = 1$. Construct the inverse of L and show that it is a linear operator.

5. If an operator L is linear and has an inverse, show that L^{-1} is also linear. [*Hint*: Use (2) and the fact that every w in the range of L can be written as $w = Lu$ for some u in the domain of L.]

6. [*Fredholm Alternative*]. Let $A: V \to W$ be a linear operator with finite-dimensional domain and codomain. It can be shown that just one of the following alternatives holds: Either $Ax = y$ has a solution for every y in W, or else the null space of A^T is nontrivial. Verify this alternative in the special case where $V \equiv \mathbf{R}^3$, $W \equiv \mathbf{R}^2$ and A is the linear operator defined by the matrix

$$\begin{bmatrix} 3 & 1 & -2 \\ 1 & -1 & -3 \end{bmatrix}$$

7. Let the $m \times n$ matrix A determine a linear operator $A: \mathbf{R}^n \to \mathbf{R}^m$ via (8). It can be shown that the equation $Ax = y$ has a solution for a given y in $\mathbf{R}^m$ if and only if $uy = 0$ for all row matrices u such that $uA = 0$. Verify this claim for the matrix in Problem 6.

8. Let the operator $L: V \to W$ be invertible, then according to (2) there is an unique operator M taking the range of L into V such that $MLu = v$ for all v in V.
(a) If the range of L is all of W, show also that $LMw = w$ for all w in W.
(b) If $L: V \to V$ and $K: V \to V$ are both invertible, show that so is KL and $(KL)^{-1} = L^{-1}K^{-1}$.

9. Find all solutions of each differential equation below. [*Hint*: Recall from Section 3.2 the technique for solving the linear equation $(D + r(t))[w] = f$ by multiplying through by the integrating factor $e^{R(t)}$, where $R(t) = \int^t r$.]

(a) $(D + 1/t)(D - 1/t)[y] = t$.
(b) $(D - 1/t)(D + 1/t)[y] = t$.
(c) $ty'' + (2 - t)y' - y = e^t$. [*Hint*: Note that $tD^2 + (2 - t)D - 1 = (D - 1)(tD + 1)$.]

10. For each of the linear operators L defined below, find all constants λ such that the equation $Lu = \lambda u$ has a nontrivial solution. For each such value of λ, what is the corresponding solution set of $Lu = \lambda u$?
(a) L has the action D^2 and domain $S = \{u \text{ in } C^2[0, 1] : u(0) = 0, u(1) = 0\}$.
(b) L has the action $-D^2$ and domain $S = \{u \text{ in } C^2[0, \pi] : u'(0) = 0, u(\pi) = 0\}$.

(c) L has the action $-D^2$ and domain

$$S = \{u \text{ in } C^2[-\pi, \pi] : u(-\pi) = u(\pi), u'(-\pi) = u'(\pi)\}$$

11. Let I be the closed interval $a \leq t \leq b$ and let $p(t)$ in $C^1(I)$, $q(t)$ in $C^0(t)$ be given. Now let L be the operator with action $Ly = (py')' + qy$ and domain $S = \{y \text{ in } C^2[a, b] : \alpha y(a) + \beta y'(a) = 0, \gamma y(b) + \delta y'(b) = 0\}$, where the given constants $\alpha, \beta, \gamma,$ and δ are such that at least one element in *each* pair (α, β) and (γ, δ) is nonzero.
(a) Show that $\int_a^b yLz \, dt = \int_a^b zLy \, dt$ for all y, z in S.
(b) Show that if u and v are nontrivial functions such that $Lu = \lambda u$, $Lv = \mu v$ with constants $\lambda \neq \mu$, then $\int_a^b uv \, dt = 0$.

12. Calculate the Green's kernel for each of the operators whose action is given below.

(a) $P(D) = D^n$.

(b) $P(D) = t^2 D^2 + 2tD - 2$, $t > 0$. [*Hint*: Look for solutions of $P(D)[y] = 0$ in the form $y = t^\alpha$ where α is a constant.]

(c) $P(D) = D^3 + 2D^2 - D - 2$.

CHAPTER 7

The Laplace Transform

In this chapter we study a special linear operator from one linear function space to another, an operator that is widely used to construct solutions of differential and integral equations. The operator involves an improper integral and "transforms" problems involving differential and integral equations into problems with a simpler structure—hence the name "integral transform."

7.1 INTRODUCTION TO THE LAPLACE TRANSFORM

An operator is defined by describing its action, its domain (the collection of elements on which it acts), and its range (the results of the action). In this section we briefly describe the *Laplace† Transform* operator and its inverse, and show how the transform may be used to solve initial value problems.

Action of the Transform

The action of the Laplace Transform operator is defined by an improper integral and acts on an appropriate class of functions defined over the half-line $0 \leq t < \infty$. This class includes all polynomials, simple exponentials, and bounded piecewise-continuous functions.

> **Action of the Laplace Transform.**   The Laplace Transform of the function $f(t)$, $0 \leq t < \infty$, is the function $L[f]$ given by
>
> $$L[f](s) = \int_0^\infty e^{-st} f(t)\, dt \tag{1}$$

† Pierre-Simon de Laplace (1749–1827) was a French applied mathematician whose work in celestial mechanics and in probability was vastly influential in the development of these subjects.

which is defined for all real $s > s_0$, where s_0 depends on f. ($s_0 = -\infty$ is a possibility.)

The Laplace Transforms of some simple functions are calculated in the following examples. In each case the reader should keep in mind that the transform $L[f]$ of a function $f(t)$ is a function in the transform variable s.

Example 7.1

Suppose that $f(t)$ is the constant function c for all $t \geq 0$ and that the transform variable s is positive. Then

$$L[f](s) = \int_0^\infty e^{-st} c \, dt = -\frac{c}{s} e^{-st} \bigg]_{t=0}^{t=\infty} = \frac{c}{s}$$

since $\lim\limits_{t \to \infty} e^{-st} = 0$ whenever $s > 0$. See Figure 7.1 for a sketch of f and $L[f]$.

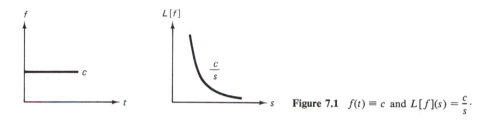

Figure 7.1 $f(t) \equiv c$ and $L[f](s) = \frac{c}{s}$.

Example 7.2

Let $f(t) = t$ for all $t \geq 0$, and suppose that $s > 0$. Then integrating by parts or using a table of integrals gives us

$$L[f](s) = \int_0^\infty e^{-st} t \, dt = -\frac{1}{s^2} e^{-st}(st+1) \bigg|_{t=0}^{t=\infty}$$

Recalling that s is assumed to be positive and using L'Hôpital's Rule, we have

$$\lim_{t \to \infty} e^{-st} t = \lim_{t \to \infty} \frac{t}{e^{st}} = \lim_{t \to \infty} \frac{1}{s e^{st}} = 0$$

Hence we have (see Figure 7.2)

$$L[f](s) = -\frac{1}{s^2} e^{-st}(st+1) \bigg|_{t=0}^{t=\infty} = 0 + \frac{1}{s^2} = \frac{1}{s^2}$$

Example 7.3

Let $f(t) = 5e^{at}$ for all $t \geq 0$, where $a \neq 0$. Then for all $s > a$, we have (see Figure 7.3)

$$L[f](s) = \int_0^\infty e^{-st} 5e^{at} \, dt = \frac{5}{a-s} e^{(a-s)t} \bigg|_{t=0}^{t=\infty} = \frac{5}{s-a}$$

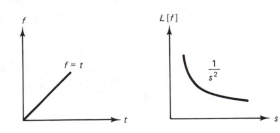

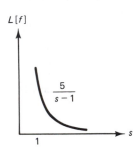
Figure 7.2 $f(t) = t$ and its transform.

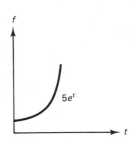

Figure 7.3 $f(t) = 5e^t$, $L[f](s) = \dfrac{5}{s-1}$.

Example 7.4

Let $f(t)$ be the piecewise-continuous function defined by

$$f(t) = \begin{cases} 1, & 0 \le t \le 1 \\ 0, & t > 1 \end{cases}$$

Then for all $s > 0$,

$$L[f](s) = \int_0^\infty e^{-st}f(t)\, dt = \int_0^1 e^{-st}\, dt = -\frac{1}{s}e^{-st}\Big]_{t=0}^{t=1} = \frac{1}{s}\left[1 - e^{-s}\right]$$

which like the previous three transforms, decays to zero as $s \to \infty$. Note that $L[f](s)$ is an infinitely differentiable function of s for $s > 0$, although the original function $f(t)$ is not continuous (see Figure 7.4).

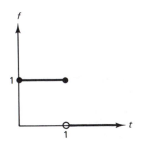

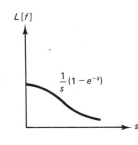

Figure 7.4 Piecewise-continuous function and its transform.

The Inverse Transform

If $L[f]$ and $L[g]$ are defined on an interval, $s > a$, so is $L[\alpha f + \beta g]$ for any real α, β. Moreover, L is a linear operator since

$$L[\alpha f + \beta g] = \alpha L[f] + \beta L[g]$$

Of course, the linear operator L would not be very useful unless a function can be recovered from its transform. The operator that does this is linear; it is denoted by L^{-1} and is called the *inverse Laplace Transform*. It is not easy to construct an explicit formula describing the action of L^{-1}, but as we shall show, we can get along without such a formula by using tables of transforms. In particular, suppose that $f(t)$ and $g(t)$ are both continuous for $t \geq 0$, and suppose that both transforms exist and are identical on a common interval $s \geq a$:

$$L[f](s) \equiv L[g](s)$$

Then it can be shown that $f(t) \equiv g(t)$, $t \geq 0$. Thus if $L[f](s)$ is, say, known to be $1/s^2$ for $s > 0$, the only continuous function f with this transform is $f(t) \equiv t$, $t \geq 0$ (Example 7.2).

We may also go back from a transform to the original function even if the original function is only piecewise continuous, as long as we are willing to tolerate ambiguities at the points of discontinuity. For example, both the function $f(t)$ of Example 7.4 and the function g defined by

$$g(t) = \begin{cases} 1, & 0 \leq t < 1 \\ 2, & t = 1 \\ 0, & t > 1 \end{cases}$$

have as their transforms

$$L[f](s) = L[g](s) = \frac{1}{s}[1 - e^{-s}]$$

An Initial Value Problem

The usefulness of the Laplace Transform in solving linear initial value problems is suggested by the following example. The central point of the example is that it shows that a differential equation can be transformed into an algebraic equation which involves no derivatives at all.

Example 7.5

Consider the initial value problem

$$y'(t) + ay(t) = f(t), \qquad y(0) = y_0 \tag{2}$$

where a is a real constant and f is, say, piecewise continuous on $[0, \infty)$. We can, of course, solve (2) by the integrating factor approach of Chapter 3. The method of Laplace Transforms may also be applied to solve (2) in the following way. Assuming that $L[y']$, $L[y]$, and $L[f]$ are defined on (s_0, ∞) for some s_0, apply L to each side of (2) and use linearity to obtain

$$L[y'](s) + aL[y](s) = L[f](s), \qquad s > s_0 \qquad (3)$$

Now, assuming that integration by parts is valid in this setting and that $\lim_{t \to \infty} e^{-st}y(t) = 0$ for each $s > s_0$, we have that

$$L[y'](s) = \int_0^\infty e^{-st}y'(t)\, dt = e^{-st}y(t)\,\Big|_{t=0}^{t=\infty} + s\int_0^\infty e^{-st}y(t)\, dt$$

$$= -y(0) + sL[y](s)$$

Thus (3) becomes the algebraic equation for the transform $L[y]$,

$$sL[y](s) - y(0) + aL[y](s) = L[f](s) \qquad (4)$$

Solving (4) algebraically for $L[y](s)$, we have that

$$L[y](s) = \frac{y(0)}{s+a} + \frac{L[f](s)}{s+a} \qquad (5)$$

At this point, we know the Laplace Transform of the solution of (2) even if we do not yet have the solution itself. Assuming that L^{-1} exists, it is linear and we can apply L^{-1} to each side of (5) to obtain

$$y(t) = y(0)L^{-1}\left[\frac{1}{s+a}\right](t) + L^{-1}\left[\frac{L[f]}{s+a}\right](t) \qquad (6)$$

The inverse terms in (6) may be determined by inspection of the Laplace Transform tables on the inside covers.

$$L^{-1}\left[\frac{1}{s+a}\right](t) = e^{-at} \qquad \text{(use number 3 of Table II with } n = 1 \text{ and } a \text{ replaced by } -a\text{)}$$

$$L^{-1}\left[\frac{L[f]}{s+a}\right](t) = \int_0^t e^{-a(t-u)}f(u)\, du \qquad \text{(use number 8 of Table I with } f_1(t) \equiv e^{-at}, f_2 \equiv f\text{)}$$

Thus we have that (6) becomes

$$y(t) = y_0 e^{-at} + \int_0^t e^{-a(t-u)}f(u)\, du \qquad (7)$$

The advantage of the transform approach is that it reduces the problem of solving an initial value problem such as (2) to the inspection of a table of transforms. In later sections of this chapter we give a number of applications of this approach.

Comments

Example 7.5 showed that the transform of a derivative does not involve a derivative. It is this single fact that makes the Laplace Transform a useful alternative method for solving linear differential equations. But the method is practical only if a sufficiently large table of transforms is available. In succeeding sections we show how some of the entries in the Laplace Transform tables on the inside covers are determined.

PROBLEMS

1. Find the Laplace Transform of each of the following functions, where a and b are real constants and n is a positive integer. Specify the largest interval $s > s_0$ on which the transform is defined and sketch the transforms.

    (a) $3t - 5$.   **(b)** t^2.   **(c)** t^n.   **(d)** $\cos bt$.   **(e)** te^{at}.   **(f)** $t \sin at$.

2. Compute the Laplace Transform for each of the following functions and determine the interval (s_0, ∞) on which the transform is defined. Sketch the transforms.

    (a) $f(t) = \begin{cases} \sin t, & 0 \leq t \leq \pi \\ 0, & t > \pi. \end{cases}$ **(b)** $f(t) = \begin{cases} 0, & 0 \leq t \leq 1 \\ t, & t > 1. \end{cases}$

    (c) $f(t) = \begin{cases} t, & 0 \leq t \leq 1 \\ 0, & t > 1. \end{cases}$

3. Solve the following initial value problems using the Laplace Transform.

    (a) $y' + 2y = 0$, $\quad y(0) = 1$.   **(b)** $y' + 2y = e^{-3t}$, $\quad y(0) = 5$.

    (c) $y' + 2y = e^{it}$, $\quad y(0) = 0$. [*Hint*: Treat i as a constant.]

7.2 THE CALCULUS OF THE TRANSFORM, I

It would not be efficient to resort to the basic definition of the Laplace Transform each time we wish to transform a function. To speed up the process we shall develop a calculus of transforms. With this calculus we may express the transform of a combination of functions in terms of the known transforms of the component functions. Thus we need to develop a list of the transforms of elementary functions and some general rules for computing Laplace Transforms and for passing back from the s-domain to the t-domain (i.e., for calculating the inverse transform). In this respect what we have done in Section 7.1 (and are about to do here) for the Laplace Transform calculus resembles the construction of the differential and integral calculus—first the underlying ideas and definitions are presented and then the formulas and rules are developed.

   By the end of this section a number of transforms will have been calculated and several rules of transform and inverse transform formulated and proved. For ease of reference, we have placed these transforms and rules, as well as many others, in the tables on the inside covers.

Smoothness Properties

From this point on we need to be more specific about the functions that the Laplace Transform acts upon.

> **Piecewise-Continuous Functions of Exponential Order.**   Let us denote by $\mathscr{E}$ the collection of all piecewise-continuous functions on $[0, \infty)$ which satisfy the following property: For each f in $\mathscr{E}$ there are positive constants M and α such that $|f(t)| \leq Me^{\alpha t}$ for all $0 \leq t < \infty$ (such functions are said to be of *exponential order*).

Unless stated otherwise, every function f considered in this chapter will be assumed to be in $\mathscr{E}$. This assumption has a very important consequence, which we state without proof.

> **Smoothness and Decay Property for $L[f]$.**   For each f in $\mathscr{E}$ the Laplace Transform $\phi(s) = L[f](s)$ is infinitely differentiable on some half-line (s_0, ∞), and $\phi(s) \to 0$ as $s \to \infty$.

Thus the Laplace Transforms of functions in $\mathscr{E}$ are very well behaved, a fact that we frequently use.

Transforms of Derivatives

As a computation in Section 7.1 suggests, the transform of a piecewise-differentiable function f with a piecewise-continuous derivative is given by $L[f'](s) = sL[f] - f(0)$. We may extend the formula above to obtain the transforms of the higher-order derivatives.

> **Transform of the nth Derivative.**   For some integer $n \geq 1$ let $f^{(n-1)}$ belong to $C^0(0,\infty)$ and $f^{(n)}$ to $\mathscr{E}$. Then $f^{(k)}$ belongs to $\mathscr{E}$, $k = 0, 1, \ldots, n - 1$, and for all sufficiently large s,
>
> $$L[f^{(n)}] = s^n L[f] - s^{n-1}f(0) - \cdots - f^{(n-1)}(0) \qquad (1)$$

Proof. The hypotheses imply that $f^{(k)}$ is in $C^0(I)$ for each $k = 0, 1, 2, \ldots, n - 1$. Writing $f^{(k)}(t) = f^{(k)}(0) + \int_0^t f^{(k+1)}(x)\, dx$ and using the fact that $f^{(k+1)}$ is in $\mathscr{E}$, we see that $f^{(k)}$ is also in $\mathscr{E}$. Thus, $f^{(k)}$ is in $\mathscr{E}$ for each $k = 0, 1, 2, \ldots, n - 1$. Formula (1) is proved by induction after integration-by-parts formula† is used to establish the induction step:

† Recall that the *integration-by-parts formula* $\int_a^b f(x)g'(x)\, dx = f(b)g(b) - f(a)g(a) - \int_a^b f'(x)g(x)\, dx$ holds for any functions f, g which are continuous on $[a, b]$, but f' and g' need only be piecewise continuous.

$$L[f^{(N)}] = \int_0^\infty e^{-st} f^{(N)}(t)\, dt = e^{-st} f^{(N-1)}(t) \Big|_{t=0}^{t=\infty} + s \int_0^\infty e^{-st} f^{(N-1)}(t)\, dt$$

$$= -f^{(N-1)}(0) + sL[f^{(N-1)}]$$

Thus if (1) holds for all $n \leq N - 1$, then the formula above shows that (1) also holds when $n = N$.

There is a dual formula in the transform calculus, a formula which tells us what transforms into $\phi^{(n)}(s)$, where $\phi = L[f]$, rather than what $f^{(n)}(t)$ transforms into.

Inverse Transform of the nth Derivative: Let f belong to $\mathscr{E}$ and let $\phi = L[f]$. Then

$$\phi^{(n)}(s) = L[(-1)^n t^n f(t)] \tag{2}$$

This is a consequence of the infinite smoothness property of the Laplace Transform. Before we see how we can apply these results, let us work out some examples.

Example 7.6

We have that for $s > 0$,

$$L[\cos at] = \int_0^\infty e^{-st} \cos at\, dt = \lim_{T \to \infty} \int_0^T e^{-st} \cos at\, dt$$

$$= \lim_{T \to \infty} \left\{ \frac{e^{-st}}{s^2 + a^2} (a \sin at - s \cos at) \Big|_0^T \right\}$$

$$= \lim_{T \to \infty} \left\{ \frac{e^{-sT}}{s^2 + a^2} (a \sin aT - s \cos aT) + \frac{s}{s^2 + a^2} \right\} = \frac{s}{s^2 + a^2}$$

Example 7.7

Let $f(t) = -(1/a) \cos at$. Now since $f'(t) = \sin at$, we have by (1) that for all sufficiently large s,

$$L[\sin at] = \frac{1}{a} + sL\left[-\frac{1}{a} \cos at \right] = \frac{1}{a} - \frac{s}{a}\frac{s}{s^2 + a^2} = \frac{a}{s^2 + a^2}$$

Example 7.8

Let $f(t) = t^n$, $n = 0, 1, 2, \ldots$; then $f^{(n)}(t) = n!$. From (1) it follows that

$$L[f] = \frac{1}{s^n} \{ L[f^{(n)}] + s^{n-1} f(0) + \cdots + f^{(n-1)}(0) \}$$

and hence we have that

$$L[t^n] = \frac{1}{s^n} \left\{ \frac{n!}{s} + 0 + \cdots + 0 \right\} = \frac{n!}{s^{n+1}}$$

Example 7.9

That $L[e^{at}] = (s - a)^{-1}$, $s > a$, follows directly from the definition of the Laplace Transform.

Example 7.10

Solve the initial value problem $y'' - y = 1$, $y(0) = 0$, $y'(0) = 1$. We know on theoretical grounds that the problem is uniquely solvable, and that the solution and its derivatives of all orders are in $\mathscr{E}$. Thus we may transform both sides of the differential equation using (1) and the initial data to obtain that

$$L[y] = \frac{1}{s(s-1)} = \frac{1}{s-1} - \frac{1}{s}, \qquad s > 1$$

Thus from Example 7.8 with $n = 0$ and Example 7.9 with $a = 1$, we have that $y(t) = e^t - 1$.

Of course, the methods of Chapter 5 could have been used to solve the initial value problem just as easily. However, we shall soon have examples where the transform technique is much more efficient than the characteristic polynomial, variation of parameter, Green's function approach of earlier chapters. Whatever method is used, the same solution will be constructed as long as the Existence and Uniqueness Theorem applies.

Another advantage of the transform approach to the solution of initial value problems is shown in Example 7.10. The transform automatically involves the initial data, unlike the situation with earlier methods, where the data were used almost as an afterthought to evaluate constants of integration. Of course, the transform approach will be successful only if we can carry the transform of the solution of the initial value problem back into the time domain.

Transforms of Integrals

We may transform an integral of a function as easily as a derivative.

Transform of an Integral. Let f belong to $\mathscr{E}$, and $a \geq 0$. Then $\int_a^t f(x)\,dx$ belongs to $\mathscr{E}$, and

$$L\left[\int_a^t f(x)\,dx\right] = \frac{1}{s}L[f] - \frac{1}{s}\int_0^a f(x)\,dx \tag{3}$$

Proof. Obviously $\int_a^t f(x)\,dx$ is a continuous function on $[0, \infty]$. Integration by parts yields (3) as follows:

$$L\left[\int_a^t f(x)\,dx\right] = \int_0^\infty e^{-st}\left(\int_a^t f(x)\,dx\right)dt$$

$$= \left[-\frac{1}{s}e^{-st}\int_a^t f(x)\,dx\right]_{t=0}^{t=\infty} + \frac{1}{s}\int_0^\infty e^{-st}f(t)\,dt$$

$$= -\frac{1}{s}\int_0^a f(x)\,dx + \frac{1}{s}L[f]$$

Example 7.11

Observe that $\int_0^t xe^x\,dx = te^t - e^t + 1$. Thus, from (3),

$$L\left[\int_0^t xe^x\,dx\right] = \frac{1}{s}L[te^t]$$

on the one hand, and $L\left[\int_0^t xe^x\,dx\right] = L[te^t] - L[e^t] + L[1]$, on the other hand. We can solve for $L[te^t]$ to find that

$$L[te^t] = \frac{1}{(s-1)^2}, \qquad s > 1$$

This result also follows immediately from (2) and Example 7.9.

Shifting Theorems: The Heaviside Function

The next three theorems are frequently used in computing transforms of functions that occur in applications.

First Shifting Theorem.   If f belongs to $\mathscr{E}$, then $e^{at}f(t)$ also belongs to $\mathscr{E}$ and

$$L[e^{at}f(t)] = L[f](s-a) \tag{4}$$

The proof is immediate from the definition of the Laplace Transform operator. We see from (4) that multiplying a function in the time domain by e^{at} *shifts* the transform variable s by the amount a.

Example 7.12

With the help of Example 7.6 we see that

$$L[e^{-2t}\cos 3t] = \frac{s+2}{(s+2)^2+9}$$

Example 7.13

Let us try to find a solution $f(t)$ to the equation

$$L[f] = (2s + 3)/(s^2 - 4s + 20)$$

We write

$$\frac{2s + 3}{s^2 - 4s + 20} = \frac{2(s - 2) + 7}{(s - 2)^2 + 16} = 2\left[\frac{s - 2}{(s - 2)^2 + 16}\right] + \frac{7}{4}\left[\frac{4}{(s - 2)^2 + 16}\right]$$

and hence it is clear from the First Shifting Theorem and Examples 7.6 and 7.7 that $f(t) = 2e^{2t} \cos 4t + \frac{7}{4}e^{2t} \sin 4t$.

One of the more useful functions in the applications and in the transform calculus is the function defined below.

Heaviside Function.   The function

$$H(t) = \begin{cases} 1, & t \geq 0 \\ 0, & t < 0 \end{cases} \tag{5}$$

is the *Heaviside* or *unit step function*.

Observe that for any real number a,

$$H(t - a) = \begin{cases} 1, & t \geq a \\ 0, & t < a, \end{cases} \qquad H(a - t) = \begin{cases} 0, & t > a \\ 1, & t \leq a \end{cases}$$

If we agree that any function f belonging to $\mathscr{E}$ is defined for all real numbers t simply by defining $f(t)$ arbitrarily for $t < 0$, then $H(t - a)f(t - a)$ is just f translated a units along the t-axis and reduced to zero for $t < a$. With this in mind, we can derive a kind of dual to the First Shifting Theorem.

Second Shifting Theorem.   Let f belong to $\mathscr{E}$, and $a \geq 0$. Then

$$e^{-as}L[f] = L[H(t - a)f(t - a)] \tag{6}$$

Proof. Observe that it is now the t-variable rather than the s-variable which is shifted. It is not hard to prove (6):

$$L[H(t - a)f(t - a)] = \int_0^\infty e^{-st}H(t - a)f(t - a)\,dt = \int_a^\infty e^{-st}f(t - a)\,dt$$

$$= \int_0^\infty e^{-s(x+a)}f(x)\,dx = e^{-as}L[f]$$

where the variables change $x = t - a$ was used in the last integral.

Example 7.14

Consider the function

$$f(t) = \begin{cases} 1, & 0 < t < 1 \\ 2 - t, & 1 < t < 2 \\ 0, & t > 2 \end{cases}$$

With the help of the Heaviside function, we write

$$f(t) = \{H(t) - H(t-1)\} + (2-t)\{H(t-1) - H(t-2)\}$$
$$= H(t) - (t-1)H(t-1) + (t-2)H(t-2)$$

Thus we have that

$$L[f] = L[H(t)] - L[(t-1)H(t-1)] + L[(t-2)H(t-2)]$$
$$= \frac{1}{s} - e^{-s}\frac{1}{s^2} + e^{-2s}\frac{1}{s^2} = \frac{s - e^{-s} + e^{-2s}}{s^2}, \qquad s > 0$$

The Second Shifting Theorem may be recast in another useful form.

Third Shifting Theorem.   Let f belong to $\mathscr{E}$, and $a \geq 0$. If we set $g(t-a) = f(t)$, then

$$e^{-as}L[g(t)] = L[H(t-a)f(t)] \tag{7}$$

Note that $g(t) = f(t+a)$.

Example 7.15

According to the Third Shifting Theorem,

$$L[H(t-a)] = L[H(t-a) \cdot 1] = e^{-as}\frac{1}{s} \tag{8}$$

which could also be obtained easily enough by direct integration.

Incidentally, it is useful to have a corresponding expression for $H(a-t)$. By integration we have, for $a > 0$,

$$L[H(a-t)] = \int_0^a e^{-st}\,dt = \frac{1 - e^{-as}}{s}, \qquad s > 0 \tag{9}$$

Example 7.16

According to the Third Shifting Theorem, we have that

$$L[H(t-a)\sin t] = e^{-as}L[\sin(t+a)] = e^{-as}L[\cos a \sin t + \sin a \cos t]$$

$$= e^{-as}\left\{\frac{\cos a}{s^2+1} + \frac{s \sin a}{s^2+1}\right\}$$

where we also used the results of Examples 7.6 and 7.7.

The Effect of a Thirty-Day Harvest on a Population

Suppose that the growth of a population may be modeled by a Malthusian law of exponential change. What will be the effect of "harvesting" the population at a constant rate for a fixed span of time? Refer to the problem set of Section 3.3 for a similar problem but with perpetual harvesting.

Let $y(t)$ denote the size of the population at time t, where time is measured, say, in days, and suppose that $y(0) = A$, a positive constant. Let us suppose that the rate constant is another positive constant, k, measured in (days)$^{-1}$, and that the harvest is carried out at the rate of h "individuals" per day over a 30-day period. A model of this is given by the initial value problem

$$y'(t) = ky(t) - hH(30-t)$$
$$y(0) = A$$

(10)

Our goal is to find an expression for $y(t)$ for all $t \geq 0$.

Applying the Laplace Transform, we have that

$$sL[y] - A = kL[y] - \frac{h}{s}(1 - e^{-30s})$$

$$L[y] = \frac{A}{s-k} - \frac{h}{s(s-k)}(1 - e^{-30s})$$

(11)

$$= \frac{A}{s-k} + \frac{h}{k}\left(\frac{1}{s} - \frac{1}{s-k}\right)(1 - e^{-30s}), \qquad s > k$$

where we have used the identity

$$\frac{1}{s(s-k)} = -\frac{1}{k}\left(\frac{1}{s} - \frac{1}{s-k}\right)$$

Rearranging the terms on the right of (11) so that we may find the inverse transforms more easily, we have that

$$L[y] = \frac{A - h/k}{s-k} + \frac{h}{k}\frac{1}{s}(1 - e^{-30s}) + \frac{h}{k}e^{-30s}\frac{1}{s-k}$$

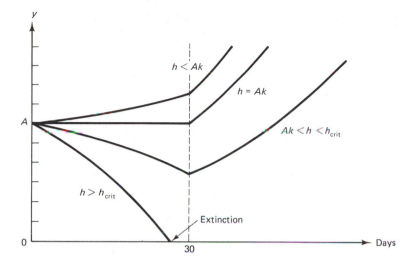

Figure 7.5   Harvested population.

Using Example 7.9 and formulas (6) and (9), we may move back into the t-domain:

$$y(t) = \left(A - \frac{h}{k}\right) e^{kt} + \frac{h}{k} H(30 - t) + \frac{h}{k} H(t - 30)e^{k(t-30)} \tag{12}$$

The last term in (12) is used for $t > 30$ (but not for $t = 30$) so that $y(t)$ is continuous at $t = 30$. The equations represent the changing population only for $y(t) > 0$. Thus, the solution makes sense only for those values of the parameters A, h, and k for which $y(t)$ remains positive for all $t \geq 0$. If $h \leq Ak$, there clearly is no difficulty. If $h > Ak$, it may happen that at some time while harvesting is going on, the population becomes extinct. It may be shown that extinction within 30 days occurs only if the harvesting rate and the initial population satisfy the inequality

$$h \geq h_{crit} = \frac{kA}{1 - e^{-30k}} \tag{13}$$

In this case, the model is not valid after the time of extinction. In Figure 7.5 we have sketched the population curves for various harvesting rates.

PROBLEMS

The following problem(s) may be more challenging: 1(j), 6, and 8.

1. Find the Laplace Transform of each of the following functions.

(a) $\sinh at$.   (b) $\cosh at$.   (c) $t^2 e^{at}$.

(d) $(1 + 6t)e^{at}$.   (e) $te^{2t}f'(t)$.   (f) $(D^2 + 1)f(t)$.

(g) $(t + 1)H(t - 1)$.   (h) $(t - 2)[H(t - 1) - H(t - 3)]$.

(i) $e^{at}[H(t - 1) - H(t - 2)]$.   (j) $te^{at}H(t - 1)$.

2. Use the Laplace Transform to solve each of the following initial value problems.
   (a) $y'' - y' - 6y = 0$, $y(0) = 1$, $y'(0) = -1$.
   (b) $y'' + y = \sin t$, $y(0) = 0$, $y'(0) = 1$.
   (c) $y'' - 2y' + 2y = 0$, $y(0) = 0$, $y'(0) = 1$.
   (d) $y'' + 4y' + 4y = e^t$, $y(0) = 1$, $y'(0) = 1$.
   (e) $y'' - 2y' + y = H(t - 1)$, $y(0) = 1$, $y'(0) = 0$.
   (f) $y'' + 2y' - 3y = H(1 - t)$, $y(0) = 1$, $y'(0) = 0$.

3. A sample of a radioactive element y decays at a rate proportional to the amount present. Suppose that at some time $t = a$, a constant flow of y is allowed to enter the sample from outside and at a later time, $t = b$, this flow is stopped.
   (a) Show that this system can be modeled by

   $$y' = -k_1 y + k_2[H(t - a) - H(t - b)], \qquad 0 < a < b$$

   where k_1 is the radioactive decay rate constant and k_2 is the amount of y coming into the sample per unit time for $a \le t < b$.
   (b) Use the Laplace Transform to find $y(t)$ when $y(0) = y_0 > 0$. Sketch the graph of $y(t)$.

4. Consider the basic LC circuit whose source of electromotive force can be turned on and off by a switch.
   (a) Use the Laplace Transform to find the charge $q(t)$ if the switch is turned on at time $t = a$:

   $$L\frac{d^2 q}{dt^2} + \frac{1}{C} q = E_0 H(t - a), \qquad q(0) = 0, \quad q'(0) = 0$$

   (b) Find $q(t)$ if the switch is turned on at $t = a$ and off at $t = b$:

   $$L\frac{d^2 q}{dt^2} + \frac{1}{C} q = E_0[H(t - a) - H(t - b)], \qquad q(0) = 0, \quad q'(0) = 0$$

5. (a) Repeat Problem 4(a) for the RLC circuit:

   $$L\frac{d^2 q}{dt^2} + R\frac{dq}{dt} + \frac{1}{C} q = E_0 H(t - a), \qquad q(0) = 0, \quad q'(0) = 0$$

   (b) Repeat Problem 4(b) for the RLC circuit:

   $$L\frac{d^2 q}{dt^2} + R\frac{dq}{dt} + \frac{1}{C} q = E_0[H(t - a) - H(t - b)], \qquad q(0) = 0, \quad q'(0) = 0$$

6. [*Maintenance of a Game Species*]. In the model of the 30-day harvest, find a relation among A, h, and k which will ensure that exactly 330 days after the end of the harvest, the population will once more be at the level of A.

7. Use the transform tables on the inside covers to find $L[f]$ if f is given, and f if $L[f]$ is given. Identify the formula used.
   (a) $f(t) = \begin{cases} 0, & 0 \le t < 2, t > 5 \\ 3, & 2 \le t \le 5 \end{cases}$.
   (b) $f(t) = \begin{cases} 2 \sin t, & 0 \le t \le \pi \\ 0, & t > \pi \end{cases}$.
   (c) $f(t) = t^{12} e^{5t}$.
   (d) $f(t) = 6t \sin 3t$.
   (e) $f(t) = J_0(t)$.
   (f) $f(t) = \sqrt{s + 2} - \sqrt{s}$.

(g) $L[f] = \ln \dfrac{s+2}{s+10}$.

(h) $L[f] = e^{-\sqrt{s}}$.

(i) $L[f] = \dfrac{1}{s} g\left(\dfrac{1}{s}\right)$, $g = L[h]$.

(j) $L[f] = \dfrac{s^2+s+1}{s(s^2-1)(s+2)}$.

8. [*Difference Equations*]. Solve the difference equation $3x(t) - 4x(t-1) = 1$, where $x(t) = 0$ if $t \leqq 0$. [*Hint*: Use III.9 in the tables.]

7.3 THE CALCULUS OF THE TRANSFORM, II

The driving force of a mechanical or electrical system often has the form of a periodic function. Any practical calculus of transforms must have a transform formula adequate to handle periodic functions ranging from smoothly contoured sinusoids to the angular pulses of a square or triangular wave train. Fortunately, there is such a formula, and we shall derive it and then apply it to determine the response of an inductor–capacitor circuit to a train of triangular pulses of voltage.

We have focused our attention mostly on techniques for calculating the transform of a function defined in the t-domain. However, this is only half of what is needed when using the Laplace transform to solve, say, an initial value problem. Once the transform of the solution has been found, we still have to invert the transform to bring the solution back into the t-domain. One of the more useful techniques for "preparing" a function in the s-domain before inverting the transform is the method of partial fractions, a method also used in the integral calculus to simplify integration. As we shall show, after a quotient of polynomials in the s-variable is decomposed into its partial fractions, the inverse Laplace Transform may be calculated quite easily.

We shall begin, however, with a review of a series method for preparing functions in the s-domain before their inverse transforms are calculated.

Geometric Series

If $|x| < 1$, then

$$\frac{1}{1-x} = 1 + x + \cdots + x^n + \cdots = \sum_{n=0}^{\infty} x^n$$

The series is said to be the *geometric series expansion* of the function $(1-x)^{-1}$. We shall use the expansion primarily with $x = \pm e^{-as}$, where a and s are both positive. Under these circumstances, we have that

$$\frac{1}{1-e^{-as}} = \sum_{n=0}^{\infty} e^{-ans} \tag{1}$$

$$\frac{1}{1+e^{-as}} = \sum_{n=0}^{\infty} (-1)^n e^{-ans} \tag{2}$$

Transform of a Periodic Function

The transform formula for a periodic function is surprisingly simple.

Transforming Periodic Functions.   Let f be piecewise continuous and periodic with period p. Then

$$L[f] = \frac{1}{1 - e^{-ps}} \int_0^p e^{-st} f(t)\, dt \tag{3}$$

Proof. We have that

$$L[f] = \int_0^p e^{-st} f(t)\, dt + \int_p^{2p} e^{-st} f(t)\, dt + \cdots$$

Making the translation $t = x + np$ in the nth integral, $n = 0, 1, 2, \ldots$, we get

$$L[f] = \sum_{n=0}^{\infty} \left\{ \int_0^p e^{-sx} f(x)\, dx \right\} e^{-psn} = \frac{\int_0^p e^{-sx} f(x)\, dx}{1 - e^{-ps}}$$

since $(1 - e^{-ps})^{-1} = \sum_{n=0}^{\infty} e^{-psn}$ by geometric series.

Example 7.17

Let us see what the transform of the square wave train f of Figure 7.6 is like. In this case the period is 2 and we have that

$$\int_0^2 e^{-st} f(t)\, dt = \int_0^1 e^{-st}\, dt - \int_1^2 e^{-st}\, dt = \frac{1}{s}(1 - 2e^{-s} + e^{-2s}) = \frac{1}{s}(1 - e^{-s})^2$$

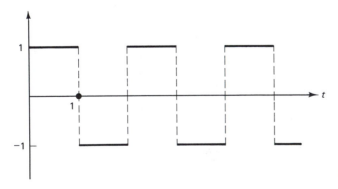

Figure 7.6 Square wave train.

Hence, by the theorem above,†

$$L[f] = \frac{1}{1 - e^{-2s}} \frac{1}{s} (1 - e^{-s})^2$$

$$= \frac{(1 - e^{-s})^2}{s(1 - e^{-s})(1 + e^{-s})} = \frac{1 - e^{-s}}{s(1 + e^{-s})}, \qquad s > 0 \qquad (4)$$

It is sometimes useful to expand the term $(1 + e^{-s})^{-1}$ in a geometric series. We then have that the Laplace Transform of the square wave train of Figure 7.6 is given by

$$L[f] = \frac{1 - e^{-s}}{s} \sum_{n=0}^{\infty} (-1)^n e^{-ns} = \frac{1}{s} (1 - e^{-s})(1 - e^{-s} + e^{-2s} - e^{-3s} + \cdots)$$

$$= \frac{1}{s} (1 - 2e^{-s} + 2e^{-2s} - \cdots) \qquad (5)$$

$$= \frac{1}{s} \left[1 + 2 \sum_{n=1}^{\infty} (-1)^n e^{-ns} \right]$$

Partial Fractions

Quotients of polynomials in the s-variable occur quite frequently as Laplace Transforms. The method of partial fractions may be used to decompose such a quotient into a linear combination of simple quotients (or "fractions"), each of which may be inverted more easily than the original quotient. The process of partial fractions is based on the unique factorization of a polynomial. Recall that any polynomial

$$Q(s) = s^n + a_{n-1}s^{n-1} + \cdots + a_1 s + a_0$$

may be factored in just one way (up to rearrangement of the order of the factors) as

$$Q(s) = (s - r_1)^{m_1} \cdots (s - r_p)^{m_p} = \prod_{k=1}^{p} (s - r_k)^{m_k}$$

where $r_1, \ldots, r_p$ are the distinct roots of $P(z)$ of respective multiplicities m_1, $\ldots, m_p$, and $m_1 + \cdots + m_p = n$. Some or all of the roots may be complex, but if the coefficients of the polynomial are real, the complex roots occur in conjugate pairs.

---

† The transform may be shown to be the function $(1/s) \tanh (s/2)$ by using the definition of the hyperbolic tangent:

$$\tanh x = \frac{\sinh x}{\cosh x} = \frac{e^x - e^{-x}}{e^x + e^{-x}}$$

The decomposition of a quotient into its partial fractions applies to quotients of polynomials in which the degree of the numerator is less than that of the denominator. Specifically, we have the following algebraic result.

Partial Fractions.   Let $P(s)$ and $Q(s)$ be polynomials, where $\deg P < \deg Q$ and $Q(s) = \prod_{k=1}^{p} (s - r_k)^{m_k}$. Then there exist unique constants a_{jk} such that

$$\frac{P(s)}{Q(s)} = \sum_{k=1}^{p} \left\{ \sum_{j=1}^{m_k} \frac{a_{jk}}{(s - r_k)^j} \right\} \qquad (6)$$

The terms $a_{jk}/(s - r_k)^j$ are the *partial fractions*.

The summation form of (6) indicates its utility in the transform calculus. For, by the linearity of L^{-1} (assuming that L^{-1} exists),

$$L^{-1}\left[\frac{P(s)}{Q(s)}\right] = \sum_{k=1}^{p} \left\{ \sum_{j=1}^{m_k} a_{jk} L^{-1}\left[\frac{1}{(s - r_k)^j}\right]\right\}$$

$$= \sum_{k=1}^{p} \left\{ \sum_{j=1}^{m_k} a_{jk}\left[\frac{t^{j-1}}{(j - 1)!}e^{r_k t}\right]\right\}$$

and we have the exact form of the inverse transform.

The following examples illustrate the techniques by which the coefficients a_{jk} of the partial fractions may be found.

Example 7.18

Since the cubic polynomial $s^3 - 2s^2 - s + 2$ has the simple roots -1, 1, and 2, we have that

$$\frac{2s^2 - 7s + 3}{s^3 - 2s^2 - s + 2} = \frac{a}{s + 1} + \frac{b}{s - 1} + \frac{c}{s - 2}$$

$$= \frac{a(s - 1)(s - 2) + b(s + 1)(s - 2) + c(s + 1)(s - 1)}{s^3 - 2s^2 - s + 2}$$

Canceling the denominators on each side and multiplying out the products in the numerator on the right, we have that

$$2s^2 - 7s + 3 = (a + b + c)s^2 + (-3a - b)s + (2a - 2b - c)$$

Since the equality must hold for all s, the coefficients of like powers of s on the two sides must match. Hence

$$2 = a + b + c, \qquad -7 = -3a - b, \qquad 3 = 2a - 2b - c$$

and $a = 2$, $b = 1$, $c = -1$. Thus

$$\frac{2s^2 - 7s + 3}{s^3 - 2s^2 - s + 2} = \frac{2}{s + 1} + \frac{1}{s - 1} - \frac{1}{s - 2}$$

Multiple or complex roots cause no difficulties, as the following examples show.

Example 7.19

$$\frac{6s^2 + 8s - 5}{s^3 + 3s^2 + 3s + 1} = \frac{6s^3 + 8s - 5}{(s + 1)^3} = \frac{a}{s + 1} + \frac{b}{(s + 1)^2} + \frac{c}{(s + 1)^3}$$

for some constants a, b, and c. The method of Example 7.18 may be used to show without difficulty that $a = 6$, $b = -4$, and $c = -7$.

Example 7.20

The quadratic polynomial $s^2 + 2s + 2$ has complex roots $-1 - i$ and $-1 + i$. Thus

$$\frac{1}{s(s^2 + 2s + 2)} = \frac{1}{s(s + 1 + i)(s + 1 - i)} = \frac{a}{s} + \frac{b}{s + 1 + i} + \frac{c}{s + 1 - i}$$

Hence we must have that

$$1 = a(s + 1 + i)(s + 1 - i) + bs(s + 1 - i) + cs(s + 1 + i)$$
$$= [a + b + c]s^2 + [2a + (1 - i)b + (1 + i)c]s + 2a$$

Thus

$$0 = a + b + c, \qquad 0 = 2a + (1 - i)b + (1 + i)c, \qquad 1 = 2a$$

and $a = \frac{1}{2}$, $b = -(1 + i)/4$, $c = -(1 - i)/4$. Hence

$$\frac{1}{s(s^2 + 2s + 2)} = \frac{1/2}{s} - \frac{(1 + i)/4}{s + 1 + i} - \frac{(1 - i)/4}{s + 1 - i}$$

The appearance of complex quantities causes no more problems here than it did in Chapters 5 and 6. The next example shows just how the complex terms drop out when calculating the inverse transform if the original quotient contains only real polynomials. We shall make use of Euler's identity for $e^{(a+bi)t}$.

Example 7.21 [*Continuation of Example 7.20*]

$$L^{-1}\left[\frac{1}{s(s^2 + 2s + 2)}\right] = \frac{1}{2}L^{-1}\left[\frac{1}{s}\right] - \frac{1 + i}{4}L^{-1}\left[\frac{1}{s + 1 + i}\right]$$

$$- \frac{1 - i}{4}L^{-1}\left[\frac{1}{s + 1 - i}\right]$$

$$= \frac{1}{2} - \frac{1 + i}{4}e^{-(1+i)t} - \frac{1 - i}{4}e^{(-1+i)t}$$

$$= \frac{1}{2} - \frac{1}{2}e^{-t}(\sin t + \cos t)$$

A Driven Circuit

The methods given above may be used to determine the current in the circuit sketched in Figure 7.7. where the driving voltage has the form of a triangular pulse train. The model for the response current $I(t)$ of the circuit to the input voltage $E(t)$ is given by the initial value problem

$$I''(t) + \omega^2 I(t) = E'(t)/L$$
$$I(0) = 0, \qquad I'(0) = 0 \tag{7}$$

where we have set $\omega^2 = 1/LC$. We assume that I is measured in amperes, the inductance L in henries, C in farads, time in seconds, and E in volts. Using Example 7.17 and other transform formulas of this and the preceding section, we have that

$$s^2 L[I] + \omega^2 L[I] = 2 \frac{1 - e^{-s}}{s(1 + e^{-s})}, \qquad s > 0$$

or, solving for the transform and using (5),

$$L[I] = \frac{2}{s(s^2 + \omega^2)} \left\{ 1 + 2 \sum_{n=1}^{\infty} (-1)^n e^{-ns} \right\}$$

$$= \frac{2}{s(s^2 + \omega^2)} + 4 \sum_{n=1}^{\infty} (-1)^n \frac{e^{-ns}}{s(s^2 + \omega^2)} \tag{8}$$

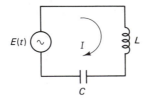

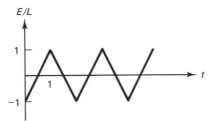

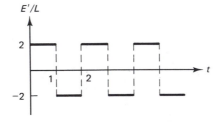

Figure 7.7 Inductor-capacitor circuit.

Thus if we determine $L^{-1}[1/s(s^2 + \omega^2)]$, we may use the Second Shifting Theorem to find $I(t)$.

Let us decompose $1/s(s^2 + \omega^2)$ into its partial fractions:

$$\frac{1}{s(s^2 + \omega^2)} = \frac{1}{s(s + \omega i)(s - \omega i)} = \frac{1/\omega^2}{s} - \frac{1/2\omega^2}{s + \omega i} - \frac{1/2\omega^2}{s - \omega i}$$

Hence

$$L^{-1}\left[\frac{1}{s(s^2 + \omega^2)}\right] = \frac{1}{\omega^2}L^{-1}\left[\frac{1}{s}\right] - \frac{1}{2\omega^2}L^{-1}\left[\frac{1}{s + \omega i}\right] - \frac{1}{2\omega^2}L^{-1}\left[\frac{1}{s - \omega i}\right] \quad (9)$$

$$= \frac{1}{\omega^2} - \frac{1}{2\omega^2}e^{-\omega it} - \frac{1}{2\omega^2}e^{\omega it} = \frac{1}{\omega^2} - \frac{1}{\omega^2}\cos \omega t$$

From (8), (9), and the Second Shifting Theorem and assuming that $L^{-1} \Sigma = \Sigma L^{-1}$, we have that

$$I(t) = 2L^{-1}\left[\frac{1}{s(s^2 + \omega^2)}\right] + 4\sum_{n=1}^{\infty}(-1)^n L^{-1}\left[\frac{e^{-ns}}{s(s^2 + \omega^2)}\right]$$

$$= \frac{2}{\omega^2}(1 - \cos \omega t) + \frac{4}{\omega^2}\sum_{n=1}^{\infty}(-1)^n H(t - n)[1 - \cos \omega(t - n)] \quad (10)$$

The form of the solution is somewhat daunting, but for any given positive value of t, there are only a finite number of terms in the sum, the remaining terms being annihilated by the Heaviside functions. See Figure 7.8 for a sketch of $I(t)$, where $\omega = 20$. Note that

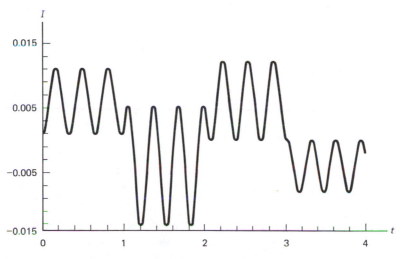

Figure 7.8 $I(t)$ versus t.

$$I(t) = \begin{cases} \dfrac{2}{\omega^2}(1 - \cos \omega t), & 0 \leq t < 1 \\[3mm] \dfrac{2}{\omega^2}(1 - \cos \omega t) - \dfrac{4}{\omega^2}(1 - \cos \omega(t - 1)), & 1 \leq t < 2 \\[3mm] \dfrac{2}{\omega^2}(1 - \cos \omega t) - \dfrac{4}{\omega^2}(1 - \cos \omega(t - 1)) + \dfrac{4}{\omega^2}(1 - \cos \omega(t - 2)), & 2 \leq t < 3 \end{cases}$$

PROBLEMS

The following problem(s) may be more challenging: 1(h), 4(k), 4(l), and 5.

1. Find the Laplace Transform of each of the following functions.
   (a) $e^t \sin t$.   (b) $\sin^2 t$.   (c) $\cos^2 t$.
   (d) $t \sin at$.   (e) $t^2 \sin at$.   (f) $\cos^3 t$.
   (g) $e^{-3t} \cos(2t + \pi/4)$.   (h) $t^2 e^t \cos t$.

2. Find the inverse Laplace Transform of each of the following.
   (a) $\dfrac{1 + e^{-s}}{s}$.   (b) $\dfrac{3e^{-2s}}{3s^2 + 1}$.   (c) $\dfrac{1}{(s - a)^n}$.

   (d) $\dfrac{1 - e^{-s}}{s^2}$.   (e) $\dfrac{(s - 1)e^{-s} + 1}{s^2}$.

   (f) $\ln \dfrac{s + 3}{s + 2}$ $\left[Hint: \dfrac{d}{ds}\left(\ln \dfrac{s + 3}{s + 2} \right) = \dfrac{-1}{(s + 2)(s + 3)}. \right]$

3. Use partial fractions to find the inverse Laplace Transform of each of the following:
   (a) $\dfrac{1}{s(s + 1)}$.   (b) $\dfrac{1}{s(s + 2)^2}$.   (c) $\dfrac{1}{(s - a)(s - b)}, a \neq b$.

   (d) $\dfrac{s^2 + 3}{(s - 1)^2(s + 1)}$.   (e) $\dfrac{3s + 1}{(s^2 + 2s + 2)(s - 1)}$.   (f) $\dfrac{s + 1}{(s - 2)(s^2 + 9)}$.

4. Use the Laplace Transform to solve the following initial value problems. Sketch the solutions of (a), (b), (i).
   (a) $y'' + 6y' + 5y = t$, $y(0) = y'(0) = 0$.
   (b) $y'' + 2y' + y = e^t$, $y(0) = y'(0) = 0$.
   (c) $y'' - 2y' + 2y = \sin t$, $y(0) = y'(0) = 0$.
   (d) $y'' - 4y' + 4y = 2e^t + \cos t$, $y(0) = 3/25, \quad y'(0) = -4/25$.
   (e) $y'' - 2y' + y = e^t \sin t$, $y(0) = y'(0) = 0$.
   (f) $y'' + 2y' + y = te^{-t}$, $y(0) = 1, \quad y'(0) = -2$.
   (g) $y''' + y'' + 4y' + 4y = -2$, $y(0) = 0, \quad y'(0) = 1, \quad y''(0) = -1$.
   (h) $y''' - y'' + 4y' - 4y = -3e^t + 2e^{2t}$, $y(0) = 0, \quad y'(0) = 5, \quad y''(0) = 3$.
   (i) $y'' + 2y' + y = H(1 - t)$, $y(0) = y'(0) = 1$.
   (j) $y'' - 2y' + 2y = (t - 1)H(1 - t) + 1$, $y(0) = 0, \quad y'(0) = 1$.
   (k) $y'' + 4y = f(t)$, where $f(t)$ is twice the function given in III.2 in the table of transforms [with $a = 1$], and $y(0) = 0, y'(0) = 0$.

(l) $y'' + 4y' + 4y = f(t)$, where $f(t)$ is the sawtooth wave function of period 1 in III.5 of the table of transforms, $y(0) = 0$, $y'(0) = 0$.

5. (a) Verify formula III.1 in the table of transforms.
   (b) Verify formula III.5 in the table of transforms.

6. Prove that $L^{-1}\left[\dfrac{1}{(s-r)^{n+1}}\right] = \dfrac{1}{n!}\, t^n e^{rt}$, $n = 0, 1, 2, \ldots$

7. Prove that $L^{-1}\left[\dfrac{Bs+C}{((s-a)^2+b^2)^n}\right] = e^{at}L^{-1}\left[\dfrac{Bs+aB+C}{(s^2+b^2)^n}\right]$

for any real B, C, a, b, $n = 1, 2, \ldots$

7.4 A MODEL FOR CAR-FOLLOWING

In heavy traffic or in constricted passages such as bridges or tunnels, cars follow one another with little or no lane changing or passing. Rear-end collisions are a common occurrence in this setting. Bad driving is the cause of many of these accidents, but some seem to occur for no apparent reason. In recent years mathematical models of car-following have been created and solved in an attempt to understand the phenomenon and, subsequently, to design traffic controls and driving codes that promote safe and efficient traffic flow. We construct and analyze a simple car-following model in this section, and then specialize it to the situation of a line of cars stopped at a red light. What happens when the light changes and the lead car accelerates? Under what circumstances will there be a rear-end collision somewhere down the line as each car accelerates in turn? We give some tentative answers to these questions.

Stimulus and Response

Many stimuli may cause the driver in a line of cars to respond with an acceleration or deceleration of his or her own car—a speeding car to the rear, a changing traffic light ahead, a variation in the relative velocity of the car directly in front. Of course, the driver's response does not occur instantaneously. The driver must first detect the stimulus and then decide whether to ease up on or to depress the accelerator or the brake pedal. The car itself does not respond instantaneously and hence contributes to the time lag in the response of the car–driver system. We shall use the word "car" to denote the car–driver system, and refer to the delay in the response of the "car" as the *response time* of the car.

   We shall assume that the dominant stimulus to the car is the difference between the velocity of the car ahead and its own velocity. The car's response, we shall assume, is an acceleration if the car ahead has a positive relative velocity, and a deceleration if the relative velocity is negative. The magnitude of the car's acceleration or deceleration (after the appropriate response time) will be assumed proportional to the magnitude of the relative velocity. The constant of proportionality (assumed positive) is called the *sensitivity* of the car to the stimulus.

Our first goal is to quantify these somewhat vague notions of sensitivity and response time and then include them with the more sharply defined concepts of velocity and acceleration in a mathematical model. Hereafter, we shall use only the word "acceleration"—deceleration is simply a negative acceleration.

A Mathematical Model

Let us suppose that there are N cars in a line, with no passing or lane changing possible. Let $v_j(t)$ and $a_j(t), j = 1, \ldots, N$, denote the velocity and the acceleration at time t of the jth car in line. Then the simplest mathematical model of the stimulus–response system above is given by the system of *differential-delay equations*

$$v_1(t) = f(t),$$
$$a_j(t + T_j) \equiv v_j'(t + T_j) = \lambda_j [v_{j-1}(t) - v_j(t)], \qquad j = 2, \ldots, N, \tag{1}$$

where $f(t)$ is the velocity profile of the leader, λ_j is the *sensitivity* coefficient of the jth car, and T_j is the *response time* of the jth car.† We shall assume that λ_j and T_j are positive. We shall measure distance in feet, time in seconds, and sensitivity in (seconds)$^{-1}$. Note that if the *relative velocity*, $v_{j-1} - v_j$, is positive, the $(j - 1)$st car is moving faster than the jth car, which responds by accelerating after the time delay T_j, as we would expect. The most distinctive aspects of (1) are the time delays T_j and the sensitivities λ_j.

System (1) does not contain enough information to be very useful in treating a traffic-flow problem. For instance, nothing has been said about the initial data for each car. Moreover, we are more interested in the locations of the cars than in their velocities. We shall append this information to (1) while setting up a model of a specific situation.

Cars at a Stoplight

Let us suppose that there are N identical cars in a line stopped at a red light. The light turns green and after 2 s the first car accelerates. Three cars back and 7 s later there is a rear-end collision (see Figure 7.9). Why? We shall show that events such as this may be predicted by the model.

First, let us measure time forward from the moment the lead car accelerates, which is 2 s after the light turns green. The acceleration will be taken to a positive constant a_0 ft/s^2. We shall assume that all cars in the line are at rest for all earlier times. Next we shall assume that each car is 15 ft long and that there is a gap of 5 ft between successive cars in the line. The location of the jth car will be measured from the stoplight to its front bumper (Figure 7.9).

Assuming identical cars (and drivers) and thus common delay time T and

† System (1) is a *velocity-control* model of car-following since the stimulus that controls the response is proportional to the relative velocity. For more on these models see the article by Robert L. Baker in *Modules in Applied Mathematics*, vol. 1, W. Lucas, M. Braun, C. Coleman, and D. Drew, eds. (New York: Springer-Verlag, 1983).

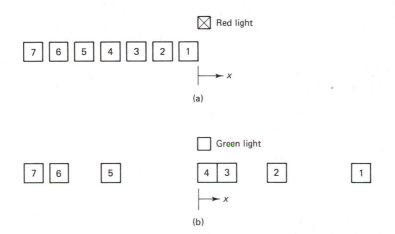

Figure 7.9 Cars at a light: (a) cars at a stoplight; (b) collision after the light turns.

sensitivity λ, we obtain the specific mathematical model involving differential-delay equations:

$$
\begin{aligned}
v_1(t) &= a_0 t, & & & t &\geq 0 \\
v_j'(t+T) &= \lambda[v_{j-1}(t) - v_j(t)], & j &= 2, \dots, N, & t &\geq 0 \\
v_j(t) &= 0, & j &= 2, \dots, N, & t &\leq 0
\end{aligned}
\tag{2}
$$

The equations of position are

$$
\begin{aligned}
x_1(t) &= \tfrac{1}{2} a_0 t^2, & t &\geq 0 \\
x_j(t) &= -20(j-1) + \int_0^t v_j(\tau)\, d\tau, & j &= 2, \dots, N, & t &\geq 0
\end{aligned}
\tag{3}
$$

Our goal is to find the functions $x_j(t)$ and, in particular, to determine if a collision occurs, that is, if for some t and j, $x_j(t) = x_{j-1}(t) - 15$, which corresponds to the jth and $(j-1)$st car colliding.

Solving the Equations of the Model.   The Laplace Transform is ideally suited for solving a linear system with time delays. Applying the transform to (2), we have

$$
L[v_1] = \frac{a_0}{s^2}
\tag{4}
$$

$$
L[v_j'(t+T)] = e^{Ts} s L[v_j] = \lambda\{L[v_{j-1}] - L[v_j]\}, \qquad j = 2, \dots, N
$$

where we have used the formula

$$
L[v'(t+T)] = L[H(t+T)v'(t+T)] = e^{Ts} L[v'(t)] = e^{Ts} s L[v]
\tag{5}
$$

which follows from the Second Shifting Theorem and the Derivative Transform. We have that for $j = 2, \dots, N$,

$$L[v_j] = \frac{\lambda}{\lambda + se^{Ts}} L[v_{j-1}] = \cdots = \left\{\frac{\lambda}{\lambda + se^{Ts}}\right\}^{j-1} L[v_1] = \left\{\frac{\lambda}{\lambda + se^{Ts}}\right\}^{j-1} \frac{a_0}{s^2} \quad (6)$$

where the equality is obtained from the second line of (4) by solving for $L[v_j]$ in terms of $L[v_{j-1}]$ and then using induction to express $L[v_j]$ in terms of $L[v_1] = a_0/s^2$.

The velocity of the jth car in line may be found by applying L^{-1} to (6). First, however, we shall replace the bracketed term by a binomial series (a special case of the binomial expansion; see Appendix D) to obtain

$$L[v_j] = a_0\lambda^{j-1} \left\{ \frac{e^{-(j-1)Ts}}{s^{j+1}} - \frac{\lambda(j-1)e^{-jTs}}{s^{j+2}} + \frac{\lambda^2(j-1)je^{-(j+1)Ts}}{2!s^{j+3}} - \cdots \right\} \quad (7)$$

Assuming that $L^{-1}\{\cdots\} = L^{-1}[\cdot] + L^{-1}[\cdot] + \cdots$ and making use of the Second Shifting Theorem, we have from (7) that

$$v_j(t) = a_0\lambda^{j-1} \left\{ \frac{1}{j!} H[t - (j-1)T][t - (j-1)T]^j - \lambda\frac{j-1}{(j+1)!} H[t - jT][t - jT]^{j+1} \right.$$

$$\left. + \lambda^2 \frac{(j-1)j}{2!(j+2)!} H[t - (j+1)T][t - (j+1)T]^{j+2} - \cdots \right\} \quad (8)$$

Integrating (8), we have that

$$x_j(t) = -20(j-1) + \frac{a_0\lambda^{j-1}}{(j-2)!} \sum_{n=0}^{\infty} \frac{(-\lambda)^n H(t - T_{nj})(t - T_{nj})^{n+j+1}}{n!(n+j-1)(n+j)(n+j+1)} \quad (9)$$

where $t \geq 0$ and $T_{nj} = (n+j-1)T$, $j = 2, \ldots, N$. Of course, we also have that the location of the lead car is given by

$$x_1(t) = \frac{1}{2} a_0 t^2, \qquad t \geq 0 \quad (10)$$

Formula (9) looks formidable, but appearances are deceiving and it is easy to use. For a fixed time $t > 0$, there are only a finite number of nonvanishing terms in each formula, the remaining terms being annihilated by the Heaviside functions. In fact, the summation in (9) has effective range from $n = 0$ only to $n = [t]$, the greatest integer not larger than t. For example, we have for the second car in line that its location at time t is given by

$$x_2(t) = \begin{cases} -20, & 0 \leq t < T \\ -20 + \dfrac{a_0\lambda}{6}(t - T)^3, & T \leq t < 2T \\ -20 + \dfrac{a_0\lambda}{6}(t - T)^3 - \dfrac{a_0\lambda^2}{24}(t - 2T)^4, & 2T \leq t < 3T \end{cases} \quad (11)$$

See Figure 7.10 for a sketch of the moving positions of the first two cars in line. In Figures 7.10 and 7.12 we have added $15(j - 1)$ to $x_j(t)$, $j = 2, \ldots, N$. The effect is to reduce the physical cars to points which are initially 5 ft apart on the

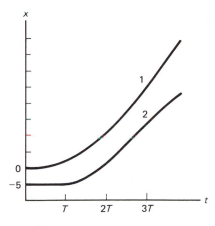

Figure 7.10 First two cars.

x-line. With this reduction a collision of the jth and $(j+1)$st cars is graphically represented by the intersection of the graphs of the functions $x_j(t)$ and $x_{j+1}(t)$.

We have reached the stage of model building and testing at which we need to select specific values for the parameters and interpret the results. Do the functions $x_j(t)$, $j = 1, 2, \ldots, N$, give reasonable estimates for the locations of the car as they move away from the light? Will there be collisions?

Bad Driving and a Rear-End Collision.   Suppose that the second car in line is a high-performance model, but driven by a drunk. It will take the driver a long time to respond to the acceleration of the lead car, but when he does, he overreacts. Thus T and λ may both be "large" in a sense that will become apparent.

Let us suppose that the lead car accelerates at $a_0 = 6$ ft/s^2. Then the distance between the first two cars in line can be obtained by using $x_1(t) = 3t^2$, $t \geq 0$, and $x_2(t)$ from (11):

$$x_1(t) - x_2(t) = \begin{cases} 3t^2 + 20, & 0 \leq t \leq T \\ 3t^2 + 20 - \lambda(t-T)^3, & T \leq t \leq 2T \quad (12) \\ 3t^2 + 20 - \lambda(t-T)^3 + \lambda^2(t-2T)^4/4, & 2T \leq t \leq 3T \end{cases}$$

If $x_1(t) - x_2(t) = 15$, the first and second cars collide at time t. This cannot happen in the first time span of T seconds, but will during the second span of T seconds if

$$x_1(2T) - x_2(2T) = 12T^2 + 20 - \lambda T^3 \leq 15$$

that is, if

$$5 \leq T^2(\lambda T - 12) \tag{13}$$

Inequality (13) suggests that T and λT are the important parameters, rather than T and λ. In Figure 7.11 we have sketched the curve defined by $5 = T^2(\lambda T - 12)$, using T and λT axes. If the point $(t, \lambda T)$ lies anywhere in the shaded region, the second car hits the first in the second span of T seconds (Figure 7.12).

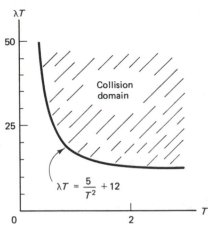

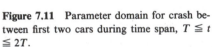

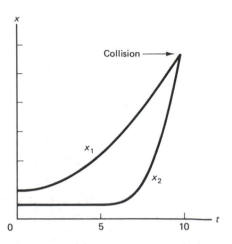

Figure 7.11 Parameter domain for crash between first two cars during time span, $T \leq t \leq 2T$.

Figure 7.12 Collision between first two cars, $\lambda T = 15$, $T = 5$ sec.

Observe that a response time of 5 s and sensitivity $\lambda = 3$ s^{-1} leads to a crash, while the same response time but the lower sensitivity of $\lambda = 1$ s^{-1} does not [at least not during the first 15 s; see (12)]. One might imagine a lower sensitivity to be inherently more dangerous. In fact, just the opposite is true since a time lag is also involved. The response does not occur at the time t an observation is made, but T seconds later, at which time the situation may have changed. High sensitivity may lead to a large response to a "deviation" that is no longer there when the response takes effect.

Comments

Mathematical models of traffic flow are of relatively recent vintage and have not yet acquired the acceptance of, say, Newton's laws or the laws of electrical circuits. There is too much of the unknown, unknowable, and random in traffic flow—as in most human systems—for there to be a single comprehensive model. There are a good many other models of car-following, models which include other stimuli, such as the separation distance between cars or the relative velocity or separation of the car behind as well as the car ahead. Some of these models are given in the problem set.

PROBLEMS

1. (a) Write out a car-following model in which the stimulus to the $(j + 1)$st driver is the separation from the car ahead. Assume a response time of T seconds and a sensitivity of λ seconds^{-1}.

(b) Repeat part (a) but in a situation where the stimulus is a linear combination of the separation distance with sensitivity λ_1 seconds^{-1} and the relative velocity with sensitivity λ_2 seconds^{-1}.

(c) For the model in part (a), assume that the lead car accelerates from a stop at the constant rate a_0 ft/s^2. Find the motion of the following car. Are there parameter values that will result in a rear-end collision? If so, find some.

2. Using the velocity control model of the text, which of the following sets of values of T and λT will result in a collision between the first two cars within the second span of T seconds?

    (a) $T = 1, \lambda T = 20$.   (b) $T = 5, \lambda T = 12.3$.   (c) $T = 2, \lambda T = 14$.

3. Using the velocity control model of the text, show that if $T = 1$ s, and $\lambda T = 2$, the third and fourth cars collide at some time t, $6 \leqq t \leqq 7$. Sketch the graph of $x_3(t) - x_4(t) - 15$, $0 \leq t \leq 7$.

4. [Reciprocal Spacing Model]. Suppose that the acceleration of the $(j + 1)$st car in a line, $j \geq 1$, is directly proportional to the relative velocity of the car with respect to that of the car ahead and inversely proportional to the separation distance. Assume a response time of T seconds and a sensitivity of λ seconds^{-1}.

(a) Justify the model equation

$$v'_{j+1}(t + T) = \lambda \frac{v_j(t) - v_{j+1}(t)}{x_j(t) - x_{j+1}(t)}$$

(b) Let l be the *effective length* of a car (i.e., the reciprocal of the *jam concentration*), which in turn is the number of cars per unit of length of road when the traffic is completely jammed. Experimental evidence suggests that l is approximately 23 ft. Show that

$$v_{j+1}(t + T) = \lambda \ln \left[\frac{x_j(t) - x_{j+1}(t)}{l} \right]$$

(c) Let $m > 1$ and consider the model equation

$$v'_{j+1}(t + T) = \lambda \frac{v_j(t) - v_{j+1}(t)}{[x_j(t) - x_{j+1}(t)]^m}$$

Show that

$$v_{j+1}(t + T) = \frac{\lambda}{m-1} \left\{ l^{1-m} - [x_j(t) - x_{j+1}(t)]^{1-m} \right\}$$

(d) The *steady-state velocity* is that of steady flow at constant velocity—hence $v_j(t) = v_0, j = 1, 2, \ldots$, for all t in this case. Find the steady-state velocity for the models in parts (b) and (c).

7.5 CONVOLUTION

When the Laplace Transform is used to solve an initial value problem, it often happens that the transform of the solution involves a product of functions. Although the Shifting Theorems allow us to find the inverse transform of certain special products, we do not yet have a general procedure for inverting a product. The far-reaching notion of the convolution product enables us to fill this gap. At the same time the convolution casts a new light on the meaning of the Green's function.

The Convolution Product

Suppose that $F(s)$ and $G(s)$ are the Laplace Transforms of the known functions $f(t)$ and $g(t)$, respectively. We shall derive an elegant formula for $L^{-1}[F(s)G(s)]$, but it is *not* $f(t)g(t)$.

First we define a new kind of product.

> **Convolution.** The *convolution product $f * g$* of f and g is given by the formula
>
> $$(f * g)(t) = \int_0^t f(t - u)g(u) \, du \tag{1}$$
>
> where f and g are assumed to be of exponential order.

The basic properties of the convolution product are simple enough.

Properties of Convolution. Let f, g, and h belong to $\mathscr{E}$. Then

(a) $f * g$ belongs to $\mathscr{E}$.
(b) (Commutativity) $f * g = g * f$.
(c) (Associativity) $(f * g) * h = f * (g * h)$.
(d) (Distributivity) $(f + g) * h = f * h + g * h$.

Proof. We shall prove only (b); the rest are left to the reader. To show (b), we set $v = t - u$ in the integral of the definition of the convolution product and obtain, for each fixed $t \geqq 0$,

$$(f * g)(t) = \int_0^t f(t - u)g(u) \, du = \int_t^0 f(v)g(t - v)(-dv)$$

$$= \int_0^t f(v)g(t - v) \, dv = (g * f)(t)$$

As the reader may have guessed, the convolution product is the sought-for inverse transform of a product.

Convolution Theorem. Let f and g belong to $\mathscr{E}$, and let F and G be their respective transforms. Then

$$L[f * g] = FG \tag{2}$$

or, equivalently,†

$$L^{-1}[FG] = f * g \tag{3}$$

---

† It is assumed that $L^{-1}[FG]$ and $f * g$ are continuous.

Proof. We shall verify (2). Let R be the region in the tu-plane defined by $\{(t, u) : 0 \leq u \leq t, 0 \leq t\}$ (see Figure 7.13). We have that

$$L[f * g] = \int_0^\infty e^{-st} \left\{ \int_0^t f(t - u)g(u)\, du \right\} dt \tag{4}$$

$$= \int_R \int e^{-st} f(t - u)g(u)\, du\, dt$$

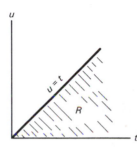

Figure 7.13 Convolution domain.

But the condition that f and g be of exponential order implies that the double integral in (4) may be evaluated by an iterated integral in either order—in the order of the iterated integral of (4), or in the reverse order. Hence

$$L[f * g] = \int_0^\infty g(u) \left\{ \int_u^\infty e^{-st} f(t - u)\, dt \right\} du \tag{5}$$

Making the change of variable $v = t - u$ in (5), we have that

$$L[f * g] = \int_0^\infty g(u) \left\{ \int_0^\infty e^{-s(v+u)} f(v)\, dv \right\} du = \int_0^\infty g(u) e^{-su} \left\{ \int_0^\infty e^{-sv} f(v)\, dv \right\} du$$

$$= \left\{ \int_0^\infty e^{-su} g(u)\, du \right\} \left\{ \int_0^\infty e^{-sv} f(v)\, dv \right\} = GF$$

and the theorem is proved since $GF = FG$.

With the Convolution Theorem in hand, we may find inverse transforms of a wide array of functions.

Example 7.22

Since $L[1] = 1/s$ and $L[\sin t] = (s^2 + 1)^{-1}$,

$$L^{-1} \left[\frac{1}{s(s^2 + 1)} \right] = L^{-1} \left[\frac{1}{s} \right] * L^{-1} \left[\frac{1}{s^2 + 1} \right]$$

$$= \int_0^t 1 \cdot \sin u\, du = 1 - \cos t$$

a formula found earlier by the lengthier method of partial functions.

Convolution and Green's Functions

An advantage of the convolution product is that it allows us to construct the Green's function of a constant-coefficient differential operator very quickly. Before stating and proving a general result to that effect, we shall illustrate the process by an example.

Example 7.23

Let $h(t)$ have exponential order, and consider the initial value problem

$$y'' + y' - 6y = h(t)$$
$$y(0) = y'(0) = 0 \tag{6}$$

Applying the Laplace Transform, we have that $(s^2 + s - 6)L[y] = L[h]$ and hence $L[y] = L[h]/(s^2 + s - 6)$. Taking inverse transforms and using the Convolution Theorem, we have that

$$y = h * L^{-1}\left[\frac{1}{s^2 + s - 6}\right] \tag{7}$$

Now since by the method of partial fractions,

$$\frac{1}{s^2 + s - 6} = \frac{1}{5}\left[\frac{1}{s - 2} - \frac{1}{s + 3}\right]$$

it follows that

$$L^{-1}\left[\frac{1}{s^2 + s - 6}\right] = \frac{1}{5}e^{2t} - \frac{1}{5}e^{-3t}$$

Hence (7) becomes

$$y(t) = \int_0^t \left\{\frac{1}{5}e^{2(t-u)} - \frac{1}{5}e^{-3(t-u)}\right\} h(u)\, du \tag{8}$$

Since the function $y(t)$ in (7) is necessarily continuous, we conclude that it must be the unique solution of the initial value problem (6).

Example 7.23 gives us a hint as to what to expect in the general case of an nth-order initial value problem with vanishing data. In operator form we have $P(D)y \equiv (D^2 + D - 6)y = h$. Denoting the (continuous) inverse transform of $1/P(s)$ by $g(t)$, we see that (8) may be written as

$$y(t) = \int_0^t g(t - u)h(u)\, du = g * h \tag{9}$$

Comparing (9) with the results of Sections 5.3 and 6.4, we observe that the function $g(t - u)$ is the Green's function of the initial value problem (6).

This reasoning still holds good in solving the initial value problem

$$P(D)[y] \equiv y^{(n)} + a_{n-1}y^{(n-1)} + \cdots + a_0y = h$$
$$y(0) = y'(0) = \cdots = y^{(n-1)}(0) = 0$$

$$(10)$$

where the a_i are real constants. Indeed, transforming (10), we easily obtain

$$L[y] = \frac{L[h]}{P(s)}$$

where $P(s)$ is the characteristic polynomial for the differential operator $P(D)$ in (10). Thus if we set $g(t) = L^{-1}[1/P(s)]$, then (9) gives the solution of initial value problem (10), where $g(t-u)$ is the Green's function for the problem. We have proved the following theorem.

Convolution and the Green's Function.   The Green's function for the operator $P(D)$ of initial value problem (10) is $g(t) = L^{-1}[1/P(s)]$. The solution of (10) is

$$y = g * h(t) = \int_0^t g(t-u)h(u) \, du \qquad (11)$$

The advantages of the Green's function approach to initial value problems have been argued in earlier chapters. The new result of this section is that the Green's function of the operator may be calculated by means of the transform calculus and then convolved with the driving force of the initial value problem to obtain the solution.

PROBLEMS

The following problem(s) may be more challenging: 3(f).

1. Find the Laplace Transform of each of the following.

   (a) $\int_0^t (t-1)e^t \, dt$.

   (b) $\int_0^t (t^2 - 2t) \, dt$.

   (c) $\int_0^t \sin(t - \pi/4)e^t dt$.

   (d) $\int_0^t e^{(t-a)} \cos t \, dt$.

2. Find the inverse Laplace Transform of each of the following by using the Convolution Theorem. Write answers as convolution products.

   (a) $\dfrac{s}{(s^2+9)^3}$.

   (b) $\dfrac{s^2+4s+4}{(s^2+4s+13)^2}$.

   (c) $\dfrac{s}{(s+1)(s+2)^3}$.

   (d) $\dfrac{s}{(s^2+10)^2}$.

(e) $\dfrac{1}{s^2(s+1)}$.

(f) $\dfrac{s}{(s^2+1)^2}$.

(g) $\dfrac{L[f]}{s^2+1}$.

(h) $\dfrac{e^{-3s}L[f]}{s^3}$.

3. Solve the following initial value problems using the methods of this section. Sketch solutions.

(a) $y'' - y = tH(t-1)$, $y(0) = y'(0) = 0.$

(b) $y'' + y = e^t - 1$, $y(0) = y'(0) = 0.$

(c) $2y'' + y' - y = \sin t$, $y(0) = y'(0) = 0.$

(d) $y'' + y = te^t$, $y(0) = y'(0) = 0.$

(e) $4y'' - 4y' + 37y = e^{t/2}\cos 3t$, $y(0) = 0, \quad y'(0) = 0.$

(f) $y'' + 2y' + 2y = \begin{cases} 0, & 0 \leq t \leq 1, \\ t-1, & 1 \leq t \leq 2, \\ 1, & t > 2 \end{cases}, \quad y(0) = y'(0) = 0.$

4. Let $D = d/dt$. Use the Convolution Theorem to construct Green's functions for each of the following differential operators.

(a) $D^2 + 6D + 13.$ (b) $D^2 + \frac{1}{3}D + \frac{1}{36}.$ (c) $D^3 + 1.$

5. Use the results in Problem 4 to solve the following initial value problems.

(a) $y'' + 6y' + 13y = f(t)$, $y(0) = 0, \quad y'(0) = 0.$

(b) $y'' + \frac{1}{3}y' + \frac{1}{36}y = f(t)$, $y(0) = 0, \quad y'(0) = 0.$

(c) $y''' + y = t$, $y(0) = 0, \quad y'(0) = 0, \quad y''(0) = 0.$

6. Show that $f * (g + h) = f * g + f * h, f * (g * h) = (f * g) * h.$

7. Show that $e^{at}(f * g) = e^{at}f * e^{at}g.$

7.6 CONVOLUTION AND THE DELTA FUNCTION

The convolution may also be used in solving the somewhat different problem of finding the response of a dynamical system to a sudden force of enormous amplitude but short duration (i.e., to an impulsive force). Impulsive forces may be modeled by the Dirac delta "function" (which turns out not to be a function at all). Green's function appears once more, this time as the response of the dynamical system to the delta function.

We shall begin on a hypothetical level, "defining" the delta function by a property we want it to possess. We then see how far we can develop the theory and the applications of the delta function from this somewhat shaky beginning. Next we take up the question of whether the delta function even exists. Finally, we conclude with an interpretation in terms of "window" functions and impulsive forces.

The Dirac Delta Function and Its Properties

We have been working up to this point with elements in the space $\mathscr{E}$ of functions of exponential order on the interval $[0, \infty)$. We shall extend the domains of all the functions f in $\mathscr{E}$ to the entire real line by taking $f(t)$ to vanish identically for all negative t. The new function space is denoted by $\mathscr{E}_0$. We may now give the definition of a very strange object, which Dirac called a "function."

> **Dirac Delta Function.**   Suppose there is an element δ in the space $\mathscr{E}_0$ such that for every f in $\mathscr{E}_0$,
>
> $$\int_{-\infty}^{\infty} \delta(t-u)f(u)\,du = f(t), \qquad t > 0 \tag{1}$$
>
> if f is continuous at t. Any such element δ is called a *Dirac delta function*.†

For the next several paragraphs we shall assume that a Dirac delta function exists in $\mathscr{E}_0$ and that (1) holds.

Properties of the Delta Function.   We have that

$$\int_{-\infty}^{\infty} \delta(t)\,dt = 1 \tag{2}$$

$$L[\delta](s) = 1 \tag{3}$$

$$L[\delta(t-u)] = e^{-us} \tag{4}$$

Proof. The reader may have doubts about the development of the calculus of a function that may not exist, but we shall plunge right ahead anyway. To prove (2), let $f(t) = H(t)$, the Heaviside function. From (1) we have that

$$H(t) = \int_{-\infty}^{\infty} \delta(t-u)H(u)\,du = \int_{-\infty}^{\infty} \delta(v)H(t-v)\,dv$$

$$= \int_{-\infty}^{t} \delta(v)\,dv \qquad\qquad [\text{since } H(t-v) = 0 \text{ if } v > t]$$

Thus $\int_{-\infty}^{t} \delta(v)\,dv = 1$ for all $t > 0$, and the desired result follows.

We leave the proof of (4) to the reader and the problem set. We prove (3) as follows. By the Convolution Theorem, we have that

$$L[\delta * H] = L[\delta]L[H] \tag{5}$$

On the other hand,

$$L[H] = L[1] \tag{6}$$

---

† Introduced by P. A. M. Dirac in the 1930s for use in quantum mechanics.

We also have that

$$L[\delta * H] = L\left[\int_0^t \delta(t-u)H(u)\, du\right] = L\left[\int_0^t \delta(v)H(t-v)\, dv\right]$$

$$= L\left[\int_{-\infty}^\infty \delta(v)H(t-v)\, dv\right] = L[1] \tag{7}$$

where we have used identities in the proof of (2) and the facts that $\delta(v) = 0$, $v < 0$, and $H(t-v) = 0$, $v > t$. From (5), (6), and (7) we conclude that $L[\delta] \equiv 1$.

One of the many applications of δ lies in its use in solving initial value problems. Suppose that $P(D) = D^n + a_{n-1}D^{n-1} + \cdots + a_1D + a_0$, where $a_{n-1}, \ldots, a_0$ are real.

Solving Initial Value Problems. The solution of the initial value problem

$$P(D)[y] = h, \qquad y^{(k)}(0) = 0, \quad k = 0, 1, \ldots, n-1 \tag{8}$$

where h lies in $\mathscr{E}_0$, is given by

$$y = \int_{-\infty}^\infty G(t, u)h(u)\, du \tag{9}$$

The function $G(t, u)$ is the solution of the problem

$$P(D)z = \delta(t-u), \qquad z^{(k)}(0) = 0, \quad k = 0, 1, \ldots, n-1 \tag{10}$$

for each value of u.

Proof. This may be shown by applying the transform to (10) and using properties of δ to obtain $L[G(t, u)] = e^{-us}/P(s)$. From the Second Shifting Theorem and the discussion concerning the Green's function g given in the preceding section, we have that $G(t, u) = H(t-u)g(t-u)$ and hence

$$y = \int_0^t g(t-u)h(u)\, du = \int_{-\infty}^t g(t-u)h(u)\, du$$

$$= \int_{-\infty}^\infty H(t-u)g(t-u)h(u)\, du = \int_{-\infty}^\infty G(t, u)h(u)\, du$$

Thus we have come up with yet another way for finding the unique solution of an initial value problem. However, this is nothing really new since $G(t, u) = H(t-u)g(t-u)$, and we have already seen in Section 7.5 how the Green's function g is used to solve the problem.

Does the Dirac Delta Function Exist?

So far we have not shown that the delta function actually exists as an element of $\mathscr{E}_0$. Indeed, we shall now show that there is no function in $\mathscr{E}_0$ having property (1). Let $t_0 > 0$ be a point of continuity for δ and assume that $\delta(t_0) > 0$. Choose any

$T, 0 < T < t_0$, such that δ is continuous and does not change sign on the interval $t_0 - T \leq t \leq t_0$. Define the function f_0 in $\mathscr{E}_0$ as follows:

$$f_0(t) = \begin{cases} 1, & 0 \leq t \leq T \\ 0, & \text{all other real } t \end{cases}$$

Now from (1) we must have that

$$0 = f_0(t_0) = \int_{-\infty}^{\infty} \delta(t_0 - u) f_0(u)\, du = \int_0^T \delta(t_0 - u)\, du$$

Hence from the conditions placed on δ we conclude that

$$\delta(t_0 - u) = 0 \qquad \text{for all } 0 \leq u \leq T$$

and, in particular, $\delta(t_0) = 0$. Thus we have proven that δ vanishes at *every* point of continuity. Thus the left-hand side of (1) vanishes for all t and hence the delta "function" can be nothing more than a convenient device for constructing the function $G(t, u)$ in (9).

Since the time of Oliver Heaviside, "functions" such as δ have been very important in the applications. In advanced treatments of modern applied mathematics a rigorous theory is constructed which contains objects, called *distributions* or *generalized functions*, that behave just like the "delta function" should.† We can do no more here than give a tiny glimpse of how the "delta function" idea can be embedded in a larger framework.

δ *as a Limit*

We might try to salvage the "delta function" identity (1) in the following way. Consider the function $\phi(t, \tau)$ defined for *positive* τ and all real t by

$$\phi(t, \tau) = \begin{cases} \dfrac{1}{\tau} & 0 < t < \tau \\ 0, & \text{otherwise} \end{cases} \tag{11}$$

Observe that $\displaystyle\int_{-\infty}^{\infty} \phi(t, \tau)\, dt = 1$ for all $\tau > 0$ and that $\phi \geq 0$. Hence $\{\phi(t, \tau) : \tau > 0\}$ is a family of density functions for computing average values of other functions. For example, let f belong to $\mathscr{E}_0$ and let f be continuous at $t_0 > 0$. From (11) we see that

$$\int_{-\infty}^{\infty} \phi(t - t_0, \tau) f(t)\, dt = \frac{1}{\tau} \int_{t_0}^{t_0 + \tau} f(t)\, dt \tag{12}$$

Hence ϕ does indeed act as a weighted average of f with the weight entirely concentrated in the interval $J = [t_0, t_0 + \tau]$.

---

† The mathematical theory of distributions was initiated by Laurent Schwartz in his book *Théorie des distributions*, Vols. 1 and 2 (Paris: Actualités Scientifiques et Industrielles, Hermann & Cie, 1957, 1959).

Now the Mean Value Theorem for Integrals asserts that if f is continuous on J, there is a number ξ in J such that

$$\int_{t_0}^{t_0+\tau} f(t)\,dt = f(\xi)\tau \tag{13}$$

From (12) and (13) we see that

$$\lim_{\tau \to 0} \int_{-\infty}^{\infty} \phi(t - t_0, \tau) f(t)\,dt = f(t_0) \tag{14}$$

which shows that the family $\{\phi(t, \tau) : \tau > 0\}$ "behaves like the delta function" in the limit. See Figures 7.14 and 7.15 for sketches of ϕ and of ϕf. Observe that the graph of each function ϕ is a "window" through which a segment of the graph of f may be seen but in distorted form.

Green's Function as a Limit

In view of the preceding remarks, we might try to define the solution of (10) in the following way. Denote by $z(t, u, \tau)$ the unique solution of the problem

$$P(D)z = \phi(t - u, \tau), \qquad z^{(k)}(0) = 0, \quad k = 0, 1, \ldots, n - 1$$

Then we shall define the solution $z(t, u)$ of (10) to be the limit of the family $z(t, u, \tau)$ as $\tau \to 0$; that is,

$$z(t, u) = \lim_{\tau \to 0+} z(t, u, \tau)$$

We shall show that this procedure leads to the same result as we had earlier for the solution of (10): namely, that $z(t, u) = H(t - u)g(t - u)$, where $g =$

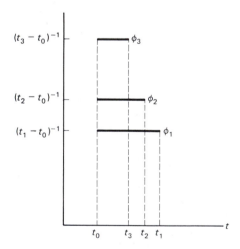

Figure 7.14 $\phi_i = \phi(t - t_0, t_i - t_0)$.

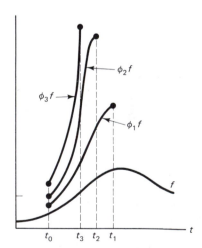

Figure 7.15 Graphs of $\phi_i f$.

$L^{-1}[1/P(s)]$. First observe that (see Figure 7.14 again)

$$z(t, u, \tau) = \int_0^t g(t - v)\phi(v - u, \tau)\, dv$$

$$= \begin{cases} 0, & t \le u \\[2mm] \dfrac{1}{\tau}\displaystyle\int_u^t g(t - v)\, dv, & u \le t \le u + \tau \\[4mm] \dfrac{1}{\tau}\displaystyle\int_u^{u+\tau} g(t - v)\, dv, & u + \tau \le t \end{cases}$$

Now let $t_0 > u$ be a fixed value of t; then for all sufficiently small positive τ, we have by the Mean Value Theorem for Integrals that

$$z(t_0, u, \tau) = \frac{1}{\tau}\int_u^{u+\tau} g(t - v)\, dv = g(t - \xi)$$

where $u < \xi < u + \tau$. Hence

$$z(t_0, u, \tau) \longrightarrow g(t - u) \qquad \text{as } \tau \to 0+$$

and the desired result is proven.

Thus the main use of the "δ-function" in this context is operational. Its use allows us to replace rather lengthy limit arguments by very brief symbolic "proofs." We remind the reader that a rigorous treatment of *distributions*, of which δ is an example, has been developed and is in widespread use in contemporary work in differential equations.†

Impulsive Forces

Another way to view the δ-function is in connection with a sudden sharp force acting on a dynamical system. For example, suppose that the weighted end of an oscillating spring is struck a hard blow (see Figure 7.16). How does the system respond? If we use the parameters of Section 5.5 for the motion of a spring, we may model the system by

$$x''(t) + \omega^2 x(t) = A\,\delta(t - T)$$
$$x(0) = \alpha \tag{15}$$
$$x'(0) = \beta$$

where $\omega^2 = k/m$, A is a positive constant, T is the time when the spring is struck, $T > 0$, and α and β are the initial data. The "force" $A\,\delta(t - T)$ is called an *impulsive force*, while its integral over time is the resulting *impulse*. Taking the transform of (15), we have that

---

† See, e.g., I. M. Gelfand and G. E. Shilov, *Generalized Functions and Operations* (New York: Academic Press, 1964).

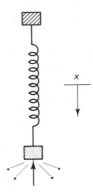

Figure 7.16 Blow to a spring.

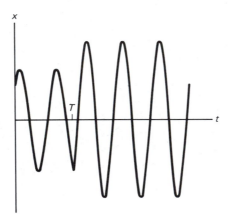

Figure 7.17 Response to blow at time t.

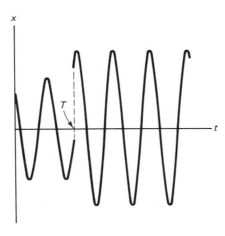

Figure 7.18 Response of the velocity to blow.

$$(s^2 + \omega^2)L[x] - s\alpha - \beta = Ae^{-Ts}$$

$$L[x] = \frac{s\alpha + \beta}{s^2 + \omega^2} + \frac{Ae^{-Ts}}{s^2 + \omega^2} \qquad (16)$$

Inverting (16), we have that

$$x(t) = \alpha \cos \omega t + \frac{\beta}{\omega} \sin \omega t + \frac{A}{\omega} H(t - T) \sin \omega(t - T) \qquad (17)$$

Note that $x(t)$ is continuous for all t, but $x'(t)$ has a jump discontinuity at $t = T$ (see Figures 7.17 and 7.18).

PROBLEMS

The following problem(s) may be more challenging: 4, 6, and 7.

1. Use the Laplace Transform to find the solution of each initial value problem.
   (a) $y'' + 2y' + 2y = \delta(t - \pi)$, $y(0) = 0$, $y'(0) = 0$.
   (b) $y'' + 4y = \delta(t - \pi) - \delta(t - 2\pi)$, $y(0) = y'(0) = 0$.
   (c) $y'' + 3y' + 2y = \sin t + \delta(t - \pi)$, $y(0) = 0$, $y'(0) = 0$.
   (d) $y'' + y = \delta(t - \pi) \cos t$, $y(0) = 0$, $y'(0) = 0$.
   (e) $y'' + y = e^t + \delta(t - 1)$, $y(0) = y'(0) = 0$.

2. Assume that $\delta(t)$ exists. Show that $L[\delta(t - u)](s) = e^{-us}$. [*Hint*: Use the Second Shifting Theorem.]

3. Show that $\delta(at) = (1/|a|)\delta(t)$, $a \neq 0$.

4. Show that if $ad \neq bc$, then

$$\delta(at + b\tau)\delta(ct + d\tau) = \frac{1}{|ad - bc|} \delta(t)\,\delta(\tau)$$

5. Let $f(t)$ be of exponential order; prove that $\delta(t) * f(t) = f(t)$.

6. A spring (spring constant k) supporting an object of mass 1 at equilibrium is subjected to a force $f(t) = A \sin \omega t$, $\omega^2 \neq k$, for $t \geq 0$. The mass is given a sharp blow at $t = 2$ that gives an impulse of 2 units. Use the Laplace Transform to solve for the motion of the system if the mass is at rest at its natural equilibrium at $t = 0$. Sketch the solution when $k = A = 1$.

7. Suppose that an LC circuit is subject to constant electromotive force E_0. At time $t = 1$ the circuit is dealt a sharp burst of EMF of size $2E_0$. Find the charge as a function of time if $q(0) = q'(0) = 0$.

CHAPTER 8

Linear Differential Equations with Nonconstant Coefficients

Linear differential equations that arise in the modeling of real systems often have nonconstant coefficients, and hence if the order is 2 or greater, the nonnumerical solution techniques of the previous chapters no longer apply. When the coefficient functions are well-enough behaved the solution of an initial value problem for such an equation can be expressed as an infinite series closely related to power series. New functions are introduced into mathematical physics in this way—as the sum of a series. In this chapter we systematize the use of such series to solve nonconstant-coefficient linear differential equations.

8.1 INTRODUCTION: AGING SPRINGS

The differential equations to be considered in this chapter are of the form

$$a_2(x)y''(x) + a_1(x)y'(x) + a_0(x)y(x) = 0 \tag{1}$$

in other words, linear homogeneous second-order equations whose coefficients may not be constant. We shall assume that the coefficient functions are either polynomials or convergent power series in x, as is the case in most of the applications.

Aging Springs

The spring "constant" of a coiled spring will gradually but inexorably diminish with advancing time until the spring stretches like a piece of chewing gum and finally snaps (see Figure 8.1). Springs are the central elements in a host of devices, such as openers, closers, dampers, and controllers, and their failure may have disastrous

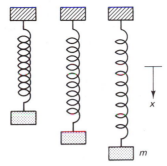

Figure 8.1 Aging spring.

consequences. For this reason, good models for the motion of a spring are of considerable importance. In one mathematical model for an aging spring we begin with the dynamical equation of an "ageless" spring,

$$mx''(t) + kx(t) = 0$$

and replace the elastic coefficient k by a decaying function of time, $ke^{-\epsilon t}$:

$$mx''(t) + ke^{-\epsilon t}x(t) = 0 \qquad (2)$$

What are solutions $x(t)$ of (2) like? Intuitively, we would expect $x(t)$ to oscillate as long as the elasticity $ke^{-\epsilon t}$ remains significant. Eventually, however, the energy of motion will overwhelm the feeble restoring force and $x(t)$ will increase unboundedly until the spring breaks or the model is no longer valid.†

All of this may be made precise if we knew more about the properties of solutions of (2) or had solution formulas. Although (2) is a linear differential equation, the techniques of Chapters 5 through 7 do not apply directly, since the coefficients are no longer constants. In this chapter we show how to use the form of the differential equation itself to deduce whether or not solutions oscillate, decay, or blow up. We also use a method of undermined coefficients to find solutions of equations such as (2) in the form of infinite series in powers of the independent variable:

$$x(t) = a_0 t^r + a_1 t^{r+1} + a_2 t^{r+2} + \cdots + a_n t^{r+n} + \cdots \qquad (3)$$

After carrying the method of undermined coefficients through the theory of Frobenius, we shall apply the techniques developed to Bessel's equations and functions.

Series Approach to Solving (1)

Notice that the coefficients in the equation of the aging spring can be written as the sum of convergent power series. With coefficients of this kind it often happens that the solutions of Eq. (1) may themselves be written as convergent series closely

---

† See Chapter 5 for the derivation of the equation of a vibrating spring. Displacement x may be measured in feet, mass m in pounds, time t in seconds, the coefficient k in pounds per second per second, and ϵ in seconds^{-1}.

related to power series. In later sections we determine just when and how this may be done, but the following simple example gives a rough idea of how power series can be used to solve linear differential equations.†

Example 8.1

Consider the initial value problem

$$I''(t) + I(t) = 0$$
$$I(0) = 1 \tag{4}$$
$$I'(0) = 0$$

for $I(t)$. We already know that the solution of this problem is $I = \cos t$, but we shall show how series can be used to solve the problem. Since (4) has a unique solution, the series solution constructed here must converge to $\cos t$. We begin by assuming no knowledge at all of the solution $I(t)$ of (4), except that it is a power series of the form

$$I(t) = \sum_{n=0}^{\infty} a_n t^n = a_0 + a_1 t + \cdots + a_n t^n + \cdots$$

convergent on an open interval containing $t = 0$. Our aim is to find the coefficients of the series and the interval of convergence.

Since the series is a solution of the differential equation, we must have that for all t inside the convergence interval,

$$\frac{d^2}{dt^2} \left(\sum_{n=0}^{\infty} a_n t^n \right) + \sum_{n=0}^{\infty} a_n t^n = 0$$

Since power series may be differentiated term by term inside the convergence interval, we have that

$$\sum_{n=2}^{\infty} n(n-1)a_n t^{n-2} + \sum_{n=0}^{\infty} a_n t^n = 0$$

where we have omitted the first two terms of the first sum since they have coefficients which vanish. Reindexing the first sum, we have that

$$\sum_{n=0}^{\infty} (n+2)(n+1)a_{n+2} t^n + \sum_{n=0}^{\infty} a_n t^n = 0$$

The two series may now be added term by term:

$$\sum_{n=0}^{\infty} [(n+2)(n+1)a_{n+2} + a_n] t^n = 0$$

According to the Identity Theorem, we must then have that

$$(n+2)(n+1)a_{n+2} + a_n = 0, \qquad n = 0, 1, 2, \ldots$$

† In Appendix D we outline the needed definitions and properties of power series.

or

$$a_{n+2} = - \frac{a_n}{(n+2)(n+1)}, \qquad n = 0, 1, 2, \ldots \qquad (5)$$

a so-called *recursion formula* for the as yet undetermined coefficients of the series solution.

Leaving (5) temporarily, let us see what effect the initial data of (4) have on the coefficients of the series:

$$I(0) = 1 = a_0 + a_1 \cdot 0 + a_2 \cdot 0^2 + \cdots = a_0$$
$$I'(0) = 0 = [a_1 + 2a_2 t + 3a_3 t^2 + \cdots]_{t=0} = a_1$$

Hence

$$a_0 = 1, \qquad a_1 = 0$$

Returning to the recursion formula (5) and using the values of a_0 and a_1, we have that

$$a_2 = - \frac{a_0}{2 \cdot 1} = - \frac{1}{2}, \qquad a_4 = - \frac{a_2}{4 \cdot 3} = \frac{1}{4 \cdot 3 \cdot 2}, \ldots$$

$$a_{2k} = - \frac{a_{2k-2}}{2k(2k-1)} = \frac{a_{2k-4}}{2k(2k-1)(2k-2)(2k-3)} = \cdots$$
$$= \frac{(-1)^k}{(2k)!}, \qquad \text{for } k = 1, 2, \ldots$$

and

$$a_3 = - \frac{a_1}{3 \cdot 2} = 0, \qquad a_5 = - \frac{a_3}{5 \cdot 4} = 0, \ldots$$

$$a_{2k+1} = - \frac{a_{2k-1}}{(2k+1)(2k)} = \cdots = \frac{(-1)^k}{(2k+1)!} a_1 = 0, \qquad \text{for } k = 1, 2, \ldots$$

Hence, inside the interval of convergence, the solution of the initial value problem is

$$I(t) = 1 - \frac{1}{2} t^2 + \frac{1}{24} t^4 - \cdots + \frac{(-1)^k}{(2k)!} t^{2k} + \cdots \qquad (6)$$

which is the Maclaurin series for cos t, as expected.

Thus under the assumption that the initial value problem (4) has a solution in the form of a power series that converges on some nontrivial interval, we have found the coefficients of the series. We may use the ratio test to determine the interval of convergence:

$$\lim_{k \to \infty} \left| \frac{(-1)^{k+1} t^{2(k+1)} / (2(k+1))!}{(-1)^k t^{2k} / (2k)!} \right| = \lim_{k \to \infty} \left| \frac{t^2}{(2k+2)(2k+1)} \right| = 0, \qquad \text{all } t$$

The interval of convergence is **R** itself in this case.

Comments

Although solutions in the form of infinite series may be unfamiliar to the reader, they do have a number of advantages. First, they are easy to use to estimate values of the solution. For example, we may use the first three terms of the power series expansion of the solution of the initital value problem (4) to approximate the solution:

$$I(t) \simeq 1 - \tfrac{1}{2}t^2 + \tfrac{1}{24}t^4$$

This approximation is quite good for small values of t (e.g., $|t| < 0.1$). In fact, from the error estimate for an alternating series, we have that

$$|I_{\text{approx}}(t) - I(t)| \leq \left| \frac{t^6}{720} \right| \leq 1.4 \times 10^{-9}$$

if $|t| < 0.1$. Of course, for t much larger than zero, we may have to use many terms in the series to achieve a reasonably accurate estimate. Because of the convergence properties of power series, we would expect the approximate solution to be most useful near the base point of expansion ($t_0 = 0$ in the example), and less useful far away from that point.

In the next section we give conditions that the solution of an initial value problem such as (4) have a power series solution which converges on a nontrivial interval. Knowing that there is a solution in series form, we may use methods like those above to determine the coefficients.

PROBLEMS

The following problem(s) may be more challenging: 1.

1. Create a model of an aging spring for which the elastic coefficient vanishes after a finite time. What would you expect to happen to the spring after that time? (Include suitable initial conditions in your model.) Compare your expectation with the actual solutions of a model equation with a vanishing elastic coefficient. How can the damping effect of friction be included in the model of an aging spring?

2. Determine the interval of convergence for the following series.

   (a) $\displaystyle\sum_{n=0}^{\infty} \frac{x^n}{n!}$.   (b) $\displaystyle\sum_{n=0}^{\infty} \frac{nx^n}{2^n}$.   (c) $\displaystyle\sum_{n=1}^{\infty} \frac{x^{2n+1}}{(2n+1)!}$.

3. Use the formula for a Taylor series to expand each function about the given point, and determine the interval of convergence.

   (a) e^x, $x_0 = 0$.   (b) $\sin x$, $x_0 = 0$.   (c) $\dfrac{1}{1+x}$, $x_0 = 0$.   (d) $\sqrt{x}$, $x_0 = 1$.

4. Reindex each of the following as a series of form $\Sigma(\cdot)x^n$.

   (a) $\displaystyle\sum_{n=0}^{\infty} \frac{2(n+1)}{n!} x^{n+1}$.   (b) $\displaystyle\sum_{n=2}^{\infty} n(n-1)a_n x^{n-2}$.   (c) $\displaystyle\sum_{n=1}^{\infty} (-1)^{n-1} \frac{x^{n+1}}{n(n+1)}$.

5. Use the Maclaurin series of e^x, $\sin x$, $\cos x$, and $1/(1 + x)$ to obtain the Maclaurin series expansions of each of the following functions.

   (a) $e^x + e^{-x}$.   (b) $\sin x - \cos x$.   (c) $\sin x \cos x$.   (d) $e^x/(1 + x)$.

6. Use the Identity Theorem to find a_n explicitly in terms of n, a_0, and a_1.

   (a) $\displaystyle\sum_{n=1}^{\infty} (na_n - n + 2)x^n = 0$.   (b) $\displaystyle\sum_{n=1}^{\infty} [(n + 1)a_n - a_{n-1}]x^n = 0$.

   (c) $\displaystyle\sum_{n=0}^{\infty} [(n + 2)(n + 1)a_{n+2} - a_n]x^n = 0$.

7. Show that $y = \displaystyle\sum_{0}^{\infty} \frac{1}{n!} x^n$ is the solution of $y'' - y = 0$, $y(0) = y'(0) = 1$ by direct substitution. How is this solution expressed in terms of elementary functions?

8. Use series to solve the problem $y'' - 4y = 0$, $y(0) = 2$, $y'(0) = 2$.

9. Use the Identity Theorem to solve the initial value problem $y'' - y = \displaystyle\sum_{n=0}^{\infty} x^n$, $y(0) = 0$,

   $y'(0) = 0$, where $|x| < 1$. $\left[Hint: y = \displaystyle\sum_{2}^{\infty} a_n x^n. \right]$

10. If you have access to a computer, use it to find the coefficient a_{100} in the series $\displaystyle\sum_{n=0}^{\infty} a_n x^n$

    if it is known that $\displaystyle\sum_{n=0}^{\infty} [(n + 1)^2 a_{n+2} - n^2 a_{n+1} + (n - 1)a_n]x^n = 0$ and that $a_0 = a_1 = 1$.

8.2 SERIES SOLUTIONS NEAR ORDINARY POINTS

Dividing by the leading coefficient, the differential equation (1) of the preceding section can be written in the normalized form

$$y'' + P(x)y' + Q(x)y = 0 \tag{1}$$

a more convenient form for expressing the results of this section. In this section we present a method for finding power series solutions of the form $\sum a_n(x - x_0)^n$ for (1) when the coefficients $P(x)$ and $Q(x)$ can be written as convergent power series of the same form. At the end of this section we use the method to find a series solution for the equation of the aging spring.

Ordinary Points

Before we begin it is useful to recall the definition of *real analyticity* (see Appendix D):

> **Real Analyticity.**   A real-valued function f defined on an interval I is said to be *real analytic* at a point x_0 in I if it is the sum of a power series $\sum a_n(x - x_0)^n$ with a nonzero radius of convergence.

We shall be interested in finding power series solutions of (1) in powers of $x - x_0$, where x_0 is an ordinary point of the equation. An ordinary point is defined as follows:

> **Ordinary Point.** x_0 is an *ordinary point* of equation (1) if both $P(x)$ and $Q(x)$ are real analytic at x_0.

The opposite of "ordinary" is "singular":

> **Singular Point.** x_0 is a singular point of (1) if $P(x)$ or $Q(x)$ is not real analytic at x_0.

If x_0 is an ordinary point, $P(x)$ and $Q(x)$ have Taylor series

$$P(x) = \sum_0^\infty p_n (x - x_0)^n, \qquad Q(x) = \sum_0^\infty q_n (x - x_0)^n$$

which converge inside a common interval centered at x_0. The coefficients p_n and q_n of these Taylor series are given by

$$p_n = \frac{P^{(n)}(x_0)}{n!}, \qquad q_n = \frac{Q^{(n)}(x_0)}{n!}$$

The identification of the ordinary points of Eq. (1) is the first step in constructing series solutions.

Example 8.2

Every point x_0 is an ordinary point for the constant coefficient equation $y'' + p_0 y' + q_0 y = 0$. Every point x_0 is also an ordinary point if $P(x)$ and $Q(x)$ are polynomials in x. For instance, the polynomial $P(x) = 1 - x^2$ can be written as a finite Taylor series about x_0:

$$\begin{aligned} P(x) &= P(x_0) + P'(x_0)(x - x_0) + \tfrac{1}{2} P''(x_0)(x - x_0)^2 \\ &= 1 - x_0^2 - 2x_0(x - x_0) - (x - x_0)^2 \end{aligned}$$

Example 8.3

Every point x_0 is an ordinary point of $y'' + e^{-x} y = 0$ since $Q(x) \equiv e^{-x}$ has a convergent Taylor series in powers of $x = x_0$ given by

$$e^{-x} = \sum_{n=0}^\infty \frac{1}{n!} (-1)^n e^{-x_0} (x - x_0)^n$$

a series that converges to e^{-x} for all x.

Example 8.4

0 is an ordinary point of $y'' + [1/(1 - x)]y = 0$ since $1/(1 - x)$ has a convergent geometric series expansion,

$$\frac{1}{1-x} = \sum_{n=0}^{\infty} x^n, \qquad \text{where } |x| < 1$$

However, 1 is a singular point since $1(1-x)$ is not defined at $x = 1$.

A second-order equation of the form

$$a_2(x)y'' + a_1(x)y' + a_0(x)y = 0 \tag{2}$$

may be converted to one of form (1) by dividing by $a_2(x)$, although one must exercise some care at points where $a_2(x)$ vanishes.

Example 8.5

The equation $(1 - x)y'' + y = 0$ is equivalent to the equation $y'' + [1/(1 - x)]y = 0$ of Example 8.4 on any interval not containing the point $x = 1$, but the two equations may not be equivalent on intervals containing 1.

Series Solutions Near an Ordinary Point: An Example

Let x_0 be an ordinary point of (1). The Null Space of $P(D)$ Theorem (Section 5.2) guarantees that the solution space of (1) in a neighborhood of x_0 is two-dimensional and is spanned by any pair of independent solutions. We shall actually show that all solutions of (1) near x_0 have the form of power series, $\sum_0^{\infty} a_n(x - x_0)^n$, and hence are real analytic at x_0. For the moment we assume that solutions of this form do exist and find necessary conditions on the coefficients a_n which must be satisfied, as in the following example.

Example 8.6

The equation to be solved is

$$y'' - xy' + y = 0 \tag{3}$$

In this simple equation $P(x) = -x$ and $Q(x) = 1$. Evidently, all points x_0 are ordinary and we look for series solutions of the form $\sum a_n(x - x_0)^n$. For simplicity we set $x_0 = 0$ and find solutions of the form $y = \sum_{n=0}^{\infty} a_n x^n$. We must find the undetermined coefficients a_n, $n = 0, 1, 2, \ldots$. Differentiating the series term by term, we have that inside the interval of convergence,

$$y' = \sum_{n=1}^{\infty} na_n x^{n-1} \quad \text{and} \quad y'' = \sum_{n=2}^{\infty} n(n-1)a_n x^{n-2}$$

Hence

$$\sum_{n=2}^{\infty} n(n-1)a_n x^{n-2} - x \sum_{n=1}^{\infty} na_n x^{n-1} + \sum_{n=0}^{\infty} a_n x^n = 0 \tag{4}$$

Since

$$\sum_{n=2}^{\infty} n(n-1)a_n x^{n-2} \equiv \sum_{n=0}^{\infty} (n+2)(n+1)a_{n+2} x^n$$

and

$$-x \sum_{n=1}^{\infty} na_n x^{n-1} \equiv \sum_{n=1}^{\infty} (-n)a_n x^n \equiv \sum_{n=0}^{\infty} (-n)a_n x^n$$

we may rewrite (4) as

$$\sum_{n=0}^{\infty} [(n+2)(n+1)a_{n+2} - na_n + a_n]x^n = 0$$

By the Identity Theorem,

$$(n+2)(n+1)a_{n+2} - na_n + a_n = 0, \qquad n = 0, 1, 2, \ldots \qquad (5)$$

Thus we have the *recursion formula* for the coefficients:

$$a_{n+2} = \frac{(n-1)a_n}{(n+2)(n+1)}, \qquad n = 0, 1, 2, \ldots \qquad (6)$$

Given the first two coefficients a_0 and a_1, we may use the recursion formula to determine all others. For example,

$$a_2 = \frac{-a_0}{2 \cdot 1} = \frac{-a_0}{2}, \quad a_4 = \frac{a_2}{4 \cdot 3} = -\frac{1}{24} a_0, \quad a_6 = \frac{3a_4}{6 \cdot 5} = \frac{-a_0}{240}, \quad \cdots$$

$$a_3 = \frac{0a_1}{3 \cdot 2} = 0, \quad a_5 = \frac{2a_3}{5 \cdot 4} = 0, \quad \cdots, \quad a_{2n+1} = 0, \quad \cdots, \quad n \geqq 1$$

The vanishing of all but the first of the odd-indexed coefficients is a fortunate occurrence here, but such simplicity is not common.

Solutions of (3) may be written as

$$y = a_0[1 - \tfrac{1}{2}x^2 - \tfrac{1}{24}x^4 - \tfrac{1}{240}x^6 - \cdots] + a_1 x \qquad (7)$$

where a_0 and a_1 are arbitrary constants. In fact, $a_0 = y(0)$ and $a_1 = y'(0)$, which might be taken as initial data. We have shown that the solution space of Eq. (3) is spanned by the particular solutions

$$y_1 = 1 - \tfrac{1}{2}x^2 - \tfrac{1}{24}x^4 - \tfrac{1}{240}x^6 - \cdots, \qquad y_2 = x \qquad (8)$$

There remain two unresolved questions about the solutions of this problem: What is the interval of convergence? What is the specific value of the nth coefficient a_n? Both questions may be answered quite simply.

Example 8.7 (*Interval of Convergence*)

Continuing with Example 8.6 we find the interval of convergence of the bracketed series in (7). This series has the form $\sum_{k=0}^{\infty} a_{2k}x^{2k}$, where

$$a_{2k+2} = \frac{(2k-1)a_{2k}}{(2k+2)(2k+1)}, \qquad k = 0, 1, \ldots \qquad (9)$$

which we obtain from the recursion formula (6) by setting $n = 2k$. Applying the Ratio Test, we have that

$$\lim_{k \to \infty} \left| \frac{a_{2k+2}x^{2k+2}}{a_{2k}x^{2k}} \right| = x^2 \lim_{k \to \infty} \left| \frac{2k-1}{(2k+2)(2k+1)} \right| = 0, \qquad \text{all } x$$

Thus the series in (7) converges for all x. The reader may show by differentiation of (7) (which is now justified since the series converges everywhere) that (7) does indeed define solutions of (3).

Occasionally, it is feasible to "solve" a recursion formula for the coefficients of the power series expansion of a solution:

Example 8.8 (*Solving the Recursion Formula*)

Continuing further with Example 8.6 we have from (9) that

$$
\begin{aligned}
a_{2k+2} &= \frac{(2k-1)a_{2k}}{(2k+2)(2k+1)} = \frac{2k-1}{(2k+2)(2k+1)} \cdot \frac{(2k-3)a_{2k-2}}{2k(2k-1)} = \cdots \\
&= \frac{(2k-1)(2k-3) \cdots 5 \cdot 3 \cdot 1 \cdot (-1)a_0}{(2k+2)(2k+1) \cdots 4 \cdot 3 \cdot 2 \cdot 1} \\
&= -\frac{(2k-1)(2k-3) \cdots 5 \cdot 3 \cdot 1}{(2k+2)!} a_0 \\
&= -\frac{(2k)!}{2^k k!(2k+2)!} a_0 = -\frac{a_0}{2^k k!(2k+2)(2k+1)}, \qquad k = 0, 1, 2, \ldots
\end{aligned}
$$

since

$$
\begin{aligned}
(2k-1)(2k-3) & \cdots 5 \cdot 3 \cdot 1 \\
&= \frac{(2k)(2k-1)(2k-2)(2k-3) \cdots 5 \cdot 4 \cdot 3 \cdot 2 \cdot 1}{(2k)(2k-2) \cdots 4 \cdot 2} = \frac{(2k)!}{2^k \cdot k!}
\end{aligned}
$$

Hence the general solution of (3) may be written in exact form as

$$y = a_0 \left\{ 1 - \sum_{k=0}^{\infty} \frac{1}{2^k k!(2k+2)(2k+1)} x^{2k+2} \right\} + a_1 x \qquad (10)$$

Although it has not been difficult to "solve" the recursion formula for the coefficients in this case, it is not always useful to do so, even when easy. The "unsolved" recursion formula (6) is already in good form for a computer program to evaluate as many coefficients as desired. The computer must be provided with a_0 and a_1, the formula itself, and a range for n (e.g., $n = 2$ to 100). The computer will do all the work simply by iterating (6) for successive even values of n, beginning with $n = 2$.

Each value of a_0 and a_1 in (10) yields a new function, a function defined by the series. Each of these functions is real analytic for all x; moreover, the functions are easy to differentiate or integrate—just differentiate or integrate the series term by term. Thus the initial value problem $y'' - xy' + y = 0$, $y(0) = a_0$, $y'(0) = a_1$ is a source of new real analytic functions. This equation is an alternate form of *Hermite's equation* of order 1 which appears in quantum mechanics (see Problem 5).

Convergence

We have carried out the details of the example above to show how a method of undetermined coefficients works in practice. Of course, all the work would be in vain if the constructed series failed to converge to the solutions, or perhaps did not converge at all. The following result assures that the method always succeeds. It is stated for the initial value problem

$$y'' + P(x)y' + Q(x)y = 0$$
$$y(0) = a_0, \qquad y'(0) = a_1$$

(11)

Convergence Theorem.   Let $P(x) = \sum\limits_{n=0}^{\infty} p_n x^n$, $Q(x) = \sum\limits_{n=0}^{\infty} q_n x^n$, where the series converge on a common interval $J = (-\rho, \rho)$, $\rho > 0$. Then (11) has the unique solution $y = \sum\limits_{n=0}^{\infty} a_n x^n$, where the coefficients are determined by the *recursion formula,*

$$(n + 2)(n + 1)a_{n+2} + \sum_{k=0}^{n} [p_{n-k}a_{k+1}(k + 1) + q_{n-k}a_k] = 0 \qquad (12)$$

The series $\sum\limits_{n=0}^{\infty} a_n x^n$ is real analytic on J.

Observe that the recursion formula expresses a_{n+2} in terms of a_k, $k < n + 2$, and thus (12) may be used to find all coefficients once a_0 and a_1 are given. In practice, (12) is rarely used for a specific equation. It is much easier to insert the unknown series $\sum a_n x^n$ and its respective derivative series into the differential equation, group the coefficients of like powers of x as we did above, and derive the recursion formula directly. The proof of the convergence theorem is omitted.

Series Solutions of the Equation of the Aging Spring

A differential equation for the vibrations of an undamped but aging spring (Figure 8.2) is $mx'' + ke^{-\epsilon t}x = 0$, where m, k, and ϵ are positive constants. Suppose that m and k have the same numerical value. Since the exponential coefficient is real

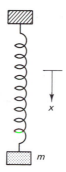

Figure 8.2   Aging spring.

analytic on the entire t-axis, we may replace that coefficient by its Maclaurin series. The equation of motion meets all conditions of the convergence theorem with, say, $t_0 = 0$ as the ordinary point about which solutions will be expanded in power series. Suppose that the spring is stretched 1 unit from its equilibrium position and then released:

$$x''(t) + e^{-\epsilon t}x(t) \equiv x''(t) + \left[\sum_0^\infty \frac{(-\epsilon t)^n}{n!} \right] x(t) = 0 \tag{13}$$

$$x(0) = 1, \qquad x'(0) = 0$$

The solution is thus a power series, $\sum_0^\infty a_n t^n$, which converges for all t. Thus

$$\sum_{n=2}^\infty n(n-1)a_n t^{n-2} + \left[\sum_{n=0}^\infty \frac{(-\epsilon)^n}{n!} t^n \right]\left[\sum_{n=0}^\infty a_n t^n \right] = 0$$

Reindexing the first series and applying the product formula to the second and third yields

$$\sum_{n=0}^\infty (n+2)(n+1)a_{n+2}t^n + \sum_{n=0}^\infty \sum_{k=0}^n \left[\frac{(-\epsilon)^k}{k!} a_{n-k} \right] t^n = 0$$

Upon adding the two series term by term, we have

$$\sum_{n=0}^\infty \left[(n+2)(n+1)a_{n+2} + \sum_{k=0}^n \frac{(-\epsilon)^k}{k!} a_{n-k} \right] t^n = 0$$

Applying the Identity Theorem, we obtain the recursion formula

$$a_{n+2} = -\frac{1}{(n+2)(n+1)} \sum_{k=0}^n \frac{(-\epsilon)^k}{k!} a_{n-k} \tag{14}$$

which expresses a_{n+2} in terms of $a_0, a_1, \ldots, a_n$. This recursion formula cannot easily be solved for a_{n+2} in terms of a_0 or a_1, but as we noted before this does not restrict its usefulness.

To find the solution of (13), we set $a_0 = 1$ and $a_1 = 0$ [since we require that

$x(0) = 1$ and $x'(0) = 0$]. We may use (14) to find the first few coefficients of the series solution:

$$n = 0: \quad a_2 = -\frac{1}{2}\frac{(-\epsilon)^0}{0!}\,a_0 = -\frac{1}{2}$$

$$n = 1: \quad a_3 = -\frac{1}{3\cdot 2}\sum_{k=0}^{1}\frac{(-\epsilon)^k}{k!}\,a_{1-k} = -\frac{1}{6}[a_1 - \epsilon a_0] = \frac{\epsilon}{6}$$

$$n = 2: \quad a_4 = -\frac{1}{4\cdot 3}\sum_{k=0}^{2}\frac{(-\epsilon)^k}{k!}\,a_{2-k} = -\frac{1}{12}\left[a_2 - \epsilon a_1 + \frac{\epsilon^2}{2}a_0\right]$$

$$= \frac{1}{24} - \frac{\epsilon^2}{24}$$

Thus the solution is

$$x(t) = 1 - \frac{1}{2}t^2 + \frac{\epsilon}{6}t^3 + \left(\frac{1}{24} - \frac{\epsilon^2}{24}\right)t^4 + \cdots \tag{15}$$

The reader may note that if we set $\epsilon = 0$, then (13) has the obvious solution $x = \cos t$ and (15) is just the Maclaurin series for $\cos t$.

Comments

This example and Example 8.7 illustrate both the advantages and the defects of solving initial value problems by series. On the one hand, we do have solution formulas which may be used to evaluate the solutions, at least approximately, for specific values of the independent variable. On the other hand, although the series solutions may converge quite rapidly near the ordinary point at which the expansions occur, the convergence may be very slow farther away. Second, solutions may not even be expandable as power series if one of the coefficient functions $P(x)$ or $Q(x)$ is not real analytic. For example, the method fails to produce a solution of the equation $y''(x) + |x|y(x) = 0$ on any interval containing $x = 0$ since the coefficient function $|x|$ is not real analytic at 0. Nevertheless, power series solutions are useful and appear frequently in both theory and applications.

PROBLEMS

The following problem(s) may be more challenging: 4, 5, and 6.

1. Determine if the given point x_0 is ordinary.
   (a) $y'' + k^2 y = 0$, x_0 arbitrary (k is a constant).
   (b) $y'' + \dfrac{1}{1+x}y = 0$, $x_0 = -1$.   (c) $y'' + \dfrac{1}{x}y' - y = 0$, $x_0 = 0$.

2. Find all singular points of the following equations.

(a) $y'' + (\sin x)y' + (\cos x)y = 0.$ (b) $y'' - (\ln|x|)y = 0.$
(c) $y'' + |x|y' + e^{-x}y = 0.$ (d) $(1-x)y'' + y' + (1-x^2)y = 0.$
(e) $(1-x^2)y'' + xy' + y = 0.$ (f) $x^2y'' + xy' + y = 0.$

3. Verify that $x = 0$ is an ordinary point and find all solutions as power series in x. Determine the interval of convergence of the series. Sketch graphs.

   (a) $y'' + 4y = 0.$ (b) $y'' + xy' + 3y = 0.$
   (c) $y'' - 4y = 0.$ (d) $(x^2+1)y'' - 6y = 0, \quad y(0) = y'(0) = 1.$

4. [*Airy's Equation*]. Find all solutions of the equation $y'' + xy = 0.$ (Certain solutions of the equation are called *Airy functions* and are used in modeling the phenomenon of the diffraction of light.)

5. [*Hermite's Equation*]. $y'' - 2xy' + ny = 0,$ where n is a nonnegative integer, is *Hermite's equation of order n*. Show that if n is an even integer, solutions $y(x)$ of Hermite's equation that satisfy $y(0) = 0$ are polynomials of degree $n/2$. Find all polynomial solutions when $n = 0, 2, 4, 6.$

6. [*Chebyshev's Equation*]. $(1-x^2)y'' - xy' + \alpha^2y = 0$ is *Chebyshev's equation of order α*. Show that if α is a nonnegative integer n, the equation has polynomial solutions of degree n. Find all polynomial solutions when $n = 0, 1, 2, 3.$

7. [*Mathieu's Equation*]. Calculate the first four nonvanishing coefficients in the power series expansion of the solution of the Mathieu equation, $y'' + (1 + 2 \cos x)y = 0,$ with initial conditions $y(0) = 1, y'(0) = 0.$

8. Assume a solution of the form $y = \sum_0^\infty a_n x^n$ and calculate the first three nonvanishing coefficients.

   (a) $y' = 1 + xy^2, \quad y(0) = 0.$
   (b) $y' = x^2 + y^2, \quad y(0) = 0.$

9. Calculate the first four coefficients in the power series expansion of the general solution of the equation $y'' + (1 - e^{-x^2})y = 0.$

10. Estimate $y(1)$ to three decimal places if $y(x)$ is the solution of $y'' + x^2y' + 2xy = 0,$ $y(0) = 1, y'(0) = 0.$

8.3 LEGENDRE POLYNOMIALS

The equation

$$(1 - x^2)y'' - 2xy' + \alpha(\alpha + 1)y = 0 \tag{1}$$

where α is a nonnegative constant, is known as *Legendre's equation of order α*.† It arises in connection with a useful technique for calculating gravitational and electrical potentials as well as in many other areas of applied mathematics. We outline below some of the properties of the solutions to this equation, especially when α is an integer.

Observe that $x_0 = 0$ is an ordinary point of Legendre's equation for any number α since the coefficients

---

† Adrien Marie Legendre (1752–1833) studied Eq. (1) and its solutions in connection with the gravitational potential of a solid sphere.

$$P(x) = \frac{-2x}{1-x^2}, \qquad Q(x) = \frac{\alpha(\alpha+1)}{1-x^2}$$

are real analytic at $x_0 = 0$. Since $P(x)$ and $Q(x)$ have power series based at $x_0 = 0$ which converge on the interval $I = (-1, +1)$, we see that every solution of Eq. (1) is real analytic on I. In particular, every series solution $\sum a_n x^n$ of (1) converges at least on the interval I. Let $y = \sum a_n x^n$ be a solution of (1) with coefficients a_n to be determined. Substituting directly into (1), we have

$$(1-x^2) \sum_0^\infty n(n-1)a_n x^{n-2} - 2x \sum_0^\infty na_n x^{n-1} + \alpha(\alpha+1) \sum_0^\infty a_n x^n = 0$$

$$\sum_0^\infty [(n+2)(n+1)a_{n+2} + (\alpha+n+1)(\alpha-n)a_n]x^n = 0$$

where we have reindexed to obtain the last equation. The recursion relation is

$$a_{n+2} = \frac{-(\alpha+n+1)(\alpha-n)}{(n+2)(n+1)} a_n, \qquad n = 0, 1, 2, \ldots$$

Using induction and noticing how the index skips a place, we obtain the two formulas

$$a_{2n} = (-1)^n \frac{(\alpha+2n-1)(\alpha+2n-3) \cdots (\alpha+1)(\alpha)(\alpha-2) \cdots (\alpha-2n+2)}{(2n)!} a_0 \qquad (2)$$

$$a_{2n+1} = (-1)^n \frac{(\alpha+2n)(\alpha+2n-2) \cdots (\alpha+2)(\alpha-1)(\alpha-3) \cdots (\alpha-2n+1)}{(2n+1)!} a_1 \qquad (3)$$

The solution of (1) together with initial conditions $y(0) = y_0$ and $y'(0) = 0$ is

$$y_1 = y_0 \left(1 + \sum_1^\infty a_{2n} x^{2n}\right) \qquad (4)$$

where a_{2n} is defined by (2) with $a_0 = 1$. Similarly, the solution of (1) with initial conditions $y(0) = 0$ and $y'(0) = y_0'$ is

$$y_2 = y_0' \left(x + \sum_1^\infty a_{2n+1} x^{2n+1}\right) \qquad (5)$$

where a_{2n+1} is defined by (3) with $a_1 = 1$. The ratio test shows that the series for (4) and (5) converge at least on $(-1, 1)$. Indeed, if α is nonintegral, we have for any n that

$$\left| \frac{a_{2n+2} x^{2n+2}}{a_{2n} x^{2n}} \right| = \left| \frac{(\alpha+2n+1)(\alpha-2n)}{(2n+2)(2n+1)} \right| |x|^2$$

$$\left| \frac{a_{2n+1} x^{2n+1}}{a_{2n-1} x^{2n-1}} \right| = \left| \frac{(\alpha+2n)(\alpha-2n+1)}{(2n+1)(2n)} \right| |x|^2 \qquad (6)$$

Since both quotients tend to $|x|^2$ as $n \to \infty$, both (4) and (5) have $I = (-1, +1)$ as their interval of convergence. Thus $\{y_1, y_2\}$ is a basis of the solution space of Legendre's equation of order α on $(-1, 1)$ [y_1 and y_2 are independent since their Wronskian does not vanish when $x = 0$]. The case where α is an integer is treated below.

Observe that we could also find series solutions of Legendre's equation in powers of $x - x_0$ since any x_0 with $|x_0| \neq 1$ is an ordinary point of the equation. These series converge at least in the interval $(x_0 - \beta, x_0 + \beta)$, where $\beta = |1 - |x_0||$.

Properties of Legendre Polynomials

It follows from (2)–(5) that Legendre's equation of order α has a nontrivial polynomial solution if and only if α is an integer. Observe that polynomial solutions of (1) for integral α have the same parity as α (i.e., the polynomial solution is even if α is even, and odd if α is odd). Moreover, when α is an integer all polynomial solutions of (1) are scalar multiples of any single one. To be specific, let $\alpha = n$, a nonnegative integer, and denote by $P_n(x)$ that polynomial solution of (1) such that $P_n(1) = 1$. This can be done by adjusting the arbitrary constants y_0 and y_0' in (4) and (5). This leads to the following formula:

$$P_n(x) = 2^{-n} \sum_{k=0}^{[n/2]} \frac{(-1)^k (2n - 2k)!}{k!(n-k)!(n-2k)!} x^{n-2k}, \qquad n = 0, 1, 2, \ldots \qquad (7)$$

where $[n/2]$ denotes the greatest integer not larger than $n/2$. The polynomial $P(x)$, $n = 0, 1, 2, \ldots$, is known as the *Legendre polynomial of order n*. Legendre polynomials have many important properties; we list some of them below, but omit many of the proofs.

Equation (7) is not a very efficient way to calculate $P_n(x)$, but $P_n(x)$ can be calculated easily, once $P_{n-2}(x)$ and $P_{n-1}(x)$ are known.

Recursion Theorem.   For any $n = 2, 3, \ldots$ we have the recursion relation

$$nP_n(x) - (2n - 1)xP_{n-1}(x) + (n - 1)P_{n-2}(x) = 0 \qquad (8)$$

The simplest way to verify that recursion relation (8) holds is by direct substitution from (7). Using the fact that $P_0(x) = 1$ and $P_1(x) = x$, we can use (8) to generate $P_2, P_3, \ldots$, recursively. Other recursion relations satisfied by the P_n are listed in Table 8.1.

Next we have the

Orthogonality Property for P_n.   For any two distinct nonnegative integers n and m,

$$\int_{-1}^{+1} P_n(x) P_m(x)\, dx = 0 \qquad (9)$$

TABLE 8.1 LEGENDRE POLYNOMIALS

Rodrigues's formula	$P_n(x) = \dfrac{1}{2^n n!} \dfrac{d^n}{dx^n} [(x^2 - 1)^n]$
Polynomial expansion†	$P_n(x) = 2^{-n} \displaystyle\sum_{k=0}^{[n/2]} \dfrac{(-1)^k (2n - 2k)!}{k!(n-k)!(n-2k)!} x^{n-2k}$
First few polynomials	$P_0(x) = 1 \qquad\qquad P_1(x) = x$
	$P_2(x) = \dfrac{1}{2}(3x^2 - 1) \qquad P_3(x) = (5x^3 - 3x)/2$
	$P_4(x) = \dfrac{1}{8}(35x^4 - 30x^2 + 3) \quad P_5(x) = \dfrac{1}{8}(63x^5 - 70x^3 + 15x)$
Roots and magnitudes	P_n has n distinct real roots, all lying in $(-1, 1)$; $\|P_n(x)\| \leq 1$, x in $[-1, 1]$;
	$P_n(1) = 1, \quad P_n(-1) = (-1)^n, \quad P_{2n+1}(0) = 0, \quad P_{2n}(0) = (-1)^n \dfrac{(2n)!}{2^{2n}(n!)^2}$
Identities	$P'_{n+1} - xP'_n = (n+1)P_n \qquad\qquad xP'_n - P'_{n-1} = nP_n$
	$P_n = xP_{n-1} + \dfrac{x^2 - 1}{n} P'_{n-1} \qquad (x^2 - 1)P'_n = nxP_n - nP_{n-1}$
Differential equation	$(1 - x^2)y'' - 2xy' + n(n+1)y = 0$
Recursion formula	$nP_n(x) - (2n - 1)xP_{n-1}(x) + (n - 1)P_{n-2}(x) = 0$
Orthogonality	$\displaystyle\int_{-1}^{+1} P_n(x)P_m(x)\,dx = \begin{cases} 0, & n \neq m \\[2mm] \dfrac{2}{2n+1}, & n = m \end{cases}$

† $[n/2]$ denotes the greatest integer $\leq n/2$.

Proof. Since

$$(1 - x^2)P''_n - 2xP'_n + n(n+1)P_n = 0$$
$$(1 - x^2)P''_m - 2xP'_m + m(m+1)P_m = 0$$

if the first equation is multiplied by P_m and the second by P_n, subtraction yields

$$(1 - x^2)[P''_n P_m - P''_m P_n] - 2x[P_m P'_n - P_n P'_m] + [n(n+1) - m(m+1)]P_n P_m = 0 \tag{10}$$

The first two terms of (10) are

$$\frac{d}{dx}[(1 - x^2)(P'_n P_m - P'_m P_n)]$$

and integrating both sides of (10) from -1 to $+1$, we have that

$$(1 - x^2)(P'_n P_m - P'_m P_n)\,\Big|_{-1}^{+1} = [m(m+1) - n(n+1)]\int_{-1}^{+1} P_n(x)P_m(x)\,dx$$

Since the left-hand side of this equality vanishes and $n \neq m$, (9) follows.

Finally,

Roots and Boundedness Theorem

(a) $P_n(x)$, $n > 0$, has n distinct real roots, all lying in the interval $(-1, +1)$ (see Figure 8.3).

(b) For any $n = 0, 1, 2, \ldots,$

$$|P_n(x)| \leq 1 \qquad \text{for all} - 1 \leq x \leq +1 \tag{11}$$

Proof. (a) Since $\int_{-1}^{1} P_n(x)\,dx = \int_{-1}^{1} P_n(x)P_0(x)\,dx = 0$ if $n > 0$ [by (9)], P_n has at least one root in $(-1, 1)$. Let $x_1, \ldots, x_k$ be the real roots of P_n in $(-1, 1)$. These roots are simple. For if x_j were a multiple root, then $P_n(x_j) = P_n'(x_j) = 0$. However, by the Existence and Uniqueness Theorem, these conditions determine a unique solution of Legendre's equation. Since the trivial solution is one solution of this initial value problem and P_n is another, we must have $P_n(x) \equiv 0$. This contradiction shows that the roots are simple. Let $Q_k(x)$ be the polynomial of degree k, $Q_k(x) = (x - x_1) \cdots (x - x_k)$. Since $P(x)$ and $Q(x)$ each change sign at each x_j, we have that $P(x)Q(x)$ has fixed sign throughout $[-1, 1]$ and hence $\int_{-1}^{+1} P_n(x)Q(x)\,dx \neq 0$. However, if Q has degree less than n, using the result of Problem 3 in the problem set we see that $\int_{-1}^{+1} P_n(x)Q(x)\,dx = 0$. Hence $Q(x)$ has degree n. Consequently, P has n distinct roots in $(-1, 1)$.

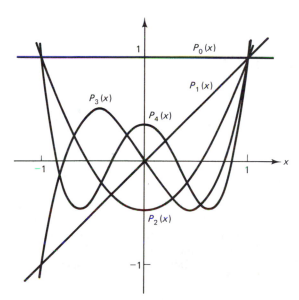

Figure 8.3 Graphs of $P_n(x)$ for $n = 0, 1, 2, 3, 4$.

(b) Squaring the identity (see Table 8.1) $P'_n = nP_{n-1} + xP'_{n-1}$, multiplying through by $(1 - x^2)/n^2$, and adding it to the square of the identity $P_n = xP_{n-1} + [(x^2 - 1)/n]P'_{n-1}$ (see Table 8.1 again) we obtain, for $|x| \leq 1$,

$$0 \leq \frac{1 - x^2}{n^2}(P'_n)^2 + P_n^2 < \frac{1 - x^2}{(n-1)^2}(P'_{n-1})^2 + P_{n-1}^2 \tag{12}$$

Since $[(1 - x^2)/n^2](P'_n)^2 + P_n^2 = 1$ when $n = 1$, we conclude from (12) that for all $n \geq 1$, $[(1 - x^2)/n^2](P'_n)^2 + P_n^2 \leq 1$. Inequality (11) follows.

Comments

Table 8.1 lists the more important properties of Legendre polynomials, but these polynomials have many other properties not listed here.

PROBLEMS

The following problem(s) may be more challenging: 3.

1. For all $n = 0, 1, 2, \ldots$ show that

$$P_{2n+1}(0) = 0, \qquad P_n(-1) = (-1)^n, \qquad P_{2n}(0) = (-1)^n \frac{(2n)!}{2^{2n}(n!)^2}$$

2. Show that x and

$$\frac{1}{2}x \ln\left(\frac{1+x}{1-x}\right) - 1$$

are solutions of Legendre's equation of order 1. Find all solutions of this equation without using infinite series. [*Hint*: Use a reduction-of-order method from the problem set of Section 5.2.]

3. For any $n > 0$, show that

$$\int_{-1}^{1} P_n(x)Q(x)\, dx = 0 \qquad \text{for all polynomials } Q \text{ with } \deg Q < n$$

[*Hint*: Since $\deg P_n = n$ for all $n = 0, 1, 2, \ldots$, note that there are constants b_j such that $Q(x) = \sum_{j=0}^{n-1} b_j P_j(x)$. Then use (9).]

4. Legendre's equation arises when solving the equation for electric potential in spherical coordinates (Laplace's equation) and it appears in the form $z''(\phi) + (\cot \phi)\, z'(\phi) + n(n+1)z(\phi) = 0$, $0 < \phi < \pi$. Show that the change of variable $x = \cos \phi$ and $y = z(\arccos x)$ leads to the Legendre equation of integer order n in y and x.

8.4 SINGULAR POINTS: EULER EQUATIONS

The method of undetermined coefficients is successful in constructing convergent power series solutions of a second-order linear differential equation on an interval centered at an ordinary point of the equation. If the point is not ordinary, but singular, the method may fail. Since a number of equations that come up in modeling phenomena of wave motion, heat flow, or electric potential are of interest primarily in the neighborhood of singular points, new methods of finding solutions are needed. In the 100 years or so since the need was recognized, methods have been found that will handle most of the singularities likely to occur in practice, at least if the singularities are "regular" in a sense to be defined in this section. The simplest class of equations with regular singularities may be solved explicitly without series, and we shall do so here. Solutions behave strangely near a regular singularity, as might be expected from the name, but not too strangely (i.e., "regular").

Regular Singular Points

We are interested in solving the equation

$$a_2(x)y'' + a_1(x)y' + a_0(x)y = 0 \tag{1}$$

near a point x_0 where the leading coefficient $a_2(x)$ has an isolated zero. It is assumed here and throughout the section that $a_0(x)$, $a_1(x)$, and $a_2(x)$ are real analytic at x_0. If one or both of the limits

$$\lim_{x \to x_0} \frac{a_1(x)}{a_2(x)}, \qquad \lim_{x \to x_0} \frac{a_0(x)}{a_2(x)}$$

fails to exist, we say that x_0 is a *singular point* of (1). There are several types of singularities, but the simplest are

> **Regular Singularities.**   The point x_0 is a *regular singularity* of Eq. (1) if x_0 is an isolated zero of $a_2(x)$ and the limits
>
> $$\lim_{x \to x_0} \frac{(x - x_0)a_1(x)}{a_2(x)}, \qquad \lim_{x \to x_0} \frac{(x - x_0)^2 a_0(x)}{a_2(x)}$$
>
> both exist.

If the two limits do exist, the functions

$$P(x) = \frac{(x - x_0)a_1(x)}{a_2(x)}, \qquad Q(x) = \frac{(x - x_0)^2 a_0(x)}{a_2(x)} \tag{2}$$

may be shown to be real analytic at x_0. Thus if x_0 is a regular singular point, we may multiply equation (1) by $(x - x_0)^2/a_2(x)$ and write the equation in the form

$$(x - x_0)^2 y'' + (x - x_0)P(x)y' + Q(x)y = 0 \qquad (3)$$

where P and Q are real analytic at x_0. Equation (3) is said to be in *standard form*.

As the following examples show, it is usually not hard to locate the singular points of an equation and to classify each as regular singular, or as *irregular singular*.

Example 8.9

$x = 0$ is a singular point of $x^2 y'' - 2xy' + 2y = 0$ since $\lim_{x \to 0} (-2x)/x^2$ does not exist. The point is regular singular since $\lim_{x \to 0} x(-2x/x^2)$ and $\lim_{x \to 0} x^2(2/x^2)$ both exist. The equation is already in standard form with $P(x) = -2$ and $Q(x) = 2$.

Example 8.10

The equation $x^3 y'' + 2y = 0$ has a singular point at 0 since $\lim_{x \to 0} 2/x^3$ does not exist. The point is an irregular singular point since $\lim_{x \to 0} x^2(2/x^3)$ does not exist.

Example 8.11 (*Legendre Equation*)

$x = 1$ is a singular point of $(1 - x^2)y'' - 2xy' + \alpha(\alpha + 1)y = 0$, where α is a constant, since $\lim_{x \to 1} -2x/(1 - x^2)$ does not exist. It is a regular singular point since

$$\lim_{x \to 1} (x - 1)\frac{-2x}{1 - x^2} = \lim_{x \to 1} \frac{2x}{1 + x} = 1$$

and

$$\lim_{x \to 1} (x - 1)^2 \frac{\alpha(\alpha + 1)}{1 - x^2} = \lim_{x \to 1} \frac{(x - 1)\alpha(\alpha + 1)}{-1 - x} = 0$$

In this case, the standard form at the singular point $x = 1$ is obtained by multiplying the Legendre equation by $(x - 1)^2/(1 - x^2) = (1 - x)/(1 + x)$, and we obtain

$$(x - 1)^2 y'' + (x - 1)\frac{2x}{1 + x} y' - \frac{\alpha(\alpha + 1)(x - 1)}{1 + x} y = 0$$

with $P(x) = (2x)/(1 + x)$ and $Q(x) = \alpha(\alpha + 1)(1 - x)/(1 + x)$. The point $x = -1$ is also a regular singular point of the Legendre equation, but we leave the verification to the reader.

All other points in the examples above are ordinary points, and the methods of Section 8.2 may be used to find real analytic solutions in the form of power series convergent in the neighborhood of such a point. Our goal here and in the rest of the chapter is to solve Eq. (1) near a regular singular point. We shall suppose from now on that 0 is a regular singular point of (1). If $x_0 \neq 0$ is regular singular, it may be translated to 0 by the change of variable $\bar{x} = x - x_0$.

Euler Equations

The simplest kind of equation of form (1) which has a regular singular point at 0 is a *Euler* (or *Cauchy*, or *Cauchy–Euler*, or *equidimensional*) equation,

$$x^2 y'' + p_0 x y' + q_0 y = 0 \tag{4}$$

where p_0 and q_0 are real constants. Euler equations can be solved completely in terms of elementary functions.

Since 0 is a singular point of (4) and we anticipate peculiar behavior of solutions at 0, let us begin by requiring that $x > 0$. The variable change $x = e^s$, or $s = \ln x$, converts (4) into an equation with constant coefficients. For, using the Chain Rule, we have

$$\frac{dy}{dx} = \frac{dy}{ds}\frac{ds}{dx} = \frac{dy}{ds}\frac{1}{x}$$

$$\frac{d^2 y}{dx^2} = \frac{d}{dx}\left[\frac{dy}{ds}\frac{1}{x}\right] = \frac{d}{ds}\left[\frac{dy}{ds}\right]\frac{ds}{dx}\frac{1}{x} + \frac{dy}{ds}\frac{d}{dx}\left(\frac{1}{x}\right) = \frac{d^2 y}{ds^2}\frac{1}{x^2} - \frac{dy}{ds}\frac{1}{x^2}$$

Inserting these expressions into (4), we have that

$$x^2\left(\frac{d^2 y}{ds^2}\frac{1}{x^2} - \frac{dy}{ds}\frac{1}{x^2}\right) + p_0 x \left(\frac{dy}{ds}\frac{1}{x}\right) + q_0 x = 0$$

that is,

$$\frac{d^2 y}{ds^2} + (p_0 - 1)\frac{dy}{ds} + q_0 y = 0 \tag{5}$$

which is an equation with constant coefficients.

The characteristic polynomial $r^2 + (p_0 - 1)r + q_0$ of (5) is known as the *indicial polynomial* of the original Euler equation. (Remember that the coefficient is $p_0 - 1$, not p_0.) Its roots r_1 and r_2 are said to be the *characteristic exponents* of the equation, for a reason that is apparent from the form of the solutions given below. The solutions of (5) are given by

$$y = \begin{cases} c_1 e^{r_1 s} + c_2 e^{r_2 s} & \text{if } r_1 \neq r_2 \\ c_1 e^{r_1 s} + c_2 s e^{r_1 s} & \text{if } r_1 = r_2 \end{cases} \tag{6}$$

Since $e^s = x$, the solutions of (4) for $x > 0$ are given by

$$y = \begin{cases} c_1 x^{r_1} + c_2 x^{r_2} & \text{if } r_1 \neq r_2 \\ c_1 x^{r_1} + c_2 x^{r_1} \ln x & \text{if } r_1 = r_2 \end{cases} \tag{7}$$

We may handle the case $x < 0$ in a similar fashion, replacing x by $|x|$ throughout (7). That this is indeed the case may be most easily shown by first making the variable change $x = -z$ in (4) and rewriting the equation in zy-variables.

It is not clear what the solutions are like if the characteristic exponents are complex conjugates. On the other hand, complex exponents cause no difficulty when the s-variable is used, as in (6). For example, we might have $r_1 = 2 + 3i = \bar{r}_2$. Then, using methods of Chapter 5, we would have $y = e^{2s}[c_1 \cos 3s + c_2 \sin 3s]$, where c_1 and c_2 are arbitrary constants, as the general solution in the s-variable. Since $s = \ln x$, $x > 0$, this implies that $y = x^2[c_1 \cos (3 \ln x) + c_2 \sin (3 \ln x)]$.

Summarizing all this analysis, the solutions of the Euler equation (1) are

$$y = \begin{cases} c_1|x|^{r_1} + c_2|x|^{r_2} & \text{if } r_1 \neq r_2, \quad r_1, r_2 \text{ real} \\ c_1|x|^{r_1} + c_2|x|^{r_1} \ln |x| & \text{if } r_1 = r_2 \\ |x|^a[c_1 \cos (b \ln |x|) + c_2 \sin (b \ln |x|)] & \text{if } r_1 = \bar{r}_2 = a + ib \end{cases} \tag{8}$$

where $x \neq 0$.

In the following examples we not only solve specific Euler equations but also sketch some of the solutions in the xy-plane near the regular singularity at $x = 0$.

Example 8.12 (*Example 8.9 Revisited*)

The equation $x^2y'' - 2xy' + 2y = 0$ is a Euler equation with indicial polynomial $r^2 - 3r + 2$ and characteristic exponents 1 and 2. The general solution of the equation is $y = c_1x + c_2x^2$, and all solutions are defined for all x, including the regular singularity at the origin. The general solution appears harmless enough, but there is one peculiarity: Every solution goes through the origin of the xy-plane. Thus the initial value problem

$$\begin{aligned} x^2y'' - 2xy' + 2y &= 0 \\ y(0) = \alpha, \qquad y'(0) &= \beta \end{aligned} \tag{9}$$

has no solutions at all if $\alpha \neq 0$, but infinitely many if $\alpha = 0$. The Existence and Uniqueness Theorem clearly fails to apply in this case. See Figure 8.4 for sketches of solution curves of (9) near the singularity. Each value of β will result in curves similar to those of Figure 8.4.

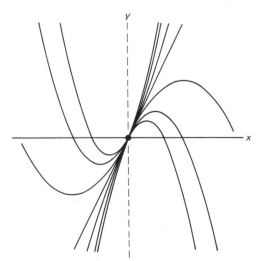

Figure 8.4 Solution curves of (9), $\alpha = 0$, $\beta = 2$.

Example 8.13

The Euler equation $x^2y'' + (1 - 2a)xy' + (1 + a^2)y = 0$ has the indicial polynomial $r^2 - 2ar + 1 + a^2$ and characteristic exponents $a + i$ and $a - i$. The general solution is $y = |x|^a[c_1 \cos (\ln |x|) + c_2 \sin (\ln |x|)]$, $x \neq 0$. The behavior of solutions near $x = 0$ is strange since the trigonometric terms have no limit at $x = 0$, but oscillate between $+1$ and -1 more and more rapidly as x approaches 0. These oscillations are magnified by the factor $|x|^a$ if $a < 0$, left unchanged if $a = 0$, and damped out if $a > 0$. See Figure 8.5 for sketches near 0 in each case.

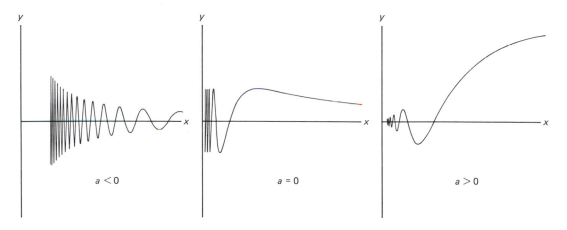

Figure 8.5   Graphs of $y = x^a \cos (\ln (x))$, $x > 0$.

Although the behavior of solutions of a Euler equation near the regular singularity at $x = 0$ is unusual, elsewhere there are no difficulties. If $x_0 \neq 0$, the initial value problem, $x^2y'' + p_0xy' + q_0y = 0$, $y(x_0) = \alpha$, $y'(x_0) = \beta$, has a unique solution for each α and β since all the conditions of the Existence and Uniqueness Theorem are met.

PROBLEMS

The following problem(s) may be more challenging: 7.

1. Determine if the given point is a regular or an irregular singular point for the corresponding equation.

   (a) $x^2y'' + xy' + y = 0$, $x_0 = 0$.   (b) $xy'' + (1 - x)y' + xy = 0$, $x_0 = 0$.

   (c) $x(1 - x)y'' + (1 - 2x)y' - 4y = 0$, $x_0 = 1$.   (d) $x^2y'' + \dfrac{2}{x}y' + 4y = 0$, $x_0 = 0$.

2. Find and classify all singular points of the following equations.

   (a) $(1 - x^2)y'' - xy' + 2y = 0$.   (b) $y'' + \left(\dfrac{x}{1 - x}\right)^2 y' + (1 + x)^2y = 0$.

(c) $x(1 - x^2)^3 y'' + (1 - x^2)^2 y' + 2(1 + x)y = 0$.

(d) $x^3(1 - x^2)y'' - x(x + 1)y' + (1 - x)y = 0$.

3. Find the general solution for $x > 0$ of each of the following equations. Sketch some solutions.

(a) $x^2 y'' - 6y = 0$. (b) $x^2 y'' + xy' - 4y = 0$.

(c) $x^2 y'' + xy' + 9y = 0$. (d) $x^2 y'' + \frac{1}{2} xy' - \frac{1}{2} y = 0$.

(e) $xy'' - y' + (5/x)y = 0$. (f) $x^2 y'' + 7xy' + 9y = 0$.

4. Find the general solution for $x > 0$ of each of the following equations.

(a) $x^2 y'' - 4xy' + 6y = 2x + 5$. (b) $x^2 y'' - 6y = 2x^2$.

5. Show that every solution $y(x)$ of $x^2 y'' + p_0 xy' + q_0 y = 0$, p_0, q_0 real, satisfies $|y(x)| \to 0$ as $x \to \infty$ if and only if $p_0 > 1$ and $q_0 > 0$.

6. [*The nth-Order Euler Equation*]. The *nth-order Euler equation* is

$$x^n y^{(n)} + a_{n-1} x^{n-1} y^{(n-1)} + \cdots + a_1 xy' + a_0 y = 0, \quad x > 0 \qquad (*)$$

where $a_0, a_1, \ldots, a_{n-1}$ are real constants. The *indicial polynomial* is $p(r) = r(r - 1) \cdots (r - n + 1) + r(r - 1) \cdots (r - n + 2)a_{n-1} + \cdots + ra_1 + a_0$.

(a) Show that $y = x^{r_1}$, $x > 0$, is a solution of (*) if and only if r_1 is a root of $p(r)$.

(b) Use part (a) to find three independent solutions of

$$x^3 y''' + 4x^2 y'' - 2y = 0, \quad x > 0$$

7. [*Aging Spring*]. The Comparison Theorem outlined in the problem set of Section 5.2 may be used to come to conclusions about the behavior of solutions of the equation of the aging spring, $x'' + Ae^{-\epsilon t} x = 0$, where A and ϵ are positive constants.

(a) Show that for all t large enough, $3t^{-2}/16 > Ae^{-\epsilon t}$.

(b) Solve the Euler equation $u'' + \frac{3}{16} t^{-2} u = 0$ and prove that each nontrivial solution has at most one zero in the interval $(0, \infty)$.

(c) Use the Comparison Theorem and parts (a) and (b) to show that each nontrivial solution of $x'' + Ae^{-\epsilon t} x$ is nonoscillatory on $(0, \infty)$. What does this imply about the long-term behavior of the aging spring? [*Hint*: No solution is oscillatory on a finite interval, and for all t large enough the Euler equation of part (b) and the equation of the aging spring may be compared.]

8.5 SERIES SOLUTIONS NEAR REGULAR SINGULAR POINTS, I

Equations with regular singular points are pervasive in the mathematical modeling of physical phenomena. If the equation has Euler form, it may be solved explicitly in terms of powers, logarithms, or trigonometric functions of logarithms of the independent variable. If the equation has a regular singularity but is not of Euler form, we must look elsewhere in our search for formulas of solutions. As it turns out, a combination of the power series of Section 8.2 and the Euler solutions of the preceding section gives us just what we want, solutions of the form

$$x^r \sum_0^\infty a_n x^n, \quad x^r \ln x \sum_0^\infty a_n x^n, \quad \text{or} \quad x^r \cos(b \ln x) \sum_0^\infty a_n x^n$$

In this section we focus our attention on solutions of the first type, leaving solutions of the second type to Section 8.7 and not considering the trigonometric solutions at all since they do not come up very often in applications. The approach we shall use is a special case of techniques developed during the nineteenth century by Sturm, Fuchs, Frobenius, and many others. Later the theory was extended by Weyl to irregular singular points. The subject is by now almost completely settled—and still of central importance in applied mathematics.†

Indicial Polynomials, Frobenius Series: An Example

Let us suppose that $x = 0$ is a regular singular point of the equation $a_2(x)y'' + a_1(x)y' + a_0(x)y = 0$. Multiplying the equation by x^2/a_2, we obtain the equation in standard form

$$x^2y'' + xP(x)y' + Q(x)y = 0 \tag{1}$$

where $P = xa_1/a_2$ and $Q = x^2a_0/a_2$. P and Q are assumed to be real analytic at 0 and have power series expansions of the form

$$P(x) = \sum_0^\infty p_n x^n, \qquad Q(x) = \sum_0^\infty q_n x^n \tag{2}$$

which converge on a common interval J centered at 0.

> **Indicial Polynomial.** The indicial polynomial of (1) is given by $r^2 + (p_0 - 1)r + q_0$. Its roots r_1 and r_2 are the *characteristic exponents*.

Observe that if all the coefficients of the series expansions of $P(x)$ and $Q(x)$ vanish (except possibly p_0 and q_0), then (1) is a Euler equation and the definition above agrees with that given in the preceding section. Our goal is to find solutions of (1) in the form of a *generalized power series*, or *Frobenius series*,

$$y = x^r \sum_0^\infty a_n x^n \tag{3}$$

As we shall see, r will turn out to be r_1 or r_2 and the coefficients $a_0, a_1, \ldots$ may be found by an extension of the method of undetermined coefficients used in Section 8.2.

Before stating the convergence theoem of Frobenius, we shall show by an example how to find solutions having the form of a Frobenius series. Consider the equation

$$4xy'' + 2y' + y = 0 \tag{4}$$

Multiplying by $x/4$ to get the equation into the standard form of (1), we have

---

† See E. A. Coddington and N. Levinson, *Theory of Ordinary Differential Equations* (New York: McGraw-Hill, 1955), for an advanced treatment of singular point theory.

$$x^2 y'' + \frac{x}{2} y' + \frac{x}{4} y = 0 \qquad (5)$$

from which we see that $P(x) = \frac{1}{2}$, $Q(x) = x/4$. Thus 0 is a regular singular point. The indicial polynomial is $r^2 - r/2$ since $p_0 = \frac{1}{2}$, $q_0 = 0$. The characteristic exponents are 0 and $\frac{1}{2}$. In the following it will be more convenient to use (4) rather than the standard form. Let x be positive in the analysis below.

Ignoring convergence questions for the time being, we assume that (4) has a solution of the form $y = x^r \sum_0^\infty a_n x^n = \sum_0^\infty a_n x^{n+r}$, insert into (4), and use a method of undetermined coefficients to calculate the constants r and a_n, $n = 0, 1, 2, \ldots$. Then

$$4x \sum_0^\infty (n+r)(n+r-1)a_n x^{n+r-2} + 2 \sum_0^\infty (n+r)a_n x^{n+r-1} + \sum_0^\infty a_n x^{n+r} = 0$$

that is,

$$\sum_0^\infty 4(n+r)(n+r-1)a_n x^{n+r-1} + \sum_0^\infty 2(n+r)a_n x^{n+r-1} + \sum_0^\infty a_n x^{n+r} = 0 \quad (6)$$

Reindexing the last sum, we have $\sum_0^\infty a_n x^{n+r} = \sum_1^\infty a_{n-1} x^{n+r-1}$. Isolating the first term ($n = 0$) in each of the other two sums so that all three begin at $n = 1$, we may replace (6) by

$$4(0+r)(0+r-1)a_0 x^{r-1} + 2(0+r)a_0 x^{r-1}$$

$$+ \sum_1^\infty [4(n+r)(n+r-1)a_n + 2(n+r)a_n + a_{n-1}]x^{n+r-1} = 0$$

or

$$x^{r-1} \left\{ 2r(2r-1)a_0 + \sum_1^\infty [2(n+r)(2n+2r-1)a_n + a_{n-1}]x^n \right\} = 0$$

By the Identity Theorem for power series, this implies that the coefficient of x^n, $n = 0, 1, 2, \ldots$, must vanish. We have the following recursion relations:

$$n = 0: \quad r(r - \tfrac{1}{2})a_0 = 0 \qquad (7a)$$

$$n \geq 1: \quad 2(n+r)(2n+2r-1)a_n + a_{n-1} = 0 \qquad (7b)$$

From the first relation we conclude that either $r = 0$ or $r = \frac{1}{2}$ or $a_0 = 0$. Observe that $r(r - \frac{1}{2})$ is the indicial polynomial and 0 and $\frac{1}{2}$ are the characteristic exponents. If we set $r = 0$ or $r = \frac{1}{2}$, a_0 is an arbitrary constant. The remaining coefficients are functions of r and a_0; they may be determined from the general *recursion formula*

$$a_n(r) = \frac{-a_{n-1}(r)}{2(n+r)(2n+2r-1)}, \qquad n = 1, 2, \ldots \qquad (8)$$

which is a rearrangement of (7b).

If r is 0 or $\frac{1}{2}$, the denominator in (8) never vanishes. Using (8) to obtain a_n in terms of a_0, we have

$$
\begin{aligned}
a_n(r) &= \frac{-a_{n-1}(r)}{2(n+r)(2n+2r-1)} \\[2mm]
&= \frac{a_{n-2}(r)}{2(n+r)(2n+2r-1)2(n-1+r)(2n-2+2r-1)} = \cdots \\[2mm]
&= \frac{(-1)^n a_0}{2^n(n+r)(n+r-1)\cdots(r+1)(2n+2r-1)(2n+2r-3)\cdots(2r+1)}
\end{aligned}
\tag{9}
$$

Observe that $a_0 = 0$ produces either the trivial solution $y \equiv 0$ or else the recursion process breaks down if a factor of the denominator in (9) vanishes. For $r = 0$ or $\frac{1}{2}$ this difficulty cannot occur. For $r = 0$ and $r = \frac{1}{2}$, we have that

$$
a_n(0) = \frac{(-1)^n a_0}{2^n n!(2n-1)(2n-3)\cdots 1} = \frac{(-1)^n a_0}{(2n)!}
$$

$$
\begin{aligned}
a_n(1/2) &= \frac{(-1)^n a_0}{2^n(n+1/2)(n-1/2)\cdots(3/2)(2n)(2n-2)\cdots 2} \\[2mm]
&= \frac{(-1)^n a_0}{(2n+1)(2n-1)\cdots 3(2n)(2n-2)\cdots 2} \\[2mm]
&= \frac{(-1)^n a_0}{(2n+1)!}
\end{aligned}
$$

We have constructed a pair of independent solutions of (4)

$$
y_1 = \sum_0^\infty \frac{(-1)^n x^n}{(2n)!}, \qquad y_2 = x^{1/2} \sum_0^\infty \frac{(-1)^n x^n}{(2n+1)!}
\tag{10}
$$

where we have taken $a_0 = 1$. More precisely, the solutions (10) are defined for $x > 0$, or for $x < 0$ if we replace $x^{1/2}$ by $|x|^{1/2}$. The first solution is defined for all x, but $y_2(x)$ is not a solution at $x = 0$, since the term $x^{1/2}$ is not differentiable there.

The two power series in (10) converge everywhere (use the Ratio Test). It is straightforward now to justify all our earlier steps and check that the two functions defined in (10) actually are solutions of (4). This can be done by direct insertion of y_1 and y_2 into (4). Of course, the recursion relations were chosen specifically so that y_1 and y_2 are solutions. Since the power series parts of y_1 and y_2 converge on **R**, it is all right to differentiate term by term. The functions y_1 and y_2 are independent since $y_1(0) = 1$, while $|y_2(x)| \to 0$ as $x \to 0$. Thus $\{y_1, y_2\}$ is a basis of the solution space of (4).

Observe that even if we had not determined the indicial polynomial and characteristic exponents before beginning the process of determining the coefficients, that polynomial (or a multiple of it) will appear as the coefficient of a_0 in (7a), and we could at that point choose r to be a root of the polynomial.

Method of Frobenius: Theory

The method of undetermined coefficients used in the example above is called the *Method of Frobenius*, and the series solutions obtained is the *Frobenius series*. A Frobenius series may be a standard power series [see y_1 in (10)] or involve x to a power that is not a nonnegative integer [see y_2 in (10)]. The technique will always generate at least one nontrivial solution defined near (but perhaps not at) the regular singular point. Many times the process will also produce a second independent solution as in the example above—but not always. In any case we have the following basic theorem which underlies the procedure.

Frobenius' Theorem I.   Suppose that the indicial polynomial $f(r) = r^2 + (p_0 - 1)r + q_0$ has real roots r_1 and r_2, $r_2 \leqq r_1$. Then Eq. (1) has a solution of the form

$$y_1 = |x|^{r_1} \sum_0^\infty a_n(r_1)x^n, \qquad x > 0 \quad \text{or} \quad x < 0 \tag{11}$$

where the coefficients are determined by the recursion relation,

$$f(r_1 + n)a_n(r_1) = -\sum_{k=0}^{n-1} [(k + r_1)p_{n-k} + q_{n-k}]a_k(r_1), \qquad n \geqq 1 \tag{12}$$

where $a_0 = 1$. The series in (11) converges to a real analytic function on the interval J. Moreover, if $r_1 - r_2$ is not an integer, a second independent solution is given by the analytic function on J.

$$y_2 = |x|^{r_2} \sum_0^\infty a_n(r_2)x^n, \qquad x > 0 \quad \text{or} \quad x < 0 \tag{13}$$

where the coefficients are determined by (12), r_2 replacing r_1.

The proof of the theorem is omitted. It must be emphasized that usually the most efficient way to solve a specific equation is to proceed by direct substitution and the method of undetermined coefficients as in the earlier example, rather than to use the recursion relations of (12) directly.

When $r_1 - r_2$ is an integer, some ingenuity must be used to construct a solution that is independent of y_1. One can see from (12) just where the recursion formula breaks down in this case. For example, suppose that $r_1 - r_2 = 3$. Then the indicial polynomial $f(r) = r^2 + (p_0 - 1)r + q_0$ factors into $(r - r_2)(r - r_2 - 3)$, and we see that $f(r_2 + 3) = 0$. Hence the recursion formula for $n = 3$ becomes $0 \cdot a_3(r_2) = -\sum_{k=0}^{2} [\cdots]a_k$, and we cannot in general determine a_3. We resolve this difficulty in Section 8.7.

Comments

We have assumed that $x_0 = 0$ is a regular singularity of the differential equation, but the process above would also apply if $x_0 \neq 0$ is a regular singularity. We would replace x by $x - x_0$ in each step.

We have seen that the construction of a Frobenius series solution $y = x^r \sum_0^\infty a_n x^n$ of the differential equation consists of the following steps.

(a) Check that the differential equation may be written in the form

$$x^2 y'' + x P(x) y' + Q(x) y = 0$$

where $P = \sum_0^\infty p_n x^n$ and $Q = \sum_0^\infty q_n x^n$, each series converging on an open interval containing 0. The point $x = 0$ is then a regular singularity and the method of Frobenius will apply. This step must not be omitted since you will need the numbers $p_0 = P(0)$ and $q_0 = Q(0)$.

(b) Write out the indicial polynomial $f(r) = r^2 + (p_0 - 1)r + q_0 = 0$ and find its roots, the characteristic exponents r_1 and r_2.

(c) Clear any denominators from the differential equation and cancel common factors. For example, rewrite $x^2 y'' + xy' + [x^3/(1 + x)]y = 0$ as $x(1 + x)y'' + (1 + x)y' + x^2 y = 0$. The resulting equation need not have the form $x^2 y'' + xP(x)y' + Q(x)y = 0$.

(d) Assuming that r_1 and r_2 are real and that $r_1 \geq r_2$, insert the series $\sum_0^\infty a_n x^{n+r_1}$ into the equation found in part (c).

(e) Reindex and rearrange the resulting series until you have

$$f(r_1)a_0 x^{r_1} + \sum_{n=1}^\infty [\cdots]x^{n+r_1} = 0$$

In reindexing and rearranging, be particularly careful not to "lose" terms at the head of the series (e.g., terms corresponding to $n = 0$ or 1). Equating the bracketed expression to zero gives the recursion formula for the coefficients. The coefficient of a_0 must be 0 since $f(r_1) = 0$; thus a_0 is arbitrary.

(f) Solve the recursion formula if it is simple enough to do so, and find the coefficient a_n in terms of a_0. Otherwise, use the recursion formula to find the first few coefficients $a_1, a_2, \ldots, a_N$ in terms of a_0.

(g) If $r_1 - r_2$ is not an integer, repeat the process to find a solution $\sum_0^\infty b_n x^{n+r_2}$.

(h) Obtain specific solutions, if desired, by setting a_0 and b_0 equal to specific constants. Otherwise, a_0 and b_0 may be considered to be the arbitrary constants of a general solution.

(i) If $r_1 - r_2$ is an integer, refer to Section 8.7 for a method for finding a second independent solution corresponding to r_2.

We have assumed that 0 is a regular singular point of the equation. If it is an ordinary point, the series methods of Section 8.2 should be used instead. If 0 is an irregular singular point, series methods may or may not work, but are worth a try.

PROBLEMS

The following problem(s) may be more challenging: 2, 3, 4, 6, and 7.

1. Use the method of Frobenius to find the general solution of each of the following equations as a series of powers of x. Sketch graphs of two independent solutions.
   (a) $9x^2y'' + 3x(x + 3)y' - (4x + 1)y = 0$.   (b) $4x^2y'' + x(2x + 9)y' + y = 0$.
   (c) $x^2y'' + x(1 - x)y' - 2y = 0$.   (d) $x^2(1 - x^2)y'' + x(x - 1)y' + \frac{8}{9}y = 0$.

2. Find the first four terms of a power series solution of the following equation:
$$x^2y'' + x(1 - x)y' - 2y = \ln(1 + x)$$

3. Solve the equation $3x^2y'' + 5xy' - e^xy = 0$ by expanding e^x as a Maclaurin series and recalling the formula for the product of two series (Appendix D). You need only find the first few terms in the Frobenius series.

4. Find a solution of $xy'' - y' - 4x^3y = 0$ for which $y(0) = y'(0) = 0$, $y(x) \not\equiv 0$. Why does this not contradict the Uniqueness Theorem?

5. Show that the equation $x^3y'' + y = 0$ has no nontrivial solutions of the form $y = x^r \sum_{n=0}^{\infty} a_n x^n$. Why does this not contradict the results of this section?

6. [*Laguerre's Equation*]. (a) Show that 0 is a regular singular point of $xy'' + (1 - x)y' + \alpha y = 0$, *Laguerre's equation of order* α.
   (b) Find one nontrivial solution.
   (c) Show that if α is a nonnegative integer, then there are polynomial solutions.

7. [*Legendre's Equation*]. (a) Find the indicial equation and the characteristic exponents of the Legendre equation of order α, $(1 - x^2)y'' - 2xy' + \alpha(\alpha + 1)y = 0$, at the regular singularity $x = 1$.
   (b) Find a series solution in powers of $x - 1$. [*Hint*: Let $s = x - 1$ and rewrite the equation in sy-variables.]

8.6 BESSEL FUNCTIONS

The most important second-order linear differential equation with a regular singular point at the origin is *Bessel's equation of order* p,

$$x^2y'' + xy' + (x^2 - p^2)y = 0 \tag{1}$$

where p is any nonnegative constant.† Nontrivial solutions of (1) are known as *Bessel functions*. Next to the trigonometric and exponential functions and their inverses, Bessel functions are the most important transcendental functions of applied mathematical analysis. Arising as they do in studies of vibrational, thermal, and elastic phenomena, their properties have been extensively studied and their values tabulated.‡

In this section we use the Frobenius Method to define some of the Bessel functions, in particular the Bessel functions of the first kind and the Bessel functions of the second kind of noninteger order. We shall also see how much Bessel functions resemble decaying sinusoids. In a way, this is not surprising since if (1) is divided by x^2, we obtain the equation $y'' + (1/x)y' + (1 - p^2/x^2)y = 0$, which for large x is very close to the simple harmonic oscillator equation, $y'' + y = 0$, with its sinusoidal solutions. For simplicity we shall assume that $x > 0$ throughout this section, although we could treat the case $x < 0$ easily enough in the manner suggested in Section 8.5. Some of the solutions of (1) are also defined at $x = 0$, as will be clear from the formulas.

Solving Bessel's Equation

Using the notation of Section 8.5, we observe that the origin is a regular singular point for (1), that $P(x) = 1$, that $Q(x) = -p^2 + x^2$, that the indicial polynomial is $f(r) = r^2 - p^2$, and that the characteristic exponents are $r_1 = p$ and $r_2 = -p$. Frobenius's Theorem I guarantees the existence of a nontrivial solution of (1) having the form $y = x^p \sum_0^\infty a_n x^n$ for $x > 0$.

To determine the coefficients a_n, we insert $y = \sum_0^\infty a_n x^{n+p}$ and the corresponding derivative series into Bessel's equation, obtaining

$$x^2 \sum_0^\infty (n+p)(n+p-1)a_n x^{n+p-2} + x \sum_0^\infty (n+p)a_n x^{n+p-1}$$

$$+ x^2 \sum_0^\infty a_n x^{n+p} - p^2 \sum_0^\infty a_n x^{n+p} = 0$$

If we move the respective factors x^2, x, and x^2 inside the first three summations, factor x^p from all four, and combine the first, second, and fourth summations, we have

$$x^p \left\{ \sum_0^\infty [(n+p)(n+p-1) + (n+p) - p^2]a_n x^n + \sum_0^\infty a_n x^{n+2} \right\} = 0$$

---

† Friedrich William Bessel (1784–1846) was a German mathematician, astronomer, and celestial mechanist who used Eq. (1) in studying perturbations of planetary orbits.

‡ See, e.g., M. Abramowitz and I. Stegun (eds.), *Handbook of Mathematical Functions* [Washington, D.C.: National Bureau of Standards, 1964; Dover (reprint), New York 1965]; G. N. Watson, *A Treatise on the Theory of Bessel Functions,* 2nd ed. (Cambridge: Cambridge University Press, 1941), is a theoretical study of the Bessel equation and its solutions.

Reindexing the final summation to the form $\sum_0^\infty a_{n-2}x^n$ and separating out the first two summands of the initial summation (the terms corresponding to $n = 0$ and $n = 1$) so that all have a common range from $n = 2$ to ∞, we have that

$$x^p \left\{ [(0+p)(0+p-1) + (0+p) - p^2]a_0 + [(1+p)(1+p-1) + (1+p) - p^2]a_1 x \right\}$$

$$+ x^p \sum_2^\infty \left\{ [(n+p)(n+p-1) + (n+p) - p^2]a_n + a_{n-2} \right\} x^n = 0$$

or

$$(1+2p)a_1 x + \sum_2^\infty [n(n+2p)a_n + a_{n-2}]x^n = 0$$

The coefficient of a_0 vanishes, as it must, since the coefficient is $f(p) = p^2 - p^2 = 0$. We have the *recursion formulas* for Bessel's equation,

$$(1+2p)a_1 = 0 \quad \text{or} \quad a_1 = 0 \qquad \text{since } p \geq 0$$

$$a_n = \frac{-a_{n-2}}{n(n+2p)}, \qquad n = 2, 3, \ldots \tag{2}$$

Hence $a_1 = a_3 = a_5 = \cdots = a_{2n+1} = \cdots = 0$, and

$$a_{2k} = \frac{-a_{2k-2}}{2k(2k+2p)} = \frac{a_{2k-4}}{2k(2k+2p)(2k-2)(2k+2p-2)}$$

$$= \cdots = \frac{(-1)^k a_0}{2^{2k}k!(k+p)(k-1+p) \cdots (1+p)}$$

Thus we have the solution

$$y_1 = a_0 x^p \left\{ 1 + \sum_{k=1}^\infty \frac{(-1)^k x^{2k}}{2^{2k}k!(1+p) \cdots (k+p)} \right\}, \qquad x > 0 \tag{3}$$

Equation (3) defines a solution of the Bessel equation of order p for each value of the constant a_0, but of course varying a_0 will not give a second independent solution. However, if the difference $p - (-p) = 2p$ of the characteristic exponents is not an integer, then according to Frobenius's Theorem we may replace p by $-p$ in the recursion formula and obtain a second independent solution,

$$y_2 = a_0 x^{-p} \left\{ 1 + \sum_{k=1}^\infty \frac{(-1)^k x^{2k}}{2^{2k}k!(1-p) \cdots (k-p)} \right\} \tag{4}$$

In fact, (4) defines a second independent solution when $p = \frac{1}{2}, \frac{3}{2}, \frac{5}{2}, \ldots$ (i.e., in the *half-integer* case) even though the recursion formula breaks down. This can be shown by first observing that (4) is defined when p is a half-integer, and then inserting y as defined by (4) directly into Bessel's equation and verifying that the equation is satisfied.

This leaves only the case when p is an integer. If $p = 0$, (4) coincides with (3) and thus represents no new solution. If $p = 1, 2, 3, \ldots$, the terms in (4) are not defined. The problem of finding a second, independent solution when p is an integer is solved in Section 8.7.

Frobenius's Theorem implies that the power series in (3) and in (4) converge for all x. This may also be shown directly by using the ratio test. The "normalizing" constant a_0 plays a curiously important role. As we shall see, a judicious choice of a_0 (a different value for each value of p) will result in some remarkable formulas. We shall begin with the case where p is an integer. We shall, for the time being, work only with the "first" solution defined by (3).

Bessel Functions of the First Kind of Integral Order

Let us suppose that the order p is the integer $n \geq 0$. We shall set $a_0 = 1/(2^n n!)$ in (3). After some simple algebra, we obtain a solution which has acquired a special name.

Bessel function of the first kind of integer order n is

$$J_n(x) = \left(\frac{x}{2}\right)^n \sum_{k=0}^{\infty} \frac{(-1)^k}{k!(k+n)!} \left(\frac{x}{2}\right)^{2k} \tag{5}$$

Since n is an integer, we may use (5) to define $J_n(x)$ for all real x, not just for $x > 0$. From (5), we have that

$$J_0(0) = 1, \quad J_n(0) = 0, \qquad n = 1, 2, 3, \ldots$$
$$J_1'(0) = \tfrac{1}{2}, \quad J_n'(0) = 0, \qquad n = 0, 2, 3, \ldots$$

For concreteness we write out the full formulas for J_0 and J_1:

$$J_0(x) = 1 - \frac{x^2}{4} + \frac{x^4}{64} - \frac{x^6}{2304} + \cdots + (-1)^k \frac{x^{2k}}{(k!)^2 2^{2k}} + \cdots$$

$$J_1(x) = \frac{x}{2} \left\{ 1 - \frac{x^2}{8} + \frac{x^4}{192} - \frac{x^6}{9216} + \cdots + (-1)^k \frac{x^{2k}}{k!(k+1)!2^{2k}} + \cdots \right\}$$

Since the series alternate, the error introduced by using a truncation of the series as an approximation is no more in magnitude than the magnitude of the first term omitted from the truncation. For example, $J_0(\frac{1}{10}) \approx 1 - \frac{1}{400}$ with an error of no more than $10^{-4}/64 = 0.0000016$. Series such as those above may be effectively used to estimate the values of $J_n(x)$ for small x, but they are not very useful for large x since we would have to go far out in the series before the denominator involving factorials begins to dominate the term x^{2k} in the numerator. See Figure 8.6 for sketches of J_0, J_1, and J_2.

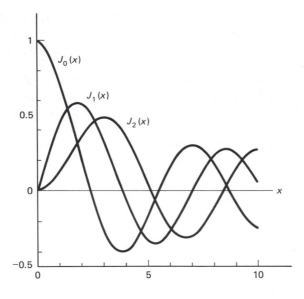

Figure 8.6 $J_0(x)$, $J_1(x)$, $J_2(x)$.

Bessel Functions of the First Kind of Nonintegral Order

Bessel functions of the first kind of integer order $p = n$ have been defined by (3) with $a_0 = 1/(2^n n!)$. But what do we do when p is not an integer? What does $p!$ mean in such a case? Euler solved the problem of constructing a useful "interpolated factorial" function by defining what is now known as the gamma function.

> **Gamma Function.**   The *Gamma function* is defined by
>
> $$\Gamma(z) = \int_0^\infty x^{z-1} e^{-x}\, dx, \qquad z > 0 \tag{6}$$

The improper integral in (6) may be shown to converge for all positive values of z. The gamma function has the following remarkable properties, the first two justifying its alternate name, the *interpolated factorial*:

$$\Gamma(n+1) = n!, \qquad\qquad n = 0, 1, 2, \ldots \tag{7}$$

$$\Gamma(z+1) = z\,\Gamma(z), \qquad\qquad \text{all } z > 0 \tag{8}$$

$$\Gamma(n + 1/2) = \frac{(2n)!}{2^{2n}n!}\,\pi^{1/2}, \qquad n = 0, 1, 2, \ldots \tag{9}$$

$$\Gamma(z) \text{ is differentiable to all orders} \tag{10}$$

See the problem set for proofs of these properties.

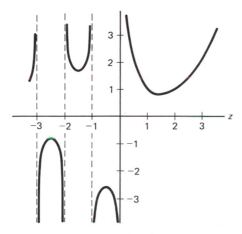

Figure 8.7 $\Gamma(z)$.

Property (8) in the form $\Gamma(z) = \Gamma(z + 1)/z$ may be used to extend the definition of the gamma function to the interval $-1 < z < 0$ using the values of the function on the interval $(0, 1)$ given by (6). Using the just-defined values of the function on $(-1, 0)$, (8) may be used again to define $\Gamma(z)$, $-2 < z < -1$, and so on. Thus the gamma function may be defined for all z except the negative integers. The resulting function is differentiable to all orders wherever it is defined. See Figure 8.7 for its graph.

After this digression on the gamma function we set $a_0 = [2^p \Gamma(p + 1)]^{-1}$ in Eq. (3) with $p \geqq 0$.

> **Bessel Functions of the First Kind.**   *Bessel function of the first kind of order p is*
>
> $$J_p(x) = \left(\frac{x}{2}\right)^p \sum_{k=0}^{\infty} \frac{(-1)^k}{k!\,\Gamma(k + p + 1)} \left(\frac{x}{2}\right)^{2k}, \qquad x > 0 \qquad (11)$$

If $x < 0$, the factor before the summation is replaced by $|x/2|^p$. Formula (11) is obtained from (3) with the foregoing choice of a_0 by observing that $\Gamma(k + p + 1) = \Gamma(p + 1)(1 + p) \cdots (k + p)$, a consequence of repeated application of property (8). Observe that (11) reduces to (3) when $p = n$, an integer.

Now that all the Bessel functions of the first kind have been defined by suitable choice of the constant factor a_0, we shall look into the "calculus" of these functions.

Recursion Formulas

Bessel functions satisfy a number of identities that relate the functions of various orders and their derivatives.

Recursion Formulas.   Bessel functions of the first kind satisfy the following identities:

$$[x^p J_p]' = x^p J_{p-1}, \qquad p \geq 1 \qquad\qquad (12)$$

$$[x^{-p} J_p]' = -x^{-p} J_{p+1}, \qquad p \geq 0 \qquad\qquad (13)$$

$$J_{p+1} = \frac{2p}{x} J_p - J_{p-1}, \qquad p \geq 1 \qquad\qquad (14)$$

$$J_{p+1} = -2 J_p' + J_{p-1}, \qquad p \geq 1 \qquad\qquad (15)$$

Proof. To verify (12), we have from (11) that

$$
\begin{aligned}
[x^p J_p]' &= \frac{d}{dx} \left\{ \sum_{k=0}^{\infty} \frac{(-1)^k 2^p}{k! \Gamma(k+p+1)} \left(\frac{x}{2}\right)^{2k+2p} \right\} \\
&= \sum_{k=0}^{\infty} \frac{(-1)^k (k+p) 2^p}{k! \Gamma(k+p+1)} \left(\frac{x}{2}\right)^{2k+2p-1} \\
&= x^p \left(\frac{x}{2}\right)^{p-1} \sum_{k=0}^{\infty} \frac{(-1)^k}{k! \Gamma(k+p)} \left(\frac{x}{2}\right)^{2k} \\
&= x^p J_{p-1}
\end{aligned}
$$

where we have used the fact that $\Gamma(k+p+1) = (k+p)\Gamma(k+p)$. This proves (12); the proof of (13) is similar. Recursion formula (14) is obtained by multiplying each side of (12) by x^{-p}, each side of (13) by x^p, and subtracting the two:

$$x^{-p}[x^p J_p]' - x^p[x^{-p} J_p]' = J_{p-1} + J_{p+1}$$

$$x^{-p}[px^{p-1} J_p + x^p J_p'] - x^p[-px^{-p-1} J_p + x^{-p} J_p'] = J_{p-1} + J_{p+1}$$

from which (14) follows after a little algebraic rearrangement. Formula (15) is derived similarly.

The recursion formulas enable us to develop a set of differentiation and integration formulas for Bessel functions which turn out to be useful in advanced applications. We shall give a single illustration of what may be done.

Example 8.14

$\int x^5 J_2(x)\, dx = x^5 J_3(x) - 2x^4 J_4(x)$, since we have that

$$
\begin{aligned}
\int x^5 J_2(x)\, dx &= \int x^2 (x^3 J_2)\, dx = \int x^2 (x^3 J_3)'\, dx & \text{[use (12)]} \\
&= x^2(x^3 J_3) - 2\int x(x^3 J_3)\, dx & \text{(integration by parts)} \\
&= x^5 J_3 - 2\int x^4 J_3\, dx = x^5 J_3 - 2\int (x^4 J_4)'\, dx & \text{[use (12) again]} \\
&= x^5 J_3 - 2x^4 J_4
\end{aligned}
$$

Bessel Functions of the First Kind of Half-Integer Order

The function $J_{1/2}(x)$ is not nearly as formidable as its definition via (11) makes it appear. In fact, we have that

$$J_{1/2}(x) = \left(\frac{2}{\pi}\right)^{1/2} \frac{\sin x}{x^{1/2}} \tag{16}$$

We see that this is so by observing from (11) and (9) that

$$J_{1/2}(x) = \left(\frac{x}{2}\right)^{1/2} \sum_{k=0}^{\infty} \frac{(-1)^k}{k!\Gamma(k+3/2)} \left(\frac{x}{2}\right)^{2k} = \left(\frac{x}{2}\right)^{1/2} \sum_{k=0}^{\infty} \frac{(-1)^k 2^{2k+2}(k+1)!}{k!(2k+2)!\pi^{1/2}} \left(\frac{x}{2}\right)^{2k}$$

$$= \left(\frac{2}{\pi x}\right)^{1/2} \sum_{k=0}^{\infty} \frac{(-1)^k}{(2k+1)!} x^{2k+1} = \left(\frac{2}{\pi x}\right)^{1/2} \sin x$$

since the last series is the Maclaurin series for $\sin x$. In the second step of the derivation above we have used Eq. (9) to replace $\Gamma(k+3/2)$, while the third series is obtained from the second by rearranging and canceling some terms.

By the same kind of argument it may be shown that

$$J_{3/2}(x) = \left(\frac{2}{\pi}\right)^{1/2} \left[-\frac{\cos x}{x^{1/2}} + \frac{\sin x}{x^{3/2}}\right]$$

Recursion formula (14) may be used to define the remaining *half-integer Bessel functions of the first kind*, $J_{n+1/2}$, $n = 0, 1, 2, \ldots$. For example, by (14), we have

$$J_{5/2} = \frac{3}{x} J_{3/2} - J_{1/2}$$

Oscillation and Decay

The Bessel functions of half-integer order oscillate like sinusoids, but with amplitudes that decay like $x^{-1/2}$ as $x \to \infty$. In fact, nontrivial solutions of Bessel equations of all orders possess these properties, but first we need a definition (see also the problem set of Section 5.2).

> **Oscillatory Solutions.**   A nontrivial solution $y(x)$ of $y'' + P(x)y' + Q(x)y = 0$ is **oscillatory** on an interval (a, b) if there is an increasing sequence $\{x_n\}_1^{\infty}$, of distinct zeros of $y(x)$ in that interval [i.e., $y(x_n) = 0$, $x_n < x_{n+1}$, $n = 1, 2, \ldots$].

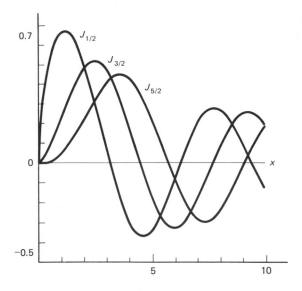

Figure 8.8 $J_{1/2}, J_{3/2}, J_{5/2}$.

Oscillation Theorem.   Every nontrivial solution of Bessel's equation of order p, $p \geqq 0$, is oscillatory on $(0, \infty)$.

Decay.   Every nontrivial solution $y(x)$ of Bessel's equation of order p, $p \geqq 0$, has the form

$$y(x) = \left(\frac{2}{\pi}\right)^{1/2} \left\{ \frac{A \sin (x + \phi)}{x^{1/2}} + \frac{h(x)}{x^{3/2}} \right\} \tag{17}$$

where A and ϕ are constants and $h(x)$ is a bounded function on $[1, \infty)$.

We shall not prove the decay property. The oscillation property for $p = 0$ was shown in the problem set of Section 5.2. See the problem set of this section for $p > 0$.

The character of decaying sinusoids is clear in the graphs of Figure 8.8. *Caution*: The zeros of a solution of Bessel's equation are not evenly spaced (with the exception of the zeros of $J_{1/2}$). However, it may be shown that the spacing of consecutive zeros approaches π (see Problem 8).

The locations of the first 20 or so zeros of J_n, $n = 0, \ldots, 8$, have been calculated to an accuracy of several decimal places.† The importance of the zeros lies in their coincidence with the nodal sets of physical systems modeled by Bessel functions (e.g., the dark rings of a diffraction pattern, the quiet lines on the membrane of a tympanum, the equilibrium points of an aging spring).

† See, e.g., Abramowitz and Stegun, op. cit.

Comments

We may easily extend the definition of the Bessel function of the first kind of any order to the entire real line:

$$J_p(x) = \left|\frac{x}{2}\right|^p \sum_{k=0}^{\infty} \frac{(-1)^k}{k!\,\Gamma(k+p+1)} \left(\frac{x}{2}\right)^{2k}, \qquad p \geqq 0, \quad \text{all} \quad x \qquad (18)$$

Note that this even function is a solution of Bessel's equation of order p on the x-interval $(0, \infty)$ and on the interval $(-\infty, 0)$, but need not be a solution at $x = 0$ since $J_p'(0)$ may not exist. In most of the applications the only x-domain of interest is $(0, \infty)$, and that is why we have confined most of our definitions and theorems to that interval.

We may extend the formula for $J_p(x)$ to all negative values of p as well. If p is not a negative integer, we have already done so, at least implicitly, by extending the domain of the gamma function from the positive half-line to the entire line with the negative integers deleted. We define $J_{-n}(x)$ for each negative integer $-n$ by removing from the range of the index k in (18) the integers $k = 0, 1, 2, \ldots,$ $n - 1$. It is not obvious that the diminished summation still defines a solution, but the reader may verify that it does by direct substitution into Bessel's equation of order n. With this alteration in the case of a negative integer, we may consider (18) to define a solution of Bessel's equation of order p even when we replace p by $-p$. The recursion formulas (12)–(15) may also be shown to hold for all p in much the same way as we proved them earlier for $p \geqq 0$.

If p is not an integer, J_{-p} is a second solution of Bessel's equation of order p which is independent of J_p. Thus $\{J_p, J_{-p}\}$ is a basis of the solution space. However, if $p = n$, an integer, $J_{-n}(x) = (-1)^n J_n(x)$ and we have not found a second, independent solution. We take up the question of second, independent solutions (i.e., of Bessel functions of the second kind) in the next section.

PROBLEMS

The following problem(s) may be more challenging: 5(d), 5(e), 6, 7, and 8.

1. Write out the first three nonvanishing terms of the series and sketch the graphs of $J_2(x)$ and $J_3(x)$.

2. Use Abel's Formula to show that the general solution of Bessel's equation of order p can be written in the form

$$y = c_1 J_p(x) + c_2 J_p(x) \int^x \frac{ds}{s\,[J_p(s)]^2}, \qquad c_1 \text{ and } c_2 \text{ any constants}$$

3. Show that the series for $J_p(x)$ converges for all x.

4. Assuming that the improper integral in the definition (6) of the gamma function converges for all $z > 0$, prove the following properties.

(a) $\Gamma(z + 1) = z\Gamma(z)$, $z > 0$.

(b) $\Gamma(n + 1) = n!$, $n = 0, 1, 2, \ldots$.

(c) $\Gamma(\tfrac{1}{2}) = \pi^{1/2}$.

(d) Use parts (a) and (c), and induction to show that $\Gamma(n + \tfrac{1}{2}) = [(2n)!/2^{2n}n!]\pi^{1/2}$, $n = 0, 1, 2, \ldots$.

5. [*Modified Bessel Equations*].

   (a) Show that if we replace the independent variable x in Bessel's equation of order p by $-it$, we obtain the *modified Bessel equation of order p*:

   $$t^2 y'' + ty' - (t^2 + p^2)y = 0$$

   (b) Show that

   $$I_0(t) = \sum_{n=0}^{\infty} \frac{1}{(n!)^2}\left(\frac{t}{2}\right)^{2n}$$

   is a solution to the modified Bessel equation of order 0. Sketch.

   (c) Show that

   $$I_1(t) = \frac{t}{2}\sum_{n=0}^{\infty} \frac{1}{n!(n+1)!}\left(\frac{t}{2}\right)^{2n}$$

   is a solution to the modified Bessel equation of order 1, $t > 0$.

   (d) [*Modified Bessel Function of the First Kind*]. Show that

   $$I_p(t) = \left(\frac{t}{2}\right)^p \sum_{n=0}^{\infty} \frac{1}{n!\,\Gamma(n+p+1)}\left(\frac{t}{2}\right)^{2n}, \qquad t > 0$$

   is a solution of the modified Bessel equation of order $p \geqq 0$. I_p is the *modified Bessel function of the first kind of order p*.

   (e) Show that no solution of the modified Bessel equation, $p \geq \tfrac{1}{2}$, has more than one zero on the interval $I = [0, \infty)$. [Note that I_0 never vanishes on $[0, \infty)$, while I_p vanishes on $[0, \infty)$ only at the origin if $p > 0$.] [*Hint*: See Problems 19 and 20 in the problem set of Section 5.2. Write the modified Bessel equation in normal form. Then compare with a suitably chosen equation.

6. [*Interlace*]. Show that the positive zeros of J_p and J_{p+1} interlace (i.e., between any two successive zeros of J_p there is a zero of J_{p+1} and between any two successive zeros of J_{p+1} there is a zero of J_p).

7. [*Orthogonality Property of Bessel Functions*]. The following steps show that if λ and μ are any two distinct nonnegative zeros of $J_p(x)$, then $\int_0^1 xJ_p(\lambda x)J_p(\mu x)\,dx = 0$, called the *orthogonality property* of Bessel functions.

   (a) Show that $y = J_p(\lambda x)$ satisfies the differential equation

   $$x^2\frac{d^2 y}{dx^2} + x\frac{dy}{dx} + (\lambda^2 x^2 - p^2)y = 0$$

   [*Hint*: Use the Chain Rule to calculate the derivatives; for example,

   $$\frac{d}{dx}J_p(\lambda x) = \frac{d}{dz}J_p(z)\frac{dz}{dx} = \lambda\frac{d}{dz}J_p(z) \quad \text{if} \quad z = \lambda x.]$$

(b) Denoting the differential equation in part (a) satisfied by $J_p(\lambda x)$ by $LJ_p(\lambda x) = 0$, write out $LJ_p(\mu x)$ and show that

$$\frac{1}{x} [J_p(\mu x)LJ_p(\lambda x) - J_p(\lambda x)LJ_p(\mu x)]$$

$$\equiv \frac{d}{dx}\left\{x[J_p(\mu x)J_p'(\lambda x) - J_p'(\mu x)J_p(\lambda x)]\right\} + x(\lambda^2 - \mu^2)J_p(\lambda x)J_p(\mu x) = 0$$

(c) Integrate the last equation in part (b) from $x = 0$ to $x = 1$ to obtain the desired result.

8. (a) Verify that every nontrivial solution of Bessel's equation of order $p > 0$ is oscillatory on $(0, \infty)$.

(b) Prove that if x_n and x_{n+1} are the nth and $(n + 1)$st consecutive zeros of a nontrivial solution of a Bessel's equation, then $x_{n+1} - x_n \to \pi$ as $n \to \infty$.

9. Verify recursion formulas (13) and (15).

8.7 SERIES SOLUTIONS NEAR REGULAR SINGULAR POINTS, II

Bessel functions of integer order were constructed by the method of generalized power series early in the nineteenth century, but the method failed to produce a second independent solution of Bessel's equation. The puzzle of the missing second solution remained unresolved until 1867, when C. G. Neumann (1832–1925) published his pioneering work on the matter. Frobenius extended Neumann's methods to a complete treatment of all solutions near a regular singular point of any linear equation of the second order. The extended theory of Frobenius fills in the gap noted in Section 8.5—the problem of the second solution when the roots of the indicial polynomial differ by an integer. In this section, we outline the extended theory, define Bessel functions of the second kind, and represent all solutions of the problem of the aging spring in terms of Bessel functions.

The Extended Method of Frobenius

To review briefly, we wish to find a pair of independent solutions in a neighborhood of the regular singular point $x = 0$ of the second-order linear equation

$$x^2 y'' + xP(x)y' + Q(x)y = 0 \tag{1}$$

where $P(x)$ and $Q(x)$ are real analytic on an interval J centered at the origin, $P(x) = \sum_0^\infty p_n x^n$, $Q(x) = \sum_0^\infty q_n x^n$. Associated with (1) are the indicial polynomial $f(r) = r^2 + (p_0 - 1)r + q_0$ and its roots (or characteristic exponents) r_1 and r_2. We suppose that r_1 and r_2 are real and that $r_2 \leq r_1$. The method of Frobenius outlined in Section 8.5 always yields a nontrivial solution of (1) of the form

$$y_1 = |x|^{r_1} \sum_0^\infty a_n x^n, \qquad x > 0 \text{ or } x < 0$$

However, the method may not always yield a second solution when $r_1 = r_2$, or even when $r_2 - r_1$ is a positive integer. In the former case our experience with Euler equations suggests that the second solution should involve a logarithm, but the case where $r_1 - r_2$ is a positive integer is a mystery. What form do the second solutions have in that situation, and how may they be found? The following complete version of Frobenius' Theorem clears up all the questions. It is stated in a somewhat repetitious form for reference purposes.

Frobenius' Theorem II. The method of undetermined coefficients will produce a nontrivial solution y_1 of Eq. (1):

$$y_1(x) = |x|^{r_1} \sum_0^\infty a_n x^n, \qquad x > 0 \text{ or } x < 0, \tag{2}$$

where a_0 is arbitrary and the a_n's are determined in terms of a_0 by the recursion formula

$$f(r_1 + n) a_n = -\sum_{k=0}^{n-1} [(k + r_1) p_{n-k} + q_{n-k}] a_k, \qquad n = 1, 2, \ldots \tag{3}$$

A second independent solution y_2 of (1) may be determined by a method of undetermined coefficients. It has one of the forms described in (a), (b), or (c) and is defined for $x > 0$ or $x < 0$ (all power series below converge in J):
(a) If $r_1 - r_2$ is not an integer, then

$$y_2 = |x|^{r_2} \left(1 + \sum_1^\infty b_n x^n \right) \tag{4}$$

where the b_n's are determined by the same recursion formula (3) as the a_n's, with r_2 replacing r_1.
(b) If $r_1 = r_2$, then

$$y_2 = y_1 \ln |x| + |x|^{r_1} \sum_0^\infty c_n x^n \tag{5}$$

(c) If $r_1 - r_2 = m$, a positive integer, then

$$y_2 = \alpha y_1 \ln |x| + |x|^{r_2} \left(1 + \sum_1^\infty d_n x^n \right) \tag{6}$$

where α is a constant (possibly 0).

It is clear from (3) just where the difficulty lies. First, in case (b), (3) can only lead to a single solution or its constant multiples. In case (c) we will have that $f(r_2 + m) = f(r_1) = 0$, and hence we cannot be sure that (3) is solvable for a_m. The proof of Frobenius' Theorem II is omitted.

The coefficients c_n, d_n, and α may be determined by substituting the appropriate form for y_2 into Eq. (1) and matching coefficients of like powers of x to obtain recursion relations.

Examples of the Difficult Cases of Frobenius' Theorem

Example 8.15

The differential equation $Ly \equiv xy'' + y' + 2y = 0$ has a regular singular point at 0. Its characteristic exponents are $r_1 = r_2 = 0$ and we are in case (b) of Frobenius' Theorem. The solution space is generated by a Frobenius series $y_1 = \sum_0^\infty a_n x^n$ and a function with a logarithmic term $y_2 = y_1 \ln |x| + \sum_0^\infty c_n x^n$.

As usual, in the calculation of the coefficients a_n and c_n we need only treat the case $x > 0$. We have, after appropriate reindexing,

$$Ly_1 = x \sum_0^\infty n(n-1)a_n x^{n-2} + \sum_0^\infty na_n x^{n-1} + 2 \sum_0^\infty a_n x^n$$

$$= \sum_1^\infty [n(n-1)a_n + na_n + 2a_{n-1}]x^{n-1} = 0$$

which gives the recurrence relation $n^2 a_n + 2a_{n-1} = 0$. Hence

$$a_n = \frac{-2a_{n-1}}{n^2} = \frac{4a_{n-2}}{n^2(n-1)^2} = \cdots = \frac{(-1)^n 2^n a_0}{(n!)^2}$$

We have, setting $a_0 = 1$,

$$y_1 = 1 + \sum_1^\infty \frac{(-1)^n 2^n x^n}{(n!)^2} \tag{7}$$

Turning to the calculation of y_2, we have for $x > 0$,

$$Ly_2 = L(y_1 \ln x) + L\left(\sum_0^\infty c_n x^n\right) = x\left(y_1'' \ln x + \frac{2y_1'}{x}\right) + \left(y_1' \ln x + \frac{y_1}{x}\right)$$

$$+ 2y_1 \ln x + y_1\left(\frac{-1}{x}\right) + \sum_1^\infty (n^2 c_n + 2c_{n-1})x^{n-1} = 0$$

Collecting the terms including $\ln x$, we have

$$Ly_2 = (Ly_1) \ln x + 2y_1' + \sum_1^\infty (n^2 c_n + 2c_{n-1})x^{n-1}$$

$$= 2y_1' + \sum_1^\infty (n^2 c_n + 2c_{n-1})x^{n-1} \qquad \text{(since } Ly_1 = 0\text{)}$$

$$= \sum_{1}^{\infty} \frac{(-1)^n 2^{n+1} x^{n-1}}{(n-1)!n!} + \sum_{1}^{\infty} (n^2 c_n + 2c_{n-1})x^{n-1} = 0$$

where it is only in the final step that we need to use the series (7) for y_1. This leads to the recursion relation

$$n^2 c_n + 2c_{n-1} + \frac{(-1)^n 2^{n+1}}{(n-1)!n!} = 0, \qquad n = 1, 2, \ldots \qquad (8)$$

It is somewhat difficult to solve this set of recursion relations to find c_n in terms of c_0. Instead, we write out the first few equations.

$$c_1 + 2c_0 - 4 = 0, \qquad c_1 = 2(2 - c_0)$$

$$4c_2 + 2c_1 + \frac{8}{2} = 0, \qquad c_2 = -3 + c_0$$

$$9c_3 + 2c_2 - \frac{16}{12} = 0, \qquad c_3 = \frac{22}{27} - \frac{2c_0}{9}$$

In this case we might as well let $c_0 = 0$. Thus our second solution is

$$y_2 = y_1 \ln |x| + \sum_{1}^{\infty} c_n x^n = y_1 \ln x + 4x - 3x^2 + \tfrac{22}{27}x^3 + \cdots$$

where the c_n's are defined recursively by (8) with $c_0 = 0$. Note once more that for all practical purposes an unsolved recursion relation is just as useful as a solved relation. If a computer is being used to calculate the coefficients, the unsolved form is to be preferred since it is almost always simpler. Initializing with c_0 and then iterating will do the trick in this case.

The following example illustrates case (c) of Frobenius' Theorem.

Example 8.16

The differential equation $Ly \equiv xy'' - y = 0$ has a regular singular point at 0. The indicial polynomial has roots 1 and 0. Thus the equation has a Frobenius series solution

$$y_1 = x \sum_{0}^{\infty} a_n x^n = \sum_{0}^{\infty} a_n x^{n+1}$$

where the a_n's may be found in the usual way. Suppressing the computations, we have that with $a_0 = 1$,

$$y_1 = \sum_{0}^{\infty} \frac{x^{n+1}}{(n+1)!n!}$$

A second solution, independent of y_1, has the form

$$y_2 = \alpha y_1 \ln |x| + \sum_{0}^{\infty} d_n x^n, \qquad d_0 = 1$$

Let x be positive and apply L to y_2. We have, after reindexing,

$$Ly_2 = \alpha L(y_1 \ln x) + L\left(\sum_0^\infty d_n x^n\right)$$

$$= \alpha(xy_1'' - y_1)\ln x + \alpha\left(2y_1' - \frac{y_1}{x}\right) + \sum_0^\infty n(n+1)d_{n+1} - d_n]x^n$$

$$= \alpha \sum_0^\infty \frac{2n+1}{(n+1)!n!}x^n + \sum_0^\infty [n(n+1)d_{n+1} - d_n]x^n = 0$$

since $xy_1'' - y_1 = 0.$† This gives the recurrence relation

$$n(n+1)d_{n+1} - d_n = \frac{-\alpha(2n+1)}{(n+1)!n!}, \qquad n = 0, 1, 2, \ldots \qquad (9)$$

The first few equations are

$$-d_0 = -\alpha, \qquad 2d_2 - d_1 = \frac{-3\alpha}{2}, \qquad 6d_3 - d_2 = \frac{-5\alpha}{12}$$

Since $d_0 = 1$, then $\alpha = 1$. Letting $d_1 = 0$ (or any other convenient value), we can then use (9) to determine recursively d_n, $n = 2, 3, \ldots$. Once again, this is quite adequate for computational use.

Although in this example $\alpha \neq 0$, it may happen that α vanishes. In such a case, the second solution would be a generalized power series like the first and there would be no logarithmic term.

Bessel Functions of the Second Kind

Let us apply Frobenius' Theorem II to the problem of finding a second solution of Bessel's equation of integer order n.

$$x^2 y'' + xy' + (x^2 - n^2)y = 0$$

a solution independent of the Bessel function $J_n(x)$ of the first kind found in Section 8.6. There are two cases. In the first, $n = 0$ and we are in case (b) of Frobenius' Theorem. Thus a second solution of Bessel's equation of order 0 independent of J_0 has the form $y_2 = y_1 \ln|x| + \sum_0^\infty c_n x^n$, and the c_n's must be determined by the method of undetermined coefficients. This can be done, but it takes a certain amount of ingenuity and patience. In practice, the second solution is taken to be a certain linear combination of $J_0(x)$ and y_2. This is the *Bessel function of the second kind* (or *Weber function*) *of order* 0 and is defined by

$$Y_0(x) = \frac{2}{\pi}\left(\gamma + \ln\left|\frac{x}{2}\right|\right)J_0(x) - \frac{2}{\pi}\sum_{k=0}^\infty \frac{(-1)^k h(k)}{(k!)^2}\left(\frac{x}{2}\right)^{2k}$$

† As in Example 8.15, there is no need to use the series form of the solution y_1 until the last step.

where $h(k) = 1 + \frac{1}{2} + \cdots + 1/k$ and γ is *Euler's constant*,†

$$\gamma = 1 + \sum_{2}^{\infty} \left(\frac{1}{n} + \ln \frac{n-1}{n} \right) = \lim_{k \to \infty} [h(k) - \ln k]$$

A second solution of Bessel's equation for $n > 0$, independent of J_n, may be obtained after a long analysis. Its usual form is given by

$$Y_n(x) = \frac{2}{\pi} \left(\gamma + \ln \left| \frac{x}{2} \right| \right) J_n(x) - \frac{1}{\pi} \left| \frac{x}{2} \right|^{-n} \sum_{k=0}^{n-1} \frac{(n-k-1)!}{k!} \left(\frac{x}{2} \right)^{2k}$$

$$- \frac{1}{\pi} \left| \frac{x}{2} \right|^{n} \sum_{k=0}^{\infty} \frac{(-1)^k [h(k) + h(k+n)]}{k!(n+k)!} \left(\frac{x}{2} \right)^{2k}$$

It is called a *Bessel function of the second kind* (or a *Weber function*) of order n.

Although the first and third terms in the expression for Y_n remain bounded near 0, the second does not, and $Y_n(x) \to -\infty$ as $x \to 0^+$. See Figure 8.9 for the graphs of the first three functions of integer order Y_0, Y_1, and Y_2. The oscillatory character of Y_n, and indeed of any nontrivial solution of Bessel's equation, has been shown in Section 8.6.

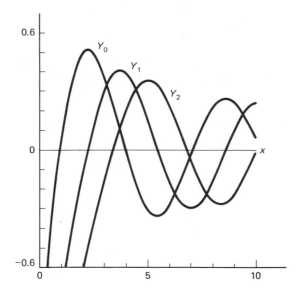

Figure 8.9 Y_0, Y_1, Y_2.

If p is not an integer, $p > 0$, then we noted in the preceding section that $J_{-p}(x)$ is a second solution independent of $J_p(x)$, and we have none of the complications encountered with the logarithmic terms. Nevertheless, it is customary to define the Bessel function of the second kind, Y_p, in this case as the linear combination of J_p and J_{-p}:

† Euler's constant is 0.5772156649 . . . ; it is not known whether the number is rational, algebraic, or transcendental.

$$Y_p(x) = \frac{\cos(p\pi)J_p(x) - J_{-p}(x)}{\sin p\pi}$$

from which it may be shown that $Y_n(x) = \lim_{p \to n} Y_p(x)$. It should be noted that the functions $Y_p(x)$, $p \geq 0$, also satisfy the recursion formulas of Section 8.6 and that $Y_p(x) \to -\infty$ as $x \to 0^+$.

Equations Reducible to Bessel Equations: The Aging Spring

It often happens that an equation superficially quite unlike any Bessel equation may be reduced to a Bessel equation by a judiciously chosen change of variable. This is true for the equation

$$mx''(t) + ke^{-\epsilon t}x(t) = 0 \tag{10}$$

the equation of the aging spring. Let us introduce a new measure of time, $s = \alpha e^{\beta t}$, where α and β are constants to be chosen later, and see what happens when we change from tx-variables to sx-variables in the equation. [An exponential measure of time is suggested by the form of the coefficient in (10).] We have by the chain rule that

$$\frac{dx}{dt} = \frac{dx}{ds}\frac{ds}{dt} = \frac{dx}{ds}\beta s$$

$$\frac{d^2x}{dt^2} = \frac{d}{dt}\left(\frac{dx}{dt}\right) = \frac{d}{ds}\left(\frac{dx}{dt}\right)\frac{ds}{dt} = \frac{d}{ds}\left(\frac{dx}{ds}\beta s\right)\frac{ds}{dt} = \frac{d^2x}{ds^2}(\beta s)^2 + \frac{dx}{ds}\beta^2 s$$

since $ds/dt = \beta\alpha e^{\beta t} = \beta s$. Hence

$$\frac{d^2x}{dt^2} + \frac{k}{m}e^{-\epsilon t}x = \beta^2 s^2 \frac{d^2x}{ds^2} + \beta^2 s\frac{dx}{ds} + \frac{k}{m}\left(\frac{s}{\alpha}\right)^{-\epsilon/\beta}x = 0$$

since $(s/\alpha)^{-\epsilon/\beta} = e^{-\epsilon t}$. Dividing by β^2 and choosing α and β so that

$$-\frac{\epsilon}{\beta} = 2, \qquad \frac{k}{m\beta^2\alpha^{-\epsilon/\beta}} = 1 \tag{11}$$

we obtain the Bessel equation of order 0 in s and x,

$$s^2\frac{d^2x}{ds^2} + s\frac{dx}{ds} + s^2 x = 0 \tag{12}$$

Thus $\beta = -\epsilon/2$ and $\alpha = (2/\epsilon)(k/m)^{1/2}$, which are obtained from (11), are the right choices for the constants in the change of variables.

The general solution of (12) is

$$x = c_1 J_0(s) + c_2 Y_0(s) \tag{13}$$

where c_1 and c_2 are arbitrary constants. In terms of the time variable t,

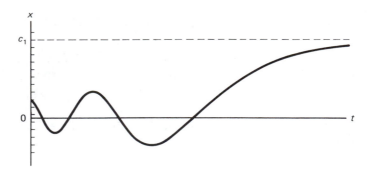

Figure 8.10   Aging spring; $x = c_1 J_0 \left[(2a/\epsilon) e^{-\epsilon t/2} \right]$.

$$ x = c_1 J_0 \left(\frac{2a}{\epsilon} e^{-\epsilon t/2} \right) + c_2 Y_0 \left(\frac{2a}{\epsilon} e^{-\epsilon t/2} \right), \qquad a = (k/m)^{1/2} \qquad (14) $$

There are two quite distinct possibilities for the behavior of the aging spring. If $c_2 = 0$, then as t tends to infinity the displacement from the rest position of the mass at the end of the spring asymptotically approaches the value $c_1 J_0(0) = c_1$ (see Figure 8.10).

It is more likely that neither c_1 nor c_2 vanish, and thus that as time goes on the displacement x approaches $c_1 J_0(0) + c_2 Y_0(0)$. Since $Y_0(s)$ tends to $-\infty$ as s approaches 0^+, the spring eventually stops oscillating and stretches without bound. Of course, what actually happens in this case is that the spring stretches and breaks (see Figure 8.11).

It should be noted in both cases that the aging spring can only pass through its equilibrium position a finite number of times before it begins to stretch without bound. This reflects the fact that the s-variable of (13) runs over a finite domain; any nontrivial solution of Bessel's equation has only finitely many zeros on a finite domain.

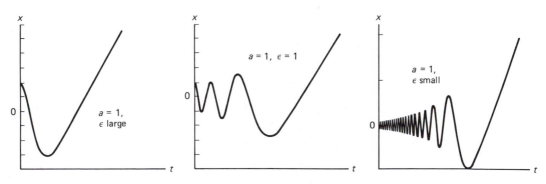

Figure 8.11   Aging spring; $x = c_1 J_0 \left(2a/\epsilon \; e^{-\epsilon t/2} \right) + c_2 Y_0 [(2a/\epsilon) e^{-\epsilon t/2}]$, $c_2 \neq 0$.

PROBLEMS

The following problem(s) may be more challenging: 1(d), 3, 4, and 6.

1. Check that 0 is a regular singular point of each equation and find a basis for the solution space on the interval $(0, \infty)$. Sketch.

   (a) $xy'' + (1 + x)y' + y = 0$.

   (b) $x^2y'' + x(x - 1)y' + (1 - x)y = 0$.

   (c) $xy'' - xy' + y = 0$.

   (d) $xy'' - x^2y' + y = 0$.

2. Suppose that $x = J_0(6e^{-t})$ is the solution of a model of an aging spring. Use the first three terms of the series expansion for J_0 and sketch x as a function of t, $0 \leq t \leq 6$.

3. (a) Show that the general solution of the general model of an aging spring, $x'' + (b^2 + a^2e^{-\epsilon t})x = 0$, is

$$x = C_1J_p(s) + C_2Y_p(s)$$

where $p = 2ib/\epsilon$ and $s = (2ae^{-\epsilon t/2})/\epsilon$, and J_p and Y_p are defined formally by replacing p by $2ib/\epsilon$ in the expressions for J_p and Y_p given in this section and in Section 8.6. Sketch the solutions for $b = 1$, $a^2 = 10$, $\epsilon = 1$, $x(0) = 1$, $x'(0) = 0$, $-5 \leq t \leq 5$.

   (b) Show that if $x(t)$ is any nontrivial solution of the differential equation in part (a), $t > 0$, then consecutive zeros of $x(t)$ are no more than π/b apart and no less than $\pi(a^2 + b^2)^{-1/2}$ apart. [*Hint*: Use the Oscillation Theorem of the problem set of Section 5.2.]

4. (a) If $w(s)$ is the general solution of $s^2w'' + sw' + (s^2 - p^2)w = 0$, show that $y(x) \equiv e^{ax}w(bx)$ is the general solution of the equation

$$x^2y'' + x(1 - 2ax)y' + [(a^2 + b^2)x^2 - ax - p^2]y = 0$$

where a and b are real constants, $b \neq 0$.

   (b) Find the general solution of $x^2y'' + x(1 - 2x)y' + (2x^2 - x - 1)y = 0$.

5. If p is not an integer, use the formula

$$Y_p = \frac{\cos(p\pi)J_p - J_{-p}}{\sin p\pi}$$

to show that Y_p satisfies the recursion formulas (12)–(15) of Section 8.6.

6. (a) Show that if $y(x)$ is a solution of Bessel's equation of order p, then $w(z) \equiv z^{-c}y(az^b)$ is a solution of

$$z^2w'' + (2c + 1)zw' + [a^2b^2z^{2b} + (c^2 - p^2b^2)]w = 0$$

   (b) [*Airy's Equation*]. Use part (a) to show that the general solution of Airy's equation, $y'' - xy = 0$, $x < 0$, is

$$y = |x|^{1/2}\left[C_1J_{1/3}\left(\frac{2|x|^{3/2}}{3}\right) + C_2J_{1/3}\left(\frac{2|x|^{3/2}}{3}\right)\right]$$

[*Hint*: First introduce a new variable $x = -s$.]

   (c) Find the general solution of each equation by using part (a).

   (i) $x^2y'' + (1/8 + x^4)y = 0$.   (ii) $y'' + x^4y = 0$).

CHAPTER 9

Introduction to Systems

The phenomena of nature are complex and interrelated. Few dynamical systems are isolated and complete in themselves, although we may imagine them to be so when we attempt to create mathematical models of their behavior. In earlier chapters that is just what we have done—we have modeled the oscillations of a single weighted spring, the transient current in a simple circuit, the solitary growth or decline in the numbers of one species. We now broaden our scope to include interacting systems: a predator species and its prey, voltages and currents in a multiloop circuit, the transitions from one element to another in the long radioactive chain from uranium to lead. The corresponding mathematical models consist of coupled differential equations in the state variables of the systems. In this chapter the underlying language and theory of first-order systems are sketched, together with the numerical and computational implementation of the theory (Sections 9.1, 9.3, and 9.4). Specific models of physical and biological systems are taken up in Sections 9.2 and 9.5, and to a lesser extent in Sections 9.1 and 9.3.

9.1 FIRST-ORDER SYSTEMS

A physical system evolves in time according to certain "laws." A mathematical model of the system represents the "laws of evolution" by equations in suitable variables. Time is a natural independent variable to be used in the equations of the model. The other variables in the equations, the dependent variables or *state variables*, are functions that depend on time and whose values portray the essential characteristics of the system. The equations themselves involve the state variables, their rates of change, and time. Systems of equations that involve the first derivatives of the state variables, but no higher derivatives, are called *first-order systems*. This section, this chapter, and Chapters 10 to 12 treat first-order systems, their sources, their properties, and their solutions.

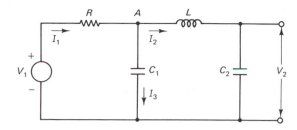

Figure 9.1 Communications circuit.

A Communications Circuit

The electrical circuit of Figure 9.1 is commonly used in communications devices. Voltage $V_1(t)$ is regarded as the known input and $V_2(t)$ is the response voltage or output. The currents $I_1(t)$ and $I_2(t)$ through the resistor and inductor are state variables together with $V_2(t)$. The circuit principles outlined in Section 5.4 may be used to derive the mathematical model of the circuit. From Coulomb's Law it follows that $V_2' = I_2/C_2$. From Kirchhoff's Voltage Law applied to the outer loop, we have that $V_1 = RI_1 + LI_2' + V_2$. Kirchhoff's Current Law at node A implies that the current I_3 equals $I_1 - I_2$. Finally, Kirchhoff's Voltage Law applied to the left loop gives $V_1 = RI_1 + (1/C_1) \int_0^t [I_1(s) - I_2(s)] \, ds$, or, in differentiated form $V_1' = RI_1' + (1/C_1)(I_1 - I_2)$. With a little rearrangement of these relations, we have derived the following model equations for the curcuit.

$$V_2' = \frac{1}{C_2} I_2$$

$$I_1' = -\frac{1}{RC_1} I_1 + \frac{1}{RC_1} I_2 + \frac{V_1'}{R} \tag{1}$$

$$I_2' = -\frac{1}{L} V_2 - \frac{R}{L} I_1 + \frac{V_1}{L}$$

Given $V_1(t)$, the circuit constants R, L, C_1, and C_2, and the initial voltages and currents, $V_2(0)$, $I_1(0)$, and $I_2(0)$, we would like to determine the state variables $V_2(t)$, $I_1(t)$, and $I_2(t)$ from (1). Of particular interest is $V_2(t)$. However, the equations of (1) are coupled, one to another, and it is not clear how they may be solved. We show how the solution of (1) can be accomplished in Section 10.5. In that section we also show that the circuit "filters out" high-frequency input signals ($V_1 = a_0 e^{i\omega t}$, $|\omega|$ large) in the sense that the amplitude of the response voltage V_2 is a small fraction of $|a_0|$. In contrast, low-frequency voltages pass through the circuit almost unchanged.

States, Systems, Solutions, Orbits, and Graphs

The mathematical model of the communications circuit contains three fairly simple rate equations in the three state variables of the circuit. More complex phenomena may lead to more equations, more state variables, and more intricate relations between the rates and the variables. An astounding number and variety of phenomena may be modeled by *first-order systems* of the form

$$x_1' = f_1(t, x_1, x_2, \ldots, x_n)$$
$$x_2' = f_2(t, x_1, x_2, \ldots, x_n)$$
$$\vdots \tag{2}$$
$$x_n' = f_n(t, x_1, x_2, \ldots, x_n)$$

in the *state variables* $x_1, \ldots, x_n$ and the independent variable t, which usually represents time. All derivatives are with respect to time, and the rate functions f_1, $f_2, \ldots, f_n$ are assumed to be known functions of time and state. *Initial conditions* of the form $x_1(t_0) = b_1$, $x_2(t_0) = b_2, \ldots, x_n(t_0) = b_n$, may also be prescribed. A *solution* is a set of functions, $x_1 = x_1(t)$, $x_2 = x_2(t), \ldots, x_n = x_n(t)$, all defined on a common interval I of time containing t_0, which satisfy the equations of (2) for all t in I, and which satisfy any given initial conditions.

The definitions above may be considerably abbreviated by the use of vector notation. Let the *state vector* x, the *initial state* x^0, and the *vector rate function* f be defined as column vectors,

$$x = \begin{bmatrix} x_1 \\ \cdot \\ \cdot \\ \cdot \\ x_n \end{bmatrix}, \qquad x^0 = \begin{bmatrix} b_1 \\ \cdot \\ \cdot \\ \cdot \\ b_n \end{bmatrix}, \qquad f = \begin{bmatrix} f_1 \\ \cdot \\ \cdot \\ \cdot \\ f_n \end{bmatrix}$$

Then the problem stated above for a first-order system takes the form

> **Initial Value Problem for a First-Order System.**   Find a state vector function $x(t)$, defined on a nontrivial interval I containing t_0, such that
>
> $$x'(t) = f(t, x(t)) \qquad \text{for all } t \text{ in } I$$
> $$x(t_0) = x^0 \tag{3}$$
>
> where t_0 is the initial time and x^0 is the initial state.

It is assumed that I is the largest interval on which the system and the solution $x(t)$ are both defined (i.e., that the solution is *maximally extended*). The state $x(t)$ is a *forward solution* for (3) if it is defined for $t \geq t_0$; a forward solution is associated with the future evolution of the state of the system. A *backward solution* for (3) is defined for $t \leq t_0$ and depicts the history of the system. It is usually clear in each case whether it is a matter of a forward, or a backward, or a *two-sided solution* (defined on an interval containing t_0 in its interior).

A solution of an initial value problem determines several curves whose nature reflects the properties of the solution.

> **Orbits and Solution Curves.**   Let $x = x(t)$ be a solution of $x' = f(t, x)$, where t lies in an interval I. The point $x(t)$ traces out the *orbit* of the solution in the *state* (or *phase*) *space* of the x-variables as t increases through I. Arrows on orbits are sometimes used to indicate the direction in which the state evolves with increasing time. A collection of several orbits is a *portrait* of the orbits. As t increases through I, $(t, x(t))$ traces out the *solution curve* in the tx-space of time and state variables. The projection of a solution curve onto the tx_j-plane is the *tx_j-component graph*.

Figures 9.2 and 9.3 illustrate these definitions. Observe that the projection of a solution curve onto the state space yields the orbit. If the state space is two-dimensional, we speak of "a planar portrait" in the "state plane."

Example 9.1

The system $x_1' = -x_1$, $x_2' = -2x_2$ is uncoupled and easily solved. The solutions are given by $x_1 = b_1 e^{-t}$ and $x_2 = b_2 e^{-2t}$, $-\infty < t < \infty$, where b_1 and b_2 are arbitrary constants. We may interpret b_1 and b_2 as the respective values of x_1 and x_2 at the initial time $t_0 = 0$. If $b_1 \neq 0$, we see that $x_2 = (b_2/b_1^2)x_1^2$, which is the equation of a parabola in the state plane. Corresponding orbits lie on these parabolas. Observe that $x_1(t)$ and $x_2(t)$ both approach zero asymptotically as time approaches infinity. Observe also that there are "ray" solutions along the axes in the state plane. See Figure 9.2 for sketches of orbits

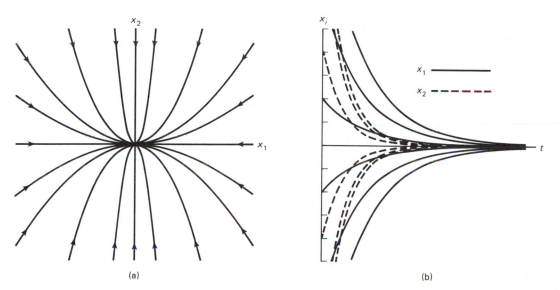

Figure 9.2   Solutions of $x_1' = -x_1$, $x_2' = -2x_2$: (a) portrait of the orbits; (b) component graphs.

and component graphs. The arrowheads on the orbits show the direction of time's increase. Note that different solutions may produce the same orbit. For example, $x_1 = e^{-t}$, $x_2 = e^{-2t}$, $-\infty < t < \infty$, and $\hat{x}_1 = 2e^{-t}$, $\hat{x}_2 = 4e^{-2t}$, $-\infty < t < \infty$, generate the same orbit, $x_2 = x_1^2$, $x_1 > 0$. The difference between the two solutions is just a matter of a translation of time, for $\hat{x}_1(t) = x_1(t - \ln 2)$ and $\hat{x}_2(t) = x_2(t - \ln 2)$. Note also the constant solution, $x_1(t) \equiv 0$, $x_2(t) \equiv 0$, and the corresponding point orbit $(0, 0)$.†

Example 9.2

The system $x_1' = -x_1 + x_2$, $x_2' = -x_1 - x_2$ has a solution, $x_1 = e^{-t} \sin t$, $x_2 = e^{-t} \cos t$, where $-\infty < t < \infty$. Another solution is the constant solution, $x_1(t) \equiv 0$, $x_2(t) \equiv 0$, called an *equilibrium solution*. The solution curve of this constant solution is a line, and its orbit is the equilibrium point at the origin. The solution curve of the first solution given above is a corkscrew-like curve that spirals around and toward the time axis (Figure 9.3). The functions $\sin t$ and $\cos t$ give the solution curve its oscillating character, while the factor e^{-t} pulls the curve toward the t-axis with the advance of time. The corresponding orbit in the state plane is a tightening spiral that winds onto the equilibrium point as t approaches infinity.

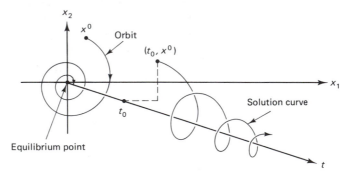

Figure 9.3 Some solution curves and orbits of the system $x_1' = -x_1 + x_2$, $x_2' = -x_1 - x_2$.

From a Scalar Equation to a System

A first-order system of differential equations may arise directly from the modeling of some physical phenomenon, as we saw in the model of the communications circuit. Systems may also be introduced to obtain insight into the behavior of the solutions of scalar differential equations. Consider the nth-order scalar differential equation

$$y^{(n)} = g(t, y, y', \ldots, y^{(n-1)}) \tag{4}$$

where the derivatives are taken with respect to t. Now introduce the state variables $x_1, \ldots, x_n$ as follows:

† Point orbits are called *equilibrium points* (see Section 9.3).

$$x_1 = y, \quad x_2 = y', \ldots, x_{n-1} = y^{(n-2)}, \quad x_n = y^{(n-1)} \tag{5}$$

Differentiating with respect to t, we have from (5) that

$$x_1' = y' = x_2, \quad x_2' = y'' = x_3, \quad \ldots, \quad x_{n-1}' = y^{(n-1)} = x_n \tag{6}$$

and from (4) and (5) that

$$x_n' = y^{(n)} = g(t, y, y', \ldots, y^{(n-1)}) = g(t, x_1, x_2, \ldots, x_n) \tag{7}$$

Thus we have from (6) and (7) that

$$
\begin{aligned}
x_1' &= x_2 \\
x_2' &= x_3 \\
&\ \ \vdots \\
x_{n-1}' &= x_n \\
x_n' &= g(t, x_1, x_2, \ldots, x_n)
\end{aligned}
\tag{8}
$$

which is a first-order system of n rate equations in the state variables and t. In vector notation, (8) has the form

$$x' = f(t, x) \tag{9}$$

where

$$
x = \begin{bmatrix} x_1 \\ x_2 \\ \cdot \\ \cdot \\ \cdot \\ x_n \end{bmatrix} \quad \text{and} \quad f = \begin{bmatrix} x_2 \\ x_3 \\ \cdot \\ \cdot \\ \cdot \\ g(t, x) \end{bmatrix}
$$

Thus from the scalar equation (4), we have created a vector system.

Example 9.3

The scalar equation $y'' + 3y' - 4y = 0$ may be converted to a system by setting $y = x_1$ and $y' = x_2$. Thus we have that

$$
\begin{aligned}
x_1' &= x_2 \\
x_2' &= 4x_1 - 3x_2
\end{aligned}
$$

There is a simple relationship between scalar solutions of (4) and vector solutions of (9). Given a solution of one, it uniquely determines a solution of the other. For suppose that $y = y(t)$, $a < t < b$, is solution of (4). Then $x_1 = y(t)$, $x_2 = y'(t)$, $\ldots$, $x_n = y^{(n-1)}(t)$, $a < t < b$, are the components of a solution of system (9) by the definition of (9). Conversely, suppose that $x_1 = x_1(t)$, $x_2 = x_2(t)$, . . . , $x_n = x_n(t)$, $a < t < b$, are the components of a solution of (9). Let $y = x_1(t)$. Then $y' = x_1' = x_2$, . . . , $y^{(n-1)} = x_{n-1}' = x_n$, $y^{(n)} = x_n' = g(t, x_1, x_2, . . . , x_n)$, and $y = x_1(t)$ is a solution of (4). This correspondence between solutions of (4) and solutions of (9) easily extends to include corresponding initial conditions:

$$y(t_0) = b_1, \quad y'(t_0) = b_2, \quad . . . , \quad y^{(n-1)}(t_0) = b_n$$

becomes

$$x_1(t_0) = b_1, \quad x_2(t_0) = b_2, \quad . . . , \quad x_n(t_0) = b_n$$

Example 9.4

The scalar equation $y'' + 3y' - 4y = 0$ has the general solution $y = c_1 e^{-4t} + c_2 e^t$. The corresponding general solution of the equivalent system $x_1' = x_2$, $x_2' = 4x_1 - 3x_2$, where $x_1 = y$ and $x_2 = y'$, is given by

$$x_1 = c_1 e^{-4t} + c_2 e^t$$
$$x_2 = -4c_1 e^{-4t} + c_2 e^t$$

Comments

There is more to solving systems than the elementary methods used in Examples 9.1 and 9.4 would indicate. We have not attempted to give any systematic treatment of solution techniques or of the theory for first-order systems in this section—that will come later. The problem set contains some special solution methods which are effective for solving special kinds of systems. One should observe that each first-order system has exactly as many equations as state variables. This makes sense since one expects one equation to determine one state variable, two equations to determine two state variables, and so on. However, without prescribed initial conditions, the state variables are not uniquely defined as functions of time by solving the equations. Generally speaking, the state variables will be uniquely defined if, in addition, there are as many prescribed initial values for these variables (at a common prescribed initial time) as there are state variables. Existence and uniqueness theorems on these matters are formulated and discussed in Section 9.3.

PROBLEMS

The following problem(s) may be more challenging: 6, 7, and 8.

1. Sometimes a coupled system can be solved one equation at a time, the solution then being used in subsequent equations. Use this method to solve the following problems.
    (a) $x_1' = -x_1$, $x_2' = x_1 - 2x_2$.
    (b) $x_1' = -x_1 + 1$, $x_2' = x_1^2 x_2$, $x_1(0) = 0$, $x_2(0) = 1$.

2. What second-order equation is equivalent to the system $x_1' = x_2$, $x_2' = -9x_1 + \cos t$? Find all solutions of the system.

3. For each scalar differential equation find an equivalent first-order system. For each system, find an equivalent scalar differential equation. In each case find the general solution of the scalar equation and the general solution of the system.
   (a) $y'' - 4y = 0$.   **(b)** $y'' + 9y = 0$.   **(c)** $y'' + 5y' + 4y = 0$.
   (d) $x_1' = x_2$, $x_2' = -2x_1 - 2x_2$.   **(e)** $x_1' = x_2$, $x_2' = -16x_1$.
   (f) $y''' + 6y'' + 11y' + 6y = 0$.   **(g)** $x_1' = x_2$, $x_2' = x_3$, $x_3' = 4x_1 - 4x_2 + x_3$.

4. In the manner of Examples 9.1 and 9.2 sketch the orbit and corresponding component graphs of the solution of the problem, $x' = y$, $y' = -2x - 2y$, $x(0) = 2$, and $y(0) = 0$.

5. Sometimes equations for orbits can be found directly without finding solutions. In each of the following problems find equations for the orbits by solving $dx_2/dx_1 = f_2/f_1$. Sketch the orbits in the state plane.
   (a) $x_1' = x_1$, $x_2' = -3x_2$. [*Hint*: Set

   $$\frac{dx_2}{dx_1} = \frac{dx_2/dt}{dx_1/dt} = \frac{-3x_2}{x_1}$$

   separate variables, and solve.]
   (b) $x_1' = x_1 x_2$, $x_2' = x_1^2 + x_2^2$. [*Hint*: Apply the method of homogeneous functions to $dx_2/dx_1 = (x_1^2 + x_2^2)/x_1 x_2$.]
   (c) $x_1' = x_2$, $x_2' = -e^{-x_1}$.

6. The solutions of the scalar equation $x' = ax$ have the form $x = x_0 e^{at}$. This suggests that solutions of $x_1' = ax_1 + bx_2$, $x_2' = cx_1 + dx_2$, where a, b, c, and d are constants, might have the form $x_1 = \alpha e^{rt} + \beta e^{st}$, $x_2 = \gamma e^{rt} + \delta e^{st}$, where r and s, α, β, γ, and δ must be determined by inserting x_1 and x_2, as given, into the differential system and matching coefficients of like exponentials. Solve the following systems.
   (a) $x_1' = x_1 + 3x_2$, $x_2' = x_1 - x_2$.
   (b) $x_1' = 2x_1 - x_2$, $x_2' = 3x_1 - 2x_2$.

7. Laplace transforms may be used to solve some systems. For example, if $x_1' = ax_1 + bx_2 + f_1(t)$, $x_2' = cx_1 + dx_2 + f_2(t)$, a, b, c, and d real constants, then applying the transform to each side of each equation, we have that $sL[x_1](s) - x_1(0) = aL[x_1](s) + bL[x_2](s) + L[f_1](s)$, and a similar equation for $sL[x_2](s) - x_2(0)$. These two linear equations may be solved for $L[x_1]$ and $L[x_2]$. Transform tables may be used to find $x_1(t)$ and $x_2(t)$. Solve the following problems this way. Let $\hat{x}(s)$ denote the Laplace transform of the function $x(t)$.
   (a) $x_1' = 3x_1 - 2x_2 + t$, $x_2' = 5x_1 - 3x_2 + 5$; $x_1(0) = x_2(0) = 0$.
   (b) $x_1' = x_1 + 3x_2 + \sin t$, $x_2' = x_1 - x_2$; $x_1(0) = 0$, $x_2(0) = 1$.

8. Find all solutions of the following system. Sketch solutions and component graphs.

   $$y_1' + 2y_2' = 3y_1 - 4y_2$$
   $$2y_1' + y_2' = -2y_1 + y_2$$

9. Suppose that the input voltage for the circuit of Figure 9.1 is the constant V_1^0. Show that $V_2 = V_1^0$, $I_1 = I_2 = 0$, defines a constant orbit. Are there any other constant orbits?

10. **(a)** Model the circuit sketched below by using as the state variables the current I through the coil and the output voltage V_2. Assume that the input voltage V_1 is given and

that L, C, and R are positive constants. [*Hint:* Show that $V_1 = LI' + V_2$ and $I = V_2/R + CV_2'$].

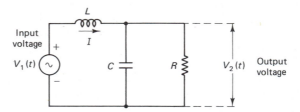

(b) Find a single scalar second-order equation in V_2 equivalent to the system in part (a). [*Hint:* Find I' from the second equation of the system in part (a) and substitute in the first.]

(c) Find $I(t)$ and $V_2(t)$ if $L = R = 1$, $C = \frac{1}{2}$, $V_1(t) = \cos t$, $I(0) = 0$, $V_2(0) = 0$.

11. Two masses are coupled to one another and to rigid supports by coiled springs as shown in the sketch. The variables x_1 and x_2 denote displacement from the equilibrium positions of the two masses. The positive constants k_1, k_2, and k_3 are the spring constants.

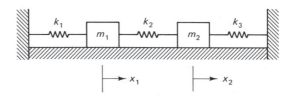

(a) Suppose that horizontal motion of the masses is frictionless. Explain why Newton's Second Law applied to the mass m_1 results in the second-order equation $m_1 x_1'' = -k_1 x_1 + k_2(x_2 - x_1)$.

(b) Suppose that frictional forces oppose motion and have magnitude proportional to the speed. Explain why the dynamic equation of motion of mass m_1 is $m_1 x_1'' = -k_1 x_1 + k_2(x_2 - x_1) - c_1 x_1'$, where c_1 is a positive constant.

(c) Write out a similar dynamical equation for the second mass, assuming a frictional force of a nature like that acting on m_1.

(d) Reduce the pair of coupled second-order equations of parts (b) and (c) to a system of four coupled first-order equations by introducing two auxiliary state variables, $x_1' = v_1$ and $x_2' = v_2$. [*Hint:* $x_1'' = f(x_1, x_2, x_1')$ becomes the pair $x_1' = v_1$, $v_1' = f(x_1, x_2, v_1)$.]

9.2 CASCADES: THE BONES OF OLDUVAI AND THE URANIUM SERIES

The rate equations of a first-order system may be so strongly coupled to one another that it is hard to solve any one of them without simultaneously solving them all. At the other extreme, there are systems where the coupling is weak enough that the equations may be solved separately, but not necessarily independently. Cascade systems are of the latter type. Specifically, a *cascade system* has the following form:

$$x'_1 = f_1(t, x_1)$$
$$x'_2 = f_2(t, x_1, x_2)$$
$$\begin{matrix} . \\ . \\ . \end{matrix} \qquad\qquad (1)$$
$$x'_n = f_n(t, x_1, x_2, \ldots, x_n)$$

where the rate function at each level may involve variables from that and preceding levels, but not variables from following levels. Solving proceeds from top down. The first equation is solved to obtain $x_1(t)$, which is then substituted for x_1 in the rate function $f_2(t, x_1, x_2)$, and so on. At each step only a single first-order rate equation in one dependent variable needs to be solved.

Linear Cascades

A cascade system is *linear* if each rate function f_j, $j = 1, \ldots, n$, has the form

$$f_j = a_{j1}(t)x_1 + \cdots + a_{jj}(t)x_j + g_j(t)$$

In the simplest linear cascades the coefficients a_{jk} are constants. The radioactive decay of an element into one or more new elements, which in turn decay into still other elements, is the most noteworthy example of a constant-coefficient linear cascade. The remainder of this section is devoted to models of this type.

The construction of a linear cascade decay model is based on the Balance Law of Section 3.1 (or the Radioactive Decay Law of Section 1.2):

$$\text{Net rate of change of mass} = \text{production rate} - \text{decay rate} \qquad (2)$$

Specifically, let the mass of substance i at time t be denoted by $x_i(t)$, and let the rate constants be denoted by positive constants k_j. Three prototypes of cascades are given in Table 9.1 and Figure 9.4. In each case there is at least one *final product* at the bottom of the cascade.

TABLE 9.1   RATE EQUATIONS OF THE CASCADES OF FIGURE 9.4

(a) Direct	(b) Branching–joining	(c) Joining–branching
$x'_1 = -k_1 x_1$	$x'_1 = -(k_1 + k_2)x_1$	$x'_1 = -k_1 x_1$
$x'_2 = k_1 x_1 - k_2 x_2$	$x'_2 = k_1 x_1 - k_3 x_2$	$x'_2 = -k_2 x_2$
$x'_3 = k_2 x_2$	$x'_3 = k_2 x_1 - k_4 x_3$	$x'_3 = k_1 x_1 + k_2 x_2 - (k_3 + k_4)x_3$
	$x'_4 = k_3 x_2 + k_4 x_3$	$x'_4 = k_3 x_3$
		$x'_5 = k_4 x_3$

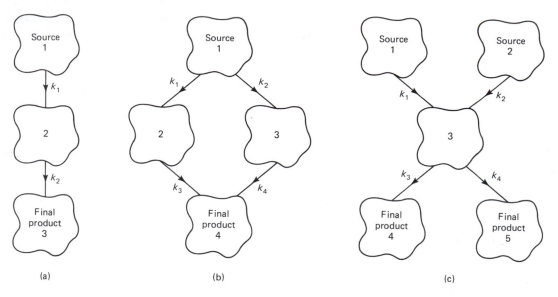

Figure 9.4   Cascades: (a) direct; (b) branching–joining; (c) joining–branching.

Observe that in each of the three models the total mass, $x_1(t) + \cdots + x_n(t)$, never changes since $x_1'(t) + \cdots + x_n'(t) = 0$ for all t. Thus mass is conserved in these models. It may also be shown that as $t \to +\infty$, all the substances end up as final products.

Example 9.5

The process by which a cascade model is solved may be illustrated by solving the direct cascade of (a) in Table 9.1. Set $x_1 = x$, $x_2 = y$, $x_3 = z$ to avoid confusion with subscripts, and assume that at time $t = 0$, $x(0) = x_0$, $y(0) = y_0$, $z(0) = z_0$. Integrating factors may be used to solve the equations one after the other, using at each stage the solution of the preceding equation. In this way the following formulas are obtained for a three-element direct decay cascade with $k_1 \neq k_2$.

$$\frac{dx}{dt} = -k_1 x \tag{3}$$

$$x(t) = x_0 e^{-k_1 t}$$

$$\frac{dy}{dt} + k_2 y = k_1 x_0 e^{-k_1 t}$$

$$y(t) = e^{-k_2 t} y_0 + \int_0^t e^{-k_2 t + (k_2 - k_1)s} k_1 x_0 \, ds \tag{4}$$

$$= e^{-k_2 t} y_0 + \frac{k_1 x_0}{k_2 - k_1} (e^{-k_1 t} - e^{-k_2 t})$$

$$\frac{dz}{dt} = k_2 y(t) = k_2 \left[e^{-k_2 t} y_0 + \frac{k_1 x_0}{k_2 - k_1} (e^{-k_1 t} - e^{-k_2 t}) \right]$$

$$z(t) = y_0 (1 - e^{-k_2 t}) + \frac{k_2 k_1 x_0}{k_2 - k_1} \left(\frac{1 - e^{-k_1 t}}{k_1} - \frac{1 - e^{-k_2 t}}{k_2} \right) + z_0 \tag{5}$$

The reader may show that $x(t) \to 0$, $y(t) \to 0$, and $z(t) \to M = x_0 + y_0 + z_0$ as $t \to \infty$, while $x(t) + y(t) + z(t) = M$ for all t. Thus x decays to y, which in turn decays to z. The orbits defined by the solutions of (3)–(5) illustrate this behavior. Some orbits are sketched in Figure 9.5 on the part of the plane of constant mass, $x + y + z = M$, which lies in the orthant $x \geq 0$, $y \geq 0$, $z \geq 0$. Observe that each orbit on the plane is asymptotic to the point orbit $(0, 0, M)$ on the z-axis.

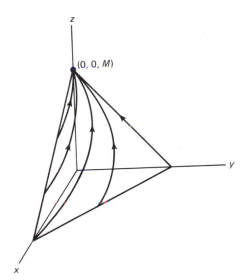

Figure 9.5   Orbits on a plane of constant mass.

We shall not carry out the detailed solution of the branching–joining or joining–branching processes. The two models discussed below illustrate particular occurrences of these processes in nature.

The Bones of Olduvai

Olduvai Gorge, in Kenya, cuts through volcanic flows, tuff (volcanic ash), and sedimentary deposits. It is the site of bones and artifacts of early hominids, considered by some to be precursors of man. In 1959, Mary and Louis Leakey uncovered in the gorge a fossil hominid skull and primitive stone tools of obviously great age, older by far than any hominid remains found up to that time. Excited by the find after nearly 30 years of searching, the Leakeys named the skull "Dear Boy" and attempted to determine its age. Carbon-14 dating methods being inappropriate for a specimen

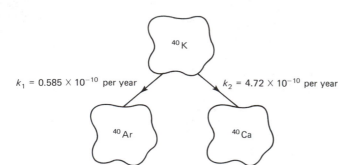

Figure 9.6   Potassium–argon and potassium–calcium decay.

of that age and nature, dating had to be based on the ages of the underlying and overlying geological strata. The method used was that of potassium–argon decay.

The potassium–argon clock is an accumulation clock, in contrast to carbon-14 dating, which is based on the disappearance of radioactive "parent" atoms (see Chapter 1). The potassium–argon method depends on measuring the accumulation of "daughter" argon atoms, which are decay products of radioactive potassium atoms. Specifically, potassium-40 (^{40}K) decays to argon-40 (^{40}Ar) and to calcium-40 (^{40}Ca) by the branching cascade illustrated in Figure 9.6. Potassium decays to calcium by emitting a β particle (i.e., an electron). Some of the potassium atoms, however, decay to argon by capturing an extranuclear electron and emitting a gamma ray.

The rate equations for this decay process may be written in terms of the masses $K(t)$, $A(t)$, and $C(t)$ of potassium, argon, and calcium in a sample of rock:

$$K' = -(k_1 + k_2)K$$
$$A' = k_1 K \tag{6}$$
$$C' = k_2 K$$

Setting $k = k_1 + k_2$, $K(0) = K_0$, $A(0) = 0$, $C(0) = 0$† and solving, we have that

$$K(t) = K_0 e^{-kt}$$
$$A(t) = \frac{k_1}{k} K_0 (1 - e^{-kt}) \tag{7}$$
$$C(t) = \frac{k_2}{k} K_0 (1 - e^{-kt})$$

where the potassium–argon–calcium clock is assumed to start at time $t = 0$ when the volcanic strata were laid down (i.e., at approximately the time when Dear Boy was alive). Observe that $K + A + C = K_0$ and that $K(t) \to 0$, $A(t) \to k_1 K_0/k$, and $C(t) \to k_2 K_0/k$ as $t \to \infty$.

The age of the volcanic strata (i.e., the current value of the time variable t)

† It would be more accurate to set $C(0) = C_0 > 0$, but doing so has no effect on the dating process given by the first two equations of (7) and by Eq. (8).

may be calculated by measuring the ratio of argon to potassium in a sample. In particular, we have from (7) that

$$\frac{A}{K} = \frac{k_1}{k} \left(e^{kt} - 1 \right)$$

Solving for t, we have that

$$\text{age of sample (in years)} = \frac{1}{k} \ln \left(\frac{kA}{k_1 K} + 1 \right) \tag{8}$$

In the case of potassium–argon–calcium decay, k_1 is known to be 0.585×10^{-10} per year, and $k_2 = 4.72 \times 10^{-10}$ per year, and k to be 5.305×10^{-10} per year (recall that $k = k_1 + k_2$). Thus a measurement of the ratio A/K of argon to potassium in a sample, or rather the average ratio in several samples, is enough to determine the age.

The measurements of the Olduvai samples were made at the University of California. After adjusting for environmental contamination of the sample, an approximate age of 1.75 million years was determined. Since stone tools were found together with the bones, it is reasonable to suppose that hominids were making tools almost 2 million years ago, a far earlier date than expected.

It should be noted that the possibility of using potassium–argon decay as a dating process depends on the fact that the argon gas is trapped in the lattices of a cool rock, but escapes if the rock is molten. It was fortunate that the strata around Dear Boy were largely volcanic material which had formed and then cooled quickly (in geological time) about 1.75 million years ago, thereby starting the clock. It was also fortunate that such layers were both above and below Dear Boy's skull.

The argon/potassium ratio in a sample is used rather than the calcium/potassium ratio because calcium has many sources, not just as a decay product from potassium. Argon, on the other hand, comes almost exclusively from the decay process. As noted earlier, there is some argon in the environment which affects the dating process, but there are techniques for taking this into account. See Figure 9.7 for the decay curve of potassium and the corresponding accumulation curves of argon and calcium.

The Uranium Series

Potassium–argon–calcium decay is a simple branching process of radioactive disintegration, a process used to date relatively young samples (in geologic time). At the other extreme there are radioactive decay cascades which began at the creation of the earth and still continue today. The cascade beginning with uranium-238 (^{238}U) has 14 levels and includes three branching–joining steps. The elements involved are uranium (U), thorium (Th), proactinium (Pa), radium (Ra), radon (Rn), polonium (Po), lead (Pb), astatine (At), bismuth (Bi), and thallium (Tl). At each decay step, α-particles (atomic mass 4) or β-particles (atomic mass effectively 0) will be emitted.

In Figure 9.8 the half-lives are given in seconds (s), minutes (m), days (d), and years (y). Half-lives are used rather than decay constants. Recall from Chapter

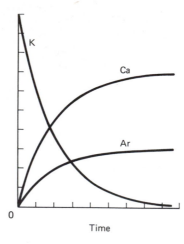

Figure 9.7   Potassium–argon–calcium branching cascade.

1 that the half-life T of a simple, nonbranching decay is given in terms of the decay constant k by the formula

$$k = \frac{1}{T} \ln 2 \tag{9}$$

Decay constants and half-lives are also simply related in a branching process. Suppose that A has half-life T and branches to B and C, $b\%$ of the decay going to B and $c\%$ to C ($b + c = 100$). Then we have

$$k = \frac{1}{T} \ln 2$$

$$k_B = \frac{b}{100} k \tag{10}$$

$$k_C = \frac{c}{100} k$$

With this in mind, we give the complete uranium series from the slowly decaying ^{238}U formed over 4.5 billion years ago to stable lead ^{206}Pb, which gradually accumulates over that long span of time. In each case we have indicated whether alpha or beta decay is involved, and in the case of branching, which percentages of decaying atoms go one way and which the other.

We shall not write out the complete set of model rate equations for the uranium series, involving as it does the 18 state variables ranging from ^{238}U to ^{206}Pb and the corresponding 17 rate constants determined by the half-lives. However, we shall write out the first two rate equations and those of one of the branches to illustrate the process. Using the data from Figure 9.8 and converting half-lives to years and then to decay constants in years^{-1} by (9) and (10), we may write out rate equations.

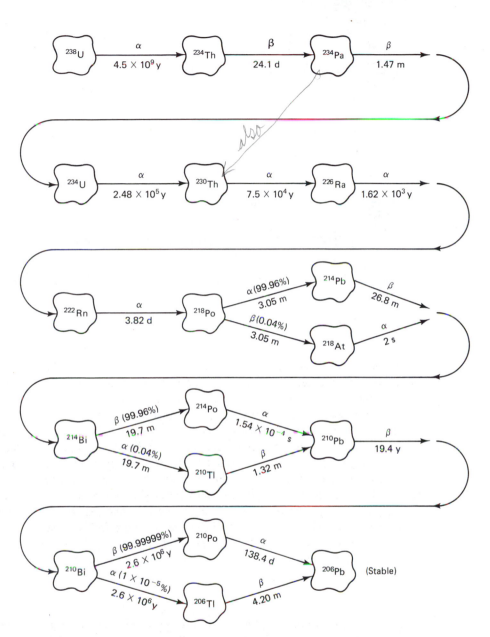

Figure 9.8 Uranium series. (A branch from ^{234}Pa to ^{230}Th involving an isomeric transition has been omitted.)

These equations are

$$\frac{d\ ^{238}U}{dt} = -1.540 \times 10^{-10}\ ^{238}U$$

$$\frac{d\ ^{234}Th}{dt} = 1.540 \times 10^{-10}\ ^{238}U - 10.498\ ^{234}Th$$

$$\cdot$$
$$\cdot$$
$$\cdot$$

$$\frac{d\ ^{214}Bi}{dt} = 1.359 \times 10^4\ ^{214}Pb + 1.093 \times 10^7\ ^{218}At - 1.849 \times 10^4\ ^{214}Bi$$

$$\frac{d\ ^{214}Po}{dt} = 1.848 \times 10^4\ ^{214}Bi - 1.419 \times 10^{11}\ ^{214}Po$$

$$\frac{d\ ^{210}Tl}{dt} = 0.001 \times 10^4\ ^{214}Bi - 2.760 \times 10^5\ ^{210}Tl$$

$$\frac{d\ ^{210}Pb}{dt} = 1.419 \times 10^{11}\ ^{214}Po + 2.760 \times 10^5\ ^{210}Tl - 3.573 \times 10^{-2}\ ^{210}Pb$$

$$\cdot$$
$$\cdot$$
$$\cdot$$

The entire set of equations may be solved by starting at the first equation for the decay of ^{238}U and working one's way from equation to equation, solving as one goes, and using the solutions of the previous equations as needed. Computer solutions speed up the process considerably, but one must use a "stiff equation" solver to handle the extreme variation in the rate constants associated with half-lives differing from one another by many orders of magnitude. A system of rate equations is *stiff* if the decay constants $k_1, \ldots, k_n$ differ by several orders of magnitude. The usual differential equations solvers on a computer may "lose" a component that decays rapidly while "tracking" a component that decays slowly.† See Figure 9.9 for some computer-produced graphs of the solutions of the first five rate equations. The time scale in Figure 9.9 is different for each of the five elements and isotopes. It is assumed that ^{238}U was present when the earth began but that none of the other substances in the uranium series existed at that time.

Comments

Cascade models of dynamical phenomena are quite popular. They may be adapted to many one-way processes, although the inherent dynamics must be linear to use the methods outlined in this section. Other methods are needed for the solution of a nonlinear cascade, but the principle of starting at the top and working one's way

---

† See C. W. Gear, *Numerical Initial Value Problems in Ordinary Differential Equations* (Englewood Cliffs, N.J.: Prentice-Hall, 1971), for a discussion of suitable numerical techniques.

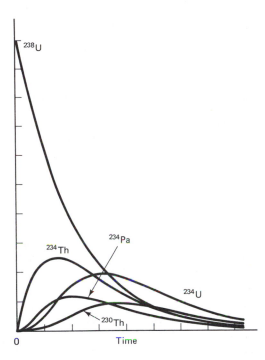

Figure 9.9 Growth and decay of the first elements and isotopes in the uranium series (the time scales are different for each substance).

down through the equations still applies. The cascade model of the bones of Olduvai is complete (i.e., it includes the two final products, argon and calcium). Thus $K(t) + C(t) + A(t) = K(0)$, $t \geq 0$. The first five steps of the uranium series are incomplete since the final products are not included and the ^{234}U of the fifth step is itself decaying. In this case, all five substances eventually decay away.

PROBLEMS

The following problem(s) may be more challenging: 3(e) and 8.

1. Solve the cascade problems below, draw a flow diagram for each, as in Figure 9.4, and find the limiting value of each state variable as $t \to \infty$. Which are source state variables, and which are final products?
   (a) $x_1' = -3x_1$, $x_2' = x_1$, $x_3' = 2x_1$; $x_1(0) = 10$, $x_2(0) = x_3(0) = 0$.
   (b) $x_1' = -2x_1$, $x_2' = 2x_1 - x_2$, $x_3' = x_2$; $x_1(0) = A$, $x_2(0) = x_3(0) = 0$.
   (c) $x_1' = -x_1$, $x_2' = x_1 - 3x_2$; $x_1(0) = A$, $x_2(0) = B$.
   (d) $x_1' = -2x_1$, $x_2' = -3x_2$, $x_3' = 2x_1 + 3x_2$; $x_1(0) = A$, $x_2(0) = B$, $x_3(0) = 0$.
   (e) $x_1' = -2x_1$, $x_2' = -x_2$, $x_3' = 2x_1 + x_2 - 3x_3$, $x_4' = x_3$, $x_5' = 2x_3$; $x_1(0) = A$, $x_2(0) = B$, $x_3(0) = x_4(0) = x_5(0) = 0$.

2. Show that in the cascade, $x_1' = -ax_1$, $x_2' = ax_1 - bx_2$, where a and b are positive constants, and $x_1(0) = A > 0$, $x_2(0) = 0$, there is a unique positive time at which $x_1(t) = x_2(t)$, if $a > b - 1$.

3. [*Example 9.5 Revisited*]. Suppose that x_0, y_0, and z_0 are nonnegative constants and that $x_0 + y_0 + z_0 = M > 0$.
   (a) Use formulas (3)–(5) to show that $x(t) \rightarrow 0$, $y(t) \rightarrow 0$, and $z(t) \rightarrow M$ as $t \rightarrow \infty$.
   (b) Prove that $x(t) + y(t) + z(t) = M$ for all t.
   (c) Prove that $x = 0$, $y = 0$, $z = M$ is a point orbit.
   (d) Prove that the orbit $x = x(t)$, $y = y(t)$, $z = z(t)$ lies on the plane $x + y + z = M$ and that $(x(t), y(t), z(t)) \rightarrow (0, 0, M)$ as $t \rightarrow \infty$.
   (e) Show that $x(t)$, $y(t)$, and $z(t)$ each lie in the interval $[0, M]$ for all $t \geq 0$.
   (f) Use formulas (3)–(5) and the results above to sketch the orbits on the plane $x + y + z = 10$ corresponding to $t \geq 0$ and the following triples of initial data for x_0, y_0, z_0: 0, 0, 10; 5, 5, 0; 10, 0, 0; 5, 0, 5; 3, 4, 3.

4. (a) Using the Maclaurin expansion of $\ln(1 + x)$ about $x = 0$, show that $\ln(1 + x) \approx x$ for $|x|$ sufficiently small.
   (b) Use part (a) and the age equation (8) to show that the simplified age equation is $t = A/(k_1 K)$.
   (c) The dating of the bones of Olduvai was not easy to accomplish because some of the samples of basalt and tuff were contaminated by earlier stream action and by careless handling. What value of A/K leads to an age of 1.75 million years? What uncertainty in the age would result if the determination of A/K had a 20% uncertainty?

5. (a) Solve the potassium–argon–calcium decay cascade assuming that $K(0) = K_0 > 0$, $A(0) = 0$, $C(0) = C_0 > 0$, and show that the assumption of positive C_0 has no effect on the potassium–argon method of dating.
   (b) What effect would a positive value of $A(0) = A_0$ have on the potassium–argon method of dating?

6. The potassium–argon method has recently been used to date the occasional reversals of the earth's magnetic field over geologic time. This is accomplished by dating samples of lava containing reversely magnetized minerals. There was an era of reversed magnetism from 1 million to 2 million years ago. What range of the fraction of argon to potassium in a sample would lead to this conclusion?

7. (a) Write out the rate equations for the uranium series from ^{238}U through ^{226}Ra, calculating the rate constants in years^{-1}.
   (b) Repeat part (a) for the series from ^{218}Po to ^{214}Bi (do not neglect the source flow into the ^{218}Po pool).

8. [*Computer Project*]. Model the entire uranium series and use a computer to find the amounts of each substance at intervals of 5×10^7 years from time 0 to time 4.5×10^9 years (roughly the age of the earth). You should use a "stiff solver" since the variation in the rate constants is extreme.

9.3 PROPERTIES OF SYSTEMS: THE SIMPLE PENDULUM

The normal form of the initial value problem for a first-order differential system is

$$x' = f(t, x), \qquad x = x^0 \quad \text{when} \quad t = t_0 \tag{1}$$

where x, x^0, and f are n-vectors. Sometimes solutions can be found explicitly, but for any system which models a physical phenomenon of any complexity, that is rarely the case. The following fundamental result provides assurance that under certain conditions on the rate function f, the initial value problem (1) is solvable uniquely for each initial point (t_0, x^0) in some region of tx-space. Moreover, the solution is well-behaved in the sense that it may be extended to the edge of the domain of f and is a continuous function of time, state variables, and initial data. The proof is omitted since the basic ideas are those of the proof given in Appendix A for the corresponding scalar problem.

Existence and Uniqueness.   Consider the initial value problem (1) where the functions f_i and $\partial f_i / \partial x_j$ $(i, j = 1, \ldots, n)$ are continuous throughout a region R of $(n + 1)$-dimensional tx-space,† and (t_0, x^0) is a point of R.

(*Existence*) (1) has a solution $x = x(t, t_0, x^0)$ defined on a t-interval I, t_0 in I.

(*Uniqueness*) (1) has no other solution defined on I.

(*Extension*) The maximally extended solution of (1) is defined on an open t-interval with the property that the solution curve tends to the boundary of R (which may be "at infinity") as t tends to either endpoint of the interval.

(*Continuity*) The solution is a continuous function of t, t_0, x^0 and any other parameters on which the rate function f depends continuously.

From now on, we consider only systems for which the Existence and Uniqueness Theorem applies. Usually, the solution of (1) will be written as $x = x(t)$, unless there is some reason to emphasize its dependence on the initial data, in which case we shall write out $x = x(t, t_0, x^0)$. To avoid ambiguity, we always assume that the solution of (1) is maximally extended. Observe that such a solution curve cannot "disappear" inside the region R where $f(t, y)$ is defined, but must approach the boundary of R as t approaches either endpoint of the t-interval on which the maximally extended solution is defined. The following example illustrates the conclusions of the Existence and Uniqueness Theorem.

Example 9.6

The uncoupled system

$$x_1' = x_1^2, \qquad x_1(t_0) = b_1$$
$$x_2' = 2x_2, \qquad x_2(t_0) = b_2 \tag{2}$$

meets the conditions of the Existence and Uniqueness Theorem in the region $\mathbf{R}^3$. The unique solution of (2) may be obtained by separating variables in each rate equation and solving. For example, the solution formula for x_2 is found in this way to be

---

† A *region* is an open and connected set. For example, the interior of a rectangle is a region in the plane.

$$x_2(t) = b_2 e^{2(t-t_0)}, \qquad -\infty < t < \infty \tag{3}$$

Solving the rate equation for x_1 is slightly more complicated. First note that $x_1(t) \equiv 0$ is a solution corresponding to $b_1 = 0$. If $x_1 \neq 0$, the variables may be separated. After integrating, using the initial data, and solving for x_1 in terms of t, we have that

$$x_1(t) = \frac{b_1}{1 - b_1(t - t_0)}, \qquad \text{where} \begin{cases} t < t_0 + \dfrac{1}{b_1} & \text{if } b_1 > 0 \\[2mm] t > t_0 + \dfrac{1}{b_1} & \text{if } b_1 < 0 \end{cases} \tag{4}$$

Observe that (4) also gives the solution $x_1(t) \equiv 0$, $-\infty < t < \infty$, if $b_1 = 0$. The change in the t-interval depending on the sign of b_1 is dictated by the need to keep t_0 inside the interval. In either case, the solution has a *finite escape time* since $x_1(t)$ becomes unbounded as t tends to the finite value $t_0 + 1/b_1$. This restriction on the t-interval of definition of $x_1(t)$ forces the same restriction on the interval of definition of $x_2(t)$ since the full solution of (2) requires both $x_1(t)$ and $x_2(t)$ to be defined on a common interval. Equation (2) has exactly one maximally-extended solution, although the solution formula and the t-interval depend on the values of t_0 and b_1. Moreover, this solution is seen through (3) and (4) to be a continuous function of t, t_0, b_1, and b_2. Small changes in any or in all of these parameters lead only to small changes in the functions $x_1(t)$ and $x_2(t)$.

Autonomous Systems

The rate functions of a first-order system may change with time, or they may be independent of time (i.e., *autonomous*). Thus the system $x' = f(x)$ is autonomous. The corresponding initial value problem is

$$x' = f(x), \qquad x = x^0 \quad \text{when} \quad t = t_0 \tag{5}$$

where the conditions of the Existence and Uniqueness Theorem are assumed to hold in a region R of tx-space. Since f does not depend on t, we may assume that R is a "cylinder with cross-sectional region S in the x-space (Figure 9.10). Let x^0 be in S and let t_0 be any real number. The four conclusions of the Existence and Uniqueness Theorem may now be interpreted in terms of S instead of R. For example, the Maximal Extension Property may be recast in the following form.

Maximal Extension (Autonomous Systems). Let the functions $f_i(x)$ and $\partial f_i / \partial x_j$, i and $j = 1, \ldots, n$, be continuous in the region S of $\mathbf{R}^n$, and suppose x^0 is in S. The solution of (5) may be extended to a unique maximal open interval (a, b) with the following property: Either $b = \infty$ $(a = -\infty)$ or else b is finite (a is finite) and $x(t)$ tends to the boundary of S as $t \to b^-$ $(t \to a^+)$.

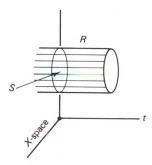

Figure 9.10   Regions for an autonomous system.

Example 9.7

Set $t_0 = 1$, $b_1 = 1$, $b_2 = 1$ in system (2). The solution is $x_1 = (2 - t)^{-1}$, $x_2 = e^{2(t-1)}$, $t < 2$ [use formulas (3) and (4)]. Observe that the maximal interval is $(-\infty, 2)$. The point (x_1, x_2) tends to the "infinite boundary" of $S = \mathbf{R}^2$ as $t \to 2^-$.

Time is only an incidental variable in an autonomous system, and the following result shows the extent to which this is so.

> **Translation in Time.**   Suppose that the function $f(x)$ satisfies the hypotheses of the Existence and Uniqueness Theorem and that $x = x(t)$, $a < t < b$, is a solution of the equation $x' = f(x)$. Then, for each T, $y(t) = x(t - T)$, $a + T < t < b + T$, is also a solution, and the orbits of the two solutions coincide.

Proof. To verify this property, set $y(t) = x(t - T)$ and observe that

$$\frac{dy(t)}{dt} = x'(t - T) = f(x(t - T)) = f(y(t)) \tag{6}$$

The orbits of the two solutions $x(t)$, $a < t < b$, and $y(t)$, $a + T < t < b + T$, coincide since $y(t + T) = x(t + T - T) = x(t)$, $a < t < b$. [Observe that the time translation property may fail for a nonautonomous system $x' = f(t, x)$ since $f(t - T, x(t - T))$ need not equal $f(t, y(t))$ as required by (6).]

A consequence of the above is that we might as well set $t_0 = 0$ if the orbit of (5) is of interest, but there is no particular reason to identify its "starting point." The following property is closely related to the Time Translation Property.

> **Successive Translations.**   Let $x(t, x^0)$, t in I, denote the solution of $x' = f(x)$, $x = x^0$ when $t = 0$, while the conditions of the Existence and Uniqueness Theorem are assumed to be satisfied. For all t and s in I for which $t + s$ belongs to I, we have that $x(t + s, x^0) = x(t, x(s, x^0))$, where the latter denotes the solution for which $x = x(s, x^0)$ when $t = 0$.

Informally, this implies that one may travel along an orbit from the point x^0 at time 0 to the point $x(t + s, x^0)$ at time $t + s$ by first moving s units of time to the point $x(s, x^0)$, and then t units of time beyond to the point $x(t, x(s, x^0))$. The property is verified by observing that $x(t + s, x^0)$ and $x(t, x(s, x^0))$ are both solutions of the problem

$$x' = f(x), \qquad x = x(s, x^0) \quad \text{when} \quad t = 0$$

and using uniqueness to conclude that the two solutions coincide.

An orbit of an autonomous system is insensitive to translations in the time variable, but even more is true when different orbits are considered.

Separation of Orbits.   Suppose that $f(x)$ satisfies the hypotheses of the Existence and Uniqueness Theorem. Let $x^1(t)$ and $x^2(t)$ be solutions of $x' = f(x)$. The corresponding orbits either never touch or else they coincide.

Proof. Suppose that for some t_1 and t_2, we have that $x^1(t_1) = x^2(t_2) = P$, a common point on the two orbits. By the Translation Property, $y^1(t) \equiv x^1(t + t_1)$ and $y^2(t) = x^2(t + t_2)$ are also solutions of $x' = f(x)$ whose orbits coincide, respectively, with the orbits of $x^1(t)$ and $x^2(t)$. Since $y^1(0) = P = y^2(0)$, the uniqueness property implies that $y^1(t) \equiv y^2(t)$, and, hence the orbits of $x^1(t)$ and $x^2(t)$ coincide, if they touch at all.

Example 9.8

The orbits of system (2) may be sketched directly on the $x_1 x_2$-state plane by using their parametric equations (3) and (4). Alternatively, since the system is autonomous, equations for the orbits may be found by dividing the two rate equations to obtain

$$\frac{x_2'}{x_1'} = \frac{dx_2/dt}{dx_1/dt} = \frac{dx_2}{dx_1} = \frac{2x_2}{x_1^2}$$

Separating variables and solving, we have $\frac{1}{2} \ln |x_2| = -(1/x_1) + c$, where c depends on initial data and $x_1 \neq 0$, $x_2 \neq 0$. Setting $K = \pm e^{2c}$, x_2 may be expressed in terms of x_1, and we have the orbital equations

$$x_2 = K e^{-2/x_1}, \qquad x_1 \neq 0$$

where $K = 0$ gives the orbits on $x_2 \equiv 0$ ($x_1 \equiv 0$ is also an orbital equation).

The orbits can now be traced and the portrait of the orbits emerges in Figure 9.11. The arrows indicate the direction of time's increase and are placed on an orbit only after an inspection of the signs of the rate functions of (2) to determine the direction of motion.

The apparent meeting of orbits at the origin is an illusion. The origin is a

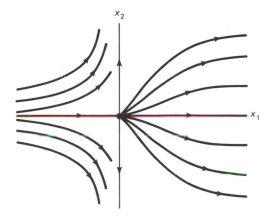

Figure 9.11   Orbits in the state plane.

point orbit corresponding to the constant solution $x_1(t) \equiv 0$, $x_2(t) \equiv 0$ for all t, while the orbits in the right half-plane approach the origin as $t \to -\infty$. An orbit on the negative x_1-axis tends to the origin as $t \to +\infty$. However, none of these orbits reaches the origin in finite time, and thus there is no actual touching of orbits.

The method used in the example to find orbits may be extended to any *planar autonomous system*

$$x_1' = f_1(x_1, x_2), \qquad x_2' = f_2(x_1, x_2) \tag{7}$$

Dividing the second equation of (7) by the first, a scalar first-order equation is obtained:

$$\frac{dx_2}{dx_1} = \frac{f_2(x_1, x_2)}{f_1(x_1, x_2)} \tag{8}$$

It may be possible to solve (8) by using one or more of the techniques of Chapters 1 to 4. Solutions of (8) define orbits of (7) since such solutions are equations in the state variables x_1 and x_2 and do not involve t. Example 9.8 illustrates how this may be done in a simple case. Observe that in passing from (7) to (8) there is no longer any need to find the solutions as functions of time.

Equilibrium Orbits

Equilibrium points and cycles are special types of orbits of an autonomous system which appear to play a fundamental organizational role in the portrait of the orbits.

> **Equilibrium Orbits.**   A point x^0 is an *equilibrium point* of $x' = f(x)$ if $x(t) \equiv x^0$, $-\infty < t < \infty$, is a constant solution. The orbit of a nonconstant periodic solution of $x' = f(x)$ is a *cycle*.

Since an equilibrium point is a constant solution, its rate of change is zero. The equilibrium points of $x' = f(x)$ are the zeros of f [i.e., the points x^0 for which

$f(x^0) = 0$]. For example, system (2) has a single equilibrium point, the point $x_1 = x_2 = 0$.

Periodic solutions and the corresponding cycles cannot be found as easily as constant solutions and equilibrium points since the rate function f neither vanishes nor remains constant along a periodic solution.

Example 9.9

The system $x_1' = x_2$, $x_2' = -4x_1$ reduces to the scalar equation $dx_2/dx_1 = -4x_1/x_2$ upon division. Separating variables and solving, we have an equation for the orbits, $4x_1^2 + x_2^2 = c$, where c is any nonnegative constant. $c = 0$ gives the single equilibrium point at the origin, while $c > 0$ determines an elliptical cycle. The corresponding periodic solutions may be found by noting that the system under consideration is equivalent to the scalar second-order equation $x_1'' + 4x_1 = 0$, whose general solution is $x_1 = A \sin(2t - \delta)$, where A and δ are arbitrary constants. Since $x_2 = x_1'$, $x_2 = 2A \cos(2t - \delta)$. Thus all nonconstant solutions are periodic of period π (see Figure 9.12).

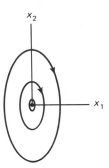

Figure 9.12   Cycles and an equilibrium point.

The Simple Pendulum

The equation of motion of a simple pendulum moving without friction or air resistance is

$$mL\theta'' + mg \sin\theta = 0 \tag{9}$$

where m is the mass, L the length of the supporting wire, g the gravitational constant, and θ the angular deviation from a downward vertical. [See Section 5.5 for the derivation of (9).] The corresponding system is

$$x_1' = x_2, \qquad x_2' = -\omega^2 \sin x_1, \qquad \omega^2 = \frac{g}{L} \tag{10}$$

The orbital equations may be found by dividing the two rate equations of (10), separating variables, and solving:

$$\tfrac{1}{2} x_2^2 = \omega^2 \cos x_1 + c \tag{11}$$

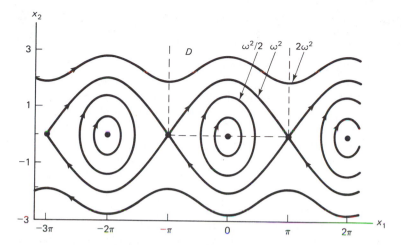

Figure 9.13 Orbits of the simple pendulum.

where c is a constant of integration. Since the left side of (11) is nonnegative, we see that c must be chosen $\geqq -\omega^2$. See Figure 9.13 for a portrait of the orbits. The equilibrium points of (9) are $(n\pi, 0)$, $n = 0, \pm1, \pm2, \ldots$; they correspond to the value $c = -\omega^2$. To sketch the orbits, notice that the graph of each orbit is symmetric about the x_1-axis, and about every vertical line $x_1 = n\pi$. Thus it is only necessary to plot the orbits in the domain $D: |x_1| < \pi$, $x_2 > 0$. The following additional observations are of use in constructing the portrait of the orbits.

1. For each c, $|c| < \omega^2$, there is a unique cycle enclosing the origin.
2. For each $c > \omega^2$, the orbit is oscillatory but not periodic; that is, x_2 is defined by (11) for all x_1, x_2 is periodic in x_1, but x_1 is not periodic in t.
3. If $c = \omega^2$, (11) defines a family of *separating orbits* (or *separatrices*), each of which tends to an equilibrium point as $t \to -\infty$, and to another equilibrium point as $t \to +\infty$. For example, the separating orbit in D tends to $(-\pi, 0)$ as $t \to -\infty$ and to $(\pi, 0)$ as $t \to +\infty$.

The arrows of time are placed on the orbits by using the first equation of (10). Observe the reversal of the arrow as one moves from the upper to the lower half-plane.

The cycles correspond to the familiar back-and-forth motions of a pendulum. An oscillating orbit corresponds to the motion of a pendulum with so much energy that it goes "over the top." A separating orbit occurs when an upended pendulum is displaced infinitesimally to one side of its unstable equilibrium. The repetitive patterns in the portrait are connected with the fact that the angles x_1 and $x_1 + 2\pi n$ are geometrically the same but correspond to distinct locations on the x_1-line.

The orbit analysis above has been carried out in the absence of solution formulas. Formula (11) for the orbits could, in principle, lead to a solution formula since x_2

is dx_1/dt, and the variables could then be separated. See the problem set for what happens when this is attempted.

Comments

Autonomous systems model the behavior of many physical systems. The cycles and equilibrium points represent the "steady states" of the system and consequently, are of considerable physical significance. Most of the models in the remainder of the book are limited to two state variables, as were the examples in this section. Although the theory of autonomous systems holds for any number of state variables, the portraits of the orbits may be quite complex in higher-dimensional state spaces. Even in $\mathbf{R}^3$ the orbits may wind around each other in such convoluted ways that the portrait may appear to be nothing more than a tangle of filaments without structure or form. Numerical approximations to orbits and corresponding computer-produced graphs of orbital projections onto various state planes may be useful in sorting out multidimensional complexities.

PROBLEMS

The following problem(s) may be more challenging: 4(b), 8, and 9.

1. Verify that each initial value problem satisfies the hypotheses of the Existence and Uniqueness Theorem throughout state space. Solve the problem and sketch the orbit or orbits. If $x_1(0) = a$ and $x_2(0) = b$, sketch the nine orbits corresponding to all possible combinations of a, $b = -1$, 0, 2. Identify any orbits that are equilibrium points or cycles.
   (a) $x_1' = x_2$, $x_2' = -x_1 - 2x_2$; $x_1(0) = 1$, $x_2(0) = 1$. [*Hint*: Observe that $x_1'' + 2x_1' + x_1 = 0$.]
   (b) $x_1' = x_2$, $x_2' = -x_1 - 2x_2$; $x_1(0) = a$, $x_2(0) = b$.
   (c) $x_1' = 2x_1$, $x_2' = -4x_2$; $x_1(0) = a$, $x_2(0) = b$.
   (d) $x_1' = x_2$, $x_2' = -9x_1$; $x_1(0) = a$, $x_2(0) = b$.
   (e) $x_1' = x_2^3$, $x_2' = -x_1^3$; $x_1(0) = a$, $x_2(0) = b$.
   (f) $x_1' = -x_1^3$, $x_2' = -x_2$; $x_1(0) = 1$, $x_2(0) = 1$.
   (g) $x_1' = x_2$, $x_2' = -26x_1 - 2x_2$, $x_3' = x_3/2$; $x_1(0) = x_2(0) = x_3(0) = 1$. Sketch the projection of the orbit in each coordinate plane.

2. Find all equilibrium points.
   (a) $x_1' = x_1 - x_2^2$, $x_2' = x_1 - x_2$.   (b) $x_1' = x_2 \sin x_1$, $x_2' = x_1 x_2$.
   (c) $x_1' = 2 + \sin(x_1 + x_2)$, $x_2' = x_1 - x_2^3 + 27$.   (d) $x_1' = x_2 + 1$, $x_2' = \sin^2 3x_1$.
   (e) $x_1' = 3(x_1 - x_2)$, $x_2' = x_2 - x_1$.

3. Find all solutions. Verify that if $x_1 = x_1(t)$, $x_2 = x_2(t)$ is a solution, then so is $x_1 = x_1(t - T)$, $x_2 = x_2(t - T)$, where T is any constant. Specify the interval on which each solution is defined.
   (a) $x_1' = 3x_1$, $x_2' = -x_2$.   (b) $x_1' = 1/x_1$, $x_2' = -x_2$.
   (c) $x_1' = -x_1^3$, $x_2' = 1$.   (d) $x_1' = x_1^2(1 + x_2)$, $x_2' = -x_2$.

4. Sometimes polar coordinates are a help in solving a planar autonomous system.
   (a) If $x_1' = f_1(x_1, x_2)$ and $x_2' = f_2(x_1, x_2)$, let $x_1 = r \cos \theta$, $x_2 = r \sin \theta$, $r^2 = x_1^2 + x_2^2$, $\theta = \arctan(x_2/x_1)$, and show that $r' = \cos \theta\, f_1(r \cos \theta, r \sin \theta) + \sin \theta\, f_2(r \cos \theta, r \sin \theta)$, $r\theta' = \cos \theta\, f_2(r \cos \theta, r \sin \theta) - \sin \theta\, f_1(r \cos \theta, r \sin \theta)$.

(b) Show that the polar form of the system

$$x_1' = x_1 - x_2 - x_1(x_1^2 + x_2^2)$$
$$x_2' = x_1 + x_2 - x_2(x_1^2 + x_2^2)$$

is $r' = r(1 - r^2)$, $\theta' = 1$. Use partial fractions or integral tables to find r as a function of t. Show that $x^2 + y^2 = r^2 = 1$ is a cycle Γ of period 2π. [*Note*: $\theta = t + c$]. Show that $(0, 0)$ is an equilibrium point. Show that the portrait of the orbits is as shown.

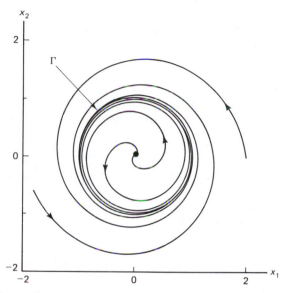

(c) Sketch the approximate orbits of $r' = r(1 - r^2)(4 - r^2)$, $\theta' = 1$ in the x_1x_2-plane, where $x_1 = r\cos\theta$, $x_2 = r\sin\theta$.

5. An orbit of a nonautonomous system may intersect itself without being a cycle. Solve the system $x_1' = \pi \cos \pi t$, $x_2' = x_2 \cos t$, $x_1(0) = 0$, $x_2(0) = 1$. Show that the orbit is as sketched here.

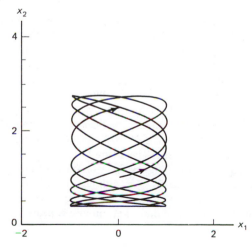

6. Sketch a portrait of the orbits of the simple pendulum in the case $\omega^2 = 1$.

7. The equation of the orbits of the simple pendulum may be derived directly from the equation of motion as follows.

   (a) Multiply each side of Eq. (9) by θ' and observe that $mL\theta'\theta'' + mg\theta' \sin \theta = [\frac{1}{2} mL(\theta')^2 - mg \cos \theta]'$. Show that $\frac{1}{2} mL(\theta)^2 - mg \cos \theta = c$, where c is a constant. Show that this equation reduces to (11).

   (b) The kinetic energy of the moving pendulum is $\frac{1}{2} m(L\theta')^2$ and the potential energy of position is $-mgL (1 - \cos \theta)$. Show that given θ and θ' at time 0, the total energy at time t is the same as that at time 0.

8. [*Periods of the Cycles of a Pendulum*]. The following problems take up the question of the periods of the cycles of a simple pendulum and the corresponding periods of the cycles of the linearized pendulum.

   (a) For small angles, $\sin \theta$ is approximately θ, where θ is the first term in the Maclaurin expansion of $\sin \theta$. The equation of the linearized pendulum is $mL\theta'' + mg\theta = 0$. Show that all nonconstant solutions are periodic of period $2\pi\sqrt{L/g}$. Show that the corresponding orbits in the $\theta\theta'$-state plane are elliptical cycles of the same period. Sketch the portrait of the orbits.

   (b) Let $\theta(0) = \theta_0$, $\theta'(0) = 0$, where $0 < \theta_0 < \pi$ for the model of the simple, nonlinearized pendulum. Show that $\theta' = -\sqrt{2g/L} \sqrt{\cos \theta - \cos \theta_0}$. [*Hint*: Recall the derivation of (11). The minus sign is used since $\theta(t)$ initially decreases as time increases from zero.]

   (c) Show that the motion under the conditions of part (b) is periodic.

   (d) Let T be the period. Show that

   $$T = 4 \sqrt{\frac{L}{2g}} \int_0^{\theta_0} \frac{d\theta}{\sqrt{\cos \theta - \cos \theta_0}}$$

   (The integral is improper since the integrand becomes infinite at $\theta = \theta_0$. It may be shown that T is finite if $0 < \theta_0 < \pi$.)

   (e) Show that the changes of variables $k = \sin \theta_0/2$, $\sin \phi = (1/k) \sin \theta/2$, give

   $$T = 4 \sqrt{\frac{L}{g}} \int_0^{\pi/2} \frac{d\phi}{\sqrt{1 - k^2 \sin^2 \phi}}$$

   The integral is an *elliptic integral of the first kind*. Its approximate values have been tabulated for various values of k [see, e.g., M. Abramowitz and I. A. Stegun, (eds.), *Handbook of Mathematical Functions* (Washington, D.C.: National Bureau of Standards, 1964; Dover (reprint), New York, 1965]. For example, if $\theta_0 = 2\pi/3$, and hence $k = \sqrt{3}/2$, the value of the integral is approximately 2.157. The corresponding period is approximately $8.628\sqrt{L/g}$, quite different from the period of the linearized pendulum, $2\pi\sqrt{L/g} \approx 6.283\sqrt{L/g}$. Why would you expect the period of the nonlinear pendulum to be greater than the period of the linearized pendulum? [*Hint*: Use the trigonometric identities $\cos \theta = 1 - 2 \sin^2 (\theta/2)$ and $\cos \theta_0 = 1 - 2 \sin^2 (\theta_0/2)$ to reduce the integral of part (d) to that given here.]

   (f) Argue that $T \to 2\pi\sqrt{L/g}$ as $\theta_0 \to 0$. Why would one expect this to be true? It is known that $T \to \infty$ as $\theta_0 \to \pi$, although a complete mathematical proof of this fact was not given here. Why is this result expected on physical grounds? [*Hint*: Think of the motion of a pendulum that initially is nearly vertical.]

9. Suppose that the conditions of the Existence and Uniqueness Theorem are met for the system $x' = f(x)$. Properties of equilibrium points and cycles are given in parts (a) and (b). The converses are true but not proved here.

   (a) Suppose that P is an equilibrium point in the region S and $x(t)$ is a nonconstant solution for which $x(t) \to P$ as $t \to T$. Show that $|T| = \infty$. [*Hint*: Since $x(t)$ is a continuous function of t, $x(T)$ must be P if T is finite. This violates the Uniqueness Principle. (Why?) Use the Extension Principle to show that P lies on the boundary of S or else $|T| = \infty$.]

   (b) Suppose that $x = x(t)$ is a nonconstant periodic solution. Show that the corresponding orbit is a simple closed curve. [*Hint*: Suppose that the least period of $x(t)$ is T and that $x(t_1) = x(0)$ for some t_1, $0 < t_1 \leq T$. Then $y(t) = x(t + t_1)$ is also a solution (why?) and $y(t) \equiv x(t)$ since $y(0) = x(0)$. Hence $x(t)$ has period t_1. Show that $t_1 = T$.]

9.4 NUMERICAL METHODS AND COMPUTED ORBITS

First-order systems of ordinary differential equations model the dynamics of electrical filters, uranium decay, simple pendula, and a host of other phenomena. The systems themselves contain much information about the phenomena being modeled, but solutions of the systems may reveal even more. Sometimes the solutions may be found as explicit functions of time, system parameters, and initial data (e.g., the cascade models), and sometimes only the orbits can be found (e.g., the model of the simple pendulum). Frequently, neither orbits nor solutions can be identified by explicit formulas, and we turn to an existence and uniqueness theorem (such as that in Section 9.3) to be sure that solutions do exist and to numerical methods and a computer to find approximations to the solutions. We outline three widely used numerical methods for solving initial value problems for systems, but we say little about the rationale behind the methods. Nor do we point out the advantages and pitfalls of each since the discussion of scalar initial value problems applies virtually verbatim to systems (refer to Chapter 2 and Appendix B).

The initial value problem under consideration is

$$x' = f(t, x), \qquad x = x^0 \quad \text{when} \quad t = t_0 \tag{1}$$

where f is assumed to satisfy the conditions of the existence and uniqueness theorem in a region R containing (t_0, x^0). For simplicity, we consider only the forward initial value problem of solving (1) for $t \geq t_0$. Let h be the *step size* used in the approximation, $t_j = t_0 + jh$, and denote by x^j the estimate for $x(t_j)$ furnished by the numerical method being used. If we allow j to range from 1 to N, we say that the *global discretization error* is $E_N = \|x(t_N) - x^N\|$. The error (and the method of approximation) is of *order p* in the step size if there is a constant M such that $E_N \leq Mh^p$. The methods to be given here are all fourth order. Thus cutting the step size in half leads (in principle) to a 16-fold improvement in accuracy. Moreover, the methods are stable in the sense that the inevitable round-off "errors" of the computer calculations at each stage of the approximation do not propagate as j increases, but tend to die out or at least remain bounded. Actually, all one-step methods are stable in this regard, but multistep methods have to be chosen with some care to avoid the perils of instability. The multistep Adams methods given below are stable, and that is a major reason they are so popular.

A One-Step Method

In a *one-step method* the estimate x^{j+1} is found in terms of h, x^j, and the rate function $f(t, x)$ evaluated at various points (t, x) in the rectangular "parallelepiped"† $[t_j, t_{j+1}] \times [x_1^j, x_1^{j+1}] \times \cdots \times [x_n^j, x_n^{j+1}]$, where $x^j = (x_1^j, \ldots, x_n^j)$ and $x^{j+1} = (x_1^{j+1}, \ldots, x_n^{j+1})$. The method is "one-step" in the sense that x^{j+1} depends only on x^j and not on x^{j-1}, x^{j-2}, ... The most widely used one-step method is the explicit *fourth-order Runge–Kutta approximation*:

$$x^{j+1} = x^j + \frac{h}{6}[k^1 + 2k^2 + 2k^3 + k^4], \qquad j = 0, \ldots, n-1 \qquad (2)$$

where the four vectors k^1, k^2, k^3, and k^4 are given by

$$k^1 = f(t_j, x^j)$$

$$k^2 = f\left(t_j + \frac{h}{2}, \quad x^j + \frac{h}{2}k^1\right)$$

$$k^3 = f\left(t_j + \frac{h}{2}, \quad x^j + \frac{h}{2}k^2\right) \qquad (3)$$

$$k^4 = f(t_{j+1}, \quad x^j + hk^3)$$

As with every one-step method, this one is easy to use, easy to program, and does not require much computer memory since there is no need to store any but the immediately preceding approximation. It is, however, somewhat slow, as the number of separate function evaluations at each step may be quite large. The method is stable, like all one-step algorithms, and it is fourth order in the step size if f has continuous derivatives through order 4.

Multistep Methods

In a one-step method the approximation at each step is calculated in terms of that obtained at the preceding step, but the approximations of two or more steps back are ignored. Multistep methods use these earlier approximations as well. The explicit *Adams–Bashforth four-step method* uses the previous four approximations:

---

† Recall that the *Cartesian Product* of the sets $A_1, \ldots, A_m$, is the set of m-tuples $A_1 \times A_2 \times \cdots \times A_m = \{(y_1, y_2, \ldots, y_m); y_1 \text{ in } A_1, \ldots, y_m \text{ in } A_m\}$.

$$x^{j+1} = x^j + \frac{h}{24} [55f(t_j, x^j) - 59f(t_{j-1}, x^{j-1}) + 37f(t_{j-2}, x^{j-2})$$

$$- 9f(t_{j-3}, \quad x^{j-3})], \qquad j = 3, \ldots, N-1 \tag{4}$$

The starting values x^1, x^2, and x^3 may be calculated from the initial value x^0 by an application of the fourth-order Runge–Kutta method.

Once x^0, x^1, x^2, and x^3 have been calculated, formula (4) may be used to calculate x^4 and the subsequent approximations. Formula (4) is stable and fourth order in the step size (again assuming that f is four times continuously differentiable).

The explicit *Adams–Bashforth–Moulton four-step predictor–corrector method* uses (4) to predict a value for x^{j+1}, but then "corrects" that prediction before going on to the next step:

$$(predictor) \quad x^{j+1}_{\text{pred}} = x^j + \frac{h}{24} [55f(t_j, x^j) - 59f(t_{j-1}, x^{j-1}) + 37f(t_{j-2}, x^{j-2})$$

$$- 9f(t_{j-3}, x^{j-3})]$$

$$\tag{5}$$

$$(corrector) \quad x^{j+1} = x^j + \frac{h}{24} [9f(t_{j+1}, x^{j+1}_{\text{pred}}) + 19f(t_j, x^j) - 5f(t_{j-1}, x^{j-1})$$

$$+ f(t_{j-2}, x^{j-2})]$$

where $j = 3, \ldots, N-1$. As with (4), one must use some other method to estimate x^1, x^2, and x^3. Algorithm (5) is both stable and fourth order in the step size.

Multistep methods are relatively fast since previously calculated function values may be used for several subsequent steps. However, more storage is required.

Example 9.10

The initial value problem below was solved numerically by using a packaged differential equations solver on a computer. The solver used a variable-step Adams–Bashforth–Moulton algorithm, but began with a Runge–Kutta starting algorithm. The tables of the numerical values of the approximate solution are omitted since the computer-produced graphs of the approximate solution are more revealing. The problem is

$$x'_1 = -x_1 + 10x_2, \qquad x_1(0) = 1$$
$$x'_2 = -10x_1 - x_2, \qquad x_2(0) = 0 \tag{6}$$

whose exact solution is $x_1 = e^{-t} \cos 10t$, $x_2 = -e^{-t} \sin 10t$. Thus we expect to see an orbit which is a tightening spiral about the equilibrium point at the origin. See Figure 9.14 for the approximate orbit and component graphs produced by the computer.

A Gallery of Pictures in State Space

A thorough understanding of a dynamical system, its solutions, and its orbits usually rests on a foundation of mathematical theory, solution formulas, numerical methods, and hand-drawn or computer-produced graphs of approximate orbits, solutions, and

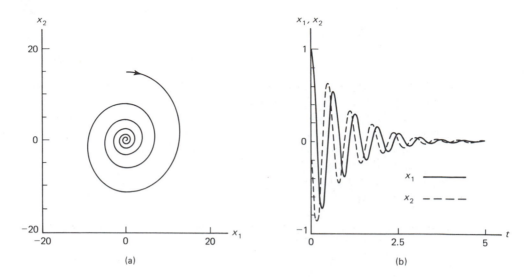

Figure 9.14   (a) Spiral orbit; (b) component graphs.

component graphs. If the differential equations of the system are intricate, the first step may well be to sketch (or have a computer sketch) some orbits or solution graphs. That is what we do here, where we have used computer graphics to generate a "gallery" of sketches of orbits and solution graphs of dynamical systems. The computer implementation of the differential equation solver employed both the Runge–Kutta and Adams–Bashforth–Moulton methods.

The dynamical systems behind these portraits range from simple linear systems with exact solution formulas to complex nonlinear systems for which no explicit solution formulas are known. Each sketch is accompanied by the corresponding dynamical system and a brief comment or two. No theorems are proved, none are stated, no solution formulas are given, no analysis or theoretical treatment is indicated. A number of the systems are treated more extensively elsewhere in the text. The reader may verify some of the features of the portraits without recourse to computer graphics by using direction field methods [replace $x' = f(x, y)$, $y' = g(x, y)$ by $dy/dx = g/f$] or by using solution techniques suggested earlier in this chapter.

Of what value are the pictures? Pictures may not prove, but they do suggest. Picasso remarked that "art is the lie by which we see the truth." That is the function of these pictures, even with the omissions, distortions, and enhancements to reality inherent in any portrait.

The Pictures.   Each picture contains the course of one or more orbits or component graphs of a dynamical system. Every system meets the conditions of the existence and uniqueness theorem of the preceding section. Thus if the system is autonomous (and most of those portrayed are), no two distinct orbits in state space can ever intersect. The apparent meeting of orbits in some of the pictures is deceptive and is an indication only of the limitations of the computer. Of course, an orbit of a nonautonomous system may intersect another orbit, or even itself, the intersections occurring at different times (see Figure 9.17). Two component graphs are shown, and these show that component graphs may contain information that is either missing or obscure in the portrait of the orbits. In all the pictures the arrowheads point in the direction of time's increase.

We have confined our attention to planar systems because the graphics are simpler, but one could deal with higher-dimensional systems by having the computer sketch portraits of any one variable plotted against any other. In this case the "picture" of the orbits or component graphs would consist of a series of projections onto various state planes or state-time planes.

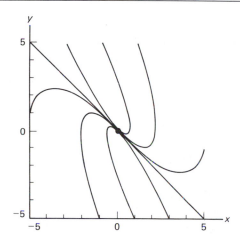

Figure 9.15

$$x' = y$$
$$y' = -2x - 3y$$

The only equilibrium solution of the system is an equilibrium point at the origin. The system is equivalent to the scalar equation $x'' + 3x' + 2x = 0$, whose solutions are known. Note that all orbits approach (but never reach) the origin as $t \to \infty$.

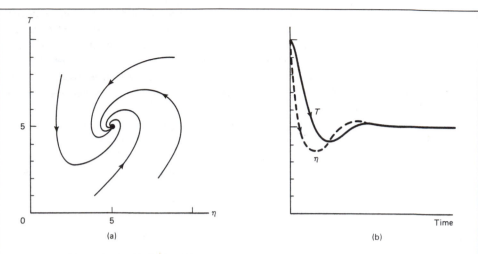

Figure 9.16   (a) Orbits; (b) component graphs corresponding to one orbit.

Orbits and component graphs for the neutron density η and the temperature T of a component of a nuclear power reactor (see Section 3.4)

$$\eta' = (\alpha - c_1 T - k_1 \eta)\eta, \qquad T' = -aT + b\eta$$

where $\alpha = a = b = 10$, $c_1 = 0.02$, and $k_1 = 1.98$ for the particular system whose orbits and solutions are portrayed. There is an equilibrium point at $(5, 5)$ in the ηT-state space and another at the origin.

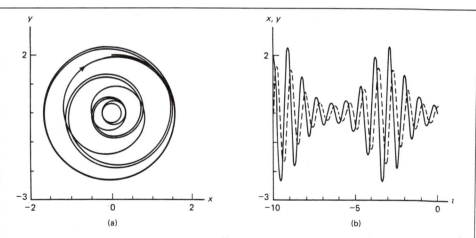

Figure 9.17   (a) Single orbit; (b) component graphs: $x(t)$, solid line; $y(t)$, dashed line.

An orbit and corresponding component graphs of the nonautonomous system

$$x' = x \sin t + 5y, \qquad y' = -10x + y \sin t$$

The orbit spirals in or out according as $\sin t$ is negative or positive, and the crossing points of the orbit correspond to different values of t.

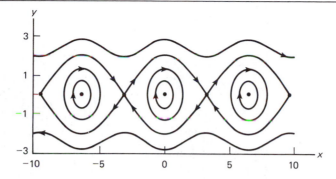

Figure 9.18 Orbits of simple pendulum.

Orbits of the simple pendulum, whose motion is modeled by

$$x' = y$$
$$y' = -\sin x$$

There are equilibrium points at $(n\pi, 0)$, $n = 0, \pm 1, \pm 2, \ldots$, and a "band" of equilibrium cycles inside each "eye." The equivalent scalar equation is $x'' + \sin x = 0$.

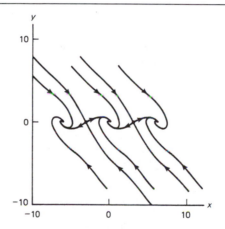

Figure 9.19 Damped simple pendulum.

Orbits of the damped simple pendulum whose dynamical system is

$$x' = y$$
$$y' = -y - \sin x$$

Damping has been added to the dynamic system of Figure 9.18. The equilibrium cycles disappear and orbits decay to the equilibrium points. The scalar equation is $x'' + x' + \sin x = 0$.

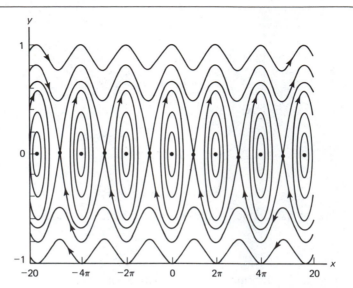

Figure 9.20 Orbits of simple pendulum.

The orbits of the simple pendulum again, but with a smaller restoring force:

$$x' = y$$
$$y' = -\tfrac{1}{10} \sin x$$

Just as in Figure 9.18 there are infinitely many equilibrium points on the line $y = 0$ and each "eye" is filled with equilibrium cycles orbiting an equilibrium point.

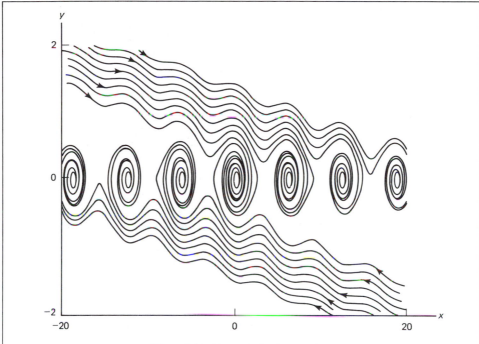

Figure 9.21   Damped simple pendulum.

A small damping term is added to the system of Figure 9.20:

$$x' = y$$
$$y' = -\tfrac{1}{20}y - \tfrac{1}{10}\sin x$$

Once more the equilibrium cycles disappear and every orbit winds down to an equilibrium point.

Remark: The systems whose orbits are pictured in Figures 9.18 and 9.20 differ only slightly ($-\sin x$ appears in one case and $-\tfrac{1}{10}\sin x$ in the other), yet the pictures look different. The difference is an illusion created mostly by changes in the scales and in the number of orbits sketched. The same comments apply to the marked contrasts between Figures 9.19 and 9.21.

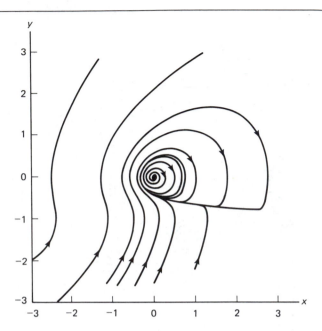

Figure 9.22   "Parrot" system.

The system for the "parrot" is

$$x' = y + y^2$$
$$y' = -x + \tfrac{1}{5}y - xy + \tfrac{6}{5}y^2$$

There is an equilibrium point at the origin and a single equilibrium cycle enclosing the origin. All other orbits spiral toward the cycle as time advances.

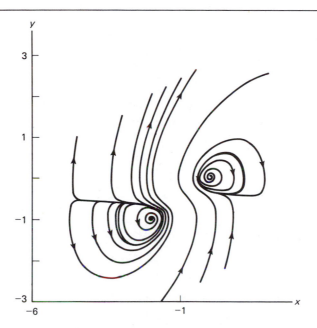

Figure 9.23 "Two-eyed monster."

$$x' = y(1+y)$$
$$y' = -\tfrac{1}{2}x + \tfrac{1}{5}y - xy + \tfrac{6}{5}y^2$$

There are two equilibrium points, $(0, 0)$ and $(-2, -1)$, and a single equilibrium cycle around each point. Here and in the preceding example, the systems are said to be *quadratic* since the right hand sides are polynomials in x and y of degree 2. There is a curious conjecture concerning such systems: A quadratic system either has infinitely many cycles or else no more than six.

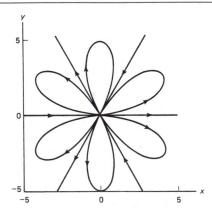

Figure 9.24 "Daisy."

The system for the "daisy" is

$$x' = 3x^4 - 12x^2y^2 + y^4$$
$$y' = 6x^3y - 10xy^3$$

The origin is an equilibrium point, while there are straight-line orbits along $y = 0$, $y = \sqrt{3}x$, and $y = -\sqrt{3}\ x$.

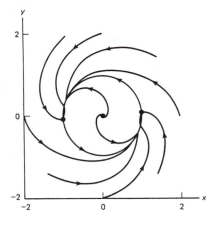

Figure 9.25 "Yin-yang."

The planar system (in polar coordinates) is

$$r' = r(1 - r^2)$$
$$\theta' = r^2 \sin^2 \theta + (r^2 - 1)^2$$

The three equilibrium points are at $r = 0$, $r = 1$, and $\theta = 0$, and $r = 1$ and $\theta = \pi$. The unit circle $r = 1$ is composed of four orbits: Two of them are equilibrium points, while each of the other two leaves one of these equilibrium points as t increases from $-\infty$ and approaches the other as t tends to $+\infty$.

Comments. The portraits in this gallery are one aspect of the *geometric* or *qualitative approach* to dynamical systems. This approach contrasts with *quantitative* studies, which are concerned with numbers, formulas, and analytic properties of solutions. The graphical capability of a computer system makes a geometric study of the orbits simple and quick. The resulting pictures may give us the first inkling of strange or unexpected behavior of the solutions of a system. Of course, the pictures only point out possibilities—they prove nothing. In addition, the limitations of numerical methods and their implementation on a computer may mean that certain delicate features of the orbits are overlooked. Another problem is that we may view the orbits of high-dimensional systems only in projections. But in spite of all these drawbacks, the pictures do suggest the directions and even the conclusions of a subsequent detailed analysis. Many of the systems given in this section are taken up elsewhere in the book, the models derived and explained, and theorems about the orbits stated and proved.

Sometimes a dynamical system is little understood, and there are no theorems outlining the specific nature and properties of the solutions. In such a case we must be content with computer-generated graphs of orbits and solution curves—at least until the underlying theory of the system is discovered.

PROBLEMS

1. Each of these problems is to be done with a programmable calculator or a computer. Either use a packaged routine or write a differential equations solver of your own. Output may be lists of approximations to the state variables at a selection of time steps, approximate graphs of the orbits, or approximate component graphs. Step size, the number of steps, and the approximation method used are left to the reader but should be clearly identified. In each case solve approximately on the interval $0 \leq t \leq T$. If you know the exact solution, compare its values with the approximations.
   (a) $x_1' = -2x_1$, $x_2' = -4x_2$; $x_1(0) = 10$, $x_2(0) = 10$, $T = 5$.
   (b) $x_1' = -2x_1$, $x_2' = 4x_2$; $x_1(0) = 10$, $x_2(0) = 1$, $T = 5$.
   (c) $x_1' = -2x_1 + 16x_2$, $x_2' = -16x_1 - 2x_2$; $x_1(0) = 0$, $x_2(0) = 1$, $T = 5$.
   [*Hint:* For the exact solution try functions of the form $x_1 = e^{at} \sin bt$ and $x_2 = e^{at} \cos bt$.]
   (d) $x_1' = x_2$, $x_2' = -16x_1$; $x_1(0) = 1$, $x_2(0) = 0$, $T = 3$.
   (e) $x_1' = 10x_2$, $x_2' = -10x_1 - 10(x_1^2 - 1)x_2$; $x_1(0) = -1$, $x_2(0) = 0$, $T = 2$.
   (f) $x_1' = -x_1$, $x_2' = -x_2 + x_3$, $x_3' = -x_2 - x_3$; $x_1(0) = 1$, $x_2(0) = 1$, $x_3(0) = 0$, $T = 10$.
   (g) $x_1' = x_1 \sin t + 5x_2$, $x_2' = -5x_1 + x_2 \cos t$; $x_1(0) = 0$, $x_2(0) = 1$, $T = 10$.
   (h) $x_1' = -x_1 + 5x_2 \sin t$, $x_2' = -5x_1 \sin t - x_2$; $x_1(0) = 0$, $x_2(0) = 1$, $T = 5$.

2–12. (a) Find the equilibrium points of the systems corresponding to Figures 9.15 to 9.25.
   (b) Using a differential equations solver and computer graphics, verify the portraits of the orbits.

13. Consider the system

$$x' = x \sin z + 5y$$
$$y' = -10x + y \sin z$$
$$z' = \ln(1 + |z|)$$

Use a differential equations solver and computer graphics package to plot orbits of the system either in three dimensions (if the option is available) or in two-dimensional projections on each of the three state planes. The projection of a single orbit pictured crosses itself. Why does this not violate the uniqueness theorem? Compare this system to that in Figure 9.17a and explain the similarity between the orbits.

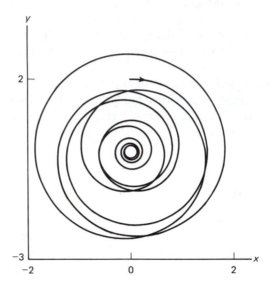

9.5 PREDATOR AND PREY

The following words of Charles Darwin set the subject for this section:

A struggle for existence inevitably follows from the high rate at which all organic beings tend to increase. Every being, which during its natural lifetime produces several eggs or seeds, must suffer destruction during some period of its life, and during some season or occasional year, otherwise, on the principle of geometrical increase, its numbers would quickly become so inordinately great that no country could support the product. Hence, as more individuals are produced than can possibly survive, there must in every case be a struggle for existence, either one individual with another of the same species, or with the individuals of distinct species, or with the physical conditions of life. It is the doctrine of Malthus applied with manifold force to the whole animal and vegetable kingdoms; for in this case there can be no artificial increase of food, and no prudential restraint from marriage. Although some species may be now increasing, more or less rapidly, in numbers, all cannot do so, for the world would not hold them. . . .

The amount of food for each species of course gives the extreme limit to which each can increase; but very frequently it is not the obtaining food, but the serving as prey to other animals, which determines the average numbers of a species.†

In the dynamics of life no species is alone. The survival or the flourishing of a species rests on strategies of cooperation, competition, and predation. We may begin to understand a small part of the biological universe by considering the interactions of just two species. The relationship may be one of mutual aid, of competition for limited resources, or of predation of the one upon the other. It is the predator–prey relationship that we shall explore. We construct a simple mathematical model of a predator–prey community. Then we determine the effect of an indiscriminate harvester or poison that kills both predator and prey. Harvesting of this type has a surprising consequence with important ecological implications (Volterra's Principle), and we conclude with a discussion of these implications.

We begin with a general discussion of the mathematical modeling of two-species interactions.

Rate Equations for Two Interactive Species

Let $x(t)$ and $y(t)$ denote the respective *populations* (i.e., the number of individuals) at time t of two species. Sometimes *population densities* (i.e., numbers per unit area or volume of habitat) are used instead, but we shall not do this. The values of $x(t)$ and $y(t)$ are integers and change by integer amounts as time goes on. However, for large populations an increase by one or two over a short time span is "infinitesimal" relative to the total, and we may think of the populations as changing continuously rather than by discrete jumps. Once we assume that $x(t)$ and $y(t)$ are continuous, we might as well smooth off any corners on the graphs of $x(t)$ and of $y(t)$ and assume that both functions are differentiable. Figure 9.26 shows this smoothing process for $x(t)$.

To say much about the changing populations of a pair of interacting species, we must know something about the "laws" of birth and mortality for each species.

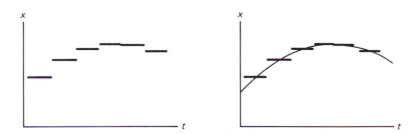

Figure 9.26   Smoothing a population curve.

† From Charles Darwin, "Struggle for Existence," Chap. 3 in *The Origin of Species*, new ed. from the 6th English ed. (New York: Appleton, 1882).

Typically, these laws are expressed in terms of rates of change. Let us see what form a rate law might take, say, for the x-species. Averaged over time and over all classes of age, sex, and fertility, a "typical" x-individual makes a net contribution, r, to the growth rate of its species. The rate of change of the total population $x(t)$ is, then,

$$x'(t) = \begin{pmatrix} \text{individual contribution} \\ \text{to the rate} \end{pmatrix} \cdot \begin{pmatrix} \text{number of} \\ \text{individuals} \end{pmatrix} = rx(t) \tag{1}$$

The *rate coefficient* r will differ from species to species, but it always has the meaning given above.

Finding $x(t)$ depends on solving (1), but this cannot be done until the form of r is known. If r is constant, $x(t)$ grows or decays exponentially, $x(t) = x_0 e^{rt}$, but there is no interaction with the y-species at all. Since we assume that the two species do interact, the coefficient r must be a function of y. It may also depend on x. In the absence of any information about the form of the dependence of r on x and y, we recall the analyst's adage, "when in doubt, linearize!" Suppose that r is the linear polynomial

$$r = a + bx + cy \tag{2}$$

where a, b, and c are constants. The rate equation for the x-species is

$$x' = (a + bx + cy)x = (a + bx)x + cxy \tag{3}$$

There is a similar equation for the y-species.

The derivation of the rate equation has been theoretical, but the terms have biological interpretations. The constant a in (2) is the *natural growth* coefficient of the x-species, while bx measures the effect of the size of the population on an individual's contribution. If b is negative, that effect is negative, and b is the *overcrowding* or *self-limiting* coefficient. [See Chapter 3 for a discussion of the logistic equation, $x' = (a + bx)x$, $a > 0$ and $b < 0$.] If b is positive, b is the *mutualism* coefficient, and bx indicates that an increase in the individual fertility occurs with growth in the population.

The coefficient c is a measure of the effect of species y on the growth rate of the x-population. If c is negative the x-species "loses" in any encounter between the two species, and x is the *prey* of the *predator* y. An alternative interpretation is that both species are in competition for the same resources and an increase in the numbers of either lowers the rate of growth (one would expect b as well as c to be negative in this case). If c is positive, the predator–prey relationship is reversed, or else the y-species contributes in some other way to the "well-being" of the x-population. The terms in the rate equation for the y-species have similar interpretations. Whatever the signs of the coefficients may be, rate terms such as bx^2, cxy, or dy^2 are called *social* or *mass-action* terms, involving as they do inter- or intraspecific effects of the total population on the individual growth rate.

From this point on the coefficients will be taken to be nonnegative, and a minus sign will be introduced when a negative rate is intended.

Example 9.11

Suppose that the rate equations for the populations of a pair of interacting species are

$$x' = (a - bx - cy)x$$
$$y' = (-d + ex)y \tag{4}$$

where the coefficients are positive. What types of biological dynamics are modeled by these equations? The first species has natural exponential growth (corresponding to the rate term ax), but this is tempered by overcrowding (through the term $-bx^2$) and by predation by the y-species (modeled by the term $-cxy$). The second species will die out exponentially if there is no food (corresponding to the term $-dy$), but it converts individuals of the x-species into increased fertility ($+exy$) and may survive, and even thrive, if its conversion efficiency e is high enough.

Example 9.12

The rate equations

$$x' = (a + by)x$$
$$y' = (c + dx)y \tag{5}$$

model a pair of interacting species, each with a natural, exponentially growing population (the consequence of the terms ax and cy), whose mutual association is beneficial to both (modeled by the rate terms $+byx$ and $+dxy$). Equations such as (5) cannot model real populations for any great length of time since the consequent explosive growth quickly exceeds reasonable bounds.

Solutions $x(t)$ and $y(t)$ of the rate equations for two interacting species are of interest only in the first quadrant (the *population quadrant*), $x \geqq 0$ and $y \geqq 0$, since negative populations and population densities have no meaning. The curve in the population quadrant defined parametrically by $x = x(t)$ and $y = y(t)$ is a *population orbit*. The vanishing of $x(t)$ or of $y(t)$ for some t means that the corresponding species vanishes. In reality, our models lose their significance if either $x(t)$ or $y(t)$ becomes small. When this occurs, individual behavior determines the fate of the species, and the averaging used to justify rate equation (1) is meaningless.

Predator and Prey

The simplest models of a predator and prey association include only natural growth and decay and the predator–prey interaction itself. All other forms of intra- or interspecific relationships are negligible and are omitted. We assume that the prey species would expand exponentially in the absence of predation, while the predator species would decline exponentially if there were no prey to consume. The predator–prey interaction is modeled by mass-action terms proportional to the product of the two populations. The rate equations for such a simplified model are

$$x' = (a - by)x \tag{6}$$
$$y' = (-c + dx)y$$

where x is the prey and y is the predator and all coefficients are positive.

The equilibrium points correspond to the simultaneous solutions of the equations

$$(a - by)x = 0$$
$$(-c + dx)y = 0$$

There are just two equilibrium points, the extinction point $(0, 0)$ and the point $P = (c/d, a/b)$, which lies inside the population quadrant. A line-element field may be constructed for the rate equations (6), and we see that the population orbits appear to wheel counterclockwise about the equilibrium point P (Figure 9.27). We cannot tell from the line-element fields whether the population orbits spiral toward or away from the point P, whether they are periodic, returning precisely to their starting point after one turn around P, or whether they exhibit some other behavior. We do know, of course, that no two distinct orbits may intersect since intersection would violate the uniqueness property of orbits.

Fortunately, system (6) is so simple that we may actually solve it. First note that (6) has the particular solutions

$$\begin{array}{ccc} x(t) = x_0 e^{at} & & x(t) = 0 \\ y(t) = 0 & \text{and} & y(t) = y_0 e^{-ct} \end{array}$$

solutions that lie along the axes and isolate the population quadrant from the other quadrants. The uniqueness theorem then implies that any orbit passing through a

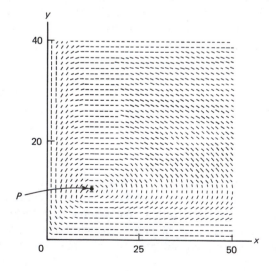

Figure 9.27 Line element field for a predator–prey model.

point in the interior of the population quadrant lies entirely inside the quadrant (otherwise, the orbit would intersect the x or y axis). Thus we may as well assume that x and y are positive, divide the second equation of (6) into the first, and obtain

$$\frac{dy}{dx} = \frac{(-c + dx)y}{x(a - by)} \tag{7}$$

which is separable. We have from (7) that

$$\frac{a - by}{y}\frac{dy}{dx} = \frac{-c + dx}{x}$$

which integrates to

$$a \ln y - by + c \ln x - dx = \tilde{K}$$

for some constant $\tilde{K}$. Upon exponentiating we have that

$$(y^a e^{-by})(x^c e^{-dx}) = e^{\tilde{K}} = K \tag{8}$$

where K is a positive constant. The orbit Γ defined by (8) has the following properties:

(i) Γ is a simple closed convex curve surrounding $P = (c/d, a/b)$ if $0 < K < K_0 = (c/de)^c(a/be)^a$.

(ii) Γ is the single point P if $K = K_0$.

(iii) Γ does not exist if $K > K_0$.

Moreover, if (x_0, y_0) is any point inside the population quadrant, there is a unique curve Γ of type (i) passing through (x_0, y_0) [type (ii) if $(x_0, y_0) = P$].

The proof that the population orbits have properties (i)–(iii) is outlined in the problem set. Aside from the equilibrium point P, every orbit inside the population quadrant is periodic (i.e., every orbit is a cycle). The periods are not constant from orbit to orbit, but increase monotonically with the "distance" of the orbit from point P.† In fact, as orbits shrink onto P, the corresponding periods decrease to $2\pi(ac)^{-1/2}$, while as orbits expand without bound the periods tend to infinity. See Figure 9.28 for a sketch of the cycles in the population quadrant and a sketch of corresponding component graphs of a periodic solution. Note that the peak of the predator population lags behind that of the prey, as expected.

These mathematical results are simple, interesting, and curious, but do they have anything to do with the populations of real species?

Volterra's Principles

In 1926, Humberto D'Ancona, an Italian biologist, completed a statistical study of the changing populations of various species of fish in the upper Adriatic Sea over

---

† The distance of an orbit Γ from a given point P is taken to be the minimum of distances from P to the points of Γ.

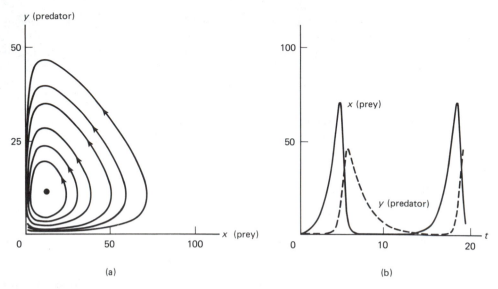

Figure 9.28 Predator–prey cycles and a periodic solution: (a) cycles; (b) component graphs of a periodic solution.

the time span 1910 to 1923. His estimates of the populations were based on the numbers of each species sold on the fish markets of the three ports, Trieste, Fiume, and Venice. D'Ancona assumed, as we shall, that the relative numbers of the various species in the markets reflected the relative abundance of those species in the marine community. A part of the data is given in Table 9.2.

TABLE 9.2 PERCENTAGES OF PREDATOR SPECIES (SHARKS, SKATES, RAYS, ETC.) OUT OF THE TOTAL FISH CATCH

Port	1914	1915	1916	1917	1918	1919	1920	1921	1922	1923
Fiume	12%	21%	22%	21%	36%	27%	16%	16%	15%	11%
Trieste	14%	7%	16%	15%	—	18%	15%	13%	11%	10%

As often happens, the data do not provide overwhelming support for any particular theory of changing fish populations. D'Ancona observed, however, that the percentages of predator species were generally higher during and immediately after World War I (1914–1918). Fishing was drastically curtailed during the war years as the fishermen abandoned their nets to fight in the war, and D'Ancona concluded that the decline in fishing caused the change in the proportions of predator to prey. He formulated the hypothesis that during the war the predator–prey community was close to its natural state of a relatively high proportion of predator fish, while the more intensive fishing of the pre- and postwar years disturbed that equilibrium to the advantage of the prey species. Unable to give a biological or ecological reason for the phenomenon, D'Ancona asked his father-in-law, the noted Italian mathematician Vito Volterra, if there was a mathematical model that might cast some light on the matter. Within a few months Volterra had outlined a series of models for the interactions of two or more species. We have given the simplest of these models in this section.†

Volterra summarized his results for a predator–prey community of two species in the form of three principles, the first of which was partly verified above.

> **The Law of the Periodic Cycle.**   The fluctuations of the populations of the predator and its prey are periodic, the period depending only on the coefficients of growth and interaction and on the initial conditions.

It is not so much that the numbers of predator and prey fluctuate periodically, but that the average numbers remain constant. This is Volterra's second principle.

> **The Law of the Conservation of Averages.**   The averages of the numbers of the predator and the prey species over the period of a cycle have the constant values of the equilibrium levels as long as the parameters of growth, decay, and interaction remain constant.

Proof. This second principle is hardly evident from the model, but it is not difficult to show. Suppose that $(x(t), y(t))$ is a periodic solution of the predator–prey equations (6) and that the period is the positive number T. From the first rate equation of (6), we have that

$$\frac{1}{T}\int_0^T \frac{x'(t)}{x(t)}\,dt = \frac{1}{T}\int_0^T \frac{(a - by(t))x(t)}{x(t)}\,dt = \frac{1}{T}\int_0^T (a - by(t))\,dt \tag{9}$$

On the other hand, since the Chain Rule implies that

$$\frac{d}{dt}\,[\ln(x(t))] = \frac{x'(t)}{x(t)}$$

we see that

---

† Volterra eventually wrote a book on his theories, *Leçons sur la théorie mathématique de la lutte pour la vie* (Paris: Gauthier-Villars, 1931; reproduced by University Microtexts, Ann Arbor, Mich.: 1976). The Soviet biologist G. F. Gause tested Volterra's theories by carrying out numerous laboratory experiments with various competing and predatory microorganisms. He published his results in *The Struggle for Existence* (Baltimore: Williams & Wilkins, 1934; reissued by Hafner, New York, 1964). D'Ancona defended Volterra's work in a book which again used the memorable phrase of Malthus and Darwin, *The Struggle for Existence* (Leiden: E. J. Brill, 1954). A. J. Lotka, an American biologist and, later in life, an actuary, arrived at many of the same conclusions independently of Volterra; see his book, *Elements of Physical Biology* (Baltimore: Williams & Wilkins, 1925).

$$\frac{1}{T}\int_0^T \frac{x'(t)}{x(t)}\,dt = \frac{1}{T}\left[\ln\left(x(T)\right) - \ln\left(x(0)\right)\right] = 0 \tag{10}$$

where we have used the fact that $x(t)$ has period T [i.e., that $x(T) = x(0)$]. From (9) and (10) we conclude that

$$\frac{1}{T}\int_0^T (a - by(t))\,dt = a - \frac{b}{T}\int_0^T y(t)\,dt = 0 \tag{11}$$

Denoting the average $(1/T)\int_0^T y(t)\,dt$ of the predator y-population over a cycle by $\bar{y}$, we have from (11) that $\bar{y} = a/b$, which is the y-coordinate of the equilibrium point of (6). A similar analysis of the second equation of (6) shows that the average of the prey population $x(t)$ over a cycle is $\bar{x} = c/d$, which is the x-equilibrium level.

Volterra's Law of the Conservation of the Averages carries us closer to real populations since averages are easy to calculate from numerical counts over time. In fact, D'Ancona's data from the fish markets are averages, although not necessarily taken over the same period as that of the presumed population cycles.

Finally, Volterra introduced the effect of fishing on the predator–prey community. The simplest model is that of *constant-effort* harvesting, in which the amount caught per week, say, is proportional to the number of fish present.† Consequently, we have Volterra's model of a predator–prey community of fish subjected to *indiscriminate, constant-effort harvesting*, in which the fishermen pull out of the sea whatever is trapped in the nets. The equations of the model are

$$\begin{aligned} x' &= (a - by)x - Hx = (a - H - by)x \\ y' &= (-c + dx)y - Hy = (-c - H + dx)y \end{aligned} \tag{12}$$

where H is a positive constant, the *harvesting coefficient*. The equilibrium point inside the population quadrant is now shifted from its original location at $(c/d, a/b)$ to the new point $((c + H)/d, (a - H)/b)$. It is assumed that $a > H$, for otherwise the average population size of the harvested predator is not a positive number. The analysis is precisely the same as before, and the harvested populations continue their periodic oscillations, but now about the new equilibrium (see Figure 9.29). Since $(c + H)/d > c/d$ and $(a - H)/b < a/b$, this leads to the third principle.

> **Volterra's Principle on the Disturbances of the Averages.** Constant-effort harvesting produces an increase in the average number of the prey and a decrease in the average number of the predator.

Before the data of Table 9.2 can be compared with the theory, the third principle must be reformulated in percentages rather than actual numbers.

† See the problem set of Section 3.3 for another harvesting model, constant rate harvesting.

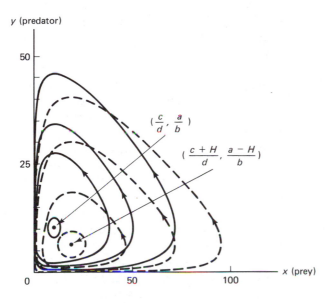

y (predator)

$(\frac{c}{d}, \frac{a}{b})$

$(\frac{c+H}{d}, \frac{a-H}{b})$

x (prey)

Figure 9.29   Effect of constant effort harvesting. The unharvested cycles are indicated by solid lines, the harvested cycles by dashed lines.

Disturbance of the Percentages.   Constant effort harvesting produces an increase in the average percentage of prey and a decrease in the percentage of predator.

Proof. The predator fractions of the total average catch are given by

$$\frac{a/b}{a/b + c/d} = \frac{ad}{ad + bc} \qquad \text{(no harvesting)}$$

$$\frac{(a-H)/b}{(a-H)/b + (c+H)/d} = \frac{ad - dH}{ad + bc + H(b-d)} \qquad \text{(harvesting)}$$

But if we think of H as a variable and differentiate the predator fraction (in the harvesting case) with respect to H, we have that

$$\left[\frac{ad - dH}{ad + bc + H(b-d)}\right]' = \frac{-bcd - abd}{[ad + bc + H(b-d)]^2} \tag{13}$$

as straightforward use of the rule for differentiating a quotient shows. Since the right-hand side of (13) is negative, the fraction (and hence the percentage) of predators must decrease as H increases. A similar analysis shows that the fraction of prey increases with an increase in the harvesting rate.

Volterra's third principle, restated in terms of percentages, answers D'Ancona's question. If Volterra's model represents conditions in the upper Adriatic during the early years of the twentieth century, the predator population should indeed increase with a drop in fishing and decline when fishing resumes, and conversely with the prey population. Volterra's models have been both challenged and supported many times in the half-century since their formulation, but the model continues to be the starting point for most serious attempts to understand just how predator–prey communities evolve.

One dramatic confirmation of the general validity of Volterra's third principle occurred when the insecticide DDT was applied to control the cottony cushion scale insect which infested American citrus orchards. The scale insect had been accidentally introduced from Australia in 1868. Its numbers were controlled (but not eliminated) by subsequent importation from Australia of the scale insect's natural predator, a particular kind of ladybird beetle. When DDT was first introduced as a pesticide, it was hoped that the scale insect could be completely wiped out. But DDT acts indiscriminately, killing all insects it touches. The consequence of the widespread application of DDT to the orchards was that the numbers of the ladybirds drastically dropped, while the population of the scale insects, freed from extensive ladybird predation, increased.

It has since been observed that the effect of insecticides such as DDT is even more drastic than Volterra's model suggests. DDT accumulates in the environment at the upper trophic levels and does more damage to the inhabitants at those levels. Generally speaking, predators are higher on the trophic scale than their prey. Thus DDT has more effect on the predators than on their prey, the pests themselves. Second, the regeneration time is usually more rapid for the prey than for the predator, and the prey evolve resistance to pesticides much faster than do their predators. For all these reasons, there is now a more cautious approach to the use of pesticides such as DDT than there once was.

Comments

Darwin's observation cited at the beginning of this section that it is "the serving as prey to other animals which determines the average numbers of a species" is supported by Volterra's model and apparently, by the data of the fish catches in the Adriatic. The model has its flaws: No account is taken of the delay in time between an action and its effect on population numbers, the averaging over all categories of age, fertility, and sex is dubious, the parameters of the model probably depend on time, and so on. Nonetheless, Ockham's Razor applies here as it does to most models: "What can be accounted for by fewer assumptions is explained in vain by more."

PROBLEMS

The following problem(s) may be more challenging: 6.

1. Using the methods of Examples 9.11 and 9.12, explain the biological dynamics modeled by the following systems. Identify predators, prey, competitors, and cooperators. The param-

eters are assumed to be positive. Which species are being harvested, which restocked from outside, and at what rate?

(a) $x' = (a - by)x,\quad y' = (c - dx)y + e.$

(b) $x' = (a - bx - cy)x,\quad y' = (d - ex - fy)y.$

(c) $x' = (a + by)x,\quad y' = (-c + dx - ey)y.$

(d) $x' = (a - bx - cy)x,\quad y' = (-d + ex - fz)y,\quad z' = (-g + hx - iy)z + 2 + \cos t.$

2. Locate all the equilibrium points in Problems 1(a)–(b) and find conditions on the coefficients which ensure that there is an equilibrium point inside the population quadrant.

3. Find the average value over one cycle of the product $x(t)y(t)$ of the populations of Volterra's model (6). [*Hint:* $\displaystyle\int_0^T x'(t)\,dt = a\int_0^T x(t)\,dt - b\int_0^T x(t)y(t)\,dt$; show that the product of the averages is the average of the product in this case.]

4. Show that the average percentage of prey in the average total catch increases with an increase in the harvesting coefficient H.

5. (a) Explain why the model $x' = r(1 - x/K)x - axy,\ y' = (-b + cx)y$, where $r,\ K,\ a,\ b,$ and c are positive constants, represents a predator–prey model, where the prey has a natural carrying capacity K.

 (b) Find all the equilibrium points of the model.

 (c) Show that there is no equilibrium state inside the population quadrant if $cK \leqq b$.

 (d) Let $cK > b$ and denote the equilibrium state inside the population quadrant by $(\hat{x}, \hat{y})$. Show that constant effort, indiscriminate harvesting leads to an increase in $\hat{x}$ and a drop in $\hat{y}$, thus verifying Volterra's Law of the Disturbance of the Averages in this case. What value of the harvesting coefficient leads to $\hat{y} = 0$ and thus to predator extinction?

6. Follow the steps below to show that for $0 < K \leqq K_0$ the graph of (8) is a simple closed curve in the interior of the population quadrant unless $K = K_0 = a^a(be)^{-a}c^c(de)^{-c}$, when the graph is the equilibrium point $(c/d, a/b)$.

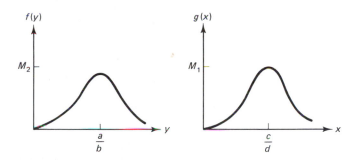

(a) Show that $f(y) = y^a e^{-by}$ and $g(x) = x^c e^{-dx}$ have the graphs indicated, each with a unique maximum $[M_1 = \max g = c^c(de)^{-c}, M_2 = \max f = a^a(be)^{-a}]$ (see the graphs).

(b) Show that Eq. (8) has no positive solutions if $K > M_1 M_2 \equiv K_0$ but that it has the unique solution $(c/d, a/b)$ if $K = K_0$.

(c) Let Δ be a positive number, $\Delta < M_1$. Show that $g(x) = \Delta$ has two solutions, $x_1 < c/d < x_2$. Show that $f(y) = \Delta M_2/g(x)$ has no solution y if $x > x_2$ or $x < x_1$, exactly one solution, $y = a/b$, if $x = x_2$ or if $x = x_1$, and two solutions, $y_1(x) <$

a/b and $y_2(x) > a/b$, if $x_1 < x < x_2$. Show that $y_1(x) \to a/b$ and $y_2(x) \to a/b$ if $x \to x_1$ or $x \to x_2$.

(d) Show that (8) defines a simple closed curve inside the population quadrant if $0 < K < K_0$.

Each system in Problems 7 to 12 is a model of interacting populations. Explain the model, identifying predator, prey, or competitor and stating why the model represents the given biological interactions. Find all equilibrium points inside or on the edge of the population quadrant. If computer graphics is available, generate portraits of the orbits like those given.

---

7. The system is

$$x' = x\left(1 - \frac{1}{100}y\right)$$

$$y' = y\left(-1 + \frac{1}{100}x\right)$$

where x is the population of a species being eaten by a species whose population is y. The nonconstant orbits inside the first quadrant are cycles that turn counterclockwise about the point $(100, 100)$ as sketched below.

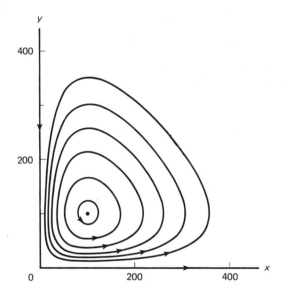

8. Let x and y denote population densities and suppose now that population x preys on population y, but that its appetite satiates. In isolation from the x species the y species follows a logistic pattern of growth, as shown at the top of page 401. A model system for this interaction is

$$x' = x\left(-\frac{1}{2} + \frac{2y}{1 + 2y}\right)$$

$$y' = y\left(1 - y - \frac{2x}{1 + 2y}\right)$$

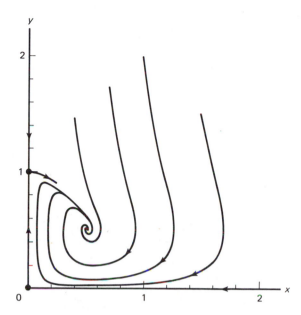

9. Write a new model for the system given in Problem 7 if constant-effort harvesting of both species is included with $H = 0.1$. Identify the new equilibrium points and compare their locations with those of the original model. What are the nonconstant orbits like?

10. (a) Suppose that the species of the system of Problem 8 undergo constant-effort harvesting with $H = 0.1$. Find the equilibria and compare with the original.

   (b) Suppose that the system of Problem 8 undergoes constant-effort harvesting with H restricted by $0 \leq H < \frac{1}{2}$. Show that the coordinates of the equilibrium point inside the population quadrant are

$$\hat{x} = \frac{1}{2} - \frac{2H}{(1-2H)^2}, \qquad \hat{y} = \frac{1+2H}{2-4H}$$

   Show that as H increases, $\hat{x}$ decreases while $\hat{y}$ increases. [*Hint:* Show that $d\hat{x}/dH < 0$ and $d\hat{y}/dH > 0$.] Does harvesting "help" the predator or the prey?

11. The system is

$$x' = x\left(-\frac{1}{2} + \frac{4y}{1+4y}\right)$$

$$y' = y\left(1 - y - \frac{4x}{1+4y}\right)$$

Although the system appears to be much like that in Problem 8, the behavior of the orbits has changed, as indicated at the top of page 402. Now there is a single equilibrium cycle enclosing the point $(\frac{9}{8}, \frac{1}{4})$ and all nonconstant orbits inside the first quadrant spiral toward it with increasing time.

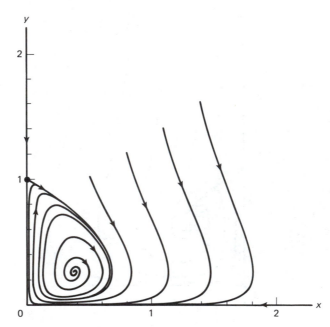

12. When two species compete for common resources, the more efficient of the two may so dominate the habitat that the other species dies out. The following system models this situation:

$$x' = x(1 - x - y)$$
$$y' = y(1 - 2x - 2y)$$

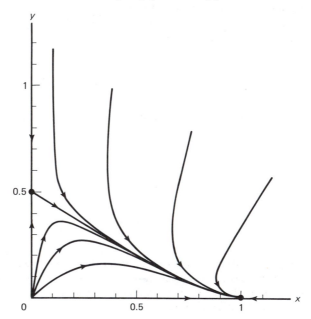

CHAPTER 10

Linear Systems
of Differential Equations

Coupled systems of linear differential equations arise in the modeling of electrical circuits, interconnected systems of shock absorbers, diffusion through a multicellular system, and a host of other phenomena. The simplest of these systems have constant coefficients. Matrices and linear algebraic concepts are basic to the characterization of the solutions of these systems. These ideas are developed in Sections 10.1 and 10.2. Section 10.3 takes up the geometry of the orbits of a constant-coefficient system, and Section 10.4 takes a theoretical approach based on linear spaces and linear operators. Sections 10.5 and 10.6 contain applications to circuits and to compartmental systems.

10.1 THE EIGENVALUES AND EIGENVECTORS OF A MATRIX

The equations to be solved in this chapter are *linear differential systems* in an unknown state vector. The general form of such systems is

$$x'_1 = a_{11}x_1 + \cdots + a_{1n}x_n + F_1(t)$$

$$\vdots \tag{1}$$

$$x'_n = a_{n1}x_1 + \cdots + a_{nn}x_n + F_n(t)$$

where $x_1, \ldots, x_n$ are the unknown *state variables*, t (usually interpreted as time) is the independent variable, the coefficients a_{ij} are real constants or real-valued continuous functions of t on a common interval I, and the *input functions* $F_1(t), \ldots, F_n(t)$ are real-valued continuous functions of t on I. In vector-matrix form (1) may be written as

$$x'(t) = A(t)x(t) + F(t)$$

or, in more compact form,

$$x' = Ax + F \tag{2}$$

where

$$A = A(t) = \begin{bmatrix} a_{11} & \cdots & a_{1n} \\ \cdot & & \cdot \\ \cdot & & \cdot \\ \cdot & & \cdot \\ a_{n1} & \cdots & a_{nn} \end{bmatrix} = [a_{ij}], \quad x = x(t) = \begin{bmatrix} x_1 \\ \cdot \\ \cdot \\ \cdot \\ x_n \end{bmatrix}, \quad F = F(t) = \begin{bmatrix} F_1 \\ \cdot \\ \cdot \\ \cdot \\ F_n \end{bmatrix}$$

The vector x is the *state vector*, A is the *system* or *coefficient matrix*, and F is the *input vector* or *driving force*. The communications circuit model of Section 9.1 is an example of a linear differential system.

There are two immediate tasks associated with (2). The first is to characterize its set of solutions, and the second to show that the related initial value problem

$$x' = Ax + F, \quad x(t_0) = x^0 \tag{3}$$

where t_0 belongs to I and x^0 lies in $\mathbf{R}^n$, has exactly one solution. The second problem is resolved by appealing to a result in Section 9.3.

Existence, Uniqueness, Extension.   Initial value problem (3) has exactly one maximally extended solution $x(t)$, and this solution is defined for all t in I.

The existence and uniqueness part follows from the corresponding theorem in Section 9.3 since the vector function $f = A(t)x + F(t)$ and all the first partial derivatives $\partial f_i / \partial x_j = a_{ij}(t)$, $i, j = 1, \ldots, n$, are continuous for t in I and x in $\mathbf{R}^n$ by assumption. The proof that the solution $x(t)$ can be extended over all of I is omitted.

The result above does not give a formula or process for finding the unique solution of (3) or for finding all solutions of (2). Our goal in this chapter is to show how this may be done, at least if A is a matrix of constants.

Constant-Coefficient Matrix and Zero Input Vector

The simplest case is when F is identically zero (no input) and A is a matrix of constants. It is this case that we address first since it turns out that special sets of scalars and vectors associated with A (the eigenvalues and eigenvectors) largely determine the solutions.

An example illustrates the nature of these sets and how they are used to construct solutions of linear differential systems.

Example 10.1

In analogy with the scalar linear case, the linear system

$$x_1' = 5x_1 + 3x_2$$
$$x_2' = -6x_1 - 4x_2$$

might be expected to have solutions of the form $x_1 = ae^{\lambda t}$, $x_2 = be^{\lambda t}$, for constants a, b, and λ. In other words, if the given system is written in the equivalent matrix form

$$x' = Ax, \quad \text{with} \quad A = \begin{bmatrix} 5 & 3 \\ -6 & -4 \end{bmatrix} \tag{4}$$

then we seek solutions of (4) of the form

$$x = \begin{bmatrix} x_1 \\ x_2 \end{bmatrix} = \begin{bmatrix} ae^{\lambda t} \\ be^{\lambda t} \end{bmatrix} = e^{\lambda t} \begin{bmatrix} a \\ b \end{bmatrix} \tag{5}$$

Substituting (5) into (4) we see that a, b, and λ must satisfy the condition

$$\lambda \begin{bmatrix} a \\ b \end{bmatrix} = \begin{bmatrix} 5 & 3 \\ -6 & -4 \end{bmatrix} \begin{bmatrix} a \\ b \end{bmatrix} = A \begin{bmatrix} a \\ b \end{bmatrix} \tag{6}$$

Using the 2×2 identity matrix E, the linear algebraic system (6) can be written as the homogeneous linear system

$$(A - \lambda E) \begin{bmatrix} a \\ b \end{bmatrix} = \begin{bmatrix} 0 \\ 0 \end{bmatrix} \tag{7}$$

Now (7) clearly has the "trivial" solution $a = b = 0$, but this corresponds to the "trivial" solution of (4) with $x \equiv 0$. As we saw in Section 6.2, the equation (7) will have nontrivial solutions if and only if the determinant of the coefficient matrix vanishes. This means that λ must satisfy the condition

$$\det (A - \lambda E) = \det \begin{bmatrix} 5 - \lambda & 3 \\ -6 & -4 - \lambda \end{bmatrix} = \lambda^2 - \lambda - 2 = 0$$

This condition is true for just two values of λ; namely, $\lambda_1 = -1$ and $\lambda_2 = 2$. First setting $\lambda = \lambda_1$ we see that (7) then reduces to the single equation $6a + 3b = 0$. All solutions of this equation can be constructed if b is assigned the value $-2c_1$, where c_1 is an arbitrary constant, and then compute the value $a = c_1$. Thus

$$x = c_1 \begin{bmatrix} 1 \\ -2 \end{bmatrix} e^{-t}, \quad c_1 \text{ an arbitrary constant}$$

are all the solutions of (4) which have the form (5) with $\lambda = -1$. Similarly we discover that

$$x = c_2 \begin{bmatrix} 1 \\ -1 \end{bmatrix} e^{2t}, \quad c_2 \text{ an arbitrary constant}$$

are all the solutions of (4) which have the form (5) with $\lambda = 2$. Direct substitution shows that

$$x = c_1 \begin{bmatrix} 1 \\ -2 \end{bmatrix} e^{-t} + c_2 \begin{bmatrix} 1 \\ -1 \end{bmatrix} e^{2t}, \qquad c_1, c_2 \text{ arbitrary constants}$$

is actually a solution of (4), and we shall show in a later section that (4) has no other solutions.

Example 10.1 suggests that all solutions of a constant coefficient linear system of first order differential equations with no input can be found by purely matrix-algebraic means. This is true, as we shall show in Section 10.2. We turn our attention now to the nature of the constants such as λ_1, λ_2, a, and b associated with the system matrix, since it is these constants which generate the solutions of the differential system. For the rest of this section we consider only matrix-algebraic questions, delaying their use in solving linear differential systems till the next section.

Eigenvalues and Eigenvectors

In Example 10.1 observe that (7) implies that

$$A \begin{bmatrix} 1 \\ -2 \end{bmatrix} = (-1) \begin{bmatrix} 1 \\ -2 \end{bmatrix} \qquad A \begin{bmatrix} 1 \\ -1 \end{bmatrix} = 2 \begin{bmatrix} 1 \\ -1 \end{bmatrix} \tag{8}$$

In other words, the matrix A applied to the vectors $[1 \quad -2]^T$, and $[1 \quad -1]^T$ produces those same vectors again but multiplied by the scalars -1 and 2, respectively.[†] Vectors and associated scalars with this property play a very important role in the solution of linear equations (as we have seen).

> **Eigenvalues and Eigenvectors.**   Let A be a square $n \times n$ matrix of constants. A scalar λ is an eigenvalue of the matrix A if there is a *nonzero* vector v such that
>
> $$Av = \lambda v \tag{9}$$
>
> The vector v is called an *eigenvector* of A corresponding to the *eigenvalue* λ. For any eigenvalue λ of A the set V_λ of all solutions of the equation $Av = \lambda v$ is called the *eigenspace* of A corresponding to λ.

It is easy to verify that any eigenspace V_λ is actually a subspace of $\mathbf{R}^n$ (or $\mathbf{C}^n$, if appropriate), and that any *nonzero* element in V_λ is an eigenvector of A corresponding to λ. Indeed, if v_1 and v_2 are any two elements in the eigenspace V_λ of a matrix A then $Av_1 = \lambda v_1$, and $Av_2 = \lambda v_2$. Thus for any constants c_1 and c_2 we have that

$$A(c_1 v_1 + c_2 v_2) = c_1 A v_1 + c_2 A v_2 = (c_1 v_1 + c_2 v_2) \tag{10}$$

[†] $[a \quad b]^T$ denotes the transpose of the row vector $[a \quad b]$; i.e., the column vector $\begin{bmatrix} a \\ b \end{bmatrix}$.

and hence $c_1v_1 + c_2v_2$ is in V_λ, proving that V_λ is a subspace. Also note that two eigenspaces V_λ and V_μ of a matrix A can only have the zero vector in common if $\lambda \neq \mu$.

Example 10.1 (*Continued*)

The equations (8) show that $\lambda = -1$ and $\mu = 2$ are eigenvalues of the matrix A in Example 10.1, and that moreover $v = [1 \quad -2]^T$, and $w = [1 \quad -1]^T$ are eigenvectors corresponding λ and μ, respectively. From the calculations in Example 10.1 it is also clear that

$$V_\lambda = \text{Span} \left([1 \quad -2]^T\right), \quad V_\mu = \text{Span} \left([1 \quad -1]^T\right)$$

and that A has no other eigenvalues.

Even though the entries of A are real, the eigenvalues and eigenvectors may be complex [and that is why in the above we used $\mathbf{C}^n$ rather than $\mathbf{R}^n$].

Example 10.2

To find the eigenvalues and eigenvectors of $A = \begin{bmatrix} 1 & 2 \\ -2 & 1 \end{bmatrix}$ we need to find scalars λ for which there are non-zero vectors v such that

$$(A - \lambda E)v = \begin{bmatrix} 1-\lambda & 2 \\ -2 & 1-\lambda \end{bmatrix} \begin{bmatrix} v_1 \\ v_2 \end{bmatrix} = \begin{bmatrix} 0 \\ 0 \end{bmatrix}$$

This is possible exactly when λ is chosen so that $\det (A - \lambda E)$ vanishes:

$$\det \begin{bmatrix} 1-\lambda & 2 \\ -2 & 1-\lambda \end{bmatrix} = \lambda^2 - 2\lambda + 5 = 0$$

The values of λ which satisfy this condition are $\lambda_1 = 1 + 2i$, and $\lambda_2 = \bar{\lambda}_1 = 1 - 2i$. The eigenspace V_{1+2i} corresponding to λ_1 is the solution set of the equation $(A - \lambda_1 E)v = 0$. Observe that $v = [a \quad b]^T$ is a solution of $(A - \lambda_1 E)v = 0$ if and only if

$$0 = (1 - \lambda_1)a + 2b = -2ia + 2b$$

or simply, $b = ia$. Thus $v = [a \quad ia]^T$, where a is an arbitrary complex number, is the solution set we seek and hence $V_{1+2i} = \text{Span} \left([1 \quad i]^T\right)$. Similarly, we obtain that $V_{1-2i} = \text{Span} \left([1 \quad -i]^T\right)$.

It is no accident that complex eigenvalues and their corresponding eigenvectors occur in conjugate pairs if A is itself real.

Complex Eigenvalues of a Real Matrix.   Let A be a real matrix with a nonreal eigenvalue λ and a corresponding eigenvector v. Then v must have at least one nonreal component, and $\bar{v}$ is an eigenvector of A with $\bar{\lambda}$ its corresponding eigenvalue.

Proof. Let $Av = \lambda v$. Then $\overline{Av} = \overline{\lambda v}$, and since the conjugate of products is the product of conjugates, we have that $\overline{A}\,\overline{v} = \overline{\lambda}\,\overline{v}$. Since A is real, $\overline{A} = A$, so $\overline{\lambda}$ is an eigenvalue of A with corresponding eigenvector $\overline{v}$. Now if v had only real components, $v = \overline{v}$, so $\overline{\lambda} v = \overline{\lambda}\,\overline{v} = A\overline{v} = \lambda v$. Thus $(\lambda - \overline{\lambda})v = 0$. Since $v \neq 0$, it follows that $\lambda = \overline{\lambda}$, a contradiction, and hence v must have a nonreal component.

A collection of eigenvectors, each corresponding to distinct eigenvalues, forms an independent set. In fact, the following useful result holds.

Eigenspace Property.   Let V_i, $i = 1, \ldots, p$ be eigenspaces of an $n \times n$ matrix A corresponding to the distinct eigenvalues λ_i, $i = 1, \ldots, p$. For $i = 1, \ldots, p$ let B_i be an independent subset of V_i. Then the set consisting of all the elements of $B_i, \ldots, B_p$ is an independent set.

The full proof is by mathematical induction on p and is outlined in the problem set.

The important consequence of the Eigenspace Property is that if A has n distinct eigenvalues, $\lambda_1, \ldots, \lambda_n$, there is a set of n independent eigenvectors. To show this, let each B_i consist of a single eigenvector of V_{λ_i}, $i = 1, \ldots, n$, and apply the Eigenspace Property.

Example 10.3

The matrix A of Example 10.1 has eigenvectors $[1 \ \ -2]^T$ and $[1 \ \ -1]^T$ corresponding, respectively, to the eigenvalues -1 and 2. The set $\{[1 \ \ -2]^T, [1 \ \ -1]^T\}$ is an independent subset of $\mathbf{C}^2$ (and of $\mathbf{R}^2$). Similarly, the set $\{[1 \ \ i]^T, [1 \ \ -i]^T\}$ of eigenvectors corresponding to the eigenvalues $1 + 2i$ and $1 - 2i$, respectively, of the matrix of Example 10.2 is an independent subset of $\mathbf{C}^2$ (but not of $\mathbf{R}^2$ in this case).

The Characteristic Polynomial

Examples 10.1 and 10.2 and the definition of an eigenvalue suggest a method for finding all the eigenvalues of a matrix. Let A be an $n \times n$ matrix with real or complex entries. If λ is an eigenvalue of A, there must be a nontrivial column vector v such that $Av = \lambda v$. Thus the matrix equation $(A - \lambda E)v = 0$ has a nontrivial solution v, and furthermore, all solutions of this equation form the eigenspace V_λ. If we denote the columns of the matrix $A - \lambda E$ by the column vectors $w^1, \ldots,$ w^n and let the vector $v = [b_1 \ \cdots \ b_2]^T$, then the equation $(A - \lambda E)v = 0$ implies that $b_1 w^1 + \cdots b_n w^n = 0$, and hence the set of vectors $\{w^1, \ldots, w^n\}$ is dependent. This is true if and only if det $[w^1 \ \cdots \ w^n] = $ det $(A - \lambda E)$ vanishes.† This determinant is a polynomial $p(\lambda)$ of degree n in λ:

---

† For clarity, sets of vectors in this and later chapters will be indexed with superscripts. The vertical bars introduced in Chapter 6 for separating column vectors of a matrix will not be used.

$$p(\lambda) = \det (A - \lambda E) = \det \begin{bmatrix} a_{11} - \lambda & a_{12} & \cdots & a_{1n} \\ a_{21} & a_{22} - \lambda & \cdots & a_{2n} \\ \cdot & \cdot & \cdot & \cdot \\ \cdot & \cdot & \cdot & \cdot \\ a_{n1} & a_{n2} & \cdots & a_{nn} - \lambda \end{bmatrix} \quad (11)$$

$$= (-1)^n \lambda^n + (-1)^{n-1}(a_{11} + \cdots + a_{nn})\lambda^{n-1} + \cdots + \det A$$

The particular form of the coefficients of λ^n and λ^{n-1} may be established by expanding the determinant in (11) by cofactors and using induction on n. That the constant term of $p(\lambda)$ is $\det A$ follows from the definition if λ is equated to 0. The polynomial p has a special name.

Characteristic Polynomial.   The polynomial $p(\lambda) = \det (A - \lambda E)$ given in (11) is called the *characteristic polynomial* of the square matrix A.

On the basis of this definition and the discussion above, we have the following result.

Characteristic Polynomial Theorem.   The scalar λ is an eigenvalue of an $n \times n$ matrix A if and only if λ is a root of the characteristic polynomial $p(\lambda) = \det (A - \lambda E)$.

Thus the calculation of the eigenvalues of an $n \times n$ matrix reduces to the problem of finding the roots of a polynomial of degree n. If $n = 2$, the quadratic formula may be used, but for $n > 2$ it may be a difficult problem to find the roots. Numerous computer software routines exist, however, for finding approximations to the eigenvalues of a matrix, and for $n > 2$ this may be the only practical method available. Sometimes, however, the factoring techniques of algebra allow the determination of all the eigenvalues even if $n > 2$.

Example 10.4

The characteristic polynomial of

$$A = \begin{bmatrix} -3 & 1 & -2 \\ 0 & -1 & -1 \\ 2 & 0 & 0 \end{bmatrix}$$

is given by

$$p(\lambda) = \det (A - \lambda E) = \det \begin{bmatrix} -3 - \lambda & 1 & -2 \\ 0 & -1 - \lambda & -1 \\ 2 & 0 & -\lambda \end{bmatrix}$$

$$= -\lambda^3 - 4\lambda^2 - 7\lambda - 6$$

Substitution of small integers for λ quickly shows that -2 is a root of $p(\lambda)$. Thus $p(\lambda) = -(\lambda + 2)q(\lambda)$ where the quadratic $q(\lambda)$ is found by synthetic division, or by straightforward calculation, to be $\lambda^2 + 2\lambda + 3$. The roots of $q(\lambda)$ are found by the quadratic formula to be the complex conjugates $-1 + \sqrt{2}i$ and $-1 - \sqrt{2}i$. Thus the eigenvalues of A are -2, $-1 + \sqrt{2}i$, and $-1 - \sqrt{2}i$.

A polynomial may have a multiple root and this leads us to the following definition.

> **Multiplicity of an Eigenvalue.**   A scalar λ_0 is said to be an eigenvalue of A of *multiplicity* k if λ_0 is a root of the characteristic polynomial $p(\lambda) = \det(A - \lambda E)$ of multiplicity k. Eigenvalues of multiplicity 1 are called *simple* eigenvalues.

Example 10.5

Let

$$A = \begin{bmatrix} 2 & 2 & 1 \\ 1 & 3 & 1 \\ 1 & 2 & 2 \end{bmatrix}$$

Then we have that

$$p(\lambda) = \det[A - \lambda E] = \det \begin{bmatrix} 2-\lambda & 2 & 1 \\ 1 & 3-\lambda & 1 \\ 1 & 2 & 2-\lambda \end{bmatrix}$$

$$= -\lambda^3 + 7\lambda^2 - 11\lambda + 5 = -(\lambda - 1)^2(\lambda - 5)$$

The roots are $\lambda_1 = 1$, $\lambda_2 = 1$, $\lambda_3 = 5$ with $\lambda = 1$ an eigenvalue of multiplicity 2, and $\lambda = 5$ a simple eigenvalue. To find the eigenspace V_1, we must find all solutions of

$$(A - E)x = \begin{bmatrix} 1 & 2 & 1 \\ 1 & 2 & 1 \\ 1 & 2 & 1 \end{bmatrix} \begin{bmatrix} x_1 \\ x_2 \\ x_3 \end{bmatrix} = \begin{bmatrix} 0 \\ 0 \\ 0 \end{bmatrix}$$

Thus $x_1 + 2x_2 + x_3 = 0$, which is the equation of a plane through the origin. Hence V_1 is a two-dimensional subspace spanned by eigenvectors $x = [1, 0, -1]^T$ and $w = [2, -1, 0]^T$, for example. The reader may show that the one-dimensional eigenspace V_5 is spanned by $[1 \quad 1 \quad 1]^T$.

It is *not* always the case that the multiplicity of an eigenvalue equals the dimension of its eigenspace. The matrix $A = \begin{bmatrix} 3 & 1 \\ 0 & 3 \end{bmatrix}$ is a counterexample. The reader should verify that this matrix has double eigenvalue 3, while the corresponding eigenspace has dimension 1. We do have the following result, which we shall not prove, concerning the dimension of an eigenspace and the multiplicity of the corresponding eigenvalue.

Eigenspace Dimension Theorem.   Let A be an $n \times n$ matrix and λ an eigenvalue of A of multiplicity m. Then the dimension of the eigenspace V_λ is no larger than m, but at least 1.

Thus if an eigenvalue is simple, its eigenspace is one-dimensional. Note that an $n \times n$ matrix A has exactly n eigenvalues if an eigenvalue of multiplicity m is counted m times. This follows from the fact that the characteristic polynomial of A has degree n, hence exactly n roots, counting multiplicities.

Comments

The eigenvalues and eigenvectors of a square constant matrix A may be used to construct solutions of the linear differential system $x' = Ax$. The eigenvalues and eigenvectors themselves are determined purely algebraically, without considering the differential system at all. This process of finding solutions of a constant-coefficient linear differential system by algebraic methods somewhat resembles the corresponding process outlined in Chapter 6 for solving an nth-order scalar constant-coefficient homogeneous differential equation. In fact, the term "characteristic polynomial" is used in both settings.

PROBLEMS

The following problem(s) may be more challenging: 4, 9, and 10.

1. Find the eigenvalues, their multiplicities, and a basis for each eigenspace of the following matrices.

(a) $\begin{bmatrix} 1 & 0 \\ 2 & 1 \end{bmatrix}$.　(b) $\begin{bmatrix} 1 & 1 \\ 1 & 1 \end{bmatrix}$.　(c) $\begin{bmatrix} 0 & -3 \\ 3 & 0 \end{bmatrix}$.　(d) $\begin{bmatrix} 6 & -7 \\ 1 & -2 \end{bmatrix}$.

(e) $\begin{bmatrix} 3 & -5 \\ 5 & 3 \end{bmatrix}$.　(f) $\begin{bmatrix} 1 & 0 & 1 \\ 0 & 1 & 0 \\ 1 & 0 & 1 \end{bmatrix}$.　(g) $\begin{bmatrix} 5 & -6 & -6 \\ -1 & 4 & 2 \\ 3 & -6 & -4 \end{bmatrix}$.

(h) $\begin{bmatrix} \cos\theta & -\sin\theta \\ \sin\theta & \cos\theta \end{bmatrix}$.　(i) $\begin{bmatrix} -1 & 36 & 100 \\ 0 & -1 & 27 \\ 0 & 0 & 5 \end{bmatrix}$.　(j) $\begin{bmatrix} a & b \\ -b & a \end{bmatrix}$, a, b real numbers.

2. Show that the eigenvalues of a triangular matrix are the diagonal entries, the multiplicity of each eigenvalue equaling the number of times the eigenvalue appears on the diagonal.

3. Find all solutions of $x' = Ax$ where $x = [x_1 \ x_2]^T$ and A is the matrix of Problem 1(b), (c), and (g). Use the format of Example 10.1 to express your answers.

4. Show that if λ is an eigenvalue of A, then λ^k is an eigenvalue of A^k. If μ is an eigenvalue of A^k, is $\mu^{1/k}$ always an eigenvalue of A?

5. Show that a square matrix A and its transpose A^T have the same characteristic polynomial.

6. Let $\lambda_1, \ldots, \lambda_n$ be the eigenvalues of a matrix A, each repeated according to its multiplicity. Show that $\det A = (\lambda_1) \cdots (\lambda_n)$. [*Hint*: Write $p(\lambda)$ as $(-1)^n(\lambda - \lambda_1) \cdots (\lambda - \lambda_n)$, expand, and compare with the form of $p(\lambda)$ in (11).]

7. [*Trace*]. The *trace* of an $n \times n$ matrix A is the sum of its diagonal entries: $\operatorname{tr} A = a_{11} + \cdots + a_{nn}$.
   (a) Show that $\operatorname{tr} A$ is the sum of the eigenvalues of A. [*Hint*: See hint in Problem 6.]
   (b) Use part (a) to show that the matrix $\begin{bmatrix} 1 & 5 & 3 \\ 6 & -7 & 10 \\ 37 & 56 & 2 \end{bmatrix}$ must have an eigenvalue with a negative real part.

8. Show that the square matrix A is nonsingular if and only if all its eigenvalues are nonzero. [*Hint*: See the hint in Problem 6.]

9. The "characteristic polynomial" of the nth-order constant-coefficient equation $y^{(n)} + a_{n-1}y^{(n-1)} + \cdots + a_1y' + a_0y = 0$ is $r^n + a_{n-1}r^{n-1} + \cdots + a_1r + a_0$.
   (a) Show that if $x_1 = y$, $x_2 = y'$, $\ldots$, $x_n = y^{(n-1)}$, the equation above becomes the system $x' = Ax$, where

$$A = \begin{bmatrix} 0 & 1 & 0 & & & 0 \\ 0 & 0 & 1 & & & \\ \cdot & & & \cdot & & \cdot \\ \cdot & & & & & \cdot \\ \cdot & & & & 0 & 1 \\ -a_0 & -a_1 & -a_2 & \cdots & -a_{n-2} & -a_{n-1} \end{bmatrix}$$

   (b) Show that the characteristic polynomial $p(\lambda)$ of matrix A is $(-1)^n(\lambda^n + a_{n-1}\lambda^{n-1} + \cdots + a_1\lambda + a_0)$. [*Hint*: Carry out a cofactor expansion of $\det (A - \lambda E)$ along the bottom row.]

10. [*Proof of the Eigenspace Property*]. Induction on p may be used to show that the collection of all vectors in $B_1, \cdots, B_p$ is an independent set if each B_i is an independent subset of V_i, where V_i is an eigenspace of an $n \times n$ matrix A corresponding to the eigenvalue λ_i, $i = 1, \ldots, p$, and these eigenvalues are distinct.
    (a) First assume that each B_i contains a single vector v_i. Verify that the eigenspace property holds for the case $p = 1$.
    (b) Suppose that the eigenspace property holds for singleton subsets B_i whenever $p \leq k - 1$ (the induction hypothesis). Show that the property holds for $p = k$.
    (c) Show that the general case where not all the sets B_i are singleton sets reduces to the case considered above.

10.2 SOLVING CONSTANT-COEFFICIENT LINEAR DIFFERENTIAL SYSTEMS

As we saw in Section 10.1, the eigenvalues and eigenvectors of the $n \times n$ constant matrix A can be used to construct families of solutions of the constant-coefficient linear system without input

$$x' = Ax \tag{1}$$

Can we obtain *all* solutions of (1) by using these eigenvalues and eigenvectors? As it happens, that is not always possible, but there is an alternative construction which does give all solutions even if the eigenvalue/eigenvector approach fails. These matters are explained in this section.

Solving x' = Ax

The following example illustrates the fact that, at least for some matrices A, all solutions can be obtained by using the eigenvalue/eigenvector approach. The basic idea is to make a change of state variables leading to an uncoupled system which can be solved one equation at a time.

Example 10.6

The matrix of coefficients of the linear system

$$\begin{aligned} x_1' &= 5x_1 + 3x_2 \\ x_2' &= -x_1 + x_2 \end{aligned} \tag{2}$$

is $A = \begin{bmatrix} 5 & 3 \\ -1 & 1 \end{bmatrix}$. The characteristic polynomial of A is $\lambda^2 - 6\lambda + 8$. The eigenvalues and a corresponding pair of eigenvectors of A are given by

$$\begin{aligned} \lambda_1 &= 2, & \lambda_2 &= 4 \\ v &= [1 \quad -1]^T, & w &= [3 \quad -1]^T \end{aligned} \tag{3}$$

Suppose now that the components of the eigenvectors are used to change the dependent variables from x to y before attempting to solve (1):

$$x = Py, \quad \text{where } P = \begin{bmatrix} 1 & 3 \\ -1 & -1 \end{bmatrix}$$

Note that the columns of P form an independent set of eigenvectors of A.

Proceeding symbolically, we have from $x = Py$ that $x' = Py'$, and so

$$y' = P^{-1}x' = P^{-1}Ax = P^{-1}APy \tag{4}$$

That P^{-1} exists follows from the columns of P being independent. In fact, $P^{-1} = \begin{bmatrix} -\frac{1}{2} & -\frac{3}{2} \\ \frac{1}{2} & \frac{1}{2} \end{bmatrix}$. Now a straightforward calculation of the matrix product shows that $P^{-1}AP$ is the diagonal matrix $\begin{bmatrix} 2 & 0 \\ 0 & 4 \end{bmatrix}$ with the eigenvalues of A on the diagonal. Thus the system for y' in (4) becomes

$$\begin{aligned} y_1' &= 2y_1 \\ y_2' &= 4y_2 \end{aligned} \tag{5}$$

which is completely uncoupled and easily solved one equation at a time. We have that $y_1 = c_1 e^{2t}$, $y_2 = c_2 e^{4t}$, where c_1 and c_2 are arbitrary constants. Since $x = Py$, all solutions x of (2) have the form

$$\begin{aligned} x &= \begin{bmatrix} 1 & 3 \\ -1 & -1 \end{bmatrix} y = \begin{bmatrix} 1 & 3 \\ -1 & -1 \end{bmatrix} \begin{bmatrix} c_1 e^{2t} \\ c_2 e^{4t} \end{bmatrix} \\ x &= c_1 \begin{bmatrix} 1 \\ -1 \end{bmatrix} e^{2t} + c_2 \begin{bmatrix} 3 \\ -1 \end{bmatrix} e^{4t} \end{aligned} \tag{6}$$

Thus, at least in this case, we may use the eigenvalues and eigenvectors of A to construct *all* solutions of the linear differential system (2).

Note that one could proceed directly from (3) to (6) without bothering with the actual change of variable x to y and back again. The example suggests that the underlying question is: Given the matrix A, is there a matrix P such that $P^{-1}AP$ is diagonal?

Diagonable Matrices

We begin with a definition.

> **Diagonable Matrix.** An $n \times n$ matrix A is *diagonable* if there is a nonsingular $n \times n$ matrix P such that $P^{-1}AP$ is a diagonal matrix D.

Thus the matrix $A = \begin{bmatrix} 5 & 3 \\ -1 & 1 \end{bmatrix}$ of Example 10.6 is diagonable by means of a matrix of eigenvectors $P = \begin{bmatrix} 1 & 3 \\ -1 & -1 \end{bmatrix}$. The following result tells us exactly which matrices are diagonable and even tells us exactly what the diagonal matrix D is.

Characterization of Diagonable Matrices.   The $n \times n$ matrix A is diagonable if and only if A has a set $V = \{v^1, \ldots, v^n\}$ of n independent eigenvectors. If the elements of V are used as the columns of a matrix $P = [v^1 \cdots v^n]$, then

$$P^{-1}AP = D = \begin{bmatrix} \lambda_1 & & & 0 \\ & \cdot & & \\ & & \cdot & \\ & & & \cdot \\ 0 & & & \lambda_n \end{bmatrix}$$

where v^i is an eigenvector of A corresponding to the eigenvalue λ_i. Each λ_i appears as often as its multiplicity.

Proof. Suppose first that A has a set of n independent eigenvectors $V = \{v^1, \ldots, v^n\}$ corresponding to the respective eigenvalues $\lambda_1, \ldots, \lambda_n$ of A (the λ_i's are *not* assumed to be distinct). Let P be the matrix $[v^1 \cdots v^n]$ whose columns are $v^1, \ldots, v^n$. Then

$$AP = [Av^1 \cdots Av^n] = [\lambda_1 v^1 \cdots \lambda_n v^n]$$

On the other hand,

$$PD = [v^1 \cdots v^n] \begin{bmatrix} \lambda_1 & & & 0 \\ & \cdot & & \\ & & \cdot & \\ & & & \cdot \\ 0 & & & \lambda_n \end{bmatrix} = [\lambda_1 v^1 \cdots \lambda_n v^n]$$

Thus $AP = PD$, or $P^{-1}AP = D$, where the invertibility of P is assured by the independence of the set of its column vectors. That each λ_i appears as often as its multiplicity is a consequence of the Eigenspace Property and the Eigenspace Dimension Theorem of Section 10.1. Conversely, suppose that there is a matrix P such that $P^{-1}AP$ is the diagonal matrix of eigenvalues of A. Then

$$AP = P \begin{bmatrix} \lambda_1 & & & 0 \\ & \cdot & & \\ & & \cdot & \\ & & & \cdot \\ 0 & & & \lambda_n \end{bmatrix}$$

If P is partitioned as $[v^1 \quad \cdots \quad v^n]$, then, the rules of matrix multiplication imply that $Av^j = \lambda_j v^j$, and we are done.

Solving Diagonable Systems.   Let A be an $n \times n$ diagonable matrix of constants. Then the family of solutions of $x' = Ax$ is given by

$$x = c_1 v_1 e^{\lambda_1 t} + \cdots + c_n v_n e^{\lambda_n t} \tag{7}$$

where $c_1, \ldots, c_n$ are arbitrary constants, $\{v^1, \ldots, v^n\}$ is any set of independent eigenvectors of A, and $\lambda_1, \ldots, \lambda_n$ are the corresponding eigenvalues. The function x as given by (7) is the *general solution* of $x' = Ax$.

Proof. The proof is a direct extension of the techniques illustrated in Example 10.6. Construct the matrix $P = [v^1 \quad \cdots \quad v^n]$, make the linear change of variable $x = Py$, write out the uncoupled system $y' = P^{-1}APy$ (i.e., $y'_i = \lambda_i y_i$, $i = 1, \ldots, n$) in the new variable y, solve each of these scalar equations, and return to the original state variable x to obtain (7). Observe that if A is diagonable, the general solution (7) may be constructed directly from the eigenvalues and a corresponding independent set of eigenvectors without actually carrying out the diagonalization process.

Example 10.7

Consider the system $x' = Ax$, where

$$A = \begin{bmatrix} 1 & -2 & -2 \\ -2 & 1 & 2 \\ 2 & -2 & -3 \end{bmatrix}$$

The characteristic polynomial of A is $p(\lambda) = (1 + \lambda)^2 (1 - \lambda)$, and the eigenvalues are -1 (a double eigenvalue) and 1 (a simple eigenvalue). Direct calculation (e.g., using Gaussian elimination) shows that V_{-1} is two-dimensional and is spanned by $\{[1 \quad 1 \quad 0]^T, [1 \quad 0 \quad 1]^T\}$ while V_1 is spanned by $\{[1 \quad -1 \quad 1]^T\}$. Thus A is diagonable since $\{[1 \quad 1 \quad 0]^T, [1 \quad 0 \quad 1]^T, [1 \quad -1 \quad 1]^T\}$ is a set of three independent eigenvectors. Hence the general solution of $x' = Ax$ is

$$x = c_1 \begin{bmatrix} 1 \\ 1 \\ 0 \end{bmatrix} e^{-t} + c_2 \begin{bmatrix} 1 \\ 0 \\ 1 \end{bmatrix} e^{-t} + c_3 \begin{bmatrix} 1 \\ -1 \\ 1 \end{bmatrix} e^t$$

Unfortunately, not all matrices are diagonable, as the following example shows.

Example 10.8

Let

$$A = \begin{bmatrix} 1 & -2 & -2 \\ -2 & 2 & 3 \\ 2 & -3 & -4 \end{bmatrix}$$

The characteristic polynomial is $P(\lambda) = (1 + \lambda)^2(1 - \lambda)$ (exactly as in Example 10.7), and the eigenvalues are $\lambda_1 = -1$ (double) and $\lambda_2 = 1$ (simple), However, unlike the matrix of Example 10.7, the eigenspace V_{-1} is only one-dimensional. For we have that the system

$$[A - \lambda_1 E] \begin{bmatrix} a \\ b \\ c \end{bmatrix} = \begin{bmatrix} 2 & -2 & -2 \\ -2 & 3 & 3 \\ 2 & -3 & -3 \end{bmatrix} \begin{bmatrix} a \\ b \\ c \end{bmatrix} = \begin{bmatrix} 0 \\ 0 \\ 0 \end{bmatrix}$$

has the solution, $a = 0$, b arbitrary, $c = -b$. Hence V_{-1} is spanned by $\{[0 \quad 1 \quad -1]^T\}$ and is one-dimensional. V_1 is spanned by $\{[1 \quad -1 \quad 1]^T\}$. Hence there does *not* exist an independent set of three eigenvectors of A, and so A is not diagonable.

A final note on diagonable matrices. If the eigenvalues of A are distinct, A is diagonable. For in this case there are n distinct eigenvalues, each eigenvalue is simple, and each eigenspace must be one-dimensional (since, otherwise, the sum of the dimensions of the eigenspaces would exceed n) and there is an independent set of n eigenvectors of A. Thus the only problematic matrices are those with multiple eigenvalues, and as Example 10.7 shows, even then it may still be possible to diagonalize the matrix.

We settle the solvability question for systems with nondiagonable matrices below.

The Triangular Form

Although diagonal systems of linear differential equations are particularly easy to solve, not all systems are diagonable. There are other "not quite" diagonal systems which can be solved fairly simply. These are systems with a triangular coefficient matrix.

Example 10.9

Consider the following system with an upper triangular coefficient matrix.

$$\begin{aligned} y_1' &= y_1 - 2y_3 \\ y_2' &= -y_2 + y_3 \\ y_3' &= -y_3 \end{aligned} \qquad A = \begin{bmatrix} 1 & 0 & -2 \\ 0 & -1 & 1 \\ 0 & 0 & -1 \end{bmatrix}$$

The system can be solved by solving the last equation first, inserting the solution $y_3(t)$ into the second equation, solving the resulting first-order linear differential equation $y_2' + y_2 = y_3(t)$, and then repeating the process with the first equation. The complete set of components of solutions obtained this way are $y_3 = c_3 e^{-t}$, $y_2 = (c_2 + c_3 t)e^{-t}$, $y_1 = c_1 e^t + c_3 e^{-t}$, where c_1, c_2, and c_3 are any constants. The family of solutions (i.e., the general solution) may be written in vector form as

$$y = c_1 \begin{bmatrix} 1 \\ 0 \\ 0 \end{bmatrix} e^t + c_2 \begin{bmatrix} 0 \\ 1 \\ 0 \end{bmatrix} e^{-t} + c_3 \begin{bmatrix} 1 \\ t \\ 1 \end{bmatrix} e^{-t} \qquad (8)$$

Observe that the triangular coefficient matrix A in this example has simple eigenvalue 1 and double eigenvalue -1, since these are the entries on the diagonal. The eigenspace V_{-1}, however, is only one-dimensional, as may be shown by a direct calculation. Thus A cannot be diagonalized, and it is not possible to use the eigenvalue–eigenvector technique at the beginning of this section to find all solutions—although, of course, each eigenvalue–eigenvector pair does generate some solutions. Note the appearance of te^{-t} in (8). Terms of the form $t^k e^{\lambda t}$ always make an appearance if λ is an eigenvalue of multiplicity $m > 1$, the corresponding eigenspace has dimension d, $1 \le d < m$, and k is an integer no larger than $m - d$. In the example λ is -1, $m = 2$, $d = 1$, and $k = 0$ or 1.

We have the following general definition.

> **Triangulable Matrices.**   An $n \times n$ matrix A is *triangulable* if there is a nonsingular $n \times n$ matrix Q such that $Q^{-1}AQ$ is upper triangular.

Every matrix is triangulable, regardless of the nature of its eigenvalues and eigenspaces.

Triangularization.   Let the $n \times n$ matrix A have eigenvalues $\lambda_1, \ldots, \lambda_n$ (not necessarily distinct). Then A is triangulable by a nonsingular $n \times n$ matrix Q, and the diagonal entries of $Q^{-1}AQ$ are the numbers $\lambda_1, \ldots, \lambda_n$ in some order.

Proof. The proof is by induction on n. The case $n = 1$ is evident. Suppose that the result holds for all $(k - 1) \times (k - 1)$ matrices (the induction hypothesis). Let A be a $k \times k$ matrix, $\lambda_1, \ldots, \lambda_k$ its eigenvalues and v an eigenvector corresponding to λ_1. Let R be the $k \times k$ matrix whose columns are $v, w^2, \ldots, w^k$, where $\{v, w^2, \ldots, w^k\}$ is any basis of $\mathbf{C}^k$. A straightforward calculation shows that

$$R^{-1}AR = \begin{bmatrix} \lambda_1 & \cdots \\ 0 & \\ \vdots & A_1 \\ 0 & \end{bmatrix}$$

where A_1 is a $(k - 1) \times (k - 1)$ matrix. We have that

$$\det (R^{-1}AR - \lambda E) = \det (R^{-1}(A - \lambda E)R)$$
$$= (\det R^{-1}) \det (A - \lambda E) \det R \qquad (9)$$
$$= \det (A - \lambda E) = (\lambda_1 - \lambda) \det (A_1 - \lambda E)$$

where we have used the facts that the determinant of a product is the product of the determinants and that $\det R^{-1} = 1/\det R$. From (9) we see that the eigenvalues of $R^{-1}AR$ are exactly the eigenvalues $\lambda_1, \ldots, \lambda_k$ of A, while the eigenvalues of A_1 are $\lambda_2, \ldots, \lambda_k$. By the induction hypothesis there is a nonsingular $(k - 1) \times (k - 1)$ matrix Q_1 such that $Q_1^{-1}A_1Q_1$ is upper triangular with diagonal entries $\lambda_2, \ldots, \lambda_k$. Let Q be defined by

$$Q = R \begin{bmatrix} 1 & 0 & \cdots & 0 \\ 0 & & & \\ . & & Q_1 & \\ . & & & \\ 0 & & & \end{bmatrix}$$

It is straightforward to show that Q is invertible and that

$$AQ = Q \begin{bmatrix} \lambda_1 & & \cdots & & & . \\ & & & & & . \\ 0 & \lambda_2 & & & & . \\ . & . & . & & & \\ & & & . & & \\ & & & & . & . & . \\ 0 & .. & & & 0 & \lambda_k \end{bmatrix}$$

Thus $Q^{-1}AQ$ is triangular with the eigenvalues down the diagonal. Thus the induction step is complete, and the result is proved.

With the Triangularization Theorem in hand, all solutions of $x' = Ax$ can be constructed even if A is not diagonable. The problem in the nondiagonable case is to find a matrix Q which triangularizes A. Suppose that such a matrix has been found. Let y be a new state variable related to x by $y = Q^{-1}x$. Then $y' = Q^{-1}x' = Q^{-1}Ax = Q^{-1}AQy$. Since $Q^{-1}AQ$ is upper triangular, all the solutions of $y' = Q^{-1}AQy$ may be found by the "bottom-up" process illustrated in Example 10.9. Note that this process introduces n arbitrary constants of integration, one at each step. Then set $x = Qy$ to obtain the *general solution* of $x' = Ax$.

Example 10.10

The matrix A of Example 10.8 cannot be diagonalized, as noted earlier. However, we can find the general solution of $x' = Ax$ by first triangularizing A. Let Q be the matrix whose first two columns are independent eigenvectors of A (see

Example 10.8) and whose third column is any vector independent of the first two (e.g., $[0 \quad 0 \quad 1]^T$ will do):

$$Q = \begin{bmatrix} 1 & 0 & 0 \\ -1 & 1 & 0 \\ 1 & -1 & 1 \end{bmatrix}, \qquad Q^{-1} = \begin{bmatrix} 1 & 0 & 0 \\ 1 & 1 & 0 \\ 0 & 1 & 1 \end{bmatrix}, \qquad Q^{-1}AQ = \begin{bmatrix} 1 & 0 & -2 \\ 0 & -1 & 1 \\ 0 & 0 & -1 \end{bmatrix}$$

The general solution $y(t)$ of $y' = Q^{-1}AQy$ is given by (8) in Example 10.9. Thus the general solution of $x' = Ax$ is

$$
x = Qy = c_1 Q \begin{bmatrix} 1 \\ 0 \\ 0 \end{bmatrix} e^t + c_2 Q \begin{bmatrix} 0 \\ 1 \\ 0 \end{bmatrix} e^{-t} + c_3 Q \begin{bmatrix} 1 \\ t \\ 1 \end{bmatrix} e^{-t}
$$

$$
= c_1 \begin{bmatrix} 1 \\ -1 \\ 1 \end{bmatrix} e^t + c_2 \begin{bmatrix} 0 \\ 1 \\ -1 \end{bmatrix} e^{-t} + c_3 \begin{bmatrix} 1 \\ -1 + t \\ 2 - t \end{bmatrix} e^{-t}
$$

(10)

where c_1, c_2, and c_3 are any constants. The first two terms in (10) are expected since 1 and -1 are eigenvalues of A with respective eigenvectors $[1 \quad -1 \quad 1]^T$ and $[0 \quad 1 \quad -1]^T$.

Comments

The techniques introduced in this section may be used to find the general solution of any linear system $x' = Ax$ once the eigenvalues and corresponding eigenspaces of A have been calculated. Diagonable $n \times n$ matrices A are the simplest to handle, but as Example 10.10 shows, even the nondiagonable case is straightforward, at least if n is small. Note that all solutions are defined for all time, $-\infty < t < \infty$.

Often, however, one wants to find, not all possible solutions, but that solution which satisfies a given initial condition. That is, the problem to be solved is

$$x' = Ax, \qquad x(0) = x^0 \tag{11}$$

where x^0 is a given vector. The Existence and Uniqueness Theorem of Section 10.1 implies that (11) has exactly one solution. That solution may be determined by imposing the initial condition on the general solution of $x' = Ax$ found by one of the methods outlined in this section.

Example 10.11

The solution of

$$x_1' = 5x_1 + 3x_2, \qquad x_1(0) = 1$$
$$x_2' = -x_1 + x_2, \qquad x_2(0) = 2$$

may be found using the general solution [see (6) in Example 10.6] and determining the constants c_1 and c_2 through the initial conditions:

$$x = c_1 \begin{bmatrix} 1 \\ -1 \end{bmatrix} e^{2t} + c_2 \begin{bmatrix} 3 \\ -1 \end{bmatrix} e^{4t} \qquad \text{(general solution)}$$

$$x(0) = \begin{bmatrix} 1 \\ 2 \end{bmatrix} = c_1 \begin{bmatrix} 1 \\ -1 \end{bmatrix} + c_2 \begin{bmatrix} 3 \\ -1 \end{bmatrix} \qquad \text{(initial condition)}$$

Thus $1 = c_1 + 3c_2$, $2 = -c_1 - c_2$. Hence $c_1 = -\frac{7}{2}$ and $c_2 = \frac{3}{2}$. The desired solution is

$$x = -\frac{7}{2} \begin{bmatrix} 1 \\ -1 \end{bmatrix} e^{2t} + \frac{3}{2} \begin{bmatrix} 3 \\ -1 \end{bmatrix} e^{4t}$$

PROBLEMS

The following problem(s) may be more challenging: 3.

1. Solve $x' = Ax$ if A is as given, using initial conditions if given.

   (a) $A = \begin{bmatrix} 5 & -2 \\ -2 & 8 \end{bmatrix}$.

   (b) $A = \begin{bmatrix} 2 & -1 \\ 1 & 2 \end{bmatrix}$, $x(0) = \begin{bmatrix} 1 \\ 1 \end{bmatrix}$.

   (c) $A = \begin{bmatrix} 2 & 2 & 1 \\ 1 & 3 & 1 \\ 1 & 2 & 2 \end{bmatrix}$.

   (d) $A = \begin{bmatrix} 3 & 0 & 0 \\ 1 & 3 & 0 \\ 0 & 1 & 3 \end{bmatrix}$, $x(0) = \begin{bmatrix} 1 \\ 0 \\ -2 \end{bmatrix}$.

2. Solve the linear differential equation

$$x' = \begin{bmatrix} 1 & -2 & -2 \\ -2 & 1 & 2 \\ 2 & -2 & -3 \end{bmatrix} x$$

3. [*Similarity*]. Two $n \times n$ matrices A and B are *similar* if there is an $n \times n$ invertible matrix Q such that $B = Q^{-1}AQ$.

   (a) Show that if A is diagonable, A is similar to the diagonal matrix of the eigenvalues of A.

   (b) Show that every matrix A is similar to an upper triangular matrix whose diagonal entries are the eigenvalues of A.

   (c) Show that similar matrices have the same eigenvalues. [*Hint*: Show that det $[A - \lambda E]$ $\equiv$ det $[B - \lambda E]$ if $B = Q^{-1}AQ$.]

   (d) Let $x' = Ax$ and $y' = By$, where B is similar to A, $B = Q^{-1}AQ$. Show that $y(t)$ is a solution of the second equation if and only if $x(t) = Qy(t)$ is a solution of the first.

4. Reduce each of the following matrices A to upper triangular form. Find all solutions of $x' = Ax$.

   (a) $\begin{bmatrix} 1 & -1 \\ 4 & -3 \end{bmatrix}$.

   (b) $\begin{bmatrix} 2 & -4 \\ 1 & -2 \end{bmatrix}$.

5. (a) Solve the initial value problem $x' = Ax$, $x(0) = [1 \ 2]^T$, where A is as in Problem 4(a).

   (b) Repeat part (a) with A given in Problem 4(b).

6. [*Real Solutions*]. Let A be a real $n \times n$ matrix with a pair of complex conjugate eigenvalues $\lambda = a + ib$, $\bar{\lambda} = a - ib$, $b \neq 0$, and corresponding complex-conjugate eigenvectors v and $\bar{v}$. Then $e^{\lambda t}v$ and $e^{\bar{\lambda} t}\bar{v}$ are independent complex solutions of the real system $x' = Ax$. The following steps show how to find corresponding real solutions.

   (a) Show that Re $(e^{\lambda t}v)$ and Im $(e^{\lambda t}v)$ are independent real solutions of $x' = Ax$.

   (b) Find all real solutions of $x'_1 = -2x_1 + x_2$, $x'_2 = -x_1 - 2x_2$. [*Hint*: Recall that $e^{(a+ib)t} = e^{at}\cos bt + ie^{at}\sin bt$.]

   (c) Find all real solutions of $x'_1 = -3x_1 + x_2 - 2x_3$, $x'_2 = -x_2 - x_3$, $x'_3 = 2x_1$.

7. Consider the parallel circuit sketched, where C, R, and L are positive constants. Let the currents and voltage drops across each of the three branches (from left to right) be denoted by I_1, V_1, I_2, V_2, and I_3, V_3.

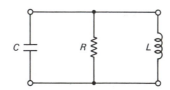

   (a) Use Kirchhoff's Laws to show that $V_1 = V_2 = V_3$ and $I_1 + I_2 + I_3 = 0$.

   (b) Use the circuit laws to show that $I_1 = C_1 V'_1$, $V_2 = RI_2$, and $V_3 = LI'_3$.

   (c) Use parts (a) and (b) to eliminate I_1, I_2, V_2, and V_3, obtaining the system

$$V'_1 = -\frac{1}{CR} V_1 - \frac{1}{C} I_3$$

$$I'_3 = \frac{1}{L} V_1$$

   (d) Show that the eigenvalues of the coefficient matrix of the system in part (c) are real and distinct if and only if $L > 4R^2C$, complex conjugates if and only if $L < 4R^2C$, real and equal if and only if $L = 4R^2C$.

   (e) Explain why all voltages and currents tend to 0 as $t \to +\infty$. [*Hint*: First show that every component of $[V_1(t) \quad I_3(t)]^T$ consists of terms of the form $e^{at}\cos bt$ or $e^{at}\sin bt$, where a and b are real numbers and $a < 0$.]

   (f) Find all the voltages and currents if $C = 0.5$ F, $R = 1\ \Omega$, $L = 1$ H, $V_1(0) = 1$ V, and $I_3(0) = 1$ A.

10.3 PORTRAITS OF THE ORBITS OF A LINEAR SYSTEM

The formulas for the solutions of a linear system define curves in state space, the orbits of the system. A collection of orbits is a portrait in state space of that system: The more orbits there are in the collection, the more detailed the portrait. These portraits provide a visual counterpart to the abstractions of the solution formulas.

We shall sketch representative portraits of affine, constant-coefficient planar systems, affine systems being slightly more general than homogeneous. Once we get out of the plane, of course, graphical depiction of the orbits becomes difficult. Nevertheless, we include a portrait of the orbits of one linear homogeneous system in 3-space as an example of what can be done in higher dimensions.

The system

$$x' = ax + by + e$$
$$y' = cx + dy + f \tag{1}$$

where the coefficients are real constants, is an *affine planar system with constant coefficients*. If $e = f = 0$, the system is homogeneous. We shall take up the case where system (1) has a unique equilibrium point, $P_0(x_0, y_0)$. In this case the straight lines $ax + by + e = 0$ and $cx + dy + f = 0$ are not parallel and intersect at P_0. We may introduce new variables u and v by setting $x = u + x_0$ and $y = v + y_0$, which has the effect of translating the origin of coordinates to P_0. In u, v variables, system (1) becomes

$$u' = au + bv$$
$$v' = cu + dv \tag{2}$$

We shall determine the nature of the orbits of system (2) and then translate the picture back to xy-variables and sketch a portrait of the orbits.

Since (2) is assumed to have a unique equilibrium point at the origin, we must have that $ad - bc \neq 0$, a condition which implies that the eigenvalues of the system matrix $A = \begin{bmatrix} a & b \\ c & d \end{bmatrix}$ are not zero. The components of the real solutions of (2) are linear combinations of functions of the respective forms $e^{\lambda_1 t}$ and $e^{\lambda_2 t}$, $e^{\lambda_1 t}$ and $te^{\lambda_1 t}$, or $e^{\alpha t}\cos \beta t$ and $e^{\alpha t}\sin \beta t$ according as λ_1 and λ_2 are real and distinct, $\lambda_1 = \lambda_2$ and the eigenspace is one-dimensional, or $\lambda_1 = \alpha + i\beta = \bar{\lambda}_2$. Thus it is easy to determine the limiting values of solutions as $t \rightarrow +\infty$ and as $t \rightarrow -\infty$. For example, if $\lambda_2 \leqq \lambda_1 < 0$, solutions must approach the origin as $t \rightarrow +\infty$. On the other hand, if $0 < \lambda_2 \leqq \lambda_1$, solutions approach the origin as $t \rightarrow -\infty$. We shall carry out the full analysis of the orbits for one case and leave the remaining cases to the reader.

Suppose that the eigenvalues of A are distinct and negative, $\lambda_2 < \lambda_1 < 0$. Let P be the nonsingular matrix of a pair of corresponding eigenvectors. As was shown in Section 10.2, the change of variable

$$\begin{bmatrix} u \\ v \end{bmatrix} = P \begin{bmatrix} r \\ s \end{bmatrix} \tag{3}$$

converts (2) to the uncoupled system

$$r' = \lambda_1 r, \qquad s' = \lambda_2 s \tag{4}$$

where the r and the s axes lie along the eigenvectors of A. Solutions of (4) are given by $r = r_0 e^{\lambda_1 t}$, $s = s_0 e^{\lambda_2 t}$ and decay to 0 as $t \rightarrow +\infty$. Moreover, the quotient

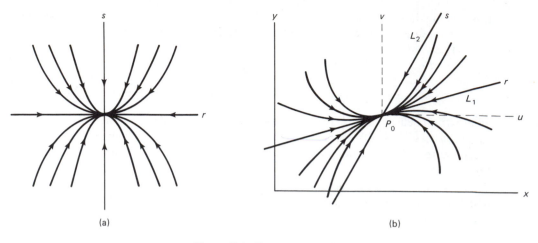

(a)   (b)

Figure 10.1   Improper node, $\lambda_2 < \lambda_1 < 0$.

$$\frac{ds}{dr} = \frac{s'}{r'} = \frac{\lambda_2 s}{\lambda_1 r} = \frac{\lambda_2 s_0 e^{\lambda_2 t}}{\lambda_1 r_0 e^{\lambda_1 t}} = \left(\frac{\lambda_2 s_0}{\lambda_1 r_0}\right) e^{(\lambda_2 - \lambda_1)t}$$

tends to 0 as $t \to \infty$ if $r_0 \neq 0$. Since ds/dr determines the orbital tangent line, orbits of (4) for which $r_0 \neq 0$ tend to the origin tangent to the r-axis as $t \to \infty$. Orbits of (4) are sketched in Figure 10.1a, while those of the original system appear in Figure 10.1b. The lines L_1 and L_2 in Figure 10.1b lie along the eigenspaces of A corresponding to the respective eigenvalues λ_1 and λ_2. The orbital distortions in going from r, s to x, y coordinates are due to the transformation matrix P.

  P_0 is a *node* if the eigenvalues of the system matrix are real, nonzero, and of a common sign. If the eigenvalues are negative, each orbit approaches the node as $t \to \infty$. The orbits approach the node tangent to the line through the node parallel to an eigenvector corresponding to the larger eigenvalue (λ_1 in Figure 10.1), with the exception of the ray solutions along the line parallel to an eigenvector associated with the smaller eigenvalue. Nodes come in varieties dependent on the signs of the eigenvalues, whether the eigenvalues are equal, and (if the eigenvalues are equal) whether the eigenspace is one- or two-dimensional. Figures 10.1 to 10.4 illustrate the varieties of nodes, but the analysis is left to the reader. (Nodes corresponding to $\lambda_1 = \lambda_2 > 0$ may be obtained from Figures 10.3 and 10.4 by reversing the arrows of time.)

  P_0 is a *saddlepoint* if the eigenvalues are real, nonzero, and of opposite sign (see Figure 10.5), a *focus* if the eigenvalues are complex conjugates with nonvanishing real part (Figures 10.6 and 10.7). The remaining case is that of a *center* (Figure 10.8) corresponding to a pair of pure imaginary conjugates as eigenvalues. The ray solutions in Figure 10.5 lie on lines parallel to eigenvectors of the system matrix. Note that these are the only solutions that tend to the equilibrium point with increasing or decreasing time. All other solutions approach P_0 but then turn away and become

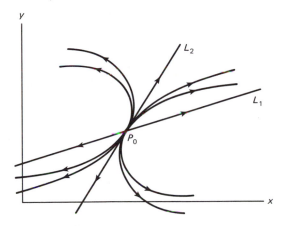

Figure 10.2 Improper node, $0 < \lambda_2 < \lambda_1$.

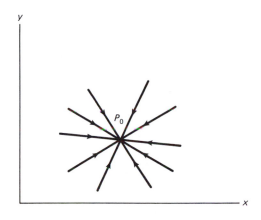

Figure 10.3 Star node, $\lambda_1 = \lambda_2 < 0$ (two-dimensional eigenspace).

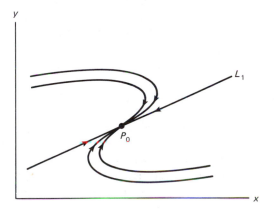

Figure 10.4 Proper node, $\lambda_1 = \lambda_2 < 0$ (one-dimensional eigenspace).

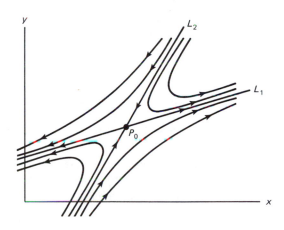

Figure 10.5 Saddlepoint, $\lambda_2 < 0 < \lambda_1$.

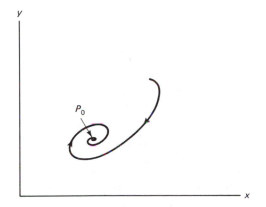

Figure 10.6 Focus, $\lambda_1 = \alpha + i\beta = \bar{\lambda}_2$, $\alpha < 0$, $\beta \neq 0$.

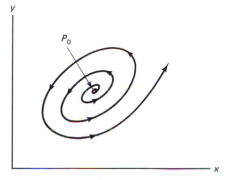

Figure 10.7 Focus, $\lambda_1 = \alpha + i\beta = \bar{\lambda}_2$, $\alpha > 0$, $\beta \neq 0$.

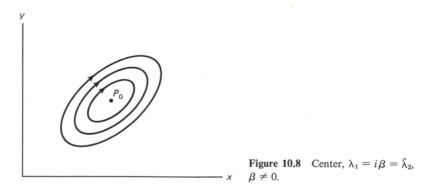

Figure 10.8   Center, $\lambda_1 = i\beta = \bar{\lambda}_2$, $\beta \neq 0$.

unbounded as time increases. There are no real eigenvectors when the eigenvalues are complex, and this fact implies a quite different kind of behavior near P_0, as we see in Figures 10.6 to 10.8. The "oscillatory" character of the orbits in the case of complex eigenvalues is produced by the trigonometric function associated with the imaginary part of the eigenvalues, while the inward or outward motion of the parametrized orbit has to do with the sign of the real part of the eigenvalues [recall that $e^{(\alpha + i\beta)t} = e^{\alpha t} (\cos \beta t + i \sin \beta t)$]. The calculations needed to verify the orbits of Figures 10.5 to 10.8 are left to the reader.

The four types of equilibrium points are node, saddlepoint, focus, and center (see Table 10.1). There are no other types if system (1) has a single equilibrium point. There is another way to view these points and the corresponding system, and that is in terms of the behavior of orbits in asymptotic time, $t \rightarrow +\infty$. Three types of behavior may be distinguished. First, all orbits may approach the equilibrium point as $t \rightarrow \infty$ (*asymptotic stability*). Second, all orbits may be bounded for all t, but nonconstant orbits do not approach the equilibrium point as $t \rightarrow \infty$ (*neutral*

TABLE 10.1   TYPES OF ISOLATED EQUILIBRIUM POINTS

Eigenvalues	Type[a]
$\lambda_1 = \bar{\lambda}_2 = \alpha + i\beta, \quad \alpha < 0$	AS, focus
$\lambda_1 = \lambda_2 < 0$	AS, star or proper node
$\lambda_2 < \lambda_1 < 0$	AS, improper node
$\lambda_1 = \bar{\lambda}_2 = i\beta$	NS, center
$\lambda_2 < 0 < \lambda_1$	UNS, saddle point
$\lambda_1 = \bar{\lambda}_2 = \alpha + i\beta, \quad \alpha > 0$	UNS, focus
$\lambda_1 = \lambda_2 > 0$	UNS, star or proper node
$0 < \lambda_2 < \lambda_1$	UNS, improper node

[a] AS, asymptotically stable; NS, neutrally stable; UNS, unstable.

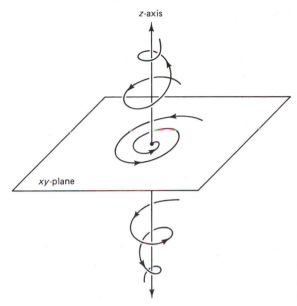

z-axis

xy-plane

Figure 10.9 Tightening helices and spirals of the system $x' = -x - y$, $y' = x - y$, $z' = z$ ($\lambda_1 = -1 + i = \bar{\lambda}_2$, $\lambda_3 = 1$).

stability). Finally, there may be at least one orbit which becomes unbounded as $t \to \infty$ (*instability*). The signs of the real parts of the eigenvalues λ_1 and λ_2 of the system matrix $A = \begin{bmatrix} a & b \\ c & d \end{bmatrix}$ are the distinguishing indicators for these three types of asymptotic behavior.

A deeper analysis of the stability and instability properties of linear and nonlinear systems is made in Chapter 11.

Orbits in Three Space

The reader may try to analyze three-dimensional affine systems, obtaining portraits like that sketched in Figure 10.9 for the system $x' = -x - y$, $y' = x - y$, $z' = z$. Even in the case of a unique critical point, there will be many more distinct portraits in 3-space than there are for planar systems. The procedure remains the same, however: Consider the various possibilities for equilibrium points and for the eigenvalues and eigenspaces of the system matrix.

PROBLEMS

The following problem(s) may be more challenging: 6 and 7.

1. Convert the *RLC* circuit equation $LI'' + RI' + (1/C)I = 0$ to the planar system $x' = y$, $y' = -L^{-1}C^{-1}x - L^{-1}Ry$, $L > 0$, $C > 0$, $R \geqq 0$. Show that the corresponding state plane portrait of the orbits has

(a) An asymptotically stable improper node, proper node, or focus if $R > 0$ and $CR^2 - 4L$ is, respectively, positive, zero, or negative.

(b) A neutrally stable center if $R = 0$.

(c) Sketch the orbits for each of the three cases in part (a), and also for part (b).

2. Sketch portraits of the orbits of the system in Problem 1 if $L = 1$ and:

    (a) $R = 5$, $C = \frac{1}{4}$.   (b) $R = 5$, $C = \frac{4}{25}$.   (c) $R = 4$, $C = \frac{4}{25}$.

    (d) $R = 0$, $C = \frac{1}{4}$.

3. Sketch a portrait of the orbits in each case, identifying the nature of the stability and the equilibrium type.

    (a) $x' = -x$, $y' = -5y$.   (b) $x' = -3x$, $y' = 5y + 1$.

    (c) $x' = 2x$, $y' = 2y$.   (d) $x' = x + y$, $y' = y$.

    (e) $x' = -x + y$, $y' = -x - y$.   (f) $x' = y$, $y' = -4x - 1$.

    (g) $x' = x + 4y$, $y' = x + y$.   (h) $x' = -14x + 20y$, $y' = -10x + 14y$.

4. Verify Figure 10.9 by solving the corresponding system.

5. Verify the entries in Table 10.1.

6. [*Project*]. Classify and sketch the portraits of the orbits of the planar affine systems $x' = ax + by + e$, $y' = cx + dy + f$, which have an infinity of equilibrium points or none at all (i.e., $ad - bc = 0$).

7. [*Project*]. Classify and sketch portraits of the orbits of $x' = Ax$, where A is a real 3×3 constant matrix with distinct nonzero eigenvalues.

10.4 THE THEORY OF LINEAR DIFFERENTIAL SYSTEMS: VARIATION OF PARAMETERS

The general linear system has the form

$$x' = A(t)x + F(t) \tag{1}$$

where the system matrix $A(t)$ is an $n \times n$ matrix of real-valued scalar functions $a_{ij}(t)$ which are continuous on an interval I, and $F(t)$ is a vector function also continuous on I. Although many of the calculations of the earlier sections are no longer valid since the system matrix is not constant, the main theoretical ideas can be extended to (1). Define the linear operator $L = D - A(t)$, which acts on $C^1(I)$. Then (1) becomes $Lx = F$. The solution set of (1) consists of all vector functions $x = v(t) + x_p(t)$, where $v(t)$ belongs to the null space of L and $x_p(t)$ is a particular solution of (1). To solve (1), then, the null space of L must first be found, that is, the solution space of the homogeneous equation

$$x' = A(t)x \tag{2}$$

Next, a particular solution $x_p(t)$ of (1) must be constructed. The null space of L and a particular solution x_p may both be characterized by a special matrix function of t called the transition matrix.

The Transition Matrix

Since (2) satisfies the hypotheses of the Existence and Uniqueness Theorem of Section 10.1, the system has a unique solution $x(t)$, defined for all t in the interval I, for which $x(t_0) = x^0$. The space of solutions of (2) has the following dimensionality property.

Solution Space Dimension.   The linear space of solutions of (2) is an n-dimensional subspace of $C^1(I)$.

Proof. This result is shown in the following way. For $j = 1, \ldots, n$, let $x^j(t)$ be the unique solution of the problem: $x' = A(t)x$, $x(t_0) = e^j$, where e^j is the jth unit basis vector of $\mathbf{R}^n$. Let $S = \{x^j(t) : j = 1, \ldots, n\}$. We claim that S is an independent subset of $C^1(I)$. For suppose that $\Sigma c_j x^j(t_1) = 0$ for some t_1 in I. Then the problem $x' = A(t)x$, $x(t_1) = 0$, has the two solutions $x = \Sigma c_j x^j(t)$ and $x \equiv 0$. By uniqueness, these two solutions must be equal on I, which can be true at $t = t_0$ only if $c_j = 0$, $j = 1, \ldots, n$. Thus S is an independent set in $C^1(I)$. Next we show that S spans the solution set of (1). Let $x(t)$ be any solution of $x' = A(t)x$ on I, $x(t_0) = x^0$. Then there exists constants $c_1, \ldots, c_n$ such that $x^0 = \Sigma c_j e^j$. We have that $x(t)$ and $\Sigma c_j x^j(t)$ are each solutions of the problem $x' = A(t)x$, $x(t_0) = x^0$. By uniqueness, these two solutions must be identical, and hence $x(t) \equiv \Sigma c_j x^j(t)$. Hence S spans the solution space of (2) on I and it follows that S is a basis for the solution set of (2). The theorem is proved.

Remark: The proof above also shows that any set of solutions of (2), $\{x^1, \ldots, x^k\}$, generates an independent subset $\{x^1(t), \ldots, x^k(t)\}$ of vectors of $\mathbf{R}^n$ (or $\mathbf{C}^n$) for each t in I if it does so for a single t_1 in I. To span the solution space of (2), it is only necessary to find n solutions which reduce to an independent set of vectors for some t_1. The techniques of Section 10.2 may be used if A is a matrix of constants.

Example 10.12

The matrix

$$A = \begin{bmatrix} 5 & 3 \\ -6 & -4 \end{bmatrix}$$

of the system

$$x_1' = 5x_1 + 3x_2$$
$$x_2' = -6x_1 - 4x_2$$

has eigenvalues -1 and 2 and respective eigenvectors $[1 \quad -2]^T$ and $[1 \quad -1]^T$. These two vectors are independent, and hence the solution vectors $e^{-t}[1 \quad -2]^T$ and $e^{2t}[1 \quad -1]^T$ are a basis of the solution space of the system.

Solutions of (2) may be made the columns of a matrix. In particular, a *solution matrix* of (2) is an $n \times n$ matrix $X(t)$ whose columns are solutions of (2). Thus a solution matrix satisfies the *matrix differential equation*

$$X' = A(t)X \tag{3}$$

since $(x^j)' = A(t)x^j$ for each column x^j of X.

Example 10.13

$$\begin{bmatrix} e^{-t} & e^{2t} \\ -2e^{-t} & -e^{2t} \end{bmatrix}, \qquad \begin{bmatrix} e^{-t} & 3e^{-t} \\ -2e^{-t} & -6e^{-t} \end{bmatrix}, \qquad \begin{bmatrix} e^{-t} + 2e^{2t} & e^{-t} - e^{2t} \\ -2e^{-t} - 2e^{2t} & -2e^{-t} + e^{2t} \end{bmatrix}$$

are all solution matrices of the system of Example 10.12.

Certain types of solution matrices $X(t)$ are of special significance. The solution matrix $X(t)$ of (3) is *fundamental* if $X(t)$ is nonsingular for all t in I. If $X(t_0)$ is nonsingular for some t_0 in I, it is nonsingular for all t in I (see the remark above). Moreover, if B is any $n \times n$ nonsingular matrix of constants and $X(t)$ is fundamental, $X(t)B$ is again fundamental since it is nonsingular and its columns are linear combinations of the columns of $X(t)$.

One type of fundamental matrix is basic to our needs.

Transition Matrix. The *transition matrix* $\Phi(t, t_0)$, t, t_0 in I, is the fundamental matrix $[x^1(t) \ \ldots \ x^n(t)]$ for which each column $x^j(t)$ is the solution of (2) satisfying $x^j(t_0) = e^j$, $j = 1, \ldots, n$ [i.e., $\Phi(t_0, t_0) = E$]. Alternatively,

$$\Phi(t, t_0) = X(t)X^{-1}(t_0), \qquad t, t_0 \text{ in } I \tag{4}$$

where $X(t)$ is any fundamental matrix.

Formula (4) is a consequence of uniqueness since $\Phi(t_0, t_0) = E = X(t_0)X^{-1}(t_0)$. The transition matrix has a number of special properties.

Properties of the Transition Matrix. Let $\Phi(t, t_0)$ be the transition matrix of (2), with t, t_0 in I. Then

(a) $\Phi(t_0, t_0) = E$.

(b) $\Phi(t, t_0)B$ is a (fundamental) solution matrix of (2) for every constant $n \times n$ (invertible) matrix B.

(c) The unique solution of

$$x' = A(t)x, \qquad x(t_0) = x^0 \tag{5}$$

is $x = \Phi(t, t_0)x^0$.

(d) For all t_0, t_1, t in I, $\Phi(t, t_0) = \Phi(t, t_1)\Phi(t_1, t_0)$, and $\Phi^{-1}(t_1, t_0) = \Phi(t_0, t_1)$.

Proof. Property (a) follows from the definition of Φ, while property (b) is true for any fundamental matrix. Property (c) follows from the definition of Φ and the fact that $\Phi(t_0, t_0) = E$. To show property (d), observe that $X = \Phi(t, t_0)$ is a solution of (3) for which $X(t_1) = \Phi(t_1, t_0)$. But $\tilde{X} = \Phi(t, t_1)\Phi(t_1, t_0)$ is also a solution of (3) since $\tilde{X}$ is a fundamental matrix of (2). Since $\tilde{X}(t_1)$ also equals $\Phi(t_1, t_0)$, uniqueness of solution implies that $\Phi(t, t_0) = \Phi(t, t_1)\Phi(t_1, t_0)$. Finally, set $t = t_0$ to obtain $E = \Phi(t_0, t_0) = \Phi(t_0, t_1)\Phi(t_1, t_0)$, which implies that $\Phi^{-1}(t_1, t_0) = \Phi(t_0, t_1)$.

Example 10.14

The transition matrix for the system of Example 10.12 may be found by applying (4) to any fundamental matrix. For instance,

$$
\Phi(t, t_0) = \begin{bmatrix} e^{-t} & e^{2t} \\ -2e^{-t} & -e^{2t} \end{bmatrix} \begin{bmatrix} e^{-t_0} & e^{2t_0} \\ -2e^{-t_0} & -e^{2t_0} \end{bmatrix}^{-1}
$$

$$
= \begin{bmatrix} e^{-t} & e^{2t} \\ -2e^{-t} & -e^{2t} \end{bmatrix} \cdot \frac{1}{e^{t_0}} \cdot \begin{bmatrix} -e^{2t_0} & -e^{2t_0} \\ 2e^{-t_0} & e^{-t_0} \end{bmatrix}
$$

(6)

Since the system is autonomous, it is customary to set $t_0 = 0$ (see Section 9.3). In this case, the formula above reduces to

$$
\Phi(t, 0) = \begin{bmatrix} -e^{-t} + 2e^{2t} & -e^{-t} + e^{2t} \\ 2e^{-t} - 2e^{2t} & 2e^{-t} - e^{2t} \end{bmatrix}
$$

Example 10.15

The system

$$
x' = \frac{1}{1-t^2} \begin{bmatrix} -t & 1 \\ 1 & -t \end{bmatrix} x, \qquad x(t_0) = x^0, \quad -1 < t, t_0 < 1
$$

has a solution matrix $X(t) = \begin{bmatrix} 1 & t \\ t & 1 \end{bmatrix}$, as can be shown by direct calculation. The transition matrix $\Phi(t, t_0)$ is, then, $X(t)X^{-1}(t_0)$:

$$
\Phi(t, t_0) = \begin{bmatrix} 1 & t \\ t & 1 \end{bmatrix} \begin{bmatrix} 1 & t_0 \\ t_0 & 1 \end{bmatrix}^{-1} = \frac{1}{1-t_0^2} \begin{bmatrix} 1 - tt_0 & t - t_0 \\ t - t_0 & 1 - tt_0 \end{bmatrix}
$$

Thus the solution of the initial value problem is

$$
x = \Phi(t, t_0)x^0 = \frac{1}{1-t_0^2} \begin{bmatrix} 1 - tt_0 & t - t_0 \\ t - t_0 & 1 - tt_0 \end{bmatrix} \begin{bmatrix} x_1^0 \\ x_2^0 \end{bmatrix}
$$

It is usually a difficult process to find the transition matrix since there are no general techniques for solving (2) when the system matrix is not constant. However, we have in principle solved the first problem posed in the introduction—that is, the problem of characterizing the null space of L, the solution space of (2).

> **Solution of a Homogeneous System.** The general solution of (2) is $x = \Phi(t, t_0)c$, where t_0 is any element of I, and c is any constant vector.

Variation of Parameters

The second problem has to do with constructing a particular solution $x_p(t)$ of (1).

> **Variation of Parameters for Systems.** Let $\Phi(t, t_0)$ be the transition matrix of (2). Then the vector function
>
> $$x_p(t) = \Phi(t, t_0) \int_{t_0}^t \Phi(t_0, s)F(s)\, ds = \int_{t_0}^t \Phi(t, s)F(s)\, ds \qquad (7)$$
>
> is the unique solution of (1) for which $x_p(t_0) = 0$.

Proof. First observe that the second equality in (7) is a consequence of the product property of transition matrices: $\Phi(t, t_0)\Phi(t_0, s) = \Phi(t, s)$. To verify that (7) defines a solution, we proceed as follows.

$$x_p' = \frac{d}{dt}\left\{ \Phi(t, t_0) \int_{t_0}^t \Phi(t_0, s)F(s)\, ds \right\} = \left\{ \Phi'(t, t_0) \int_{t_0}^t \Phi(t_0, s)F(s)\, ds \right\}$$

$$+ \Phi(t, t_0)\Phi(t_0, t)F(t)$$

(by the vector version of the Fundamental Theorem of Calculus)

$$= \left\{ A(t)\Phi(t, t_0) \int_{t_0}^t \Phi(t_0, s)F(s)\, ds \right\} + \Phi(t, t)F(t)$$

$$= A(t)x_p(t) + F(t)$$

where we have used several properties of the transition matrix. The origin of the title "Variation of Parameters" is outlined in Problem 10.

The general solution of (1) and the solution of the corresponding initial value problem may now be given.

> **Solution of a Linear System.** Let $\Phi(t, t_0)$ be the transition matrix for $x' = A(t)x$. Then the general solution of $x' = A(t)x + F(t)$, t in I, is
>
> $$x(t) = \Phi(t, t_0)c + \int_{t_0}^t \Phi(t, s)F(s)\, ds \qquad (8)$$
>
> where c is any constant vector and t_0 is any point in I. The solution of the initial value problem $x' = A(t)x + F(t)$, $x(t_0) = x^0$, t_0 in I, is

$$x(t) \quad = \quad \Phi(t, t_0)x^0 \quad + \quad \int_{t_0}^{t} \Phi(t, s)F(s)\, ds \qquad (9)$$

$$\underset{\substack{total \\ response}}{} \qquad \underset{\substack{response\ to \\ initial\ data}}{} \qquad \underset{\substack{response\ to\ driving \\ force}}{}$$

Example 10.16

Add the driving force $F = \begin{bmatrix} F_1(t) \\ F_2(t) \end{bmatrix}$ to the initial value problem of Example 10.15. The solution is given by

$$x = \Phi(t, t_0)\begin{bmatrix} x_1^0 \\ x_2^0 \end{bmatrix} + \int_{t_0}^{t} \Phi(t, s)\begin{bmatrix} F_1(s) \\ F_2(s) \end{bmatrix} ds$$

where

$$\Phi(t, s) = \frac{1}{1 - s^2}\begin{bmatrix} 1 - ts & t - s \\ t - s & 1 - ts \end{bmatrix}$$

The Matrix Exponential, e^{tA}

The transition matrix has particularly simple properties if the system matrix is constant. These properties are so much like those of a scalar exponential function that we use the same notation.

> **Matrix Exponential e^{tA}.**   Let A be an $n \times n$ matrix of real constants. The transition matrix for $x' = Ax$ at $t_0 = 0$ is called the *matrix exponential* and is denoted by e^{tA} where $e^{tA} \equiv \Phi(t, 0)$.

The following result justifies this suggestive notation.

> **Properties of e^{tA}.**   Let A be an $n \times n$ matrix of real constants. Let $e^{tA} = \Phi(t, 0)$. Then
> (a) $e^{(t-t_0)A}$ is the transition matrix for $x' = Ax$ at t_0.
> (b) $e^{tA}e^{sA} = e^{(t+s)A}$,   all t and s.
> (c) $e^{tA}e^{-tA} = E$, $(e^{tA})^{-1} = e^{-tA}$,   all t.
> (d) $\dfrac{d}{dt}(e^{tA}) = e^{tA}A = Ae^{tA}$,   all t.
>
> (e) $e^{Dt} = \begin{bmatrix} e^{\lambda_1 t} & & & 0 \\ & \cdot & & \\ & & \cdot & \\ & & & \cdot \\ 0 & & & e^{\lambda_n t} \end{bmatrix}$ if $D = \begin{bmatrix} \lambda_1 & & & 0 \\ & \cdot & & \\ & & \cdot & \\ & & & \cdot \\ 0 & & & \lambda_n \end{bmatrix}$

(f) $e^{tA} = E + tA + \dfrac{t^2}{2!} A^2 + \cdots + \dfrac{t^n}{n!} A^n + \cdots$,   all t.†

(g) e^{tA} is real.

(h) The solution of $x' = Ax + F(t)$, $x(0) = x^0$, is

$$x = e^{tA} x^0 + \int_0^t e^{(t-s)A} F(s) \, ds \qquad (10)$$

Proof. Property (a) is a consequence of the autonomy of the system $x' = Ax$, which implies that $x(t - t_0)$ is a solution for all t and t_0 if $x(t)$ is a solution (see Section 9.3). Thus $\Phi(t, t_0) \equiv \Phi(t - t_0, 0)$ since both are solution matrices and both reduce to E at $t = t_0$.

Property (b) is a consequence of a general property of autonomous systems: the solution $x(t, x^0)$ of $x' = F(x)$, $x(0) = x^0$, has the property that $x(t + s, x^0) = x(t, x(s, x^0))$. Thus $e^{(t+s)A} x^0 = e^{tA}(e^{sA} x^0) = e^{tA} e^{sA} x_0$ for all vectors x^0. Hence $e^{(t+s)A} = e^{tA} e^{sA}$. Property (c) follows from (b) with $s = -t$.

To show property (d) observe that $Y(t) = e^{tA} A$ is a solution matrix for $y' = Ay$ since $Y'(t) = (e^{tA})' A = (Ae^{tA}) A = AY(t)$. Observe that $Z(t) = Ae^{tA}$ is also a solution matrix of $y' = Ay$ since $Z'(t) = A(e^{tA})' = A(Ae^{tA}) = AZ(t)$. Now since $Y(0) = Z(0)$, it follows from the Uniqueness Theorem that $Y = Z$, as asserted. Property (e) is immediate since $x' = Dx$ consists of n uncoupled scalar equations in this case. The proof of property (f) is omitted, while property (g) follows from (f). Property (h) is a restatement of (9) with $\Phi(t, 0) = e^{tA}$.

Remark. It is easy to calculate e^{tA} if A is diagonal [see property (e)], but in other cases more elaborate techniques must be used. Properties (b)–(g) all reduce to familiar properties of the scalar exponential if A is a 1×1 matrix, the scalar a. *Warning:* If the system matrix $A(t)$ is nonconstant, the transition matrix $\Phi(t, 0)$ for $x' = A(t)x$ does *not* in general have the properties listed above. In particular, $e^{tA(t)}$ as given by the series in property (f) is *not* a solution matrix of $x' = A(t)x$.

Example 10.17

Referring to Examples 10.12 to 10.14, we see that if $A = \begin{bmatrix} 5 & 3 \\ -6 & -4 \end{bmatrix}$, then

$$e^{tA} = \Phi(t, 0) = \begin{bmatrix} -e^{-t} + 2e^{2t} & -e^{-t} + e^{2t} \\ 2e^{-t} - 2e^{2t} & 2e^{-t} - e^{2t} \end{bmatrix}$$

Note that $e^{tA} \neq \begin{bmatrix} e^{5t} & e^{3t} \\ e^{-6t} & e^{-4t} \end{bmatrix}$. Thus the solution to $x' = Ax + F(t)$, $x(0) = x^0$, may be written as

† Convergence of a matrix series means the convergence of the corresponding scalar series for each component.

$$x(t) = e^{tA} \begin{bmatrix} x_1^0 \\ x_2^0 \end{bmatrix} + \int_0^t e^{(t-s)A} \begin{bmatrix} F_1(s) \\ F_2(s) \end{bmatrix} ds$$

or

$$x(t) = e^{tA} \begin{bmatrix} x_1^0 \\ x_2^0 \end{bmatrix} + e^{tA} \int_0^t e^{-sA} \begin{bmatrix} F_1(s) \\ F_2(s) \end{bmatrix} ds$$

where e^{tA} is given above.

Example 10.18

If some power A^k of A is the zero matrix, the series expansion of property (f) may be the simplest way to find e^{tA}. For example, if

$$A = \begin{bmatrix} 0 & 1 & 0 \\ 0 & 0 & 1 \\ 0 & 0 & 0 \end{bmatrix}$$

then

$$A^2 = \begin{bmatrix} 0 & 0 & 1 \\ 0 & 0 & 0 \\ 0 & 0 & 0 \end{bmatrix}$$

while A^3 is the zero matrix. Hence

$$e^{tA} = A + tA + \frac{t^2}{2!} A^2 = \begin{bmatrix} 1 & t & \dfrac{t^2}{2!} \\ 0 & 1 & t \\ 0 & 0 & 1 \end{bmatrix}$$

PROBLEMS

The following problem(s) may be more challenging: 6, 8, 9, 12, and 13.

1. Find e^{tA}, A as given. Then solve $x' = Ax$, $x(0) = [1 \quad 2]^T$.

(a) $\begin{bmatrix} 1 & 3 \\ 1 & -1 \end{bmatrix}$. (b) $\begin{bmatrix} 0 & 1 \\ -9 & 0 \end{bmatrix}$. (c) $\begin{bmatrix} -1 & 1 \\ 0 & -1 \end{bmatrix}$.

2. Find e^{tA}, A given below. Then solve $x' = Ax$, $x(0) = [1 \quad 2 \quad 3]^T$.

(a) $\begin{bmatrix} 0 & 1 & 1 \\ 0 & 0 & 1 \\ 0 & 0 & 0 \end{bmatrix}$. (b) $\begin{bmatrix} 2 & 0 & 0 \\ 0 & -3 & 0 \\ 0 & 0 & 7 \end{bmatrix}$. (c) $\begin{bmatrix} 5 & 4 & 0 \\ -1 & 0 & 0 \\ 0 & 0 & 1 \end{bmatrix}$.

3. **(a)** Show that if A is the block matrix $\begin{bmatrix} B & 0 \\ 0 & C \end{bmatrix}$, then e^{tA} is the block matrix $\begin{bmatrix} e^{tB} & 0 \\ 0 & e^{tC} \end{bmatrix}$.

   [*Hint*: The system $x' = Ax$ uncouples to two subsystems $y' = By$, $z' = Cz$.]

   (b) Use part (a) to find e^{tA} if

$$A = \begin{bmatrix} 1 & 3 & 0 \\ 1 & -1 & 0 \\ 0 & 0 & 2 \end{bmatrix}, \quad \text{and} \quad A = \begin{bmatrix} 1 & 3 & 0 & 0 \\ 1 & -1 & 0 & 0 \\ 0 & 0 & 0 & 1 \\ 0 & 0 & 0 & 0 \end{bmatrix}$$

4. Solve the following problems.

   (a) $x' = \begin{bmatrix} 0 & 2 \\ -2 & 0 \end{bmatrix} x + \begin{bmatrix} 1 \\ 0 \end{bmatrix}$, $\quad x(0) = \begin{bmatrix} a \\ b \end{bmatrix}$.

   (b) $x' = \begin{bmatrix} 2 & -1 \\ 3 & -2 \end{bmatrix} x + \begin{bmatrix} 3e^t \\ t \end{bmatrix}$, $\quad x(0) = \begin{bmatrix} 1 \\ 2 \end{bmatrix}$. $\cdot$ Sketch the component graphs.

   (c) $x' = \begin{bmatrix} 2 & -5 \\ 1 & -2 \end{bmatrix} x + \begin{bmatrix} \cos t \\ 0 \end{bmatrix}$, $\quad x(0) = \begin{bmatrix} a \\ b \end{bmatrix}$.

   (d) $x' = \begin{bmatrix} -1 & -4 \\ 1 & -1 \end{bmatrix} x + \begin{bmatrix} e^{-3t} \\ 1 \end{bmatrix}$, $\quad x(0) = \begin{bmatrix} 0 \\ 0 \end{bmatrix}$. $\cdot$ Sketch the component graphs.

   (e) $x' = \begin{bmatrix} 3 & -1 & 1 \\ 2 & 0 & 1 \\ 1 & -1 & 2 \end{bmatrix} x + \begin{bmatrix} f_1(t) \\ f_2(t) \\ f_3(t) \end{bmatrix}$, $\quad x(0) = \begin{bmatrix} a \\ b \\ c \end{bmatrix}$.

5. Sometimes a particular solution $\tilde{x}_p(t)$ of $x' = Ax + F$ can be found by a method of undetermined coefficients. Although $\tilde{x}_p(t)$ need not be the particular solution given by Variation of Parameters, it can be used in the same way in the representation of the solution set of $x' = Ax + F$. Find the solution set in each case.

   (a) $x' = \begin{bmatrix} 2 & -1 \\ 5 & -2 \end{bmatrix} x + \begin{bmatrix} e^t \\ 1 \end{bmatrix}$. [*Hint*: assume a particular solution of the form $\tilde{x}_1 = ae^t$
   $+ b$, $\tilde{x}_2 = ce^t + d$ and find the undetermined coefficients a, b, c, d by inserting x_1
   and x_2 into the differential equations and matching the coefficients of corresponding
   terms. Then add this particular solution to the general solution of the undriven system.]

   (b) $x' = \begin{bmatrix} 1 & 3 \\ 1 & -1 \end{bmatrix} x + \begin{bmatrix} \cos t \\ 2t \end{bmatrix}$. [*Hint*: Assume a particular solution of the form $x_1 =$
   $a_1 \cos t + b_1 \sin t + c_1 + d_1 t$, $x_2 = a_2 \cos t + b_2 \sin t + c_2 + d_2 t$, where the
   coefficients a_j, b_j, c_j, and d_j are constants.]

6. **(a)** Show that $e^{tA}e^{tB}$ need not be either $e^{t(A+B)}$ or $e^{tB}e^{tA}$ by calculating all three where
   $$A = \begin{bmatrix} 0 & 1 \\ 0 & 0 \end{bmatrix} \text{ and } B = \begin{bmatrix} 1 & 0 \\ 0 & 0 \end{bmatrix}.$$

   (b) Suppose that $AB = BA$. Show that $e^{t(A+B)} = e^{tA}e^{tB}$ for all t. [*Hint*: Show that if
   $P(t) = e^{t(A+B)}e^{-tA}e^{-tB}$, then $P'(t) = 0$ for all t.]

7. **(a)** Show that $X(t) = \begin{bmatrix} t^2 & t \\ 2t & 1 \end{bmatrix}$ is a fundamental solution matrix for $x_1' = x_2$, $x_2' =$
   $-2x_1/t^2 + 2x_2/t$, $t > 0$.

   (b) Find $\Phi(t, t_0)$ and verify that $\Phi(t_0, t_0) = E$, $\Phi^{-1}(t, t_0) = \Phi(t_0, t)$, $t_0 > 0$.

   (c) Show that the system in part (a) is equivalent to the Euler equation, $t^2 x'' - 2tx' + 2x = 0$.

8. **(a)** Let the $n \times n$ nonconstant matrix $A(t)$ have period T, $A(t + T) = A(t)$ for all t. Let $\Phi(t, t_0)$ be the transition matrix for $x' = A(t)x$. Show that there is a constant nonsingular matrix B such that $\Phi(t + T, t_0) = \Phi(t, t_0)B$ for all t. [*Hint*: Show that $\Phi(t + T, t_0)$ is a fundamental solution matrix by calculating

$$\frac{d}{dt} \Phi(t + T, t_0) = \frac{d}{d(t + T)} \Phi(t + T, t_0)]$$

(b) Find $\Phi(t, t_0)$, the period T, and the matrix B. Does $\Phi(t, t_0)$ have period T in t?
(i) $x'_1 = x_1 \sin t$.　　(ii) $x'_1 = a(t)x_1$,　$a(t + T) = a(t)$.
(iii) $x'_1 = x_1 \sin t$,　$x'_2 = x_1 + x_2 \sin t$.

9. **(a)** Let $X(t)$ be a fundamental solution matrix for $X' = A(t)X$. Show that $X^{-1}(t)$ satisfies the *adjoint matrix equation* $Y' = -YA(t)$.

(b) If $A(t)$ is real, $x'(t) = A(t)x(t)$ and $(z'(t))^T = -z^T(t)A(t)$, show that $\sum_1^n x_i(t)z_i(t) \equiv$ $x(t)^Tz(t)$ is constant for all t. [*Hint*: $\dfrac{d}{dt} (x^Tz) = x'^Tz + x^Tz' = 0$.]

(c) Let A be *skew symmetric* $[A = -A^T]$. Show that $x(t)^Tx(t)$ is constant for all t if $x'(t) = A(t)x(t)$; that is, show that each orbit of this differential equation lies on a "sphere" centered at the origin in state space.

(d) Illustrate part (c) by solving the system $x'_1 = (\sin t)x_2$, $x'_2 = -(\sin t)x_1$.

10. Prove the Variation of Parameters Theorem by assuming that $x' = A(t)x + F$, $x(t_0) = 0$ has a solution of the form $x = \Phi(t, t_0)c(t)$, where Φ is the transition matrix for $x' = A(t)x$ and $c(t)$ is the unknown "varied parameter" vector. [*Hint*: Insert Φc for x in the equation, use properties of Φ, and show that $c' = \Phi(t_0, t)F(t)$, from which $c(t) = \displaystyle\int_{t_0}^t \Phi(t_0, s)F(s) \, ds$ if $c(t_0) = 0$.]

11. **(a)** Let $X(t)$ be a solution matrix of $x' = A(t)x$, where the components of $A(t)$ are continuous for t in I. The *Wronskian* $W(t)$ of X is the determinant $\det [X(t)]$. Show that $W(t) = W(t_0) \exp \left[\displaystyle\int_{t_0}^t \operatorname{tr} A(s) \, ds \right]$ for t, t_0 in I. [*Hint*: Write $X(t) =$ $[r^1 \cdots r^n]^T$, where $r^1, \ldots, r^n$ are the rows of X. Then use the fact that $(\det [r^1 \cdots r^n]^T)' = \det [(r^1)' r^2 \cdots r^n]^T + \cdots + \det [r^1 \cdots r^{n-1} (r^n)']^T$.]

(b) Use part (a) to give another proof that a solution matrix is either singular for all t in I, or else it is nonsingular for all t in I.

(c) Use part (a) to show that at least one component of one solution of the system $x'_1 = (7 + t^{-1})x_1 + x_2$, $x'_2 = (\cos t)x_1 + (e^{-t})x_2$, $t \geq 1$, must become unbounded as $t \to +\infty$.

12. [*Laplace Transforms for Systems*]. The following steps outline the method of Laplace Transforms for solving

$$x' = Ax + F, \qquad x(0) = x^0 \tag{*}$$

where A is a matrix of constants. The Laplace Transform $L[x(t)]$ of a vector function is the vector of the Laplace Transform of the individual components. The definitions and results of Chapter 7 are assumed.

(a) Apply L to (*) to obtain $[sE - A]L[x] = x^0 + L[F]$.

(b) Show that the matrix function of s, $sE - A$, is invertible for all complex s of large enough magnitude. [*Hint*: $sE - A$ is singular precisely when s is an eigenvalue of A. Recall that A has a finite number of eigenvalues.]

(c) Show that the solution of (*) is $x = L^{-1}[(sE - A)^{-1}x^0] + L^{-1}[(sE - A)^{-1}L[F]]$. Compare with (9) and show that $L^{-1}[(sE - A)^{-1}] = e^{tA}$, $L^{-1}[(sE - A)^{-1}L[F]] = e^{tA} * F(t)$, the convolution of e^{tA} and $F(t)$.

(d) Use the Laplace Transform to solve the systems of Problem 1(b) and (c).

13. [*Coupled Oscillators*]. Let A be an $n \times n$ diagonal matrix with positive diagonal entries. Let B be a symmetric matrix with positive eigenvalues. If x is an n-vector, the system

$$Ax'' + Bx = 0 \qquad (*)$$

is said to be the *coupled oscillator system*, A the *mass matrix*, and B the *stiffness matrix* [see below for the reasons behind the terminology]. The second-order coupled oscillator system may be reduced to a first-order system in $2n$ state variables, $x' = y$, $y' = -A^{-1}Bx$; however, it is usually easier to work with the coupled oscillator system directly.

(a) Show that $x = e^{\lambda t}v$, where λ is a scalar and v is an n-vector, is a solution of (*) if and only if $-\lambda^2$ is an eigenvalue of $A^{-1}B$ and v is a corresponding eigenvector.

(b) If A and B have the properties listed above, then it can be shown that the eigenvalues of $A^{-1}B$ are positive. Show that the numbers λ in part (a) are pure imaginary, $\lambda = \pm i\omega$, ω positive.

(c) It can be shown that $A^{-1}B$ has n independent real eigenvectors $v^1, \ldots, v^n$, called *normal modes*, which correspond to the numbers $i\omega_1, \ldots, i\omega_n$ in part (b). The numbers $\omega_1, \ldots, \omega_n$ are the *natural frequencies*. Show that $x = (a_1 \cos \omega_1 t + b_1 \sin \omega_1 t)v^1 + \cdots + (a_n \cos \omega_n t + b_n \sin \omega_n t)v^n$, where a_i, b_i, $i = 1, \ldots, n$, are constants, is a solution of $Ax'' + Bx = 0$. All solutions have this form and may be interpreted as a superposition of coupled oscillators each with its own frequency.

(d) The system $mx_1'' + 2kx_1 - kx_2 = 0$, $2mx_2'' + 2kx_2 - kx_1 = 0$ models the displacements from rest of two bodies (of mass m and $2m$, respectively) joined by springs to one another and to fixed walls (see Problem 11 of Section 9.1). The stiffness coefficient of the springs is assumed to be k. Find the normal modes and the natural frequencies.

(e) The corresponding sinusoidally driven system is $Ax'' + Bx = w\cos \alpha t$, where w is a constant n-vector and α is a positive constant. Show that the driven system has the unique periodic solution of frequency α given by $x = [-\alpha^2 A + B]^{-1}w\cos \alpha t$ if $\alpha \neq \omega_j$, $j = 1, \ldots, n$. Such a periodic solution is called a *forced oscillation*.

10.5 DECAY, STEADY STATE, AND A LOW-PASS FILTER

A well-designed physical system should be relatively impervious to disturbances. A disturbance may drive the system away from its equilibrium state, but not far away if the disturbance is small. Once the disturbance ends, one would expect the system to return to equilibrium. In this section we formulate the mathematical equivalents of these ideas for a system with a linear, constant-coefficient model

$$x' = Ax + F(t), \qquad x(0) = x^0 \qquad (1)$$

The theory is then applied to analyze the steady-state response of an electrical circuit that acts as a low-pass filter.

Attraction and Decay

We shall begin with the undriven system

$$x' = Ax \qquad (2)$$

where A is a nonsingular $n \times n$ matrix of constants. Then the origin 0 is the unique equilibrium state. We may view $x^0 \neq 0$ as the initial disturbance, and we seek conditions on A under which the solution $x = e^{tA}x^0$ of

$$x' = Ax, \qquad x(0) = x^0 \tag{3}$$

tends to the origin 0 as $t \to \infty$ regardless of the value of x^0. If this occurs, the origin is a *global attractor*. It is easy to characterize those systems (2) for which 0 is a global attractor.

Linear Attractors.   The origin 0 is a global attractor for (2) if and only if every eigenvalue of A has a negative real part.

Proof. According to the constructions of Section 10.2, the components of the solutions of (2) are linear combinations of terms of the form $t^k e^{\beta t} \cos \gamma t$ and $t^k e^{\beta t} \sin \gamma t$, where $\lambda = \beta + i\gamma$ is an eigenvalue of A and k is a nonnegative integer smaller than the multiplicity of λ. L'Hôpital's Rule applied k times to $t^k / e^{-\beta t}$ shows that these terms tend to zero as t tends to infinity if and only if $\beta < 0$. This proves the Linear Attractors Theorem.

If the origin is a global attractor for (2), then the orbits are said to *decay* or be *transient* [equivalently, the system (2) itself is said to *decay* or be *transient*].

Example 10.19

The parameters a and b of the system $x' = ax + y$, $y' = bx - 3y$ are to be chosen so that the system decays. The system matrix $\begin{bmatrix} a & 1 \\ b & -3 \end{bmatrix}$ has eigenvalues which are roots of the characteristic polynomial $\lambda^2 + (3 - a)\lambda - 3a - b$. But the roots of the quadratic have negative real parts if and only if all the coefficients are positive. Hence we must have $a < 3$ and $b < -3a$.

The following result, stated without proof, gives more information about the decay of solutions of (3).

Decay Estimates.   Let all eigenvalues of A have real parts less than $-\alpha$, where α is a positive constant. Then there is a positive constant M such that the solution $x = x(t, x^0)$ of (3) satisfies

$$|x(t, x^0)|_{\max} \leq M e^{-\alpha t} |x^0|_{\max}$$

for all $t \geq 0$ and each x^0 in $\mathbf{R}^n$.†

The Routh Array

The difficulty with all the results above is that one must know in advance that all the eigenvalues of A have negative real parts. The following tests give information

† For x in $\mathbf{R}^n$, $|x|_{\max} = \max \{|x_j| : j = 1, 2, \ldots n\}$.

about the signs of the real parts of the roots of a polynomial (e.g., the eigenvalues as roots of the characteristic polynomial of A) without the necessity of actually finding the roots.

Coefficient Test.   Let $P(r) = r^n + a_{n-1}r^{n-1} + \cdots + a_0$, where the coefficients are real. If any coefficient of $P(r)$ is either zero or negative, at least one root has a nonnegative real part.

Proof. We shall prove the contrapositive. Suppose that all roots are negative or have negative real parts. Thus $P(r)$ is a product of factors of the form $(r + r_j)$ and $(r + \beta_k + i\gamma_k)(r + \beta_k - i\gamma_k) = (r^2 + 2\beta_k r + \beta_k^2 + \gamma_k^2)$, where r_j and β_k are positive (i.e., $-r_j$ and $-\beta_k \pm i\gamma_k$ are roots of the type assumed). Hence $P(r)$ has only positive coefficients. The contrapositive is proved, and hence so is the Coefficient Test.

Unfortunately, as the example $r^3 + r^2 + r + 1 = (r + 1)(r^2 + 1) = (r + 1)(r + i)(r - i)$ shows, the positivity of all the coefficients of a polynomial does not necessarily guarantee that all the roots have negative real parts. However, the Routh Array will handle this case, and, in fact, every polynomial. The proof is quite difficult and is omitted.

Routh Array.   Let the coefficients of $P(r) = r^n = a_{n-1}r^{n-1} + \cdots + a_0$ be real. Then all roots of $P(r)$ have negative real parts if and only if all entries in the first column of the $(n + 1)$-rowed *Routh Array* are positive:

$$\begin{bmatrix} 1 & a_{n-2} & a_{n-4} & \cdots \\ a_{n-1} & a_{n-3} & a_{n-5} & \cdots \\ b_{11} & b_{12} & b_{13} & \cdots \\ b_{21} & b_{22} & b_{23} & \cdots \\ \cdot & \cdot & \cdot & \\ \cdot & \cdot & \cdot & \\ \cdot & \cdot & \cdot & \\ b_{n-1,1} & b_{n-1,2} & b_{n-1,3} & \cdots \end{bmatrix}$$

where $a_n = 1$, $a_j = 0$ for $j < 0$, and

$$b_{1j} = \frac{a_{n-1}a_{n-2j} - a_n a_{n-2j-1}}{a_{n-1}}, \qquad b_{2j} = \frac{b_{11}a_{n-2j-1} - a_{n-1}b_{1,j+1}}{b_{11}}$$

$$b_{ij} = \frac{b_{i-1,1}b_{i-2,j+1} - b_{i-2,1}b_{i-1,j+1}}{b_{i-1,1}}, \qquad i > 2$$

The algorithm for constructing the array looks formidable, but for any particular polynomial it is easy to use. The rows are constructed from the top down and from left to right. In the first place, the Routh Array need only be used for a polynomial with positive coefficients since the Coefficient Test handles the other cases. Next, if a zero or a negative number ever appears in the left-hand column of the Routh Array during its construction, there is no point in proceeding further since the criterion has already been violated and the polynomial must have at least one root with nonnegative real part.

Example 10.20

The Coefficient Test implies that the polynomials $r^3 + 3r + 10$ and $r^3 - r^2 + 5r + 7$ each have at least one root with nonnegative real part. That test does not apply to the polynomial $r^3 + 2r^2 + 3r + 5$, but the Routh Array does apply. The Routh Array in this case is a 4-rowed matrix:

$$\begin{bmatrix} 1 & 3 & 0 \\ 2 & 5 & 0 \\ \frac{1}{2} & 0 & 0 \\ 5 & 0 & 0 \end{bmatrix} \begin{matrix} \Big\}\text{(coefficient rows)} \\ \\ \Big\}(b_{ij}\text{ rows}) \end{matrix}$$

Since the entries in the left-hand column are positive, all roots of the polynomial have negative real parts.

Driven Systems: The Steady State

According to Section 10.4, the general solution of

$$x' = Ax + F(t) \tag{4}$$

may be written in the form

$$x(t) = e^{tA}c + e^{tA} \int^t e^{-sA}F(s)\, ds \tag{5}$$

where c is any constant vector, $F(t)$ is continuous for all t, and the indefinite integral denotes any fixed antiderivative of $e^{-tA}F(t)$. If in addition the eigenvalues of A have negative real parts and $F(t)$ is periodic of period T, then system (4) has a unique periodic solution (called a *forced oscillation*) and this solution also has period T (see Problem 14 for an outline of the proof). In this case we have the following definition and theorem.

> **Steady State.**   If the eigenvalues of A have negative real parts and $F(t)$ is periodic of period T, then the forced oscillation of (4) is called the *steady-state* solution of (4). The steady-state solution has the form
>
> $$x_s(t) = e^{tA} \int^t e^{-sA}F(s)\, ds \tag{6}$$
>
> for some particular antiderivative of $e^{-tA}F(t)$.

The reason for this terminology becomes apparent in the following theorem.

Approach to Steady State.   Let system (4) have the steady-state solution $x_s(t)$. Then every solution of (4) approaches $x_s(t)$ as $t \to \infty$.

Proof. Let $x(t)$ be any solution of (4), $x_s(t)$ the steady-state solution. Then from (5), $x(t) = e^{tA}c + x_s(t)$ for some constant vector c. Hence, $\lim\limits_{t \to \infty} [x(t) - x_s(t)] = \lim\limits_{t \to \infty} e^{tA}c = 0$.

Example 10.21

The general solution of $x_1' = -x_1 + \cos t$, $x_2' = -2x_2$ is $x_1 = ae^{-t} + \frac{1}{2}(\sin t + \cos t)$, $x_2 = be^{-2t}$. The steady-state solution is obtained by discarding the terms that decay to 0 as $t \to \infty$: $x_1 = \frac{1}{2}(\sin t + \cos t)$, $x_2 = 0$.

If the input $F(t)$ is constant, even more can be said. As usual, the eigenvalues of A are assumed to have negative real parts.

Constant Steady State.   Let the input F be the constant vector F_0. Then the constant vector $x_s = -A^{-1}F_0$ is the steady-state solution of (4).

Suppose on the other hand all that is known about the input $F(t)$ is that it is continuous and bounded, i.e., for some constant M, $\|F(t)\| \le M$ for all $t \ge 0$. We have the following general result in this case.

Bounded Input–Bounded Response.   Let the input $F(t)$ be bounded for $t \ge 0$. Then every solution of (4) is bounded for $t \ge 0$.

The proof of the Constant Steady-State Theorem is immediate since $x_s = -A^{-1}F_0$ is evidently a constant solution of (4) and all solutions approach it. The Bounded Input–Bounded Response Theorem follows from the decay estimates given earlier, but the details of the proof are omitted. See Section 3.4 for a similar result in the scalar case.

A Low-Pass Filter

An electrical filter is a network of coils, capacitors, and resistors designed to allow signals in certain bands of frequencies to pass through relatively unchanged (*passbands*), while attenuating or suppressing signals in other frequency bands (*stopbands*). A *low-pass* filter passes low-frequency signals but stops those with high frequency.

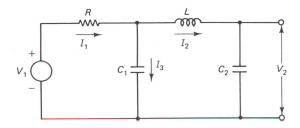

Figure 10.10   Low-pass filter.

The circuit of Figure 10.10 is a low-pass filter commonly used in communication devices. The equations of the circuit were derived in Section 9.1 and take the form (4) with

$$x = \begin{bmatrix} V_2 \\ I_1 \\ I_2 \end{bmatrix}, \qquad A = \begin{bmatrix} 0 & 0 & \dfrac{1}{C_2} \\ 0 & \dfrac{-1}{RC_1} & \dfrac{1}{RC_1} \\ \dfrac{-1}{L} & \dfrac{-R}{L} & 0 \end{bmatrix}, \qquad F = \begin{bmatrix} 0 \\ \dfrac{V_1'}{R} \\ \dfrac{V_1}{L} \end{bmatrix} \qquad (7)$$

We shall show that the undriven system $x' = Ax$ decays and that the steady-state response voltage $V_2(t)$ to an oscillating input voltage $V_1(t) = a_0 e^{i\omega t}$ of frequency ω has amplitude comparable to $|a_0|$ if $|\omega|$ is small, but amplitude close to 0 if $|\omega|$ is large.

The characteristic polynomial of A, $p(\lambda)$, is

$$p(\lambda) = -\left[\lambda^3 + \frac{1}{RC_1}\lambda^2 + \frac{1}{L}\left(\frac{1}{C_1} + \frac{1}{C_2}\right)\lambda + \frac{1}{LRC_1C_2} \right]$$

Since the circuit parameters R, L, C_1, C_2 are assumed to be positive, we may proceed directly to the Routh Array to verify that the roots of $-p(\lambda)$ [hence of $p(\lambda)$] have negative real parts. The reader may show that the four entries in the left-hand column of the Routh Array are, respectively, 1, $1/RC_1$, $1/LC_2$, and $1/LRC_1C_2$. Hence, as expected on physical grounds, the eigenvalues of A have negative real parts.

Since the input voltage $V_1 = a_0 e^{i\omega t}$ is bounded, so is the input vector,

$$F = \begin{bmatrix} 0 \\ \dfrac{V_1'}{R} \\ \dfrac{V_1}{L} \end{bmatrix} = \alpha e^{i\omega t}, \qquad \text{where } \alpha = \begin{bmatrix} 0 \\ \dfrac{a_0 i\omega}{R} \\ \dfrac{a_0}{L} \end{bmatrix}$$

Thus all responses of the circuit are also bounded and approach the steady-state response. We expect the steady-state response to have the form $x_s = \beta e^{i\omega t}$ for some constant vector β. To find β, insert x_s into (4) for this system and solve for β.

$$i\omega\beta e^{i\omega t} = A\beta e^{i\omega t} + \alpha e^{i\omega t}, \qquad \beta = [i\omega E - A]^{-1}\alpha \qquad (8)$$

The inverse matrix exists since A cannot have the pure imaginary number $i\omega$ as an eigenvalue (no eigenvalue of A has zero real part).

We are interested in the first component $V_2(t)$ of x_s. A straightforward calculation using (8) shows that

$$V_2(t) = \frac{a_0}{p(i\omega)LRC_1C_2}\, e^{i\omega t}$$

To show the filtering character of the circuit most directly, we give the constants of the circuit values that simplify the calculations: $L = \frac{4}{3}$, $R = 1$, $C_1 = \frac{1}{2}$, $C_2 = \frac{3}{2}$. Then the response voltage is

$$V_2(t) = \frac{a_0}{-i\omega^3 - 2\omega^2 + 2i\omega + 1}\, e^{i\omega t}$$

and its amplitude is

$$|V_2(t)| = \frac{|a_0|}{(\omega^6 + 1)^{1/2}}$$

Thus if $|\omega|$ is small, the amplitude of the response is comparable to the amplitude $|a_0|$ of the input voltage. If $|\omega|$ is large, the response is drastically attenuated, and the high-frequency input signal is effectively stopped. See Figure 10.11 for some representative graphs of the response voltage $V_2(t)$ versus t.

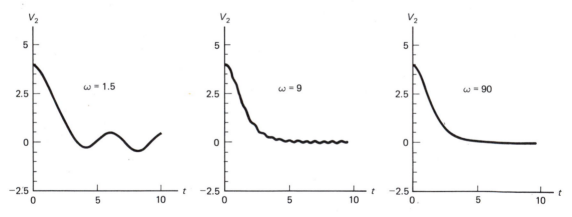

Figure 10.11   Response of a low-pass filter to the input $\sin \omega t$.

Comments

A system such as $x' = Ax$ is commonly said to be *stable*, or to have *practical stability* or *engineering stability*, if it responds to bounded input and bounded initial data in a bounded fashion. Thus, we have shown that a sufficient condition for the system to have this type of stability is that all eigenvalues of A have negative real parts. As outlined in the problem set, the condition is also necessary in the sense that if one or more eigenvalues have nonnegative real parts, there is always some bounded driving force which results in an unbounded response (Problem 6).

PROBLEMS

The following problem(s) may be more challenging: 10, 11, 12, and 14.

1. Is the origin a global attractor for $x' = Ax$ if A is as given?

(a) $\begin{bmatrix} -10 & 1 & 2 \\ -3 & 4 & 5 \\ -7 & -6 & 7 \end{bmatrix}$.   (b) $\begin{bmatrix} -10 & 1 & 2 \\ -3 & 4 & -2 \\ -7 & -6 & -40 \end{bmatrix}$.   (c) $\begin{bmatrix} -10 & 1 & 2 \\ 0 & -1 & 100 \\ 0 & 0 & -1 \end{bmatrix}$.

2. Find all values of the real constant α for which all solutions of $x' = Ax$ decay to 0 as $t \to \infty$ if A is as given.

(a) $\begin{bmatrix} \alpha & 2 \\ 3 & -4 \end{bmatrix}$.   (b) $\begin{bmatrix} 0 & 1 & 0 \\ 0 & 0 & 1 \\ \alpha & -1 & -1 \end{bmatrix}$.   (c) $\begin{bmatrix} 0 & 1 & 0 \\ 0 & 0 & 1 \\ -2 & -3 & \alpha \end{bmatrix}$.

3. The system $x_1' = x_2$, $x_2' = -bx_1 - ax_2 + f(t)$ is equivalent to the scalar equation $x'' + ax' + bx = f(t)$. Show that every solution of the system is bounded for $t \geq 0$ if $a > 0$, $b > 0$, and $f(t)$ is continuous and bounded for $t \geq 0$.

4. Show that the system $x_1' = x_2$, $x_2' = -x_1 + \cos t$ has unbounded solutions, but that this does not contradict the Bounded Input–Bounded Response Theorem.

5. Show that the system $x_1' = -x_1 + e^{3t}x_2$, $x_2' = -x_2$ has unbounded solutions for $t \geq 0$ even though the eigenvalues of the system matrix are negative constants. Why is this not a contradiction of the Bounded Input–Bounded Response Theorem?

6. (a) Let the matrix A have an eigenvalue λ with zero real part. Show that if v is an eigenvector corresponding to λ, the system $x' = Ax + e^{\lambda t}v$ has the unbounded solution $te^{\lambda t}v$, $t \geq 0$.

(b) Show that $x' = Ax$ has unbounded solutions if A has an eigenvalue with positive real part.

7. Let $P(r) = r^n + a_{n-1}r^{n-1} + \cdots + a_0$ have positive coefficients. Use the Routh Array to show that the roots have negative real parts if:

(a) $n = 2$.   (b) $n = 3$ and $a_2a_1 > a_0$.

(c) $n = 4$ and $a_3a_2a_1 > a_3^2a_0 + a_1^2$, and $a_3a_2 > a_1$.

8. (a) Apply the Routh Criterion to each of the following polynomials and decide which have all roots with negative real parts.

(i) $z^3 + z^2 + 2z + 1$.   (ii) $z^4 + z^3 + 2z^2 + 2z + 3$.   (iii) $z^6 + 10z^4 + z + 1$.

(iv) $z^6 + z^5 + 6z^4 + 5z^3 + 11z^2 + 6z + 6$.   (v) $z^5 + 6z^4 + 12z^3 + 12z^2 + 11z + 6$.

(b) Show that the roots of $z^4 + 2z^3 + 3z^2 + z + a$ have negative real parts if and only if $0 < a < 1.25$.

9. Find the steady-state solution of the following systems. Sketch component graphs of the steady-state and of several other solutions.

(a) $x_1' = x_2$, $x_2' = x_1 - 2x_2 + \cos t$.

(b) $x_1' = -x_1 + x_2 + \cos t$, $x_2' = -x_1 - x_2$.

10. Show that if the eigenvalues of A have negative real parts, all solutions of $x' = Ax + F(t)$ are unbounded as $t \to \infty$ if one of them has this property. [*Hint*: If $x^1(t)$ and $x^2(t)$ are solutions, $x^1(t)$ bounded and $x^2(t)$ unbounded, then $x^1 - x^2$ is an unbounded solution of $y' = Ay$.]

11. Explain why the circuit of Figure 10.10 acts as a low-pass filter for all positive values of L, R, C_1, and C_2. [*Hint*: Show that for $|\omega|$ small, $|V_2(t)|$ is approximately $|a_0|$, while if $|\omega|$ is large, $|V_2(t)|$ is approximately $|a_0|/\omega^3 RLC_1C_2$.]

12. Solve the low-pass filter system of the text by using Laplace Transforms. Assume that $V_2(0)$, $I_1(0)$, and $I_2(0) = 0$, while $V_1(t) = a_0 e^{i\omega t}$.

13. Show that the steady-state currents in the circuit sketched are $I_1 = 3$ A and $I_2 = 0$ if $E(t) = 3$ V. When is $I_2(t)$ at maximal amplitude?

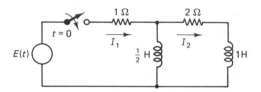

14. [*Forced Oscillations*]. A nonconstant solution $x(t)$ of period T of the system $x' = Ax + F(t)$, where A is a matrix of real constants, and $F(t)$ has period T, is a *forced oscillation* (see Problem 13 in Section 10.4). Show that the system has a unique forced oscillation if *no* solution of the homogeneous system $y' = Ay$ has period T. [*Hint*: Use (10) in Section 10.4.]

10.6 LEAD IN THE HUMAN BODY: COMPARTMENTAL MODELS

The concentration of lead in the human body may be quite high due to industrial wastes and automobile exhaust. Lead enters the body by inhalation and by eating and drinking. From the lungs and the gut lead is taken up by the blood and rapidly distributed to the liver and kidneys. It is slowly absorbed by other soft tissues and very slowly by the bones. Lead is excreted from the body primarily through the urinary system and hair, nails, and sweat (see Figure 10.12).

A mathematical model for the flow of lead through the body may be based on the conservation principle (see Section 3.1):

Net rate of change of a substance = rate in − rate out in each compartment   (1)

Let $x_i(t)$ be the amount of lead in compartment i at time t. We shall assume that the transfer rate of lead from compartment j into compartment i, $j \neq i$, is proportional

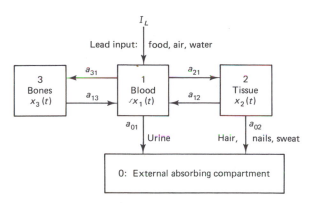

Figure 10.12 Lead ingestion, distribution, and excretion.

to x_j, while the reverse flow rate from i to j is proportional to x_i. The coefficient of proportionality from j into i is a_{ij} and from i into j is a_{ji}. The two coefficients need not be equal. A flow coefficient a_{ij} is positive unless flow is not possible from j into i, in which case $a_{ij} = 0$. With (1) and these conventions in mind, we have the following mathematical model:

$$\text{(blood)}\quad x_1' = -(a_{01} + a_{21} + a_{31})x_1 + a_{12}x_2 + a_{13}x_3 + I_L$$
$$\text{(tissue)}\quad x_2' = a_{21}x_1 - (a_{02} + a_{12})x_2 \tag{2}$$
$$\text{(bones)}\quad x_3' = a_{31}x_1 - a_{13}x_3$$

where $I_L(t)$ is the ingestion rate of lead into the bloodstream. The external absorbing compartment 0 is usually called the *environmental compartment*. Its rate equation is omitted since it does not send the substance being tracked (lead in this case) back into the system. Note that the diagonal entry in the jth column of the coefficient matrix in (2) is the negative of the sum of the other entries in the column diminished by the transfer coefficient from compartment j to the environment. This is a characteristic property of this type of model.

The coefficients a_{ij} are assumed to be constants. Given the coefficients and the ingestion rate, the techniques of this chapter may be used to determine the changing levels of lead in the parts of the body. In fact, the Routh Array Test may be applied directly to the characteristic polynomial of the coefficient matrix of (2). After the straightforward but lengthy calculations first to find the characteristic polynomial and then to construct the Routh Array, it may be seen that for all positive values of the constants a_{ij} in (2) all three roots of the characteristic polynomial have negative real parts. Thus, as expected, the eigenvalues of the coefficient matrix have negative real parts. By the results of the preceding section, all solutions of (2) approach the steady-state solution, which is bounded if the input $I_L(t)$ is bounded. None of this is surprising, but what does it have to do with the specifics of real people ingesting contaminated food and drink and breathing in air laced with industrial pollution and exhaust fumes from cars running on leaded gasoline?

A Case Study

To answer this question in a quantitative way, Rabinowitz, Wetherill, and Kopple did a careful study of the lead intake and excretion of a healthy male volunteer in an industrial urban setting (Los Angeles).† Based on the measurements of this 6-month-long study, and under the reasonable assumption that the ingestion rate I_L had reached equilibrium, all the parameters in (2) may be calculated:

$$\text{(blood)} \quad x_1' = -\frac{65}{1800}x_1 + \frac{1088}{87,500}x_2 + \frac{7}{200,000}x_3 + 49.3$$

$$\text{(tissue)} \quad x_2' = \frac{20}{1800}x_1 - \frac{20}{700}x_2 \tag{3}$$

$$\text{(bones)} \quad x_3' = \frac{7}{1800}x_1 - \frac{7}{200,000}x_3$$

where the rates x_1', x_2', x_3', and $I_L = 49.3$ are in units of micrograms (μg) of lead per day. The relatively small coefficient of x_1 in the last rate equation of (3) is a measure of the slow absorption of lead from the blood by the skeleton, while the very small coefficient of x_3 in the same equation shows how slowly lead diffuses out of the bones and back into the blood.

Since the input vector $[49.3 \quad 0 \quad 0]^T$ is a constant, the steady state is also a constant. In fact, the steady state is easily found by equating the rates in (3) to zero and solving for x_1, x_2, and x_3. The equilibrium amounts are

$$\bar{x}_1 = 1800, \qquad \bar{x}_2 = 701, \qquad \bar{x}_3 = 200,010$$

micrograms of lead in the blood, tissues, and bones, respectively. See Figure 10.13 for sketches of the buildup of lead in the three compartments. Noteworthy is the very slow rise toward equilibrium in the bones, while the lead in the blood and tissues is close to equilibrium in less than a year. The reader with access to a computer solver may observe the results of doubling the input rate from 49.3 to 98.6 μg of lead per day, or of removing all lead from the environment (change 49.3 to 0). The disturbing aspect of the latter is that although the blood and tissues rid themselves of most of the lead in a few months, the skeleton releases its lead so slowly that it would take more than a lifetime to reach insignificant levels.

The Aleutian Carbon Ecosystem

There is a somewhat similar model for the carbon cycle in the vicinity of the Aleutian Islands of Alaska. The human inhabitants of the region are at the top of a complex natural food chain, which may be partially understood by tracing the passage of carbon through its compartments. The nine compartments are portrayed in Figure

---

† Their work was reported in *Science 182* (1973), 725–727, and later extended by Batschelet, Brand, and Steiner in *J. Math. Biol. 8* (1979), 15–23.

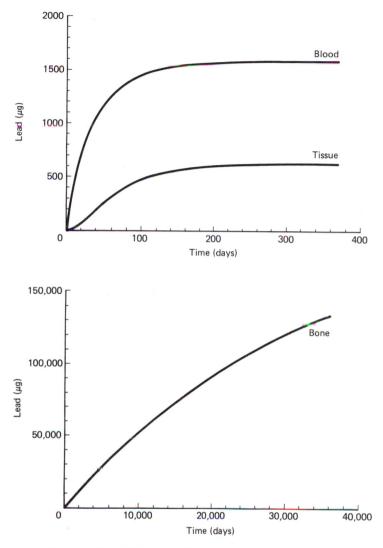

Figure 10.13   Buildup of lead in the compartmental model (3).

10.14 as nodes, while the connecting arrows indicate a transfer of carbon from one compartment to another.

The variables in the Aleutian system are the amounts $x_1(t), \ldots, x_9(t)$ of carbon in the compartments at time t. We shall assume that the transfer rate of carbon from compartment j into compartment i is proportional to x_j. The coefficient of proportionality a_{ij}, $i \neq j$, is a positive constant if transfer is possible (i.e., if there is a direct curve from j to i in Figure 10.14); otherwise, a_{ij} is zero. We shall write out two of the rate equations to illustrate the process:

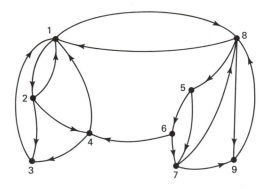

Figure 10.14 Aleutian carbon ecosystem: 1, atmosphere; 2, land plants; 3, dead land matter; 4, man; 5, phytoplankton; 6, zooplankton and marine animals; 7, dead organic marine matter; 8, surface water; 9, the deep sea.

(atmospheric carbon) $\quad x'_1 = -(a_{21} + a_{81})x_1 + a_{12}x_2 + a_{13}x_3 + a_{14}x_4 + a_{18}x_8$

(human carbon) $\qquad\quad x'_4 = a_{42}x_2 - (a_{14} + a_{34})x_4 + a_{46}x_6$ $\hspace{2cm}$ (4)

Observe that there is no external absorbing compartment in the system, nor is there a source.

Observe that in both systems (2) and (4) the "diagonal" coefficients are negative; that is, the coefficient a_{ii} of x_i in the ith rate equation is negative. For example, $a_{44} = -(a_{14} + a_{34})$ in (4), while $a_{11} = -(a_{01} + a_{21} + a_{31})$ in (2). The negativity of a_{ii} reflects the fact that compartment i is not absorbing, but transfers material to other compartments or to the absorbing compartment 0. From here on we shall merge all absorbing compartments (if there are any) into an *environmental compartment* labeled 0, but we shall not write out the rate equation for the environment. Note that we have already followed this convention in the body lead system.

Compartmental Systems

The two models suggest the following definition.

> **Compartmental System.** For $i = 0, \ldots, n$ and $j = 1, \ldots, n$, let a_{ij}, $i \neq j$, be a nonnegative constant and let $a_{jj} = -\sum\limits_{k \neq j}^{n} a_{kj}$, a nonpositive constant. Let $F(t)$ be a continuous n-vector function of t, $t \geqq 0$. Then the $n \times n$ matrix $A = [a_{ij}]$, $1 \leqq i, j \leqq n$, is a *compartment matrix*, while
>
> $$x' = Ax \hspace{2cm} (5)$$
>
> is a *homogeneous compartmental system*. The corresponding *driven compartmental system* is
>
> $$x' = Ax + F \hspace{2cm} (6)$$

The Aleutian carbon system is homogeneous; the body lead system is driven. The lead and carbon systems also suggest the next two definitions:

Open to the Environment.   Compartment i of a compartmental system is *open to the environment* if there is a sequence of compartments $j_1, \ldots, j_r$ for which $a_{j_1 i}, a_{j_2 j_1}, \ldots,$ and $a_{0 j_r}$ are positive.

Particles of the substance being tracked may thus move out of an open compartment through a chain of successive compartments and into the absorbing environment. Observe that all compartments of the body lead system are open to the environment.

Closed Systems.   A compartmental system and compartmental matrix are each said to be *closed* if the n column sums $\sum\limits_{i=1}^{n} a_{ij}$ of A, $j = 1, \ldots, n$, vanish.

Consideration of the Aleutian model shows that that system is closed.

The following two theorems, whose proofs are omitted, support the basic principles of open and closed compartmental systems.

Washout.   Suppose that compartment k of a homogeneous compartmental system (5) is open to the environment. Then $x_k(t) \to 0$ as $t \to \infty$ (i.e., the substance being tracked "washes out"). If the input rate F in the driven system (6) is a constant, $x_k(t)$ approaches a constant as $t \to \infty$.

Closed systems behave somewhat differently. First note that for a closed homogeneous compartmental system $x_1' + \cdots + x_n' = 0$ since the column sums of the coefficients all vanish. Thus $[1 \quad \cdots \quad 1]A = [0 \quad \cdots \quad 0]$, which implies that $A^T[1 \quad \cdots \quad 1]^T$ is the zero vector. Hence 0 is an eigenvalue of A^T. Since A and A^T have the same eigenvalues, 0 is also an eigenvalue of A. Suppose that v is a corresponding eigenvector. Then v defines a line of equilibrium points, quite unlike the case of a homogeneous open compartment system with its isolated equilibrium at 0.

Equilibrium.   Let $x(t)$ be a solution of a closed homogeneous compartmental system, $x(0) \geqq 0$. Then $\lim\limits_{t \to \infty} x(t) = b \geqq 0$, where b is an equilibrium point of the system and $\sum\limits_{1}^{n} b_j = \sum\limits_{1}^{n} x_j(0)$.

See Figure 10.15 for a sketch of the orbits of a closed system. Note that since the sum of the rates is zero, we must have that $x_1(t) + \cdots + x_n(t) = x_1(0) + \cdots + x_n(0)$ for all t, that is, each orbit $x = x(t)$ remains for all time on a hyperplane $S : x_1 + \cdots + x_n = $ const., tending to an equilibrium point in that hyperplane as $t \to \infty$. Thus any redistribution of the carbon in the Aleutian system away from the equilibrium levels would tend to return in time to the equilibrium levels.

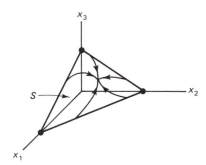

Figure 10.15 Orbits of a closed system.

PROBLEMS

The following problem(s) may be more challenging: 7.

1. (a) [*Tetracycline in the Body*]. The antibiotic tetracycline is prescribed for ailments ranging from acne to acute infections. The drug is taken orally and absorbed through the intestinal wall into the bloodstream, and eventually, is removed from the blood by the kidneys and excreted in the urine. If the compartments of the system are the intestinal tract, the bloodstream, and the urinary system, diagram the system as in Figure 10.12 and derive the compartmental model, $x_1' = -a_{21}x_1 + I_1$, $x_2' = a_{21}x_1 - a_{32}x_2$, $x_3' = a_{32}x_2$, and explain the meaning of each term [$x_i(t)$ denotes the amount of tetracycline in milligrams per cubic centimeter at time t]. Why may the urinary system be considered an absorbing or environmental compartment?

   (b) It has been experimentally shown that $a_{21} = 0.72$ per hour and $a_{32} = 0.15$ per hour. Show that the amount of tetracycline in the bloodstream reaches a maximum about 2.75 h after absorption from the intestine [assuming that $x_1(0) > 0$, but that ingestion has ceased]. Sketch the graph of $x_2 = x_2(t)$ if $x_1(0) = 0.0001$, $x_2(0) = x_3(0) = 0$.

   (c) Show that the amount in the bloodstream declines to about 20% of the maximum after about 15 h.

2. (a) [*The Tracer Inulin*]. Water molecules move back and forth between blood plasma (compartment 1) and intercellular areas (compartment 2) and from the plasma into the urinary system (compartment 3). This motion may be tracked by injecting the tracer inulin into the system. Molecules of inulin attach to molecules of water and may be followed by x-ray. Diagram the system and derive the equations $x_1' = -(a_{21} + a_{31})x_1 + a_{12}x_2 + I_1$, $x_2' = a_{21}x_1 - a_{12}x_2$, $x_3' = a_{31}x_1$.

   (b) Explain why, after the injection of inulin ceases, the amounts of inulin in the plasma and the intercellular compartment tend to 0 as $t \to \infty$.

3. Write out the rate equations for the Aleutian carbon ecosystem and show from the equations that the system is closed.

4. Solve the following compartment models. First diagram each and identify compartments open to the environment. Is the system closed? In each case let $x_1(0) = 1$, $x_i(0) = 0$, $i > 1$. Sketch the component graphs.

   (a) $x_1' = -x_1 + x_2$, $x_2' = x_1 - x_2 + x_3$, $x_3' = -x_3$.

   (b) $x_1' = -x_1$, $x_2' = x_1 - x_2$, $x_3' = x_2 - x_3$.

5. Give an example of a homogeneous compartmental system with two compartments open to the environment, two others that are not open, and such that every compartment has direct access to or from every other compartment. [*Hint*: Draw a diagram first.]

6. Explain how a system that is not closed becomes closed by including the environmental compartment.

7. [*Computer Project*]. Solve body lead system (3) for the equilibrium levels $\bar{x}_1$, $\bar{x}_2$, $\bar{x}_3$. Use the computer to show that the eigenvalues of the compartmental matrix are $\lambda_1 = -4.469 \times 10^{-2}$, $\lambda_2 = -2.000 \times 10^{-2}$, and $\lambda_3 = -3.064 \times 10^{-5}$. Show that if $x_1(0) = x_2(0) = x_3(0) = 0$, $x_1(t) \rightarrow \bar{x}_1$ and $x_2(t) \rightarrow \bar{x}_2$ within a fraction of a year, but the approach of $x_3(t)$ to $\bar{x}_3$ is very slow (measured in tens of years, i.e., a life span).

Nonlinear Systems and Stability

Queries about differential systems and their solutions fall into two categories:

(A)
Is there any solution at all?
If there are solutions, how many are there?
Is there a formula for the solutions?

(B)
Is the system stable?
What are the dominant features of the portrait of the orbits?
What are the equilibrium orbits?

It is questions of type B, and others like them, that dominate the work of this chapter and the next. There are two reasons for the move from the specific and detailed representation of solutions to the broader qualitative point of view, and from the linear to the nonlinear. First, these are the questions that arise in the design and analysis of real systems. Second, it is much harder to come by adequate representations of the solutions of general nonlinear systems since nonlinear systems have little in common except their nonlinearity. Although the systems of this chapter are termed "nonlinear," that does not exclude linear systems. A more accurate term might be "not necessarily linear" systems. For simplicity, we shall work throughout only with autonomous systems.

11.1 STABILITY, CONSERVATION, AND INTEGRALS

The behavior of the orbits of a dynamical system is strongly affected by the location and the nature of the equilibrium points and cycles, which correspond to constant and periodic solutions. Equilibrium points and their stability characteristics are studied in this chapter; cycles are taken up in Chapter 12. We begin with the central idea of the stability of a system at an equilibrium point, an idea based on what we already know of the behavior of the orbits of a constant-coefficient homogeneous linear system.

Stability

We consider the autonomous system

$$x' = F(x) \tag{1}$$

where x and F are real n-vectors and F belongs to $C^1(R)$ for a region R of $\mathbf{R}^n$. Under these conditions, the initial value problem

$$x' = F(x), \qquad x(0) = x^0 \text{ in } R \tag{2}$$

has a unique maximally extended solution $x = x(t, x^0)$. For now it is assumed that R contains the origin, and $F(0) = 0$ [i.e., the origin is an equilibrium point of (1)]. These conditions and notations are assumed throughout the chapter.

The word "stability" has virtually the same technical meaning as it has in common usage. A system is stable at an equilibrium point if it responds to small displacements from equilibrium with only small changes in its subsequent states. We have the following exact definition.

> **Stability.** System (1) is *stable* at the equilibrium point 0 if for each positive number ϵ there is a positive number δ such that $\|x(t, x^0)\| < \epsilon$ whenever $t \geqq 0$ and $\|x^0\| < \delta$.

Thus, stability means that as time increases, each orbit remains near 0 (i.e., $\|x(t, x^0)\| < \epsilon$) whenever the initial displacement (i.e., x^0) is sufficiently close to 0 (i.e., $\|x^0\| < \delta$). Given ϵ, the number δ is not unique; $a\delta$ is suitable if δ is, where $0 < a \leqq 1$. The central idea of stability is that for *every* positive ϵ there is a δ with the desired property, although it usually happens that the choice of δ may change as ϵ changes.

Stability may be *local* in the sense that the orbit defined by $x(t, x^0)$ may wander away from the neighborhood of the origin if $\|x^0\|$ is too large, or even become unbounded, while only orbits with small $\|x^0\|$ stay near 0 with advancing time. A system that is not stable at 0 is said to be *unstable* at 0. See Figures 11.1 and 11.2 for sketches of the orbits of some stable and some unstable systems.

Explicit verification of stability or instability is illustrated in the following two examples.

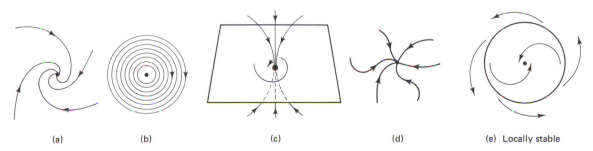

(a)   (b)   (c)   (d)   (e) Locally stable

Figure 11.1 Orbits of stable systems.

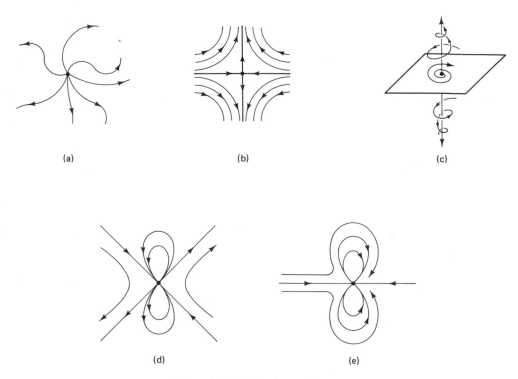

Figure 11.2 Orbits of unstable systems.

Example 11.1

The general solution of $x_1' = 4x_2$, $x_2' = -x_1$ is $x_1 = 2a \sin (2t + b)$, $x_2 = a \cos (2t + b)$, where a and b are real constants. The orbits are ellipses, $E_a : x_1^2 + 4x_2^2 = 4a^2$, whose minimal distance from the origin is a and maximal distance is $2a$. Any solution whose initial state x^0 satisfies $\|x^0\| < a$ satisfies $\|x(t)\| < 2a$ for all t. Thus the system is stable at the origin (let $a = \epsilon$ in the definition of stability and take for δ any positive number less than $\epsilon/2$). See Figure 11.3.

Example 11.2

The general solution of $x_1' = x_1$, $x_2' = -2x_2$ is $x_1 = ae^t$, $x_2 = be^{-2t}$, where a and b are real constants. The system is unstable since the solution $x = (ae^t, 0)$ becomes unbounded for every $a \neq 0$ as $t \to \infty$ (see Figure 11.4 for some orbits).

Asymptotic stability is a strong kind of stability possessed by many physical systems.

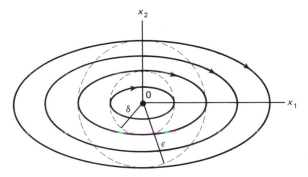

Figure 11.3 Elliptical orbits and stability.

Attractors and Asymptotic Stability.   Suppose that for some positive constant r, $\lim_{t \to \infty} x(t, x^0) = 0$ if $\|x^0\| < r$. Then the origin 0 is an *attractor* for the system (1). It is a *global attractor* if $r = \infty$. In either case, the *domain of attraction* of 0 is $\{x^0 : \lim_{t \to \infty} x(t, x^0) = 0\}$. If (1) is stable at 0 and if 0 is an attractor, then (1) is *asymptotically stable* at 0, and *globally asymptotically stable* if 0 is a global attractor.

The systems whose orbits are sketched in Figures 11.1a, c, and d are globally asymptotically stable, while that of Figure 11.1e is only locally asymptotically stable. The system of Figure 11.1b is stable but not asymptotically stable. Asymptotic stability requires stability as well as an attractor. Thus the system of Figure 11.2e is not asymptotically stable even though the origin is an attractor (see Problem 7).

Example 11.3

The general solution of $x_1' = -x_1$, $x_2' = -2x_2$ is $x_1 = ae^{-t}$, $x_2 = be^{-2t}$, where a and b are constants. The origin is stable since $x_1^2(t) + x_2^2(t) = a^2 e^{-2t} + b^2 e^{-4t} < a^2 + b^2 = x_1^2(0) + x_1^2(0)$ for all $t > 0$ (hence, for each $\epsilon > 0$, we may take $\delta = \epsilon$ in the definition of stability). The origin is a global attractor since $x_1^2(t) + x_2^2(t) \to 0$ as $t \to \infty$. Hence the system is globally asymptotically stable at the origin (Figure 11.5).

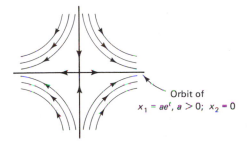

Orbit of
$x_1 = ae^t, a > 0; x_2 = 0$

Figure 11.4 Unstable system.

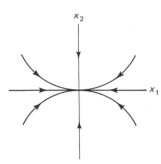

Figure 11.5 Global asymptotic stability.

The stability properties of the system of Example 11.1 are weaker than those of the system of Example 11.3. There is a term for that weak kind of stability.

> **Neutral Stability.**   System (1) is *neutrally stable* at 0 if it is stable at 0 but not asymptotically stable.

From now on the terminology will frequently be simplified or inverted if there is no danger of ambiguity. For example, the statement "system (1) is stable at the equilibrium point at the origin" may be truncated to "system (1) is stable" or inverted to "the origin is stable."

Constant-Coefficient Linear Systems and Stability

It is no accident that Examples 11.1 to 11.3 pertain to constant-coefficient homogeneous linear systems since such systems provide the simplest examples of the varieties of stability and instability. The following result is a consequence of the analysis in Sections 10.2 and 10.5.

Linear Systems and Stability.   Let A be an $n \times n$ matrix of real constants. If λ is an eigenvalue of A, let m_λ and d_λ denote, respectively, the multiplicity of λ and the dimension of its eigenspace. The system $x' = Ax$ is

(a) globally asymptotically stable (at 0) if every eigenvalue of A has negative real part;

(b) neutrally stable if every eigenvalue of A has nonpositive real part, and if there is at least one zero or pure imaginary eigenvalue, and if $m_\lambda = d_\lambda$ for every zero or pure imaginary eigenvalue;

(c) unstable if some eigenvalue of A has positive real part or if there is at least one zero or pure imaginary eigenvalue λ for which $m_\lambda > d_\lambda$.

Proof. We shall only sketch the proof. If all eigenvalues of A have negative real parts, then by Decay Theorem of Section 10.5 there are positive constants α and M such that for all $t \geqq 0$, $\|x(t, x^0)\| \leqq M\|x^0\|e^{-\alpha t}$ for every solution

$x(t, x^0)$ of $x' = Ax$. In the definition of stability, we may take $\delta = \epsilon/M$ for each $\epsilon > 0$ since $\|x(t, x^0)\| \leq M\|x^0\|e^{-\alpha t} < M(\epsilon/M)e^{-\alpha t} \leq \epsilon$ for $t \geq 0$ if $\|x^0\| < \epsilon/M$. Thus the origin is stable. The origin is also a global attractor since $e^{-\alpha t} \to 0$ as $t \to \infty$. Hence the system is globally asymptotically stable. Turning to (b) and (c) of the theorem, we observe that $x' = Ax$ is unstable if A has an eigenvalue $\lambda = \alpha + i\beta$ with positive real part α since the system has a real solution of the form $(e^{\alpha t} \cos \beta t)v$, where v is a constant vector, and $e^{\alpha t} \to \infty$ as $t \to \infty$. If $\alpha = 0$ and $m_\lambda < d_\lambda$, there are solutions with components of the form $(t^k \cos \beta t)v$; thus the system has solutions that are unbounded as $t \to \infty$. The system is unstable in this case.

Conservation and Integrals

Testing the orbital properties of a general system is not easy if the process depends on having exact formulas for the solutions of the system. There are indirect ways of determining the behavior of the orbits which do not require solution formulas. Most of these methods use a scalar function of the state variables. The values of the function are noted over a span of time, and inferences are made about the behavior of the orbits.

We have already seen one example of this indirect approach. Consider the planar system,

$$\frac{dx}{dt} = N(x, y), \qquad \frac{dy}{dt} = -M(x, y) \tag{3}$$

and the corresponding scalar equation $dy/dx = -M/N$ or

$$M(x, y) + N(x, y)\frac{dy}{dx} = 0 \tag{4}$$

According to Section 4.5, if $K(x, y)$ is an exactness function for (4), the graph of a solution $y = f(x)$ of (4) lies on the level set $K(x, y) = K(x_0, y_0)$, where $y_0 = f(x_0)$. Thus the orbit of (3) defined by $x = x(t, x_0, y_0)$, $y = y(t, x_0, y_0)$ lies on the level set $K(x, y) = K(x_0, y_0)$. Observe that $K(x(t), y(t))$ remains constant on the orbit as t varies (i.e., K is conserved along an orbit). Thus knowledge of a scalar exactness function is a considerable help in a study of orbital structure. The examples and figures of Section 4.5 illustrate what can be done along these lines for a planar system.

The idea of an exactness function may be extended to higher dimensions. The system under consideration is (1), but it is no longer required that R contain the origin nor even that (1) have an equilibrium point in R.

> **Conservation and Integrals.**   Let $K(x)$ be a continuous real-valued scalar function belonging to $C^1(R)$. K is an *integral* of (1) on R if $K(x(t, x^0))$ is constant on each solution $x = x(t, x^0)$ of (1), but K is nonconstant on every open subset of R. If an integral K exists on R, (1) is *conservative* on R.

The term "integral" is used because K is often obtained by an integration, while the word "conservation" alludes to the constancy of K on an orbit (i.e., K is "conserved"). If an integral of (1) on R exists at all, it is not unique since cK, for constant $c \neq 0$, is also an integral. On the basis of this definition and discussion, the following result is not surprising.

Integrals and Orbits.   Suppose that (1) has an integral $K(x)$ on R. Then each orbit of (1) in R lies on some level set of K, and every level set of K is a union of orbits.

Proof. Let x^0 be in R and suppose that $K(x^0) = K_0$. Then $K(x(t, x^0)) = K_0$ for all t for which $x(t, x^0)$ is defined. Hence the orbit of the solution $x = x(t, x^0)$ lies on the level set $K(x) = K_0$. Similarly, the level set must be a union of orbits.

Example 11.4

$K(x_1, x_2) = x_1^2 x_2$ is an integral for the unstable conservative system of Example 11.2 since $K(x_1(t), x_2(t)) = (ae^t)^2 (be^{-2t}) = a^2 b$. Observe that $K(x_1, x_2)$ is an exactness function for the equivalent scalar equation $2x_2 + x_1(dx_2/dx_1) = 0$ (see Figure 11.6).

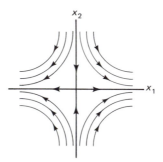

Figure 11.6   Level sets and instability.

Example 11.5

$K = x_1^2 + 4x_2^2$ is an integral for the neutrally stable system of Example 11.1 since $x_1^2 + 4x_2^2 \equiv 4a^2$ for all t along the orbit corresponding to the solution $x_1 = 2a \sin(2t + b)$, $x_2 = a \cos(2t + b)$.

Examples 11.4 and 11.5 illustrate conservative systems which are, respectively, unstable and neutrally stable. One might expect an example of an asymptotically stable conservative system, but no such system exists.

Conservation, but No Attractors.   If (1) is conservative on the region R, then R contains no attractors.

Proof. Let K be an integral of (1) on R and suppose to the contrary that R contains an attractor which we take to be the origin 0. For some $r > 0$, $x(t, x^0) \to 0$ as $t \to \infty$ if $\|x^0\| < r$. Since $K(x)$ is continuous, $K(x(t, x^0)) \to K(0)$ as $t \to \infty$. Since $K(x(t, x^0))$ remains constant as t changes, $K(0)$ must be that constant value. But this reasoning applies to every orbit whose initial point x^0 satisfies $\|x^0\| < r$. Hence $K(x)$ has the constant value $K(0)$ on the open set $\|x\| < r$ and this violates the condition of the nonconstancy of an integral on open sets. Therefore, R cannot contain an attractor.

The existence of an integral is a counter indicator for asymptotic stability, but is consistent with neutral stability or instability. The problem sets of this section and Section 11.2 give several examples of conservative systems. The final example is an instance of one common source of integrals—the total energy of a physical system.

Example 11.6

The equation of motion of the simple pendulum is $mL\theta'' + mg \sin \theta = 0$. Its total energy is $K = $ kinetic $+$ potential energy $= \frac{1}{2}m(L\theta')^2 - mgL(1 - \cos \theta)$. It is easy to show that $dK/dt = 0$ along an orbit of the equivalent system, $x_1' = x_2$, $x_2' = -\omega^2 \sin x_1$, where $\omega^2 = g/L$. Hence K is an integral since K is clearly not constant on open planar sets. See Figure 9.13 for sketches of the level sets of K. Observe that in this setting of conserved total energy there are both neutrally stable and unstable equilibrium points.

Comments

The analysis above illustrates how a scalar function may be used to obtain information about the orbits of a system. However, the integral is not always of assistance in deciding stability questions, nor is an integral easy to find, and, of course, may not even exist over R. Consequently, in the remainder of the chapter we turn generally to a different type of scalar function (the test functions of Lyapunov) to determine the stability properties of a system.

PROBLEMS

The following problem(s) may be more challenging: 6 and 7.

1. Why is each of the following systems asymptotically stable at the origin? Sketch a portrait of the orbits.

   (a) $x_1' = -x_1$.   (b) $x_1' = -x_1^3$.   (c) $x_1' = -\sin x_1$.
   (d) $x_1' = -4x_1$, $x_2' = -3x_2$.   (e) $x_1' = x_1 - 3x_2$, $x_2' = 4x_1 - 6x_2$.
   (f) $x_1' = -x_1 + 4x_2$, $x_2' = -3x_1 - 2x_2$.   (g) $x_1' = -2x_1^3$, $x_2' = -5x_2$.
   (h) $x_1' = -x_1 + x_2 + x_3$, $x_2' = -2x_2$, $x_3' = -3x_3$.   (i) $x_1' = -x_1^3$, $x_2' = -x_2^3$.

2. Show that each of the following systems is neutrally stable at the origin. Sketch a portrait of the orbits.
   (a) $x_1' = 2x_2$, $x_2' = -8x_1$.   (b) $x_1' = 0$, $x_2' = -x_2^3$.

3. The following systems are unstable at the origin. Explain why and sketch the orbits.
   (a) $x_1' = x_1^2$.   (b) $x_1' = \sin x_1$.   (c) $x_1' = |x_1|$.
   (d) $x_1' = 3x_1 - 2x_2$, $x_2' = 4x_1 - x_2$.   (e) $x_1' = 3x_1 - 2x_2$, $x_2' = 2x_1 - 2x_2$.
   (f) $x_1' = -4x_1 - 3x_2$, $x_2' = 4x_1 + 3x_2$.   (g) $x_1' = x_1^3$, $x_2' = -3x_2$.

4. Find an integral and sketch the orbits. What are the stability properties of the equilibrium points? [*Hint*: Write each as $dx_2/dx_1 = F_2/F_1$, and solve to find an integral.]
   (a) $x_1' = 3x_1$, $x_2' = -x_2$.   (b) $x_1' = -x_2$, $x_2' = 25x_1$.
   (c) $x_1' = x_2^3$, $x_2' = -x_1^3$.   (d) $x_1' = x_1$, $x_2' = x_2 + x_1^2 \cos x_1$.

5. Show that the linear system $x_1' = -x_1$, $x_2' = -2x_2$, (x_1, x_2) in $\mathbf{R}^2$, is asymptotically stable at the origin and that the function $K = x_2 x_1^{-2}$ is constant on each orbit not touching the x_2-axis. Why doesn't this contradict the theorem that a conservative system has no attractor? [*Hint*: The integral K is not continuous on any region containing the attractor at the origin. Observe that the system *is* conservative on any region disjoint from the x_2-axis. One must specify a region before determining whether a system is conservative.]

6. (a) Show that the system $x_1' = x_2$, $x_2' = -g(x_1)$ [equivalent to $x'' + g(x) = 0$] is conservative on $\mathbf{R}^2$ if g is in $C^1(\mathbf{R})$. Show that the system possesses no attractors.
   (b) [*Hard Spring*]. Find an integral and sketch the orbits of $x_1' = x_2$, $x_2' = -10x_1 - x_1^3$. What are the stability properties of the origin?
   (c) [*Soft Spring*]. Repeat part (b) for $x_1' = x_2$, $x_2' = -10x_1 + x_1^3$ and discuss the stability properties of each of the three equilibrium points.
   (d) Prove that the system $x_1' = f(x_2)$, $x_2' = g(x_1)$, where f, g belong to $C^1(\mathbf{R})$, is conservative on $\mathbf{R}^2$.

7. (a) Separate variables and solve the system, $r' = r(1 - r^2)$, $\theta' = 1 - \cos \theta$. Let $x_1 = r \cos \theta$, $x_2 = r \sin \theta$ $(r > 0)$ and show that the equivalent Cartesian system is

$$x_1' = x_1 - x_2 - x_1^3 - x_1^2 x_2 + x_1(x_1^2 + x_2^2)^{-1/2}$$
$$x_2' = x_1 + x_2 - [x_1^2 x_2 + x_2^3 + x_1^2(x_1^2 + x_2^2)^{-1/2}]$$

   (b) Show that the point $x_1 = 1$, $x_2 = 0$ is an unstable global attractor for the system in part (a). (Thus, as noted in the text, an attractor need not be stable.)

8. (a) Show that if the real $n \times n$ matrix A is *skew symmetric* (i.e., $A^T = -A$), the system $x' = Ax$ is conservative on $\mathbf{R}^n$ and neutrally stable at the origin. [*Hint*: Let $K = x^T x \equiv \|x\|^2$.]
   (b) Prove that if all nonzero eigenvalues of the skew-symmetric matrix A are pure imaginary numbers, every solution of $x' = Ax$ is a superposition of periodic solutions (not necessarily of the same period) and each solution lies on the surface of a "sphere," $\|x\|^2 = $ constant, in $\mathbf{R}^n$.

11.2 THE STABILITY TESTS OF LYAPUNOV

The stability of a physical system may be determined by observing how a scalar function of the state variables of the system changes in time. For example, the physiological condition of a person is correlated with the way in which body temperature changes. The Soviet mathematician A. M. Lyapunov abstracted from this common-place idea three mathematical tests for the determination of the stability properties of a system modeled by differential equations.† The first test relates to stability itself but does not distinguish between neutral and asymptotic stability. The second test has to do with asymptotic stability, and the third with instability. Each test is associated with a corresponding scalar test function. In the first and third cases (but not the second), the test function may be an integral of the system.

The autonomous system to be considered is

$$x' = F(x) \qquad (1)$$

with an equilibrium point at the origin $[F(0) = 0]$. As usual, F belongs to $C^1(R)$ for some region R containing 0, and the problem

$$x' = F(x), \ x(0) = x^0, \qquad x^0 \ \text{in} \ R \qquad (2)$$

has a unique maximally extended solution, $x = x(t, x^0)$.

Two definitions are needed before Lyapunov's tests are given.

> **Definiteness.** Let $V(x)$ be a real-valued scalar function belonging to $C^0(S)$ for some region S of $\mathbf{R}^n$ containing 0. Suppose that $V(0) = 0$. V is *positive (negative) definite* on S if $V(x) > 0 \ [V(x) < 0]$ for $x \neq 0$, *positive (negative) semidefinite* if $V(x) \geq 0 \ [V(x) \leq 0]$, and *indefinite* otherwise.

Example 11.7

$V = ax_1^2 + 2bx_1x_2 + cx_2^2$ is positive definite (positive semidefinite) on $\mathbf{R}^2$ if and only if $a > 0$ and $b^2 < ac$ $(a \geq 0$ and $b^2 \leq ac)$. V is negative definite or negative semidefinite if and only if $-V$ is positive definite or positive semidefinite. Otherwise V is indefinite.

In the subsequent analysis we need to calculate the rate of change of a scalar function along an orbit of (1) (indeed, we have already done so in Example 11.6). The following general definition and formula is useful in this regard.

> **Total Derivative.** Let $V(x)$ be a real-valued scalar function belonging to $C^1(R)$. The *total derivative* of V with respect to an orbit $x = x(t, x^0)$ of (1) is
>
> $$V' \equiv \frac{d}{dt} V(x(t, x^0)) = \sum_{i=1}^{n} \frac{\partial V}{\partial x_i} x_i'(t, x^0) = (\nabla V(x))^T F(x) \qquad (3)$$
>
> where ∇V denotes the gradient of V.

---

† Alexander Mikhailovich Lyapunov (1857–1918) developed his ideas on stability in his doctoral thesis in 1890. It is no exaggeration to say that Lyapunov's thesis and the work of the French mathematician Henri Poincaré (see Section 12.2) in the 1870s are the foundation of most of the contemporary developments in the theory and applications of ordinary differential equations.

The second and third equalities of (3) follow from the Chain Rule, the definition of the gradient, and the fact that $x_i' = F_i$. Note that $V'(x)$ may be calculated without knowing the orbit of (1) through the point x.

Example 11.8

Let $x_1' = x_2$, $x_2' = -x_1 - x_2$. Let $V = x_1^2 + x_2^2$. Then $V' = 2x_1x_1' + 2x_2x_2'$ $= 2x_1(x_2) + 2x_1(-x_1 - x_2) = -2x_1^2$. Observe that V is positive definite and V' is negative semidefinite on the x_1x_2-plane.

A Test for Stability

Our aim is to find a test for the stability of (1) at the equilibrium point 0 which makes use of a scalar "test" function $V(x)$ with the property given in the following definition.

> **Stability Functions.**   A real-valued scalar function $V(x)$ which belongs to $C^1(R)$ is a *stability function* for (1) at the origin 0 if V is positive definite on R, while V' is negative semidefinite on R.

Thus the function V in Example 11.8 is a stability function for the system of the example. Observe that if a stability function V exists at all for a system, it cannot be unique since cV, c a positive constant, is also a stability function.

The existence of a stability function is intimately connected with the stability of (1) as one would expect from the name.

> **Lyapunov's Test for Stability.**   System (1) is stable at 0 if and only if there is a stability function for (1) at 0.

Proof. We shall prove only the sufficiency of the test. Let V be a stability function for (1) at 0. It will be shown that for each positive number ϵ there is a positive number δ such that $\|x(t, x^0)\| < \epsilon$ for all $t \geq 0$ whenever $\|x^0\| < \delta$. Let $\epsilon > 0$ be given and suppose that B is a closed ball centered at 0 of some radius r, $r < \epsilon$, and contained in R. The stability function $V(x)$ is continuous and positive on the closed, bounded "spherical" boundary S of B. By the Minimum Value Theorem for continuous functions, $V(x)$ has a positive minimum value V_0 on S:

$$0 < V_0 \leqq V(x), \qquad \text{all } x \text{ in } S \tag{4}$$

Since $V(x)$ is continuous and has value 0 at $x = 0$, there is an open ball C of radius δ, say, which is centered at 0 and lies entirely inside B and for which

$$V(x) < \tfrac{1}{2}V_0 \qquad \text{for all } x \text{ in } C$$

See Figure 11.7 for a sketch of the sets B, S, and C in two dimensions, where balls become disks and their boundaries become circles. Let x^0 belong to C (i.e.,

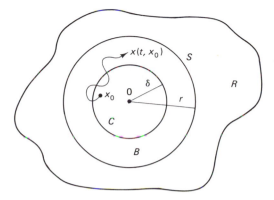

Figure 11.7 Sets of the stability test.

$\|x^0\| < \delta$). Since $V(x^0) \leqq \frac{1}{2}V_0$ and since V does not increase along an orbit as time increases, we have that

$$V(x(t, x^0)) \leqq \tfrac{1}{2}V_0 \tag{5}$$

for all $t \geqq 0$ for which $x(t, x^0)$ is defined. Suppose that the interval of definition of $x(t, x^0)$ is $0 \leqq t < T$. First we shall show that $x(t, x^0)$ lies in B, $0 \leqq t < T$ (i.e., that $\|x(t, x^0)\| < r < \epsilon$) and then we shall show that $T = +\infty$, thus proving that (1) is stable at 0.

Suppose that for some t_1, $0 < t_1 < T$, the point $x(t_1, x^0)$ lies outside the region B. Then we have for $t = 0$ and $t = t_1$ that

$$0 < \|x(0, x^0)\| = \|x^0\| < \delta \leqq r < \|x(t_1, x^0)\|$$

Since $\|x(t, x^0)\|$ is a continuous function of t, the Intermediate Value Theorem implies that there is a number $\bar{t}$ between 0 and t_1 such that

$$\|x(\bar{t}, x^0)\| = r$$

Thus at time $\bar{t}$ the orbit defined by $x(t, x^0)$ lies on the surface S of the region B. From (4) and (5) we are led to the impossible conclusion that

$$V_0 \leqq V(x(\bar{t}, x^0)) < \tfrac{1}{2}V_0 \tag{6}$$

Hence the earlier supposition that the point $x(t_1, x^0)$ is outside B is wrong and $\|x(t, x^0)\| \leqq r < \epsilon$ for $0 \leqq t < T$. Now T must be $+\infty$, for if that were not so, $\|x(t, x^0)\|$ becomes unbounded as $t \to T^-$, a consequence of the Extension Property of Section 9.3. Thus $\|x(t, x^0)\|$ exceeds r for some values of t, $0 < t < T$. This cannot be so, and the sufficiency of the Stability Test is verified.

Application of the Stability Test to (1) depends on finding a stability function. If (1) models a mechanical or other physical system for which total energy can be defined, energy is a reasonable candidate for such a function.

Example 11.9

The system $x_1' = x_2$, $x_2' = -k^2 x_1$, where $k^2 > 0$, models a unit mass oscillating without friction at the end of a spring. The total energy is $V = \frac{1}{2}x_2^2 + \frac{1}{2}k^2 x_1^2$, which is positive definite. We also have $V' = x_2 x_2' + k^2 x_1 x_1' = x_2(-k^2 x_1) + k^2 x_1 x_2 \equiv 0$, which is trivially negative semidefinite. Thus the system is stable at the equilibrium point at the origin. In this case the elliptical level sets of V [defined by $V(x_1, x_2) = $ constant] are the orbits. Hence the system is stable, but not asymptotically stable (i.e., it is neutrally stable). Observe that V is an exactness function or integral for the equation $x_2'/x_1' = dx_2/dx_1 = -k^2 x_1/x_2$ if it is written as $x_2 \, dx_2 + k^2 x_1 \, dx_1 = 0$ (see Figure 11.8).

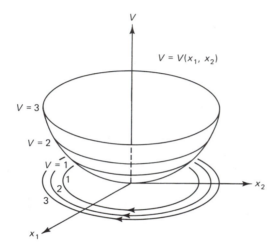

Figure 11.8 Stability function and its level sets. The system is neutrally stable and orbits lie along the level sets.

A Test for Asymptotic Stability

If a system is stable at an equilibrium point, nearby orbits remain nearby with increasing time, but they do not necessarily approach the equilibrium point. If the system is asymptotically stable, however, all nearby orbits decay with increasing time to the equilibrium point. This more stringent kind of stability is associated with a decrease in "energy" along the orbits. Following Lyapunov, we formulate the following definition and test.

> **Asymptotic Stability Function.** Let V be a continuously differentiable stability function for (1) at the origin 0. It is an *asymptotic stability function* if V' is negative definite on some open set containing 0.

Thus the total derivative of V with respect to (1), $V' = (\nabla V(x))^T F(x)$, is negative near 0 except at 0 itself where $V' = 0$. Every asymptotic stability function of (1) at 0 is also a stability function, but the converse need not be true (e.g., the V function of Example 11.8).

> **Test for Asymptotic Stability.**   System (1) is asymptotically stable at the origin 0 if and only if there is an asymptotic stability function for (1) at 0.

Thinking of an asymptotic stability function as a measure of "energy" one can imagine the negative definiteness of V' as implying that the system is losing or dissipating energy throughout some region containing 0. It is then reasonable to expect orbits in such a region gradually to "slow down" and to approach the equilibrium position at 0 where the energy is minimal. The formal proof of the asymptotic stability test is omitted. For physical systems asymptotic stability is often more significant than stability.

Example 11.10

The system $x_1' = -x_1 + x_2 + x_1(x_1^2 + x_2^2)$, $x_2' = -x_1 - x_2 + x_2(x_1^2 + x_2^2)$ has an equilibrium point at the origin 0. The positive definite function $V = x_1^2 + x_2^2$ is an asymptotic stability function at 0, but only if we restrict V to the open ball defined by $x_1^2 + x_2' < 1$. For we have that for $r^2 = x_1^2 + x_2^2$,

$$V' = 2x_1x_1' + 2x_2x_2' = -2(x_1^2 + x_2^2) + 2(x_1^2 + x_2^2)^2 = -2r^2(1 - r^2)$$

Thus V' is negative definite for $r < 1$ and positive for $r > 1$. The system is asymptotically stable at 0, and the domain of attraction is the open disk, $r < 1$. See Figure 11.9 for a sketch of the orbits. Note that $x_1^2 + x_2^2 = 1$ is an orbit that separates the domains of stability and instability.

Geometrically, the gradient vector ∇V of the asymptotic stability function and the field vector F of the differential system form an obtuse angle at each point (see Figure 11.9). The gradient vector points outward toward higher values of V, but the field vector directs the orbit inward across lower V levels toward the rest state at the origin, where V has its minimal value.

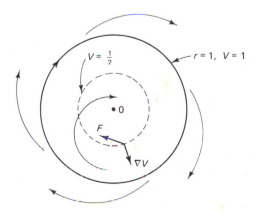

Figure 11.9 Local asymptotic stability.

Instability

Lyapunov not only found tests for stability, but he also gave a test for instability. As might be expected, the test involves a scalar "energy function," but now the energy increases along orbits, or at least along some of them.

> **Instability Function.**   Let the scalar real-valued function $V(x)$ belong to $C^1(R)$. Then V is an *instability function* for (1) at the origin 0 if $V(0) = 0$, if V is positive at at least one point in every region containing 0, and if V' is positive definite on R.

An orbit starting at a point where V is positive is carried away from the origin with increasing time since V' is positive definite, and this is the idea behind Lyapunov's third test.

> **Test for Instability.**   System (1) is unstable at the origin 0 if there is an instability function for (1) at 0.

The proof is omitted, but the next example illustrates the main point.

Example 11.11

The system to be considered is $x_1' = -x_1 - x_1^2 x_2$, $x_2' = x_2 + 2x_1^3$. The function $V = x_2^2 - x_1^2$ is positive on the region $|x_2| > |x_1|$. We have that

$$V' = 2x_2(x_2 + 2x_1^3) - 2x_1(-x_1 - x_1^2 x_2)$$
$$= 2r^2(1 + 3r^2 \cos^3 \theta \sin \theta) \geqq 2r^2(1 - 3r^2)$$

where $x_1 = r\cos \theta$, $x_2 = r\sin \theta$, and the inequality $\cos^3 \theta \sin \theta \geqq -1$ has been used. Thus V' is positive definite on the region $R : r^2 < 1/3$. The Instability Test applies, and the system is unstable at the origin.

Test Functions

The success of the Lyapunov method depends on choosing an appropriate test function V. Energy functions, integrals, and exactness functions are frequently used for this purpose. Many of these functions are *quadratic forms* in the state variables:

$$V = \sum_{i,j=1}^{n} b_{ij} x_i x_j$$

where the coefficients b_{ij} are real constants. There is no loss of generality if it is assumed that $b_{ji} = b_{ij}$ since $b_{ij}x_i x_j + b_{ji}x_j x_i = (b_{ij} + b_{ji})x_i x_j$ and we may set $\bar{b}_{ij} = \bar{b}_{ji} = \frac{1}{2}(b_{ij} + b_{ji})$ in the sum without changing the values of V. Observe that if B is the symmetric matrix $[b_{ij}]$ of coefficients, V may be written as $V = x^T B x$.

Quadratic forms have definiteness properties which are quite easy to determine.

We observed these properties in Example 11.7 for two variables while the following theorem applies to quadratic forms in any number of variables.

Quadratic Forms and Definiteness.   Let $V = x^T Bx$, where B is a real symmetric matrix. Then V is positive definite, positive semidefinite, negative semidefinite, or negative definite if and only if all eigenvalues of B are, respectively, positive, nonnegative, nonpositive, or negative. Otherwise, V is indefinite. [Note: Real symmetric matrices only have real eigenvalues.]

If one is uncertain about the stability properties of a planar system, it sometimes helps to begin with a test function of the form $V = ax_1^2 + cx_2^2$, a and c unspecified. After calculating V', it may be possible to choose values of a and c so that the resulting definiteness properties of V and V' reveal the nature of the stability. If this process is inconclusive, a term $2bx_1x_2$ may be added to V and the calculations repeated.

Example 11.12

To test the stability properties of the system $x_1' = x_2 - x_1^3$, $x_2' = -x_1 - x_2^3$ at the origin, set $V = ax_1^2 + cx_2^2$ and calculate V'. We have that $V' = 2(a - c)x_1x_2 - 2ax_1^4 - 2cx_2^4$. Let $a = c$, $c > 0$. Then V is positive definite and V' is negative definite. The system is asymptotically stable at the origin.

PROBLEMS

The following problem(s) may be more challenging: 5, 6, 7, and 8.

1. Derive the definiteness properties given in Example 11.7 for the planar quadratic form $V = ax_1^2 + 2bx_1x_2 + cx_2^2$. [Hint: V vanishes if and only if either $a \neq 0$ and $x_1 = x_2(-b \pm \sqrt{b^2 - ac})/a$, or $a = 0$ and either $x_2 = 0$ or $2bx_1 + cx_2 = 0$. Show that V is definite if and only if $b^2 < ac$.]

2. Test the stability properties of each system at the origin by using a function of the form $V = ax_1^2 + cx_2^2$, adding the mixed term $2bx_1x_2$ only if necessary. Distinguish between neutral and asymptotic stability if possible.
   (a) $x_1' = -4x_2 - x_1^3$, $x_2' = 3x_1 - x_2^3$.   (b) $x_1' = x_2$, $x_2' = -9x_1$.
   (c) $x_1' = -x_1 + 3x_2$, $x_2' = -3x_1 - x_2$.   (d) $x_1' = 2x_1 + 2x_2$, $x_2' = 5x_1 + x_2$.
   (e) $x_1' = (-x_1 + x_2)(x_1^2 + x_2^2)$, $x_2' = -(x_1 + x_2)(x_1^2 + x_2^2)$.
   (f) $x_1' = -2x_1 - x_1e^{x_1x_2}$, $x_2' = -x_2 - x_2e^{x_1x_2}$.
   (g) $x_1' = -x_1^3 + x_1^3x_2 - x_1^5$, $x_2' = x_2 + x_2^3 + x_1^4$.

3. Sketch level sets of the function $V = 3x_1^2 + 4x_2^2$. Sketch orbits of the system in Problem 2(a) and note that orbits cut across level sets and tend to the origin.

4. Use $V = ax_1^{2m} + bx_2^{2n}$, m and n positive integers, to determine the stability properties of the following systems at the origin.
   (a) $x_1' = -2x_2^3$, $x_2' = 2x_1 - x_2^3$.
   (b) $x_1' = -x_1^3 + x_2^3$, $x_2' = -x_1^3 - x_2^3$.

5. [*Gradient Systems*]. Let $W(x)$ be a real-valued scalar function belonging to $C^2(\mathbf{R}^n)$. Then $x' = -\nabla W(x)$ is said to be a *gradient system*.

   (a) Show that a gradient system is (asymptotically) stable at the origin if W has an (isolated) local minimum value at the origin. [*Hint*: Let $V = W(x) - W(0)$ be the test function. Then

   $$V' = \sum_{i=1}^{n} \frac{\partial W}{\partial x_i} x_i' = -\|\nabla W(x)\|^2]$$

   (b) Show that the planar system $x_1' = F_1$, $x_2' = F_2$ is a gradient system if F_1 and F_2 belong to $C^1(\mathbf{R}^2)$ and $\partial F_1/\partial x_2 = \partial F_2/\partial x_1$. [*Hint*: Recall from multivariable calculus that the latter condition implies the existence of a scalar function V such that $\partial V/\partial x_i = F_i$, $i = 1, 2$.]

   (c) Show that $x_1' = -4x_1^3 - x_2\sin(2x_1x_2)$, $x_2' = -2x_2 - x_1\sin(2x_1x_2)$ is a gradient system which is asymptotically stable at the origin.

   (d) Give an example of an unstable gradient system.

6. (a) Let $x^T F(x)$ be negative semidefinite (definite) on $\mathbf{R}^n$ and let $F(0) = 0$. Show that $x' = F(x)$ is (asymptotically) stable at 0. [*Hint*: Let $V = x^T x$.]

   (b) What is the geometric interpretation of the condition $x^T F(x) \leqq 0$?

7. [*Volterra's Predator–Prey Model*]. The system $x' = (a - by)x$, $y' = (-c + dx)y$, where the coefficients are positive, models the interactions of a predator–prey community (see Section 9.5). Show that the system is neutrally stable at the equilibrium point $(c/d, a/b)$ by using the integral $W = (y^a e^{-by})(x^c e^{-dx})$. [*Hint*: set $V = W(c/d, a/b) - W(x, y)$.]

8. Parts (a)–(d) below refer to the *Hamiltonian system*

   $$x_i' = \frac{\partial H}{\partial y_i}, \qquad y_i' = -\frac{\partial H}{\partial x_i}, \qquad i = 1, \ldots, k \qquad (*)$$

   where the *Hamiltonian function* H is a real-valued scalar function belonging to $C^2(R)$, R is a subset of $\mathbf{R}^{2k}$, and H is nonconstant on every open subset of R.

   (a) Show that H is an integral of (*).

   (b) Show that Hamiltonian systems cannot be asymptotically stable. [*Hint*: By a result in Section 11.1, no system with an integral has an attractor.]

   (c) Show that the system $x_1' = x_2$, $x_2' = -g(x_1)$ (equivalent to $x'' + g(x) = 0$, the generalized undamped mass–spring or simple pendulum equation) has no asymptotically stable equilibrium points. [*Hint*: The system has Hamiltonian $H = \frac{1}{2}x_2^2 + \int_0^{x_1} g(s)\,ds$, the "total energy" of the system.]

   (d) What does the result in part (c) imply about the possibility of constructing a "restoring force" $-g(x)$ such that a physical system modeled by $x'' = -g(x)$ tends toward equilibrium for any initial position and velocity.

9. Show that the system $x_1' = (x_1 - x_2)^2(-x_1 + x_2)$, $x_2' = (x_1 - x_2)^2(-x_1 - x_2)$ is neutrally stable at the origin. [*Hint*: Use $V = x_1^2 + x_2^2$. Note the line of equilibrium points.]

11.3 STABILITY AND THE LINEAR APPROXIMATION

The Lyapunov tests provide ways to settle stability questions for a differential system without first having to find the solutions. This is a distinct advantage both in the analysis of a given physical system modeled by differential equations and in the design of a dynamical system to operate in or near the equilibrium state. In this section we show how the Lyapunov method may be used to treat a system whose rate functions consist of an asymptotically stable linear part plus a "higher-order" nonlinear perturbation.

The Linear Approximation to a System

Suppose that a physical system with an equilibrium state can be modeled by the differential system in n state variables,

$$x' = F(x) \tag{1}$$

where F belongs to $C^2(R)$, R a region of $\mathbf{R}^n$ containing the origin 0, and $F(0) = 0$. There is no way to determine the stability of (1) at 0 unless we have more information about F. A rule frequently followed in engineering and applied analysis comes to mind: When in doubt, linearize.† In the current context, this suggests expanding $F(x)$ by the vector version of Taylor's Theorem and looking at the linear terms. We have

$$x' = F(0) + \frac{\partial F}{\partial x}\bigg|_{x=0} x + P(x) = Ax + P(x) \tag{2}$$

where $F(0) = 0$, $\partial F / \partial x|_{x=0} \equiv A$ is the real $n \times n$ Jacobian matrix of the first-order partial derivatives $\partial F_i / \partial x_j$ evaluated at $x = 0$, and $P(x)$ is the remainder term in the Taylor expansion of $F(x)$. Since F is twice continuously differentiable, $P(x)$ is *second order* in x; that is, there is a constant c such that

$$\|P(x)\| \leqq c\|x\|^2 \tag{3}$$

for all x in some region $\tilde{R}$ which contains the origin.

$P(x)$ may be considered to be a higher-order disturbance of the *linear system of first approximation*

$$x' = Ax \tag{4}$$

System (4) is asymptotically stable at the origin if and only if every eigenvalue of A has a negative real part, which we assume to be the case from this point on. The domain of attraction of the origin for the asymptotically stable linearized system (4) is the entire state space $\mathbf{R}^n$.

There are two questions here: First, can one construct a "simple" Lyapunov asymptotic stability function for (4), and second, can one use this function to show that the original system (1) is asymptotically stable at 0?

---

† We have used this rule implicitly several times. For example, in the study of the simple pendulum equation, $mL\theta'' + gL \sin \theta = 0$, we replaced $\sin \theta$ by θ, the first term in its Maclaurin expansion, and obtained the more tractable linear equation of the harmonic oscillator. Of course, ignoring all but linear terms means that some information is altered in content or even lost. In this example of the "linearization process" all nontrivial solutions of the harmonic oscillator equation are bounded and periodic with a common period, whereas this is definitely not the case for the simple pendulum equation. In fact, some solutions of the latter equation are unbounded and the periodic solutions do not share a common period. Care is thus needed in ascribing properties of a linearized system to the original nonlinear system.

Asymptotic Stability in the First Approximation

The asymptotic stability of (4) at the origin carries over to the original system (2) although, as later examples show, the domain of attraction may be restricted. The following theorem also shows how to construct a Lyapunov asymptotic stability function for (4) which may be used for (2) as well.

Asymptotic Stability in the First Approximation. Suppose that A is a real $n \times n$ matrix whose eigenvalues have negative real parts. Let $P(x)$ be a vector function satisfying (3) for all x in some open set R containing 0. Then (2) is asymptotically stable at 0, and the function $V = x^T B x$, where $B = \int_0^\infty e^{A^T t} e^{At} \, dt$, is a Lyapunov asymptotic stability function for systems (2) and (4) at 0.

Proof. First it must be shown that the improper integral defining B converges. This is a consequence of the fact that every entry in the matrix e^{At} (hence also in the product matrix $e^{A^T t} e^{At}$) is a linear combination of terms of the form $p(t) e^{-at} k(t)$, where $p(t)$ is a polynomial, a is a negative real number, and $k(t)$ is $\sin bt$ or $\cos bt$ for some real constant b. The antiderivatives of such terms have the same form, and by L'Hôpital's Rule these antiderivatives tend to 0 as $t \to +\infty$. Hence the matrix B is obtained by evaluating at $t = 0$ the antiderivatives of the entries of the matrix $e^{A^T t} e^{At}$.

The matrix B has two other properties to be used in the proof below. First, $V(x) = x^T B x$ is positive definite on $\mathbf{R}^n$. For we have that

$$x^T B x = \int_0^\infty x^T e^{A^T t} e^{At} x \, dt = \int_0^\infty y^T(t) \, y(t) \, dt = \int_0^\infty \|y(t)\|^2 \, dt$$

where $y(t) = e^{At} x$. Observe that $y(t)$ is the zero vector if and only if $x = 0$ since the matrix e^{At} is nonsingular. Second, $A^T B + BA$ is the matrix $-E$ since

$$A^T B + BA = \int_0^\infty (A^T e^{A^T t} e^{At} + e^{A^T t} e^{At} A) \, dt = \int_0^\infty (e^{A^T t} e^{At})' \, dt$$

$$= [e^{A^T t} e^{At}]_{t=0}^{t=\infty} = -E \tag{5}$$

where we have used the previously noted fact that the entries in the matrix $e^{A^T t} e^{At}$ all vanish at $t = \infty$, while $e^Z = E$, Z the $n \times n$ zero matrix. B is also symmetric, but we leave the proof of that fact to the problem set.

The properties of B may now be used to prove the Asymptotic Stability Theorem. Calculating the total derivative of $V(x)$ along the orbits of system (2), we have that

$$V'(x) = (x')^T Bx + x^T Bx' = (x^T A^T + P^T)Bx + x^T B(Ax + P)$$
$$= x^T(A^T B + BA)x + P^T Bx + x^T BP = x^T(-E)x + g(x) \tag{6}$$
$$= -\|x\|^2 + g(x)$$

where $g(x) = P^T Bx + x^T BP$. We may use (3) to estimate the quantity $|g(x)|$:

$$|g(x)| = |P^T Bx + x^T BP| \leqq |P^T Bx| + |x^T BP| = 2|P^T Bx| \tag{7}$$
$$\leqq 2\|P\| \cdot \|Bx\| \leqq 2c\|x\|^2 \|Bx\| \leqq k\|x\|^3$$

where k is a positive constant.† From (6) and (7) we have that

$$V'(x) \leqq -\|x\|^2[1 - k\|x\|] \tag{8}$$

If x is restricted to the open subset S of R inside the "sphere" defined by $\|x\| = 1/k$, (8) implies that V' is negative definite on that subset. $V(x)$ is an asymptotic stability function in S for system (2). It is also an asymptotic stability function in all of $\mathbf{R}^n$ for system (4) since the restriction $\|x\| < 1/k$ is no longer needed for that system. The theorem is proved.

The domain of attraction of the origin for system (4) is $\mathbf{R}^n$, but that may not be true for the perturbed system (2). The linear terms Ax dominate near the origin (hence both systems are asymptotically stable at the origin), but far away from the origin the nonlinear terms $P(x)$ become significant and may act as destabilizers. In fact, the domain of attraction of the origin may be quite limited, as the following example shows.

Example 11.13

The scalar equation $x' = -x + ax^3$, where a is a positive constant, has three equilibrium points, $x = 0, \pm a^{-1/2}$. Although the domain of attraction of the origin for the linearized system, $x' = -x$, is the entire real line, that for the original equation is only the interval $(-a^{-1/2}, a^{-1/2})$.

The theorem above provides a technique for constructing a Lyapunov asymptotic stability function $V = x^T Bx$ from the linear part of the rate function if the linear system matrix A has eigenvalues with negative real parts. However, the method requires the calculation of e^{At}, $e^{A^T t}[= (e^{At})^T]$, and $e^{A^T t}e^{At}$ followed by a subsequent antidifferentiation. It may be easier to calculate B directly from the equation $A^T B + BA = -E$. It is known that if the eigenvalues of A have negative real parts, this matrix equation has a unique matrix solution B, which must be the matrix given by the integral in the theorem. This result is called *Lyapunov's Lemma*. The following example shows how the matrix B can be found.

---

† The triangle inequality $|a + b| \leqq |a| + |b|$ has been used, and also the Cauchy–Schwarz inequality for vectors v, w: $|v^T w| \leqq \|v\| \cdot \|w\|$. The Cauchy–Schwarz inequality extends to a vector-matrix product, $\|Bv\| \leq b\|v\|$, where $b = \sum_{i,j} |b_{ij}|$ is the *matrix norm* of $B = [b_{ij}]$.

Example 11.14

The system $x'_1 = x_2$, $x'_2 = -bx_1 - ax_2 + p(x_1, x_2)$ is a system equivalent to the scalar equation $x'' + ax' + bx = p(x, x')$. Suppose that a and b are positive constants and $p(x, x')$ is of "higher order," $|p(x, x')| \le c(x^2 + (x')^2)$ for some constant c for all points (x, x') for which $x^2 + (x')^2 \le 1$, say. The equation $A^T B + BA = -E$ becomes

$$A^T B + BA = \begin{bmatrix} 0 & -b \\ 1 & -a \end{bmatrix} \begin{bmatrix} b_{11} & b_{12} \\ b_{12} & b_{22} \end{bmatrix} + \begin{bmatrix} b_{11} & b_{12} \\ b_{12} & b_{22} \end{bmatrix} \begin{bmatrix} 0 & 1 \\ -b & -a \end{bmatrix} = \begin{bmatrix} -1 & 0 \\ 0 & -1 \end{bmatrix}$$

for the symmetric matrix $B = [b_{ij}]$. We have the following linear equations for the three unknown entries of B:

$$-bb_{12} - bb_{12} = -1, \qquad -bb_{22} + b_{11} - ab_{12} = 0, \qquad 2b_{12} - 2ab_{22} = -1$$

The unique solutions are

$$b_{11} = \frac{1}{2ab}(a^2 + b^2 + b), \qquad b_{12} = \frac{1}{2b}, \qquad b_{22} = \frac{1}{2ab}(b + 1)$$

Hence

$$V = x^T Bx = \frac{1}{2ab}(a^2 + b^2 + b)x_1^2 + \frac{1}{b}x_1 x_2 + \frac{1}{2ab}(b + 1)x_2^2$$

is a Lyapunov asymptotic stability function both for the system and for its linear approximation, $x'_1 = x_2$, $x'_2 = -bx_1 - ax_2$.

Perturbed Planar Systems

Consider the planar system

$$x' = ax + by + P(x, y), \qquad y' = cx + dy + Q(x, y) \tag{9}$$

and its linearization,

$$x' = ax + by, \qquad y' = cx + dy, \qquad ad - bc \ne 0 \tag{10}$$

where P and Q are at least second order in (x, y) in a neighborhood of the origin. The previous theorem shows that if (10) is asymptotically stable at the origin, so is (9). In fact, much more is true. Generally speaking, the orbits of (9) near the origin resemble those of (10). For example, if the orbits of (9) spiral onto a focal point at the origin, so do the orbits of (10), at least those that start out close enough to the origin. There are two cases where the disturbance terms alter the configuration of the orbits in a substantial way. If the origin is a star or proper node of (10), it may become an improper node or even a focal point for (9). Second, if the origin is a neutrally stable center for (10), the disturbance terms may result in an unstable or a stable focus for (9). Table 11.1 (which follows the problem set) summarizes all the possibilities, but the proofs are not at all simple and are omitted. [See Section 10.3 for the full analysis of (10).]

PROBLEMS

The following problem(s) may be more challenging: 4, 9, and 10.

1. Use the method of Example 11.14 to find a Lyapunov asymptotic stability function for each system at the origin.
   (a) $x_1' = -x_1 - x_2$, $x_2' = x_1 - x_2$.
   (b) $x_1' = x_1 - 3x_2$, $x_2' = 4x_1 - 6x_2$.
   (c) $x_1' = -3x_2$, $x_2' = 2x_1 - 5x_2 + x_1^3 x_2$.
   (d) $x_1' = -7x_1 + x_1^3 x_2$, $x_2' = -x_2 + x_1^{10}$.

2. Determine the stability properties of each system at the origin.
   (a) $x_1' = -x_1 + x_2^2$, $x_2' = -8x_2 + x_1^2$.
   (b) $x_1' = 2x_1 + x_2 + x_1^4$, $x_2' = x_1 - 2x_2 + x_1^3 x_2$.
   (c) $x_1' = 2x_1 - x_2 + x_1^2 - x_2^2$, $x_2' = x_1 - x_2$.
   (d) $x_1' = e^{x_1 + x_2} - \cos(x_1 - x_2)$, $x_2' = \sin x_1$.

3. Find the equilibrium points and discuss the stability of the system at each point. [*Hint*: If x^0 is an equilibrium point of $x' = F(x)$, calculate the eigenvalues of the Jacobian matrix $[\partial F / \partial x]_{x = x^0}$ to determine the character of stability at x^0.]
   (a) $x_1' = x_2$, $x_2' = -6x_1 - x_2 - 3x_1^2$.
   (b) $x_1' = x_2^2 - x_1$, $x_2' = x_1^2 - x_2$.
   (c) $x_1' = -x_1 - x_1^3$, $x_2' = x_2 + x_1^2 + x_2^2$.
   (d) $x_1' = -x_2 - x_1(x_1^2 + x_2^2)$, $x_2' = x_1 - x_2(x_1^2 + x_2^2)$.
   (e) $x_1' = -x_1 + x_2^2$, $x_2' = x_1 + x_2$.
   (f) $x_1' = x_1 + x_1 x_2^2$, $x_2' = x_1$.

4. Sketch the general appearance of the orbits of systems in Problem 3(a)–(e) in the neighborhood of each equilibrium point. [*Hint*: Find the linear approximation at the point.]

5. Show that the linear approximation at the origin to the system $x_1' = -x_1$, $x_2' = ax_2^3$ has eigenvalues -1 and 0. Show that the system is asymptotically stable at the origin if a is a negative constant, unstable if a is a positive constant. What if $a = 0$? Sketch the orbits in all three cases. Why does the theorem proved in this section not apply?

6. Show that the equilibrium points of the system of the damped pendulum, $x_1' = x_2$, $x_2' = -ax_2 - b\sin x_1$, a and b positive constants, are alternating stable foci and unstable saddle points (see Figure 9.19).

7. Show that the system $x_1' = x_2$, $x_2' = -2x_1 - x_2 - 3x_1^2$ is asymptotically stable at the origin, but is not globally asymptotically stable.

8. Show that the matrix $B = \int_0^\infty e^{A^T t} e^{At}\, dt$ is symmetric if the improper integral converges.

9. The equation of motion of a damped simple pendulum with constant external torque is $x'' + ax' + b\sin x = L$, where a, b, and L are positive constants. If $0 \le L < b$, give a complete analysis of the character of stability of every equilibrium point of the equivalent system, $x_1' = x_2$, $x_2' = -b\sin x_1 - ax_2 + L$.

10. Find functions $P_i(x_1, x_2)$, $i = 1, 2$, each of the form $(x_1^2 + x_2^2)z_i$, where z_i is 0, $\pm x_1$, or $\pm x_2$, such that the system $x_1' = -x_2 + P_1$, $x_2' = x_1 + P_2$ is
    (a) Globally asymptotically stable.
    (b) Unstable.
    (c) Neutrally stable.
    (d) The perturbation terms P_i may affect the stability properties of the linearized system. Why is this not a contradiction to the main theorem of the section?

TABLE 11.1 EFFECTS OF DISTURBANCE ON STABILITY[a]

$p \equiv a + d$ $q \equiv (a-d)^2 + 4bc$ $\Delta \equiv ad - bc \neq 0$	Linear system $x' = ax + by$ $y' = cx + dy$	Perturbed system[b] $x' = ax + by + P(x, y)$ $y' = cx + dy + Q(x, y)$
1. $p < 0, q < 0$	AS, focus	AS, focus
2. $p < 0, q = 0$	AS, star or proper node	AS, node or focus
3. $p < 0, q > 0$, and $\Delta > 0$	AS, improper node	AS, improper node
4. $p = 0, q < 0$	NS, center	Stability not known

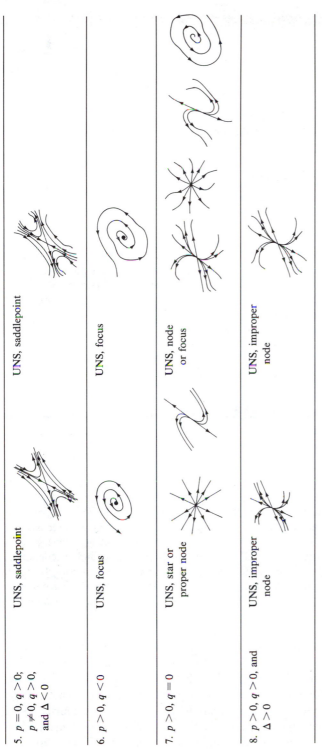

5. $p = 0$, $q > 0$;   UNS, saddlepoint   UNS, saddlepoint
$p \neq 0$, $q > 0$,
and $\Delta < 0$

6. $p > 0$, $q < 0$ UNS, focus   UNS, focus

7. $p > 0$, $q = 0$ UNS, star or   UNS, node
proper node   or focus

8. $p > 0$, $q > 0$, and   UNS, improper   UNS, improper
$\Delta > 0$ node   node

[a] AS, asymptotically stable; NS, neutrally stable; UNS, unstable.
[b] P, Q at least of second order near origin.

CHAPTER 12

Cycles
and Bifurcations

The simplest equilibrium states of a dynamical system are its equilibrium points. These points correspond to static behavior, the system at rest. In this chapter another type of equilibrium state is studied. These are the cycles, corresponding to periodic solutions of the dynamical equations, the system not at rest, but endlessly turning in the same simple closed paths. Of special importance are the attracting cycles, cycles with the property that nearby orbits wind down onto the cycles in asymptotic time. Most of the chapter is devoted to the study of cycles and the physical systems possessing cyclical behavior.

12.1 THE VAN DER POL CYCLE

The aeolian harp, a pneumatic hammer, the scratching noise of a knife on a plate, the waving of a flag in the wind, the humming noise sometimes made by a water-tap, the squeaking of a door, the tetrode multivibrator . . . , the intermittent discharge of a condensor through a neon tube, the periodic reoccurrence of epidemics and of economical crises, the periodic density of an even number of species of animals living together and the one species serving as food for the other, the sleeping of flowers, the periodic recurrence of showers behind a depression, the shivering from cold, menstruation, and finally, the beating of a heart.

In this quotation, Balthazar van der Pol [*Phil. Mag.*, 6 (ser. 7) (1928), 763–775] has grouped the most varied phenomena into a single category of systems that exhibit periodic oscillations, or cycles, even though there are no external periodic forces causing such oscillations. (One wonders, however, about some of van der Pol's systems; for example, the sleeping of flowers is surely a response to the diurnal cycle of darkness and light.) According to van der Pol, most of these systems can be modeled more or less accurately by a pair of differential equations of the form

$$x' = f(x, y)$$
$$y' = g(x, y)$$
(1)

The symbols x and y denote the magnitudes of the principal quantities being modeled. For example, x might denote the voltage across the condensor of a neon tube and y the rate at which that voltage changes. Or x and y might denote the population densities of a predator and its prey. For the present, however, let us simply assume that the differential system (1) is given and that we are interested in any cycles that may be present in the orbit set.

Limit Cycles

The cycles of system (1), if there are any at all, may form a continuous band filling up the state plane (except for an equilibrium point at the center), or there may be a spectrum of disjoint bands separated by noncyclical spiral orbits, or there may be several "eyes" surrounded by bands, or the cycles may be solitary and isolated from one another, or the portrait of the orbits may include combinations of bands of cycles and solitary cycles. By an "isolated" cycle we mean a cycle that has an annular neighborhood containing no other cycles. Figure 12.1 illustrates some possibilities.

Whatever the overall portrait in state space may be, each cycle has the following two distinctive properties:

1. It divides the plane into a bounded interior region and an unbounded exterior.
2. There is at least one equilibrium point in the interior region.

The first property seems obvious but is difficult to prove, while the second is not obvious but is easier to show. We shall use both, but prove neither.†

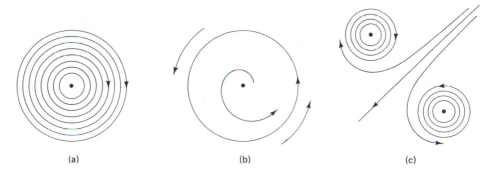

(a)   (b)   (c)

Figure 12.1   Planar cycles: (a) band of cycles; (b) isolated cycle; (c) two bands and eyes.

† Property (a) is a special case of the Jordan curve theorem. See *Elements of the Topology of Plane Sets of Points*, (2nd ed. Cambridge: Cambridge University Press, 1951), M. H. A. Newman, for a proof. A proof of property (b) may be found in M. W. Hirsch and S. Smale, *Differential Equations, Dynamical Systems, and Linear Algebra* (New York: Academic Press, 1974).

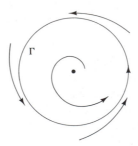

Figure 12.2   Attracting limit cycle.

Although bands of cycles appear in several important models (e.g., the linearized harmonic oscillator, the simple pendulum, Volterra's predator-prey model), they are not of interest to us here. Small disturbances of such a band may drastically alter the behavior of the orbits and destabilize what had previously been a stable system. It is the isolated attracting cycles that are likely to persist.

> **Limit Cycle.**   Let Γ be a cycle of (1). Γ is a *limit cycle* if there is a point Q not on Γ such that the orbit of the solution through Q approaches Γ either as $t \rightarrow +\infty$ or as $t \rightarrow -\infty$.

> **Attracting Limit Cycle.**   Let Γ be a limit cycle of (1). Γ is *attracting* if there is an annular neighborhood of Γ with the property that any orbit which penetrates that neighborhood approaches Γ as $t \rightarrow +\infty$.

Repelling limit cycles have a similar definition, but nearby orbits tend to the limit cycles as $t \rightarrow -\infty$. See Figure 12.2 for a sketch of an attracting limit cycle and its domain of attraction. Examples 12.1 to 12.3 show that limit cycles may arise in relatively simple nonlinear equations.

Example 12.1

The system $x' = x - y - x(x^2 + y^2)$, $y' = x + y - y(x^2 + y^2)$ has a single equilibrium point at the origin and a unique attracting limit cycle Γ of period 2π defined by $x^2 + y^2 = 1$. The simplest way to see this is to introduce polar coordinates $r^2 = x^2 + y^2$, $\tan \theta = y/x$, obtaining, after differentiating and simplifying, the system $r' = r(1 - r^2)$, $\theta' = 1$, whose general solution is $r(t) = r_0[r_0^2 + (1 - r_0^2)e^{-2t}]^{-1/2}$, $\theta(t) = t + \theta_0$. [$r(t)$ is found by solving the Bernoulli equation $r' - r = -r^3$.] A solution corresponding to $r_0 = 1$ is $r \equiv 1$, $\theta = t$; in xy-coordinates, we have $x = \cos t$, $y = \sin t$, a solution of period 2π. Since $\lim_{t \to \infty} r(t) = 1$ if $r_0 > 0$, the unit circle Γ is an attracting limit cycle (see Figure 12.2).

Example 12.2

The following system is written directly in polar coordinates: $\theta' = 1$, $r' = r(1 - r^2)(4 - r^2)^2(9 - r^2)$. It may be shown that the system has three limit cycles: one attracts ($r = 1$), another repels ($r = 3$), and a third ($r = 2$) attracts from

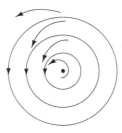

Figure 12.3   Three limit cycles.

one side ($2 < r < 3$) and repels from the other ($1 < r < 2$) (see Figure 12.3). These results are suggested by the changes of sign in r' on the three circles $r = 1, 2, 3$.

Example 12.3

The system $x' = y(1 + y)$, $y' = -\frac{1}{2}x + \frac{1}{5}y - xy + \frac{6}{5}y^2$ has a pair of limit cycles, one attracting and the other repelling, but we shall not prove this. The system was listed earlier in the gallery of portraits in Section 9.4 (see Figure 12.4).

Limit cycles cannot occur in linear systems since any constant multiple of a solution of a linear system is again a solution. Thus limit cycles represent distinctly nonlinear behavior.

With these examples and definitions in mind, we return to one of the physical phenomena listed by van der Pol and possessing an attracting limit cycle.

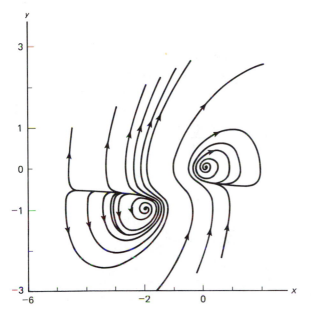

Figure 12.4   Two-eyed monster.

Oscillator Circuits: The van der Pol Limit Cycle

Van der Pol encountered the phenomenon and the equation now known by his name in his study of the tetrode multivibrator circuit in early commerical radios. The circuit is in essence a simple RLC loop in which the passive resistor is replaced by an active element. First we shall review the equations of the passive RLC circuit, but from a state-variable and system point of view. Suppose that a source of electrical current is attached to the circuit of Figure 12.5 and then withdrawn. How do current and the voltages across the elements attenuate with the passage of time? Kirchhoff's Current Law implies that there is a single current $I(t)$ flowing through the loop. According to the voltage law, the voltage drops across the three circuit elements satisfy

$$V_{13} = V_{12} + V_{23} \qquad (2)$$

The separate voltages are related to the corresponding circuit elements by the laws

$$V_{12} = LI', \qquad V_{23} = RI, \qquad V_{13}' = -\frac{1}{C} I \qquad (3)$$

where the minus sign in the last equation follows from the relation $V_{13} = -V_{31}$. From the equations of (2) and (3), we see that the two state variables I and $V = V_{13}$ are enough to characterize the dynamics of the circuit:

$$I' = \frac{1}{L} V_{12} = \frac{1}{L} (-V_{23} + V) = \frac{1}{L} (-RI + V)$$

$$V' = -\frac{1}{C} I \qquad (4)$$

If R, L, and C are positive constants, every solution $(I(t), V(t))$ of (4) decays to the equilibrium point $(0, 0)$ of the IV-state plane.

The reason for the decaying character of the circuit is that the voltage RI across the resistor dissipates energy. In fact, defining energy by the positive-definite function

$$E = \frac{1}{2} I^2 + \frac{C}{2L} V^2 \qquad (5)$$

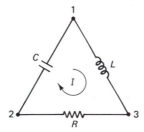

Figure 12.5   Passive circuit.

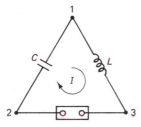

Figure 12.6   Simple circuit with an active element.

the rate of change $E'(t)$ along an orbit $(I(t), V(t))$ of (4) is given by

$$E' = I \cdot I' + \frac{C}{L} V \cdot V' = I \cdot \frac{1}{L} (-RI + V) + \frac{C}{L} V \cdot \left(-\frac{1}{C} I\right) = -\frac{R}{L} I^2 \leq 0 \quad (6)$$

$E(t)$ is a nonincreasing function of time along an orbit, a stability function in the sense of Section 11.2. Since $I(t)$ and $V(t)$ decay to 0 with increasing time, so does $E(t)$. There are no attracting limit cycles, no cycles of any type.

The circuit would be active instead of passive if energy were pumped into the circuit whenever the amplitude of the current fell below some level. One way to inject energy would be to replace the resistor by an active element which acts as a "negative resistor" at low current levels but dissipates energy at high levels. In van der Pol's time there were vacuum tubes and now there are semiconductor devices that do just this. See Figure 12.6 for a sketch of such a circuit and Figure 12.7 for a comparison of the voltage–current characteristics of a passive resistor with that of a typical semiconductor device.

The corresponding rate equations (7) are obtained from (4) of the passive circuit by replacing RI by $f(I)$, where the curve $V = f(I)$ is as sketched in Figure 12.7b:

$$I' = \frac{1}{L} (-f(I) + V)$$

$$V' = -\frac{1}{C} I$$

$$(7)$$

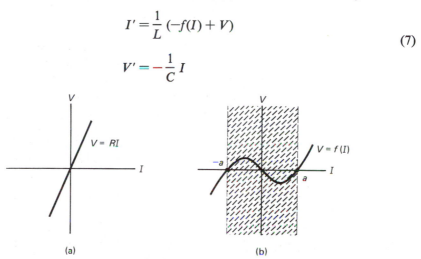

(a)   (b)

Figure 12.7   Voltage–current characteristics: (a) passive resistor with positive resistance; (b) active element: shaded region is region of negative resistance.

The rate of change of energy E along an orbit of (7) is given by

$$E' = I \cdot I' + \frac{C}{L} V \cdot V' = I \cdot \frac{1}{L} (-f(I) + V) + \frac{C}{L} V \cdot \left(-\frac{1}{C} I\right) = -\frac{1}{L} If(I)$$

Since $If(I)$ is positive for $|I| > a$ and negative for $0 < |I| < a$, energy is dissipated along those arcs of a nonconstant orbit Γ of (7) lying in the region $I > a$ or $I < -a$ of the state plane, while energy is injected into the system along arcs of Γ lying in the strip $|I| < a$. This raises the possibility of the existence of a cycle Γ for which the net energy change over one period is zero. Presumably, Γ would cut across regions of energy dissipation and zones of energy injection. Orbits lying entirely in the region $|I| < a$ would gain energy as time advances and move away from the point of minimal energy at the origin, while orbits that spend most of the time in the regions $I > a$ and $I < -a$ would suffer a net loss of energy and move inward. With this in mind, the existence of an attracting limit cycle Γ may be proved.

The van der Pol Cycle.   Let the differentiable function $f(I)$ have the following properties:
(a) $f(-I) = -f(I)$;
(b) $f(I) < 0$ for $0 < I < a$, $f(I) > 0$ for $I > a$, $a > 0$.
(c) $f(I) \to \infty$ as $I \to \infty$.
Then system (7) has a unique attracting limit cycle whose domain of attraction is the VI-state plane with the exception of the origin.

It is not mathematically difficult to prove this result, but the proof is long and we shall not give it. [See M. W. Hirsch and S. Smale, *Differential Equations, Dynamical Systems, and Linear Algebra* (New York: Academic Press, 1974) for a proof of a special case.] The mode of the proof has been suggested above. Show that the net change of energy along one turn of an orbit about the origin is negative if the orbit is far away from the origin, positive if the orbit lies close to the origin. Then somewhere there is an orbit where the net change is zero, which means that the E value after one turn is the same as the E value at the start. But this implies that the values of I and V are the same after one turn as they were at the start, and this implies that the orbit is a cycle. By a "turn about the origin" we mean that an orbit starting, say, on the positive V-axis rotates with advancing time through the quadrants clockwise and eventually returns to the positive V-axis. The hypotheses of the theorem are strong enough that it may be shown that not only does there exist a cycle but the cycle is unique and attracts all nonconstant orbits as time increases.†

The use of an energy function E in the proof is in the spirit of the Lyapunov tests in Chapter 11. We may deduce the nature of the state-space portrait from the

---

† There are extensions of this result to more general systems than (7). The extensions usually go by the name of Liénard systems, in honor of a French engineer and mathematician who extended van der Pol's work.

way a scalar-valued function changes along the orbits, even though no formulas for the orbits are known.

Van der Pol considered a particular case of the theorem above. Introduce new state and time variables by $x = L^{1/2}I$, $y = C^{1/2}V$, $T = (LC)^{-1/2}t$. In these variables system (7) becomes

$$x' = y - \mu F(x), \qquad y' = -x$$

where $\mu = C^{1/2}$ and $F(x) = f(L^{-1/2}x)$. Van der Pol then assumed that $F(x) = x^3/3 - x$ and analyzed the system

$$x' = y - \mu\left(\frac{1}{3}x^3 - x\right), \qquad y' = -x \qquad (8)$$

See Figure 12.8 for a sketch of isoclines (curves on which $dy/dx = m$ is constant) and orbits of system (8). The attracting limit cycle is the curve in the figure between the inner and outer spirals.

The xy-state plane for (8) is the Liénard plane for the equivalent scalar equation

$$x'' + \mu(x^2 - 1)x' + x = 0 \qquad (9)$$

which is known as *van der Pol's equation*. In general, the *Liénard plane* for $x'' +$

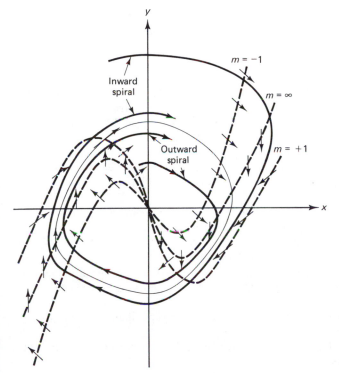

Figure 12.8 State-space portrait of van der Pol system (8). Isoclines for $m = -1$, $+1$, ∞. Inward and outward spiral orbits bracket the limit cycle.

$f(x)x' + g(x) = 0$ is the xy-plane for the corresponding system

$$x' = y - \int_0^x f(s)ds, \quad y' = -g(x)$$

PROBLEMS

The following problem(s) may be more challenging: 2(f) and 4.

1. **(a)** [*Lienard Construction*]. Verify the following method for constructing the line-element field for the system $x' = y - F(x)$, $y' = -x$. This is the *Liénard system* equivalent to the equation $x'' + f(x)x' + x = 0$, $F(x) = \int_0^x f(s)\, ds$. The xy-plane is called the "Liénard plane" in this case. To find the line element at a point P, draw a vertical through P to cut the curve C, $y= F(x)$, at a point Q; draw the horizontal through Q to cut the y-axis at R; then the line element at P is orthogonal to $\overline{RP}$ (see the sketch).

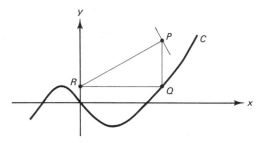

(b) Use Liénard's construction to produce a line-element field for the van der Pol system (8) with $\mu = 1$. Repeat for $\mu = 0.1$ and $\mu = 5$. In each case construct an approximation to the limit cycle by starting at $(2, 0)$. Observe that when $\mu = 0.1$, the limit cycle is essentially circular, whereas when $\mu = 5$, quite distorted.

(c) Construct a line-element field for $x' = y + \cos x$, $y' = -x$.

(d) Use a construction like Liénard's to handle the line-element field for the system $x' = y$, $y' = -x - h(y)$ [equivalent to $x'' + h(x') + x = 0$]. [*Hint*: Replace C by the curve defined by $x = -h(y)$.]

(e) Construct the line element field for the oscillator with *Newtonian damping*, $x'' + x' |x'| + x = 0$. [*Hint*: Use part (d) for the system $x' = y$, $y' = -x - y|y|$.]

(f) Solve the equation in part (e) explicitly. [*Hint*: Find an equation for $d(y^2)/dx$ and then solve.] Sketch the graphs in the line-element field of part (e).

2. **(a)** Show that the polar coordinate system $r' = r(r^2 - 1)^2$, $\theta' = 1$ has a unique limit cycle and that the cycle is semistable (i.e., an attractor on one side, a repeller on the other).

(b) Show that the system $r' = (r^2 - 1)r$, $\theta' = 1$ has a unique limit cycle and that the cycle is a repeller.

(c) Show that $r' = r \sin(1/r)$, $\theta' = 1$ has infinitely many limit cycles which are, alternately, repellers and attractors.

3. Prove that if $\mu < 0$, van der Pol's equation (9) has a unique limit cycle, which is unstable. [*Hint*: Reverse time in the equation with $\mu > 0$.]

4. Construct the line-element field in the Liénard plane (see Problem 1) for the system corresponding to $x'' + 2(|x| - 1)x' + 2x = 0$. Give an informal argument for the existence of a limit cycle.

12.2 THE ALTERNATIVES OF POINCARÉ AND BENDIXSON

Portraits of the orbits of various planar autonomous systems have been sketched in earlier sections. The variations in the portraits are extreme, but the portraits do have some common features and these are brought out in this section. In the process, tests for the existence or nonexistence of cycles will be derived, thereby solving in part a question posed in Section 12.1.

The planar autonomous system to be considered is

$$
\begin{aligned}
x' &= f(x, y) \\
y' &= g(x, y)
\end{aligned}
\tag{1}
$$

A solution $x = x(t)$, $y = y(t)$ of (1) will be denoted by the vector function $z = z(t)$ and the corresponding orbit by Γ. Γ and the solution are said to be *positively bounded* if the subset $\Gamma_T^+ = \{z(t) : t \geq T\}$ of Γ is a bounded subset of the xy-state plane for some T. The value of T in the definition makes no difference, since if $\Gamma_{T_0}^+$ is bounded for some T_0 for which $z(T_0)$ is defined, then Γ_T^+ is bounded for every T for which $z(t)$ is defined. Γ is *negatively bounded* if $\Gamma_T^- = \{z(t) : t \leq T\}$ is a bounded subset of Γ. Γ is *bounded* if it is both positively and negatively bounded. As we shall see, although a positively bounded orbit may appear to move through the state plane in an almost random way, its asymptotic behavior as $t \rightarrow +\infty$ is severely restricted. Henri Poincaré and Ivar Bendixson discovered just what the alternatives for the asymptotic behavior are. These alternatives provide the unifying features of state-plane portraits.

Limit Sets

Since we are interested in the long-term asymptotic behavior of an orbit, we shall define what are known as its limit sets.

Limit Sets.   Let Γ be an orbit of (1). The *omega limit set* of Γ, $\omega(\Gamma)$, is the set of all points Q such that for some increasing sequence of times, $t_1 < t_2 < \cdots < t_n < \cdots$, which diverges to $+\infty$, $\lim_{t \to \infty} z(t_n) = Q$. The *alpha limit set* $\alpha(\Gamma)$ is defined similarly but for decreasing sequences of times converging to $-\infty$.†

---

† The definition is not restricted to planar systems. It may be shown that the omega or alpha limit set of a bounded orbit in any dimension is a nonempty, closed, bounded, connected, and invariant set. By "invariant" we mean that any orbit intersecting the limit set stays in it for all t.

The limit sets of some orbits are easy to determine. For example, the alpha and omega limit sets of an equilibrium point and of a cycle are the point and the cycle, respectively. The following examples illustrate some of the possibilities.

Example 12.4

The system $x' = x$, $y' = -y$ has an orbit Γ_1, $x = 0$, $y = e^{-t}$; $\omega(\Gamma_1) = (0, 0)$, while $\alpha(\Gamma_1)$ is empty since $y \to \infty$ as $t \to -\infty$. For the orbit Γ_2, $x = e^t$, $y = 0$, we have $\omega(\Gamma_2)$ empty and $\alpha(\Gamma_2) = (0, 0)$. The orbit Γ_3, $x = e^t$, $y = e^{-t}$, has both limit sets empty. The orbit Γ_4 consisting of the equilibrium point at the origin is its own alpha and omega limit set. Note that Γ_1 is positively bounded, Γ_2 is negatively bounded, Γ_3 is unbounded, and Γ_4 is bounded (see Figure 12.9).

A spiraling orbit may have a limit cycle in its omega limit set (see, e.g., Example 12.1). The next example illustrates a new phenomenon.

Example 12.5

The Cartesian system

$$x' = [x(1 - x^2 - y^2) - y][(x^2 - 1)^2 + y^2]$$
$$y' = [y(1 - x^2 - y^2) + x][(x^2 - 1)^2 + y^2]$$

(2)

becomes in polar coordinates

$$r' = r[1 - r^2][(r^2 \cos^2 \theta - 1)^2 + r^2 \sin^2 \theta]$$
$$\theta' = (r^2 \cos^2 \theta - 1)^2 + r^2 \sin^2 \theta$$

(3)

These equations are obtained from the equations, $r' = r(1 - r^2)$, $\theta' = 1$, of Example 12.1 by multiplying the right-hand sides by the factor $(r^2 \cos^2 \theta - 1)^2 + r^2 \sin^2 \theta$, which vanishes at $r = 1$, $\theta = n\pi$ and is otherwise positive. Thus, $dr/d\theta = r(1 - r^2)$ for both systems. System (3), however, has the two

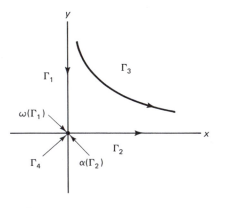

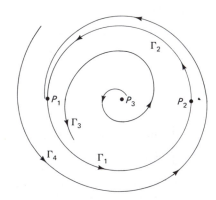

Figure 12.9 Orbits and their limit sets.   **Figure 12.10** Spiraling toward a cycle graph.

Figure 12.11   Cycle graphs.

new equilibrium points, $P_1(r = 1, \theta = \pi)$ and $P_2(r = 1, \theta = 0)$, in addition to the equilibrium point P_3 at the origin common to both systems. See Figure 12.10 for a sketch of the orbits.

The circle in Example 12.5 is not a cycle of (2) since it is not one orbit, but four. It is an example of one of the more mysterious objects in planar systems, a cycle graph.

> **Cycle Graph.**   Let CG be an ordered collection of orbits of (1) containing nonconstant orbits and a finite number of equilibrium points. CG is a *cycle graph* of (1) if its nonconstant orbits tend to its equilibrium points as $t \rightarrow \pm\infty$, and the ordering within CG is consistent with the advance of time and orients CG either clockwise or counterclockwise as a closed curve.

The ordered collection $P_1\Gamma_1P_2\Gamma_2$ of Example 12.5 is a cycle graph that is oriented counterclockwise. See Figure 12.11 for other cycle graphs (the corresponding systems are omitted).

The examples suggest that the omega limit set of a positively bounded orbit can be an equilibrium point, a cycle, or a cycle graph. Bendixson and Poincaré showed that under a certain mild restriction, there are no other possibilities.

The Threefold Way.   Let Γ be a positively bounded orbit of (1), and suppose that $\omega(\Gamma)$ contains at most a finite number of equilibrium points. Then either $\omega(\Gamma)$ is a single equilibrium point, or it is a cycle, or it is a cycle graph. Moreover, in the latter two cases Γ "spirals" toward $\omega(\Gamma)$ with the advance of time.

A similar result holds for $\alpha(\Gamma)$ if Γ is negatively bounded. The proof is omitted.†

Consequences of the Threefold Way

Now suppose that Γ is bounded, both positively and negatively. The alternatives for $\omega(\Gamma)$ may be combined with those for $\alpha(\Gamma)$ in exactly 12 ways. These are sketched in Figure 12.12.

---

† See M. W. Hirsch and S. Smale, *Differential Equations, Dynamical Systems, and Linear Algebra* (New York: Academic Press, 1974), or J. K. Hale, *Ordinary Differential Equations* (New York: Wiley-Interscience, 1969).

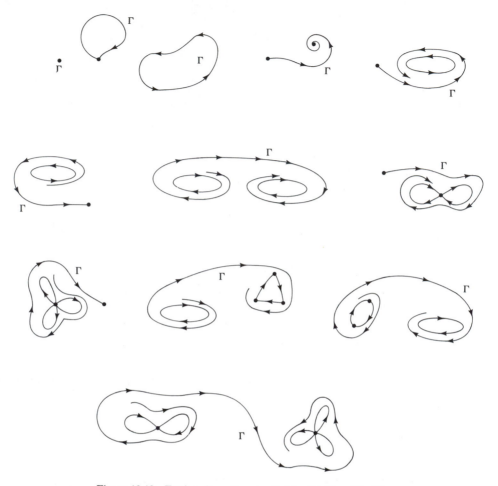

Figure 12.12 Twelve alternatives implied by the Threefold Way.

With these alternatives in mind, we see the special role that equilibrium points and cycles play in the study of the structure of orbits.

Bounded Orbits. Every bounded orbit has an equilibrium point or a cycle in each of its limit sets.

This is an immediate consequence of the Threefold Way.

Since the asymptotic behavior of an orbit is intimately connected with the equilibrium points and the cycles of the system, the first step in a general structural analysis of the portrait of the orbits would be to locate the equilibrium points and cycles. Finding the former is the algebraic problem of finding all point solutions of the simultaneous equations

$$f(x, y) = 0, \qquad g(x, y) = 0$$

The cycles are more difficult to detect. We shall give three results having to do with the existence or nonexistence of cycles.

The first is an immediate corollary of the Threefold Way.

Poincaré–Bendixson Test for Cycles.   Let Γ be a positively bounded orbit of (1). Suppose that $\omega(\Gamma)$ contains no equilibrium points. Then $\omega(\Gamma)$ is a cycle.

The test may be applied even if all that is known about Γ is that from some time on it remains within a bounded region A of the state plane, where neither A nor its boundary contains equilibrium points of (1). This implies that $\omega(\Gamma)$ has no equilibrium points and hence $\omega(\Gamma)$ is a cycle.

Example 12.6

Suppose that the vector field $(f(x, y), g(x, y))$ at each point (x, y) of the circle $C_1 : x^2 + y^2 = 1$ points outward, but it points inward at each point of the circle $C_2 : x^2 + y^2 = 2$. Suppose that the system $x' = f(x, y)$, $y' = g(x, y)$ has no equilibrium points in the ring $R : 1 \leq x^2 + y^2 \leq 2$. Then the Poincaré–Bendixson Test implies that the system has at least one cycle in R since any orbit touching a boundary circle is directed into R by the field vector and must remain in R as time advances (see Figure 12.13).

The Poincaré–Bendixson Test indicates that the absence of equilibrium points may show the presence of cycles. But there is a notable result in an opposing direction (see also Section 12.1).

Cycles and Equilibrium Points.   Let Γ be a cycle of (1). There must be at least one equilibrium point in the region bounded by Γ.

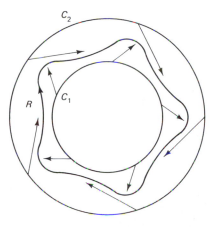

Figure 12.13   Cycle in a ring.

This result is not contradicted by Example 12.5, but it does imply that the system of the example must have equilibrium points inside C_1.

Example 12.7

The system $x' = x(1 - x - y)$, $y' = y(1 - 2x - 2y)$ may be used to model competing species (see the problem set of Section 9.5). The three equilibrium points are $(0, 0)$, $(0, \frac{1}{2})$, and $(1, 0)$. However, the x-axis and the y-axis are unions of orbits. Thus no cycle can cross either axis. Hence if the system had a cycle, it would have to lie inside one of the quadrants. Since there are no equilibrium points inside any quadrant, there are no cycles.

Bendixson's Negative Criterion.   Suppose that the divergence $\partial f/\partial x + \partial g/\partial y$ of the field vector (f, g) of (1) has a fixed sign in a region A. Then (1) has no cycle that lies, together with its interior region, entirely in A.

Proof. We shall use Green's Theorem to prove Bendixson's Test. There is no loss of generality in supposing that $\partial f/\partial x + \partial g/\partial y$ is positive in A. Suppose that there is a cycle Γ, and A contains Γ and its interior. We shall show that this supposition leads to a contradiction. Let Γ correspond to a solution $(x(t), y(t))$ of period Γ. Let S denote the union of Γ and its interior region. Then

$$0 < \int_S \int \left(\frac{\partial f}{\partial x} + \frac{\partial g}{\partial y} \right) dx \, dy \qquad \left(\text{since } \frac{\partial f}{\partial x} + \frac{\partial g}{\partial y} > 0 \text{ in } S \right)$$

$$= \oint_\Gamma (g \, dx - f \, dy) \qquad \text{(Green's Theorem)}$$

$$= \int_0^T \left(g \frac{dx}{dt} - f \frac{dy}{dt} \right) dt \qquad \text{(definition of line integral)}$$

$$= \int_0^T (gf - fg) \, dt = 0 \qquad \text{(since } x' = f \text{ and } y' = g\text{)}$$

This contradiction proves the Bendixson Test.

Since the divergence of a vector field is a measure of how much orbits tangent to field vectors spread apart (positive divergence means that the orbits move away from one another with advancing time, negative divergence means they move closer), the Bendixson Criterion makes intuitive sense. For one would expect the net divergence around a cycle to be zero. Divergence of fixed sign implies no cycles.

Example 12.8

The equation $x'' = g(x, x')$ is an extension of the equation of motion of a pendulum, or of a mass on the end of a coiled spring, or of the current in a simple circuit. Under what circumstances could we be certain that there are no nontrivial periodic solutions, that is, that the equivalent system, $x' = y$, $y' = g(x, y)$, has no cycles? The divergence of the vector field is

$$\frac{\partial}{\partial x}(y) + \frac{\partial}{\partial y}(g(x, y)) = \frac{\partial g}{\partial y}$$

Thus if $\partial g / \partial y$ has a fixed sign for all x and y, the Bendixson Negative Criterion implies that there are no cycles.

For example, the equation $x'' + ax' + G(x) = C$, where a and C are constants, $a \neq 0$, has no periodic solutions. In this case, the system is $x' = y$, $y' = C - ay - G(x)$ and the divergence of the corresponding field vector is

$$\frac{\partial}{\partial x}(y) + \frac{\partial}{\partial y}(C - ay - G(x)) = -a$$

The criterion implies that there are no periodic solutions. The physical reason we would not expect periodic solutions is that the term ax' in the original equation represents a constant energy drain if $a > 0$ and an energy source if $a < 0$. One expects the net energy change once around any cycle, however, to be neither positive nor negative, but zero.

Comments

The tests and alternatives of this section are qualitative, not quantitative. General behavior of orbits is described rather than solution formulas given or estimates made of growth or decay. Results such as these are appropriate at the early stages of an analysis of a complex system modeling some physical phenomenon. They are also useful in designing a system, if for no other purpose than to tell us what cannot be done. For example, if the aim if to construct a planar autonomous system with a cycle, we must allow for an equilibrium point.

We have restricted attention to planar systems and for good reasons. In the plane, a cycle splits the plane into two components, and no orbit may pass from one to another. In state spaces of three or more dimensions, cycles do not have this property. As a result, orbits may wind around one another, become knotted, or trace out what appears to be an almost random path which goes everywhere but settles nowhere. Current mathematical research is aimed at making sense out of the apparently limitless possibilities for limit sets in higher-dimensional state spaces. One of the more promising approaches is to give a precise mathematical definition of "chaotic" orbits and then to characterize classes of equations that possess such orbits.

PROBLEMS

The following problem(s) may be more difficult: 5, 7, 8, and 11.

1. Show that the equation of the damped simple pendulum $mL\theta'' + a\theta' + mg \sin \theta = 0$, where m, L, a, g are positive constants, has no cycles. [*Hint*: See Example 12.8.]

2. Determine whether the following systems have any cycles. Find the cycles if they exist.
   (a) $x' = e^x + y^2$, $y' = xy$ [*Hint*: Does x ever decrease?]
   (b) $x' = 3x^3y^4 + 5$, $y' = 2ye^x + x^3$ [*Hint*: Use Bendixson's Negative Criterion.]
   (c) [*Polar coordinates*]. $r' = r \sin r^2$, $\theta' = 1$.

3. Show that the alternatives sketched in Figure 12.12 are the only ones possible for a bounded orbit that lies in a region with finitely many equilibrium points.

4. Prove that the average value of the divergence $\partial f/\partial x + \partial g/\partial y$ on the region inside a cycle of $x' = f(x, y)$, $y' = g(x, y)$ must be zero. [*Hint*: Apply Green's Theorem to
$$\oint_\Gamma g \, dx - f \, dy,$$
as in the proof of Bendixson's Negative Criterion.]

5. Show that the system $x' = x - y - 5x(x^2 + y^2) + x^5$, $y = x + y - 5y(x^2 + y^2) + y^5$ has a cycle. [*Hint*: Use polar coordinates, and find circles C_1 and C_2 as in Example 12.6.]

6. Show that $x' = x - xy^2 + y^2 \sin y$, $y' = 3y - yx^2 + e^x \sin x$ has no cycles inside C : $x^2 + y^2 = 4$. [*Hint*: Use Bendixson's Negative Criterion.]

7. Suppose that the system in Example 12.6 possesses exactly three cycles in the ring. Show that at least one is an attractor on both sides, but that all three could not have this property. Generalize to any finite number of cycles.

8. Suppose that $K(x) = c$, $a \le c \le b$, $x = (x_1, x_2)$, defines a family of smooth simple closed curves, K_c, where K_{c_1} is inside K_{c_2} if $c_1 < c_2$. Suppose that the system $x' = F(x)$ has no equilibrium points in the region R between K_a and K_b, while $F \cdot \nabla K$ is positive on K_a and negative on K_b. Show that the system has a cycle inside R.

9. Prove that the planar linear system $x' = ax + by$, $y' = cx + dy$ has no limit cycle.

10. [*Dulac Criterion*]. Prove the following result. Let R be an open subset of $\mathbf{R}^2$, f, g, K in $C^1(R)$; suppose that $\partial(Kf)/\partial x + \partial(Kg)/\partial y$ has a fixed sign in R. Then the system $x' = f$, $y' = g$ cannot have a nonconstant periodic solution, the interior of whose orbit is a subset of R.

11. [*Ragozin's Negative Criterion for Interactive Species*]. Suppose that a two-species interaction is governed by the model equations $x' = xF(x, y)$, $y' = yG(x, y)$. The species are *self-regulating* if the per unit specific growth rate (i.e., F or G) decreases as the species population increases. If $\partial F/\partial x < 0$ and $\partial G/\partial y < 0$, each species is self-regulating. Suppose that $\partial F/\partial x < 0$ and $\partial G/\partial y < 0$. Use Dulac's Criterion (see Problem 10) to show that the model equations have no cycles in the population quadrant $x > 0$, $y > 0$.

12.3 BIFURCATIONS

Cycles and equilibrium points are distinctive features of the portrait of the orbits of planar autonomous systems. If the field vector of the system is changed slightly, one expects the new portrait to resemble the old. A cycle might contract a little, an equilibrium point shift somewhat, a spiral tighten or loosen, but the dominant features of the portrait would remain—or so one might anticipate. This seems reasonable, but it is not necessarily true. Small changes in a coefficient may mean the disappearance of one feature and the sudden appearance of another which is altogether different. The particular phenomenon studied in this section is the bifurcation of an asymptotically stable equilibrium point into an attracting limit cycle and an unstable equilibrium point. The loss of stability and the birth of a limit cycle occurs as a parameter passes through a critical value. The phenomenon is usually called a *Hopf bifurcation* and is now widely used to model the sudden appearance of a limit cycle and consequent destabilization of an equilibrium point.

Changing Parameters, Changing Portraits

Before taking up the Hopf bifurcation explicitly, we show by examples that small changes in parameters may lead to marked changes in the character of the orbits of a system.

Example 12.9

Let ϵ be a real constant in the linear system

$$x' = \epsilon x + y, \qquad y' = -x + \epsilon y$$

The eigenvalues of the coefficient matrix are $\epsilon \pm i$ and we expect oscillatory solutions. In polar coordinates, $r' = \epsilon r$, $\theta' = -1$. The solutions are given by $r = r_0 e^{\epsilon t}$, $\theta = -t + \theta_0$. The pictures in Figure 12.14 show what happens as ϵ increases from -0.1, through 0, to $+0.1$. Inward spiraling slows down, ceases altogether as cycles suddenly appear at $\epsilon = 0$, and then these cycles break up into outward-moving spirals as ϵ becomes positive. This is not a Hopf bifurcation since there never is a limit cycle, but the example shows how portraits may change, and change quite abruptly. Note how asymptotic stability becomes neutral stability, and then instability and the change occurs the instant ϵ moves through 0.

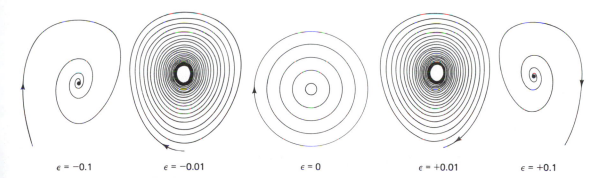

$\epsilon = -0.1$ $\qquad$ $\epsilon = -0.01$ $\qquad$ $\epsilon = 0$ $\qquad$ $\epsilon = +0.01$ $\qquad$ $\epsilon = +0.1$

Figure 12.14 Portraits for the equations of Example 12.9: from a stable focus, through a center, to an unstable focus.

Example 12.10

The equilibrium point of the linear system

$$x' = -x, \qquad y' = \epsilon y$$

changes from an isolated asymptotically stable node when $\epsilon < 0$ to a nonisolated neutrally stable point, and then to an unstable saddle point when ϵ is positive. The exact solutions are given by $x = Ae^{-t}$, $y = Be^{\epsilon t}$ (see Figure 12.15).

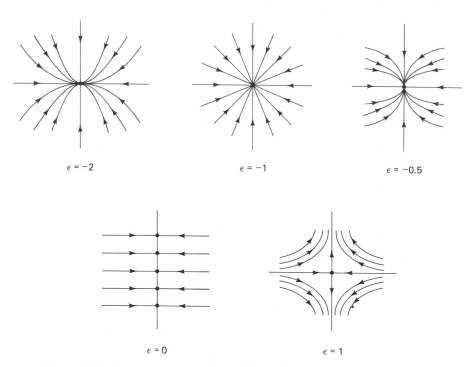

$\epsilon = -2$ $\epsilon = -1$ $\epsilon = -0.5$

$\epsilon = 0$ $\epsilon = 1$

Figure 12.15 Portraits for the equations of Example 12.10: from a stable improper node, through a star node, to a saddle point.

There is much more that can be done along these lines, but that is not our aim. We want to show that small changes in a parameter can lead to marked and sudden changes in the portrait of the orbits, and the examples illustrate this point. In imprecise terms, we say that a *bifurcation* occurs in each system when the parameter passes through a particular value at which the structure of the orbits undergoes a qualitative change.

The Hopf Bifurcation

The following simple example illustrates the particular phenomenon of the Hopf bifurcation.

Example 12.11

The linear part of the nonlinear system

$$x' = \epsilon x + y - x(x^2 + y^2)$$
$$y' = -x + \epsilon y - y(x^2 + y^2)$$

(1)

is identical to that of Example 12.9. By the results of Section 11.3 one expects the orbits of the two systems to be similar in a neighborhood of the origin. That expectation is justified, but only up to a point. In polar coordinates we have

$$r' = r(\epsilon - r^2)$$
$$\theta' = -1$$

(2)

The solution may be found by separating variables and integrating, but the nature of the orbits can be determined directly from (2) without reference to solution formulas. We have that

(a) If $\epsilon < 0$, orbits spiral clockwise into the origin as $t \rightarrow \infty$ since $r' < 0$ for $r > 0$ and $\theta = -t + \theta_0$.
(b) If $\epsilon = 0$, the same behavior as in (a) occurs, but the inward motion is not quite as rapid.
(c) If $\epsilon > 0$, then $r = \epsilon^{1/2}$, $\theta = -t + \theta_0$, is an attracting limit cycle since $r' > 0$ if $0 < r < \epsilon^{1/2}$ and $r' < 0$ if $r > \epsilon^{1/2}$ (see Figure 12.16).

The general situation of which the system of Example 12.11 is a special case may be stated in the following terms for the system

$$x' = f(x, y, \epsilon) = \alpha(\epsilon)x - \beta(\epsilon)y + P(\epsilon, x, y)$$
$$y' = g(x, y, \epsilon) = \beta(\epsilon)x + \alpha(\epsilon)y + Q(\epsilon, x, y)$$

(3)

where P and Q are at least second order in (x, y) near $x = 0$, $y = 0$, and real analytic in (x, y, ϵ) throughout $\mathbf{R}^3$, while $\alpha(\epsilon)$ and $\beta(\epsilon)$ are real analytic on an ϵ-interval centered at $\epsilon = 0$.† Thus (3) has an isolated equilibrium point at the origin $x = 0$, $y = 0$ for every ϵ. Observe that the eigenvalues of the linear part of the right-hand side of (3) are $\lambda_1(\epsilon) = \alpha(\epsilon) + i\beta(\epsilon) = \bar{\lambda}_2(\epsilon)$.

The following is a variant of a result proved by E. Hopf in 1942 and since much extended and applied in mathematics, physics, vibrational mechanics, population dynamics, and elsewhere.

Hopf Bifurcation.   Suppose that the foregoing conditions on (3) are satisfied, that $\alpha(0) = 0$, $\alpha'(0) > 0$, and $\beta(0) > 0$, and that (3) is asymptotically stable at the origin when $\epsilon = 0$. Then for each sufficiently small positive ϵ, system (3) is unstable at the origin and possesses an attracting limit cycle. The limit cycle encloses the origin and shrinks into it as $\epsilon \rightarrow 0^+$.

---

† See Appendix D for the definition of real analyticity.

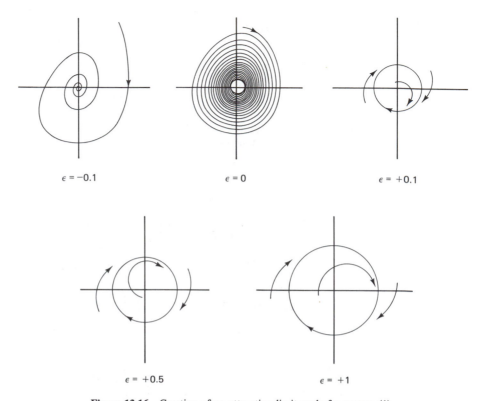

$\epsilon = -0.1$ $\epsilon = 0$ $\epsilon = +0.1$

$\epsilon = +0.5$ $\epsilon = +1$

Figure 12.16   Creation of an attracting limit cycle for system (1).

Observe that the system of Example 12.11 meets all the conditions of the Hopf Bifurcation, but that the system of Example 12.9 does not since it is only neutrally stable when $\epsilon = 0$. The Hopf phenomenon is exclusive to nonlinear systems—the terms P and Q must be present to preserve the asymptotic stability of the system at the bifurcation point $\epsilon = 0$.

We shall not prove the theorem since the proof is deep. However, the Poincaré–Bendixson Test of Section 12.2 makes the existence of a limit cycle plausible. The argument goes as follows. System (3) is assumed to be asymptotically stable at the origin when $\epsilon = 0$. By the converse of the Test for Asymptotic Stability (see Section 11.2), a result which we shall not prove, there is an asymptotic stability function V for (3). Orbits of (3) with $\epsilon = 0$ cross a level curve $V = C$, for some small C, moving inward and nontangentially across the level curve. Since the field vector of (3) is continuous in ϵ, the same thing happens for orbits of (3) (for sufficiently small positive ϵ) which intersect the level curve $V = C$ (see Figure 12.17). But the origin is unstable for $\epsilon > 0$ and repels all nearby orbits since the eigenvalues $\alpha(\epsilon) \pm i\beta(\epsilon)$ have positive real parts for $\epsilon > 0$. Thus no orbit Γ cutting across the level set $V = C$ can approach the origin as $t \to \infty$. Since Γ must remain inside the level curve as time increases

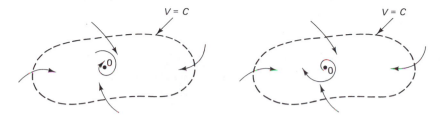

Figure 12.17   Orbits of (3): (a) $\epsilon = 0$; (b) $\epsilon > 0$.

and there are no equilibrium points of (3) other than the origin, the ω limit set of Γ must be a cycle. This is the cycle of the Hopf bifurcation.

Example 12.12

The system

$$x' = \epsilon(x - \epsilon) + y - (x - \epsilon)[(x - \epsilon)^2 + y^2]$$
$$y' = -(x - \epsilon) + \epsilon y - y[(x - \epsilon)^2 + y^2]$$

(4)

has a unique equilibrium point $P_\epsilon : x = \epsilon, y = 0$. In polar coordinates centered at P_ϵ, $x = \epsilon + r \cos \theta$, $y = r \sin \theta$, we have that

$$r' = r(\epsilon - r^2), \qquad \theta' = -1$$

We leave it to the reader to verify the correctness of the portraits in Figure 12.18. Note that the limit cycle is given by $x = \epsilon + \epsilon^{1/2} \cos t$, $y = \epsilon^{1/2} \sin t$, and that it grows in amplitude and drifts away to the right like a smoke ring.

Comments

The word "bifurcation" has the root meaning of "splitting into two." Suppose that a dynamical system has a parameter ϵ, and as that parameter is made to increase through a value ϵ_0, say, a single equilibrium point splits into two, or a single equilibrium point divides into an equilibrium point and a surrounding cycle, or one isolated cycle becomes two which bound an annulus. All of these phenomena are termed bifurcations, and ϵ_0 is the bifurcation point. Typically, there is a change of stability as well. In the Hopf case, an asymptotically stable equilibrium point transfers its stability to an attracting cycle and becomes unstable. It should be noted that the Hopf Bifurcation Theorem holds even though the linear terms do not have the special form (3) as long as the eigenvalues $\alpha(\epsilon) \pm i\beta(\epsilon)$ of the linear part meet all the other conditions.

Contemporary research in this area suggests that mysterious oscillations in the concentrations of reactants and products in a chemical reactor may be explained by a dynamical system model possessing a Hopf bifurcation triggered, for example, by an increase in the input rate of one of the reactants. A Hopf bifurcation has also been proposed as a model for one of the most basic of all biological phenomena—the life cycles of a cell.

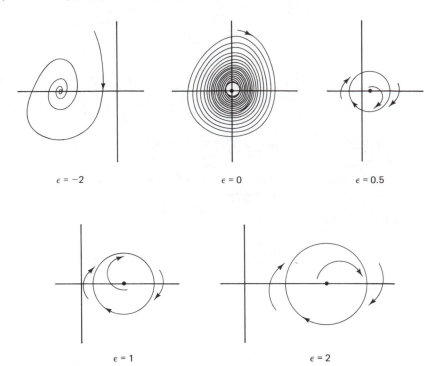

Figure 12.18 Birth, expansion, and migration of a Hopf cycle for system (4).

PROBLEMS

The following problem(s) may be more challenging: 2(c), 4, 5, and 7.

1. [*Pitchfork Bifurcation*]. A stable equilibrium point of a simple scalar rate equation may split into a triad of two stable and one unstable equilibrium points as a parameter in the rate function changes. The graph of the locations of the equilibrium points as functions of the parameter looks like a pitchfork, hence the name.

    (a) Let ϵ_0 and b be positive constants, ϵ a real parameter, in the rate equation $x' = x(\epsilon - \epsilon_0 - bx^2)$. Show that if $\epsilon \leq \epsilon_0$ there is a single equilibrium point $x = 0$, which is asymptotically stable, but if $\epsilon > \epsilon_0$ there are three equilibrium points, $x = 0$ and $x = \pm \sqrt{(\epsilon - \epsilon_0)/b}$. Show that in the latter case $x = 0$ is unstable, but the other two equilibrium points are stable.

    (b) Sketch the graphs of the location of the equilibrium points as functions of ϵ, and justify the name "pitchfork bifurcation."

    (c) Solve the system explicitly and sketch the solutions in the tx-plane, where $\epsilon_0 = 2$, $b = 1$, in the three cases $\epsilon = 0, 2, 4$.

2. Show that each of the following systems exhibits a Hopf bifurcation as ϵ increases through 0. Sketch the orbits for $\epsilon = -1, 0, +0.1, +0.5$, and $+1$.

(a) $x' = \epsilon x + 2y - x(x^2 + y^2)$, $y' = -2x + \epsilon y - y(x^2 + y^2)$.

(b) $x' = \epsilon x - 3y - x(x^2 + y^2)^3$, $y' = 3x + \epsilon y - y(x^2 + y^2)^3$.

(c) $x' = y - x^3$, $y' = -x + \epsilon y - y^3$.

3. Show that the system $x' = -\epsilon x + y + x(x^2 + y^2)$, $y' = -x - \epsilon y + y(x^2 + y^2)$ bifurcates as ϵ increases through 0 from an unstable focus to a stable focus surrounded by a repelling limit cycle.

4. Show that the standard system equivalent of *Rayleigh's equation*, $x'' + (x')^3 - \epsilon x' + x = 0$, has a Hopf bifurcation at $\epsilon = 0$. Sketch the portraits of the orbits for values of ϵ near 0.

5. [*A Satiable Predator*]. Consider the system

$$x' = x\left(-\frac{1}{2} + \frac{ay}{1 + ay}\right), \qquad y' = y\left(1 - y - \frac{ax}{1 + ay}\right)$$

where a is a constant, $a > 1$.

(a) Explain why x may be interpreted as the predator of the prey y. Show that the coefficient $a/(1 + ay)$ may be interpreted as meaning that the appetite of the predator satiates as the number y of prey increases.

(b) Show that $x = 2(a - 1)a^{-2}$, $y = a^{-1}$ is an equilibrium point.

(c) Using a computer to plot orbits, show that as a increases from 2 to 4, the equilibrium point in part (b) changes from stable to unstable and an attracting cycle appears. This gives computer evidence of a Hopf bifurcation.

6. Describe in words how the orbits change as the parameter ϵ increases through 0. Sketch the orbits for various values of ϵ.

(a) $x' = \epsilon x, y' = x - \epsilon y$.

(b) $x' = \epsilon x, y' = x - \epsilon^2 y$.

(c) $x' = \epsilon x - \epsilon^2 y, y' = \epsilon y$.

7. [*Computer Project: The Lorenz System and Chaotic Motion on a Strange Attractor*]. In 1963 the meteorologist and mathematician E. N. Lorenz published numerical studies of the solutions of a simplified model for atmospheric turbulence in an air cell beneath a thunderhead [*J. Atmos. Sci.*, v. 20 (1963), pp. 130–141]. The model equations are

$$\begin{cases} x' = -\sigma x + \sigma y \\ y' = rx - y - xz \\ z' = -bz + xy \end{cases} \qquad (L)$$

where σ, r, and b are positive constants denoting certain dimensionless physical parameters, x is the amplitude of the convective air currents in the cell, y is the temperature difference between the rising and falling currents, and z is the deviation from normal of the temperature in the cell. Lorenz observed strange, almost random behavior in the computed orbits of (L), small changes in the initial conditions leading to marked and apparently unpredictable changes in the orbits. Mathematical theories and experimental evidence to explain and support the numerical results for (L) and similar systems have developed rapidly since 1963, but much still remains a mystery. Computer graphics is at the heart of many of the studies, providing compelling visual demonstrations of the unexpected twists and turns of the orbits.

Since the limit sets of the orbits dominate their behavior, the first step is to classify the most common types of limit set and then determine those of system (L). Consistent with the definition in Chapter 11, we say that a connected union of limit sets is an *attractor* if all nearby orbits tend to the set as $t \to \infty$. According to the Three-Fold Way of Section 12.2, the only possible attractors for a planar system with bounded orbits and finitely many equilibrium points are attracting equilibrium points, limit cycles, or cycle graphs. In the three-dimensional state space of a system such as (L) we should add attracting tori (i.e., doughnuts) to this list of possible attractors, but there may be others. A closed, bounded, connected, and invariant union Z of limit sets is a *strange attractor* if Z is an attractor but is not a single point, a single cycle, or a torus. Z may be considered to be a three-dimensional generalization of a cycle graph. Motion on a strange attractor is *chaotic* if small changes in initial points lead to large (and apparently unpredictable) changes in the orbits on the attractor and if there is at least one orbit on the attractor which lies in every neighborhood of every point of the attractor infinitely often (such an orbit is said to be *dense* in the attractor). The chaotic wanderings of certain Euler approximations to the solutions of the logistic equation illustrate this behavior (see Section 2.5). There are other possible definitions of strange attractors and chaotic motion, but the somewhat imprecise definitions given in this paragraph suffice for what we have in mind.

Computer evidence and mathematical theory strongly suggest that the Lorenz system has a chaotic strange attractor. Problem 7(a)–(e) outline a few of the mathematical properties of (L), while in 7(f) the reader is asked to use computer graphics to construct orbital portraits. A 25-minute high-resolution computer-animated color movie has been made of the orbits of the Lorenz system for various values of r (B. Stewart, "The Lorenz System," Visumath Library Film Series, Aerial Press, Santa Cruz, Calif.).

(a) [*Symmetry*]. Show that if $(x(t),\ y(t),\ z(t))$ is a solution of (L), so is $(-x(t),\ -y(t),\ z(t))$.

(b) [*Equilibrium Points*]. Show that if $0 < r \leq 1$ the origin 0 is the only equilibrium point of (L), while if $r > 1$ there are two additional equilibrium points $P_1(-a,\ -a,\ r - 1)$ and $P_2(a,\ a,\ r - 1)$, where $a = (br - b)^{1/2}$.

(c) [*Stability of the Equilibrium Point*]. Denote by $J(0)$, $J(P_1)$, $J(P_2)$ the Jacobian matrix of the right-hand side of (L) evaluated at the respective three equilibrium points. Show that the eigenvalues of $J(0)$ are λ_1, $\lambda_2 = [-\sigma - 1 \pm (\sigma^2 - 2\sigma + 1 + 4\sigma r)^{1/2}]/2$, $\lambda_3 = -b$, and that 0 is asymptotically stable if $0 < r < 1$, but becomes unstable as r increases through 1. Show that the eigenvalues of $J(P_1)$ and $J(P_2)$ are the roots of $\lambda^3 + (1 + b + \sigma)\lambda^2 + (\sigma + r)b\lambda + 2\sigma b(r - 1)$. Use the Routh Array to show that all three eigenvalues at P_1 and P_2 have negative real parts if $1 < r < r_{cr} \equiv \sigma(\sigma + b + 3)/(\sigma - b - 1)$ [hence, P_1 and P_2 are locally asymptotically stable for $1 < r < r_{cr}$.] Show that if $r > r_{cr}$ then 0, P_1, and P_2 are unstable and there are no single-point attractors.

(d) [*The Lorenz Squeeze*]. Let $V(R)$ denote the volume of a bounded region R of xyz-space. Let $V(R_t)$ be the volume of the region R_t obtained from R by following each orbit of (L) beginning at a point P in R forward t units of time. Show that $V(R_t) = V(R)e^{-(1+b+\sigma)t}$, and, hence, that as $t \to \infty$ the region R_t "squeezes" into a set of zero volume. [Hint: $dV(R_t)/dt = \iint_{S_t} F \cdot n\, dA = \iiint_{R_t} \text{div } F\, dt$, where S_t is the bounding surface of R_t, n the outward pointing unit normal field on S_t, and the Divergence Theorem has been applied to the vector field F defined by the right side of (L).]

(e) [*The Lorenz Attractor*]. It may be shown that there is an ellipsoid E: $rx^2 + \sigma y^2 + \sigma(z - 2r)^2 \leq C$, where C is a certain constant, with the properties that orbits of (L) cross the boundary of E moving inward with increasing t and that all orbits "eventually" belong to E. The *Lorenz Attractor* Z is defined to be the limiting region $\lim_{t \to \infty} E_t$. By (d) Z is a region of zero volume. Show that Z contains all equilibrium points and cycles of (L).

(f) Set $\sigma = 10$ and $b = \frac{8}{3}$ (the values Lorenz used) and show that $r_{cr} = 470/19 \approx 24.74$. For $1.346 < r < r_{cr}$ it may be shown that $J(P_1)$ $[J(P_2)]$ has one negative eigenvalue and a pair of complex conjugate eigenvalues with negative real parts. As r increases through r_{cr} the complex eigenvalues move across the imaginary axis and acquire positive real parts. The points P_1 and P_2 each destabilize as this occurs by absorbing a repelling limit cycle which exists nearby for every r, $1.346 < r <$ some r^{**}. The set Z apparently is a strange attractor on which chaotic motion occurs. Plot the orbits indicated below. (It is strongly advised that before doing any of these problems, you should read the warnings following the problem statement.)

  (i) Four orbits that decay to the origin, one of them being a vertical line ($r = 0.5$).

  (ii) Two orbits with one approaching P_1 and the other approaching P_2 ($r = 10$). Plot the corresponding tx- and tz-component graphs.

  (iii) Two orbits as described in (ii), but $x_0 = y_0 = \pm(a + 50)$, where the number a is given in (b), and $z_0 = 14$ ($r = 15$). Plot the corresponding tx- and tz-component graphs.

  (iv) Two orbits which start out close together (near the origin) and diverge as t increases, i.e., one moves toward P_1 and one toward P_2 ($r = 20$). Plot the corresponding tx- and tz-component graphs.

  (v) An orbit that is very near an asymptotically stable equilibrium point, yet far enough from its domain of attraction that after a sufficiently long time the solution appears to be chaotic ($r = 24.5$). Plot the corresponding tx- and tz-component graphs.

  (vi) A chaotic orbit ($r = 40$). Plot the corresponding tx- and tz-component graphs.

  (vii) A periodic orbit ($r = 70$). Plot the corresponding tx- and tz-component graphs. (See hint below.)

  (viii) Two orbits which are symmetric through the z-axis (i.e., replacing x and y by $-x$ and $-y$) and graph their projections onto the xy-, xz-, and yz-planes ($r = 87$).

  (ix) A chaotic orbit ($r = 114$). Plot the corresponding tx- and tz-component graphs.

  (x) It would be interesting to see the orbits which correspond to the values of $r = 500$ and 1000. If your computer allows, plot some of these orbits.

Warning: The software that is normally used to solve differential equations may not be appropriate for solving the Lorenz system. If this is the case, one may use another program that solves the system using other approximation techniques. For example, the technique that can be used in this problem is Euler's method [Section 2.2], extended to a system of differential equations. Doing this yields the difference equations:

$$x_{n+1} = x_n + h(-\sigma x_n + \sigma y_n)$$
$$y_{n+1} = y_n + h(-y_n - x_n z_n + r x_n)$$
$$z_{n+1} = z_n + h(-b z_n + x_n y_n)$$

where h is the step size ($h = 0.005$ is a good step size to use).

Because computers are constructed differently, the pictures produced by different computers may not be exactly alike, even though the same initial conditions and values of r are used. The problem is compounded when solving the Lorenz system because of its sensitivity to small changes in the initial data. Sometimes, the pictures of the orbits in xyz-space are not the same if one looks at them from a different angle. This problem can be eliminated by simply changing the view position, if it is possible on your computer. Manipulating the point of view may clarify the pictures, as well. Moreover, the time interval that the problem is solved on can also make two identical graphs dissimilar. The longer the time interval, the less clear the picture appears to be because of the apparent tangle of the orbits.

[*Hint*: The periodic orbits are the most difficult to locate. The student may wish to change the value of r given in (vii). Even though r is the same, two graphs may not look alike if the initial conditions are different, and vice versa, since there are many periodic orbits when r is "large." If the student decides to use $r = 70$, he or she may want to use $x_0 = -4$ and $z_0 = 64$ and $0 < y_0 < 10$. Keep track of the range of y_0 which produces the orbits that almost repeat themselves. It may help to look at the projection onto the xz-plane, instead of the 3-D plot. Narrow down the range of y_0. Also, do not solve the system over too large a time interval, but not too small either. The time should be long enough such that if one were to look at the component graphs, at least four repeating cycles can be seen.]

CHAPTER 13

Partial Differential Equations and Fourier Series

The internal temperatures of a solid body whose boundary temperatures are controlled, voltage along a transmission line, the displacement of a vibrating string—these physical systems and many more can be modeled by partial differential equations with appropriate boundary and initial conditions. Our aim in this chapter is to study in some depth a few simple physical systems which have partial differential equations as models and to introduce the Method of Separation of Variables for solving the corresponding mathematical problems. That method rests on the theory and techniques of Fourier Series and of Sturm–Liouville boundary value problems for ordinary differential equations. Consequently, a brief introduction to these topics is also included.

13.1 INTRODUCTION TO PARTIAL DIFFERENTIAL EQUATIONS

Partial differential equations are equations that involve the partial derivatives of functions of several variables. Using the "variable subscript" notation for the partial derivatives of the real-valued function $u(x, y, t)$ we shall write u_x, u_y, u_t, u_{xx}, u_{yx}, u_{xxx}, and so on, for the partial derivatives $\partial u/\partial x$, $\partial u/\partial y$, $\partial u/\partial t$, $\partial^2 u/\partial x^2$, $\partial^2 u/\partial x\,\partial y \equiv \partial(\partial u/\partial y)/\partial x$, $\partial^3 u/\partial^3 x$, and so on, respectively.† The basic partial differential equations that occur most frequently in the applications are listed below.

$$
\begin{aligned}
&(\textit{Laplace or Potential Equation}) && u_{xx} + u_{yy} + u_{zz} = 0 \\
&(\textit{Wave Equation}) && u_{tt} = c^2\{u_{xx} + u_{yy} + u_{zz}\} \\
&(\textit{Heat or Diffusion Equation}) && u_t = K\{u_{xx} + u_{yy} + u_{zz}\}
\end{aligned}
$$

† Recall that if $\partial^2 u/\partial x\,\partial y$ and $\partial^2 u/\partial y\,\partial x$ are continuous at a point P, they have the same value at P. In this case, the order of the "indices" in u_{xy} is immaterial. Thus to avoid unpleasantness, we shall always assume enough smoothness of derivatives to make the order of differentiation unimportant.

where c and K are positive constants whose values are determined by the physical properties of the materials being modeled. The variables x, y and z are space variables and t is time. The Wave Equation arises in models involving wave propagation (and hence the name), while Laplace's equation appears in models of steady-state phenomena, and the Heat Equation is connected with diffusion processes such as heat flow. A convenient operator notation ∂_t, ∂_x, ∂_{xx}, and so on, is frequently used to denote the operations of differentiation $\partial/\partial t$, $\partial/\partial x$, $\partial^2/\partial x^2$, and so on. Thus, defining the *Laplacian* operator $\nabla^2 \equiv \partial_{zz} + \partial_{yy} + \partial_{zz}$, we observe that the Laplace, Wave, and Heat Equations can be written as

$$\nabla^2 u = 0, \qquad (\partial_{tt} - c^2\nabla^2)u = 0, \qquad (\partial_t - K\nabla^2)u = 0$$

The *order* of a partial differential equation is the order of the highest derivative appearing in the equation. All the equations above are second order. Notice that the Laplacian ∇^2 is a linear operator since $\nabla^2[\alpha u + \beta v] = \alpha\nabla^2 u + \beta\nabla^2 v$, for any constants α, β, and any functions u, v which are twice continuously differentiable. The operators ∂_t and ∂_{tt} are linear as well. Thus the basic equations above are *linear* because they have the form $L[y] = f$, where L is a linear operator (see Section 3.2 for a general discussion of operators). Since our basic equations are homogeneous (i.e., the driving term f vanishes, and the equations have the form $L[u] = 0$), it follows that a linear combination of solutions to any one of them is again a solution of that same equation. This simple but important fact will be used over and over again in this chapter (it is sometimes called the *Principle of Superposition of Solutions*).

We now try to learn something about the solvability of partial differential equations by looking at some examples. For simplicity, however, we shall consider equations with only one or two space variables.

Examples with Solutions

In contrast with the situation in the theory of ordinary differential equations, there is not yet anything like a comprehensive theory concerning basic existence or uniqueness questions for partial differential equations—not even for linear equations. The following examples indicate some situations that can occur when dealing with partial differential equations. First it will be helpful to define the continuity classes for functions of several variables.

> **Continuity Classes.**   A function $u(x, y, z)$ defined on an open (or closed) region S in $\mathbf{R}^3$ is said to belong to the *continuity class* $C^k(S)$, for some nonnegative integer k, if all partial derivatives of u of order less than or equal to k exist as continuous functions on S.† There are similar definitions for functions of n variables, $n \geq 2$.

---

† Recall that a *region* R in $\mathbf{R}^n$ is a connected (i.e., not in disjoint pieces) set consisting only of interior points. The *closed region* denoted by cl R is the set consisting of the region R together with its boundary points.

Example 13.1

The simple second-order equation $u_{yx} = 0$ for the unknown function $u(x, y)$ can be solved if we rewrite the equation in the form $(u_y)_x = 0$. Now if u is any solution of $u_{yx} = 0$ belonging to $C^2(\mathbf{R}^2)$, evidently u_y must be a C^1-function independent of x, say $c(y)$. Now the equation $u_y = c(y)$ can be considered as an ordinary differential equation in y for each fixed value of x. Thus antidifferentiation yields that $u = f(x) + g(y)$, where f and g are C^2-functions in x and y, respectively. Conversely, any function of this form for *arbitrary* C^2-functions f and g is a C^2-solution of $u_{yx} = 0$. Thus we have found *all* C^2-solutions of $u_{yx} = 0$, namely $u = f(x) + g(y)$, for arbitrary C^2-functions f and g. It is interesting to recall that the general solution of a second-order ordinary differential equation might involve two arbitrary *constants*, whereas two arbitrary *functions* seem to be involved in the general solution of second order partial differential equations. Observe that prescribing the value of u and *any* number of its derivatives at a point would not be enough to single out a specific solution of $u_{yx} = 0$.

Example 13.2

Sometimes a partial differential equation can be solved by an iteration of the integrating factor technique of Chapter 3. Say that $u(x, y)$ is a C^2-solution of the second-order equation $u_{xy} + yu_x = 0$. Putting $v(x, y) \equiv u_x(x, y)$ we see that $v_y + yv = 0$. But this equation can be treated as an ordinary differential equation in the independent variable y for each fixed value of x. With $e^{y^2/2}$ as an integrating factor, this equation becomes $(e^{y^2/2}v)_y = 0$. Thus $e^{y^2/2}v = c(x)$, where c is a C^1-function, so $v = u_x = c(x)e^{-y^2/2}$, which can be considered an ordinary differential equation in x for each fixed value of y. It follows that any C^2-solution of $u_{xy} + yu_x = 0$ has the form $u = f(x)e^{-y^2/2} + g(y)$, where f and g are C^2-functions. Conversely, for *any* choice of C^2-functions f and g, $u = f(x)e^{-y^2/2} + g(y)$ is a C^2-solution of $u_{xy} + yu_x = 0$. thus we have found *all* C^2-solutions of $u_{xy} + yu_x = 0$. (There are many other "solutions," but none of them are C^2-solutions.)

In the next example we go from an overabundance of solutions to a scarcity.

Example 13.3

Consider the first-order equation

$$xu_x + yu_y + u = 0, \qquad x, y \text{ in } \mathbf{R}^2 \tag{1}$$

We shall show that the *only* solution of (1) that belongs to $C^1(\mathbf{R}^2)$ is $u \equiv 0$. Let $u(x, y)$ be a C^1-solution of (1) and suppose that m is the minimum value of u on the square $S = [-1, 1] \times [-1, 1]$. If $u = m$ at an interior point of S, then $u_x = u_y = 0$ at that point and hence $u = -xu_x - yu_y = 0$ there. Thus $m = 0$ in this case. Now suppose that $u = m$ at a boundary point of

S, say at $(x_0, -1)$. Then certainly, $u_y(x_0, -1) \geq 0$, and the minimality of u at $(x_0, -1)$ implies that

$$u_x(x_0, -1) \begin{cases} = 0 & \text{if } |x_0| < 1 \\ \geq 0 & \text{if } x_0 = -1 \\ \leq 0 & \text{if } x_0 = 1 \end{cases} \qquad (2)$$

From (1) and (2) we have that

$$u(x_0, -1) = m = -x_0 u_x(x_0, -1) + u_y(x_0, -1) \geq 0$$

A similar argument applies if u takes on the value m anywhere on the boundary of S. Thus $u \geq m \geq 0$ on S.

Now the same argument can be applied to $-u$, which is also a solution of (1). Thus $M = \min(-u(x, y)) \geq 0$, over (x, y) in S, which implies that $u(x, y) \leq -M \leq 0$ for all (x, y) in S. Hence $m = M = 0$ and $u \equiv 0$ on S. Clearly, this extends to any square $[-a, a] \times [-a, a]$, and hence $u \equiv 0$ on $\mathbf{R}^2$.

As if this were not bad enough, it was discovered only in 1955 that there are very simple *linear* nonhomogeneous partial differential equations that have no solution at all in any neighborhood of a given point!

What have we learned from these examples? We have seen that certain partial differential equations can be solved by techniques developed for ordinary differential equations; and the solutions involve arbitrary functions, in contrast to solutions of ordinary differential equations, which involve only arbitrary constants. But the most telling difference between ordinary and partial differential equations is that apparently no simple but comprehensive description of the solution sets of partial differential equations has yet been given. It will come as no surprise at all, however, that all the partial differential equations which arise in models of physical phenomena behave "nicely," and hence we need not be unduly worried. On the other hand, it is clear that we cannot attempt anything like a theory of partial differential equations which parallels that developed in Chapters 1 to 12 for ordinary differential equations.

We next take up some useful "tricks" for finding exponential solutions and separated solutions for linear homogeneous equations, and we end this section with a discussion of initial value problems for partial differential equations of second order.

Exponential Solutions

The second-order linear homogeneous equation with constant real coefficients

$$A u_{xx} + B u_{xy} + C u_{yy} + D u_x + E u_y + F u = 0 \qquad (3)$$

always has complex-valued exponential solutions of the form $e^{\alpha x} e^{\beta y}$ for complex numbers α, β. Indeed, substituting $e^{\alpha x} e^{\beta y}$ for u into (3) we see that for any solution pair (α, β) of the quadratic equation

$$A\alpha^2 + B\alpha\beta + C\beta^2 + D\alpha + E\beta + F = 0$$

the exponential $e^{\alpha x + \beta y}$ is a (possibly complex-valued) solution of (3). Since (3) has real coefficients, the real and imaginary parts of the exponential solution $e^{\alpha x + \beta y}$ are *real*-valued solutions of (3).

Example 13.4

What are the exponential solutions $e^{\alpha t}e^{\beta x}$ of the wave equation $u_{tt} = u_{xx}$? By substitution we see that the condition $\alpha^2 = \beta^2$ must be satisfied. Thus $\beta = \pm\alpha$, so $e^{\alpha t}e^{\pm \alpha x}$ is a solution for any constant α, real or complex. If we seek a periodic solution of the wave equation, we must take $\alpha = i\omega$ for some real ω. Then $\beta = \pm i\omega$, so

$$e^{i\omega t}e^{\pm i\omega x} = e^{i\omega(t \pm x)} \equiv \cos \omega(t \pm x) + i \sin \omega(t \pm x)$$

whose real and imaginary parts are *real*-valued periodic solutions of $u_{tt} = u_{xx}$.

Example 13.5

Direct substitution shows that $e^{\alpha t}e^{\beta x}$ is a solution of the heat equation $u_t = u_{xx}$ if $\alpha = \beta^2$. Thus $e^{\beta^2 t}e^{\beta x}$ is a solution of $u_t = u_{xx}$ for any real or complex constant β. If we want a periodic solution of the diffusion equation, we need only choose β such that $\beta^2 = i\omega$ for any positive ω. Thus β must be one or the other of the complex numbers $\pm(\omega/2)^{1/2}(1 + i)$ (see Appendix C for a technique to compute square roots of complex numbers), so if we put $\gamma = (\omega/2)^{1/2}$, then

$$e^{i\omega t}e^{\pm\gamma(1+i)x} = e^{\pm\gamma x}e^{i(\omega t \pm \gamma x)} = e^{\pm\gamma x}\cos(\omega t \pm \gamma x) + ie^{\pm\gamma x}\sin(\omega t \pm \gamma x)$$

whose real and imaginary parts are real-valued solutions of $u_t = u_{xx}$.

Separated Solutions

There is another approach to finding solutions of (3) which generalizes the exponential solution approach and sometimes is successful even when the coefficients in (3) are nonconstant. In this approach we use a *Method of Separation of Variables* to find so-called *separated solutions* of (3) [i.e., solutions having the form $X(x)Y(y)$]. An example will explain the simple procedure.

Example 13.6

Let us find all solutions of the equation $x^2 u_{xx} + xu_x = u_{yy}$ which have the "separated" form $X(x)Y(y)$. Substituting $X(x)Y(y)$ for u in the equation and then dividing by $X(x)Y(y)$, we can write the resulting equation in the "variables separated" form

$$\frac{x^2 X''(x) + xX'(x)}{X(x)} = \frac{Y''(y)}{Y(y)} \tag{4}$$

The left-hand side of (4) depends only on the variable x, whereas the right-hand side depends only on y; thus the equation can hold only if each side is equal to the *same* constant, say λ, called the *separation constant*. Hence $X(x)$, $Y(y)$ must satisfy the pair of *ordinary* differential equations

$$x^2 X'' + xX' - \lambda X = 0 \tag{5}$$

$$Y'' - \lambda Y = 0 \tag{6}$$

for the *same* value of λ. Observe that (5) is an Euler equation whose characteristic roots satisfy the equation $m^2 - \lambda = 0$ (see Section 8.4). For example, if $\lambda = 1$, the general solution of (5) is $X(x) = Ax + Bx^{-1}$. For $\lambda = 1$, (6) has the general solution $X(y) = Ce^y + De^{-y}$. Hence

$$X(x)Y(y) \equiv (Ax + Bx^{-1})(Ce^y + De^{-y})$$

is a separated solution of $x^2 u_{xx} + xu_x = u_{yy}$ for all values of the constants A, B, C, D.

Initial Value Problems

When a partial differential equation has many solutions, what sort of conditions can be imposed which will "select" a unique solution from among the class of all solutions? Our experience with ordinary differential equations suggests that initial conditions can be used for this purpose, but what form do such conditions take for partial differential equations? Since there is no natural "normal" form for a partial differential equation, initial conditions must be expressed in the following general fashion. If L is a second-order partial differential operator in the three independent variables (x, y, z), an *initial manifold* is a smooth surface S in $\mathbf{R}^3$ (see Figure 13.1). For a smooth unit normal vector field $\mathbf{n}$ to S, the *initial value problem* for L on S is given by

$$
\begin{aligned}
L[u] &= h & &\text{for all } (x, y, z) \text{ "near" } S \\
u(x, y, z) &= F(x, y, z), & &\text{all } (x, y, z) \text{ on } S \\
\left.\frac{du}{d\mathbf{n}}\right|_{(x, y, z)} &= G(x, y, z), & &\text{all } (x, y, z) \text{ on } S
\end{aligned}
\tag{7}
$$

where the "driving term" h and the *initial data* F and G are given and $du/d\mathbf{n}$ (also written as $\partial u/\partial n$) denotes the directional derivative of u in the direction of $\mathbf{n}$. Thus there is a great deal of choice for the initial manifold S in problem (7). As we shall see in the example below, the properties of the initial value problem (7) for a given partial differential operator L depend crucially on the choice of S. Of course, there is no analog of this fact in ordinary differential equations.

Example 13.7

As we saw in Example 13.2, the general C^2-solution of $u_{xy} + yu_x = 0$ is $u = f(x)e^{-y^2/2} + g(y)$, where f and g are arbitrary C^2-functions of the single variables

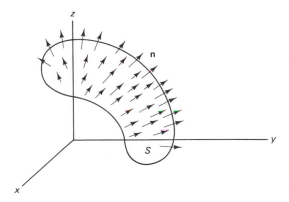

Figure 13.1   Geometry for an initial manifold.

indicated. First we take the x-axis as the initial manifold S. Then (7) takes the form

$$u_{xy} + yu_x = 0$$
$$u(x, 0) = F(x), \qquad u_y(x, 0) = G(x)$$

$$(8)$$

where we have set the driving term h equal to zero. Using the general solution, we see that (8) is solvable if and only if f and g can be found such that $f(x) + g(0) = F(x)$ and $g'(0) = G(x)$. Thus (8) is not solvable unless G is a constant function, say c, and in that case $g(y) \equiv cy + d$, for an arbitrary constant d, and $f(x) = F(x) - d$. Hence (8) in this case has infinitely many solutions

$$u = (F(x) - d)e^{-y^2/2} + (cy + d), \qquad d \text{ an arbitrary constant}$$

Next we consider the initial value problem for $u_{xy} + yu_x = 0$, where the initial manifold S is the line $x = y$. Thus $\mathbf{n} = (1/\sqrt{2}, -1/\sqrt{2})$ is a unit normal field to S, and the problem becomes

$$u_{xy} + yu_x = 0$$
$$u(x, x) = F(x), \qquad u_x(x, x) - u_y(x, x) = \sqrt{2}G(x)$$

$$(9)$$

Recalling that $u = f(x)e^{-y^2/2} + g(y)$ is the general solution to $u_{xy} + yu_x = 0$, we must choose f and g such that

$$f(x)e^{-x^2/2} + g(x) = F(x)$$
$$f'(x)e^{-x^2/2} + xf(x)e^{-x^2/2} - g'(x) = \sqrt{2}G(x)$$

Differentiating the first relation and adding to the second, we obtain $2f'(x)e^{-x^2/2} = F'(x) + \sqrt{2}G(x)$. Thus we see that the initial value problem (9) has a *unique* solution with $f(x)$ chosen such that $f'(x) \equiv \frac{1}{2}e^{x^2/2}(F'(x) + \sqrt{2}G(x))$ and $g(y) = F(y) - f(y)e^{-y^2/2}$. Comparing problems (8) and (9), we see that the choice of an initial manifold can seriously affect the solvability of an initial value problem for a given partial differential equation.

PROBLEMS

The following problem(s) may be more challenging: 5.

1. Find all exponential solutions of the following equations.
   (a) $u_{tt} = u_{xx} - 4u_x + 2u$.
   (b) $u_{xx} + u_{yy} = 0$.
2. Find all separated solutions of the following equations.
   (a) $u_{tt} = c^2 u_{xx}$.
   (b) $u_t = K u_{xx}$.
   (c) $u_{xx} + u_{yy} = 0$.
   (d) $x u_x + y^2 u_{yy} = 0$.
   (e) $u_{xx} - y u_y + u = 0$.
   (f) $u_{yy} = a^2 u_{xx} - 2h u_y$; a, h positive constants.
3. Find all C^2-solutions $u(x, y)$ of the equation $u_{xy} + (yu)_y + xu_x + xyu = 0$. [*Hint*: Show that for $v = u_x + yu$ the equation assumes the form $v_y + xv = 0$.]
4. Find all solutions (or discuss the solvability) of the initial value problems that arise from the equation $u_{xy} + y u_x = 0$ when the initial conditions are chosen as follows:
   (a) $u(0, y) = F(y)$, $u_x(0, y) = G(y)$.
   (b) $u(x, x) = x^2$, $u_x(x, x) - u_y(x, x) = 0$.
5. [*Picone*]. Let a, b, c be functions in $C^0(S)$, where $S = [-1, 1] \times [-1, 1]$, and assume that for $|x| \leq 1$ and $|y| \leq 1$, $a(-1, y) \geq 0$, $a(1, y) \leq 0$, $b(x, -1) \geq 0$, $b(x, 1) \leq 0$, $c(x, y) < 0$. Using the techniques of Example 13.3, show that $u \equiv 0$ is the only solution of $a u_x + b u_y + c u = 0$ defined for all (x, y) in S.

13.2 WAVE MOTION: VIBRATIONS OF A STRING

Periodic disturbances and oscillations play an important role in many diverse areas in science and engineering. Examples are (a) the motion of a pendulum in a gravitational field, (b) the motion of celestial bodies, and (c) oscillations in an electrical circuit. When oscillations travel in space, such phenomena are known under the collective title of "wave motion." Some familiar examples of wave motion are (a) the motion of ripples on the surface of a pond, (b) sound waves, and (c) transverse vibrations on a string.

We shall analyze the simplest and most intuitive system that involves wave motion: the transverse vibrations of a taut, flexible string. This special case is not as restrictive as it appears since many other physical phenomena can be treated by the same mathematical model (e.g., the longitudinal vibration of acoustical waves in a narrow column of gas). After constructing the linearized model for the motion of a vibrating string, we give a characterization of solutions of the model equations from which many properties of wave motion can be deduced.

Linear Model for the Vibrating String

Suppose that a string is stretched between the points $x = 0$ and $x = L > 0$ of the x-axis. We shall construct a model for the motion of the string under the action of the tension when no external applied forces are acting. To simplify the problem, let us assume that the string is *flexible*; that is, that the force of tension in the string acts tangentially to the string and hence the string offers no resistance to bending. Now since the equilibrium position of the string under the action of tension alone is the interval $0 \leq x \leq L$, we may describe the motion of the string by specifying three functions, $u_1(x, t)$, $u_2(x, t)$, and $u_3(x, t)$, the spatial coordinates at time t of the point x on the string in equilibrium position. On physical grounds it is clear that the motion of the string depends not only on the initial deflection and velocity of the string but also on how the string is restrained at the boundary points $x = 0$ and $x = L$. We discuss boundary conditions in Section 13.3.

We make some further simplifying assumptions in order to obtain a simple, linear partial differential equation describing the motion of the string. Specifically, we shall assume that:

1. The motion of the string takes place in a fixed plane, and points on the string are constrained to move only in a direction transverse to the string. Thus the motion of the string may be described by a single function $u(x, t)$ defined in the region $R = \{(x, t) : 0 < x < L, t > 0\}$.

2. The motion u belongs to the classes $C^2(R)$ and $C^1(\mathrm{cl}R)$, and only "small" deflections from equilibrium are allowed in the sense that second-order terms in u_x may be ignored when compared to terms of lower order.†

3. The tension in the string $\mathbf{T}(x, t)$ is of class $C^1(\mathrm{cl}R)$. The density ρ of the string is constant.

4. There are no external transverse forces acting on the string. In particular, the weight of the string is neglected.

Now we are ready to derive the equation of motion of the string. Let us compute the forces acting on the segment $[x, x']$ of the string. The forces acting on this segment are due to tension alone. We shall find the parallel and transverse components of the tension forces acting on $[x, x']$ at each of these endpoints. Referring to Figure 13.2, we have the following computation (the labeling in Figure 13.2 defines the symbols used below).

$$\|\mathbf{T}_u(x', t)\| = \|\mathbf{T}(x', t)\| |\sin \alpha|; \qquad \|\mathbf{T}_x(x', t)\| = \|\mathbf{T}(x', t)\| |\cos \alpha|$$

$$\|\mathbf{T}_u(x, t)\| = \|\mathbf{T}(x, t)\| |\sin \beta|; \qquad \|\mathbf{T}_x(x, t)\| = \|\mathbf{T}(x, t)\| |\cos \beta|$$

---

† A function $h(z)$ is nth *order* in z, denoted by $h = O(|z|^n)$, if there is a positive constant C such that for all $|z|$ small enough, $|h(z)| \leq C|z|^n$. Thus for small $|z|$ the lower-order terms in a sum of nonnegative powers of z are dominant, and for simplicity, higher-order terms may be dropped. For example, for small $|z|$, $1 + 3z + 5z^2$ may be replaced by $1 + 3z$, or even by 1. See also Section 11.3.

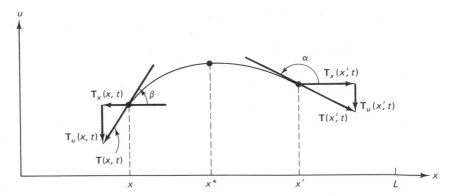

Figure 13.2 Profile on $[x, x']$ of the string at time t.

But since we have that

$$|\sin \alpha| = \frac{|u_x(x', t)|}{\sqrt{1 + |u_x(x', t)|^2}} \qquad |\cos \alpha| = \frac{1}{\sqrt{1 + |u_x(x', t)|^2}}$$

and similar formulas for $\sin \beta$, $\cos \beta$, we observe from assumption (2) above that

$$|\sin \alpha| = |u_x(x', t)|, \qquad |\sin \beta| = |u_x(x, t)|, \qquad |\cos \beta| = 1, \qquad |\cos \alpha| = 1$$

where second- and higher-order terms in u_x have been ignored.† Thus, supplying the proper orientations to the components, we see that the transverse and parallel components of the net tensile forces acting on the segment $[x, x']$ are given, respectively, by

$$\|\mathbf{T}(x', t)\| u_x(x', t) - \|\mathbf{T}(x, t)\| u_x(x, t) \qquad \text{(transverse)}$$

and

$$\|\mathbf{T}(x', t)\| - \|\mathbf{T}(x, t)\| \qquad \text{(parallel)}$$

But since we assumed that only transverse motions of the string were possible, we conclude that $\|\mathbf{T}(x', t)\| = \|\mathbf{T}(x, t)\|$, and since x and x' were arbitrary we must have that $T \equiv \|\mathbf{T}\|$ is independent of x in $[0, L]$. Now applying Newton's Second Law to the segment $[x, x']$, we have the equation

$$\rho(x' - x)u_{tt}(x^*, t) = T\{u_x(x', t) - u_x(x, t)\} \qquad (1)$$

where x^* is the x-coordinate of the center of mass of the string segment between x and x'. Dividing (1) through by $\rho(x' - x)$, denoting T/ρ by c^2, and taking limits as $x' \to x$, we obtain the *wave equation in one space dimension*,

$$u_{tt} - c^2 u_{xx} = 0 \qquad (2)$$

---

† We have expanded $(1 + |u_x|^2)^{-1/2}$ into a binomial series, $1 - \frac{1}{2}|u_x|^2 + \frac{3}{8}|u_x|^4 - \cdots$, and then kept only the first term, 1.

which must be satisfied for all points (x, t) in R. We shall assume that c^2 is constant. Note that (2) is a second-order homogeneous linear partial differential equation.

The wave equation (2) has many solutions. For example, if F and G are any C^2-functions, $u = F(x - ct) + G(x + ct)$ is a solution. This may be shown by repeated application of the Chain Rule to calculate partial derivatives; for example, if r denotes $x - ct$, then

$$F_t = \frac{dF}{dr}\frac{\partial r}{\partial t} = F'(r)(-c) \quad \text{and} \quad F_{tt} = \frac{d}{dr}(F'(r))\frac{\partial r}{\partial t}(-c) = c^2 F''(r)$$

Physical intuition for the circumstances of the vibrating string suggests that *initial conditions* such as prescribing $u(x, 0)$ and $u_t(x, 0)$, $0 < x < L$, together with *boundary conditions* such as prescribing $u(0, t)$ and $u(L, t)$, $t \geq 0$, are needed to obtain a unique and physically plausible solution. We take up the problem of the vibrating string in more detail in the next section. In the remainder of this section we solve the idealized problem of the vibrations of an infinitely long string (for example, $-\infty < x < \infty$) and derive the fundamental properties of all solutions of the wave equation (2).

Characteristics of Solutions of the Wave Equation

The following observation will be useful for finding solutions of the wave equation:

Constancy Theorem. Let Ω be a region in the xt-plane and let u in the class $C^2(\Omega)$ be any solution of (2) in Ω. Then

(a) $u_t - cu_x$ is constant on any line segment in Ω of the form $x - ct =$ constant.

(b) $u_t + cu_x$ is constant on any line segment in Ω of the form $x + ct =$ constant.

Proof. We prove only assertion (a); the proof of (b) is very similar. Observe that (2) can be written as $v_t + cv_x = 0$, where $v(x, t) \equiv u_t(x, t) - cu_x(x, t)$. But notice that $v_t + cv_x$ is proportional to the directional derivative of v in the xt-plane in the direction defined by the vector with x-coordinate c and t-coordinate unity. Thus the line $x - ct =$ constant in Ω is parallel to this direction and the function $v \equiv u_t - cu_x$ has a vanishing directional derivative in the direction of the line. The desired result follows.

Thus we see that the two families of lines $x - ct =$ constant and $x + ct =$ constant have special significance for the wave equation (2).

Characteristics. Each line of the form $x + ct =$ constant or $x - ct =$ constant is called a *characteristic* for the wave equation (2).

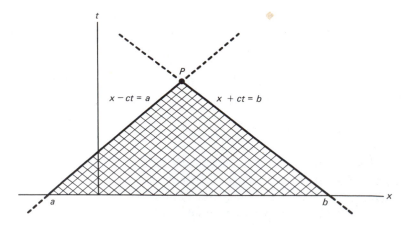

Figure 13.3   Characteristic triangle $\Delta(a, b)$ for the wave equation (2).

Characteristic Triangle.   Let $a < b$ be any two points in **R** and denote by $\Delta(a, b)$ the interior of the triangle whose base is the interval (a, b) and whose sides are segments of the characteristic lines $x - ct = a, \; x + ct = b$. The set $\Delta(a, b)$ is called a *characteristic triangle for* (2) (see Figure 13.3).

We now exploit the special property of characteristics for deriving solutions for (2). Let H be the half-space $-\infty < x < +\infty, \; t > 0$. Suppose that $u(x, t)$ is a C^2-solution of the wave equation (2) in H such that u and u_t are continuous on cl $H : -\infty < x < +\infty, \; t \geq 0$. If we put $u(x, 0) = f(x)$ and $u_t(x, 0) = g(x)$, we shall show that $u(x, t)$ is uniquely determined by the *initial data* $f(x)$ and $g(x)$. More generally, we shall show that given any choice of initial data f in $C^2(\mathbf{R})$ and g in $C^1(\mathbf{R})$, the *initial value* (or *Cauchy*) *problem*

$$u_{tt} = c^2 u_{xx}, \qquad x \text{ in } \mathbf{R}, \qquad t > 0$$
$$u(x, 0) = f(x), \qquad x \text{ in } \mathbf{R} \tag{3}$$
$$u_t(x, 0) = g(x), \qquad x \text{ in } \mathbf{R}$$

has a unique C^2-solution u in H for which u and u_t are C^0-functions on cl H. First we show uniqueness.

> **Uniqueness for Problem (3).**   Let u and v be two solutions of the initial value problem (3). Then $u \equiv v$ in the half-space H.

Proof. Let $w \equiv u - v$. Then w solves problem (3), too, with initial data $f(x) \equiv 0, \; g(x) \equiv 0$. Thus w_t and w_x both vanish for every point on the x-axis. Let P be a given point in the half-space H; we shall show that $w_t(P) = 0$ and $w_x(P) = 0$. Indeed, let $\Delta(a, b)$ be the characteristic triangle formed by following the two characteristic lines through P back to where they cross the x-axis, say at a and b, with $a < b$ (see Figure 13.3). Now the Constancy Theorem implies that

$$w_t(P) - cw_x(P) = w_t(a, 0) - cw_x(a, 0) = 0$$
$$w_t(P) + cw_x(P) = w_t(b, 0) + cw_x(b, 0) = 0$$

Therefore, $w_t(P) = w_x(P) = 0$. But P in H was arbitrary; thus $w_t \equiv 0$, $w_x \equiv 0$ in H. It follows that $w \equiv$ constant on H.† But since w vanishes on the x-axis, the constant must be zero, and hence $w \equiv 0$ on H. This implies that $u \equiv v$ on H, and we are done.

Remark: Actually, the proof above shows that if two solutions u and v of the wave equation (2) have initial data that coincide only on an interval (a, b) of the x-axis, then $u \equiv v$ on the characteristic triangle $\Delta(a, b)$.

Now we turn to the question of the existence of a solution for (3).

D'Alembert's Solution of the Initial Value Problem (3).   Let $f(x)$ be in the class $C^2(\mathbf{R})$, and $g(x)$ in the class $C^1(\mathbf{R})$. Then the function

$$u(x, t) = \frac{f(x + ct) + f(x - ct)}{2} + \frac{1}{2c}\int_{x-ct}^{x+ct} g(s)\, ds, \qquad (x, t)\ \text{in}\ H \qquad (4)$$

is the unique solution of the problem (3) with the required smoothness.

Proof. A direct calculation shows that $u(x, t)$ in (4) is indeed a solution of problem (3).‡ By the Uniqueness Theorem, (4) is the only solution and we are done.

Actually, we can show much more. We can derive (4) from the Constancy Theorem by carrying the initial data forward along characteristics. Here is how this is done. Let (x, t) be a given point in H, and denote by X_1 and X_2, $X_1 < X_2$, the intersections of the x-axis and the two characteristic lines through (x, t). Clearly, $X_1 = x - ct$, $X_2 = x + ct$. From the Constancy Theorem we conclude that

$$u_t(x, t) + cu_x(x, t) = u_t(X_2, 0) + cu_x(X_2, 0)$$
$$u_t(x, t) - cu_x(x, t) = u_t(X_1, 0) - cu_x(X_1, 0) \tag{5}$$

But since $u_t(X_2, 0) = g(X_2)$, $u_t(X_1, 0) = g(X_1)$, $u_x(X_2, 0) = f'(X_2)$, $u_x(X_1, 0) = f'(X_1)$, we see that (5) becomes

---

† The following theorem is a natural extension of the Vanishing Derivative Theorem of Section 1.1. Let R be a region in $\mathbf{R}^n$ and let f in $C^1(R)$ be such that $\partial f/\partial x_i = 0$ in R, for each index $i = 1, 2, \cdots, n$. Then $f \equiv$ constant in R.

‡ The reader is reminded that the *Leibniz Rule* for differentiating integrals with respect to parameters states that under appropriate smoothness conditions on $\alpha(t)$, $\beta(t)$, and $g(s, t)$, we have that

$$\frac{d}{dt}\int_{\alpha(t)}^{\beta(t)} g(s, t)\, ds = g(\beta(t), t)\beta' - g(\alpha(t), t)\alpha' + \int_{\alpha(t)}^{\beta(t)} g_t(s, t)\, ds$$

$$u_t(x, t) + cu_x(x, t) = g(x + ct) + cf'(x + ct)$$
$$u_t(x, t) - cu_x(x, t) = g(x - ct) - cf'(x - ct)$$

(6)

Solving the identities (6) for u_t and u_x we obtain that

$$2u_t(x, t) = [g(x + ct) + g(x - ct)] + c[f'(x + ct) - f'(x - ct)]$$
$$2u_x(x, t) = \frac{1}{c}[g(x + ct) - g(x - ct)] + [f'(x + ct) + f'(x - ct)]$$

(7)

Now observe that if we put

$$U(x, t) \equiv u(x, t) - \frac{f(x + ct) + f(x - ct)}{2} - \frac{1}{2c}\int_{x-ct}^{x+ct} g(s)\, ds$$

we can use (7) to show that $U_t \equiv 0$, $U_x \equiv 0$ in H. Thus $U \equiv$ constant in H, and since U vanishes on the x-axis, it must vanish for all (x, t) in H. Hence (4) results and we are done.

Note that the wave equation has no other solutions than those given by the form (4) for suitable "arbitrary" functions $f(\cdot)$ and $g(\cdot)$. Thus (4) is the *general solution* of the wave equation. The wave equation is one of the very few second-order partial differential equations for which a general solution can be conveniently found. The process outlined above for continuing the initial data along the characteristics to build the D'Alembert solution is particularly important in Section 13.3.

Properties of Solutions of the Wave Equation

Having the explicit solution of problem (3), we can now examine not only its mathematical properties but also its significance to the problem of the vibrating string. Examples of solutions appear in Section 13.3.

(A) *Superposition of Traveling Waves*. First observe that the function $u(x, t)$ in (4) can be written as

$$u(x, t) = P(x + ct) + Q(x - ct)$$

(8)

where

$$P(s) = \frac{1}{2}f(s) + \frac{1}{2c}\int_0^s g(s')\, ds', \qquad Q(s) = \frac{1}{2}f(s) - \frac{1}{2c}\int_0^s g(s')\, ds'$$

(9)

Thus (8) implies that to find the profile of our string at the given time T [i.e., $u(x, T)$] we need only take the graph of P on the x-axis, translate it to the *left* by cT units, and then add it to the translation of Q on the x-axis to the *right* by cT units. Thinking of the graphs of P and Q on the x-axis as *wave forms*, we see that $P(x + ct)$ is the wave form P moving uniformly in time to the *left* with velocity c. Similarly, $Q(x - ct)$ is just the wave form Q moving uniformly in time to the *right* with velocity c. Thus the solution of any initial value problem (3) is the sum of two *traveling waves*, one moving to the right, the other to the left [as is evident from (8)].

(B) *Huygens' Principle*.   Say that the string is undeflected and at rest initially except on the bounded interval J of the x-axis, where the string is deflected and given some initial velocity [i.e., f and g in (3) vanish identically everywhere but on J]. It is of interest to determine how much time must pass for a point x_0 not in J to "see" the "disturbance" on the set J. From (4) it is clear that $u(x, t)$ will vanish in any characteristic triangle $\Delta(x_0 - ct, x_0 + ct)$ whose base does not meet J. If $J = [a, b]$ and a is the nearest endpoint of J to x_0, the earliest time that x_0 can "feel" the "disturbance" on J is $t = |a - x_0|/c$. This observation agrees with (A) in that the wave forms P and Q both travel with speed c. Thus x_0 "perceives" a "sharp beginning" for the waves generated by the disturbance on J. But observe from (4) that if the initial velocity, g, is nontrivial on J, then x_0 will *not* in general "perceive" a "sharp end" to the waves generated by the "disturbance" on J, even when J is chosen *arbitrarily* small in length. Physicists express this peculiar property of the problem (3) by saying that "The Cauchy Problem (3) does *not* satisfy *Huygens' Principle*," since that principle requires both a sharp beginning and a sharp end to any disturbance passing through an "observers" location at x_0.

(C) *Cauchy Problem (3) Is Well-Set*.   The concept of a well-set (or well-posed) problem for a partial differential equation is the same as that for an ordinary differential equation (see Section 3.2). Thus the Cauchy Problem (3) is *well-set* if it has a unique solution for all appropriately smooth data, and if the solution changes "continuously" with respect to changes in the initial data f and g. This last condition needs some interpretation, just as was the case in Section 3.2. Let $u(x, t)$ be the solution of problem (3) for data $f(x)$ and $g(x)$. Choose a fixed $T > 0$, and define the strip S : $-\infty < x < \infty$, $0 \leq t \leq T$. Now suppose that the data are bounded and suppose that M_f and M_g are the smallest constants such that $|f(x)| \leq M_f$ and $|g(x)| \leq M_g$ for all x in **R**. Then according to (4), we have that

$$|u(x, t)| \leq M_f + TM_g, \qquad \text{all } (x, t) \text{ in } S \qquad (10)$$

Now suppose that f^*, g^* is another data pair which is bounded, and that $u^*(x, t)$ is the corresponding solution of problem (4). Then $u - u^*$ is the solution of problem (4) with the data pair $f - f^*$ and $g - g^*$, as is easy to verify. Thus the estimate (10) applies and we have that

$$|u(x, t) - u^*(x, t)| \leq M_{f-f^*} + TM_{g-g^*} \qquad \text{for all } (x, t) \text{ in } S \qquad (11)$$

Hence it follows from (11) that the difference $|u(x, t) - u^*(x, t)|$ may be made uniformly small over S if the differences $|f(x) - f^*(x)|$ and $|g(x) - g^*(x)|$ are made sufficiently small over **R**. Thus the Cauchy Problem (3) is well-set in the sense outlined above.

(D) *Propagation of Singularities*.   What can we say about the solvability of the Cauchy Problem (3) when the data are only piecewise smooth on **R**? The D'Alembert formula (4) still makes sense for such data. Notice that from (8) if the data are such that $P(s)$ in (9) has a discontinuity at some s_0, this discontinuity "propagates" in the half-space H along the characteristic line $x + ct = s_0$. A similar statement holds for $Q(s)$ and the characteristic line $x - ct = s_0$. Thus in particular we observe that if either $g(s) + cf'(s)$ or $g(s) - cf'(s)$ is discontinous at s_0, then u_t and u_x cannot be continuous along the characteristic lines $x \pm ct = s_0$. This situation is summarized by saying "the wave equation propagates singularities in the data into the interior along characteristic lines." Observe that the function u defined by (4) cannot satisfy the wave equation at interior singularities since the second derivatives u_{tt} and u_{xx} may not even exist at such points.

Comments

We have studied the Cauchy Problem (3) for the wave equation in some detail. The key tool was a simple trick (the Constancy Theorem) that enabled us to find the general solution of the wave equation. We saw that the Cauchy Problem (3) was a model for the motion of infinitely long flexible strings under tension. In spite of the fact that one does not often encounter infinitely long (or even *very* long) strings, we shall find this physically idealistic case useful in the next section for obtaining a solution of the more realistic problem of finite strings.

PROBLEMS

The following problem(s) may be more challenging: 4 and 5.

1. Find the D'Alembert solution of (3) with $c = 1$ for each of the following sets of initial data f, g. Sketch the profiles of the string at $t = 0$, $t = 1$, and $t = 10$.

   (a) $f(x) = \begin{cases} \cos x, & |x| < \pi/2 \\ 0, & \text{otherwise} \end{cases}$ $g(x) \equiv 0$, all x.

   (b) $f(x) = \begin{cases} 1 - |x|, & |x| < 1 \\ 0, & \text{otherwise} \end{cases}$ $g(x) \equiv 0$, all x.

   (c) $f(x) \equiv 0$, all x; $g(x) = \begin{cases} 1, & |x| \leq 1 \\ 0, & \text{otherwise.} \end{cases}$

2. Find some particular solutions of $u_{xx} + u_{yy} = 0$ other than linear polynomials in x or y. [*Hint*: Write the equation as $u_{xx} - i^2 u_{yy} = 0$, and recall that real and imaginary parts of complex-valued solutions must be real-valued solutions of $u_{xx} + u_{yy} = 0$.]

3. (a) Prove part (b) of the Constancy Theorem.

   (b) Prove part (a) of the Constancy Theorem directly by parameterizing the characteristic line segment $x - ct = k$ as $x = k + cs$, $t = s$, for $s_1 \leq s \leq s_2$ and using the Chain Rule.

4. Show that the boundary problem

$$u_{xy} = -au, \qquad u(x, 0) = 1, \qquad u(0, y) = 1, \qquad x \geq 0, \ y \geq 0$$

where a is a positive constant, has a solution $u = J_0(2(axy)^{1/2})$, where J_0 is the Bessel function of the first kind of order zero.

5. Show that the following initial value problem is not well posed:

$$u_{tt} + u_{xx} = 0, x \text{ in } \mathbf{R}, \qquad t > 0$$

$$u(x, 0) = f(x), x \text{ in } \mathbf{R}$$

$$u_t(x, 0) = g(x), x \text{ in } \mathbf{R}$$

[*Hint:* Observe that $u \equiv 0$ is a solution if $f \equiv g \equiv 0$, and observe that $u_m^* = m^{-2}(\sin mx)(\sinh mt)$ is a solution of the initial value problem with data $f_m^* \equiv 0$ and $g_m^*(x) = m^{-1} \sin mx$ for each positive integer m, but that $u_m^* \not\to 0$ as $m \to \infty$ even though f_m^* and $g_m^* \to 0$ as $m \to \infty$.]

13.3 BOUNDARY/INITIAL VALUE PROBLEMS FOR THE WAVE EQUATION

In Section 13.2 we derived the equation of transverse motion of a taut, flexible string under some simplifying assumptions. We found there that this motion is described by the second-order homogeneous linear partial differential equation

$$u_{tt} - c^2 u_{xx} = 0 \tag{1}$$

called the wave equation in one space variable. Note that any consistent set of units can be used in (1). The variables u and x carry units of length, t of time, and the constant c has units of length/time.

The effect of the boundary points of a finite string were not considered at all in Section 13.2. We saw that initial data (i.e., initial deflection and initial velocity of the string) specified for the infinite string determine a unique solution of the wave equation on the half-space $H : -\infty < x < \infty, t > 0$. We are now ready to consider the effect of various conditions at the boundaries of a finite string. It is assumed that the string has length L (i.e., that $0 \leqq x \leqq L$) and that time is measured forward from the initial time $t = 0$.

Conditions at the Ends of the String: Boundary Operators

Only transverse motions of the string are considered. Thus if the endpoints of the string (i.e., $x = 0$, $x = L$) are allowed to move, we may imagine the string to be looped around transverse rods at these two ends, but otherwise free to move (other interpretations are, of course, possible). There are three types of boundary conditions of interest. The simplest is when the transverse displacement of an endpoint is a prescribed function of time. For example, at $x = 0$ we may require that

$$u(0, t) = \alpha(t), \qquad t \geqq 0 \tag{2}$$

where $\alpha(t)$ is a given function. If $\alpha(t) = 0$, the string is fastened securely to the x-axis at $x = 0$. This special case is called a *fixed boundary condition*. A similar condition might be imposed at $x = L$.

In the second kind of boundary condition the endpoint is acted on by a transverse force, say $F = \gamma(t)$. Taking the endpoint to be $x = 0$ and applying this force and the transverse component of the tensile forces described in Section 13.2 to the segment $0 \leqq x \leqq h$ of the string, we have by Newton's Second Law that $\rho h u_{tt}(x^*, t) = T u_x(h, t) + \gamma(t)$, where ρ is the density of the string, x^* is the center of mass of the string segment $[0, h]$ and T is the (constant) tension in the string. Let h decrease to 0 in this equation. Since $u_{tt}(x, t)$ is bounded in x for each $t \geqq 0$, the left-hand side $\rho h u_{tt}(x^*, t)$ tends to 0. Hence this second kind of condition has the form

$$T u_x(0, t) + \gamma(t) = 0, \qquad t \geqq 0 \tag{3}$$

where $\gamma(t)$ is given. If $\gamma(t) \equiv 0$, it is a *free boundary condition*. A similar condition may be imposed at the endpoint $x = L$.

Finally, suppose that the left endpoint of the string is connected to the point $x = 0$ on the x-axis by a spring with spring constant k. Then the function $\gamma(t)$ in (3) must be $-ku(0, t)$. Hence we have that

$$T u_x(0, t) - ku(0, t) = 0, \qquad t \geqq 0 \tag{4}$$

which is called an *elastic boundary condition*. A similar condition could be imposed at the other endpoint.

The boundary conditions above may be described in terms of *boundary operators*. For example, the condition $u_x(0, t) - 4u(0, t) = 0$ may be written as $B[u(x, t)] = 0$, where the operator B is defined by $B[u(x, t)] = u_x(0, t) - 4u(0, t)$. Note that B may be applied to any function for which $u(0, t)$ and $u_x(0, t)$ are defined. Note also that B is a linear operator: $B[au + bv] = (au + bv)_x(0, t) - 4(au + bv)(0, t) = a[u_x(0, t) - 4u(0, t)] + b[v_x(0, t) - 4v(0, t)] = aB[u] + bB[v]$. The operators associated with the other boundary conditions are also linear. All of this may be formalized in the following way.

Boundary Conditions Expressed with Operators.   The fixed, free, and elastic boundary conditions at $x = 0$ described in (2)–(4) above may be expressed as $B_0[u] = 0$, where the boundary operator B_0 is defined respectively by:

(a) *Fixed Boundary*: $B_0[u(x, t)] \equiv u(0, t)$

(b) *Free Boundary*: $B_0[u(x, t)] \equiv u_x(0, t)$

(c) *Elastic Boundary*: $B_0[u(x, t)] \equiv T u_x(0, t) - ku(0, t)$

A similar definition holds for fixed, free, or elastic boundary conditions at $x = L$ with boundary operators B_L.

Since the boundary operators B_0 (and B_L) in the definitions above are linear operators, it follows that the null space of any one of them is closed under the operation of forming linear combinations. This fact will be of some importance later. The operators above may also be used to express nonhomogeneous boundary conditions such as (3): $B_0[u(x, t)] \equiv u_x(0, t) = -\gamma(t)/T$.

The Mixed Initial/Boundary Value Problem for the Vibrating String

As we saw in Section 13.2, the deflection of a vibrating string, $u(x, t)$ must satisfy the *wave equation* (1) in the region $R = \{(x, t) : 0 < x < L, t > 0\}$, provided that $u(x, t)$ in $C^2(R)$ satisfies some simplifying conditions. We observed above that if u is in $C^1(\text{cl } R)$, it is appropriate to demand that u and its first partial derivative with respect to x satisfy certain conditions at the boundaries $x = 0$ and $x = L$ for all $t \geq 0$ (depending on the character of the restraint of the string at the endpoints). Of course, the initial deflection $u(x, 0)$ and the initial velocity $u_t(x, 0)$ of the string may also be specified. Thus, in mathematical terms, the motion of a vibrating string with given initial and endpoint conditions will be a function u in $C^2(R)$ and $C^1(\text{cl } R)$ which satisfies all of the relations below inside the region R or on its boundaries (see Figure 13.4):

$$
\begin{array}{llll}
\text{(PDE)} & u_{tt} - c^2 u_{xx} = 0 & \text{for } 0 < x < L, & t > 0 \\[4pt]
\text{(IC)} & u(x, 0) = f(x), & 0 \leq x \leq L & \\
& u_t(x, 0) = g(x), & 0 \leq x \leq L & \text{(5)} \\[4pt]
\text{(BC)} & B_0[u] = \phi(t), & t \geq 0 & \\
& B_L[u] = \mu(t), & t \geq 0 &
\end{array}
$$

where $f(x)$, $g(x)$, $\phi(t)$, and $\mu(t)$ are given functions and the boundary operators B_0 and B_L may be (independently) any of the three types of boundary operators mentioned above. The problem (5) with *initial conditions* (IC) and *boundary conditions* (BC) can be shown not to have more than one solution (see the problem set) for any appropriately smooth *initial data* $[f(x)$ and $g(x)]$ and *boundary data* $[\phi(t)$ and $\mu(t)]$.

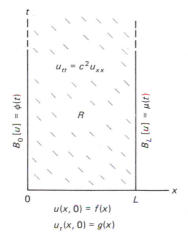

Figure 13.4 Geometry for problem (5).

We shall solve problem (5) in the particular case of fixed boundary conditions (i.e., the ends of the string are clamped).

The Method of Images

We shall show how the D'Alembert solution of the Cauchy Problem of Section 13.2 can be used to solve the mixed initial/boundary value problem (5) in the case where the boundary data vanishes [i.e., $\theta(t) \equiv \mu(t) \equiv 0$]. Called the *method of images*, an example will clarify the procedure. Let us consider the case of a taut flexible string of length L which is fastened at both endpoints. Then the motion of the string is characterized by the problem (5) with the fixed boundary conditions

$$u(0, t) = 0, \quad u(L, t) = 0 \qquad \text{for all } t \geq 0 \tag{6}$$

It is remarkable that we can construct a suitable Cauchy Problem for an infinite string whose solution will turn out to be a solution of this finite string problem as well. Let $\bar{f}(x)$ and $\bar{g}(x)$ be the initial data for the finite string problem. Observe that if initial data $f(x)$, $g(x)$ used in the Cauchy problem of Section 13.2 are periodic with period $2L$ and odd about $x = 0$ and $x = L$, the D'Alembert formula of Section 13.2 implies that the fixed conditions (6) will automatically be satisfied.† The question then is whether or not we can find such data $f(x)$, $g(x)$ on **R** which coincide with $\bar{f}(x)$, $\bar{g}(x)$ on $0 \leq x \leq L$. This may easily be done as follows. Extend $\bar{f}$, $\bar{g}$ into $[-L, 0]$ as odd functions about $x = 0$, and then define f, g to be the periodic extension of these functions on $[-L, L]$ into **R** with period $2L$ (see Figures 13.5 and 13.6.) That the data f and g constructed in this manner satisfy the stated properties is left to the reader. Now in order to guarantee the required smoothness properties for f and g, it is necessary and sufficient that

$$\bar{f} \text{ is in } C^2[0, L], \qquad \bar{g} \text{ is in } C^1[0, L]$$
$$\bar{g}(0) = \bar{g}(L) = 0 \quad \text{and} \quad \bar{f}(0) = \bar{f}(L) = \bar{f}''(0^+) = \bar{f}''(L^-) = 0$$

The proof of this fact is omitted. Using the D'Alembert formula of Section 13.2, we easily verify that the solution $u(x, t)$ of the Cauchy Problem with the data f, g constructed above does in fact solve the finite string problem with fixed endpoints.

Finite string problems but with free endpoint conditions (or one endpoint fixed and the other free) can be treated similarly (see the problem set). We may also seek

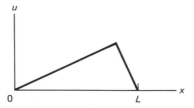

Figure 13.5   Initial profile of plucked string.

† A function $h(x)$ on **R** is *odd* about a point x_0 if $h(x_0 - r) = -h(x_0 + r)$ for all $r \geq 0$ (i.e., h takes opposite values at points which are symmetric about x_0. Note that $h(x_0) = 0$.

solutions of the finite string problem when the data $\bar{f}$ and $\bar{g}$ only satisfy the smoothness requirements piecewise. The procedure above works perfectly well in this case and it is not hard to see that discontinuities in the data propagate into the region $\{(x, t) : 0 < x < L, t > 0\}$ along characteristic line segments $x \pm ct = s_0$ (where s_0 is a point of discontinuity for the data) and their "reflections" in the lines $x = 0$, $x = L$.

Example 13.8

Suppose that a string of length L clamped at both ends is "plucked" and released from rest. What will the subsequent motion be? If the string is not plucked too vigorously then the deflection $u(x, t)$ will be a solution to the problem.

$$
\begin{aligned}
u_{tt} &= c^2 u_{xx}, & 0 < x < L, \quad t > 0 \\
u(0, t) &= u(L, t) = 0, & t \geq 0 \\
u(x, 0) &= f(x), & 0 \leq x \leq L \\
u_t(x, 0) &= 0, & 0 \leq x \leq L
\end{aligned}
\tag{7}
$$

where $f(x)$ is the function defined graphically in Figure 13.5. Let $\bar{f}(x)$ be the periodic extension into **R** of the odd extension of f into $[-L, L]$ (see Figure 13.6).

Using the D'Alembert formula, we see that

$$
u(x, t) = \tfrac{1}{2}\bar{f}(x + ct) + \tfrac{1}{2}\bar{f}(x - ct)
\tag{8}
$$

which displays the solution of (7) as a sum of two traveling waves, one moving to the left and the other to the right with velocity c. In Figure 13.7 we have used the solution formula (8) to plot a profile of the string at a time T where $cT = L/2$.

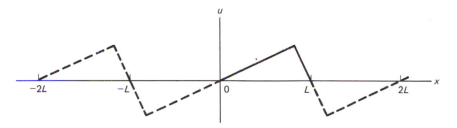

Figure 13.6 Extension of plucked initial string profile required by the method of images.

Standing Waves and Separated Solutions

There is a class of separated solutions for the wave equation (1) which have considerable significance in the theory of acoustics. These are the *standing waves* $X(x)T(t)$, where $T(t)$ is a periodic function of time. [More accurately, $X(x)$ is the standing wave

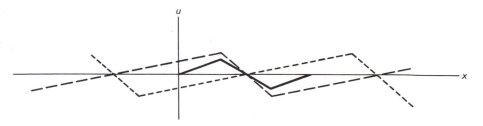

Figure 13.7   Profile of plucked string at time $t = L/(2c)$.

with amplitude modulated by $T(t)$]. Substituting $X(x)T(t)$ into (1) and separating variables, we see that all separated solutions of (1) have the form

$$X''(x) = \lambda X(x), \qquad T''(t) = c^2 \lambda T(t) \tag{9}$$

for any constant λ. Thus we see that for standing waves we must have $\lambda < 0$ for $T(t)$ to be periodic.

Example 13.9

Let us consider the special case of a guitar string of length L with tension T, density ρ, and hence $c^2 = T/\rho$. Since it is clamped at both ends, any standing wave supported by the string must satisfy the boundary conditions $X(0) = X(L) = 0$. Thus to find all standing waves, we see from (9) that we must first determine all constants λ such that the problem

$$\begin{aligned} X''(x) &= \lambda X(x) \\ X(0) &= X(L) = 0 \end{aligned} \tag{10}$$

has a nontrivial C^2-solution. This is an example of a *Sturm–Liouville Problem* (more on such problems in Section 13.5). Notice that if $X(x)$ is any C^2- function on $[0, L]$ such that $X(0) = X(L) = 0$, integration by parts shows that

$$\int_0^L X(x)X''(x)\, dx = -\int_0^L (X'(x))^2\, dx \leqq 0 \tag{11}$$

Thus if $X(x)$ is a solution of (10) for some constant λ, (11) implies that

$$\lambda \int_0^L (X(x))^2\, dx = \int_0^L X(x)X''(x)\, dx \leqq 0$$

and hence $\lambda \leqq 0$. Now clearly (10) does not have a nontrivial solution for $\lambda = 0$, so let us try $\lambda = -k^2$ for some $k > 0$. The general solution of $X'' + k^2 X = 0$ is $X(x) = A \cos kx + B \sin kx$ for arbitrary constants A and B. Thus the condition $X(0) = 0$ implies that $A = 0$, and the condition $X(L) = 0$ implies that $B \sin kL = 0$. Wishing to avoid the trivial solution, we assume that $B \neq 0$ and hence we must take $kL = n\pi$, where n is a nonnegative integer. Thus for any value $\lambda_n = -(n\pi/L)^2$, $n = 1, 2, \ldots$, problem (10)

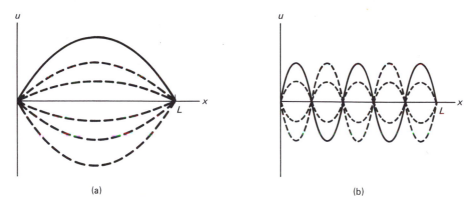

Figure 13.8   Standing waves for a guitar string: (a) $n = 1$; (b) $n = 5$.

has the corresponding solution $X_n(x) = \sin(n\pi x/L)$, up to an arbitrary multiplicative constant. Using this fact together with (9), we see that all standing waves for the guitar string have the form

$$u_n(x, t) \equiv \sin \frac{n\pi x}{L}\left(A_n \cos \frac{n\pi ct}{L} + B_n \sin \frac{n\pi ct}{L}\right), \quad n = 1, 2, 3, \ldots \quad (12)$$

where A_n and B_n are arbitrary constants. If time is measured in seconds, the *period* of the standing wave u_n is $2L/nc$ seconds (per cycle), and the standing wave recovers its initial sinusoidal profile $u_n(x, 0) = A_n \sin(n\pi x/L)$ at $t = k(2L/nc)$ seconds for every integer k. The *frequency* of u_n is the reciprocal of the period: $f_n = nc/2L$ cycles/second (or hertz). The numbers f_n are sometimes said to be the *natural frequencies supported by the string*. Observe that the points $x = 0, L/n, \ldots, (n - 1)L/n, L$, are rest points, or *nodes*, of the standing wave u_n. In Figure 13.8 we have sketched two standing waves; the solid line is the initial profile of the string, while the dashed lines show the profile of the string at later moments. Observe, finally, that the standing wave u_n is a solution of problem (5) with fixed boundary conditions and initial conditions

$$u(x, 0) = A_n \sin \frac{n\pi x}{L}$$

$$u_t(x, 0) = \frac{n\pi c}{L} B_n \sin \frac{n\pi x}{L}$$

A Musical Interlude

The various standing waves (12) and their frequencies have acquired musical names over the years because of their association with musical tones and instruments. The standing wave

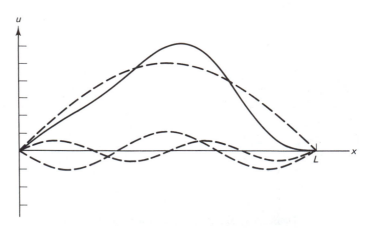

Figure 13.9   Superposition of standing waves.

$$u_1(x, t) = \left(A_1 \cos \frac{\pi ct}{L} + B_1 \sin \frac{\pi ct}{L} \right) \sin \frac{\pi x}{L}$$

is called the *fundamental* or *first harmonic*, while the standing waves of higher frequency are the *overtones* or *higher harmonics*. For example,

$$u_3(x, t) = -2 \cos \frac{3\pi ct}{L} \sin \frac{3\pi x}{L}$$

is a second overtone (or third harmonic). The waveform sketched in Figure 13.9 is the profile of a fundamental plus a third and a fourth harmonic. Each stringed instrument produces its own characteristic combinations of fundamentals and overtones and it is the variety of these combinations that distinguishes a guitar from a harp, or even one guitar from another.

The comments above seem to imply that it is the physical vibrations of the guitar string, violin string, or piano wire that we actually hear. Of course, what really happens is that these mechanical vibrations of the wire cause corresponding variations in the air pressure, variations that we hear as various musical sounds. In fact, it may be shown (but we do not) that air pressure itself is a solution of a wave equation, where the constant c^2 is a ratio of specific heats and of standard air pressures and densities. Thus what our ears actually detect are air-pressure waves set in motion by the vibrations of a string, a larynx, or some other source of sonic energy.

The Plucked Guitar String Revisited: Superposition of Standing Waves

We saw above how the plucked guitar string problem (7) could be solved using the method of images together with D'Alembert's formula. It is a remarkable fact that problem (7) can also be solved as a superposition of the standing waves $u_n(x, t)$

defined in (12), where the constants A_n, B_n are determined appropriately for all $n = 1, 2, \ldots$. As we shall see in later sections, this technique can be generalized to an important method for solving boundary/initial value problems for partial differential equations. Indeed, let us suppose that a solution $u(x, t)$ of problem (12) has the form $u(x, t) = \sum_{n=1}^{\infty} u_n(x, t)$, where the constants A_n, B_n are such that the series converges at each point in the closed region $S = \{(x, t) : 0 \leq x \leq L, t \geq 0\}$. We now describe a procedure for finding the values that the coefficients A_n, B_n must necessarily have in order that the series $\sum_{n=1}^{\infty} u_n(x, t)$ converge to a "solution" of problem (12). First note that each u_n satisfies the string endpoint conditions, and hence so does the sum $\sum u_n$. If the partial derivatives $\partial^2 u / \partial x^2$ and $\partial^2 u / \partial t^2$ can be computed by differentiation of the series $\sum u_n$ term by term, $u = \sum u_n$ satisfies the wave equation because each u_n does. Thus the constants A_n, B_n must be determined so that the series $\sum u_n$ satisfies the initial conditions. This evidently implies that

$$f(x) = u(x, 0) = \sum_{n=1}^{\infty} A_n \sin \frac{n \pi x}{L}, \qquad 0 \leq x \leq L \qquad (13)$$

$$0 = u_t(x, 0) = \sum_{n=1}^{\infty} \frac{n \pi c}{L} B_n \sin \frac{n \pi x}{L}, \qquad 0 \leq x \leq L \qquad (14)$$

Now comes a derivation which is of crucial importance in determining the constants A_n, B_n in (13) and (14). Notice that for all integers m and n we have the integral formula

$$\int_0^L \sin \frac{n \pi x}{L} \sin \frac{m \pi x}{L} \, dx = \begin{cases} 0, & m \neq n \\ L/2, & m = n \end{cases} \qquad (15)$$

Thus if we multiply both sides of (13) and (14) by $\sin (m \pi x / L)$ and integrate over $0 \leq x \leq L$ (and assume that the integration and infinite summation can be interchanged), we obtain

$$\int_0^L f(x) \sin \frac{m \pi x}{L} \, dx = A_m \left(\frac{L}{2} \right), \qquad m = 1, 2, \ldots$$

$$0 = \frac{m \pi c}{L} B_m \left(\frac{L}{2} \right), \qquad m = 1, 2, \ldots$$

which uniquely determines A_m and B_m. Hence if our assumptions are valid, we are led to the characterization of the solution of problem (7) as the series

$$u(x, t) = \sum_{n=1}^{\infty} A_n \sin \frac{n \pi x}{L} \cos \frac{n \pi c t}{L} \qquad (16)$$

where

$$A_n = \frac{2}{L} \int_0^L f(x) \sin \frac{n \pi x}{L} \, dx \tag{17}$$

We shall show later in Section 13.7 that if the data $f(x)$ are "nice" enough, then (16) with coefficients as in (17) does indeed solve problem (7), and since we know that this problem cannot have more than one solution, it must be *the* solution of problem (7).

Comments

As we hinted above, boundary/initial value problems can be solved by the technique of superposition of standing waves. Although this technique is rather limited in its scope, it is very important since it provides explicit solution formulas for a wide class of commonly occurring boundary/initial value problems. It is called the Method of Separation of Variables, and we will discuss this approach in more detail in Sections 13.7 to 13.9.

PROBLEMS

The following problem(s) may be more challenging: 6.

1. Plot profiles of the solution of the plucked guitar string problem (7) for the times
   (a) $T = 3L/4c$.
   (b) $T = L/c$.
   (c) $T = L/8c$.

2. Use the method of images to solve problem (5) with
   (a) Free boundary conditions at the endpoints. [*Hint*: Extend f and g into $[-L, 0]$ as even functions, and then into **R** periodically.]
   (b) A fixed boundary condition at $x = 0$ and a free boundary condition at $x = L$.

3. (a) Find all standing waves for the vibrating string of length L with free boundary conditions at the endpoints.
   (b) Use a superposition of standing-wave technique to solve problem (5) with free boundary conditions.

4. Verify directly that the Sturm–Liouville Problem (10) has no nontrivial solutions when either $\lambda = 0$, or $\lambda > 0$.

5. Verify the integration formulas (15). [*Hint*: Use the identity $\sin \alpha \sin \beta = \frac{1}{2}[\cos (\alpha - \beta) - \cos (\alpha + \beta)]$.]

6. [*Uniqueness Theorem for Problem* (5)]. Consider the region $R = \{(x, t) : 0 < x < L, t > 0\}$, and let u, v in the classes $C^2(R)$ and $C^1(clR)$ be solutions of the wave equation $w_{tt} = c^2 w_{xx}$ in R such that $u(x, 0) = v(x, 0)$ and $u_t(x, 0) = v_t(x, 0)$ for all $0 < x < L$, and $B_0[u] = B_0[v]$, $B_L[u] = B_L[v]$ for all $t > 0$, where B_0, B_L are any of the three boundary operators defined in this section. Show that $u \equiv v$ in R. [*Hint*: Use the Constancy Theorem (see Section 13.2) to show that $w \equiv u - v$ vanishes in R.]

13.4 SHAKING A STRING TO REST

Suppose that one end of a vibrating string is pinned down. How can the other end be shaken in order to bring the string to rest in minimal time? What is the minimal time? These two questions are answered in this section. Although control problems of this general nature have a long history, the particular problem posed above was only solved in 1973.†

The Optimal Control Problem

We shall consider a taut, flexible string of length 1. It is assumed for simplicity that the coefficient c^2 in the corresponding wave equation is also 1. The motion of the string with one end fastened and the other initially fastened and subsequently shaken may be modeled by the boundary/initial value problem

$$
\begin{array}{llll}
\text{(PDE)} & u_{tt} = u_{xx}, & 0 < x < 1, \quad t > 0 & \\
\text{(IC)} & u(x, 0) = f(x), & u_t(x, 0) = g(x), \quad 0 \le x \le 1 & \text{(1)} \\
\text{(BC)} & u(0, t) = 0, & u(1, t) = S(t), \quad t \ge 0 &
\end{array}
$$

The initial data f and g and the control function $S(t)$ are assumed to be continuous and piecewise smooth. To ensure continuity around the boundary we must have the end point conditions $f(0) = f(1) = S(0) = 0$, $g(0) = 0$, and $g(1) = S'(0)$. Such functions f, g, S are *admissible*. Other restrictions are imposed later to ensure the uniqueness of the solution $u(x, t)$ of (1). The specific optimal control problem to be solved is as follows:

Find the least positive time T such that for all admissible f and g there is an admissible control S which brings the string to rest at time T,

$$
u(x, T) = 0, \qquad u_t(x, T) = 0, \quad 0 \le x \le 1 \tag{2}
$$

Find the corresponding control function S.

It is shown in the problems that if condition (2) holds, the string is undeflected and at rest for all subsequent time $t \ge T$, assuming, of course, that the control is turned off at time T, $S(t) \equiv 0$ for $t \ge T$. Thus we need only consider (1) and (2) on $R : 0 < x < 1, 0 < t < T$ and on cl $R : 0 \le x \le 1, 0 \le t \le T$.

The problem is solved through a sequence of steps:

1. Show that the minimal time T is at least 2.
2. Assuming that $T = 2$, construct the optimal control S.
3. With the control S and $T = 2$, construct a solution $u(x, t)$ of (1) and (2).

† P. C. Parks, "On how to shake a piece of string to a standstill," in D. J. Bell (ed.), *Recent Mathematical Developments in Control* (New York: Academic Press, 1973), pp. 267–287.

4. Show that Eqs. (1) and (2) have a unique solution if $T = 2$ and S is the control constructed above (additional conditions are imposed on the data at this step).

Thus it will follow that the minimal time is indeed $T = 2$ and that the optimal control S constructed corresponds to this minimal time.

The Minimal Time T is at Least 2

Suppose, to the contrary, that there is a time T, $0 < T < 2$, such that for arbitrary admissible data f and g there is an admissible control S which brings the string to rest at time T. The corresponding space-time region is sketched in Figure 13.10. Note that $u_x \equiv 0$ on TB (as well as $u_t \equiv 0$) since $u \equiv 0$ on TB. Note also that $u_t \equiv 0$ on OT since the end $x = 0$ is permanently fastened. The characteristics for (1) are the lines $x + t = $ const. and $x - t = $ const.; along each of the former $u_x + u_t$ is fixed, while $u_x - u_t$ is unchanged along each of the latter (the Constancy Theorem of Section 13.2). Thus the value 0 of $u_x - u_t$ on TB is carried down to the left along the characteristics $x - t = $ constant to the segment CT. Since u_t vanishes on CT, we conclude that $u_x = 0$ on CT as well. An application of the Constancy Theorem along all characteristics from CT and from TB then shows that $u_x + u_t \equiv 0$ inside the triangle CBT. The Constancy Theorem may also be used to carry the values of $u_x + u_t$ from the bottom edge OA along characteristics $x + t = $ const. and into the triangle CDE: $u_x + u_t = f'(x + t) + g(x + t)$ for all (x, t)

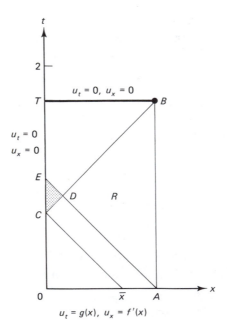

Figure 13.10 Space-time geometry and characteristics for the shaken string: $T < 2$.

inside CDE. But $u_x + u_t \equiv 0$ inside CDE by the previous argument. Hence this means that for all z in some interval $(\bar{x}, 1)$.

$$0 = f'(z) + g(z) \tag{3}$$

But it is *not* the case that (3) holds for *all* admissible data functions f and g. This contradiction shows that the minimal time T is at least 2.

The Optimal Control Function

Suppose, for now, that T is exactly 2. Let us see what the optimal control function S must be in this case. We shall construct S as a function of arbitrary admissible data f and g. This is done by a further application of the Constancy Theorem. The sketches in Figure 13.11 illustrate the process of carrying the values (both 0) of $u_x + u_t$ and $u_x - u_t$ on the top edge of R downward along characteristics to combine with known values of u_t (or u_x) on other edges and thereby deduce the value of u_x (or u_t) on these other edges.

From Figure 13.11(c), since $u_x + u_t$ has a fixed value along each characteristic line $x + t = $ constant, we have that

$$-2S'(1+t) = g(t) + f'(t), \qquad 0 \le t \le 1 \tag{4}$$

By a similar process, but moving off the top edge downward to the left along characteristics, rather than to the right, it may be shown that

$$2S'(t) = g(1-t) - f'(1-t), \qquad 0 \le t \le 1 \tag{5}$$

Setting $s = 1 + t$ in (4) and then renaming s by t, (4) may be replaced by an equivalent equation valid for $1 \le t \le 2$. Equation (5) and the new equation equivalent to (4) characterize S' for $0 \le t \le 2$:

$$S'(t) \equiv H(t) = \begin{cases} \frac{1}{2}[g(1-t) - f'(1-t)], & 0 \le t \le 1 \\ -\frac{1}{2}[g(t-1) + f'(t-1)], & 1 \le t \le 2 \end{cases} \tag{6}$$

Since $S(0) = 0$ (one of the admissibility conditions), (6) may be integrated to obtain the control function $S(t)$:

$$S(t) = \int_0^t H(s)\, ds, \qquad 0 \le t \le 2 \tag{7}$$

Formula (7) gives the control function for shaking the string to rest in 2 units of time, assuming, as we have all along, that this can indeed be accomplished (i.e., that the minimal time is indeed 2).

Solving the Boundary/Initial Value Problem

Still assuming that the control problem has a solution with $T = 2$, we may insert the control function S just calculated into (1) and solve for $u(x, t)$, subject to the additional boundary condition (2). This is done in two steps. First, the Constancy

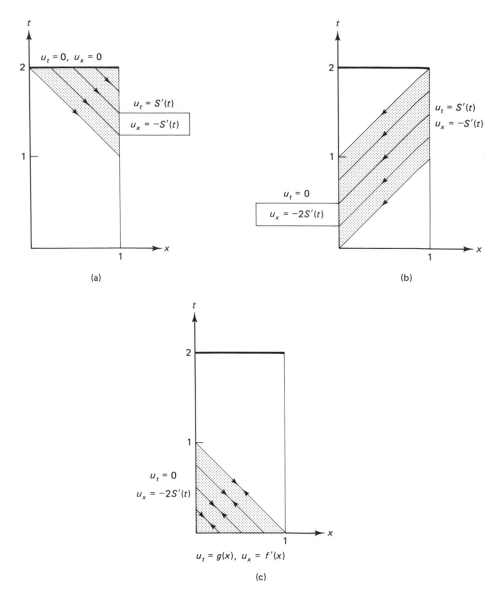

Figure 13.11 Carrying data along characteristics: (a) finding u_x on upper right; (b) finding u_x on lower left; (c) matching data on lower left and bottom.

Theorem may be used much as above to derive formulas for $u_t - u_x$ and $u_t + u_x$ valid in R:

$$u_t - u_x = \begin{cases} g(x-t) - f'(x-t), & 0 \le x - t \le 1 \\ -\{g(t-x) + f'(t-x)\}, & -1 \le x - t \le 0 \\ 0, & -2 \le x - t \le -1 \end{cases} \qquad (8)$$

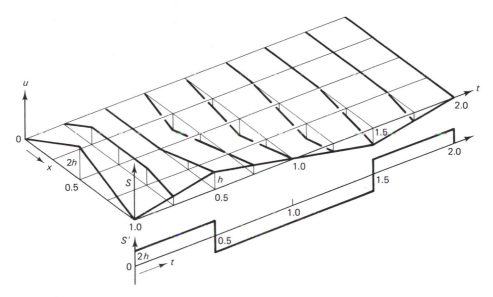

Figure 13.12   Shaking the plucked string to a standstill in two units of time. (Adapted by permission of the Institute of Mathematics and its Applications and the Editor from "On How to Shake a Piece of String to a Standstill," P. C. Parks; in *Recent Mathematical Developments in Control*, D. J. Bell, editor, Academic Press, 1973.)

$$u_t + u_x = \begin{cases} g(x+t) + f'(x+t), & 0 \le x+t \le 1 \\ 0, & 1 \le x+t \le 2 \\ 0, & 2 \le x+t \le 3 \end{cases} \tag{9}$$

The derivation of (8) and (9) is left to the problem set.

The second step consists of using (8) and (9) to express u_t and u_x in terms of the data f and g. Then $u(x, t)$ is found by integration, using the known values of u on the boundary of R. We shall not carry out the details of this process.

Figure 13.12 illustrates the solution in the particular case of a plucked string whose vibrations are stilled at $T = 2$ by the piecewise linear control S indicated. The calculations needed to determine S via (6) and (7) are quite simple for a plucked string since f' is piecewise constant and $g(x) \equiv 0$.

With all the constructions above, we have shown that a string can indeed be shaken to rest during the time span $0 \le t \le 2$, and moreover, a control function has been given that will accomplish this fact. We have now shown that (1) and (2) have a solution u under these conditions.

Additional Conditions to Ensure the Uniqueness

The uniqueness theorem require that the boundary and initial data be sufficiently smooth. In essence, to ensure that (1) and (2) have a unique solution $u(x, t)$ belonging to $C^2(R)$ and to $C^1(\text{cl } R)$, we require that the odd extension of $f(x)$ to the entire real line be twice continuously differentiable, while additional conditions are imposed to ensure that the data remain smooth around ∂R:

$$f \text{ belongs to } C^2[0, 1], \quad g \text{ to } C^1[0, 1] \tag{10a}$$

$$f(0) = f(1) = f''(0) = 0, \qquad g(0) = 0 \tag{10b}$$

$$g(1) + f'(1) = 0 \tag{10c}$$

$$g'(1) + f''(1) = 0 \tag{10d}$$

Some of these conditions represent a restatement or a strengthening of conditions imposed earlier. Others are consequences of the construction of the control function. For example, (10c) is a consequence of (5) and of the earlier requirement that $S'(0) = g(1)$.

The following result summarizes what is accomplished by these additional restrictions.

Existence and Uniqueness Theorem. Suppose that the data f and g satisfy (10a)–(10d), and that the control function is given by (7). Then (1) has a unique solution $u(x, t)$ which satisfies (2) for $T = 2$ and which lies in $C^2(R)$ and in $C^1(\text{cl } R)$.

The comments made earlier indicate the main points of the proof of this theorem, but the details are omitted. It should be noted that the example of the plucked string shown in Figure 13.12 does not meet all the conditions of the theorem since f is only piecewise smooth, yet the problem clearly has a solution even in this case.

Comments

We have shown that under reasonable restrictions on the initial deflection and velocity, a taut, flexible string fastened at one end can be brought to rest by a shaking action applied to the other end. Moreover, this can be accomplished in two units of time (assuming a string of length 1 with $c^2 = 1$) regardless of the initial deflection and velocity. The optimal shaking function $S(t)$ depends explicitly, of course, on the initial data [see (6) and (7)].

There are other curious problems of a similar nature. For example, it is possible to find a periodic control function $S(t)$ of arbitrarily small amplitude which will break the string in a finite time. Resonance is the phenomenon involved in this setting, the period of $S(t)$ being taken equal to a natural frequency of the vibrating string (see the problem set of Section 13.7).

PROBLEMS

The following problem(s) may be more challenging: 3, 4, and 5.

1. For a string of length 1 and $c^2 = 1$, find the control law $S(t)$ that will bring the string to rest if $g(x) \equiv 0$, and, for a fixed $0 < x_0 < 1$,

$$f(x) = \begin{cases} x(1 - x_0), & 0 \le x \le x_0 \\ x_0(1 - x), & x_0 \le x \le 1 \end{cases}$$

(See Figure 13.12 for the case $x_0 = 0.5$.)

2. Modify the procedure presented for shaking a string to rest so that it will apply when the string has length L, and the constant c in the wave equation is not necessarily unity.

3. Derive (8) and (9).

4. [*The Semigroup Property for the Wave Equation*]. Let $u(x, t)$ be the unique solution of the Cauchy problem (3) of Section 13.2 for a given initial data f and g. Let $T > 0$ be given and let $u*(x, t)$ be the unique solution of the Cauchy problem for the initial data $f*(x) = u(x, T)$, $g*(x) = u_t(x, T)$. Then show that $u*(x, t) \equiv u(x, t + T)$ for all x in **R** and all $t > 0$. In other words, the motion of the string that results when using the state of the string at $t = T$ as initial data is precisely the same as would have occurred had we used the original initial data and reset our clocks back T units when t reaches the time T. [*Hint*: Express $u(x, t)$ and $u*(x, t)$ each as the sum of two traveling waves (see Section 13.2).]

5. Suppose that $T > 2$. Show that for any admissible data there exists a solution of (1) which has the property that $u(x, T) = u_t(x, T) = 0$ for all $t \ge T$. Show, however, that control functions with this property are not unique. [*Hint*: The Semigroup Property in Problem 4 also holds for linear boundary/initial value problems.]

13.5 BOUNDARY VALUE PROBLEMS AND ORTHOGONAL FUNCTIONS

In Section 4.3 we defined the dot (or scalar) product for vectors in **R**³ and the consequent concept of orthogonality among such vectors. We saw that these concepts not only brought a certain geometric sense to the construction of models but significant computational simplifications as well. Problem 2 of Section 4.3 brings into focus the basic properties of the dot product which make it so useful as a computational device. In this section we begin by using some of these basic properties to define the notion of a scalar product on a linear space. Then we go on to set up specific scalar products on commonly occurring linear spaces of functions and show how orthogonality in such spaces provides a valuable tool for solving boundary/initial value problems for differential equations.

Scalar Products

It turns out that the basic properties of a scalar product are the symmetry, bilinearity, and positive-definite properties (observe from Problem 2 of Section 4.3 that all other properties of the dot product can be derived from just these three). This leads us to ask whether or not on *any* given linear space we can find a way to associate a scalar with each (ordered) pair of vectors so that these three properties are satisfied. When this is possible, we may imitate what was done for the dot product to develop notions of "orthogonality," "distance," and "convergence of a sequence" in the linear space.

Scalar Product.   Let V be a linear space, and suppose that $\langle \cdot, \cdot \rangle$ is a function which associates with every pair of vectors u, v the scalar $\langle u, v \rangle$. Then $\langle \cdot, \cdot \rangle$ is a *scalar product* if the following properties are satisifed:

$$(Symmetry) \quad \langle v, u \rangle = \overline{\langle u, v \rangle} \text{ for all } u, v \text{ in } V \tag{1a}$$

(Bilinearity) $\langle \alpha u + \beta v, w \rangle = \alpha \langle u, w \rangle + \beta \langle v, w \rangle$
for all scalars α, β and all u, v, w in V $\tag{1b}$

(Positive Definiteness) $\langle u, u \rangle \geq 0$ for all u in V
and $\langle u, u \rangle = 0$ if and only if $u = 0$ $\tag{1c}$

[*Note*: If V is a real linear space, the complex conjugate in (1a) is not needed.] A linear space V with a scalar product is called a *Euclidean space*.

There are many scalar products for the same linear space; examples follow. It is easy to check that $\langle x, y \rangle = \sum_1^n x_k y_k$ is a scalar product on $\mathbf{R}^n$; this *standard scalar product* makes $\mathbf{R}^n$ into a real Euclidean space and generalizes the dot product on $\mathbf{R}^3$ (see Problem 1 in Section 4.3). Similarly, $\langle x, y \rangle = \sum_1^n x_k \bar{y}_k$ makes $\mathbf{C}^n$ into a complex Euclidean space and is the standard scalar product on $\mathbf{C}^n$.

Since the scalar product of any vector with itself is always a nonnegative number (even for complex linear spaces), we can define the length of a vector in the same way as for the dot product.

Norm.   In a linear space V with scalar product $\langle \cdot, \cdot \rangle$ we define the *norm* (or *length*) of a vector u by

$$\|u\| = + (\langle u, u \rangle)^{1/2} \tag{2}$$

This norm is called *the norm induced by the scalar product* (other norms are possible). Vectors of length 1 are called *unit vectors*. (Any nonzero vector can be *normalized* by dividing it by its length.)

Continuing the analogy, we define

Distance between Two Vectors.   In a linear space V with scalar product $\langle \cdot, \cdot \rangle$ we define the *distance* between any two vectors u and v in V to be $\|u - v\|$, where $\|\cdot\|$ is the norm induced by $\langle \cdot, \cdot \rangle$.

Recall that a function $f(x)$ on a closed interval $[a, b]$ is said to be *piecewise continuous* if there are finitely many points $c_1, c_2, \ldots, c_k$ such that $a < c_1 < c_2 < \cdots < c_k < b$, and f is continuous on each open interval (a, c_1), (c_1, c_2), $\ldots$, (c_{k-1}, c_k), (c_k, b), and the one-sided limits of f exist at all the points $x = a, c_1$, $\ldots, c_k, b$. Examples are step functions. (*Note*: the actual values of f at the points $x = a, c_1, \ldots, c_k, b$ play no role in the definition. In fact, f need not even be defined at these points.) The set of all piecewise continuous functions on $[a, b]$ is denoted by PC$[a, b]$. Using the usual notion of addition of functions and multiplication of a function by a scalar, we see that PC$[a, b]$ is a linear space (a real linear space if the functions are real-valued and the scalars are real numbers).

Example 13.10

For the linear space PC$[a, b]$ it is easy to see that

$$\langle f, g \rangle = \int_a^b f(x)\overline{g(x)}\, dx, \qquad f, g \text{ in PC}[a, b] \tag{3}$$

is a scalar product provided that we agree to identify functions in PC$[a, b]$ which differ at only a finite set of points [otherwise (1c) fails to hold]. We still use the symbol PC$[a, b]$ for this Euclidean space. Notice that the statement $f = 0$ means that f vanishes in $[a, b]$ except at most on a finite set. The scalar product (3) is called the *standard scalar product* on PC$[a, b]$ (there are infinitely many others). Observe that since $C^n[a, b]$ can be considered a subspace of PC$[a, b]$, (3) is also a scalar product on $C^n[a, b]$. Finally, notice that the norm induced by (3) has the form

$$\|f\| = \left[\int_a^b |f|^2\, dx \right]^{1/2}, \qquad f \text{ in PC}[a, b] \tag{4}$$

Any scalar product $\langle \cdot, \cdot \rangle$ and its associated norm $\|\cdot\|$ have the following useful properties (proofs omitted):

Properties of Scalar Product.   Let V be a linear space with inner product $\langle \cdot, \cdot \rangle$ and corresponding norm $\|\cdot\|$. For all u, v in V, and scalars α the following hold:

(*Cauchy–Schwartz Inequality*) $|\langle u, v \rangle| \leq \|u\| \|v\|$.

(*Positive definiteness*) $\|u\| \geq 0$ and $\|u\| = 0$ if and only if $u = 0$.

(*Homogeneity*) $\|\alpha u\| = |\alpha| \|u\|$.

(*Triangle inequality*) $|\, \|u\| - \|v\| \,| \leq \|u + v\| \leq \|u\| + \|v\|$.

These properties are very useful in calculations and in producing error estimates in Euclidean spaces, as we shall see presently.

Orthogonality

A scalar product is used to define the orthogonality of vectors.

Orthogonality.   Two vectors u, v in a Euclidean space V are said to be *orthogonal* if and only if $\langle u, v \rangle = 0$. A subset S in V is an *orthogonal* *subset* if $\langle u, v \rangle = 0$ for any two vectors u, v in S. By convention we do not allow the zero vector to be in an orthogonal set.

Of course, the zero vector is orthogonal to all elements of V, but it is the only element with that property. This notion of orthogonality extends the familiar notion of perpendicularity in the standard Euclidean spaces $\mathbf{R}^3$ and $\mathbf{R}^2$. For instance, $(a, b) \neq (0, 0)$ in $\mathbf{R}^2$ is orthogonal to $(-b, a)$, and this coincides with the familiar fact that a pair of lines intersects orthogonally if their slopes are negative reciprocals. The notion of orthogonality in a function space such as $C^0[a, b]$ or $PC[a, b]$ has no such simple geometric interpretation.

Example 13.11

The set $\Phi = \{1, \cos x, \sin x, \ldots, \cos nx, \sin nx, \ldots\}$ is an orthogonal subset of $PC[-\pi, \pi]$ under the standard scalar product. Indeed, using a trigonometric identity we can show that for any integers m and n,

$$\langle \sin nx, \cos mx \rangle = \int_{-\pi}^{\pi} \sin nx \cos mx \, dx$$

$$= \int_{-\pi}^{\pi} \frac{1}{2} [\sin (n + m)x + \sin (n - m)x] \, dx = 0$$

The other orthogonality relations follow in a similar fashion.

Example 13.12

Let $PC[-T, T]$ consist of complex-valued functions. The set $\Psi = \{e^{ik\pi/T} : k = 0, \pm 1, \pm 2, \ldots\}$ is an orthogonal set in $PC[-T, T]$. Using the standard scalar product in $PC[-T, T]$, we have for $k \neq m$,

$$\langle e^{ik\pi x/T}, e^{im\pi x/T} \rangle = \int_{-T}^{T} e^{ik\pi x/T} e^{-im\pi x/T} \, dx = \frac{T e^{i(k-m)\pi x/T}}{i(k - m)\pi} \Big|_{-T}^{T} = 0$$

Mean Approximation

Let f be a given element in a Euclidean space E. If g in E is considered as an approximation to f, then $\|f - g\|$ is known as the *error in the sense of the mean*, or simply *error in the mean*. Suppose that $\Phi = \{\phi_1, \phi_2, \ldots\}$ is a finite or infinite orthogonal set in the Euclidean space E. Then for a given element f in E we may ask the following question: What linear combination of the first n elements of Φ is "closest" to f in the sense of the distance measure on E (i.e., *in the mean*)? In symbols: For what values of the constants $c_1, c_2, \ldots, c_n$ is $\|f - \sum_{k=1}^{n} c_k \phi_k\|$ a minimum? The answer to this question follows:

Mean Approximation Theorem. For any orthogonal set $\Phi = \{\phi_1, \phi_2, \ldots\}$ in a Euclidean space E, any f in E, and any positive integer n, the problem

$$\|f - \sum_{k=1}^{n} c_k\, \phi_k\| = \text{minimum}$$

has the unique solution

$$(c_k)_{\min} = \langle f, \phi_k \rangle / \|\phi_k\|^2, \qquad k = 1, 2, \ldots, n \tag{5}$$

Moreover,

$$\|f - \sum_{k=1}^{n} (c_k)_{\min}\phi_k\|^2 = \|f\|^2 - \sum_{k=1}^{n} |\langle f, \phi_k \rangle|^2/\|\phi_k\|^2 \tag{6}$$

Proof. For simplicity of notation we assume that E is a real Euclidean space. Now notice that by the bilinearity of the scalar product $\langle \cdot, \cdot \rangle$ in E,

$$\|f - \sum_{k=1}^{n} c_k\, \phi_k\|^2 = \langle f - \sum_{k=1}^{n} c_k\, \phi_k, f - \sum_{k=1}^{n} c_k\, \phi_k \rangle$$

$$= \|f\|^2 - 2 \sum_{k=1}^{n} c_k \langle f, \phi_k \rangle + \sum_{k=1}^{n} c_k^2 \|\phi_k\|^2 \tag{7}$$

Completing the square on the quadratic polynomial in c_k in (7), we see that

$$\|\phi_k\|^2 c_k^2 - 2\langle f, \phi_k \rangle c_k = \|\phi_k\|^2 \left(c_k - \frac{\langle f, \phi_k \rangle}{\|\phi_k\|^2} \right)^2 - \frac{|\langle f, \phi_k \rangle|^2}{\|\phi_k\|^2}$$

and hence (7) becomes

$$\|f - \sum_{k=1}^{n} c_k\, \phi_k\|^2 = \|f\|^2 + \sum_{k=1}^{n} \|\phi_k\|^2 \left(c_k - \frac{\langle f, \phi_k \rangle}{\|\phi_k\|^2} \right)^2 - \sum_{k=1}^{n} \frac{|\langle f, \phi_k \rangle|^2}{\|\phi_k\|^2} \tag{8}$$

To minimize the right-hand side of (8), it is clear that the c_k must be chosen as in (5), and when this is done, (6) results from (8), finishing the proof.

Example 13.13

It is easy to verify that $\Phi = \{\sin \dfrac{k\pi x}{L} : k = 1, 2, \ldots\}$ is an orthogonal set in PC[0, L], for any positive constant L, and that $\|\sin \dfrac{k\pi}{L}\|^2 = L/2$, for any $k = 1, 2, \ldots$. Let f be the function defined in Figure 13.13. Clearly f is in PC[0, L].

We shall find the best mean-square approximation to f in the form $c_1 \sin (\pi x/L) + c_2 \sin (2\pi x/L) + c_3 \sin (3\pi x/L)$. Using (5), we see that for any $k > 0$,

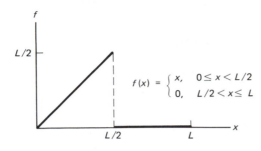

$$f(x) = \begin{cases} x, & 0 \le x < L/2 \\ 0, & L/2 < x \le L \end{cases}$$

Figure 13.13 Function f in PC[0, L].

$$(c_k)_{min} = \frac{2}{L} \int_0^{L/2} x \sin \frac{k\pi x}{L} dx = \frac{2L}{k\pi} \left(\frac{1}{k\pi} \sin \frac{k\pi}{2} - \frac{1}{2} \cos \frac{k\pi}{2} \right)$$

Thus $(c_1)_{min} = 2L/\pi^2$, $(c_2)_{min} = L/2\pi$, $(c_3)_{min} = -2L/9\pi^2$, and hence

$$g_{min} = \frac{2L}{\pi^2} \sin \frac{\pi x}{L} + \frac{L}{2\pi} \sin \frac{2\pi x}{L} - \frac{2L}{9\pi^2} \sin \frac{3\pi x}{L}$$

is the best approximation in the mean to f of the desired form. The square of the mean error (or the *mean-square error*) of this approximation is given by (5) and (6) as

$$\|f - g_{min}\|^2 = \|f\|^2 - \sum_{k=1}^{3} (c_k)^2_{min} \| \sin \frac{k\pi x}{L} \|^2 = \frac{L^3}{24} - \frac{L}{2} \left(\frac{4L^2}{\pi^4} + \frac{L^2}{4\pi^2} + \frac{4L^2}{81\pi^4} \right)$$

Convergence in the Mean

As we saw earlier, the induced norm $\|\cdot\|$ on a Euclidean space E measures the "distance" between two elements u, v in E as $\|u - v\|$. We can use this "distance" measure to define convergence of infinite sequences and series in E.

Convergence in the Mean.   Let $\|\cdot\|$ be the induced norm on a Euclidean space E. We say that the sequence $\{f_n\}$, $n = 1, 2, \ldots$ in E *converges in the mean* if and only if there exists an element f in E such that $\|f_n - f\| \to 0$, as $n \to \infty$. In symbols, we write $f_n \to f$ as $n \to \infty$. For g_k, $k = 1, 2, \ldots$, in E the series $\sum_{k=1}^{\infty} g_k$ is said to *converge in the mean* if and only if the sequence of partial sums $f_n = \sum_{k=1}^{n} g_k$, $n = 1, 2, \ldots$, converges in the mean. If $f_n \to f$ as $n \to \infty$, then we write $f = \sum_{k=1}^{\infty} g_k$.

As in ordinary convergence, a mean convergent sequence $\{f_n\}$ can have only one limit.

Mean Convergence Theorem. For any sequence $\{f_n\}$ in a Euclidean space E there is not more than one element f in E such that $\|f_n - f\| \to 0$ as $n \to \infty$.

Proof. Suppose that $f_n \to g$ and $f_n \to h$ as $n \to \infty$, where $g \neq h$. Now $\|g - h\| > 0$, and from the Triangle Inequality we have that

$$0 < \|g - h\| = \|(g - f_n) + (f_n - h)\| \leq \|f_n - g\| + \|f_n - h\|, \qquad \text{for all } n$$

Thus it is impossible that both $\|f_n - g\| \to 0$, and $\|f_n - h\| \to 0$, unless $g = h$, establishing our claim.

Mean convergence has another important property which we state without proof (see the problem set).

Continuity of the Scalar Product Theorem. Suppose that $f_n \to f$, and $g_m \to g$ as $n, m \to \infty$ in a Euclidean space. Then $\langle f_n, g_m \rangle \to \langle f, g \rangle$, as $n, m \to \infty$.

Example 13.14

Mean convergence of a sequence $\{f_n(x)\}$ in $PC[a, b]$ implies the existence of an element f in $PC[a, b]$ such that $\int_a^b (f_n(x) - f(x))^2 \, dx \to 0$ as $n \to \infty$. Consider the sequence g_k in $PC[0, 1]$, $k = 1, 2, \ldots$ defined in Figure 13.14. We claim that $\{g_k\}$ does *not* converge in the mean. Indeed, say $g_k \to g$, for some g in $P[0, 1]$. Writing $g_k = (g_k - g) + g$ and applying the Triangle Inequality, we have $\|g_k\| \leq \|g_k - g\| + \|g\|$, and hence the sequence $\{\|g_k\|\}$ is bounded. But since $\|g_k\|^2 = \int_0^1 |g_k|^2 dx = 2k/3 \to \infty$, as $k \to \infty$, it follows that $\{\|g_k\|\}$ is unbounded and hence $\{g_k\}$ cannot converge.

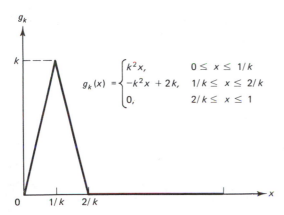

$$g_k(x) = \begin{cases} k^2 x, & 0 \leq x \leq 1/k \\ -k^2 x + 2k, & 1/k \leq x \leq 2/k \\ 0, & 2/k \leq x \leq 1 \end{cases}$$

Figure 13.14   Sequence $\{g_k\}$ in $PC[0, 1]$.

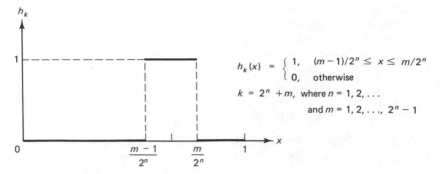

Figure 13.15 Sequence $\{h_k\}$ in PC[0, 1].

On the other hand, consider the sequence h_k in PC[0, 1], $k = 1, 2, \ldots$ defined in Figure 13.15. Note that h_k, for $k = 2^n + m$, has the value 1 inside the interval $[(m - 1)/2^n, m/2^n]$ and vanishes otherwise in [0, 1]. Thus $\int_0^1 |h_{2^n + m}|^2 \, dx = 1/2^n$ for all $n = 1, 2, \ldots$, and all $m = 1, 2, \ldots, 2^n - 1$. It follows that $\|h_k\|^2 = \|h_k - 0\|^2 \to 0$ as $k \to \infty$. Thus $\{h_k\}$ converges in the mean to the zero function on [0, 1], even though for each k there is an interval on which h_k has the value one.

Mean, Pointwise, and Uniform Convergence

Mean convergence for sequences in PC[a, b] is not to be confused with the usual notions of pointwise and uniform convergence for function sequences. We review now pointwise and uniform convergence for a sequence $\{f_n(x)\}$, $n = 1, 2, \ldots$, in PC[a, b] (see also the discussion in Appendix A). The sequence $\{f_n(x)\}$ is said to converge *pointwise* at x_0 in [a, b] if there is a value y_0 such that $f_n(x_0) \to y_0$ as $n \to \infty$ in the ordinary sense of number sequences. Restated, this means that for any interval of length $2\epsilon > 0$ centered at y_0, the values $f_n(x_0)$ "eventually" all remain in this interval, that is, for all $n \geq N$ for some positive integer N (see Figure 13.16).

Uniform convergence of the function sequence $\{f_n(x)\}$ on a closed interval

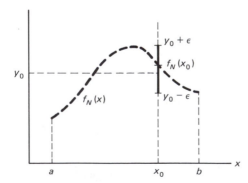

Figure 13.16 Geometry for pointwise convergence.

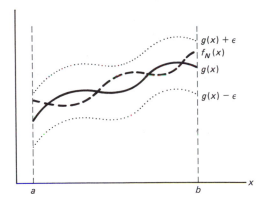

Figure 13.17   Uniform convergence.

$[a, b]$ is *more* than just saying that $\{f(x)\}$ converges pointwise at *each* x_0 in $[a, b]$. Uniform convergence of $\{f_n(x)\}$ to a function $g(x)$ on $[a, b]$ means that for any "tube" with center spine $g(x)$ and vertical diameter $2\epsilon > 0$, the graphs of $f_n(x)$ are "eventually" all contained in this tube, for all $n \geq N$ for some positive integer N (see Figure 13.17).

Example 13.15

It is easy to see that the first sequence $\{g_k\}$ in Example 13.14 converges pointwise to the zero function on $[0, 1]$. On the other hand, since $g_k(1/k) = k$ it is clear that the graphs of the $g_k(x)$ cannot eventually all lie in any tube about the zero function on $[0, 1]$. Thus $\{g_k\}$ does *not* converge uniformly on $[0, 1]$. It is true, however, that $\{g_k\}$ converges uniformly to the zero function on any interval of the form $[\delta, 1]$, where $0 < \delta < 1$. The second sequence $\{h_k\}$ considered in Example 13.14 does not converge pointwise for *any* point in $[0, 1]$. Indeed, for any x_0 in $[0, 1]$ and any positive integer K, there are integers $r > K, s > K$ such that $h_r(x_0) = 0$ and $h_s(x_0) = 1$. Thus the sequence $\{h_k(x_0)\}$ cannot converge.

Finally, it is simple to see that if the sequence $\{g_k\}$ in PC$[a, b]$ converges uniformly to g on $[a, b]$, then $g_k \to g$ in the mean (over PC$[a, b]$). This fact and Examples 13.14 and 13.15 provide Figure 13.18, summarizing the general relationship between the three modes of convergence in PC$[a, b]$ considered here. Uniform convergence implies mean convergence only if $-\infty < a < b < \infty$ (see Problem 12).

Orthogonal Series: Bases

Let $\Phi = \{\phi_1, \phi_2, \ldots\}$ be an orthogonal set in a Euclidean space E. The series $\sum_{k=1}^{\infty} c_k \phi_k$, for any scalars $c_1, c_2, \ldots$, is called an *orthogonal series* (whether or

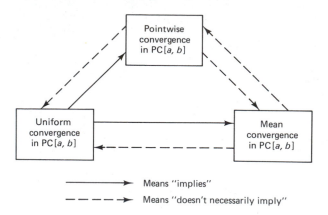

Means "implies"

Means "doesn't necessarily imply"   **Figure 13.18**   Convergence in PC[a, b].

not it converges). By definition, the series $\sum_{k=1}^{\infty} c_k \phi_k$ converges to the element f in E if and only if $\left\| \sum_{k=1}^{n} c_k \phi_k - f \right\| \to 0$ as $n \to \infty$.

In finite-dimensional linear spaces we have seen how useful it is to have a set with the property that every element in the space can be uniquely written as a linear combination over this set. There is an analog of this concept in infinite-dimensional Euclidean spaces like PC[a, b] which makes use of the convergence property to attach a meaning to the sum of an orthogonal series: An orthogonal set $\Phi = \{\phi_k : k = 1, 2, \ldots\}$ in a Euclidean space E is called a *basis* for E if and only if for every f in E there exists a unique set of scalars $\{c_k\}_1^{\infty}$ such that $f = \sum_{k=1}^{\infty} c_k \phi_k$.

The Continuity of the Scalar Product Theorem implies that if $f = \sum_{k=1}^{\infty} c_k \phi_k$, then necessarily $c_n = \langle f, \phi_n \rangle / \|\phi_n\|^2$ for each $n = 1, 2, \ldots$.

Fourier–Euler Theorem.   If for an orthogonal set $\Phi = \{\phi : k = 1, 2, \ldots\}$ in the Euclidean space E we have that $f = \sum_{k=1}^{\infty} c_k \phi_k$, then the constants c_k are necessarily given by the *Euler Formulas*

$$c_k = \frac{\langle f, \phi_k \rangle}{\|\phi_k\|^2} \qquad \text{for each } k = 1, 2, \ldots \tag{9}$$

The constant c_k given by (9) is called the kth *Fourier coefficient* of f over Φ.

Proof. Using the Continuity of Scalar Product Theorem, we see that since $\sum_{k=1}^{N} c_k \phi_k \to f$ as $N \to \infty$, it follows that $\left\langle \sum_{k=1}^{N} c_k \phi_k, \phi_n \right\rangle \to \langle f, \phi_n \rangle$, for each (fixed) $n = 1, 2, \ldots$. But by orthogonality, $\left\langle \sum_{k=1}^{N} c_k \phi_k, \phi_n \right\rangle = c_n \langle \phi_n, \phi_n \rangle$, if $N > n$, and the assertion is established.

This result and the Mean Approximation Theorem give the following equivalent (and more convenient) definition of a basis:

> **Basis.** An orthogonal set $\Phi = \{\phi_k : k = 1, 2, \ldots\}$ is a basis for the Euclidean space E if and only if
>
> $$\sum_{k=1}^{\infty} \frac{\langle f, \phi_k \rangle}{\|\phi_k\|^2} \phi_k = f \qquad \text{for each } f \text{ in } E \tag{10}$$

An important property of bases is given by the following result:

Totality Theorem. Let $\Phi = \{\phi_k : k = 1, 2, \ldots\}$ be a basis for the Euclidean space E. If f in E is such that $\langle f, \phi_k \rangle = 0$, for all $k = 1, 2, \ldots$, then $f = 0$.

Proof. Follows immediately from (10).

Fourier Series

Orthogonal series constructed in a certain way have acquired a special name.

> **Fourier Series.** Let $\Phi = \{\phi_k : k = 1, 2, \ldots\}$ be an orthogonal set in the Euclidean space E. Then for any f in E the orthogonal series
>
> $$\text{FS}[f] = \sum_{k=1}^{\infty} \frac{\langle f, \phi_k \rangle}{\|\phi_k\|^2} \phi_k \tag{11}$$
>
> is called *the Fourier Series of f over* Φ. [The symbol $\text{FS}[f]$ stands for the formal series in (11) and no implication is made about the convergence.]

The same function may have any number of Fourier Series, depending on the particular choice of an orthogonal set Φ in E. The definition of a basis can now be stated as follows: An orthogonal set Φ in a Euclidean space E is a basis for E if and only if every f in E is the sum of its Fourier Series with respect to Φ. The results below will be important later.

Let $\Phi = \{\phi_k : k = 1, 2, \ldots\}$ be an orthogonal set in the Euclidean space E. Then for any f in E, the series $\sum_{k=1}^{\infty} |\langle f, \phi_k \rangle|^2 / \|\phi_k\|^2$ converges. Moreover,

Bessel Inequality.   We have that

$$\|f\|^2 \geq \sum_{k=1}^{\infty} \frac{|\langle f, \phi_k \rangle|^2}{\|\phi_k\|^2} \qquad \text{for all } f \text{ in } E \tag{12}$$

Decay of Coefficients.   If c_k is the kth Fourier coefficient of f given by (9), and Φ is nonfinite, then

$$\|\phi_k\| \cdot |c_k| \longrightarrow 0 \qquad \text{as } k \to \infty \tag{13}$$

Parseval Relation. Φ is a basis for E if and only if

$$\|f\|^2 = \sum_{k=1}^{\infty} \frac{|\langle f, \phi_k \rangle|^2}{\|\phi_k\|^2} \qquad \text{for all } f \text{ in } E \tag{14}$$

Proof. The identity (6) in the Mean Approximation Theorem holds for all positive integers n. Since the left-hand side of (6) can never be negative, it follows that

$$\|f\|^2 \geq \sum_{k=1}^{n} \frac{|\langle f, \phi_k \rangle|^2}{\|\phi_k\|^2} \qquad \text{for all } n \tag{15}$$

Thus the sequence $\left\{ \sum_{k=1}^{n} |\langle f, \phi_k \rangle|^2 / \|\phi_k\|^2 \right\}$, $n = 1, 2, \ldots$, is monotone increasing and bounded from above, and hence the infinite series $\sum_{k=1}^{\infty} |\langle f, \phi_k \rangle|^2 / \|\phi_k\|^2$ is convergent. The inequality (12) now follows immediately from (15) by taking the limit as $n \to \infty$. Because the infinite series in (12) is convergent, the kth term converges to zero. The property (13) follows from the fact that $|\langle f, \phi_k \rangle|^2 / \|\phi_k\|^2 = \|\phi_k\|^2 |c_k|^2$, where c_k is the Fourier coefficient given by (9). Finally, the Parseval Relation property follows immediately from (5) and (6) of the Mean Approximation Theorem.

Example 13.16

Let us consider the orthogonal set Φ in PC$[-\pi, \pi]$ given in Example 13.11. We now use the definition (11) to calculate the Fourier Series of the function $f(x) \equiv x$ on $[-\pi, \pi]$. Notice first that $\int_{-\pi}^{\pi} \sin^2 nx \, dx = \int_{-\pi}^{\pi} \cos^2 nx \, dx = \pi$, for all $n \geq 1$, and that $\int_{-\pi}^{\pi} 1^2 \, dx = 2\pi$. Now FS$[f]$ has the form

$$\text{FS}[f] = A_0 + \sum_{k=1}^{\infty} (A_k \cos kx + B_k \sin kx) \tag{16}$$

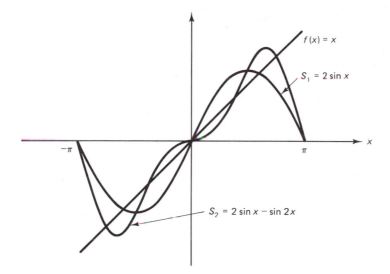

Figure 13.19   Graphs of partial sums of (17).

where $A_0 = \dfrac{1}{2\pi} \displaystyle\int_{-\pi}^{\pi} 1 \cdot x \, dx = 0$, $A_k = \dfrac{1}{\pi} \displaystyle\int_{-\pi}^{\pi} x \cos kx \, dx = 0$ (since $x \cos kx$ is an odd function on $[-\pi, \pi]$), and $B_k = \dfrac{1}{\pi} \displaystyle\int_{-\pi}^{\pi} x \sin kx \, dx = 2(-1)^{k+1}/k$.

Therefore, we have that

$$\text{FS}[f] = 2 \sum_{k=1}^{\infty} \frac{(-1)^{k+1}}{k} \sin kx \tag{17}$$

It is interesting to plot the first and second partial sums of the Fourier Series (17) and compare them with the function $f(x) = x$ itself on the interval $[-\pi, \pi]$ (see Figure 13.19).

Sturm–Liouville Problems

The results below are important for the Method of Separation of Variables technique for solving boundary/initial value problems for partial differential equations. It will be seen that orthogonality and orthogonal series play a key role in these results.

Let M be the operator defined on $C^2[a, b]$ by

$$My \equiv \frac{1}{\rho(x)} \left\{ \frac{d}{dx} \left[p(x) \frac{dy}{dx} \right] + q(x)y \right\} \tag{18}$$

where

$$\rho \text{ is in } C^0[a, b], \quad \rho \neq 0 \text{ on } [a, b], \quad q \text{ is in } C^0[a, b],$$
$$p \neq 0 \text{ on } [a, b], \quad p \text{ is in } C^1[a, b] \tag{19}$$

M in (18) is not as special as it seems. Indeed, the general normal second-order linear ordinary differential operator $P(D) \equiv a_2(x)D^2 + a_1(x)D + a_0(x)$, where $a_i(x)$ is in $C^0[a, b]$, $i = 0, 1, 2$, and $a_2(x) \neq 0$ on $[a, b]$, can be written in the form (18) if we let

$$p(x) = e^{\alpha(x)}, \qquad q(x) = \frac{a_0(x)}{a_2(x)} e^{\alpha(x)}, \qquad \rho(x) = \frac{e^{\alpha(x)}}{a_2(x)} \tag{20}$$

where $\alpha(x) = \int^x [a_1(x)/a_2(x)] \, dx$. Observe that if we use the weighted scalar product in $C^0[a, b]$, $\langle f, g \rangle = \int_a^b \rho(x) f(x) \bar{g}(x) \, dx$, then integration by parts yields the relation

$$\langle Mu, v \rangle = p(u'v - uv') \Big|_a^b + \langle u, Mv \rangle \qquad \text{for all } u, v \text{ in } C^2[a, b] \tag{21}$$

Now we define an operator L whose action is given by M in (18) and whose domain Dom (L) is a restriction of $C^2[a, b]$ defined by conditions at the endpoints $x = a$ and $x = b$.

Boundary Conditions.   For u in $C^2[a, b]$, the conditions

$$\alpha u'(a) + \beta u(a) = 0, \qquad \gamma u'(b) + \delta u(b) = 0 \tag{22}$$

are called *separated* (or *unmixed*) *boundary conditions* at $x = a$ and $x = b$, respectively, where the constants α, β, γ, δ are such that $\alpha^2 + \beta^2 \neq 0$ and $\gamma^2 + \delta^2 \neq 0$. The conditions

$$u(a) = u(b), \qquad u'(a) = u'(b) \tag{23}$$

are referred to as *periodic boundary conditions*.

Now we have the following results whose proof is immediate from (21).

Symmetric Operator Theorem.   Let L be an operator with action given by M in (18) and domain Dom (L) one of the two types:
1. Dom $(L) = \{u$ in $C^2[a, b] : u$ satisfies a separated boundary condition at each endpoint where $p \neq 0\}$.
2. Dom $(L) = \{u$ in $C^2[a, b] : u(a) = u(b), u'(a) = u'(b)$ and $p(a) = p(b)\}$.

Then L is a *symmetric operator*; that is, $\langle Lu, v \rangle = \langle u, Lv \rangle$, for all u, v in Dom (L), where $\langle \cdot, \cdot \rangle$ is the weighted scalar product in $C^0[a, b]$, $\langle f, g \rangle = \int_a^b \rho(x) f(x) \overline{g(x)} \, dx$.

Symmetric linear operators such as L defined in the theorem above are defined on a subspace Dom (L) of $C^0[a, b]$ and take values in $C^0[a, b]$. The eigenvalues and eigenspaces of such operators play an important role in applied mathematics. The problem of finding all the eigenvalues and corresponding eigenspaces of such operators is called a *Sturm–Liouville Problem*.† In general, we have the following definition:

> **Eigenvalues, Eigenspaces.**   Let S be a subspace of a linear space V and let $L: S \rightarrow V$ be a linear operator. A scalar λ (a real or complex number according as V is a real or complex linear space) is an *eigenvalue* for L if there exists a *nonzero* v in S such that $Lv = \lambda v$. That vector v is an *eigenvector* of L corresponding to the eigenvalue λ. For any eigenvalue λ of L the set V_λ, of all vectors in V that satisfy the equation $Lv = \lambda v$ is the *eigenspace* of L corresponding to λ.

For a similar definition in finite-dimensional spaces, see Section 10.1. Since our symmetric operators act on functions in $C^2[a, b]$, their eigenvectors are often called *eigenfunctions*. Notice that for any eigenvalue λ, the eigenspace V_λ is indeed a nontrivial subspace of V, and any nonzero element in V_λ is an eigenvector for L corresponding to λ. (The zero vector can never be an eigenvector of any linear operator.) Thus V_λ is comprised of all the eigenvectors of L corresponding to λ with the zero vector thrown in. Thus two eigenspaces V_λ and V_μ of a linear operator with $\lambda \neq \mu$ can only have the zero vector in common.

Eigenspaces and eigenvalues for our symmetric differential operators have several important properties, which we list below.

Let L be any differential operator defined in the Symmetric Operator Theorem. Then the following properties hold:

Orthogonality of Eigenspaces.   Any two eigenspaces of L, V_λ and V_μ, corresponding to two distinct eigenvalues λ and μ, must be orthogonal; that is, any element of V_λ is orthogonal to any element of V_μ under the scalar product

$$\langle f, g \rangle = \int_a^b \rho(x) f(x) \overline{g(x)} \, dx.$$

Simplicity of Eigenspaces.   If a separated boundary condition is used in defining Dom (L), then all eigenspaces of L are one-dimensional.

Proof. The orthogonality of eigenspaces follows immediately from the symmetry relation $\langle Lu, v \rangle = \langle u, Lv \rangle$ for all u, v in Dom (L). The only choices for the dimension of any eigenspace is one or two. Now if the dimension of the eigenspace corresponding to an eigenvalue λ is two, *all* solutions of the homogeneous equation $(p(x)y')' + $

---

† Charles Sturm (1803–1855) and Joseph Liouville (1809–1882) jointly introduced this system to resolve problems concerning the solutions of the partial differential equations of wave motion and thermal diffusion.

$q(x)y = \lambda \rho(x)y$ must satisfy the endpoint conditions. This is impossible if Dom (L) is defined with a separated condition at either $x = a$ or $x = b$. Thus the dimension of the eigenspace is one, as asserted.

Nonpositivity of Eigenvalues. Let L be any differential operator defined in the Symmetric Operator Theorem such that $q(x) \leq 0$ on $[a, b]$, and if either of the separated conditions (22) is used to define Dom (L), then $\alpha\beta \leq 0$, and $\gamma\delta \geq 0$. Then the eigenvalues of L are all nonpositive.

Proof. Use integration by parts to show that $\langle Lu, u \rangle \leq 0$, for all u in Dom (L). The assertion follows immediately from this inequality.

Example 13.17

Suppose that the operator L has action $Lu = u''$ and domain Dom (L) = $\{u$ in $C^2[0, T]: u(0) = u(T) = 0\}$. Then L is a symmetric differential operator with $a = 0$, $b = T$, $\rho(x) \equiv p(x) \equiv 1$, $q(x) \equiv 0$ on $[0, T]$, and separated boundary conditions (22) with $\alpha = 0$, $\beta = 1$, $\gamma = 0$, $\delta = 1$. Thus all the eigenvalues of L are nonpositive, and all eigenspaces are one-dimensional and mutually orthogonal with the scalar product $\langle f, g \rangle = \int_0^T f(x)g(x)\, dx$. To find the eigenvalues and eigenspaces, first try $\lambda = 0$ in the eigenvalue equation $u'' = \lambda u$. The general solution of $u'' = 0$ is $u = Ax + B$ for arbitrary constants A, B. The condition $u(0) = u(T) = 0$ in this case yields that $A = B = 0$, and hence $\lambda = 0$ cannot be an eigenvalue of L. Now try $\lambda = -k^2$ for some positive constant k. The general solution of the eigenvalue equation $u'' = -k^2u$ is $u = A \cos kx + B \sin kx$. The conditions $u(0) = u(T) = 0$ implies that $A = 0$ and $k = n\pi/T$. Thus L has the eignvalues $\lambda_n = -(n\pi/T)^2$, with corresponding eigenspaces spanned by the respective eigenfunctions $u_n = \sin n\pi x/T$, $n = 1, 2, \ldots$.

Example 13.18 (*Bessel Functions of Order 0*)

Consider the eigenvalue equation $(1/x)\{(xu')'\} = \lambda u$, where u is in $C^2[0, R_0]$, $R_0 > 0$, and $u(R_0) = 0$. Now the associated operator L has $a = 0$, $b = R_0$, $\rho \equiv x$, $p \equiv x$, $q \equiv 0$. Notice that $p(0) = 0$ and that we have a separated condition at $x = R_0$, with $\gamma = 0$, $\delta = 1$. Thus L is a symmetric differential operator, and since a separated condition was used, the eigenspaces are all one-dimensional. The eigenspaces are mutually orthogonal, with the scalar product $\langle f, g \rangle = \int_0^{R_0} xf(x)g(x)\, dx$, and the eigenvalues are nonpositive. First try $\lambda = 0$; the equation $(xu')' = 0$ has the general solution $u = A \ln x + B$. But the conditions u in $C^1[0, R_0]$ and $u(R_0) = 0$ imply that $A = B = 0$. Thus $\lambda = 0$ is not an eigenvalue. Now try $\lambda = -k^2$, for $k > 0$. The eigenvalue equation $(xu')' + k^2xu = 0$ has only one solution (up to constant multiples) which belongs to the class $C^2[0, R_0]$; it is $u = J_0(kx)$, where J_0 is the Bessel function of order zero (see Section 8.6). The condition $u(R_0) = 0$ now implies that $J_0(kR_0) = 0$. But we know from Section 8.6 that J_0 has infinitely many positive zeros: $x_1, x_2, \ldots$. Thus $k_n = x_n/R_0$, $n = 1, 2, \ldots$ so the eigenvalues of L are $\lambda_n = -(x_n/R_0)^2$, $n = 1, 2, \ldots$. The corresponding eigenspaces are spanned by the eigenfunctions $u_n = J_0(x_n x/R_0)$, $n = 1, 2, \ldots$, $\{u_n\}$ is a basis of $PC[0, R_0]$ with the weighted inner product given above, but we omit the proof.

Example 13.19 (*Legendre Polynomials*)

Consider the eigenvalue equation $((1 - x^2)u')' = \lambda u$ for u in $C^2[-1, 1]$. The associated operator L has $a = -1$, $b = 1$, $\rho \equiv 1$, $p(x) = (1 - x^2)$, $q(x) \equiv 0$, and since $p(-1) = p(1) = 0$, Dom $(L) = C^2[-1, 1]$. Thus L is a symmetric operator under the scalar product $\langle f, g \rangle = \int_{-1}^{+1} f(x)g(x)\, dx$, and moreover the eigenvalues are nonpositive. Although our earlier theorem does not guarantee it, we will see that the eigenspaces are all one-dimensional. Of course, the eigenspaces are mutually orthogonal. Observe that for $\lambda = -n(n + 1)$, $n = 0, 1, 2, \ldots$, the eigenvalue equation $((1 - x^2)u')' = -n(n + 1)u$ is Legendre's equation (see Section 8.3) whose only solution (up to constant multiples) in $C^2[-1, +1]$ is $P_n(x)$, $n = 0, 1, \ldots$, the Legendre polynomial of order n. It is a fact that $\{P_n(x) : n = 0, 1, 2, \ldots\}$ is a basis for $PC[-1, 1]$ (although we will not prove it) under the standard scalar product. We use this fact to show that L has no other eigenvalues. Then assume that $\mu \neq -n(n + 1)$, $n = 0, 1, 2, \ldots$, is an eigenvalue of L with corresponding eigenfunction $v(x)$. Now L is a symmetric operator and so for any $n = 0, 1, 2, \ldots$,

$$\mu \langle v, P_n \rangle = \langle \mu v, P_n \rangle = \langle Lv, P_n \rangle = \langle v, LP_n \rangle = -n(n + 1)\langle v, P_n \rangle$$

Thus $\langle v, P_n \rangle = 0$, for all $n = 0, 1, 2, \ldots$. Thus by the Totality Theorem, $v = 0$ and hence cannot have been an eigenfunction. Thus the eigenvalues of L are $\lambda_n = -n(n + 1)$, $n = 0, 1, 2, \ldots$, with corresponding eigenspace spanned by $P_n(x)$.

We end now with a fundamental property of Sturm–Liouville systems.

Sturm–Liouville Theorem (in the Regular Case).   Let the operator L have action given by M in (18) and (19) and domain in $C^2[a, b]$ characterized by separated conditions (22) at both $x = a$ and $x = b$. Then the eigenvalues of L form a sequence λ_n, $n = 1, 2, \ldots$, with $|\lambda_n| \to \infty$; the corresponding eigenspaces V_n are one-dimensional and mutually orthogonal under the scalar product $\langle f, g \rangle = \int_a^b \rho fg\, dx$; if Φ is any set consisting of precisely one eigenfunction from each V_n, then Φ is a basis for $PC[a, b]$; and finally, the Fourier Series over Φ of any function u in Dom (L) converges uniformly to u on $[a, b]$.

Example 13.20

The Sturm–Liouville system in Example 13.17 is in the regular case. Thus the set $\Phi = \{\sin n\pi x/T : n = 1, 2, \ldots\}$ is a basis for $PC[0, T]$ under the standard scalar product. The function $v = -(x - T/2)^2 + T^2/4$ is in the domain of the associated operator L. Thus the Fourier series of v converges uniformly to v on $[0, T]$.

The Sturm–Liouville systems in Examples 13.18 and 13.19 are not in the regular case. Such systems are said to be *singular Sturm–Liouville systems*. Nevertheless, it is true that conclusions similar to those in the regular Sturm–Liouville Theorem hold. For example, $\Phi = \{J_0(x_n x/R_0) : n = 1, 2, \ldots\}$, where x_n is the nth positive zero of J_0, is a basis for $PC[0, R_0]$ under the scalar product $\langle f, g \rangle = \int_0^{R_0} xfg \, dx$. Also, as mentioned earlier, $\Phi = \{P_n(x) : n = 0, 1, 2, \ldots\}$, is a basis for $PC[-1, 1]$ under the standard scalar product, where P_n is the Legendre polynomial of order n.

PROBLEMS

The following problem(s) may be more challenging: 2, 8, 12, and 13(i).

1. Let E be a real Euclidean space with scalar product $\langle \cdot, \cdot \rangle$ and its associated norm $\|\cdot\|$. Show that for all x, y in E,
   (a) $\|x + y\|^2 + \|x - y\|^2 = 2\|x\|^2 + 2\|y\|^2$.
   (b) $\langle x, y \rangle = \frac{1}{2}\{\|x + y\|^2 - \|x\|^2 - \|y\|^2\} = \frac{1}{4}\{\|x + y\|^2 - \|x - y\|^2\}$.

2. Let $\mathbf{R}^\infty$ be the set of all sequences of real numbers, $(x_n)_{n=1}^\infty$. Define addition ("+") and multiplication by reals ("·") in $\mathbf{R}^\infty$ termwise; that is, $(x_n) + (y_n) = (x_n + y_n)$, $\alpha \cdot (x_n) = (\alpha x_n)$. Let $l^2 = \{(x_n) \text{ in } \mathbf{R}^\infty \text{ with } \sum_{n=1}^\infty |x_n|^2 < \infty\}$. Show that

   (a) For any x, y in l^2, $\sum_{i=1}^\infty x_i y_i < \infty$.

   (b) $\langle x, y \rangle = \sum_{i=1}^\infty x_i y_i$ is a scalar product on l^2.

3. Let ρ be in $C^0[a, b]$ and such that ρ is positive on $[a, b]$ except at a finite number of points. Show that $\langle f, g \rangle = \int_a^b \rho fg \, dx$ is a scalar product on $C^0[a, b]$.

4. (a) Show that the real-valued functions $E = \{f \text{ in } C^0(\mathbf{R}) : \int_{\mathbf{R}} |f|^2 \, dx < \infty\}$ is a linear space if addition and multiplication by scalars are defined in the usual way for functions.
   (b) Show that $\langle f, g \rangle = \int_{\mathbf{R}} fg \, dx$ is a scalar product on E.

5. [*Properties of Orthogonal Sets*]. Let $A = \{u^1, \ldots, u^k\}$ be an orthogonal subset of a Euclidean space E, and let $S = \text{Span}(A)$. Show that
   (a) The set A is linearly independent and hence a basis for S.
   (b) $k \leq \dim E$.

(c) If u is in S, then $u = \sum\limits_{i=1}^{k} [\langle u, u^i \rangle / \|u^i\|^2] u^i$.

(d) u in E is orthogonal to S if and only if u is orthogonal to each u^i, $i = 1, 2, \ldots, k$.

6. [*Orthogonal Projection*]. Let E be a Euclidean space and S an n-dimensional subspace of E spanned by the orthogonal set $\{u^1, u^2, \ldots, u^n\}$. For any u in E the vector

$$\text{proj}_S(u) = \sum_{i=1}^{n} [\langle u, u^i \rangle / \|u^i\|^2] u^i \text{ is called the } orthogonal\ projection \text{ of } u \text{ onto } S. \text{ [}Note\text{:}$$

$\text{proj}_S(u)$ does not depend on the orthogonal basis of S chosen.] Show that

(a) For every u in E there is a unique decomposition $u = v + w$ with v in S and w orthogonal to S.

(b) For any u in E, $\|u - \text{proj}_S(u)\| \le \|u - v\|$, for all v in S, and equality holds if and only if $v = \text{proj}_S(u)$.

(c) In $\mathbf{R}^3$, find the distance from the point $(1, 1, 1)$ to the plane: $x + 2y - z = 0$.

7. In the Euclidean space $PC[-\pi, \pi]$, consider the subspace S_N spanned by the set $\Phi_N = \{1, \cos x, \sin x, \ldots, \cos Nx, \sin Nx\}$. Find the element in S_N that is closest to the element $f(x) = x$, $|x| < \pi$, in $PC[-\pi, \pi]$.

8. Prove the Continuity of the Scalar Product Theorem.

9. Let Φ be a basis for a Euclidean space E. Show that if two elements f and g in E have the same Fourier Series over Φ, then $f = g$.

10. Let $\Phi = \{1, \cos x, \sin x, \ldots, \cos nx, \sin nx, \ldots\}$ in the Euclidean space $PC[-\pi, \pi]$. Find the Fourier Series of the following elements of $PC[-\pi, \pi]$ with respect to Φ.

(a) *Sawtooth*

$$f(x) = \begin{cases} -A\left(1 + \dfrac{x}{\pi}\right), & -\pi \le x < 0 \\[2mm] A\left(1 - \dfrac{x}{\pi}\right), & 0 < x < \pi \end{cases}$$

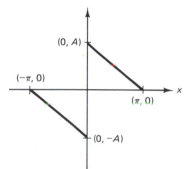

(b) *Triangular*

$$f(x) = \begin{cases} A\left(1 + \dfrac{x}{\pi}\right), & -\pi \le x \le 0 \\[2mm] A\left(1 - \dfrac{x}{\pi}\right), & 0 < x < \pi \end{cases}$$

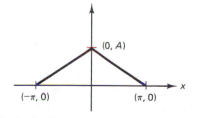

(c) *Triangular*

$$f(x) = |x|, \qquad |x| \le \pi$$

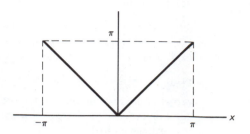

11. Consider the orthogonal set $\Phi = \{\sin kx : k = 1, 2, \ldots\}$ in $PC[0, \pi]$. Find the Fourier Series of the following elements of $PC[0, \pi]$.
    (a) $f(x) = x, 0 \le x \le \pi$.
    (b) $f(x) = 1 + x/\pi, 0 \le x \le \pi$.
    (c) $f(x) = 1, 0 \le x \le \pi$.

12. Let E be the Euclidean space in Problem 4, and consider the sequence $f_n(x) = n^{-1/2}e^{-x^2/n^2}, n = 1, 2, \ldots$.
    (a) Show that $\{f_n\}$ converges uniformly to the zero function on $\mathbf{R}$.
    (b) Show that $\{f_n\}$ does not converge in the mean in E. [*Hint*: If $\{f_n\}$ is mean convergent, the limit must be the zero function (why?), but the Continuity of the Scalar Product Theorem contradicts this.]

13. In each of the Sturm–Liouville systems below identify the operator L associated with it, find the eigenvalues and eigenspaces of L, and state (if you can) the orthogonality and basis properties of the eigenfunctions.
    (a) $y'' = \lambda y;$ $y(0) = y(\pi/2) = 0.$
    (b) $y'' = \lambda y;$ $y(0) = 0, \qquad y'(T) = 0.$
    (c) $y'' = \lambda y;$ $y'(0) = 0, \qquad y'(T) = 0.$
    (d) $y'' = \lambda y;$ $y(-T) = y(T), \quad y'(-T) = y'(T).$
    (e) $y'' = \lambda y;$ $y(0) = 0, \qquad y(\pi) + y'(\pi) = 0.$
    (f) $y'' - 4y' + 4y = \lambda y;$ $y(0) = y(\pi) = 0.$
    (g) $x^4 y'' + 2x^3 y' = \lambda y;$ $y(1) = 0, \quad y(2) = 0.$ [*Hint*: To solve the equation, try the independent variable substitution $s = 1/x$.]
    (h) $xy'' - y' = \lambda x^3 y; y(0) = 0, y(a) = 0.$ [*Hint*: To solve the equation, try the independent variable substitution $s = x^2$.]
    (i) $\dfrac{1}{x}\left[(xy')' - \dfrac{p^2}{x}y\right] = \lambda y, \quad y$ in $C^2[0, R_0], y(R_0) = 0$, where p is a positive number. [*Hint*: This is a Bessel operator of order p.]

13.6 FOURIER TRIGONOMETRIC SERIES

As we have seen in earlier sections, many techniques of applied analysis crucially depend on the possibility of writing a given real-valued function as the sum of specialized trigonometrical series over some interval. In fact, this situation arises so frequently in the applications that a considerable body of literature has been generated on this topic. The purpose of this section is to pursue these ideas in a systematic fashion.

It is not difficult to see that

$$\Phi_T = \left\{ 1, \cos\frac{\pi x}{T}, \sin\frac{\pi x}{T}, \cdots, \cos\frac{n\pi x}{T}, \sin\frac{n\pi x}{T}, \cdots \right\} \tag{1}$$

is an orthogonal set in PC[$-T$, T] under the standard scalar product $\langle f, g \rangle = \int_{-T}^{T} fg\, dx$. In fact, Φ_T turns up as a set of orthogonal eigenfunctions of the Sturm–Liouville problem

$$y'' = \lambda y, \qquad y(-T) = y(T), \qquad y'(-T) = y'(T) \tag{2}$$

Indeed, associated with (2) is the symmetric operator $Ly \equiv y''$, with domain defined by Dom $(L) = \{y$ in $C^2[-T, T] : y(-T) = y(T), y'(-T) = y'(T)\}$. From Section 13.5 we see that the eigenvalues of L are nonpositive. It is easy to see that $\lambda = 0$ is an eigenvalue and the corresponding eigenspace V_0 is spanned by the constant function 1. The other eigenvalues are $\lambda_n = -(n\pi/T)^2$, $n = 1, 2, \ldots$, and each corresponding eigenspace V_n is two-dimensional and is spanned by the orthogonal set $\{\cos(n\pi x/T), \sin(n\pi x/T)\}$. The Sturm–Liouville Problem (2) arises often in the applications.

Notice that

$$\left\| \cos\frac{k\pi x}{T} \right\|^2 = \int_{-T}^{T} \cos^2\frac{k\pi x}{T}\, dx = \begin{cases} 2T & \text{if } k = 0 \\ T & \text{if } k > 0 \end{cases}$$

$$\left\| \sin\frac{k\pi x}{T} \right\|^2 = \int_{-T}^{T} \sin^2\frac{k\pi x}{T}\, dx = T \qquad k > 0$$

Thus for any f in PC[$-\pi$, π], we see from Section 13.5 that the Fourier Series of f over Φ_T is

$$\text{FS}[f] = A_0 + \sum_{k=1}^{\infty} \left(A_k \cos\frac{k\pi x}{T} + B_k \sin\frac{k\pi x}{T} \right) \tag{3}$$

where the coefficients are given by the Euler formulas,

$$A_0 = \frac{1}{2T}\int_{-T}^{T} f(x)\, dx, \qquad A_k = \frac{1}{T}\int_{-T}^{T} f(x)\cos\frac{k\pi x}{T}\, dx$$

$$B_k = \frac{1}{T}\int_{-T}^{T} f(x)\sin\frac{k\pi x}{T}\, dx, \qquad \text{for } k > 0 \tag{4}$$

This particular orthogonal series (3) is called a *Fourier Trigonometrical Series* on [$-T$, T] and evidently consists of a superposition of a constant term and sinusoids whose frequencies are multiples (harmonics) of a basic (fundamental) frequency of $1/2T$ hertz.

Although it is not evident (and *not* covered by the Sturm-Liouville Theorem in Section 13.5), the orthogonal set Φ_T is a basis for PC[$-T$, T].

> **Fourier Trigonometrical Basis Theorem.**   The subset Φ_T given by (1) is a basis for PC[$-T$, T].

See the problem set for a proof in a special case.

Calculation of Fourier Trigonometrical Series

Given a function f in PC[$-T$, T], the process of setting up the trigonometrical series (3) with coefficients (4) is known as "expanding f into a Fourier Trigonometrical Series," and the values of A_k and B_k in (4) are "the Fourier coefficients of f." Calculating the Fourier coefficients of a function f can be a tedious exercise in integration techniques. Keep in mind, however, that two functions f and g in PC[$-T$, T] whose values differ at only a finite number of points in [$-T$, T] have the *same* Fourier coefficients. Thus, changing the values of a function at a finite number of points in [$-T$, T] will not change its Fourier coefficients.

The process of calculating Fourier coefficients can be simplified if the function has certain symmetry properties. A function f in PC[$-T$, T] is *even* if $f(-x) = f(x)$ for $|x| < T$ (except possibly a finite number of points), while f is odd if $f(-x) = -f(x)$, for $|x| < T$. It is easy to show that $f(x)g(x)$ is even if f and g are both even or both odd and $f(x)g(x)$ is odd if one factor is even and the other odd. Note that $\int_{-a}^{a} f(x)\,dx = 0$ if f is odd, and $\int_{-a}^{a} f(x)\,dx = 2\int_{0}^{a} f(x)\,dx$ if f is even. Thus if f in PC[$-T$, T] is odd, the Fourier coefficients $A_k = 0$, since $f(x)\cos(k\pi x/T)$ is odd for each $k = 1, 2, \ldots$. Similarly, $B_k = 0$ for $k = 1$, $2, \ldots$ if f is even. See the problems for more "tricks" of this kind.

Example 13.21

Let us find the Fourier (Trigonometrical) Series of the function $f(x) = |x|$, $-T \le x \le T$. Since f is even, all the coefficients B_k vanish. Now we have

$$A_0 = \frac{1}{T}\int_{0}^{T} x\,dx = \frac{T}{2}, \qquad A_k = \frac{2}{T}\int_{0}^{T} x\cos\frac{k\pi x}{T}\,dx = \frac{2T}{\pi^2 k^2}(\cos k\pi - 1)$$

for $k > 0$

Note that since $\cos k\pi = (-1)^k$, $A_k = 0$ when k is a positive even integer. Setting $k = 2m + 1$, $m = 0, 1, \ldots$, we see that $A_{2m+1} = -4T/\pi^2(2m + 1)^2$. Thus

$$\mathrm{FS}[f] = \frac{T}{2} + \sum_{m=0}^{\infty} \frac{-4T}{\pi^2(2m + 1)^2}\cos\frac{(2m + 1)\pi x}{T}$$

Convergence Properties

Since Φ_T is a basis for PC$[-T, T]$, we know that for any f in PC$[-T, T]$, the Fourier Trigonometrical Series (3) and (4) converges in the mean to f; that is,

$$\left\| A_0 + \sum_{k=1}^{n} (A_k \cos kx + B_k \sin kx) - f \right\| \to 0, \qquad \text{as } n \to \infty$$

But what about the pointwise or uniform convergence properties of Fourier Series? Let us consider a simple, but instructive, example.

Example 13.22

The function

$$f(x) = \begin{cases} 5, & 0 < x < \pi \\ -5, & -\pi < x < 0 \end{cases}$$

is in PC$[-\pi, \pi]$ and since it is an odd function, its Fourier Series contains no cosine terms (or constant term). A quick calculation shows that

$$\text{FS}[f] = \frac{20}{\pi} \sum_{k=1}^{\infty} \frac{\sin(2k-1)x}{2k-1} \tag{5}$$

In Figure 13.20, the graphs of the partial sums of the series (5) using 5, 9, and 15 terms are compared with the graph of f. Apparently, we have pointwise convergence of FS$[f]$ except at the endpoints $x = \pm\pi$, or the "jump" discontinuity $x = 0$. The convergence also appears to be uniform on any closed interval not containing any of these three exceptional points.

The behavior of the partial sums near $x = 0$ and $x = \pm\pi$ is somewhat mysterious. Apparently, the "hump" in the graph of the partial sums near these points does not diminish in size but merely "moves" over closer and closer to the exceptional points, $x = 0$ and $x = \pm\pi$, as more terms are included in the partial sums. This is *Gibbs phenomenon* and always occurs at jump discontinuities. We now examine the precise pointwise convergence properties of the Fourier Trigonometrical Series.

> **Piecewise Smoothness, PS$[-T, T]$.** A function f on $[-T, T]$ is *piecewise smooth* if it is piecewise continuous and is differentiable at all but at most a finite set of points in $(-T, T)$, and f' is piecewise continuous on $[-T, T]$. The set of all piecewise-smooth functions on $[-T, T]$ is denoted by PS$[-T, T]$. Thus f in PS$[-T, T]$ if and only if f, f' are in PC$[-T, T]$.

We will need the following few facts:

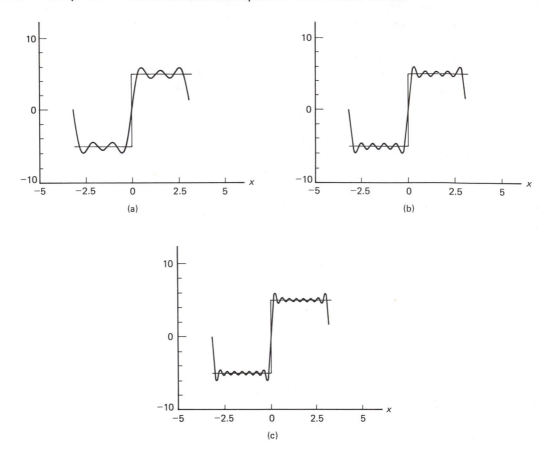

Figure 13.20   Partial sum of (5): (a) 5 terms; (b) 9 terms; (c) 15 terms.

Decay of Coefficients.   Let f be in PC$[-T, T]$, and let FS$[f]$ be the Fourier Trigonometrical Series of f given by (3) and (4). Then $A_k \to 0$, $B_k \to 0$, as $k \to \infty$.

Integration of Periodic Functions Theorem.   Let f be any periodic function on **R** with period $2T$ which is piecewise continuous over the interval $[-T, T]$. Then

$$\int_{-T}^{T} f(x)\, dx = \int_{a}^{a+2T} f(s)\, ds, \qquad \text{for any real } a \tag{6}$$

Proof. Notice $\|\cos\,(k\pi x/T)\|^2 = \|\sin\,(k\pi x/T)\|^2 = T$ for $k > 0$, and from the general Decay of Coefficients Theorem in Section 13.5 we conclude that $A_k \to 0$, $B_k \to 0$, as $k \to \infty$. The integration formula (6) is treated in the problems.

In what follows we set $T = \pi$ for convenience; it will be clear how the results can be modified for general T.

Dirichlet Representation Theorem.   Let f be in $PC[-\pi, \pi]$, and let $\tilde{f}$ be the periodic extension of f in $\mathbf{R}$ with period 2π. Let $S_n(x)$ denote the nth partial sum of $FS[f]$ as in (3) and (4); that is, $S_n(x) = A_0 + \sum_{k=1}^{n} (A_k \cos kx + B_k \sin kx)$, where A_k, B_k are as in (4). Then we have the *Dirichlet Representation* for $S_n(x)$,

$$S_n(x) = \int_{-\pi}^{\pi} \tilde{f}(x + s) \frac{\sin (n + \frac{1}{2})s}{2\pi \sin (s/2)} \, ds, \qquad x \text{ in } \mathbf{R}, \quad n = 0, 1, 2, \ldots \quad (7)$$

The function $D_n(s) = \sin((n + \frac{1}{2})s)/(2\pi \sin (s/2))$ which appears in the integrand of (7) is called the *Dirichlet kernel* and has the property that

$$\int_{-\pi}^{\pi} D_n(s) \, ds = 1, \qquad \text{for all } n = 0, 1, 2, \ldots \quad (8)$$

Proof.   Recalling the definition of the coefficients A_k, B_k given in (4) and substituting in $S_n(x)$, we obtain (after interchanging summation and integration) that

$$S_n(x) = \frac{1}{\pi} \int_{-\pi}^{\pi} f(t) \left[\frac{1}{2} + \sum_{k=1}^{n} \cos k(t - x) \right] dt, \qquad \text{for any } n = 0, 1, 2, \ldots$$

Using the identity $\frac{1}{2} + \sum_{k=1}^{n} \cos ks = \sin((n + \frac{1}{2})s)/(2 \sin (s/2))$ (see the problem set for a proof), we have that

$$S_n(x) = \frac{1}{\pi} \int_{-\pi}^{\pi} f(t) \frac{\sin (n + \frac{1}{2})(t - x)}{2 \sin \frac{1}{2}(t - x)} \, dt \qquad (9)$$

Now after replacing f by $\tilde{f}$ in (9), making the change of variables $t - x = s$, and applying the Integration Theorem for Periodic Functions, the desired formula (7) for $S_n(x)$ results. The identity (8) results immediately from application of (7) to the function $f \equiv 1$, and we are done.

Now we are ready to state a major result of this section.

Pointwise Convergence Theorem for Fourier Series.   Let f be piecewise smooth on $[-\pi, \pi]$, and $\tilde{f}$ be the periodic extension of f into $\mathbf{R}$ with period 2π. Then for each x_0 in $\mathbf{R}$, the Fourier Series (3) and (4) converges to $\frac{1}{2}[\tilde{f}(x_0{}^+) + \tilde{f}(x_0{}^-)]$.

Proof. Writing $S_n(x)$ in (7) as

$$S_n(x) = \int_{-\pi}^{0} \tilde{f}(x+s)D_n(s)\,ds + \int_{0}^{\pi} \tilde{f}(x+s)D_n(s)\,ds$$

we will show that

$$\int_{0}^{\pi} \tilde{f}(x_0+s)D_n(s)\,ds \to \frac{1}{2}\tilde{f}(x_0^+), \qquad \text{as } n \to \infty,$$

$$\int_{-\pi}^{0} \tilde{f}(x_0+s)D_n(s)\,ds \to \frac{1}{2}\tilde{f}(x_0^-) \qquad \text{as } n \to \infty \tag{10}$$

which implies the assertion. We prove only the first statement in (10); the proof of the other follows similarly. Notice that $D_n(s)$ is an even function and hence from (8), $\int_{0}^{\pi} D_n(s)\,ds = \int_{-\pi}^{0} D_n(s)\,ds = \frac{1}{2}$. Thus we may write

$$\int_{0}^{\pi} \tilde{f}(x_0+s)D_n(s)\,ds - \frac{1}{2}\tilde{f}(x_0^+) = \int_{0}^{\pi} [(\tilde{f}(x_0+s) - \tilde{f}(x_0^+)]D_n(s)\,ds$$

$$= \frac{1}{2\pi}\int_{-\pi}^{\pi} \{H(s)[\tilde{f}(x_0+s) - \tilde{f}(x_0^+)]\}\cos ns\,ds \tag{11}$$

$$+ \frac{1}{\pi}\int_{-\pi}^{\pi} \left\{ H(s)\cdot\frac{\tilde{f}(x_0+s) - \tilde{f}(x_0^+)}{s}\frac{\frac{1}{2}s}{\sin(s/2)}\cos\left(\frac{s}{2}\right) \right\}\sin ns\,ds$$

where $H(s)$ is the piecewise constant function which has the value 1 for $0 < s < \pi$, and zero for $-\pi < s < 0$. Now $H(s)[\tilde{f}(x_0 + s) - \tilde{f}(x_0^+)]$ belongs to the class $PC[-\pi, \pi]$ as a function of s, and hence the Decay of Coefficients Theorem implies that the first integral in (11) tends to zero as $n \to \infty$. Now if we can show that the function in braces $\{\cdot\}$ in the second integral in (11) is also in $PC[-\pi, \pi]$ as a function of s, the Decay of Coefficients Theorem also implies that this integral tends to zero as $n \to \infty$, finishing the proof. Observe that $H(s)$, $\frac{1}{2}s/\sin(s/2)$, and $\cos(s/2)$ are all piecewise continuous on $-\pi \le s \le \pi$. Now since $H(s) \equiv 0$ for $-\pi \le s < 0$, it remains only to show that $[\tilde{f}(x_0 + s) - \tilde{f}(x_0^+)]/s$ is piecewise continuous on $0 \le s \le \pi$. Evidently, the only trouble point is $s = 0$. Recall that f is piecewise smooth on $[-\pi, \pi]$; thus $\tilde{f}(x_0 + s)$ is also piecewise smooth for $-\pi \le s \le \pi$. Thus for all sufficiently small positive s, we have from the Mean Value Theorem that

$$\frac{\tilde{f}(x_0+s) - \tilde{f}(x_0^+)}{s} = \tilde{f}'(x_0+s^*) \qquad \text{for some } s^* \text{ with } 0 < s^* < s \tag{12}$$

Since $\tilde{f}'(x_0 + s^*) \to \tilde{f}'(x_0^+)$ as $s \to 0+$, we see from (12) that the function $[\tilde{f}(x_0+s) - \tilde{f}(x_0^+)]/s$ has passed the last test to show that it is piecewise continuous on $0 \le s \le \pi$, and hence we are done.

Example 13.23

Let us apply the Pointwise Convergence Theorem to the function in Example 13.22. For $-\pi < x < \pi$ we easily calculate that

$$\tfrac{1}{2}[\,\tilde{f}(x_0{}^+) + \tilde{f}(x_0{}^-)] = \begin{cases} 5, & 0 < x_0 < \pi \\ 0, & x_0 = 0 \\ -5, & -\pi < x_0 < 0 \end{cases}$$

But what about the endpoints? A little thought shows that $\tilde{f}(\pi^+) = f(-\pi^+)$ and $\tilde{f}(-\pi^-) = f(\pi^-)$, as is true for the periodic extension of any piecewise continuous function. Thus in our case,

$$\tfrac{1}{2}[\,\tilde{f}(x_0{}^+) + \tilde{f}(x_0{}^-)] = 0, \qquad x_0 = \pm\pi$$

Thus we see that Fourier Series of piecewise smooth functions converge pointwise for all x *in* **R**, and the sum at a point x_0 where $\tilde{f}$ is continuous is just $\tilde{f}(x_0)$, whereas at both endpoints $x = \pm\pi$ the sum has the same value, $\tfrac{1}{2}[f(\pi^-) + f(-\pi^+)]$.

If we assume a bit more smoothness for f, FS$[f]$ will converge uniformly and not just pointwise. Clearly, the function in Example 13.22 shows that something more than merely piecewise smoothness is required. First, however, we need an effective test for uniform convergence. The simplest, perhaps, is the following (presented without proof).

Weierstrass M-Test.   Let the functions $g_k(x)$, $k = 1, 2, \ldots$, be continuous on the interval I, and suppose that the positive constants M_k are such that $|g_k(x)| \le M_k$ for all x in I, and $\sum\limits_{k=1}^{\infty} M_k < \infty$. Then there exists a continuous function $g(x)$ on I such that the series $\sum\limits_{k=1}^{\infty} g_k(x)$ converges uniformly to g.

Now we can prove the

Uniform Convergence Theorem for Fourier Series.   Let f be in PS$[-\pi, \pi]$ such that $\tilde{f}$ is in $C^0(\mathbf{R})$. Then FS$[f]$ converges uniformly to $\tilde{f}$ on **R**, where $\tilde{f}$ is the periodic extension of f into **R** with period 2π.

Proof. Obviously, $\tilde{f}$ will be in $C^0(\mathbf{R})$ if and only if f is in $C^0[-\pi, \pi]$ and $f(-\pi) = f(\pi)$. The Cauchy–Schwartz Inequality in $\mathbf{R}^2$, when applied to $g_k(x) = A_k \cos kx + B_k \sin kx$, the kth-term in FS$[f]$, yields the inequality

$$|A_k \cos kx + B_k \sin kx| \le (A_k^2 + B_k^2)^{1/2} \qquad \text{for all } x \text{ in } \mathbf{R}$$

Thus if we take $M_k = (A_k^2 + B_k^2)^{1/2}$ and if $\sum\limits_{k=1}^{\infty} M_k < \infty$, the Weierstrass M-test provides the desired result. The rest of the proof is aimed toward showing that $\sum_{k=1}^{\infty} M_k < \infty$. Since both f and f' are in PC$[-\pi, \pi]$, we may write

$$\text{FS}[f] = \frac{A_0}{2} + \sum_{k=1}^{\infty} (A_k \cos kx + B_k \sin kx)$$

$$\text{FS}[f'] = \frac{A_0'}{2} + \sum_{k=1}^{\infty} (A_k' \cos kx + B_k' \sin kx)$$

where the coefficients are given by the Euler formulas (4). Integration by parts in the Euler Formulas for A_k' and B_k' implies that

$$A_0' = 0, \quad A_k' = kB_k, \quad B_k' = -kA_k, \qquad k = 1, 2, \ldots \tag{13}$$

Thus the Bessel Inequality applied to FS$[f']$ implies that $\sum k^2(A_k^2 + B_k^2) < \infty$.

Using this and the Cauchy–Schwartz Inequality, we obtain that

$$\sum_{k=1}^{n} [A_k^2 + B_k^2]^{1/2} \le \left[\sum_{k=1}^{\infty} \frac{1}{k^2} \right]^{1/2} \left[\sum_{k=1}^{\infty} k^2(A_k^2 + B_k^2) \right]^{1/2}$$

Thus the series $\sum [A_k^2 + B_k^2]^{1/2}$ converges because its partial sums form a bounded monotone sequence, and hence we are done.

Decay Estimates

Now it is time to give a precise estimate on how fast the coefficients in FS$[f]$ decay.

Decay of Fourier Coefficients of Piecewise Smooth Functions. Let g be in PS$[-\pi, \pi]$. Then the coefficients in FS$[g]$ are $0(1/k)$, as $k \to \infty$. If $\tilde{g}$, the periodic extension of g, is not in $C^0(\mathbf{R})$, then this estimate cannot be improved.

Proof. Since g is in PS$[-\pi, \pi]$ we can partition $[-\pi, \pi]$ with points $-\pi = x_0 < x_1 < x_2 < \cdots < x_N < x_{N+1} = \pi$ such that g is continuously differentiable on each subinterval $[x_n, x_{n+1}]$. Then we may integrate the coefficients of the sine terms in FS$[g']$ by parts as follows:

$$\frac{1}{\pi} \int_{-\pi}^{\pi} g'(x) \sin kx \, dx = \frac{1}{\pi} \sum_{n=0}^{N} \int_{x_n}^{x_{n+1}} g'(x) \sin kx \, dx$$

$$= \frac{1}{\pi} \sum_{n=0}^{N} g(x) \sin kx \, \Big|_{x_n^+}^{x_{n+1}^-} - \frac{k}{\pi} \sum_{n=0}^{N} \int_{x_n}^{x_{n+1}} g(x) \cos kx \, dx \tag{14}$$

$$= G - k \cdot \frac{1}{\pi} \int_{-\pi}^{\pi} g(x) \cos kx \, dx$$

Now since G is bounded and the coefficients of FS$[g']$ decay to zero as $k \to \infty$, we see that the coefficients of the cosine terms in FS$[g]$ are $0(1/k)$, as $k \to \infty$. A similar computation holds for the coefficients of the sine terms in FS[g], and hence the asserted decay rate for the coefficients in FS[g] is established. We omit the proof that this decay rate cannot be improved.

Now let us assume that f in PC[$-\pi, \pi$] is such that $\tilde{f}$ is in $C^m(\mathbf{R})$, and $f^{(m)}$ is in PS[$-\pi, \pi$]. Set

$$\text{FS}[f^{(j)}] = A_0^j + \sum_{k=1}^{\infty} (A_k^j \cos kx + B_k^j \sin kx), \qquad \text{for } j = 0, 1, \ldots, m+1$$

where j is a superscript, not a power. Using the definition of the coefficients A_k^j, B_k^j in (4) and integration by parts we have the recursion relation

$$A_k^{j+1} = kB_k^j, \quad B_k^{j+1} = -kA_k^j, \qquad k > 0, \text{ and } j = 0, \ldots, m \qquad (15)$$

Now let us assume in addition just a bit more smoothness; namely, that $f^{(m+1)}$ is *piecewise smooth*. Taking $g \equiv f^{(m+1)}$ we conclude from the previous result that the coefficients in FS[$f^{(m+1)}$] are $0(1/k)$, as $k \rightarrow \infty$. Putting this fact together with (15) produces the following estimate.

Decay Theorem for Fourier Trigonometrical Coefficients.   Let f belong to PC[$-\pi, \pi$], $\tilde{f}$ belong to $C^m(\mathbf{R})$, and $f^{(m+1)}$ belong to PS[$-\pi, \pi$] for some $m = 0, 1, 2, \ldots$. If A_k, B_k are the Fourier coefficients of f given by (4), then

$$A_k = 0(1/k^{m+2}), \quad B_k = 0(1/k^{m+2}) \qquad \text{as } k \rightarrow \infty \qquad (16)$$

Moreover, if $\tilde{f}$ is *not* in $C^{m+1}(\mathbf{R})$, the estimates (16) cannot be improved.

Proof. The decay estimates (16) were proven above. We omit the proof that these estimates are best possible.

The Decay Theorems are very useful as a check when calculating Fourier Trigonometrical Series.

Example 13.24

The "square wave" of Example 13.22 does not come directly under the Decay Theorem above since its periodic extension is not continuous. It is, however, in PS[$-\pi, \pi$], but from the first decay theorem we conclude that the Fourier coefficients of a square wave decay precisely like $O(1/k)$, which we observe is the case.

Half-Range Expansions

There are some standard ways of constructing orthogonal sets (and even bases) for the Euclidean space PC[$0, L$] with the standard scalar product.

We begin by finding bases for PC[$0, L$] which consist only of sine functions or only of cosine functions. Let f in PC[$0, L$] be given and put

$$f_{\text{even}}(x) = \begin{cases} f(x), & 0 \leq x \leq L \\ f(-x), & -L \leq x \leq 0 \end{cases}$$

$$f_{\text{odd}}(x) = \begin{cases} f(x), & 0 < x \leq L \\ -f(-x), & -L < x < 0 \end{cases}$$

Observe that (a) f_{odd} is odd about $x = 0$ and f_{even} is even about $x = 0$, and (b) f_{odd} and f_{even} are in PC[$-L$, L]. The functions f_{odd} and f_{even} are called the *odd and even extensions*, respectively, of f to the interval [$-L$, L]. Now we observe that FS[f_{odd}] contains only sine functions and FS[f_{even}] contains only cosine functions. We shall define the *Fourier Sine Series* and *Fourier Cosine Series of f*, denoted by FSS[f] and FCS[f], as follows:

$$\text{FSS}[f] \equiv \text{FS}[f_{\text{odd}}], \qquad \text{FCS}[f] \equiv \text{FS}[f_{\text{even}}]$$

Thus for f in PC[0, L], we have that

$$\text{FSS}[f] = \sum_{k=1}^{\infty} b_k \sin \frac{k\pi x}{L}, \qquad \text{where } b_k = \frac{2}{L} \int_0^L f(x) \sin \frac{k\pi x}{L} \, dx, \quad k = 1, 2, \ldots$$

Our knowledge of convergence properties of Fourier Series tells us much about Fourier Sine Series. In particular, the functions $\Phi_s = \{\sin(k\pi x/L) : k = 1, 2, \ldots\}$ form an orthogonal set in PC[0, L] since for all $k \neq m$,

$$0 = \int_{-L}^{L} \sin \frac{k\pi x}{L} \sin \frac{m\pi x}{L} \, dx = 2 \int_0^L \sin \frac{k\pi x}{L} \sin \frac{m\pi x}{L} \, dx$$

It is not difficult to show that Φ_s is a basis of PC[0, L] since the Fourier Sine Series of any f in PC[0, L] converges in the mean to f. The convergence properties of FSS[f] can be traced back to those criteria for Fourier Series. In the same fashion we see that the Fourier Cosine Series of a function f in PC[0, L] may be defined by

$$\text{FCS}[f] = \frac{a_0}{2} + \sum_{k=1}^{\infty} a_k \cos \frac{k\pi x}{L}, \qquad a_k = \frac{2}{L} \int_0^L f(x) \cos \frac{k\pi x}{L} \, dx, \quad k = 0, 1, 2, \ldots$$

Again, the convergence properties of FCS[f] are known from the behavior of FS[f_{even}]. In particular we may verify that the set $\Phi_c = \{\cos(k\pi x/L) : k = 0, 1, 2, \ldots\}$ is an orthogonal set in PC[0, L] and is a basis for that space.

The series FSS[f] and FCS[f] for f in PC[0, L] are frequently referred to as *half-range expansions* in the applications. See the problem set for examples.

Complex Fourier Trigonometrical Series

If the functions in PC[$-\pi$, π] are complex valued, it is convenient to use the orthogonal set $\Psi = \{e^{ikx} : k = 0, \pm 1, \pm 2, \ldots\}$. Recall that the scalar product in PC[$-\pi$, π] is $\langle f, g \rangle = \int_{-\pi}^{\pi} f\bar{g} \, dx$, where $\bar{g}$ denotes the complex conjugate of g. Note that $\|e^{ikx}\|^2 = 2\pi$ for all k. Thus the Fourier Series of an element f in PC[$-\pi$, π] with respect to Ψ is

$$FS[f] \equiv c_0 + \sum_{n=1}^{\infty} (c_{-n}e^{-inx} + c_n e^{inx}) \tag{17}$$

where

$$c_n = \frac{\langle f, e^{inx} \rangle}{\|e^{inx}\|^2} = \frac{1}{2\pi} \int_{-\pi}^{\pi} f(x)e^{-inx}\,dx, \qquad n = 0, \pm 1, \pm 2, \ldots \tag{18}$$

There are advantages in grouping the nth and $(-n)$th terms as shown in (17), but it is customary to write $FS[f]$ in the following equivalent two-sided form (leaving aside questions of ordering):

$$FS[f] = \sum_{n=-\infty}^{\infty} c_n e^{inx} \tag{19}$$

Observe that if f in $PC[-\pi, \pi]$ happens to be real valued, then

$$c_{-n} = \bar{c}_n \quad \text{and} \quad c_n = \tfrac{1}{2}(A_n - iB_n) \qquad \text{for all } n > 0 \tag{20}$$

and hence the series in (17) reduces to the series in (3). We have intentionally used the same symbol, $FS[f]$, to denote both the real and complex forms of Fourier Series in (3) and (17) or (19). There is little risk of confusion in this practice because of the special relation between the orthogonal sets Φ and Ψ.

Example 13.25

Let us find the complex form of $FS[f]$ when $f(x) = x$, $|x| \le \pi$. Clearly, $c_0 = 0$, and for $n \neq 0$, $c_n = \dfrac{1}{2\pi} \int_{-\pi}^{\pi} xe^{-inx}\,dx = (i/n)\cos n\pi = i(-1)^n/n$. Thus

$$FS[f] = i \sum_{n=1}^{\infty} \left[\frac{(-1)^n}{n} e^{inx} - \frac{(-1)^n}{n} e^{-inx} \right]$$

Recalling that $e^{i\theta} = \cos\theta + i\sin\theta$, we easily see that also

$$FS[f] = 2 \sum_{n=1}^{\infty} \frac{(-1)^{n+1}}{n} \sin nx$$

which we could have obtained by expanding f directly into a Fourier Series over the trigonometrical set Φ. Using (20) we see that in general no confusion can arise if the complex form of $FS[f]$ is computed when f is real valued.

Example 13.26

Let us compute $FS[f]$ for $f(x) = e^{i\omega x}$, $|x| \le \pi$, where ω is not an integer. We have

$$c_n = \frac{1}{2\pi} \int_{-\pi}^{\pi} e^{i\omega x} e^{-inx}\,dx = \frac{1}{2\pi} \frac{1}{i(\omega - n)} e^{i(\omega - n)x} \Big]_{-\pi}^{+\pi} = \frac{\sin(\omega - n)\pi}{(\omega - n)\pi}$$

Thus

$$\text{FS}[f] = \frac{\sin \omega\pi}{\omega\pi} + \sum_{n=1}^{\infty} \left[\frac{\sin(\omega - n)\pi}{(\omega - n)\pi} e^{inx} + \frac{\sin(\omega + n)\pi}{(\omega + n)\pi} e^{-inx} \right]$$

It is not difficult to see that Ψ is a basis for the complex space $\text{PC}[-\pi, \pi]$, and that the Pointwise and Uniform Convergence Theorems hold for $\text{FS}[f]$ when f is complex valued.

Application to an RLC Circuit

Consider the simple *RLC* circuit of Figure 13.21. As we saw in Section 5.4, the charge $q(t)$ on the capacitor satisfies the equation

$$Lq''(t) + Rq'(t) + \frac{1}{C} q(t) = E(t) \tag{21}$$

Although we have already solved equations of the form (21) for arbitrary piecewise continuous input voltages $E(t)$ (see Sections 5.4 and 6.4), the Fourier Series approach is more natural if $E(t)$ is periodic. Suppose, then, that $E(t)$ has period 2π and that the restriction of $E(t)$ to $[-\pi, \pi]$ is piecewise continuous.

We want to find all functions $q(t)$ of period 2π for which

$$q \text{ is in } C^1(\mathbf{R}), \qquad q'' \text{ is in } \text{PC}[-\pi, \pi] \tag{22a}$$

$$Lq'' + Rq' + \frac{1}{C}q - E = 0 \qquad \begin{array}{l} \text{on } [-\pi, \pi] \text{ except possibly} \\ \text{for a finite number of points} \end{array} \tag{22b}$$

Suppose that (21) has a solution q meeting these requirements. First put

$$\text{FS}[q] = \sum_{k=-\infty}^{\infty} c_k e^{ikt} \quad \text{and} \quad \text{FS}[E] = \sum_{k=-\infty}^{\infty} b_k e^{ikt} \tag{23}$$

where the c_k's are to be found, while the b_k's are known. Using the smoothness conditions of (22) and integration by parts, we have that

$$\text{FS}[q'] = \Sigma \, ikc_k e^{ikt}, \qquad \text{FS}[q''] = \Sigma \, (ik)^2 c_k e^{ikt} \tag{24}$$

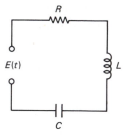

Figure 13.21 *RLC* circuit.

Thus from (21), (23), and (24) we have that

$$\sum [P(ik)c_k - b_k]e^{ikt} = 0, \qquad \text{where } P(x) = Lx^2 + Rx + \frac{1}{C} \qquad (25)$$

Since $\{e^{ikt} : k = 0, \pm 1, \pm 2, \ldots\}$ is a basis, it follows from (25) that $P(ik)c_k - b_k = 0$, $k = 0, \pm 1, \ldots$. But $P(ik) \neq 0$, $k = 0, \pm 1, \ldots$, and hence $c_k = b_k/P(ik)$, all k. Thus

$$\text{FS}[q] = \sum_{k=-\infty}^{\infty} \frac{1}{P(ik)} b_k e^{ikt} \qquad (26)$$

We omit the proof that the series in (26) converges uniformly to a function $q(t)$ satisfying (22a). By construction the Fourier Series of $Lq'' + Rq' + (1/C)\, q - E$ is the same as the Fourier Series of the zero function. Hence by the Totality Property, we have that $Lq'' + Rq' + (1/C)\, q - E \equiv 0$ on $[-\pi, \pi]$ except possibly at a finite number of points. Thus (22b) holds and we are done.

We have shown that the *RLC* circuit has a unique periodic response $q(t)$ to a periodic input voltage $E(t)$. The response $q(t)$ is known as a *forced oscillation* (see the problem sets of Sections 10.4 and 10.5). Thus *every* solution of (21) is the superposition of damped exponentials (the *transients*) and the steady-state periodic solution just constructed.

PROBLEMS

The following problem(s) may be more challenging: 10, 13, 15, and 17.

1. Use trigonometric identities, common sense, or previous results to find the Fourier Series of each of the following functions without using the Euler formulas.
   (a) $5 - 4\cos 6x - 7\sin 3x - \sin x$.
   (b) $\cos^2 7x + (\sin \frac{1}{2}x)(\cos \frac{1}{2}x) - 2\sin x \cos 2x$.
   (c) $\sin^3 x$.

2. Calculate FS[f] for each of the following functions assumed to have domain $[-\pi, \pi]$.
   (a) $2x - 2$.   (d) $\sin \pi x$.
   (b) x^2.   (e) $|x| + e^x$.
   (c) $a + bx + cx^2$.   (f) $|x(x^2 - 1)|$.
   (g) [*Rectangular Pulse Train*]

$$f(x) = \begin{cases} 0, & 2n\pi + B < x < (2n+2)\pi - B, & n \text{ an integer} \\[2mm] \dfrac{A}{2}, & x = 2n\pi \pm B, & n \text{ an integer} \\[2mm] A, & 2n\pi - B < x < 2n\pi + B, & n \text{ an integer} \end{cases}$$

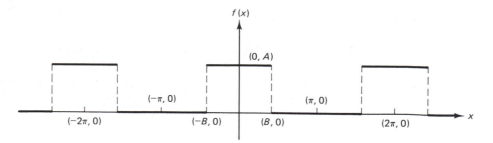

(h) [*Alternating Pulse* (or *Square Wave*)]

$$f(x) = \begin{cases} 0, & -\pi \leq x \leq -B \quad \text{or} \quad B \leq x \leq \pi \\ A, & -B < x < 0 \\ -A, & 0 < x < B \end{cases}$$

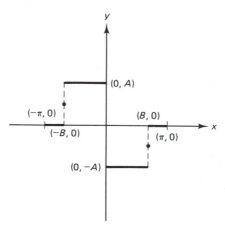

3. Find the Fourier Sine Series, FSS[f], of each of the following functions. Then find the Fourier Cosine Series, FCS[f].
   (a) $f(x) = 1, 0 < x < \pi$.
   (b) $f(x) = \sin x, 0 \leq x \leq \pi$.

4. Let f be in PC[0, c], $f(x) = f(c - x)$ for all x, $c/2 < x < c$ (i.e., f is even about $x = c/2$).
   (a) Prove that FSS[f] contains only terms of the form $\sin 2(k + \frac{1}{2})(\pi x/c)$, for $k \geq 0$.
   (b) Reflect the function $\sin x$, $0 \leq x \leq \pi/4$, about $\pi/4$ to obtain a function f defined on $[0, \pi/2]$ which is even about $x = \pi/4$. Use part (a) to find FSS[f].

5. Let f be in PC[0, c], $f(x) = -f(c - x)$, $c/2 < x < c$ (i.e., f is "odd" with respect to $x = c/2$).
   (a) Prove that FCS[f] contains only terms of the form $\cos 2(k + \frac{1}{2})(\pi x/c)$.
   (b) Use part (a) to find FCS[f] for the function f defined by reflecting x^2, $0 \leq x < 1$, "oddly" through $x = 1$.

6. Compute the Fourier Series of each of the following functions.
   (a) $f(x) = \sin x, \quad -\pi \leq x \leq \pi$.

(b) $f(x) = \sin x$, $\quad -\pi/2 \leq x \leq \pi/2$.
(c) $f(x) = \sin x$, $\quad -3\pi/2 \leq x \leq 3\pi/2$.

7. Find the complex form of the Fourier Series of each of the following functions defined on $[-\pi, \pi]$.
(a) $\cos x$.
(b) x^2.
(c) $|x| + ix$.

8. Find the steady-state charge in the capacitor in an RLC loop if $R = 10 \ \Omega$, $L = 0.5$ H, and $C = 10^{-4}$ F, and let the impressed electromotive force $E(t)$, measured in volts, be described by the figure.

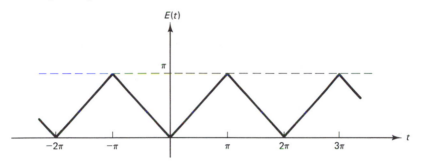

9. In a spring–mass system, let the spring constant be 1.01 newtons/meter and the weight of the mass be 1 kilogram. Suppose that while in motion, the mass is acted on by viscous damping with coefficient equal to 0.2, and also suppose that the mass is driven by the force (measured in newtons) described in the figure. Find the steady-state motion of the system. [See Problem 8 in Section 5.5.]

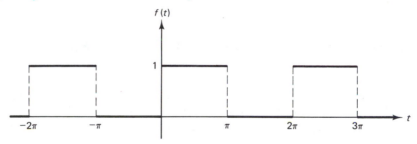

10. Show that $\int_{-T}^{T} f(x) \, dx = \int_{a}^{a+2T} f(s) \, ds$ for all real numbers a if f is periodic of period $2T$ and piecewise continuous on every finite interval.

11. Show that

$$\frac{1}{2} + \sum_{k=1}^{n} \cos ks = \frac{\sin \left(n + \frac{1}{2}\right)s}{2 \sin (s/2)}$$

[*Hint:* Use L'Hôpital's Rule if $s = 2m\pi$, m an integer. For all other s, use the identity]

$$2 \sin \frac{s}{2} \cos ks = \sin \left(k + \frac{1}{2}\right)s - \sin \left(k - \frac{1}{2}\right)s$$

12. Prove that $\|f - S_n\| \leqq \|f - S_m\|$ if $m \leqq n$, where S_n and S_m are the corresponding partial sums of FS[f].

13. Show that

$$\sum_1^\infty \frac{1}{n^{1/4}} \cos nx, \qquad \sum_1^\infty (\sin n) \sin nx, \qquad \sum_2^\infty \frac{1}{\ln (n)} \sin nx$$

are *not* the Fourier Series of any functions in PC[$-\pi, \pi$]. [*Hint*: Use the Decay Theorem or the Parseval Theorem of Section 13.5.]

14. Evaluate

$$\lim_{n \to \infty} \int_{-\pi}^{\pi} e^{\sin x}(x^5 - 7x + 1)^{52} \cos nx \, dx$$

[*Hint*: Relate the integral to a Fourier coefficient of a portion of the integrand.]

15. (a) Prove that $\{1, \cos x, \ldots, \cos nx, \ldots\}$ is not a basis of PC[$-\pi, \pi$].
   (b) Prove that $\{1, \cos x, \ldots, \cos nx, \ldots\}$ is a basis of the subspace of all even functions in PC[$-\pi, \pi$].
   (c) Formulate and prove similar statements for $\{\sin x, \ldots, \sin nx, \ldots\}$.

16. Let $f(x)$ be the function $|x|$ on $[-\pi, \pi]$.
   (a) Using FS[f] show that

$$\frac{\pi^2}{8} = \sum_{k=1}^\infty \frac{1}{(2k - 1)^2}$$

   (b) Discuss the convergence properties of FS[f] on all of **R**.

17. Let S be the linear space of functions f in PS[$-T, T$] whose periodic extension $\tilde{f}$ is in $C^0(\mathbf{R})$. Show that Φ_T in (1) is a basis for S. [*Hint*: Use the Uniform Convergence Theorem and an implication in Figure 13.18 of Section 13.5.]

13.7 SEPARATION OF VARIABLES AND EIGENFUNCTION EXPANSIONS

The method of separated variables used in Section 13.3 to construct a solution of the problem of the plucked guitar string has general application to a wide class of boundary/initial value problems for linear partial differential equations. We shall build on the theory and examples of earlier sections to outline this general method. We shall also generalize the method of eigenfunctions in Section 13.6 to solve boundary/initial value problems involving nonhomogeneous linear partial differential equations.

The Method of Separation of Variables

Although we shall work out the method only in the context of a specific problem, the method itself applies to many other problems (see Sections 13.8 and 13.9).

Consider the transverse motion of a taut, flexible string of length L. Assuming that the endpoints are held fixed, we see from Sections 13.2 and 13.3 that the deflection

$u(x, t)$ of the string at the point x and time t is the unique solution of the initial/boundary value problem (see Figure 13.22):

For $G = \{(x, t) : 0 < x < L, t > 0\}$ we seek a function u which is both in $C^2(G)$ and in $C^1(\text{cl } G)$, and such that

(PDE) $u_{tt} - c^2 u_{xx} = 0$ in G

(BC) $u(0, t) = 0, \quad u(L, t) = 0, \quad t \geq 0$ (1)

(IC) $u(x, 0) = f(x), \quad u_t(x, 0) = g(x), \quad 0 \leq x \leq L$

---

where c^2 is a constant and the functions $f(x)$ and $g(x)$ are the initial deflection and initial velocity, respectively, of the string. A solution of (1) that satisfies these smoothness conditions is called a *classical solution*.

I. Set Up Mathematical Model as a Boundary/Initial Value Problem.   We have already done this in (1). If (PDE) is not homogeneous or if the boundary conditions are not homogeneous, the techniques at the end of this section must be applied.

II. Choose Independent Variables Appropriate to Shape of Domain.   For (1), G is a region bounded by line segments on which x is constant or t is constant (see Figure 13.22). Thus the rectangular coordinates (x, t) should be used in (PDE), as they already have been. If G were, for example, a circular disk, polar coordinates should be chosen (see Section 13.9). Generally speaking, each part of the boundary of G should be a set on which one of the independent variables is constant (i.e., a coordinate level set).

III. Separate the Variables.   Suppose that $u = X(x)T(t)$ satisfies (PDE). Substituting into (PDE), we have

$$X(x)T''(t) - c^2 X''(x)T(t) = 0 \tag{2}$$

In any coordinate rectangle R in G in which $X(x)T(t)$ does not vanish, we have that

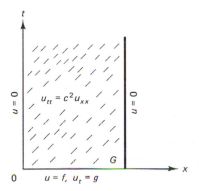

Figure 13.22   Regions for (1).

$$\frac{X''(x)}{X(x)} = \frac{1}{c^2}\frac{T''(t)}{T(t)} \qquad \text{for all } (x, t) \text{ in } R \tag{3}$$

The name *Method of Separation of Variables* derives from the fact that equation (2) can be put into the variables-separated form (3). Now let (x, t_0) be any point *inside R*. Then we see that

$$\frac{X''(x)}{X(x)} = \frac{1}{c^2}\frac{T''(t_0)}{T(t_0)}$$

for all x such that (x, t_0) lies in R. Thus the ratio $X''(x)/X(x)$ is constant in cl R; call the constant λ. Then the ratio $T''(t)/c^2T(t)$ is also equal to λ in cl R. Thus there exists a *separation constant* λ such that

$$\frac{X''(x)}{X(x)} = \frac{1}{c^2}\frac{T''(t)}{T(t)} \equiv \lambda \tag{4}$$

for all (x, t) in cl R. It can be shown that the closed region $0 \le x \le L, 0 \le t \le T$ can be covered with finitely many nonoverlapping closed coordinate rectangles such as R above. Moreover, it can also be shown that if the ratios in (4) have the constant value λ in one coordinate rectangle R, they have the *same* constant value λ in *every* coordinate rectangle, and in fact throughout G. This separation argument is the same every time the Method of Separation of Variables is used and we shall not repeat it again in this or in later sections.

IV. Set Up a Sturm–Liouville Problem. The functions $X(x)$ and $T(t)$ that solve (2) must necessarily satisfy the pair of equations

$$\begin{aligned} X'' - \lambda X &= 0, & 0 < x < L \\ T'' - \lambda c^2 T &= 0, & t > 0 \end{aligned} \tag{5}$$

for some constant λ. Conversely, any solution of Eq. (5) for any real constant λ must be such that $X(x)T(t)$ is a solution of (PDE) in G.

Now we shall demand, in addition, that the solution $X(x)T(t)$ of (PDE) generated by (5) also satisfy (BC). This will amount to a restriction on the choice of λ. If we require that $X(x)T(t)$ satisfy (BC), we must have that

$$X(0)T(t) = 0, \quad X(L)T(t) = 0, \qquad \text{all } t \ge 0$$

Now if either $X(0) \neq 0$ or $X(L) \neq 0$, we have that $T(t) \equiv 0$ and hence $v(x, t) = X(x)T(t)$ would be the trivial solution. Thus to ensure a nontrivial solution, we must take $X(0) = X(L) = 0$. But recall that $X(x)$ satisfies the differential equation $X'' - \lambda X = 0$ on $0 < x < L$; thus we consider the Sturm–Liouville Problem

$$X'' - \lambda X = 0, \qquad X(0) = X(L) = 0 \tag{6}$$

Recall from Section 13.5 that (6) is an eigenvalue problem for an operator with action d^2/dx^2 and domain $\{X \text{ in } C^2[0, L] : X(0) = X(L) = 0\}$. An eigenfunction of this operator [i.e., a solution of (6)] is often called a *normal mode*.

V. Solve the Sturm–Liouville Problem.   We studied (6) in Section 13.3 and found that it has a nontrivial solution if and only if $\lambda = \lambda_n = -(n\pi/L)^2$, $n = 1$, 2, . . . ; the λ_n are the *eigenvalues* of the operator associated with the Sturm–Liouville Problem (6). The general solution of (6) corresponding to $\lambda = \lambda_n$ is given by

$$X_n(x) = C_n \sin \frac{n\pi x}{L}, \qquad n = 1, 2, \ldots \tag{7}$$

where C_n is an arbitrary real constant. The X_n are *eigenfunctions* of the operator associated with (6). Now inserting $\lambda = \lambda_n$ in the second equation in (5) gives the equation $T'' + (n\pi c/L)^2\, T = 0$, $n = 1, 2, \ldots$, whose general solution is given by

$$T_n(t) = \alpha_n \cos \frac{n\pi ct}{L} + \beta_n \sin \frac{n\pi ct}{L}, \qquad n = 1, 2, \ldots$$

where the constants α_n and β_n are arbitrary real numbers for each n. Thus we have found all the standing waves $X(x)T(t)$ supported by the string, that is, all separated solutions of (PDE) which satisfy (BC):

$$v_n(x, t) = \sin \frac{n\pi x}{L} \left(A_n \cos \frac{n\pi ct}{L} + B_n \sin \frac{n\pi ct}{L} \right), \qquad n = 1, 2, \ldots$$

where A_n and B_n are arbitrary constants. The numbers $\lambda_n = n\pi c/L$ are the *natural frequencies*.

VI. Construct the Formal Solution.   We shall look for a solution of (1) as a linear combination of the standing waves $\{v_n\}$. That is, assume that the solution of (1) has the form

$$u(x, t) = \sum_{n=1}^{\infty} \sin \frac{n\pi x}{L} \left(A_n \cos \frac{n\pi ct}{L} + B_n \sin \frac{n\pi ct}{L} \right) \tag{8}$$

for some choice of the constants A_n and B_n. Let us now restrict ourselves to only those choices for A_n and B_n for which all the series derived by termwise differentiation of (8) up to two times with respect to t and x converge uniformly in cl G. (It would be very difficult to determine in advance the conditions on A_n and B_n to ensure the uniform convergence of the series mentioned above, but fortunately as we shall see, there is no need to do so.) Thus for *any* such choice of the A_n and B_n, $u(x, t)$ satisfies (PDE) and both boundary conditions. Hence it remains only to determine values for A_n and B_n in (8) such that $u(x, t)$ also satisfies the initial conditions. Now $u(x, t)$ will satisfy (IC) if

$$f(x) = \sum_{n=1}^{\infty} A_n \sin \frac{n\pi x}{L}, \qquad 0 \le x \le L \qquad \text{[set } t = 0 \text{ in (8)]}$$

$$g(x) = \sum_{n=1}^{\infty} \frac{n\pi c}{L} B_n \sin \frac{n\pi x}{L}, \qquad 0 \le x \le L \qquad \begin{array}{l}\text{[differentiate (8) termwise}\\ \text{and then set } t = 0\text{]}\end{array}$$

Now recall from Section 13.5 that the set of functions $\Phi_S = \{\sin n\pi x/L :$ $n = 1, 2, \ldots\}$ is a basis for $PC[0, L]$. Thus it follows that for $n = 1, 2, \ldots,$

$$A_n = \frac{\langle f, \sin (n\pi x/L)\rangle}{\|\sin (n\pi x/L)\|^2} = \frac{2}{L}\int_0^L f(x)\sin\frac{n\pi x}{L}\,dx$$

and

$$\frac{n\pi c}{L}B_n = \frac{\langle g, \sin (n\pi x/L)\rangle}{\|\sin (n\pi x/L)\|^2} = \frac{2}{L}\int_0^L g(x)\sin\frac{n\pi x}{L}\,dx$$

Thus we have that the *formal solution* of (1) by the Method of Separation of Variables is

$$u(x, t) = \sum_1^\infty \left(A_n \cos\frac{n\pi ct}{L} + B_n \sin\frac{n\pi ct}{L}\right)\sin\frac{n\pi x}{L}$$

$$A_n = \frac{2}{L}\int_0^L f(x)\sin\frac{n\pi x}{L}\,dx \tag{9}$$

$$B_n = \frac{2}{n\pi c}\int_0^L g(x)\sin\frac{n\pi x}{L}\,dx$$

where we have simply *assumed* that the series in (9) and its derived series up to second order converge uniformly. Although $u(x, t)$ obviously satisfies (BC), we need the stated uniform convergence properties to verify (PDE) and (IC).

VII. Determine If the Formal Solution is a Classical Solution.   To verify that $u(x, t)$ as defined by (9) is a classical solution, we must show that u belongs to $C^2(G)$ and to $C^1(\text{cl } G)$ and that $u(x, t)$ satisfies (PDE) and (IC). To do this, we must impose smoothness conditions on the initial data f and g. Let $\tilde{f}$ and $\tilde{g}$ denote the odd extensions of f and g into the interval $[-L, L]$ which are further extended periodically to $\mathbf{R}$. We shall use these extensions in the proof of the following theorem.

Classical Solution.   Let f and g satisfy the conditions:
(a) f is in $C^2[0, L]$, f''' is in $PC[0, L]$, $f(0) = f(L) = f''(0) = f''(L) = 0$;
(b) g is in $C^1[0, L]$, g' is in $PC[0, L]$, $g(0) = g(L) = 0$.
Then $u(x, t)$ defined by (9) is a classical solution of (1).

Proof.   To prove this, we calculate the Fourier Sine Series of f and g on $[0, L]$:

$$\text{FSS}[f](x) = \sum_{n=1}^\infty \alpha_n \sin\frac{n\pi x}{L}, \qquad \text{FSS}[g](x) = \sum_{n=1}^\infty \beta_n \sin\frac{n\pi x}{L}$$

The hypotheses imply that $\tilde{f}$ lies in $C^2(\mathbf{R})$, $\tilde{g}$ in $C^1(\mathbf{R})$, while $\tilde{f}'''$ and $\tilde{g}'$ are in $PC(I)$ for every interval I. Apply the Decay of Coefficients Theorem of Section 13.6 to the Fourier Sine Series for f and g on $[0, L]$; we have that

$$\sum_1^\infty n^2|\alpha_n| < \infty \quad \text{and} \quad \sum_1^\infty n|\beta_n| < \infty \tag{10}$$

Since A_n and $n\pi c\, B_n/L$ are Fourier Sine Series coefficients of f and g, respectively, we have from (10) that

$$\sum_1^\infty n^2|A_n| < \infty \quad \text{and} \quad \sum_1^\infty n^2|B_n| < \infty \tag{11}$$

Using (11) and the Weierstrass M-test (see Section 13.6), the series obtained by term-by-term differentiation of (9) up to second order in x and t all converge uniformly on cl G. Thus $u(x, t)$ as defined by (9) belongs to C^2(cl G) and all derivatives of u up to second order can be computed by term-by-term differentiation. Thus (9) certainly defines a classical solution of (1).

Remarks: Often it is enough to find a formal solution such as (9) without bothering to check that the initial data are smooth enough to produce a classical solution. Indeed, as we saw in Sections 13.2 and 13.3 by a different approach, some of the most interesting behavior, at least for the wave equation, occurs precisely in problems where the initial data are not smooth but have "corners" which propagate into G. In this case the formal solution cannot possibly be a classical solution since at the interior points where $u(x, t)$ has "corners," u is not even differentiable.

The Method of Eigenfunction Expansions

The Method of Separation of Variables applies if the partial differential equation and boundary conditions involved are homogeneous [see (1)], but what do we do if this is not the case? We saw how to do this in Section 13.6 in the solution of a driven electrical circuit, a solution technique that involved the expansion of the driving force in an eigenfunction series of an associated Sturm–Liouville Problem. First we describe the process for an ordinary differential equation, using our experience with the circuit problem as a guide. Then we adapt the method to partial differential equations.

Suppose that an ordinary differential operator has action given by

$$My \equiv \frac{1}{\rho(x)}\left\{\frac{d}{dx}\left[p(x)\frac{dy}{dx}\right] + q(x)y\right\}$$

where the coefficients satisfy the conditions (19) of Section 13.5. The regular Sturm–Liouville problem,

$$\begin{cases} My = \lambda y \\ B_a y \equiv \alpha_1 y(a) + \beta_1 y'(a) = 0, & \alpha_1^2 + \beta_1^2 \neq 0 \\ B_b y \equiv \alpha_2 y(b) + \beta_2 y'(b) = 0, & \alpha_2^2 + \beta_2^2 \neq 0 \end{cases} \tag{12}$$

has eigenvalues λ_i, $i = 1, 2, \ldots$ and a corresponding orthogonal collection of eigen-functions $\Phi = \{y_n(x)\}$ which is a basis of PC[a, b]. Now consider the boundary value problem

$$My = f, \qquad B_a y = 0, \qquad B_b y = 0 \tag{13}$$

where f is a given function belonging to PC[a, b]. We look for a solution to this problem in the form $y(x) = \sum\limits_{k=1}^{\infty} c_k y_k$. The function $y(x)$ always satisfies the boundary conditions for any choice of the constants c_k, so it remains just to choose the c_k's such that $My = f$ is satisfied (if we can). Expanding f in a Fourier series with respect to the basis $\{y_n\}$ of eigenfunctions of the Sturm–Liouville system (12), we have $f = \sum a_k y_k$. Substitution in the differential equation of (13) yields

$$My = M \sum_{k=1}^{\infty} c_k y_k = \sum_{k=1}^{\infty} c_k \lambda_k y_k = \sum_{k=1}^{\infty} a_k y_k$$

where we have *assumed* that we can interchange M and Σ. Now since $\Phi = \{y_n(x)\}$ is a basis for PC[a, b] it follows that $c_j \lambda_j = a_j$, $j = 1, 2, \ldots$. If no $\lambda_j = 0$, we only need take $c_j = a_j/\lambda_j$, $j = 1, 2, \ldots$, and obtain the solution of (13) in the form

$$y(x) = \sum_{k=1}^{\infty} \frac{a_k}{\lambda_k} y_k \tag{14}$$

Nonhomogeneous Partial Differential Equations

Now we shall adapt the Method of Eigenfunction Expansions to boundary/initial value problems for nonhomogeneous partial differential equations. We begin with the following problem:

$$\begin{cases} u_{tt} - c^2 u_{xx} = F(x, t), & 0 < x < L, \quad t > 0 \\ u(0, t) = u(L, t) = 0, & t \geq 0 \\ u(x, 0) = f(x), & 0 \leq x \leq L \\ u_t(x, 0) = g(x), & 0 \leq x \leq L \end{cases} \tag{15}$$

where $F(x, t)$ is in $C^0(G)$, $G = \{(x, t) : 0 < x < L, t > 0\}$ and f, g lie in PC[0, L]. Now we know that the solution of the homogeneous version of (15) with $F \equiv 0$ [i.e., system (1)] is a superposition of functions from the basis $\{\sin(n\pi x/L)\}$ with time-varying coefficients. In the *Method of Eigenfunction Expansions* for (15) with $F \not\equiv 0$, the solution $u(x, t)$ is expressed as a superposition $\sum U_n(t) \sin(n\pi x/L)$, but the coefficients $\{U_n(t)\}$ will be different. This coefficient set is determined by a technique of undetermined coefficients. We carry out the detailed calculations below.

All the data functions F, f, g may be expanded in terms of the basis functions

$\{\sin{(n\pi x/L)}\}$, obtaining Fourier Sine Series in x (assuming throughout that the functions are sufficiently smooth):

$$\left[\begin{array}{ll}
\text{FSS}[f](x) = \sum_{n=1}^{\infty} A_n \sin \dfrac{n\pi x}{L}, & A_n = \dfrac{2}{L} \int_0^L f(x) \sin \dfrac{n\pi x}{L} \, dx & \text{(16a)} \\[3mm]
\text{FSS}[g](x) = \sum_{n=1}^{\infty} B_n \sin \dfrac{n\pi x}{L}, & B_n = \dfrac{2}{L} \int_0^L g(x) \sin \dfrac{n\pi x}{L} \, dx & \text{(16b)} \\[3mm]
\text{FSS}[F](x,t) = \sum_{n=1}^{\infty} C_n(t) \sin \dfrac{n\pi x}{L}, & C_n(t) = \dfrac{2}{L} \int_0^L F(x,t) \sin \dfrac{n\pi x}{L} \, dx & \text{(16c)}
\end{array} \right.$$

Now expand the solution $u(x,t)$ of (15) in a Fourier Sine Series in x:

$$\text{FSS}[u](x,t) = \sum_{n=1}^{\infty} U_n(t) \sin \frac{n\pi x}{L}, \qquad (x,t) \text{ in } G \qquad (17)$$

where $U_n(t) = (2/L) \int_0^L u(x,t) \sin{(n\pi x/L)} \, dx$. Proceeding formally, we assume that $u(x,t)$ is the sum of its Fourier Sine Series $\text{FSS}[u](x,t)$, insert (17) into the partial differential equation of (15), and use (16) to calculate the set of coefficient functions $\{U_n(t)\}$. We have [using Eq. (16c)]

$$\sum_{n=1}^{\infty} \left\{ -c^2 \left(\frac{n\pi}{L} \right)^2 U_n(t) - U_n''(t) + C_n(t) \right\} \sin \frac{n\pi x}{L} = 0, \qquad (x,t) \text{ in } G$$

Hence since $\Phi = \{\sin n\pi x/L : n = 1, 2, \ldots\}$ is a basis for $PC[0, L]$ we have that

$$U_n''(t) + \left(\frac{n\pi c}{L} \right)^2 U_n(t) = C_n(t), \qquad t > 0, \quad n = 1, 2, \ldots \qquad (18)$$

Using (16a) to determine initial conditions for each $U_n(t)$, we have that

$$u(x,0) = f(x) = \sum_{n=1}^{\infty} U_n(0) \sin \frac{n\pi x}{L} = \sum_{n=1}^{\infty} A_n \sin \frac{n\pi x}{L}, \qquad 0 \le x \le L$$

$$u_t(x,0) = g(x) = \sum_{n=1}^{\infty} U_n'(0) \sin \frac{n\pi x}{L} = \sum_{n=1}^{\infty} B_n \sin \frac{n\pi x}{L}, \qquad 0 \le x \le L$$

Thus the initial conditions on U_n are

$$U_n(0) = A_n, \qquad U_n'(0) = B_n \qquad (19)$$

and (18) and (19) is a second-order, linear, nonhomogeneous initial value problem with solution

$$U_n(t) = A_n \cos \frac{n\pi ct}{L} + \frac{LB_n}{\pi nc} \sin \frac{n\pi ct}{L} + \frac{L}{\pi nc} \int_0^t \sin \frac{n\pi c}{L}(t-s) C_n(s) \, ds \qquad (20)$$

Thus the *formal solution* of (15) is given by (17) and (20), with A_n, B_n, and C_n given by (16). Observe that this solution can be written as

$$u(x, t) = u_1(x, t) + u_2(x, t) \tag{21}$$

where

$$u_1(x, t) = \sum_{n=1}^{\infty} \sin \frac{n\pi x}{L} \left(A_n \cos \frac{n\pi ct}{L} + \frac{LB_n}{\pi nc} \sin \frac{n\pi ct}{L} \right) \tag{22}$$

$$u_2(x, t) = \sum_{n=1}^{\infty} \frac{L}{\pi nc} \int_0^t \sin \frac{n\pi c}{L}(t - s) \sin \frac{n\pi x}{L} C_n(s)\, ds$$

Thus $u_1(x, t)$ represents the response of the string to the initial conditions, while $u_2(x, t)$ is the response to the external force $F(x, t)$.

Remark. If the "driving force" $F(x, t)$ is periodic of a frequency near a "natural frequency," one might expect resonance to occur. This is indeed the case. See the problem set and also Section 13.4 for a related problem. We omit any discussion of the conditions on $F, f,$ and g which will make the foregoing formal solution a classical solution.

Shifting the Data

The Method of Separation of Variables handles initial conditions together with homogeneous boundary conditions and a homogeneous partial differential equation. The Method of Eigenfunction Expansions extends to a nonhomogeneous partial differential equation. Now we show how to reduce a problem where *all* the conditions are nonhomogeneous to one or the other of the problems above. As always, the discussion is in terms of the wave equation on a finite interval.

Consider the vibrating string moving under the influence of a vertical external force $F(x, t)$, subject to the usual initial conditions and with time-varying boundary conditions:

$$\begin{aligned}
&\text{(PDE)} &u_{tt} - c^2 u_{xx} &= F(x, t), &0 < x < L, \quad t > 0 \\
&\text{(BC)} &u(0, t) &= \alpha(t), &u(L, t) = \beta(t), \quad t \geq 0 \\
&\text{(IC)} &u(x, 0) &= f(x), &u_t(x, 0) = g(x), \quad 0 \leq x \leq L
\end{aligned} \tag{23}$$

Observe that the function $v(x, t) = \alpha(t) + (x/L)(\beta(t) - \alpha(t))$ satisfies the conditions $v(0, t) = \alpha(t), v(L, t) = \beta(t)$. Let $w(x, t)$ be a solution of the problem

$$\begin{aligned}
w_{tt} - c^2 w_{xx} &= F(x, t) - (v_{tt} - c^2 v_{xx}) \\
w(0, t) &= w(L, t) = 0 \\
w(x, 0) &= f(x) - v(x, 0) \\
w_t(x, 0) &= g(x) - v_t(x, 0)
\end{aligned} \tag{24}$$

We have that $u(x, t) = w(x, t) + v(x, t)$ is a solution of (23).

What we have done is to introduce the *boundary function* $v(x, t)$, which satisfies both boundary conditions of (23) but none of the other conditions. The effect of letting $w = u - v$ is to introduce a boundary/initial value problem for w in which the boundary data have been *shifted* away from the endpoints $x = 0$ and $x = L$ and attached in altered form to the right-hand sides of the partial differential equation and the initial conditions. But a problem such as (24) for w can be solved by the Method of Eigenfunction Expansions outlined earlier in this section.

In many special cases of practical interest, nonzero data can be "shifted" in such a way as to make the modified problem have a homogeneous partial differential equation *and* homogeneous boundary conditions (but at the expense of altered initial data), which can then be treated directly by the Method of Separation of Variables. For illustrations of this technique, see the problem set.

Comments

The Method of Separation of Variables and the Method of Eigenfunction Expansions are based on the same idea—that of expanding functions in a series of eigenfunctions of an appropriate Sturm–Liouville Problem. The Sturm–Liouville Problem itself is inherent in the geometry and the partial differential and boundary operators of the boundary/initial value problem being solved.

Different operators lead to different Sturm–Liouville Problems, to different eigenfunctions and eigenvalues, and to different bases of different spaces. The possibilities seem unlimited. In Sections 13.8 and 13.9 we explore some other boundary/initial value problems which lead to various types of Sturm–Liouville Problems. Finally, the reader should be reminded that at least for boundary/initial value problems for the wave equation, the Method of Images may sometimes be simpler than the Method of Separation of Variables.

PROBLEMS

The following problem(s) may be more challenging: 3, 6, 8, and 9.

1. Use the Method of Separation of Variables to solve (1) under the following conditions.

   (a) $f(x) = 0$, $g(x) = 3 \sin (\pi/L)x$, $0 \leq x \leq L$.

   (b) $f(x) = g(x) = \begin{cases} x, & 0 \leq x \leq L/2 \\ L - x, & L/2 \leq x \leq L. \end{cases}$

   (c) $f(x) = x(L - x) = -g(x)$, $0 \leq x \leq L$.

2. (a) Solve the problem with free end conditions $u_{tt} - c^2 u_{xx} = 0$, $0 < x < L$, $t > 0$; $u_x(0, t) = u_x(L, t) = 0$, $t \geq 0$; $u(x, 0) = f(x)$ and $u_t(x, 0) = g(x)$, $0 \leq x \leq L$. Show that the solution is an oscillatory motion about the x-axis in the transverse direction superimposed on a uniform transverse translation.

   (b) Solve the boundary/initial value problem of part (a) but with the boundary condition $u_x(0, t) = 0$ replaced by $u(0, t) = 0$. (Thus one end is fastened and the other is free.)

3. (a) Construct the Sturm–Liouville Problem associated with the boundary/initial value problem $u_{tt} - c^2 u_{xx} = 0$, $0 < x < L$, $t > 0$; $u(0, t) = 0$, $u_x(L, t) = -hu(L, t)$, $t \geq 0$, h a positive constant; $u(x, 0) = f(x)$, $u_t(x, 0) = g(x)$, $0 \leq x \leq L$.

(b) Show that the eigenvalues of the Sturm–Liouville operator of part (a) are $\{-s_n^2/L^2:$ $n = 1, 2, \ldots\}$, where s_n is the nth consecutive positive zero of the equation $s + hL \tan s = 0$. Show that there are infinitely many zeros, s_n, but do not try to evaluate them, and find a corresponding basis of eigenfunctions for the space PC[0, L].

4. [*Damped Wave Equation*]. The equation $u_{tt} + b^2 u_t - a^2 u_{xx} = 0,\ 0 < x < L,\ t > 0$, models a vibrating string taking into account air resistance. Find the formal solution $u(x, t)$ of the boundary problem of the damped wave equation:

$$u_{tt} + b^2 u_t - a^2 u_{xx} = 0, \qquad\qquad 0 < x < L, \quad t > 0$$
$$u(0, t) = u(L, t) = 0, \qquad\qquad t \geq 0$$
$$u(x, 0) = f(x), \qquad\qquad u_t(x, 0) = g(x), \quad 0 \leq x \leq L$$

where b and a are positive constants and f and g belong to PC[0, L].

5. [*Shifting Data*]. The boundary/initial value problem

$$u_{tt} - c^2 u_{xx} = g, \qquad\qquad 0 < x < L, \quad t > 0$$
$$u(0, t) = u(L, t) = 0, \qquad\qquad t \geq 0$$
$$u(x, 0) = u_t(x, 0) = 0, \qquad\qquad 0 \leq x \leq L$$

models the vertical displacement $u(x, t)$ of a taut flexible string tied at both ends, with vanishing initial data, and acted on by gravity (g is the constant gravitational acceleration). The outline below shows how to shift the nonhomogeneity in the partial differential equation onto an initial condition.

(a) Find a $v(x, t)$ such that $v_{tt} - c^2 v_{xx} = g,\ 0 < x < L,\ t > 0;\ v(0, t) = v(L, t) = 0$, $t > 0$. [Hint: Try $v(x) = Ax + Bx^2$, where A, B are constants.] Note that $v(x)$ is the steady-state sag of the string under the force of gravity.

(b) Let $w = u - v$ and show that w satisfies the same equations as u, but with g replaced by 0 and the condition $u(x, 0) = 0$ replaced by $w(x, 0) = -gx(L - x)/2c^2$.

(c) Find $u(x, t)$.

6. [*Shifting Data*]. A string of unit length with $c^2 = 1$ is clamped at one end, driven by $\sin \pi t/2$ at the other end, and given an initial velocity. A model for the problem is $u_{tt} - u_{xx} = 0,\ 0 < x < 1,\ t > 0;\ u(0, t) = \sin \pi t/2,\ u(1, t) = 0,\ t \geq 0;\ u(x, 0) = f(x)\ u_t(x, 0) = g(x),\ 0 \leq x \leq 1$. The following steps show how to shift the boundary data $\sin \pi t/2$ onto an initial condition.

(a) Let v be any solution of $v_{tt} - v_{xx} = 0,\ v(0, t) = \sin \pi t/2,\ v(1, t) = 0,\ t \geq 0$. Let $w = u - v$. Show that w is a solution of the same problem as u except the boundary condition that $u(0, t) = \sin \pi t/2$ is replaced by $w(0, t) = 0$, and the data $f(x)$ and $g(x)$ are replaced by $f(x) - v(x, 0)$, and $g(x) - v_t(x, 0)$.

(b) Find $v(x, t)$ in the form $X(x)T(t)$.

(c) Solve the problem if $g(x) \equiv 0$.

7. Solve the following problems.

(a) $u_{xx} = u_{tt} - 6x,\quad 0 < x < 1,\quad t > 0$ **(b)** $u_{xx} = u_{tt},\quad 0 < x < L,\quad t > 0$

$u(0, t) = u(1, t) = 0,\quad t \geq 0$ $u(0, t) = \sin \dfrac{3\pi t}{2L},\quad t \geq 0$

$u(x, 0) = u_t(x, 0) = 0,\quad 0 \leq x \leq 1$ $u(L, t) = 0,\quad t \geq 0$

$\qquad\qquad\qquad\qquad\qquad\qquad\qquad u(x, 0) = u_t(x, 0) = 0,\quad 0 \leq x \leq L$

8. [*Breaking a String by Shaking It*]. Consider the problem of a string of length L with one end fastened and the other driven:

$$u_{tt} = c^2 u_{xx}, \quad 0 < x < L, \quad t > 0$$

$$u(0, t) = 0, \quad u(L, t) = \mu(t), \quad t \geq 0$$

$$u(x, 0) \neq 0, \quad u_t(x, 0) = 0, \quad 0 \leq x \leq L$$

Show that there is a periodic function $\mu(t)$ such that the string will eventually break. [*Hint*: What happens if $\mu(t) = A \cos \omega t$ when ω is near a natural frequency $n\pi c/L$? Observe that this problem is in some sense the opposite of the problem solved in Section 13.4.]

9. Use the Method of Eigenfunction Expansions to solve the problem

$$u_{tt} - c^2 u_{xx} = F(x, t), \qquad\qquad 0 < x < L, \quad t > 0$$

$$u(0, t) = u_x(L, t) = 0, \qquad\qquad t \geq 0$$

$$u(x, 0) = f(x), u_t(x, 0) = g(x), \qquad 0 \leq x \leq L$$

13.8 THE HEAT EQUATION: OPTIMAL DEPTH FOR A WINE CELLAR

The heat equation is a linear partial differential equation which models both the flow of heat and a host of other physical, chemical, and biological phenomena involving diffusion processes. We derive the heat equation as a mathematical model of thermal conduction and then solve some simple heat flow problems. Finally, we discuss the mathematical properties of solutions of the heat equation and the physical meaning of these properties.

Heat Conduction

Molecular vibrations of a material body generate energy which we feel as heat. Heat flows from warm to cool parts of the body by *conduction*, a process in which the collision of neighboring molecules transfers thermal energy from one molecule to another. It is conduction that is modeled below. Flow of heat also takes place through *convection*, molecules moving from region to region carrying their thermal energies along with them, but we shall not model convection here.

   The thermal state of a material body at each of its points P at time t is measured by the *temperature* $u(P, t)$. The units of temperature are degrees centigrade (or Celsius). Heat itself is measured in *calories* (or *joules*), 1 calorie being the energy needed to raise the temperature of 1 gram of water 1 degree (1 joule $= 0.239$ calorie). Each substance has a characteristic *specific heat* c, which is the energy required to raise the temperature of 1 gram of the substance 1 degree (thus the specific heat of water is 1). The *heat* in a portion D of a material body at time t is given by

$$H(t) = \int_D c\rho u(P, t)\, dP \tag{1}$$

where ρ is the density of the material and the integration is over D. The integral in (1) is single or multiple according to the spatial dimension of D. H as defined by (1) has indeed the units of energy, as the following unit analysis of the factors on the right of (1) shows:

$$
\underset{\dfrac{\text{energy}}{\text{mass} \cdot \text{degree}}}{c} \quad \cdot \quad \underset{\dfrac{\text{mass}}{\text{content}}}{\rho} \quad \cdot \quad \underset{\text{degree}}{u} \quad \cdot \quad \underset{\text{content}}{dP}
$$

where "content" denotes length, area, or volume according to the dimension of D. For simplicity in the derivations below we take D to be three-dimensional. The one- and two-dimensional cases are treated in the same way.

The conductive flow of heat through D is governed by the balance equation

$$
\begin{array}{c}
\text{rate of change of} \\
\text{heat in } D
\end{array}
\;=\;
\begin{array}{c}
\text{heat generated or} \\
\text{consumed inside} \\
D \text{ per unit time}
\end{array}
\;+\;
\begin{array}{c}
\text{heat moving across} \\
\text{the boundary of } D \\
\text{per unit time}
\end{array}
\tag{2}
$$

We shall show how (2) may be represented mathematically by the partial differential equation of heat flow. From (1) we see that the left side of the balance equation has the form

$$\frac{dH(t)}{dt} = \frac{d}{dt} \int_D c\rho u(P, t)\, dt = \int_D \frac{\partial}{\partial t} [c\rho u(P, t)]\, dP \tag{3}$$

where c, ρ, and u are assumed to be continuously differentiable functions of t and continuous in P.

The first term on the right of the balance equation may be expressed as

$$\int_D F(P, t)\, dP \tag{4}$$

where $F(P, t)$ is the heat generated or consumed per unit volume per unit time at the point P at time t. P is a *source* if $F(P, t)$ is positive, a *sink* if $F(P, t)$ is negative.

The last term on the right-hand side of (2) may be written as an integral over the boundary ∂D of D. The integrand is the amount of heat moving across the boundary per unit of surface area per unit of time at the point P on the boundary at time t,

$$\int_{\partial D} k \nabla u(P, t) \cdot \mathbf{n}\, dS \tag{5}$$

where ∇u is the spatial gradient of u, $\mathbf{n}$ is the unit outward normal to the boundary ∂D at P, and dS denotes integration over the boundary surface. The coefficient k in (5) is called the *thermal conductivity* and measures the ability of the body to conduct heat. Specifically, k is the time rate of change of heat through unit thickness per unit of surface area per unit of temperature. The integral in (5) is the mathematical representation of the *Euler–Fourier Law of Heat Conduction* (*Fick's Law* in the case of gas or liquid diffusion): Heat flows in the direction of maximal temperature drop per unit of distance at a rate proportional to the magnitude of that drop. Since ∇u points in the direction of maximal rise in temperature, heat flow is actually in the direction of $-\nabla u$. Note that if $\nabla u \cdot \mathbf{n}$ is positive at a point P on the boundary of D, the temperature outside D near P is higher than that at P, and hence thermal energy flows into D through P. The surface integral in (5) may be replaced by a volume integral over D:†

$$\int_{\partial D} k \nabla u \cdot \mathbf{n} \, dS = \int_{D} \operatorname{div}(k \nabla u) \, dP \tag{6}$$

In Cartesian x, y, z coordinates,

$$\nabla u = \frac{\partial u}{\partial x} \mathbf{i} + \frac{\partial u}{\partial y} \mathbf{j} + \frac{\partial u}{\partial z} \mathbf{k}$$

$$\operatorname{div}(k \nabla u) = \frac{\partial}{\partial x}\left(k \frac{\partial u}{\partial x}\right) + \frac{\partial}{\partial y}\left(k \frac{\partial u}{\partial y}\right) + \frac{\partial}{\partial z}\left(k \frac{\partial u}{\partial z}\right)$$

Inserting the expressions (3)–(6) into the balance equation (2), we have, after some rearranging, that

$$\int_{D} \left[\frac{\partial}{\partial t}(c\rho u) - F - \operatorname{div}(k \nabla u) \right] dP = 0 \tag{7}$$

If the integrand is continuous in P and in t, (7) is true for all regions D if and only if the integrand vanishes. This gives the *heat* (or *diffusion*) *equation*,

$$\frac{\partial}{\partial t}(c\rho u) - \operatorname{div}(k \nabla u) = F \tag{8}$$

at each point P inside the body for all time t for which the model is valid. The *temperature equation* would be a more precise name for (8). Note that (8) is linear in the unknown u and its derivatives.

---

† The *Divergence Theorem* (or *Gauss's Theorem*) of multivariable calculus justifies the equality in (6). If D is planar, ∂D is a curve and the integrals in (6) are, respectively, a line integral and a double integral. If D is the interval $a \leq x \leq b$, and $u = u(x, t)$, then $\nabla u = \partial u / \partial x$ and the boundary integral in (6) reduces to

$$k \frac{\partial u}{\partial x}\bigg]_{b} - k \frac{\partial u}{\partial x}\bigg]_{a}$$

If there are no internal sources or sinks, F vanishes and we have the *homogeneous* (or *source-free*) *heat* equation,

$$\frac{\partial}{\partial t}(c\rho u) - \text{div}(k\nabla u) = 0 \tag{9}$$

If the material coefficients c, ρ, and k are constants, the *diffusivity* $K = k/c\rho$ may be defined. The units of diffusivity typically are cm²/s and its values range from 0.0014 for water (a good insulator, but a poor conductor) up to 1.71 for silver (a good conductor). In this case (9) can be written in terms of the Laplacian operator ∇^2 as

$$u_t - K\nabla^2 u \equiv u_t - K(u_{xx} + u_{yy} + u_{zz}) = 0 \tag{10}$$

Boundary and Initial Conditions

On physical grounds one would not expect the heat equation to be enough to determine uniquely the temperatures $u(P, t)$ within a material body M. Initial and boundary conditions are also needed. The *initial condition* may be written as

$$u(P, 0) = f(P), \qquad \text{all } P \text{ in } M \tag{11}$$

where f is a given function. Unlike the situation with the wave equation, it is *not* necessary to prescribe $u_t(P, 0)$, since the heat equation itself does that.

There are three common types of thermal conditions imposed on the boundary of M. The first has to do with *prescribed boundary temperatures*:

$$u(P, t) = g(P, t), \qquad P \text{ in } \partial M, \quad t \geq 0 \tag{12}$$

where g is a given function. For example, the body may be submerged in an ice bath, and hence $u(P, t) = 0$ on the boundary.

Alternatively, the body may be wrapped with *thermal insulation*, which affects the flow of heat across the boundary:

$$\frac{\partial u(P, t)}{\partial n} [= \nabla u \cdot \mathbf{n}] = h(P, t), \qquad P \text{ in } \partial M, \quad t \geq 0 \tag{13}$$

where h is a prescribed function. If $h \equiv 0$, the insulation is said to be *perfect*, and there is no heat flow through the boundary.

A third type of boundary condition is given by *Newton's Law of Cooling*,

$$\frac{\partial u(P, t)}{\partial n} = r[U(t) - u(P, t)], \qquad P \text{ in } \partial M, \quad t \geq 0 \tag{14}$$

where $U(t)$ is the given *ambient temperature* outside M and r is a given *heat transfer coefficient*. If $r > 0$ and if the ambient temperature is higher than the boundary temperature, $\partial u/\partial n$ is positive, indicating a flow of heat into M across the boundary.

There are other types of boundary conditions, but these three cover most cases. Each condition may be written in terms of a linear boundary operator in u and

$\partial u / \partial n$. For example, (14) may be written as $Bu = rU$, where B is the linear operator $(\partial/\partial n) + r$ acting on the linear space of sufficiently smooth functions $u(P, t)$ defined for P in ∂M and $t \geq 0$. In a given problem part of the boundary may be subject to one condition, another part to a different condition. For example, one end of an iron bar may be immersed in an ice bath, the other end in boiling water, while the middle is covered with insulation.

With the remarks above in mind, we may formulate a typical *boundary/initial value problem for the heat equation* in a material body M of constant diffusivity and with no internal sources and sinks: Find $u(P, t)$ such that

$$\text{(PDE)} \quad u_t - K\nabla^2 u = 0, \qquad P \text{ inside } M, \quad t > 0$$

$$\text{(BC)} \quad \alpha \frac{\partial u}{\partial n} + \beta u = f, \qquad P \text{ in } \partial M, \quad t \geq 0 \tag{15}$$

$$\text{(IC)} \qquad\qquad u = g, \qquad P \text{ in } M, \quad t = 0$$

where α, β, f, and g are prescribed functions of P and t. Continuity and smoothness conditions may also be imposed on α, β, f, g and on ∂M so that (15) is well-set (see the end of this section). Rather than consider general problems, we take up and solve two simple and illustrative special cases of problem (15).

Temperature in a Rod

Suppose that a straight rod of constant diffusivity has uniform cross sections. Suppose, moreover, that the lateral surface of the rod is perfectly insulated, while the two ends are maintained at a temperature of $0°$. Suppose that at some initial time the temperature distribution is known. The problem is to determine the temperature distribution within the rod at later times.

We may take the central axis of the rod to be the x-axis. Because of the cross-sectional symmetry and the insulation on the lateral walls, we shall assume there is no temperature variation in the y and z directions. Denoting the diffusivity by K and the length of the rod by L, we see that we must solve the following boundary/initial value problem for the temperature $u(x, t)$:

$$\text{(PDE)} \qquad u_t - Ku_{xx} = 0, \qquad 0 < x < L, \quad t > 0$$

$$\text{(BC)} \qquad u(0, t) = 0, \qquad u(L, t) = 0, \quad t \geq 0 \tag{16}$$

$$\text{(IC)} \qquad u(x, 0) = f(x), \qquad 0 \leq x \leq L$$

where the initial temperature distribution f is given. See Figure 13.23 for a sketch of $R : 0 < x < L$, $t > 0$, the region in which (PDE) is to be solved. The shape of R and the nature of the problem itself suggest that the Method of Separation of Variables might be used to construct the solution.

First we look for "separated solutions" $u = X(x)T(t)$ of (PDE) and of (BC) in (16), leaving the satisfaction of (IC) to a subsequent superposition of separated solutions. Inserting $X(x)T(t)$ into (PDE), we now have that $X(x)dT(t)/dt =$

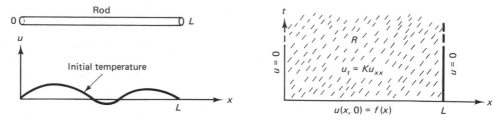

Figure 13.23 Temperature in a rod.

$KT(t)\ d^2X(x)/dx^2$. Separating variables in the usual way, we have a function of x equaling a function of time t, which can happen only if both equal a *separation constant* λ.

$$\frac{1}{X(x)}\frac{d^2X(x)}{dx^2}=\frac{1}{KT(t)}\frac{dT(t)}{dt}=\lambda \tag{17}$$

Conditions (BC) impose addition restrictions on $X(x)$ but not on $T(t)$. Combining these restrictions with the equations of (17), we see that $X(x)$ and $T(t)$ must be solutions of

$$\frac{d^2X}{dx^2}-\lambda X=0,\qquad X(0)=0,\quad X(L)=0 \tag{18a}$$

$$\frac{dT}{dt}-\lambda KT=0 \tag{18b}$$

The values of λ for which (18a) is solvable and the solutions $X(x)$ were determined in Section 13.3: $\lambda_n = -(n\pi/L)^2$, $X_n(x) = A_n \sin(n\pi x/L)$, $n = 1, 2, 3, \ldots$, where A_n is any constant. Corresponding solutions of (18b) are $T_n(t) = B_n \exp[-K(n\pi/L)^2 t]$, where B_n is any constant. Hence solutions of (PDE) and (BC) in (16) are given by

$$u_n = X_n(x)T_n(t) = C_n \sin\left(\frac{n\pi x}{L}\right)\exp\left[-K\left(\frac{n\pi}{L}\right)^2 t\right],\qquad n = 1, 2, \ldots \tag{19}$$

where $C_n \equiv A_n B_n$ is an arbitrary constant.

Since (PDE) and (BC) are homogeneous in this problem, any superposition of functions of the form given in (19) is again a solution of (PDE) and (BC). We shall determine constants C_n so that the superposition

$$u = \sum_{n=1}^{\infty} C_n \sin\left(\frac{n\pi x}{L}\right)\exp\left[-K\left(\frac{n\pi}{L}\right)^2 t\right] \tag{20}$$

is also a solution of the initial condition (IC). That is, we must choose C_n so that

$$u(x, 0) = f(x) = \sum_{n=1}^{\infty} C_n \sin\frac{n\pi x}{L},\qquad 0 \le x \le L$$

This suggests a Fourier Sine Series. The coefficients C_n may be found by the methods of Section 13.6:

$$C_n = \frac{\langle f, \sin(n\pi x/L)\rangle}{\|\sin(n\pi x/L)\|^2} = \frac{2}{L}\int_0^L f(x)\sin\frac{n\pi}{L}x\,dx, \qquad n = 1, 2, \ldots \tag{21}$$

Thus ignoring questions of convergence, the formal solution of (16) is given by (20) and (21). The series defines a classical solution, $u(x, t)$, which belongs to $C^2(R)$ and to $C^0(\text{cl } R)$ and satisfies (PDE) in R and (BC) and (IC) on ∂R if the data f are continuous, piecewise smooth, and $f(0) = f(L) = 0$ (see the end of this section for more on the classical solution).

The calculation of (21) may be carried out explicitly in the particular case of a rod of length 2, diffusivity 1, and initial temperature

$$f(x) = \begin{cases} x, & 0 \le x \le 1 \\ 2 - x, & 1 \le x \le 2 \end{cases} \tag{22}$$

In fact, $C_n = (-1)^{(n-1)/2}\,8/(n\pi)^2$ for n odd, $C_n = 0$ for n even. Thus the temperature function in this case is

$$u(x, t) = \frac{8}{\pi^2}\sum_{\text{odd } n}(-1)^{(n-1)/2}\frac{1}{n^2}\sin\left(\frac{n\pi x}{2}\right)\exp\left[\frac{-n^2\pi^2 t}{4}\right] \tag{23}$$

Several profiles of this temperature function are sketched in Figure 13.24. Note the "smoothing property" of the heat operator and note also the rapid decay of the temperature as the initial heat in the rod "leaks" out through the ends.

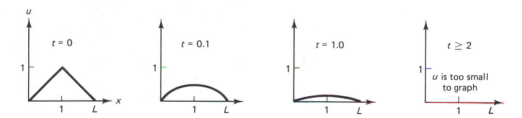

Figure 13.24   Decaying temperatures in a rod.

Optimal Depth for a Wine Cellar

One of the first applications of Fourier Series was to model heat flow through soil and rock, Fourier himself taking up this question. We shall consider a simple case here. First, however, we solve a heat problem different from (16):

(PDE) $u_t - Ku_{xx} = 0,$ $0 < x < \infty, \quad -\infty < t < \infty$

(BC) $u(0, t) = A_0 e^{i\omega t},$ $-\infty < t < \infty$ $\tag{24}$

(Boundedness) $|u(x, t)| < C,$ $0 \le x < \infty, \quad -\infty < t < \infty$

K, A_0, ω, and C are assumed to be positive constants. The xt- region, $R : 0 < x < \infty$, $-\infty < t < \infty$, and the data of the problem are sketched in Figure 13.25. The use of the complex exponential in the boundary condition of (24) is for convenience in calculation. It may be shown that (24) has no more than one solution that belongs to $C^0(\text{cl } R)$ and satisfies (PDE) throughout R. We shall find a solution, which then must be the only one.

The exponential "input" $A_0 e^{i\omega t}$ along the boundary suggests that the solution of (24) might have the separated form $u = A(x)e^{i\omega t}$. The amplitude $A(x)$ may be found by inserting u into the equation of (24):

$$i\omega A(x)e^{i\omega t} = K\frac{d^2 A(x)}{dx^2}e^{i\omega t}, \qquad A(0) = A_0, \quad |A(x)| < C \qquad (25)$$

Canceling $e^{i\omega t}$ from the first equation of (25) and solving for $A(x)$, we have that

$$A(x) = C_1 e^{\alpha(1+i)x} + C_2 e^{-\alpha(1+i)x}, \qquad \alpha = \left(\frac{\omega}{2K}\right)^{1/2} > 0$$

where C_1 and C_2 are arbitrary constants. The condition $|A(x)| < C$ implies that $C_1 = 0$ since $e^{\alpha x}$ becomes unbounded as $x \to \infty$. The condition $A(0) = A_0$ implies that $C_2 = A_0$. Thus the solution of (24) is

$$u = A_0 e^{-\alpha(1+i)x}e^{i\omega t} = A_0 e^{-\alpha x}e^{i(\omega t - \alpha x)}, \qquad \alpha = \left(\frac{\omega}{2K}\right)^{1/2}, \quad (x, t) \text{ in cl } R \quad (26)$$

Problem (24) and its solution (26) may be interpreted in terms of finding the optimal depth for locating a storage cellar. In this setting x is the depth below the surface of the earth, while $A_0 \cos \omega t$ (the real part of $A_0 e^{i\omega t}$) is a crude model of surface temperature normalized about a mean of $0°$. Suppose that the period of the surface wave is 1 year (i.e., $2\pi/\omega = 1 \text{ yr} = 3.15 \times 10^7$ s). Then, taking only the real part of $u(x, t)$ from (26), the temperature at depth x at time t is

$$u = A_0 e^{-\alpha x} \cos(\omega t - \alpha x), \qquad \omega = \frac{2\pi}{3.15 \times 10^7}, \qquad \alpha = \left(\frac{\omega}{2K}\right)^{1/2}$$

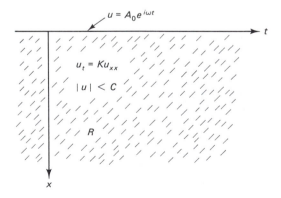

Figure 13.25   Diagram for (24).

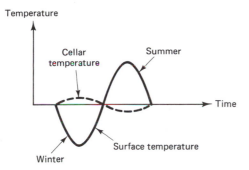

Figure 13.26 Optimal cellar temperature.

The *optimal depth* for the storage cellar is defined to be the smallest positive x at which the cellar's "seasons" are 6 months out of phase with the surface seasons. At this depth thermal convection currents from the surface tend to move the cellar temperature even closer to the mean. Thus the optimal depth satisfies $\alpha x = \pi$. Now the diffusivity K of average dry soil is 0.002 cm²/s. Hence the optimal depth is

$$x = \frac{\pi}{\alpha} = \pi \left(\frac{2K}{\omega}\right)^{1/2} = \pi (0.004)^{1/2} \left(\frac{3.25 \times 10^7}{2\pi}\right)^{1/2} \approx 445 \text{ cm} \quad \text{or} \quad 4.45 \text{ m}$$

The amplitude of the surface wave has dropped from A_0 to $A_0 e^{-x\alpha} = A_0 e^{-\pi} \approx A_0/25$ at the optimal depth. This 25-fold reduction in the amplitude coupled with the reversal of the seasons implies a nearly constant temperature in the cellar (see Figure 13.26).

Note that there is no need for an initial condition in this problem (indeed, there is no "initial time"). Note also that the model is valid near the surface of the earth in a region where there are no subsurface thermal sources or sinks.

Properties of Solutions of the Heat Equation

Boundary/initial value problems for the heat equation have physical significance only if they are *well-set* in the sense that each problem has exactly one solution and the solution changes continuously with the data. Separation of variables may often be used to construct a formal series solution as we saw above in the problem of the rod with zero boundary data. However, to show that this series solution has enough convergence properties to be a classical solution requires a detailed study of the series (see below). The other two aspects of a well-set problem (uniqueness and continuity) are somewhat easier to resolve. We shall consider the following problem in this regard:

$$
\begin{aligned}
&\text{(PDE)} && u_t - K u_{xx} = 0, && 0 < x < L, \quad t > 0 \\
&\text{(BC)} && u(0, t) = g_1(t), && u(L, t) = g_2(t), \quad t \geq 0 \\
&\text{(IC)} && u(x, 0) = f(x), && 0 \leq x \leq L
\end{aligned}
\qquad (27)
$$

where g_1 and g_2 belong to $C^0[0, \infty)$ and are bounded for $0 \leq t < \infty$, f belongs to $C^0[0, L]$, $g_1(0) = f(0)$, and $g_2(0) = f(L)$. The equations of (27) model the temperature

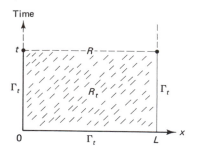

Figure 13.27 Space-time regions for (27).

in a uniform rod with perfect insulation on the lateral boundary, a rod whose endpoints have prescribed but varying temperatures. The conditions on f and g ensure the joint continuity of the boundary and initial data.

The first result refers to certain space-time regions shown in Figure 13.27. These are defined as follows: R, $0 < x < L$, $t > 0$; R_t, $0 < x < L$, $0 < \bar{t} \le t$; Γ_t, the boundary of R_t with the upper edge deleted. The following basic principle holds for each region R_t.

> **Maximum Principle for the Heat Equation.** Let $u(x, t)$ be any solution of (PDE) for which u is in $C^2(R)$ and also in $C^0(\text{cl } R)$. Then for every point (x, t) in cl R_t, $|u(x, t)| \le \max |u(\bar{x}, \bar{t})|$, where the minimum is taken over all $(\bar{x}, \bar{t})$ in Γ_t.

The proof of the Maximum Principle is omitted. The principle has a clear physical interpretation: In the absence of internal sources and sinks, the magnitudes of the temperatures in the rod at time t do not exceed the extreme magnitudes of the end temperatures up to time t or the extreme magnitudes of the initial temperatures.

The Maximal Principle implies uniqueness and continuity in the data:

> **Uniqueness.** Problem (27) has no more than one solution.

Proof. Suppose that u_1 and u_2 are both solutions of (27). Then $w = u_1 - u_2$ satisfies (PDE) and the homogeneous conditions, $w(0, t) = w(L, t) = 0$, $w(x, 0) = 0$, $t \ge 0$, $0 \le x \le L$. Since $w = 0$ on the three segments of Γ_t (see Figure 13.27), the Maximal Principle implies that $|w(x, t)| \le \max |w(\bar{x}, \bar{t})| = 0$, where $(\bar{x}, \bar{t})$ is taken over Γt. Hence $w(x, t) \equiv 0$ and $u_1(x, t) \equiv u_2(x, t)$. Thus if (27) has a solution at all with the required smoothness, it is unique.

> **Continuity in the Data.** Every solution of (27) changes continuously with respect to changes in the data f, g_1, and g_2.

Proof. Let u be a solution of (27), $\bar{u}$ a solution of (27) with $\bar{f}$, $\bar{g}_1$, $\bar{g}_2$ replacing f, g_1, g_2 throughout. Then $u - \bar{u}$ is a solution of (PDE) with respective initial and boundary data $f - \bar{f}$, $g_1 - \bar{g}_1$, $g_2 - \bar{g}_2$. Hence, by the Maximal Principle, for all $t \geq 0$ and all $0 \leq x \leq L$,

$$|u(x, t) - \bar{u}(x, t)| \leq \max_{\substack{0 \leq \bar{x} \leq L \\ 0 \leq \bar{t} \leq t}} \{|f(\bar{x}) - \bar{f}(\bar{x})|, |g_1(\bar{t}) - \bar{g}_1(\bar{t})|, |g_2(\bar{t}) - \bar{g}_2(\bar{t})|\} \tag{28}$$

Thus small changes in boundary and initial data mean at most small changes in the temperature, (28) being the mathematical version of this assertion.

For simplicity we shall now set the boundary data g_1 and g_2 equal to 0, reducing (27) to (16). The series given in (20) with coefficients defined by (21) provides a formal solution to the boundary/initial value problem (16). The following results give additional properties of that solution.

Smoothing Properties.   Suppose that the initial data belong to $PC[0, L]$. Then the formal solution $u(x, t)$ of (16) defined by (20), (21) belongs to $C^\infty(R)$ and satisfies (PDE) in R.

Proof. The proof of the Smoothing Property rests on the Weierstrass M-test. First observe from (21) that there is a positive constant A such that $|C_n| \leq A$, $n = 1, 2, \ldots$. Hence, for $N = 1, 2, \ldots$ and all (x, t) in the set S_{t_0} described by $0 \leq x \leq L$, $t \geq t_0 > 0$,

$$C_n \sin\left(\frac{n\pi}{L} x\right) \exp\left[-K\left(\frac{n\pi}{L}\right)^2 t\right] \leq A \exp\left[-K\left(\frac{n\pi}{L}\right)^2 t_0\right]$$

Now the series $\sum \exp[-K(n\pi/L)^2 t_0]$ converges by the Ratio Test. Hence by the Weierstrass M-test, the series in (20) converges uniformly in S_{t_0} to a function which, of course, we call $u(x, t)$. If each term of the series in (20) is differentiated k times (r times in x and $k - r$ times in t, say), the nth term of the derived series is no larger in magnitude than a term of the form $Bn^{k-r} \exp[-K(n\pi/L)^2 t_0]$ for all (x, t) in S_{t_0}, for some constant B (which may depend on k, but not on n). But the series $\sum_n n^{k-r} \exp[-K(n\pi/L)^2 t_0]$ also converges by the Ratio Test. Hence by the Weierstrass M-test and a basic theorem on uniform convergence, the derived series converges uniformly to a kth derivative of $u(x, t)$ (i.e., to $\partial^k u/\partial x^r \partial t^{k-r}$) on the closed region S_{t_0}. Since all of the arguments above hold for every $t_0 > 0$ and for every $k = 1, 2, \ldots$, it follows that $u(x, t)$ belongs to $C^\infty(R)$. A direct calculation involving the term-by-term differentiation of the series for u shows that (20) satisfies (PDE) in R.

That $u(x, t)$ possesses all derivatives of all orders is quite remarkable since the initial data are only required to be piecewise continuous. Visual evidence of the smoothing property may be seen in Figure 13.24, where the sharp corner in the initial data is immediately rounded off for $t > 0$. Under additional assumptions on the initial data, the series in (20) defines a function $u(x, t)$ which is a classical solution of (16), that is, satisfies the initial conditions of (16) [as well as (PDE)].

Classical Solution.   Let $f(0) = f(L) = 0$ and suppose that $f(x)$ is continuous and piecewise smooth, $0 \le x \le L$. Then the function $u(x, t)$ defined by (20) and (21) belongs to $C^\infty(R)$ and to $C^0(\text{cl } R)$. Moreover, $u(x, t)$ is a solution of (16).

The proof is an extension of the series techniques used above, but the details are omitted.

Comments

The "smoothing property" of the *heat operator* $(\partial/\partial t) - K(\partial^2/\partial x^2)$ is quite unlike any property of the wave operator, $(\partial^2/\partial t^2) - c^2(\partial^2/\partial x^2)$. In fact, smoothing gives a direction to time that the wave operator cannot. We may argue as follows in the context of (22) and (23) corresponding to piecewise linear initial data with a "corner." By the smoothing character of the heat operator, the temperature function lies in $C^\infty(R)$ for all $t > 0$. Thus the "corner singularity" in the initial data does not propagate from ∂R into R, and the process $f(x) \to u(x, t)$ is irreversible. The sequence of temperature profiles in Figure 13.24 cannot be read backward and *time has an arrow for heat conduction.*

Corner singularities in initial data do propagate with the wave operator, and in fact, are repeated periodically, as we saw in Section 13.3. The sequence of displacement profiles for a plucked guitar string can be read forward or backward in time without distinction. *Time has no arrow for the wave equation in one spatial dimension.*

There is more. As we saw from D'Alembert's formula, initial disturbances propagate with speed c under the influence of the wave operator. It may be shown, but we shall not do so here, that the *speed of propagation of an initial temperature disturbance is infinity*. In fact, suppose that the initial temperature of a uniform rod is $0°$ except for a temperature $T_0 > 0$ in some segment of arbitrarily small length at the middle of the rod. Then (with end temperatures maintained at $0°$) for any positive t, no matter how small, the solution $u(x, t)$ of the corresponding boundary/initial value problem is positive for every point x inside the rod. The thermal disturbance at the center of the rod has traveled infinitely fast and raised the temperature everywhere inside the rod.

The infinite speed of propagation for the heat operator and the periodic repetition of "corners" on initial displacements for the plucked guitar string and the wave operator show the defects of the respective mathematical models. Thermal diffusion and wave motion cannot be perfectly modeled by the boundary/initial value problems constructed in this chapter. However, the mathematical models presented here are sufficiently accurate in most regards that they continue to be the models of first choice in the treatment of simple heat and wave phenomena.

PROBLEMS

The following problem(s) may be more challenging: 4, 5, and 7.

1. Use the Separation of Variables Technique to construct a series solution of the one-dimensional heat equation $u_t - Ku_{xx} = 0$, where $0 < x < L$, $t > 0$ and the following initial and boundary conditions are imposed. Give a physical interpretation of each problem and sketch time profiles along the interval $0 \leq x \leq L$ at $t = 0.1$, $t = 1.0$, and $t = 10.0$. [*Hint*: See the subsection above on temperature in a rod.]

   (a) $u(0, t) = u(L, t) = 0$, $u(x, 0) = \sin(2\pi/L)x$.

   (b) $u(0, t) = u(L, t) = 0$, $u(x, 0) = x$.

   (c) $u(0, t) = u(L, t) = 0$, $u(x, 0) = u_0 > 0$, u_0 a constant.

   (d) $u(0, t) = u(L, t) = 0$, $u(x, 0) = \begin{cases} u_0, 0 \leq x \leq L/2. \\ 0, L/2 < x \leq L. \end{cases}$

   (e) $u(0, t) = 0, u_x(L, t) = 0$, $u(x, 0) = \sin(\pi/2L)x$.

   (f) $u(0, t) = 0, u_x(L, t) = 0$, $u(x, 0) = x$.

   (g) $u_x(0, t) = u_x(L, t) = 0$, $u(x, 0) = x$.

2. [*Constant End Temperatures*]

   (a) Find the series solution of the problem $u_t - Ku_{xx} = 0$, $0 < x < 1$, $t > 0$; $u(0, t) = 10$, $u(1, t) = 20$, $t \geq 0$; $u(x, 0) = 0$, $0 \leq x \leq 1$. [*Hint*: Write $u(x, t) = A(x) + v(x, t)$, where $A''(x) = 0$, $A(0) = 10$, $A(1) = 20$, and $v_t - Kv_{xx} = 0$, $0 < x < 1$, $t > 0$; $v(0, t) = 0$, $v(1, t) = 0$, $t \geq 0$; $v(x, 0) = -A(x)$, $0 \leq x \leq L$. Then find v in the usual way once $A(x)$ has been found.]

   (b) Show that the *steady-state solution* in part (a) is $u = A(x)$ in the sense that $v(x, t) \rightarrow 0$ as $t \rightarrow \infty$.

3. [*Variable End Temperatures*]. Let u be a solution to the problem $u_t - Ku_{xx} = 0$, $0 < x < 1$, $t > 0$; $u(0, t) = g_1(t)$, $u(1, t) = g_2(t)$, $t \geq 0$; $u(x, 0) = f(x)$, $0 \leq x \leq 1$.

   (a) If $V = g_1(t) + [g_2(t) - g_1(t)]x$ and $U(x, t)$ satisfies the equations $U_t - KU_{xx} = -KV_t$, $0 < x < 1$, $t > 0$; $U(0, t) = U(1, t) = 0$, $t \geq 0$; $U(x, 0) = f(x) - V(x, 0)$, $0 \leq x \leq 1$, show that $u = V(x, t) + U(x, t)$.

   (b) Use the Eigenfunction Expansion Method of Section 13.7 to find $U(x, t)$ if $g_1(t) = \sin t$ and $g_2(t) = 0$.

   (c) Find $u(x, t)$.

4. [*Internal Sources/Sinks*]. Use the Eigenfunction Expansion Method of Section 13.7 to solve the problem $u_t - Ku_{xx} = 3e^{-2t} + x$, $0 < x < 1$, $t > 0$; $u(0, t) = u(1, t) = 0$, $t \geq 0$; $u(x, 0) = 0$, $0 \leq x \leq 1$.

5. [*Wine Cellars*]

   (a) Let the surface temperature wave be $T_0 + A_0 \cos \omega t$, where A_0, T_0, and ω are positive constants. Find the optimal depth of the wine cellar. What is that depth if ω corresponds to 1 day instead of 1 year as in the example given in the text?

   (b) Find the temperature function $u(x_0, t)$ at the optimal depth x_0

   (c) Formulate and solve the wine cellar problem where the surface wave has the form $T_0 + A_1 \cos \omega_1 t + A_2 \cos \omega_2 t$, where ω_1 corresponds to 1 year and $\omega_2 = 365\omega_1$ corresponds to 1 day.

6. (a) Prove that the series (23) diverges for every x, $0 < x < 2$, if $t = t_0 < 0$.

   (b) Explain this in terms of time's arrow.

(c) Find an initial data function $f \neq 0$ so that (20) and (21) do give solutions defined for all time, even for $t < 0$. [*Hint*: Consider $f(x) = \sin(\pi x/L)$.]

7. Suppose that the temperature $u(x, t)$ in a rod of length 1 and diffusivity 1 satisfies the equations $u_t - u_{xx} = 0$, $0 < x < 1$, $t > 0$; $u(0, t) = te^{-t}$, $t \geq 0$; $u(1, t) = 0$, $t \geq 0$; $u(x, 0) = 0.01x(1 - x)$, $0 \leq x \leq 1$. Show that $|u(x, t)| \leq 1/e$ for $0 \leq x \leq 1$, $t \geq 0$.

13.9 LAPLACE'S EQUATION AND HARMONIC FUNCTIONS

Laplace's equation is the homogeneous linear second-order partial differential equation

$$\nabla^2 u = 0 \tag{1}$$

where ∇^2 is the Laplacian operator. Laplace's equation models steady-state temperatures in a body of constant material diffusivity. By "steady state" we mean that the temperature function u does not change with time, although it may vary from point to point within the body. Laplace's equation also models the gravitation and magnetic potentials in empty space, electric potential, and the velocity potential of ideal fluids. For these reasons, (1) is also called the *potential equation*.

The Laplacian operator in the Cartesian rectangular coordinates of 3-space is $\nabla^2 = \partial^2/\partial x^2 + \partial^2/\partial y^2 + \partial^2/\partial z^2$. The operator has the following form in other coordinate systems:

Polar coordinates:

$$u_{xx} + u_{yy} = u_{rr} + \frac{1}{r} u_r + \frac{1}{r^2} u_{\theta\theta} \tag{2}$$

Cylindrical coordinates:

$$u_{xx} + u_{yy} + u_{zz} = u_{rr} + \frac{1}{r} u_r + \frac{1}{r^2} u_{\theta\theta} + u_{zz} \tag{3}$$

Spherical coordinates:

$$u_{xx} + u_{yy} + u_{zz} = \frac{1}{\rho^2} (\rho^2 u_\rho)_\rho + \frac{1}{\rho^2 \sin \phi} (\sin \phi u_\phi)_\phi + \frac{1}{\rho^2 \sin^2 \phi} u_{\theta\theta} \tag{4}$$

Recall that Cartesian coordinates (x, y, z) in $\mathbf{R}^3$ are related to spherical coordinates (ρ, θ, ϕ) as follows (see Figure 13.28):

$$x = \rho \sin \phi \cos \theta$$
$$y = \rho \sin \phi \sin \theta$$
$$z = \rho \cos \phi$$

Warning: Some authors interchange θ and ϕ.

Laplace's equation has many solutions; for example, $u = c_1 e^{-x} \cos y + c_2 z + c_3 e^{-4z} \cos 4x$ gives solutions in Cartesian coordinates for all constants c_1, c_2, c_3, while $u = c_1 r \cos \theta + c_2 r^2 \sin 2\theta$ gives solutions in polar coordinates for all c_1 and c_2. Boundary or boundedness conditions are needed to select a unique solution. We shall solve representative problems in various coordinate systems and derive the fundamental properties of these solutions.

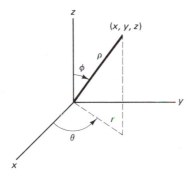

Figure 13.28   Non-Cartesian coordinates.

The Dirichlet Problem: Harmonic Functions

Let G be a region in $\mathbf{R}^2$ or $\mathbf{R}^3$, and let h be a piecewise continuous function defined on ∂G, the boundary of G. Then the *Dirichlet Problem for* G *with boundary data* h is defined as the following boundary value problem. Find a function u in $C^2(G)$ such that $\nabla^2 u = 0$ in G with the additional property that if P is a point of ∂G where h is continuous, then $\lim_{x \to P} u(x) = h(P)$, where x is in G. With this understanding we shall briefly write the Dirichlet Problem for G with boundary data h as

$$\text{(PDE)} \quad \nabla^2 u = 0 \quad \text{in } G$$
$$\text{(BC)} \qquad u = h \quad \text{on } \partial G \tag{5}$$

A solution u of (PDE) is called a *harmonic function* or a *potential function*. Observe that when h lies in $C^0(\partial G)$, a solution u of (5) belongs to $C^0(\text{cl } G)$ and $u(P) = h(P)$ for all P in ∂G. In this case (5) is called the *classical* Dirichlet Problem and u is called a *classical solution* for the Dirichlet Problem.

Steady-State Temperatures in the Unit Disk: Circular Harmonics

Let $G = \{(x, y) : x^2 + y^2 < 1\}$ and suppose that h is a given piecewise continuous function on ∂G. We shall construct a solution of the problem (5) for this region, G, and the given function, h, by using the Method of Separation of Variables. If the region G and the conditions (PDE) and (BC) are expressed in Cartesian coordinates, separation of variables will *not* provide us with a formal solution to this problem because ∂G is not composed of level curves in the Cartesian coordinate system. From this point of view, if we find a problem that is equivalent to (5) but expressed in terms of polar coordinates, separation of variables would have some chance for success. We see that (5) is equivalent to the problem

$$\text{(PDE)} \quad w_{rr} + \frac{1}{r} w_r + \frac{1}{r^2} w_{\theta\theta} = 0 \qquad \text{for } \theta \text{ in } \mathbf{R}, \quad 0 < r < 1$$
$$\text{(BC)} \qquad \qquad w(1, \theta) = f(\theta) \qquad \text{for } \theta \text{ in } \mathbf{R} \tag{6}$$

where f is the function h expressed in the polar angle θ, and $w(r, \theta) = u(r \cos \theta, r \sin \theta)$. Observe that f belongs to $PC[-\pi, \pi]$ and that w is required to be periodic in θ with period 2π and twice continuously differentiable in the "strip," $0 < r < 1$, $-\infty < \theta < \infty$. Moreover, to ensure that u is twice continuously differentiable near the origin, we must impose the condition that for any θ_0 in **R**, the limits of all derivatives of w up to second order exist as $(r, \theta) \to (0, \theta_0)$ and are independent of θ_0. We may interpret (6) as the model for steady-state temperatures in a thin homogeneous disk whose top and bottom faces are perfectly insulated and for which there are prescribed edge temperatures.

Following the method of Separation of Variables to construct a formal solution of (6), we first look for all twice continuously differentiable solutions of (PDE) which have the form $R(r)\Theta(\theta)$. After substituting $R(r)\Theta(\theta)$ into (PDE), the variables can be separated and we are led to consider the separated ordinary differential equations

$$\Theta'' - \lambda\Theta = 0, \qquad r^2 R'' + rR' + \lambda R = 0$$

where λ is the separation constant. We must demand that the solution $R(r)\Theta(\theta)$ be periodic in θ and smooth across $\theta = \pi$; thus we must necessarily have the conditions $\Theta(-\pi) = \Theta(\pi)$, $\Theta'(-\pi) = \Theta'(\pi)$. Hence we are led to consider the Sturm–Liouville problem

$$\Theta'' - \lambda\Theta = 0$$
$$\Theta(-\pi) = \Theta(\pi), \qquad \Theta'(-\pi) = \Theta'(\pi)$$

But we have considered this problem in Section 13.6 and found that it has a nontrivial solution if and only if $\lambda = \lambda_n = -n^2, n = 0, 1, 2, \ldots$. The corresponding nontrivial solutions are given by

$$\Theta_0(\theta) = 1, \qquad \Theta_n(\theta) = A_n \cos n\theta + B_n \sin n\theta, \qquad n = 1, 2, \ldots$$

where A_n and B_n are arbitrary real numbers. Replacing λ by λ_n in the other separated equation, we obtain the differential equation

$$r^2 R'' + rR' - n^2 R = 0$$

This is an Euler equation (see Section 8.4) and is easily found to have the general solution

$$R(r) = Ar^n + Br^{-n} \qquad \text{when } n = 1, 2, \ldots \tag{7}$$

and

$$R(r) = A + B \ln r \qquad \text{when } n = 0$$

But since R must be well-behaved as $r \to 0^+$, we must take $B = 0$ for any $n = 0$, $1, 2, \ldots$, and hence we have that

$$R_n = r^n, \qquad n = 0, 1, 2, \ldots$$

Thus we look for a formal solution to (6) in the form

$$w(r, \theta) = \frac{A_0}{2} + \sum_{n=1}^{\infty} r^n (A_n \cos n\theta + B_n \sin n\theta) \tag{8}$$

We compute the A_n and B_n now by imposing the boundary condition

$$f(\theta) = \frac{A_0}{2} + \sum_{n=1}^{\infty} (A_n \cos n\theta + B_n \sin n\theta) \qquad \text{for } -\pi \le \theta \le \pi$$

Recalling that $1, \cos x, \sin x, \ldots$ is a basis for $PC[-\pi, \pi]$, we have that

$$A_n = \frac{1}{\pi} \int_{-\pi}^{\pi} f(\theta) \cos n\theta \, d\theta, \qquad n = 0, 1, 2, \ldots$$

$$B_n = \frac{1}{\pi} \int_{-\pi}^{\pi} f(\theta) \sin n\theta \, d\theta, \qquad n = 1, 2, \ldots \tag{9}$$

Recall that a solution of Laplace's equation is called a harmonic function; (8) is an expansion in *circular harmonics*.

The solution (8) with coefficients defined by (9) is a formal solution to the Dirichlet problem, since nothing has been said as yet about convergence properties. In fact, just as in the preceding section, the function w defined by (8) and (9) belongs to $C^{\infty}(G)$ and satisfies (PDE) if the boundary function h (i.e., f in polar coordinates) is piecewise continuous. Thus steady-state temperature functions have the same strong smoothness properties as the time-dependent temperature functions of Section 13.8. If, in addition, h is continuous on ∂G and also piecewise smooth, (8) and (9) define a classical solution of (6).

The reader may easily show that if the disk has radius r_0, $r_0 > 0$ rather than radius 1, then (8) and (9) define the solution of the corresponding Dirichlet problem if r in (8) is replaced by r/r_0.

Properties of Harmonic Functions

Harmonic functions have a number of distinctive properties. For simplicity all of the results below are stated and proved only for planar regions.

Mean Value Property.   Suppose that $u(x, y)$ is a solution of Laplace's equation in a planar region R. Let (x_0, y_0) be a point of R, and D a disk of radius r_0 centered at (x_0, y_0) and lying entirely in R. Then

$$u(x_0, y_0) = \frac{1}{2\pi} \int_0^{2\pi} u(x_0 + r_0 \cos \theta, y_0 + r_0 \sin \theta) \, d\theta$$

That is, the value of a harmonic function at a point is the average of its values around any circle centered at the point.

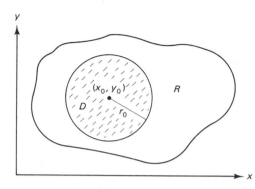

Figure 13.29   Geometry for the Mean Value Property.

Proof. Figure 13.29 illustrates the geometry. Let the polar coordinates r, θ be referred to the point P. Then using (8) with r replaced by r/r_0, we have that at $r = 0$

$$w(0, \theta) = \frac{A_0}{2}$$

where by (9) A_0 is given by

$$\frac{1}{\pi} \int_{-\pi}^{\pi} u(x_0 + r_0 \cos \theta, y_0 + r_0 \sin \theta) \, d\theta$$

and we are done.

Harmonic functions also satisfy a Maximum Principle.

Maximum Principle for Harmonic Functions.   Unless the harmonic function u is a constant function, u cannot attain either its maximum or its minimum value inside a region R of $\mathbf{R}^2$. That is, if R is closed and u is continuous on cl R, then the extreme values of a nonconstant harmonic function u are attained only on the boundary of R.

Proof. Suppose that u attains its maximum in R at some interior point P of R. Then this maximum value is the average of the values of u around the edge of any disk centered at P and lying in R. But $u(P)$ cannot be both an average and a maximum unless the values of u are constant. A similar argument applies to the minimum and leads to a minimum principle.

One of the central questions is whether a problem has exactly one solution which changes continuously with the data. Such problems are *well-set*, as we have noted before. The existence of a solution for an arbitrary Dirichlet problem is not easily shown, although for regions of simple shape solutions may be constructed by the method of Separation of Variables (e.g., the Dirichlet Problem solved earlier in this section). However, the Maximum Principle may be applied to derive the other two aspects of a well-set Dirichlet Problem (i.e., no more than one solution and continuity in the data).

Uniqueness and Continuity.   The Dirichlet Problem (5) for a region G in $\mathbf{R}^2$ has no more than one continuous solution u in cl G if the boundary data h is continuous on ∂G. Moreover, the solution (if it exists) varies continuously with respect to the data.

Proof. Suppose that $\nabla^2 u = \nabla^2 v = 0$ in G while $u = h$ and $v = h + \epsilon$ on ∂G, where h and ϵ are continuous on ∂G. We shall show that

$$\min_{P \text{ on } \partial G} \epsilon(P) \le v - u \le \max_{P \text{ on } \partial G} \epsilon(P)$$

for values of $v - u$ everywhere in cl G. Thus if $|\epsilon|$ is small, v is close to u and we have continuity in the data. Let $w = v - u$. Then $\nabla^2 w = \nabla^2 v - \nabla^2 u = 0$ in G. By the Maximum Principle (and the corresponding Minimum Principle),

$$\min_{P \text{ on } \partial G} w(P) \le w \le \max_{P \text{ on } \partial G} w(P)$$

for all values w in cl G. But $w = v - u = h + \epsilon - h = \epsilon$ on ∂G, and we are done.

Now suppose that u_1 and u_2 are both solutions of (5). Then $U = u_1 - u_2$ satisfies (PDE) in G but with $U = 0$ on ∂G. Hence by the same argument as above, $0 \le U \le 0$, for all values of U in cl G. Hence $U \equiv 0$ and $u_1 = u_2$. Thus (5) has no more than one solution.

From the results above we see that harmonic functions have strong and distinctive properties. The reader may interpret these properties in terms of steady-state temperatures within a thin plate R whose top and bottom faces are perfectly insulated and whose edge temperatures are prescribed.

Steady Temperatures in a Ball: Zonal Harmonics and Legendre Polynomials

Now let G be the interior of the unit ball in $\mathbf{R}^3 : x^2 + y^2 + z^2 < 1$, and consider the Dirichlet Problem (5) with the boundary function h defined and, say, continuous on the boundary sphere $x^2 + y^2 + z^2 = 1$. The natural coordinate system is spherical in this case. Thus we shall actually solve (4) for $u = u(\rho, \phi, \theta)$ inside G subject to the boundary condition $u = f(\theta, \phi)$ on ∂G, where f denotes the function h of (5) expressed in spherical coordinates:

$$\begin{aligned} \text{(PDE)} \quad & \nabla^2 u = 0 \text{ in } G \ (\nabla^2 \text{ in spherical coordinates}) \\ \text{(BC)} \quad & u = f(\theta, \phi) \text{ on } \partial G \end{aligned} \tag{10}$$

We shall assume further that f is independent of θ *and that the solution u has the same property.* As usual, it is assumed that u belongs to $C^2(G)$ and to $C^0(\text{cl } G)$.

In using the Method of Separation of Variables we need only look for solutions of (PDE) of the form $R(\rho)\Phi(\phi)$; the variables are easily separated to obtain the differential equations,

$$\rho^2 R'' + 2\rho R' + \lambda R = 0, \qquad 0 < \rho < 1 \tag{11}$$

$$\Phi'' + \cot \phi \, \Phi' - \lambda = 0, \qquad 0 < \phi < \pi \tag{12}$$

where λ is the separation constant. Now since we assumed that $R(\rho)\Phi(\phi)$ belongs to $C^2(G)$, we shall look for those constants λ such that Eqs. (11) and (12) have solutions R and Φ with

$$R \text{ in } C^2[0, 1] \tag{13a}$$

$$\Phi \text{ in } C^2[0, \pi] \tag{13b}$$

Observe that Eqs. (11) and (12) are *nonnormal*; thus the conditions (13) are not redundant. The Method of Separation of Variables has provided us with a singular Sturm–Liouville Problem to solve.

If we change the independent variable with the mapping $s = \cos \phi$, we see that the interval $0 \le \phi \le \pi$ is mapped one-to-one onto the interval $-1 \le s \le 1$ and that (12) and (13) become the Singular Sturm–Liouville system

$$(1 - s^2)\frac{d^2\Phi}{ds^2} - 2s\frac{d\Phi}{ds} - \lambda\Phi = 0, \qquad -1 < s < +1, \quad \Phi \text{ in } C^2[-1, +1] \tag{14}$$

Now, we have seen Eq. (14) before; indeed, if λ is replaced by $-n(n + 1)$, (14) is precisely Legendre's equation and has as one of its solutions the Legendre polynomial P_n (n can be any nonnegative integer).

According to Example 13.19, we have that $\lambda_n = -n(n + 1)$, $n = 0, 1, \ldots$, $\Phi_n(\phi) = P_n(\cos \phi)$, $n = 0, 1, \ldots$.

If we substitute λ_n for λ in (11), we arrive at the problem

$$\rho^2 R'' + 2\rho R' - n(n + 1)R = 0, \quad R \text{ in } C^2[0, 1], \qquad n = 0, 1, 2, \ldots \tag{15}$$

[recall the condition (13a)]. Now the equation in (15) is an Euler equation and is easily seen to have the general solution.

$$R(\rho) = A\rho^n + B\rho^{-n-1}, \qquad n = 0, 1, 2, \ldots$$

Since we demand that R be in $C^2[0, 1]$, we must take $B = 0$, so we have that

$$R_n(\rho) = \rho^n, \qquad n = 0, 1, 2, \ldots$$

Now we look for a solution for (10) in the form

$$u(\rho, \phi) = \sum_{n=0}^{\infty} A_n \rho^n P_n(\cos \phi) \tag{16}$$

Inserting $\rho = 1$ into (16), we have the condition

$$f(\phi) = \sum_{n=0}^{\infty} A_n P_n(\cos \phi), \qquad 0 < \phi < \pi \tag{17}$$

If we make the change of variables $x = \cos \phi$ in (17) and let g be the function on $|x| \le 1$ such that $g(\cos \phi) = f(\phi)$ for all $0 \le \phi \le \pi$, then (17) becomes

$$g(x) = \sum_{n=0}^{\infty} A_n P_n(x), \qquad |x| \le 1 \tag{18}$$

Referring to Example 13.19, we see that (18) holds if we choose A_n by using the Fourier–Euler formula appropriate to a Legendre basis of PC$[-1, 1]$:

$$A_n = \frac{\langle g, P_n \rangle}{\|P_n\|^2} = \frac{2n+1}{2} \int_{-1}^{1} g(x) P_n(x)\, dx, \qquad n = 0, 1, 2, \ldots \tag{19}$$

where $g(x) = f(\arccos x)$. Thus the series (16) with coefficients given by (19) gives a *formal* solution to (10).

Those values of ϕ for which $\cos \phi$ is a root of $P_n(\cos \phi)$ [recall that $P_n(x)$ has n distinct roots, all of which lie in $(-1, 1)$] determine *nodal latitudes* on the unit sphere. The regions between consecutive nodal latitudes are called *zones* and the functions $\rho^n P_n(\cos \phi)$ are called *zonal harmonics*. The solution of (16) is a *superposition of zonal harmonics* (see Figure 13.30). These results may be interpreted in terms of steady-state temperatures in a ball, given boundary temperatures independent of longitude θ.

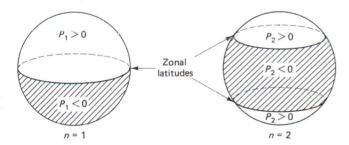

Figure 13.30   Zones on the surface of the sphere.

Steady Temperatures in a Cylinder: Cylindrical Harmonics and Bessel Functions

Let C be the cylinder described in cylindrical coordinates by $0 \le r \le 1$, $0 \le z \le a$, and let $u(r, \theta, z)$ be the steady temperature in C at the point whose cylindrical coordinates are given by (r, θ, z). We consider the problem of finding the steady temperature in C when the base of the cylinder $(z = 0)$ is maintained at the given temperature $f(r)$ degrees, independent of θ, and the rest of ∂C is maintained at $0°$. Thus u is the solution of the boundary problem (see Figure 13.31).

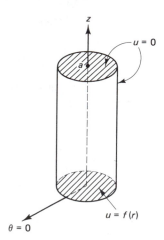

Figure 13.31 Boundary conditions on a cylinder.

$$\nabla^2 u = 0, \qquad 0 \le r < 1, \quad 0 < z < a, \quad -\pi \le \theta < \pi$$
$$u(1, \theta, z) = 0, \qquad 0 \le z \le a, \quad -\pi \le \theta < \pi$$
$$u(r, \theta, a) = 0, \qquad 0 \le r \le 1, \quad -\pi \le \theta < \pi \tag{20}$$
$$u(r, \theta, 0) = f(r), \qquad 0 \le r \le 1, \quad -\pi \le \theta < \pi$$

where $\nabla^2 u$ is expressed in cylindrical coordinates [see (3)]. Since the data are a function of the variable r only, we are led to suspect that the solution of (20) is a function of r and z only. Assuming this and separating the variables in the usual way, we obtain the equations

$$R'' + \frac{1}{r} R' - \lambda R = 0, \qquad R(0+) < \infty, \qquad R(1) = 0 \tag{21}$$

$$Z'' + \lambda Z = 0, \qquad Z(a) = 0 \tag{22}$$

The Singular Sturm–Liouville Problem (21) was treated in Example 13.18. The differential equation of (21) is an eigenvalue equation for Bessel's operator of order $p = 0$. Hence the separation constant λ may assume only the values $\lambda = \lambda_n = -k_n^2$, where the k_n are the consecutive positive zeros of $J_0(x)$, $n = 1, 2, 3, \ldots$. The corresponding eigenfunctions for (21) are given by

$$R_n(r) = J_0(k_n r), \qquad n = 1, 2, \ldots$$

Thus (22), with λ replaced by $-k_n^2$, yields

$$Z_n(z) = \sinh k_n(a - z), \qquad n = 1, 2, \ldots$$

Hence we are led to expect a solution of (20) in the form

$$u(r, z) = \sum_{n=1}^{\infty} A_n \sinh k_n(a - z) J_0(k_n r) \tag{23}$$

The remaining condition in (20) which must be satisfied now implies that

$$f(r) = u(r, 0) = \sum_{n=1}^{\infty} A_n \sinh(k_n a) J_0(k_n r)$$

This suggests an orthogonal expansion of f in terms of the functions $J_0(k_n r)$, $n = 1, 2, \ldots$ (see Example 13.18). We obtain that

$$A_n = \frac{2}{[J_1(k_n)]^2 \sinh(k_n a)} \int_0^1 rf(r)J_0(k_n r)\, dr \tag{24}$$

where we have used the formula $\int_0^1 rJ_0^2(k_n r)\, dr = \frac{1}{2} J_1^2(k_n)$ (see the problem set). The series (23) with coefficients given by (24) defines the formal solution to (20). Although we shall not do it here, it can be shown that this formal solution is indeed a classical solution if $f(r)$ is smooth enough and $f(1) = 0$. The functions $\sinh k_n(a - z)J_0(k_n r)$ are *cylindrical harmonics*.

Comments

We have analyzed a sample of Dirichlet Problems which may be solved by the Method of Separation of Variables. As always, the boundary of the region involved must be composed of level sets of the coordinate variables. In practice, this restricts the method to rectangular or boxlike regions (Cartesian coordinates), circular or "piece-of-pie" regions (polar coordinates), cylinders (cylindrical coordinates), balls (spherical coordinates), or to simple combinations of these regions. We have also showed by two examples that non-Cartesian geometry may lead to Singular Sturm–Liouville Problems.

Given the complexities of actually constructing solutions of the Dirichlet Problem, it is remarkable that properties of solutions (e.g., the Maximal Principle) can be proved independently of the particular shape of the region involved. Although we proved the properties only for planar regions, they remain true, after appropriate reformulations, in any number of dimensions.

PROBLEMS

The following problem(s) may be more challenging: 2, 4, 7, and 8.

1. **(a)** Show that the solution of the steady-state temperature problem in the rectangular plate $G : 0 < x < L, 0 < y < M$, modeled by

$$\text{(PDE)} \quad \frac{\partial^2 u}{\partial x^2} + \frac{\partial^2 u}{\partial y^2} = 0 \qquad \text{in } G$$

$$\text{(BC)}_1 \quad u(0, y) = \alpha(y), \qquad 0 \le y \le M$$

$$\text{(BC)}_2 \quad u(L, y) = 0, \qquad 0 \le y \le M$$

$$(BC)_3 \qquad u(x, 0) = 0, \qquad 0 \le x \le L$$
$$(BC)_4 \qquad u(x, M) = 0, \qquad 0 \le x \le L$$

is

$$u_\alpha(x, y) = \sum_{n=1}^{\infty} A_n \sin \frac{n\pi y}{M} \sinh \frac{n\pi}{M} (L - x)$$

where

$$A_n \sinh \frac{n\pi L}{M} = \frac{\langle \alpha, \sin (n\pi y/M) \rangle}{\| \sin (n\pi y/M) \|^2} = \frac{2}{M} \int_0^M \alpha(y) \sin \frac{n\pi y}{M} dy$$

(b) Show that the formal solution of the temperature problem in G with general prescribed boundary temperatures

$$\text{(PDE)} \quad u_{xx} + u_{yy} = 0 \qquad \text{in } G$$
$$(BC)_1 \qquad u(0, y) = \alpha(y), \qquad 0 \le y \le M$$
$$(BC)_2 \qquad u(L, y) = \beta(y), \qquad 0 \le y \le M$$
$$(BC)_3 \qquad u(x, 0) = \gamma(x), \qquad 0 \le x \le L$$
$$(BC)_4 \qquad u(x, M) = \delta(x), \qquad 0 \le x \le L$$

is $u = u_\alpha + u_\beta + u_\gamma + u_\delta$, where u_α is given in part (a) and u_β, u_γ, u_δ are defined analogously.

(c) Repeat part (a) with $(BC)_1$ replaced by the "insulation" condition $u_x(0, y) = \tilde{\alpha}(y)$, $0 \le y \le M$.

(d) Repeat part (a) with $(BC)_3$ and $(BC)_4$ replaced by the "perfect insulation" conditions, $u_y(x, 0) = 0 = u_y(x, M)$, $0 \le x \le L$.

2. Find the steady temperature in a thin annulus, $G = \{(r, \theta) : \rho < r < R, -\pi \le \theta \le \pi\}$ whose faces are insulated and whose boundaries are maintained at the temperatures $f(\theta)$ and $g(\theta)$ at $r = \rho$ and $r = R$, respectively. [*Hint:* Proceed as in the text for a disk, but keep both terms in (7). Thus, in determining the superposition constants one will have $f(\theta) = A_0 + B_0 \ln \rho + \Sigma \rho^n [A_n \cos n\theta + B_n \sin n\theta] + \Sigma \rho^{-n} (C_n \cos n\theta + D_n \sin n\theta)$ and a similar expression with f replaced by g, ρ by R.]

3. **(a)** Solve the Dirichlet Problem for steady-state temperatures in the unit disk if $f(\theta) = 3 \sin \theta$ on the edge. Find the maximum and minimum temperatures in the disk.
   (b) Repeat part (a) if $f(\theta) = \theta + \pi$ for $-\pi \le \theta < 0$, $f(\theta) = -\theta + \pi$ for $0 \le \theta < \pi$.
   (c) Repeat part (a) if $f(\theta) = 0$ for $-\pi < \theta < 0$, $f(\theta) = 1$ for $0 < \theta < \pi$.

4. Find the steady-state temperatures in a ball of unit radius if the surface temperatures f are as given.
   (a) $f = \cos 2\phi - \sin^2 \phi$.
   (b) $f(\phi) = \begin{cases} 1, & 0 < \phi < \pi/2 \\ 0, & \pi/2 < \phi < \pi. \end{cases}$
   (c) $f(\phi) = |\cos \phi|$.

5. Find the steady-state temperatures in a spherical shell, $0 < \rho_1 < \rho < \rho_2$, if the temperatures on the inner sphere are given by $f(\phi)$, on the outer sphere by $g(\phi)$. [*Hint:* Proceed as in the text, but keep both terms in $R(\rho) = A\rho^n + B\rho^{-n-1}$.]

6. **(a)** Solve (20) if $f(r) \equiv 1$, $0 \le r \le 1$.
   (b) Solve (20) if the side-wall temperature condition $u(1, \theta, z) = 0$ is replaced by the

perfect-insulation condition $u_r(1, \theta, z) = 0$, $0 \leq z \leq a$, $-\pi \leq \theta \leq \pi$. [*Hint*: Use (13) of Section 8.6 with $p = 0$.]

7. Show that $\int_0^1 r J_0^2(k_n r)\, dr = J_1^2(k_n)$, where k_n is a positive zero of J_0 and J_1 is the Bessel function of the first kind of order 1. [*Hint*: Multiply Bessel's equation of order 0 by $2 J_0'(r)$ and rewrite to obtain $[r^2(J_0')^2]' + r^2(J_0^2)' = 0$. Integrate from 0 to k_n using an integration by parts, and then use the recursion formula (13) of Section 8.6 with $p = 0$: $J_0' = -J_1$.]

8. [*Neumann Problems*]. Let G be a bounded planar region whose boundary is a simple smooth closed curve. The problem $\nabla^2 u = 0$ in G, $\partial u / \partial n = f$ on ∂G, where u belongs to $C^2(G)$ and to $C^0(\mathrm{cl}\ G)$ and f is continuous on ∂G, is a *Neumann Problem* for G.

   (a) Show that any two solutions of the Neumann Problem differ by a constant.

   (b) Show that if the Neumann Problem has a solution at all, then $\int_{\partial G} f(s)\, ds = 0$.

   (c) Solve the Neumann Problem above if G is the unit disk in the plane and if $f(\theta) = \sin\theta$, while $u = 0$ at $\theta = 0$, $r = 1$.

Basic Theory of Initial Value Problems

A.1 UNIQUENESS

The aim of Appendix A is the presentation of basic theory for the initial value problem

$$y' = f(t, y), \qquad y(t_0) = y_0 \tag{1}$$

That is, we shall prove and extend the theorems of Section 2.2.

The uniqueness question for problem (1) is in some ways the easiest to handle, so we will treat it first.

Uniqueness Principle. In the initial value problem (1) let f and $\partial f/\partial y$ be continuous on some region R containing (t_0, y_0). Then on any t-interval I containing t_0 there is at most one solution of the problem (1).

Proof. We shall prove the claim only in the case where I is the interval $t_0 \leq t \leq t_0 + a$ and R is the closed rectangle $t_0 \leq t \leq t_0 + a$, $y_0 - b \leq y \leq y_0 + b$ for some positive constants a, b. This special case can be used to establish the more general claim, but we omit the details. Now suppose that $y_1(t)$ and $y_2(t)$ are two solutions of problem (1) which remain in R over the interval $I : t_0 \leq t \leq t_0 + a$ (see Figure A.1). Then

$$[y_1(t) - y_2(t)]' = f(t, y_1(t)) - f(t, y_2(t)) \qquad \text{for } t_0 < t < t_0 + a \tag{2}$$

Observe that $y_1(t) - y_2(t)$ is continuous on I and vanishes at $t = t_0$. Integrating (2) from t_0 to some t in I, we have that

$$y_1(t) - y_2(t) = \int_{t_0}^{t} [f(s, y_1(s)) - f(s, y_2(s))] \, ds \qquad \text{for } t \text{ in } I \tag{3}$$

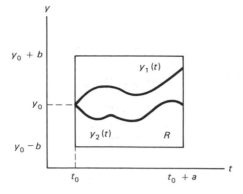

Figure A.1 Specialized geometry for the problem (1).

Now since $\partial f/\partial y$ is continuous on R, it follows via the Mean Value Theorem that there is a positive constant L such that†

$$|f(t, y_1) - f(t, y_2)| \leq L|y_1 - y_2| \qquad \text{for any points } (t, y_1), (t, y_2) \text{ in } R \qquad (4)$$

Functions $f(t, y)$ which satisfy the inequality (4) for some constant $L > 0$ are said to satisfy a *Lipschitz condition* in y on R; the constant L is called a *Lipschitz constant*. Thus taking absolute values of each side of (3), denoting $|y_1(t) - y_2(t)|$ by $w(t)$, and estimating the integral in the usual way, we obtain the integral inequality

$$0 \leq w(t) \leq L \int_{t_0}^{t} w(s)\, ds, \qquad \text{all } t \text{ in } I \qquad (5)$$

We shall show that the only non-negative solution of the inequality (5) is the trivial function $w(t) \equiv 0$. Indeed, let $v(t) \equiv \int_{t_0}^{t} w(s)\, ds$, then since $v'(t) \equiv w(t)$ on I we can rewrite (5) as the differential inequality

$$v'(t) - Lv(t) \leq 0, \qquad \text{all } t \text{ in } I \qquad (6)$$

Multiplying (6) through by the "integrating" factor $\exp[-Lt]$, we have the inequality $\{v(t) \exp[-Lt]\}' \leq 0$ for all t in I. Thus $v(t)e^{-Lt} \leq v(t_0)e^{-Lt_0}$, for all t in I. But since $v(t_0) = 0$, it follows that $v(t) \leq 0$ for all t in I. On the other hand if follows from the definition that $v(t) \geq 0$ for all t in I, and hence $v(t) \equiv 0$. This shows that $w(t) \equiv 0$ as well, so we have shown that $y_1(t) \equiv y_2(t)$ for all t in I. Thus the problem cannot have more than one solution on I, finishing our proof.

---

† L may be taken to be max $|\partial f/\partial y|$ for (t, y) in R.

PROBLEMS

The following problem(s) may be more challenging: 2.

1. Show that the function $f(t, y) = |y|$ does *not* satisfy the hypotheses of the Uniqueness Principle in any region containing all or part of the t-axis in the ty-plane. Show that $y' = |y|$, $y(t_0) = 0$ has a unique solution even so. Is there a contradiction?

2. Let m and n be positive integers without common factors (i.e., relatively prime). Consider the initial value problem

$$y' = |y|^{m/n}, \qquad y(0) = 0 \tag{*}$$

   (a) Show that (*) has the unique solution $y(t) \equiv 0$ if $m \geq n$.
   (b) Show that (*) has infinitely many solutions if $m < n$. [*Hint*: See Example 2.2.]

A.2 THE PICARD PROCESS FOR SOLVING AN INITIAL VALUE PROBLEM

We shall now establish the existence of a unique solution of the initial value problem (1) in a region R of the ty-plane containing (t_0, y_0) on which f and $\partial f / \partial y$ are both continuous functions. The method we shall use is to construct a sequence of iterate functions—called *Picard iterates*—which converges to the unique solution of (1) in an appropriate fashion.

An Equivalent Integral Equation

To approach the task of finding a solution for the problem (1), we shall first convert it into an "equivalent" integral equation. Suppose that $f(t, y)$ is continuous on the region R. If $y(t)$ is a solution of (1) on the interval I containing t_0, then integrating the relation $y'(t) = f(t, y(t))$ from t_0 to some t in the interval I and using the initial condition, we have

$$y(t) - y_0 = \int_{t_0}^{t} y'(s)\, ds = \int_{t_0}^{t} f(s, y(s))\, ds$$

Hence if $y(t)$ is a solution of (1) on the interval I containing t_0, then $y(t)$ is a solution of the integral equation

$$y(t) = y_0 + \int_{t_0}^{t} f(s, y(s))\, ds, \qquad t \text{ in } I \tag{7}$$

Conversely, if $y(t)$ on I satisfies the integral equation (7), then surely $y(t_0) = y_0$, and from the Fundamental Theorem of Calculus, $y(t)$ is differentiable at each interior point of I, and

$$y'(t) \equiv \frac{d(y(t))}{dt} = \frac{d}{dt}\left(y_0 + \int_{t_0}^{t} f(s, y(s))\, ds\right) = f(t, y(t))$$

It follows that $y(t)$ is a solution of (1) on the interval I. This is the sense in which the initial value problem (1) and the integral equation (7) are equivalent. Thus the solvability of (1) is equivalent to the solvability of the integral equation (7). We shall exploit this equivalence.

Existence of Solution

After this slight detour, we are at last ready to prove the existence of a solution to the problem (1). Recall that under the condition that f and $\partial f / \partial y$ are continuous on R, we have already shown that the problem (1) cannot have more than one solution on any interval containing t_0.

> **Existence Theorem.** If the functions f and $\partial f / \partial y$ are continuous on a region R of the ty-plane and if (t_0, y_0) is an interior point of R, the initial value problem (1) has a (unique) solution $y(t)$ on an interval I containing t_0 in its interior.

Preparation for Proof. The first step in our proof is to construct in R a closed rectangle S which has the form $t_0 \le t \le t_0 + a$, $y_0 - b \le y \le y_0 + b$, where a and b are positive numbers (Figure A.2). Such a construction is possible since (t_0, y_0) is an interior point of R. Now the assumptions of the theorem imply that f and $\partial f / \partial y$ are continuous on the closed rectangle S. Hence there are positive constants M and L (not unique) such that

$$|f(t, y)| \le M, \quad |f(t, y_1) - f(t, y_2)| \le L|y_1 - y_2|, \quad \text{all } (t, y), (t, y_1), (t, y_2) \text{ in } S \quad (8)$$

Our technique of proof will produce the unique solution, $y(t)$, of the differential equation $y' = f(t, y)$ on the interval $t_0 < t < t_0 + c$ where $c = \min\{a, b/M\}$, which is continuous on $t_0 \le t \le t_0 + c$, with $y(t_0) = y_0$. The corresponding solution curve remains in the rectangle S. It may be regarded as a solution of the forward initial value problem associated with (1). The same technique can be used to show

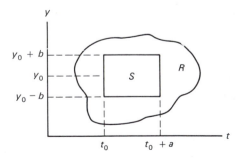

Figure A.2 Geometry for a forward initial value problem.

the existence of a solution for (1) on an interval $t_0 - d \leq t \leq t_0$ for some $d > 0$. Putting this "backward" solution together with the "forward" solution, we would then have a "two-sided" solution to problem (1) on the interval $t_0 - d \leq t \leq t_0 + c$, as asserted in the statement of the Existence Theorem. We turn now to the construction of the forward solution.

Construction of the Picard Iterates.   We begin with the equivalent formulation of the forward problem for (1) as the integral equation (7). We saw that $y(t)$ is a solution of the forward problem for (1) over an interval $t_0 \leq t \leq t_0 + c$ if and only if $y(t)$ satisfies the integral equation

$$y(t) = y_0 + \int_{t_0}^{t} f(s, y(s))\, ds, \qquad t_0 \leq t \leq t_0 + c \tag{9}$$

We shall describe below an algorithm for generating a sequence of functions $y_0(t)$, $y_1(t)$, $y_2(t)$, . . . all defined and continuous on $t_0 \leq t \leq t_0 + c$ which will be used to generate a solution $y(t)$ of (9) in a sense to be made clear later. The functions $y_n(t)$, $n = 0, 1, 2, \ldots$ constructed in this proof arise in a distinctive way and are called *Picard*† *iterates*. They are generated recursively [i.e., $y_{n+1}(t)$ is computed directly from $y_n(t)$, for each $n = 0, 1, 2, \ldots$] in the following way. Taking $y_0(t) \equiv y_0$, we define $y_1(t)$, $y_2(t)$, . . . via the recursion relation

$$y_{n+1}(t) = y_0 + \int_{t_0}^{t} f(s, y_n(s))\, ds, \qquad n = 0, 1, 2, \ldots \tag{10}$$

with $t_0 \leq t \leq t_0 + c$. In constructing the Picard iterates $y_n(t)$ via (10) we must take care that at each stage the points $(t, y_n(t))$ remain in R for all $t_0 \leq t \leq t_0 + c$ before we can go on to compute $y_{n+1}(t)$ via (10). Indeed, $y_0(t)$ belongs to S. We shall now show by induction that $y_n(t)$, $n = 1, 2, \ldots$ also satisfies this condition. For let us assume for some positive integer n that the graph of $y_n(t)$ lies in S for all $t_0 \leq t \leq t_0 + c$. Using (10), we have the estimate

$$|y_{n+1}(t) - y_0| \leq |t - t_0| M \leq cM \leq b, \qquad t_0 \leq t \leq t_0 + c \tag{11}$$

since $c = \min \{a, b/M\}$, and thus the graph of $y_{n+1}(t)$ remains in S for all $t_0 \leq t \leq t_0 + c$. Hence there is no difficulty in constructing the Picard iterates via (10) if we restrict t to the interval $[t_0, t_0 + c]$.

Uniform Convergence of a Sequence of Functions

We must interrupt our proof of the Existence Theorem to describe the sense in which the Picard iterates $y_0(t)$, $y_1(t)$, . . . are used to generate the solution of the integral equation (9). Our remarks actually apply to any sequence of functions defined over a common interval, so we shall not restrict ourselves to Picard iterates at the moment.

---

† Emile Picard (1856–1941) was an eminent French mathematician and the permanent secretary of the Paris Academy of Sciences. His mathematical work includes deep results in complex analysis and partial differential equations, as well as in ordinary differential equations.

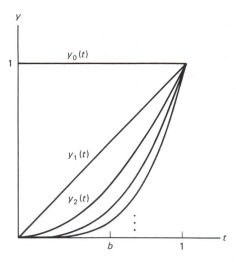

Figure A.3   Convergence of $\{t^n\}$.

Uniform Convergence of a Sequence of Functions.   A sequence of functions $\{y_n(t)\}$, $n = 0, 1, 2, \ldots$, all defined on a common interval I, is said to converge *uniformly* to the function $y(t)$ on I if given any error tolerance $E > 0$ there exists a positive integer N such that

$$|y_n(t) - y(t)| < E, \qquad \text{for all } t \text{ in } I \quad \text{and all } n \geq N \tag{12}$$

As we shall see, uniform convergence is the way we will generate the solution of the integral equation (9) from the Picard iterates.

As an example, note that the sequence $y_n(t) \equiv t^n$, $n = 0, 1, 2, \ldots$ converges uniformly to the zero function on any closed interval $[0, b]$ where $0 < b < 1$, but does *not* converge uniformly on the closed interval $[0, 1]$ (see Figure A.3). It is sometimes useful to view uniform convergence graphically as follows (see Figure A.3 again). The sequence $\{y_n(t)\}$ converges to $y(t)$ uniformly over I if for every $E > 0$ the graphs of $y_n(t)$ eventually all remain inside a "tube" of radius E about the graph of $y(t)$.

The inequality (12) is not a very useful test for uniform convergence of a sequence $y_n(t)$ because one must "guess" the limit function $y(t)$ in advance before the test can be applied. As with sequences of numbers, there is a *Cauchy Test for uniform convergence*. It has the advantage that one need not know the limit function in advance before the test can be applied; on the other hand, the Cauchy Test has the disadvantage that when it succeeds one does not precisely know the limit function.

Cauchy Test for Uniform Convergence.   The sequence $\{y_n(t)\}$ defined over a common interval I is uniformly convergent† if given any $E > 0$ there is a positive integer N such that

$$|y_n(t) - y_m(t)| < E \qquad \text{for all } t \text{ in } I, \quad \text{all } n, m \geq N \tag{13}$$

---

† That is, there is a function $y(t)$ on I such that for any $E > 0$, there is an N such that (12) holds.

Note that although the limit is not known from the Cauchy Test, it is unique and can be uniformly approximated by the sequence element $y_n(t)$ as closely as desired by choosing n large enough. Another fact we accept without proof is that the uniform limit of a sequence of continuous functions on a common interval I is also continuous on that same interval. Figure A.3 provides an example of this phenomenon on the interval $[0, \frac{1}{2}]$, say, as well as an example of what goes wrong when the convergence is not uniform (look at the closed interval $[0, 1]$).

Convergence of the Picard Iterates.   Now we show that the sequence of Picard iterates $\{y_n(t)\}$ converges uniformly to a solution of the integral equation (9) on the interval $[t_0, t_0 + c]$. First, however, recall that

$$|f(t, y_1) - f(t, y_2)| \leq L|y_1 - y_2|, \qquad \text{all } (t, y_1), (t, y_2) \text{ in } S \tag{14}$$

where L is defined via (8). Thus, using (10) and (14), we have the estimate

$$|y_{n+1}(t) - y_n(t)| \leq L \int_{t_0}^{t} |y_n(s) - y_{n-1}(s)| \, ds,$$

$$t_0 \leq t \leq t_0 + c, \quad n = 1, 2, \ldots \tag{15}$$

Now observe from (11) that

$$|y_1(t) - y_0| \leq b, \qquad \text{all } t_0 \leq t \leq t_0 + c \tag{16}$$

and hence using (15) with $n = 1$, we have

$$|y_2(t) - y_1(t)| \leq Lb(t - t_0) \qquad \text{for } t_0 \leq t \leq t_0 + c \tag{17}$$

Using (17) and (15) with $n = 2$, we see that

$$|y_3(t) - y_2(t)| \leq L^2 b \frac{(t - t_0)^2}{2} \qquad \text{for } t_0 \leq t \leq t_0 + c$$

Proceeding along in this manner, we can use (15) to establish the estimate

$$|y_{n+1}(t) - y_n(t)| \leq bL^n \frac{(t - t_0)^n}{n!}, \qquad t_0 \leq t \leq t_0 + c, \quad n = 0, 1, 2, \ldots \tag{18}$$

So for any $n > m$ we have by (18) and the triangle inequality the estimate

$$|y_n(t) - y_m(t)| \leq |y_n(t) - y_{n-1}(t)| + \cdots + |y_{m+1}(t) - y_m(t)|$$

$$\leq b \sum_{k=m}^{n-1} \frac{L^k(t - t_0)^k}{k!} \leq b \sum_{k=m}^{\infty} \frac{L^k(t - t_0)^k}{k!} \tag{19}$$

If we denote by $S_n(t)$ the nth partial sum of $\exp[L(t - t_0)]$, that is,

$$S_n(t) = \sum_{k=0}^{n-1} \frac{L^k(t - t_0)^k}{k!}$$

then (19) can be written as

$$|y_n(t) - y_m(t)| \leq b[S_n(t) - S_m(t)], \qquad t_0 \leq t \leq t_0 + c \tag{20}$$

Now since the partial sums $\{S_n(t)\}$ converge uniformly to $\exp[L(t - t_0)]$ on any interval of the t-axis, it follows from what was said earlier that $\{S_n(t)\}$ satisfies the Cauchy Test (13) and hence, from (20), so does the sequence of Picard iterates $\{y_n(t)\}$ over the interval $[t_0, t_0 + c]$. Thus it follows that $y_n(t)$ converges uniformly to a continuous function $y(t)$ over $[t_0, t_0 + c]$.

To show that $y(t)$ satisfies the integral equation (9) on $[t_0, t_0 + c]$, we need the following result.

Integral Convergence Theorem.   Let the sequence $\{y_n(t)\}$ of continuous functions converge uniformly to $y(t)$ on $t_0 \le t \le t_0 + c$ and be such that $|y_n(t) - y_0| \le b$ for all $t_0 \le t \le t_0 + c$, where b, c are defined as above. Then for $f(t, y)$ as in problem (1),

$$\int_{t_0}^{t} f(s, y_n(s))\, ds \rightarrow \int_{t_0}^{t} f(s, y(s))\, ds \qquad \text{for any } t_0 \le t \le t_0 + c \qquad (21)$$

Observe that the sequence of Picard iterates $\{y_n(t)\}$ satisfies the conditions leading to (21), and also the relation (10) over $[t_0, t_0 + c]$. Thus taking limits of each side of (10) for each fixed t in $[t_0, t_0 + c]$, we see via (21) that $y(t)$ satisfies the integral equation (9), and hence the forward initial value problem for (1), finishing the proof.

Example A.1

Consider the initial value problem

$$y' = ty + 1$$
$$y(0) = 1 \qquad (22)$$

Comparing this problem with (1), we see that $f(t, y) \equiv ty + 1$, $t_0 = 0$ and $y_0 = 1$. We may as well take the region R where f is defined to be the whole ty-plane. Let us take the rectangle S to be $0 \le t \le a$, $1 - b \le y \le 1 + b$, where a, b are arbitrary positive numbers. Now $|ty + 1| \le a(b + 1) + 1$ on S, so we may take $M = a(b + 1) + 1$, thus

$$c = \min\left\{a, \frac{b}{a(b + 1) + 1}\right\}$$

Because $f(t, y)$ is defined and well-behaved over the entire ty-plane, (10) defines the Picard iterates for all t, but our proof only guarantees convergence of the iterates over the interval $0 \le t \le c$. We have

$$y_0(t) = 1$$
$$y_1(t) = 1 + \int_0^t f(s, y_0(s))\, ds = 1 + \int_0^t [sy_0(s) + 1]\, ds$$
$$= 1 + \int_0^t (s + 1)\, ds = 1 + t + \frac{t^2}{2}$$

$$y_2(t) = 1 + \int_0^t [sy_1(s) + 1] \, ds$$

$$= 1 + \int_0^t \left[s \left(1 + s + \frac{1}{2}s^2 \right) + 1 \right] ds$$

$$= 1 + t + \frac{t^2}{2} + \frac{t^3}{3} + \frac{t^4}{8}$$

and so on. Taking $a = 1$, $b = 8$, we see that $c = \min\{1, 0.8\} = 0.8$, and hence the Picard iterates converge in the prescribed sense at least over the interval $0 \le t \le 0.8$ (see Figure A.4).

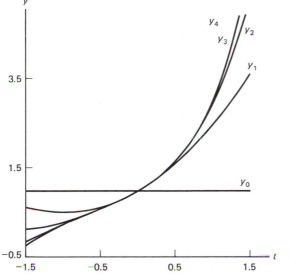

Figure A.4 Picard iterates for $y' = ty + 1$, $y(0) = 1$.

Comments

Suppose that all the conditions of the Existence and Uniqueness Theorem (EUT) for the initial value problem (1) are met. The problem has exactly one solution, $y = y(t)$, say. But how can you find it? Only for very special types of functions f is there any hope of finding an explicit formula for the solution y. Alternatively, a computer can be used to come up with an approximate graph of the solution, or a table of its approximate values for a given set of values of t. The computer will use some numerical approximation method to do this. Contemporary computer methods are remarkably robust and accurate, and are now the first choice in solving such problems. But for most users the computer is a "black box" that generates graphs or tables by magic. The Picard iteration scheme, on the other hand, both illuminates what is going on when the initial value problem is solved and lies at the heart of the proof of the EUT. These two assets of the scheme may outweigh its wild impracticality.

The problems below illustrate these points. The Picard iterates $y_n(t)$, $n = 0$, 1, 2, . . . , are defined by.

$$y_0(t) \equiv y_0, \quad . . . \quad , \quad y_{n+1}(t) = y_0 + \int_{t_0}^{t} f(s, y_n(s)) \, ds, \qquad n = 0, 1, . . . \qquad (10)$$

According to the Existence and Uniqueness Theorem, $y_n(t) \rightarrow y(t)$ (the actual, but unknown solution) as $n \rightarrow \infty$ for each t in some open interval containing t_0.

PROBLEMS

The following problem(s) may be more challenging: 3, 5, 6, and 9.

1. (a) Find the Picard iterates y_1, y_2, y_3 for the problem $y' = -y$, $y(0) = 1$.
   (b) Sketch the graphs of these iterates for $|t| < 2$.
   (c) Find a formula for the nth iterate y_n by induction.
   (d) Find the exact solution $y(t)$ and sketch its graph along with those of y_0, y_1, y_2, and y_3.
   (e) Show that $y_n(t)$ is a partial sum of the Maclaurin series for $y(t)$.

2. Repeat Problem 1 for $y' = 2ty$, $y(0) = 2$. [*Hint:* The Maclaurin series for $y(t^k)$, k a positive integer, is obtained from that for $y(t)$ by replacing t by t^k.]

3. (a) Find the Picard iterates y_1, y_2, and y_3 for $y' = y^2$, $y(0) = 1$, but do not bother to find an explicit formula for $y_n(t)$.
   (b) Find the solution $y(t)$ directly and plot y_0, y_1, y_2, y_3, and y on their maximal domains.
   (c) Prove by induction that the nth Picard iterate $y_n(t)$ is a polynomial of degree $2^n - 1$.
   (d) Show that every Picard iterate is defined for all t, but that the exact solution is only defined for $t < 1$. Then show that the Maclaurin series of the exact solution converges only for $|t| < 1$. Thus Picard iterates, the exact solution, and the Maclaurin series of the exact solution may all be defined on different intervals.

4. The Picard iteration scheme is sturdy enough to survive a bad choice for the first approximation $y_0(t)$. Show that the Picard iterates for the problem $y' = -y$, $y = 1$ when $t = 0$, converge to the exact solution e^{-t} even if the wrong starting "point" $y_0(t) \equiv 0$ is used to start the iteration scheme. [*Hint:* $y_1(t) = 1 + \int_0^t (-y_0(s)) \, ds = 1$, $y_2(t) = 1 + \int_0^t (-y_1(s)) \, ds = 1 - t$, and so on. The correct initial data, $y(0) = 1$, is used in any case.]

5. Repeat Problem 4 but with the absurd first approximation $y_0(t) = \sin t$. Show that the sequence of iterates still converges to e^{-t}.

6. (a) Find the Picard iterates y_1, y_2, and y_3 for $y' = 1 + y^2$, $y(0) = 0$.
   (b) Prove by induction that $y_n(t)$ is a polynomial of degree $2^{n+1} - 1$.
   (c) Find the exact solution $y(t)$ and identify the maximal interval on which it is defined.
   (d) Does $y_n(t) \rightarrow y(t)$ for all t on which $y_n(t)$ is defined?

7. How many Picard iterates can you calculate for $y' = \sin(t^3 + ty^5)$, $y(0) = 1$? What is the practical difficulty in finding the iterates?

8. Let $y_0(t) = 3$, $y_1(t) = 3 - 27t$, . . . , $y_n(t) = 3 - \int_0^t y_{n-1}^3(s) \, ds$, ... Find

$\lim\limits_{n\to\infty} y_n(t)$. [*Hint*: The problem is related to the Picard iteration scheme for a certain initial value problem.]

9. (a) Show that the sequence $\{(1/n)\sin nt\}_{n=1}^{\infty}$ converges uniformly to the zero function on any interval I. [*Hint*: $|(1/n)\sin nt - 0| \leq (1/n).$]

   (b) Show that the sequence $\{t^n\}_{n=1}^{\infty}$ converges uniformly to the zero function on the interval $0 \leq t \leq 0.5$, while it converges, but not uniformly, on the interval $0 \leq t \leq 1$ to the function $f(t) = 0$, $0 \leq t < 1$, $f(1) = 1$.

A.3 EXTENSION OF SOLUTIONS

A solution defined on an interval can be extended outside that interval when the conditions on the data allow it. This brings up the question of the relationship between the maximally extended solutions of $y' = f(t, y)$ and the region R where f is defined. In the proof of the Existence Theorem we showed that for any *closed* rectangle S in the ty-plane where f and $\partial f/\partial y$ are continuous, the solution curve that "enters" S at the midpoint of a vertical side will "exit" through one of the other sides. The result below shows that this behavior of solution curves is typical.

Extension Principle. Let R be a region in the ty-plane and suppose that f and $\partial f/\partial y$ are continuous on R. Let D be a closed bounded region contained in R and consider the initial value problem (1) with (t_0, y_0) a point in the interior of D. Then the solution of (1) can be extended forward and backward in t until its solution curve exits through the boundary of D.

Proof. We shall consider only the forward extension assertion since the backward assertion is proved in the same way. Now since D is bounded there is a *smallest* value $\bar{t}$ such that no point of D lies in the half-plane $t > \bar{t}$. Thus, if the forward maximally extended solution curve for (1) contains a point $(t, y(t))$ with $t \geq \bar{t}$, the solution curve must have exited D through its boundary, so there is nothing to prove. To have something to prove, let us assume that the forward maximally extended solution of (1) remains in the interior of D and is defined for all t in the interval $t_0 \leq t < \bar{\bar{t}} < \bar{t}$. We shall show that this assumption leads to a contradiction, and hence the forward solution curve cannot both lie inside D and to the left of some value $\bar{\bar{t}} < \bar{t}$ (see Figure A.5). Thus the solution curve must indeed exit D. To show this, recall that (1) is equivalent to the integral equation (7), and hence

$$y(t_2) - y(t_1) = \int_{t_1}^{t_2} f(s, y(s))\, ds \qquad t_0 \leq t_1, t_2 \leq \bar{\bar{t}} \qquad (23)$$

Since f is continuous on a closed bounded region there is a constant $M > 0$ such that $M \geq |f(t, y)|$ for all (t, y) in D. Thus (23) implies via the standard estimate for integrals that

$$|y(t_1) - y(t_2)| \leq M|t_1 - t_2|, \qquad t_0 \leq t_1, \quad t_2 < \bar{\bar{t}} \qquad (24)$$

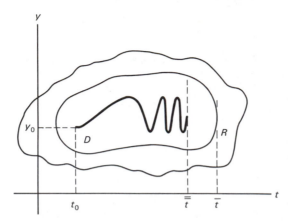

Figure A.5 Geometry for the Extension Principle.

It follows† from (24) that the limit $y(\bar{\bar{t}}^-)$ exists. Now the point $(\bar{\bar{t}}, y(\bar{\bar{t}}^-))$ cannot lie in the interior of D, for then the method described in the proof of the Existence Theorem would allow us to extend $y(t)$ beyond $t = \bar{\bar{t}}$. Thus $(\bar{\bar{t}}, y(\bar{\bar{t}}^-))$ lies on the boundary of D and we are done.

Comments

The problem $y' = y^2$, $y(0) = y_0$, where $y_0 \neq 0$, has the unique solution $y(t) = y_0(1 - ty_0)^{-1}$ defined on the interval $(-\infty, y_0^{-1})$ if $y_0 > 0$, and on the interval (y_0^{-1}, ∞) if $y_0 < 0$ (see Example 1.2). This problem illustrates the fact that it is not easy to predict from the rate function itself just how a solution will approach the boundary of the region in which the existence and uniqueness hypotheses hold; nevertheless, for *linear* problems the situation is much simpler (Problem 3).

PROBLEMS

1. Solve and find the largest t-interval on which the solution is defined. What happens to the solution as t approaches each endpoint of the interval?
   (a) $y' = -y^3$, $y(0) = 1$.   (b) $y' = -te^{-y}$, $y(0) = 2$.
2. Find a first-order equation $y' = f(t, y)$, each solution of which is defined only on a finite t-interval.
3. Show that the solution $y(t)$ of $y' + p(t)y = q(t)$, $y(t_0) = y_0$, where $p(t)$ and $q(t)$ are continuous for all t in an open interval I containing t_0, is defined for all t in I. [*Hint:* Use the solution formula of Section 3.2.]

---

† Via the Cauchy Test for convergence.

A.4. DEPENDENCE OF SOLUTIONS ON THE DATA

There remains a last question concerning the solution of (1). Loosely phrased, the question amounts to this: Is it always possible to find bounds on the determination of the data $f(t, y)$ and y_0 in (1) which will guarantee that the corresponding solution will be within given prescribed error bounds over a given t-interval? If this question can be answered in the affirmative, one consequence is that any "small" change in the data of an initial value problem produces only a "small" change in the solution. Or, stated another way: In spite of the fact that the initial value problem arising from a model has empirically determined components (and therefore can never be known precisely), the model is still nevertheless relevant and useful.

In our study of the initial value problem (1) we have been guided by the four questions posed in Section 2.2: Does the problem (1) have any solutions? How many? What are they? How do solutions respond to changes in the data? If the functions f and $\partial f / \partial y$ are continuous in some region R in the ty-plane and (t_0, y_0) is a point of R, we gave satisfactory answers to the first, second, and third questions. In this section we show that these same simple conditions on the data also lead to a satisfactory answer to the last question.

In addressing the fourth question it would be extremely helpful to have a formula for the solution of (1) in which the data appear explicitly. But for general nonlinear differential equations there rarely is a solution formula for problem (1) in which the data appear explicitly. Thus we shall have to find some other way to answer the fourth question in the general case.

To estimate the change in the solution to problem (1) as the data $f(t, y)$ and y_0 are modified, it will be very helpful to have an estimate of the type given below.

Basic Estimate.   For given constants A, B, and L with $L > 0$, suppose that $z(t)$ is a continuous function on the interval $[a, b]$ which satisfies the integral inequality

$$z(t) \leq A + B(t - a) + L \int_a^t z(s)\, ds \qquad \text{for all } a \leq t \leq b \qquad (25)$$

Then $z(t)$ satisfies the estimate

$$z(t) \leq A e^{L(t-a)} + \frac{B}{L} \left(e^{L(t-a)} - 1 \right) \qquad \text{for all } a \leq t \leq b \qquad (26)$$

Proof. Let us put

$$Q(t) = \int_a^t z(s)\, ds \qquad \text{for } a \leq t \leq b$$

Then Q is continuous on $[a, b]$ and the Fundamental Theorem of Calculus implies that $Q'(t) = z(t)$ for $a < t < b$, so (25) can be rewritten as

$$Q'(t) - LQ(t) \le A + B(t-a), \qquad a < t < b \tag{27}$$

Multiply each side of (27) by the positive factor e^{-Lt} and then use the fact that $[e^{-Lt}Q]' = e^{-Lt}Q' - e^{-Lt}LQ$ to obtain

$$[Q(t)e^{-Lt}]' \le [A + B(t-a)]e^{-Lt}, \qquad a < t < b \tag{28}$$

Since $F(t) \equiv -L^{-2}[B + LA + LB(t-a)]e^{-Lt}$ is an antiderivative to the right-hand side of (28), we see that (28) can be written as

$$[Q(t)e^{-Lt} - F(t)]' \le 0, \qquad a < t < b \tag{29}$$

The function $G(t) \equiv Q(t)e^{-Lt} - F(t)$ is continuous on the interval $[a, b]$, and hence because of (29), we see that $G(t) \le G(a)$ for $a \le t \le b$, or after simplification [note that $Q(a) = 0$],

$$A + B(t-a) + LQ(t) \le Ae^{L(t-a)} + \frac{B}{L}(e^{L(t-a)} - 1), \qquad a \le t \le b \tag{30}$$

Now from (25) we see that

$$z(t) \le A + B(t-a) + LQ(t), \qquad a \le t \le b$$

which together with (30) yields the inequality (26), and we are done.

There is also a "backward" form of the Basic Estimate, which under the same hypotheses states that the inequality

$$z(t) \le A + B(b-t) + L \int_t^b z(s)\, ds, \qquad a \le t \le b \tag{31}$$

implies that

$$z(t) \le Ae^{L(b-t)} + \frac{B}{L}(e^{L(b-t)} - 1), \qquad a \le t \le b \tag{32}$$

The proof of (32) from (31) is very similar to the foregoing proof for (26) from (25), and hence is omitted (see the problem set).

We need some estimate of the change in the solution of (1) which occurs when the data $f(t, y)$ and y_0 are modified (or, to use more suggestive terminology, perturbed).

Perturbation Estimate.   Let the function f in problem (1) be continuous along with $\partial f / \partial y$ in a rectangle R described by the inequalities $t_0 \le t \le t_0 + a$, $|y - y_0| \le b$. Now suppose that on some common interval $t_0 \le t \le t_0 + c$, with $c \le a$, the function $y(t)$ is the solution of (1) and the function $\bar{y}(t)$ is a solution of the "perturbed" problem

$$\begin{aligned} \bar{y}' &= f(t, \bar{y}) + g(t, \bar{y}) \\ \bar{y}(t_0) &= \bar{y}_0 \end{aligned} \tag{33}$$

where g and $\partial g/\partial \tilde{y}$ are continuous functions on R and $\tilde{y}_0$ is in the interval $|y - y_0| < b$ (see Figure A.6). Then we have the estimate

$$|y(t) - \tilde{y}(t)| \le |y_0 - \tilde{y}_0|e^{L(t-t_0)} + \frac{M}{L}(e^{L(t-t_0)} - 1), \qquad t_0 \le t \le t_0 + c \qquad (34)$$

where L and M are any numbers such that

$$M \ge |g(t, y)|, \quad L \ge \left|\frac{\partial f}{\partial y}\right|, \qquad \text{all } (t, y) \text{ in } R.$$

Proof. The proof is an immediate consequence of the Basic Estimate above. Using (9), we see that $y(t)$ and $\tilde{y}(t)$ must satisfy the integral equations (for $t_0 \le t \le t_0 + c$)

$$y(t) = y_0 + \int_{t_0}^{t} f(s, y(s))\, ds$$

$$\tilde{y}(t) = \tilde{y}_0 + \int_{t_0}^{t} [f(s, \tilde{y}(s)) + g(s, \tilde{y}(s))]\, ds$$

Subtracting the second equation from the first, we have for any $t_0 \le t \le t_0 + c$ that

$$y(t) - \tilde{y}(t) = y_0 - \tilde{y}_0 + \int_{t_0}^{t} [f(s, y(s)) - f(s, \tilde{y}(s))]\, ds - \int_{t_0}^{t} g(s, \tilde{y}(s))\, ds \qquad (35)$$

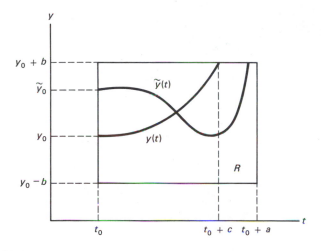

Figure A.6 Geometry for perturbation estimate.

Putting $z(t) = |y(t) - \tilde{y}(t)|$ and estimating the integrals in (35) using the inequality (4), we have the inequality

$$z(t) \leq z(t_0) + M(t - t_0) + L \int_{t_0}^{t} z(s)\, ds, \qquad t_0 \leq t \leq t_0 + c \qquad (36)$$

Comparing (36) with (35) and applying the Basic Estimate with A and B replaced by $|y_0 - \tilde{y}_0|$ and M, respectively, we obtain the desired estimated (34), concluding our proof.

Example A.2

As an illustration of the Perturbation Estimate we graph in Figure A.7 solutions of the initial value problems

$$
\begin{array}{ll}
y' = y \sin y & \tilde{y}' = \tilde{y} \sin \tilde{y} + \bar{a} \sin(\bar{b}t\tilde{y}) \\
y(0) = y_0 & \tilde{y}(0) = \tilde{y}_0
\end{array}
\qquad (37)
$$

for various values of the constants $\tilde{y}_0$, $\bar{a}$, and $\bar{b}$. The solid line is the solution curve of $y(t)$ that corresponds to the data in the table below.

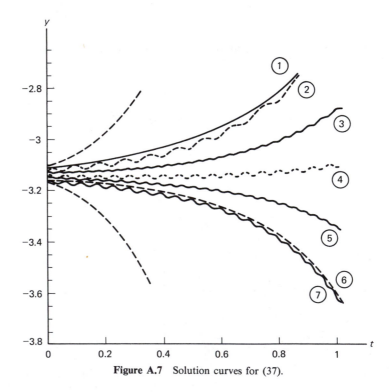

Figure A.7 Solution curves for (37).

$\tilde{y}(t)$	$\tilde{a}$	$\tilde{b}$	$\tilde{y}_0$
①	0	0	−3.11
②	0.7	24	−3.11
③	0.5	60	−3.13
④	0.7	36	−3.14
⑤	0.5	48	−3.15
⑥	0	0	−3.16
⑦	0.7	36	−3.16

The heavy dashed line is the error bound predicted by the Perturbation Estimate. Notice that the error bound is not very sharp as we move away from $t_0 = 0$, and hence is not really all that practical in estimating errors.

Now at last we are in a position to answer the last of the basic questions.

Continuity in the Data.   Let f, $\partial f/\partial y$, g, and $\partial g/\partial y$ be continuous functions of t and y on the rectangle R defined by $t_0 \le t \le t_0 + a$, $|y - y_0| \le b$. Let $E > 0$ be a given error tolerance. Then there exist positive constants $H < b$ and $c \le a$ such that the respective solutions $y(t)$ and $\tilde{y}(t)$ of

$$\text{(a)} \begin{cases} y' = f(t, y) \\ y(t_0) = y_0 \end{cases} \qquad \text{(b)} \begin{cases} \tilde{y}' = f(t, \tilde{y}) + g(t, \tilde{y}) \\ \tilde{y}(t_0) = \tilde{y}_0 \end{cases} \qquad (38)$$

satisfy the inequality

$$|y(t) - \tilde{y}(t)| \le E \qquad t_0 \le t \le t_0 + c \qquad (39)$$

for any choice of $\tilde{y}_0$ for which $|y_0 - \tilde{y}_0| \le H$.

Proof. Let H be any positive constant with $H < b$, and let $\tilde{M}$ be such that $\tilde{M} \ge \max |g(t, y)|$ for all (t, y) in R. Then according to the Existence and Uniqueness Theorem, the solution $\tilde{y}(t)$ of problem (38b) must be defined at least over the interval $t_0 \le t \le t_0 + c$, with

$$c = \min\left\{ a, \frac{b - H}{M + \tilde{M}} \right\} \qquad (40)$$

where M is a positive constant such that $M \ge |f(t, y)|$ for all (t, y) in R. Now by the Perturbation Estimate we have that

$$|y(t) - \tilde{y}(t)| \le H e^{Lc} + \frac{\tilde{M}}{L}(e^{Lc} - 1), \qquad t_0 \le t \le t_0 + c \qquad (41)$$

where $L > 0$ is any number such that $L \ge |\partial f/\partial y|$, for all (t, y) in R. The positive constants H and c can be determined to make the right-hand side of (41) less than, or equal to, E. Hence (39) holds and we are done.

Properties of Solutions

Putting together the results of this section with those of previous sections, we now have a rather comprehensive view of how solutions of initial value problems such as (1) behave. If $f(t, y)$ and $\partial f/\partial y$ are continuous on a closed region R of the ty-plane, then:

1. Through each point in the interior of R there is one and only one solution curve of the equation $y' = f(t, y)$, and this solution curve can be extended up to the boundary of R (from the Existence Theorem, the Uniqueness Principle, and the Extension Principle).

2. The change in the solution of an initial value problem (1) can be made as small as desired near the initial time t_0 if the change in the data $f(t, y)$ and y_0 is made sufficiently small (Continuity in the Data Theorem).

The first property above shows that an initial value problem is an effective tool in solving problems that involve dynamical systems. The second property indicates that models involving initial value problems are not invalidated because empirically determined elements of the model do not have their "true" values.

PROBLEMS

The following problem(s) may be more challenging: 1 and 3.

1. Prove that the "backward" estimate (32) follows from the inequality (31).

2. Use the Perturbation Estimate to approximate $|\tilde{y}(t) - y(t)|$ if $\tilde{y}'(t) = (e^{-\tilde{y}^2} + 0.1)\tilde{y}$, $\tilde{y}(0) = \tilde{a}$, and $y'(t) = 0.1y$, $y(0) = a$.

3. Let $y(t)$ be a continuously differentiable function on the real line such that $|y'(t)| \leq |y(t)|$ for all t. Show that either $y(t)$ never vanishes or else that $y(t) = 0$ for all t.

Numerical Methods

The initial value problem under consideration is

$$y' = f(t, y), \qquad y(t_0) = y_0 \tag{1}$$

where f and $\partial f / \partial y$ are continuous in a rectangle R "centered" at (t_0, y_0). As noted in Appendix A, there is a positive constant L such that the following Lipschitz condition holds:

$$|f(t, y_1) - f(t, y_2)| \leq L|y_1 - y_2| \tag{2}$$

for all points (t, y_1) and (t, y_2) in R. First we shall derive bounds for the discretization and the total errors incurred when Euler's Method is applied to the numerical approximation of the solution of (1). Then we shall show how the various Adams–Bashforth multistep algorithms are derived.

Errors of Euler's Method

Euler's method with step size h rests on the formula

$$y_{j+1} = y_j + f(t_j, y_j)h, \qquad j = 0, \ldots, N-1 \tag{3}$$

We have the following result concerning the global discretization error E_N when Euler's method is used to approximate the solution of (1) at $T = t_0 + Nh$.

Discretization Error of Euler's Method.   If $f(t, y)$ is continuously differentiable in the rectangle R, the error of the Euler algorithm (3) for solving the initial value problem (1) is first order in h: $E_N \leq Mh$ for some constant M.

Proof. Suppose that the function f is continuous and continuously differentiable in t and y in the rectangle R. Then the solution $y(t)$ has a continuous second derivative, as well as a continuous first derivative. For we have from the differential equation that

$$\frac{d^2y(t)}{dt^2} = \frac{\partial f(t, y(t))}{\partial t} + \frac{\partial f(t, y(t))}{\partial y} \frac{dy(t)}{dt}$$

$$= \frac{\partial f}{\partial t} + \frac{\partial f}{\partial y} f(t, y)$$

According to Taylor's Theorem with Lagrange form for the remainder, for each $j = 0, 1, 2, \ldots, N - 1$ there is a number c_j in $[t_j, t_{j+1}]$ such that

$$y(t_{j+1}) = y(t_j) + y'(t_j)(t_{j+1} - t_j) + \tfrac{1}{2} y''(c_j)(t_{j+1} - t_j)^2 \tag{4}$$

or

$$y(t_{j+1}) = y(t_j) + f(t_j, y(t_j))h + \tfrac{1}{2} y''(c_j)h^2$$

Subtracting (3) from (4), we have that

$$y(t_{j+1}) - y_{j+1} = y(t_j) - y_j + [f(t_j, y(t_j)) - f(t_j, y_j)]h + \tfrac{1}{2} y''(c_j)h^2 \tag{5}$$

Taking the absolute value of each side of (5) and using the triangle inequality and the Lipschitz condition (2), we have that the global discretization errors $E_j = |y(t_j) - y_j|$ satisfy the conditions

$$E_{j+1} \leqq E_j + |f(t_j, y(t_j)) - f(t_j, y_j)|h + \tfrac{1}{2}|y''(c_j)|h^2$$
$$\leqq E_j + |y(t_j) - y_j|hL + \tfrac{1}{2}|y''(c_j)|h^2$$
$$\leqq (1 + hL)E_j + Kh^2$$

where L is a Lipschitz constant for f on R, and K is the largest value of $\tfrac{1}{2}|y''(t)|$ on the interval $t_0 \leqq t \leqq T$. [The existence of K is a consequence of the fact that a continuous function (in this case, $\tfrac{1}{2}|y''(t)|$) always has a maximum value on a closed interval.]

Thus we have shown that for $j = 0, 1, \ldots, N - 1$,

$$E_{j+1} \leqq \alpha E_j + Kh^2 \tag{6}$$

where $\alpha = 1 + hL$. Since E_0 vanishes, we have from (6) that

$$E_1 \leqq Kh^2, \quad E_2 \leqq \alpha E_1 + Kh^2 \leqq (\alpha + 1)Kh^2, \quad \ldots,$$

$$E_N \leqq (\alpha^{N-1} + \cdots + \alpha + 1)Kh^2 = \frac{\alpha^N - 1}{\alpha - 1} Kh^2 = \frac{\alpha^N - 1}{hL} Kh^2 \tag{7}$$

Using the fact that $\alpha = 1 + hL = 1 + [(T - t_0)L/N]$ and the fact that for any positive u, $\{(1 + u/N)^N\}_{N=1}^{N=\infty}$ is an increasing sequence which converges to e^u, we have from (7) that

$$E_N \leqq \frac{K}{L}[e^{(T - t_0)L} - 1]h \tag{8}$$

Thus Euler's algorithm is first order in the step size, where the constant M is the factor multiplying h in (8).

Since (8) implies that $E_N \to 0$, as $h \to 0$, we see that if exact arithmetic were possible, Euler's Method provides as good an approximation of $y(T)$ as desired, provided that h is chosen small enough. But as we know, exact arithmetic is not a practical possibility, and our next result indicates that round-off may have serious consequences for the convergence of Euler's Method.

We may carry out a similar analysis of the cumulative round-off errors when Euler's method is implemented by computer or calculator, or by hand. Suppose that at each step j there are various round-off and function evaluation errors whose totality is denoted by r_j. Thus we do not actually see the numbers y_j of the theoretical Euler method, but instead we see the numbers $\tilde{y}_j$, where

$$\begin{aligned} \tilde{y}_0 &= y_0 + r_0 \\ \tilde{y}_{j+1} &= \tilde{y}_j + f(t_j, \tilde{y}_j)h + r_j, \qquad j = 0, 1, 2, \dots, N-1 \end{aligned} \tag{9}$$

For simplicity, suppose that the round-off errors are all bounded in magnitude from above by a positive constant r. Define $\epsilon_j = |y(t_j) - \tilde{y}_j|$, $j = 0, 1, \dots, N$.

Now we are ready to prove the

Bound for Total Errors in Euler's Method.   With the definition of terms above, an upper bound for the total error ϵ_N in Euler's Method is given by

$$\epsilon_N \leqq e^{(T-t_0)L}\epsilon_0 + \frac{1}{L}\left(e^{(T-t_0)L} - 1\right)\left(Kh + \frac{r}{h}\right) \tag{10}$$

Proof. The estimate (10) is proven by an argument similar to that used above for the discretization errors E_j. First, let us put $\epsilon_j' = |y_j - \tilde{y}_j|$, where the y_j and $\tilde{y}_j$ are the Euler and "rounded" Euler iterates given by (3) and (9), respectively. Observe that $\epsilon_0' = \epsilon_0$. Now subtracting (9) from (3) and using the Triangle Inequality, we obtain the estimate

$$\epsilon_{j+1}' \leq \alpha\epsilon_j' + r, \qquad j = 0, 1, 2, \dots, N-1 \tag{11}$$

where $\alpha = 1 + hL$ and r is an upper bound for the magnitudes of the round-off errors r_j. Using (11) iteratively, we find that

$$\epsilon_j' \leqq \alpha^j \epsilon_0 + (\alpha^{j-1} + \alpha^{j-2} + \cdots + 1)r \tag{12}$$

By using the facts that $1 + \alpha + \cdots + \alpha^{j-1} = (\alpha^j - 1)/(\alpha - 1)$ and that $\{(1 + u/m)^m\}_{m=1}^{\infty}$ is an increasing sequence converging to e^u, we have from (12) that

$$\epsilon_N' \leq e^{(T-t_0)L}\epsilon_0 + \frac{r}{hL}\left(e^{(T-t_0)L} - 1\right) \tag{13}$$

Using the Triangle Inequality, we have that

$$\epsilon_N \leq E_N + \epsilon_N'$$

and hence using (8) and (13), the desired estimate follows.

The term r/h in (10) clearly shows the difficulty encountered when the step size h is reduced in an attempt to decrease the upper bound on the discretization error. The upper bound on the error due to round-off may actually increase as a result. In fact, the factor $Kh + r/h$ in (10) has its minimum value for positive h when $h = h_0 = (r/K)^{1/2}$. Thus the upper bound on the error ϵ_N decreases as the step size is reduced until the step size falls below h_0. Further reductions in h may lower the upper bound on the discretization error E_N, but would have the opposite effect on the upper bound for ϵ_N. Thus it is not easy to say just what step size would minimize the total of all the errors.

Adams–Bashforth Methods: Interpolating Polynomials

Suppose that $y(t)$, $t_0 \leq t \leq T$, is the solution of the initial value problem (1). Our goal, as always, is to approximate $y(T)$. First subdivide the interval $[t_0, T]$ into N subintervals $[t_j, t_{j+1}]$ of equal length $h = (1/N)[T - t_0]$ and then move from t_0 to t_1, from t_1 to $t_2, \ldots$, from t_{N-1} to $t_N = T$, approximating $y(t_j)$ by a number y_j, for each $j = 0, 1, \ldots, N$. In particular, we shall determine a polynomial $p(t)$ which approximates $f(t, y(t))$ on the interval $[t_j, t_{j+1}]$ and then use that polynomial to replace the exact algorithm

$$y(t_{j+1}) = y(t_j) + \int_{t_j}^{t_{j+1}} y'(t)\, dt = y(t_j) + \int_{t_j}^{t_{j+1}} f(t, y(t))\, dt \tag{14}$$

by the approximating algorithm

$$y_{j+1} = y_j + \int_{t_j}^{t_{j+1}} p(t)\, dt \tag{15}$$

Let K be a nonnegative integer. We shall take for $p(t)$ the unique *interpolating polynomial* of degree no more than K which passes through the $K + 1$ points,

$$(t_{j-K}, f_{j-K}), \ldots, (t_j, f_j), \qquad \text{where } f_j = f(t_j, y_j)$$

Thus $p(t)$ is determined by the numbers $y_{j-K}, \ldots, y_j$ calculated at the $K + 1$ steps preceding the $(j + 1)$st step.† For example, if K is 0, then $p(t)$ is the constant polynomial passing through the point (t_j, f_j). Hence $p(t) = f(t_j, y_j)$ and (15) becomes

$$y_{j+1} = y_j + \int_{t_j}^{t_{j+1}} f(t_j, y_j)\, dt = y_j + f(t_j, y_j)h$$

which is Euler's method. If $K = 1$, then $p(t)$ is the linear function through the two points, (t_{j-1}, f_{j-1}) and (t_j, f_j), and we have that

---

† The proof of the existence and uniqueness of the interpolating polynomial may be found, for example, in the book by J. M. Ortega and W. G. Poole, *Numerical Methods for Differential Equations* (London: Pitman, 1981).

$$p(t) = -\frac{1}{h}(t - t_j)f_{j-1} + \frac{1}{h}(t - t_{j-1})f_j \tag{16}$$

Using this $p(t)$ in (15), we obtain the *two-step Adams–Bashforth* algorithm

$$y_{j+1} = y_j + \int_{t_j}^{t_{j+1}} p(t)\, dt = y_j + \frac{h}{2}(3f_j - f_{j-1}) \tag{17}$$

For $K \geq 2$, the coefficients of the polynomial

$$p(t) = a_0 + a_1 t + \cdots + a_K t^K$$

are found by solving the $K + 1$ equations

$$f_i = a_0 + a_1 t_i + \cdots + a_K t_i^K, \qquad i = j - K, \ldots, j \tag{18}$$

for $a_0, a_1, \ldots, a_K$. Then $p(t)$ is integrated from t_j to t_{j+1} and (15) yields y_{j+1}. When $K = 2$ and 3 we obtain the respective *three- and four-step Adams–Bashforth algorithms*,

$$y_{j+1} = y_j + \frac{h}{12}(23f_j - 16_{j-1} + 5f_{j-2}) \tag{19}$$

$$y_{j+1} = y_j + \frac{h}{24}(55f_j - 59f_{j-1} + 37f_{j-2} - 9f_{j-3}) \tag{20}$$

Adams–Bashforth–Moulton Methods: Prediction and Correction

The idea is to use the Adams–Bashforth method based on $f_{j-K}, \ldots, f_j$ to *predict* a value $y_{j+1}^{(p)}$ and then use $f_{j-K+1}, \ldots, f_j, f_{j+1}$, where $f_{j+1} = f(t_{j+1}, y_{j+1}^{(p)})$, to determine a new interpolating polynomial $\tilde{p}(t)$, which via (15) will produce a *corrected* value y_{j+1}. We shall illustrate how these methods work for some low values of K.

If $K = 1$, then as we saw in (17), the Adams–Bashforth method predicts the value

$$y_{j+1}^{(p)} = y_j + \frac{h}{2}(3f_j - f_{j-1})$$

Then set $f_{j+1}^{(p)} = f(t_{j+1}, y_{j+1}^{(p)})$ and use (16) to obtain the linear function $\tilde{p}(t)$ through the points (t_j, f_j) and $(t_{j+1}, f_{j+1}^{(p)})$,

$$\tilde{p}(t) = -\frac{1}{h}(t - t_{j+1})f_j + \frac{1}{h}(t - t_j)f_{j+1}^{(p)} \tag{21}$$

Thus inserting $\tilde{p}(t)$ from (21) into (15) and integrating, we have the Adams–Bashforth–Moulton corrected value,

$$y_{j+1} = y_j + \int_{t_j}^{t_{j+1}} \tilde{p}(t)\, dt = y_j + \frac{h}{2}(f_j + f_{j+1}^{(p)})$$

as a straightforward integration of $\tilde{p}(t)$ shows. This two-step Adams–Bashforth–Moulton predictor–corrector method is comparable to Heun's method.

The most widely use multistep predictor–corrector method of this type uses the four-step Adams–Bashforth algorithm of Eq. (20) to calculate the predicted value $y_{j+1}^{(p)}$, then finds the interpolating cubic polynomial $\tilde{p}(t)$ through the four points

$$(t_{j-2}, f_{j-2}), \quad (t_{j-1}, f_{j-1}), \quad (t_j, f_j), \quad (t_{j+1}, f_{j+1}^{(p)})$$

and calculates the corrected value of y_{j+1} by the formula

$$y_{j+1} = y_j + \int_{t_j}^{t_{j+1}} \tilde{p}(t)\, dt$$

We have the "predictor"

$$y_{j+1}^{(p)} = y_j + \frac{h}{24}\left(55f_j - 59f_{j-1} + 37f_{j-2} - 9f_{j-3}\right) \tag{22}$$

which is then corrected by,

$$y_{j+1} = y_j + \frac{h}{24}\left(9f_{j+1}^{(p)} + 19f_j - 5f_{j-1} + f_{j-2}\right) \tag{23}$$

where $f_{j+1}^{(p)} = f(t_{j+1}, y_{j+1}^{(p)})$. We omit the lengthy but straightforward derivation of the coefficients in the formulas.

Complex-Valued Functions of a Real Variable

It is customary to write *complex numbers* in the form $a + ib$, where a and b are real numbers. Complex numbers are introduced to facilitate calculation of roots of polynomials of degree 2 or higher. We review below the basic features of the complex number system. Our interest, however, goes beyond the calculation of roots of polynomials. To develop an elegant and simple solution technique for linear differential equations with constant coefficients, we shall need to develop the notion of differentiability of complex-valued functions of a single real variable. At first glance this approach may seem like a far cry from our goal of studying real-world dynamical systems, but in fact complex-valued functions are so useful that they are now widely infused into the scientific literature.

If $z = a + ib$, the real number a is called the *real part* of z and is denoted by Re[z]; the real number b is called the *imaginary part* of z and is denoted by Im[z]. Real numbers are, of course, identified with those complex numbers whose imaginary parts vanish. The complex number $a - ib$ is called the *complex conjugate* of $z = a + ib$ and is denoted by $\bar{z}$. Complex numbers are added and multiplied as linear polynomials in the variable i with real coefficients, remembering to replace i^2 by -1. This definition extends to the complex numbers the corresponding operations for real numbers. Sometimes the representation $z = a + ib$ is called the *Cartesian* form of the complex number z since we can associate with z the point (a, b) in the Cartesian plane. For this reason we also sometimes refer to the collection of all complex numbers as the *complex plane*. Observe that $\overline{z + w} = \bar{z} + \bar{w}$ and $\overline{zw} = \bar{z}\bar{w}$.

Using polar coordinates for $z = a + ib$ (see Figure C.1), we see that $a = r \cos \theta$ and $b = r \sin \theta$, where $r = (a^2 + b^2)^{1/2}$, $0 \le \theta < 2\pi$; hence we have the alternative representation $z = r(\cos \theta + i \sin \theta)$, called the *polar* form for z. For any real number θ we shall write

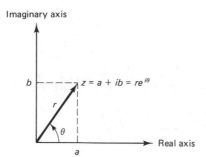

Imaginary axis

$z = a + ib = re^{i\theta}$

Real axis

Figure C.1 Complex plane.

| **(Euler's Formula)** $\quad e^{i\theta} \equiv \cos \theta + i \sin \theta$ | (1) |

and hence the polar form of a complex number z can be written as $z = re^{i\theta}$. The nonnegative real number r is called the *absolute value* or the *modulus* of z denoted also by $|z|$, and θ is called the *argument* or *angle* of z. Notice that $|z|$ agrees with the usual notion of absolute value if z is a real number. The complex number zero (denoted simply by 0) is given by $0 + i0$; also note that $z = 0$ if and only if $|z| = 0$. Observe that $z\bar{z} = |z|^2 = r^2 = a^2 + b^2$.

It is a consequence of the addition laws for the sine and cosine functions that for any complex numbers in polar form $z = re^{i\theta}$, $w = \rho e^{i\phi}$,

$$z^{-1} = \frac{1}{z} = \frac{e^{-i\theta}}{r} \qquad \text{if } z \neq 0,$$

and

$$zw = r\rho e^{i(\theta+\phi)}$$

Hence we use these identities to prove the following useful result.

| **DeMoivre's Formula.** For *any* integer n, positive, negative, or zero, $$z^n = r^n e^{in\theta} = r^n[\cos n\theta + i \sin n\theta]$$ |

Clearly, the advantage of the polar form for complex numbers is the simple geometric characterization of the product and quotient of complex numbers.

Just as for real numbers, absolute value can be used to measure the distance between complex numbers; indeed, the *distance* between the complex numbers z and w is defined to be $|z - w|$. Thus the sequence $\{z_n\}$ is said to *converge* if and only if there exists a z_0 in the complex plane such that $|z_n - z_0| \to 0$. Now suppose that f is a complex-valued function defined over an interval I on the real t-axis; that is, $w = f(t)$ is a complex number for each t in the interval I. We can define the notion of a derivative for such functions in a manner consistent with the usual notion of derivative for real-valued functions.

> **Derivative of a Complex-Valued Function.** We say that the complex-valued function $f(t)$ defined on the real t-interval I is differentiable at some t_0 in I if and only if there exists a complex number z_0 such that
>
> $$\left| \frac{f(t_0 + h) - f(t_0)}{h} - z_0 \right| \to 0 \qquad \text{as } h \to 0$$
>
> If this is so, z_0 is said to be the derivative of f at t_0 and is customarily denoted by $f'(t_0)$.

Note that if f is real-valued, the definition above reduces to the usual notion of differentiation for real-valued functions. Hence no confusion can arise if we do not know in advance whether a function is real or complex valued. Since differentiation of complex-valued functions of a real variable is defined in precisely the same formal manner as for real-valued functions of a real variable, it is not surprising that all the usual differentiation formulas (such as the Chain Rule, differentiation of sums, products, etc.) hold for these more general functions also.

We now present a useful computational device for computing derivatives of complex-valued functions of a real variable. For each t in I, let $\alpha(t)$ and $\beta(t)$ be the real and imaginary parts of $f(t)$, respectively. Then f can be uniquely written as $f(t) = \alpha(t) + i\beta(t)$, where $\alpha(t)$ and $\beta(t)$ are real-valued functions defined over I.

Differentiation Principle for Complex-Valued Functions. The complex-valued function of a real variable $f(t) = \alpha(t) + i\beta(t)$ is differentiable at t_0 if and only if $\alpha(t)$ and $\beta(t)$ are both differentiable at t_0. If $f'(t_0)$ exists, then

$$f'(t_0) = \alpha'(t_0) + i\beta'(t_0) \tag{2}$$

The proof of this result is a straightforward result of applying the definitions and will be omitted.

As one might suspect, antidifferentiation or integration of $f(t) = \alpha(t) + i\beta(t)$ is completely equivalent to performing the same operation on both $\alpha(t)$ and $\beta(t)$, and conversely.

Example C.1

The function $f(t) = e^{ikt} \equiv \cos kt + i \sin kt$, where k is a real constant, is differentiable for all t. Using (2) and the definition of e^{ikt},

$$(e^{ikt})' = -k \sin kt + ik \cos kt = ik(\cos kt + i \sin kt) = ike^{ikt} \tag{3}$$

Formula (3) is easy to remember because it says that the complex-valued function e^{ikt} follows the usual differentiation rule for exponential functions. Let $z = a + ib$ be a (fixed) complex number; then e^{zt} is defined to be a $e^{(a+ib)t} = e^{at}e^{ibt}$. Using (2) and (3) together with the product rule, we see that

$$(e^{zt})' = ae^{at}e^{ibt} + e^{at}ibe^{ibt} = (a + ib)e^{at}e^{ibt} = ze^{zt} \tag{4}$$

a result that is equally easy to remember.

Properties
of Power Series

Power series have some remarkable properties that make them especially well suited for certain types of applications. We shall list these properties below, without proof, in a form that allows easy reference.

(A) *Power Series*. The series $a_0 + a_1(x - x_0) + a_2(x - x_0)^2 + \cdots + a_n(x - x_0)^n + \cdots \equiv \sum_{n=0}^{\infty} a_n(x - x_0)^n$, where x_0 and the coefficients a_n are real, is a *power series based at* x_0. We write $\sum_{0}^{\infty} a_n x^n$ or $\sum a_n x^n$ for $\sum_{n=0}^{\infty} a_n x^n$ if there is no danger of confusion. We treat only series of the form $\sum_{n=0}^{\infty} a_n x^n$ since $\sum_{n=0}^{\infty} a_n(x - x_0)^n$ may be converted to that form by replacing $x - x_0$ by x.

(B) *Intervals of Convergence.* $\sum a_n x^n$ converges at a given point x if $\lim_{N \to \infty} \sum_{n=0}^{N} a_n x^n$ exists. Unless $\sum a_n x^n$ converges only at $x = 0$, there is a largest positive number R (possibly infinity) such that $\sum a_n x^n$ converges for all $|x| < R$. R is the *radius of convergence* and $J = (-R, R)$ is the *open interval of convergence*. The series diverges (i.e., does not converge) if $|x| > R$ and may or may not converge if $x = \pm R$. We shall only consider series for which $R > 0$ and J is nontrivial, and we shall only work with a series inside its open interval of convergence.

(C) *Ratio Test.* The *ratio test* for series is a useful way to determine the radius of convergence of a power series. For example, suppose that $\lim_{n \to \infty} |a_{n+1}/a_n| = a$. Then

$$\lim_{n \to \infty} \left| \frac{a_{n+1} x^{n+1}}{a_n x^n} \right| = a|x|$$

According to the ratio test, we have convergence if the limit $a|x| < 1$ and divergence if $a|x| > 1$. Thus the radius of convergence R is $1/a$ and $\sum a_n x^n$ converges for all $|x| < 1/a$. (Note that $R = \infty$ if $a = 0$.)

(D) *Calculus of Power Series.* The function $f(x) = \sum\limits_{n=0}^{\infty} a_n x^n$, x in J, is continuous and possesses derivatives of all orders. Moreover, $f'(x) = \sum\limits_{n=1}^{\infty} n a_n x^{n-1}$ and this series also has J as its open interval of convergence. In general, $f^{(k)}(x) = \sum\limits_{n=k}^{\infty} n(n-1) \cdots (n-k+1)a_n x^{n-k}$, where the series converges for x in J. In addition, the series

$$\sum_{0}^{\infty} \frac{a_n}{n+1} (x^{n+1} - b^{n+1})$$

converges to $\int_{b}^{x} \sum\limits_{0}^{\infty} (a_n s^n) \, ds$ for all b and x in J.

(E) *Identity Theorem.* If $\sum\limits_{n=0}^{\infty} b_n x^n = \sum\limits_{n=0}^{\infty} a_n x^n$ for all x in an open interval about 0, then $b_n = a_n$ for $n = 0, 1, 2, \ldots$. In particular, if $\sum\limits_{n=0}^{\infty} a_n x^n = 0$ for all x near 0, every coefficient a_n vanishes. This is called the *Identity Theorem* for power series.

(F) *Algebra of Power Series.* Convergent power series may be added or multiplied for all x in a common interval of convergence. The resulting *sum* or *product* series converges on that interval. Specifically,

(i) $\sum\limits_{n=0}^{\infty} a_n x^n + \sum\limits_{n=0}^{\infty} b_n x^n = \sum\limits_{n=0}^{\infty} (a_n + b_n)x^n$.

(ii) $\left(\sum\limits_{n=0}^{\infty} a_n x^n \right)\left(\sum\limits_{n=0}^{\infty} b_n x^n \right) = \sum\limits_{n=0}^{\infty} c_n x^n$,   where $c_n = \sum\limits_{k=0}^{n} a_k b_{n-k}$.

(iii) If $\sum\limits_{n=0}^{\infty} b_n x^n \neq 0$ on an interval, the quotient series,

$$\left(\sum_{n=0}^{\infty} a_n x^n \right) \Big/ \left(\sum_{n=0}^{\infty} b_n x^n \right) = \sum_{n=0}^{\infty} d_n x^n$$

converges on that interval. The coefficients d_n are given by complicated expressions in the a_n's and b_n's which we shall not write out.

(G) *Reindexing.* A series may be reindexed in any convenient way. For example,

$$\sum_{n=0}^{\infty} a_n x^n \equiv \sum_{n=2}^{\infty} a_{n-2} x^{n-2} \equiv \sum_{n=-3}^{\infty} a_{n+3} x^{n+3}$$

Writing out the first few terms of a series explicitly sometimes helps in determining the equivalence of indexing schemes. Sometimes it is convenient to introduce negative indexed coefficients, but it is understood that they have value zero. For example, we could rewrite $\sum_{n=0}^{\infty} a_n x^n + \sum_{n=2}^{\infty} b_{n-2} x^n$ either as $a_0 + a_1 x + \sum_{n=2}^{\infty} (a_n + b_{n-2}) x^n$ or as $\sum_{n=0}^{\infty} (a_n + b_{n-2}) x^n$, where it is understood that $b_{-2} = 0$ and $b_{-1} = 0$.

(H) *Real Analytic Functions*. A function $f(x)$ which has a *Taylor series expansion*,

$$\sum_{n=0}^{\infty} \frac{f^{(n)}(x_0)}{n!} (x - x_0)^n$$

convergent to $f(x)$ on an interval $(x_0 - R, x_0 + R)$, is said to be *real analytic* on that interval, or, alternatively, at x_0. If $x_0 = 0$, the Taylor series is said to be the *Maclaurin series* of f. By the Identity Theorem, a real analytic function f at x_0 has a unique power series in powers of $x - x_0$; hence this must be the Taylor (or Maclaurin) series of f. The definitions above extend to functions of several variables. For example, the Maclaurin series of $f(x, y)$ is

$$\sum_{n=0}^{\infty} \sum_{i=0}^{\infty} \binom{n}{i} \frac{1}{n!} \frac{\partial^n f(0, 0)}{\partial x^i \partial y^{n-i}} x^i y^{n-i}$$

where $\binom{n}{i}$ is the binomial coefficient, $n!/i!(n-i)!$.

(I) Some Maclaurin series are used so often that it is convenient to list them here.

(*Sine*)	$\sin x = x - \dfrac{x^3}{3!} + \dfrac{x^5}{5!} - \cdots + (-1)^n \dfrac{x^{2n+1}}{(2n+1)!} + \cdots,$	all x
(*Cosine*)	$\cos x = 1 - \dfrac{x^2}{2!} + \dfrac{x^4}{4!} - \cdots + (-1)^n \dfrac{x^{2n}}{(2n)!} + \cdots,$	all x
(*Exponential*)	$e^x = 1 + x + \dfrac{x^2}{2!} + \cdots + \dfrac{x^n}{n!} + \cdots,$	all x
(*Geometric*)	$\dfrac{1}{1-x} = 1 - x + x^2 + \cdots + (-1)^n x^n + \cdots,$	$|x| < 1$
(*Binomial*)	$(1+x)^a = 1 + ax + \dfrac{a(a-1)}{2} x^2 + \cdots$ $+ \dfrac{a(a-1) \cdots (a-n+1)}{n!} x^n + \cdots,$	$|x| < 1$

(J) *Error Estimate*. Let $\alpha_n > 0$, $n = 0, 1, 2, \ldots$. The *alternating series*, $\sum_{0}^{\infty} (-1)^n \alpha_n$ (not necessarily a power series) converges if $\lim_{n \to \infty} \alpha_n = 0$ and $\alpha_n \geqq \alpha_{n+1}$ for every n. The *error* in using the partial sum $\sum_{0}^{N} (-1)^n \alpha_n$ to approximate the true sum $\sum_0^\infty (-1)^n a_n$ has magnitude no greater than that of the first term omitted:

$$|\text{error}| \leq a_{N+1}$$

(K) *Taylor's Formula with Remainder.* Let $f(x)$ belong to $C^{N+1}(x_0 - R, x_0 + R)$. Then for all x, $x_0 - R < x < x_0 + R$,

$$f(x) = \sum_{n=0}^N \frac{f^{(n)}(x_0)}{n!}(x - x_0)^n + \frac{f^{(N+1)}(c)}{(N+1)!}(x - x_0)^{N+1} \tag{1}$$

where c is some point between x and x_0, and the last term on the right is the *Lagrange form of the remainder*.

The following examples illustrate the properties given above.

Example D.1

$\sum_0^\infty n^5(x^n/2^n)$ converges for $|x| < 2$ by the ratio test since

$$\lim_{n\to\infty} \left| \frac{(n+1)^5 x^{n+1}/2^{n+1}}{n^5 x^n/2^n} \right| = |x| \lim_{n\to\infty} \frac{(n+1)^5}{2n^5} = \frac{|x|}{2} \lim_{n\to\infty} (1 + 1/n)^5 = \frac{|x|}{2}$$

and $|x|/2 < 1$ for all $|x| < 2$.

Example D.2

$$\sum_0^\infty \frac{x^n}{n+3} + \sum_2^\infty x^n = \frac{1}{3} + \frac{1}{4}x + \sum_2^\infty \left(\frac{1}{n+3} + 1 \right) x^n \qquad \text{(reindexing and adding)}$$

Example D.3

$$-\frac{1}{(1+x)^2} = \frac{d}{dx}\left(\frac{1}{1+x} \right) = \frac{d}{dx}(1 - x + x^2 + \cdots + (-1)^n x^n + \cdots)$$

$$= -1 + 2x + \cdots + (-1)^n n x^{n-1} + \cdots$$

if $|x| < 1$ (differentiating a geometric series within the interval of convergence).

Example D.4

$$\left(\sum_{n=0}^\infty n x^n \right)\left(\sum_{n=0}^\infty \frac{x^n}{n+1} \right) = \sum_{n=0}^\infty \left(\sum_{k=0}^n k \frac{1}{n-k+1} \right) x^n$$

$$= \quad 0 \quad + \left(\sum_{k=0}^1 \frac{k}{1-k+1} \right)x + \left(\sum_{k=0}^2 \frac{k}{2-k+1} \right)x^2 + \cdots$$

$$= x + \frac{5}{2}x^2 + \cdots$$

The product series converges to the product of the two factor series for $|x| < 1$ since each factor series converges for $|x| < 1$.

Answers and Hints
for Selected
Even-Numbered Problems

CHAPTER 1

Section 1.1

2. (b), (c), (d) are linear.
4. (a) $ty' = y$. (b) $y'' = y$. (c) $ty' = 2y$. (d) $(y')^2 = 9ty$. (e) $y'' - 4y' + 4y = 0$.
(f) $y'' + 2y' + 5y = 0$. (g) $y = ty' + \sin y'$.

Section 1.2

2. 93%   **4.** 15.34 months.   **6.** 7.40 s.   **8.** 278.5 days.
10. Doubling time is $69.315/r$; 50% increase time is $40.547/r$. In both cases the bank keeps money longer than needed to earn the desired return.

Section 1.3

2. $y = +\sqrt{1 + t^2}$.   **4.** 4103 years old.

CHAPTER 2

Section 2.1

2. (b) $y = ct + \ln c$.   (c) $y = ty' - (y')^2/4$, $y = t^2$ is a solution that is not a straight line. The equations of the tangent lines to the graph of the function and the equation of the function all define solutions of the differential equation.

A33

4. (a) $y' = y$, $y'' = y' = y$; slope is positive and concavity is upward if y is positive, negative and downward if y is negative.

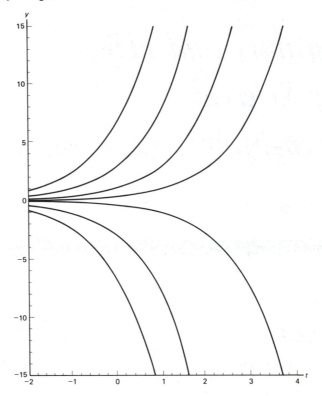

(b) $y' = 2y/t$, $y'' = 2y'/t - 2y/t = 2y/t^2$. Solution curves rise (fall) in first and third quadrants (second and fourth); curves turn up (down) where $y > 0$ ($y < 0$).

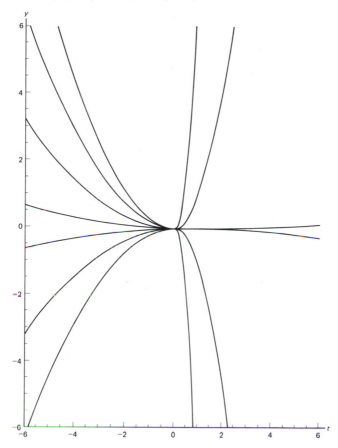

(c) $y'' = (y + 3)(y - 2)(2y + 1)$. Solution curves rise (fall) where $y < -3$ or $y > 2$ $(-3 < y < 2)$; turn up (down) where $y > 2$ or $-3 < y < -\frac{1}{2}$ $(y < -3$ or $-\frac{1}{2} < y < 2)$.

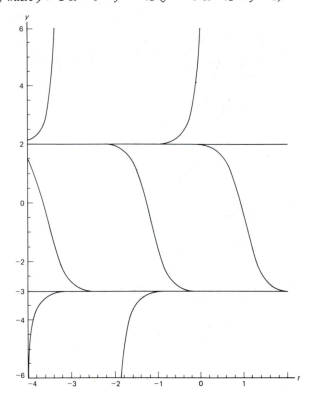

(d) $y' = 2t - y$, $y'' = 2 - y' = 2 - 2t + y$. Thus slope is positive if $y < 2t$, concavity is upward if $y > 2t - 2$. Similarly, slope is negative if $y > 2t$ and concavity is downward if $y < 2t - 2$.

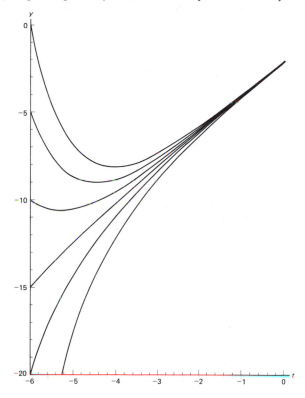

(e) $N'' = r^2N (1 - N/K)(1 - 2N/K)$. Solution curves rise (fall) where $0 < N < K$ ($N < 0$ or $N > K$) and turn up (down) where $0 < N < K/2$ or $N > K$ ($N < 0$ or $K/2 < N < K$). Inflection points on the horizontal line $N = K/2$.

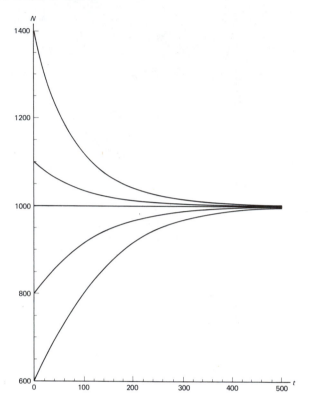

(f) $y'' = yt(t - 2)$. Solution curves rise (fall) where $t < 1$ and $y > 0$ or where $t > 1$ and $y < 0$ ($t < 1$ and $y < 0$ or where $t > 1$ and $y > 0$); turn up (down) where $y > 0$ and $t > 2$, or $y > 0$ and $t < 0$, or $y < 0$ and $0 < t < 2$. Inflection points are at $t = 0, 2$, while $y \equiv 0$ is a solution. Elsewhere solution curves turn down.

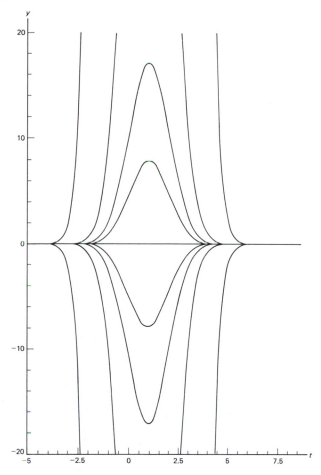

(g) $y'' = y - t^2 - 2t$. Solution curves rise (fall) if $y > t^2$ ($y < t^2$), turn up (down) if $y > t^2 + 2t$ ($y < t^2 + 2t$). Inflection points on the parabola $y = t^2 + 2t$.

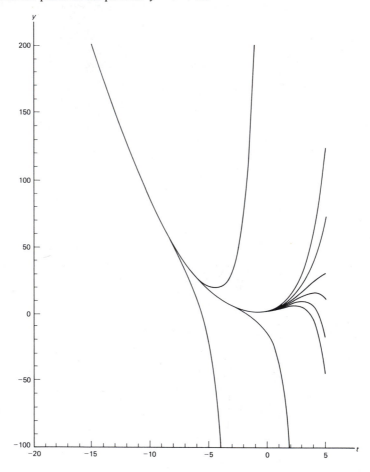

6. (a) $y' = 2t$
$y(2) = 2.$

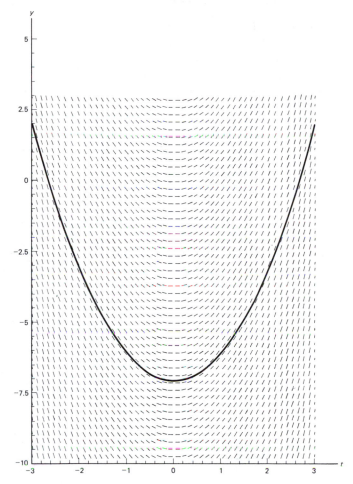

(b) $y' = t + y$
    $y(0) = 2.$

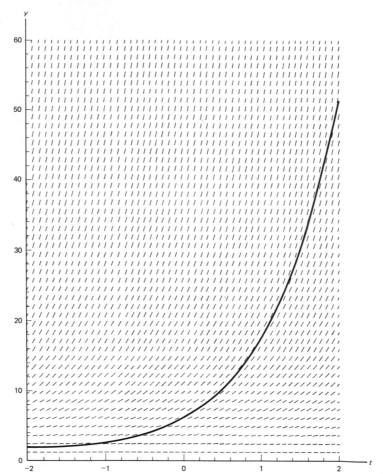

(c) $y' + ty = t^2$
$\quad y(1) = 1.$

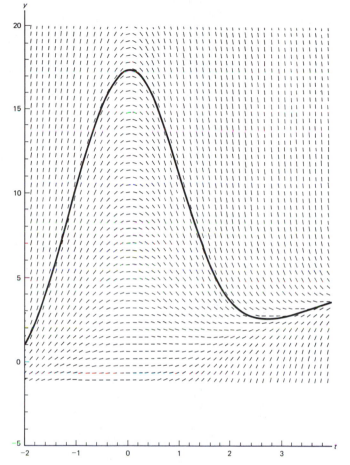

(d) $ty' = 3y$
$y(0) = 0.$

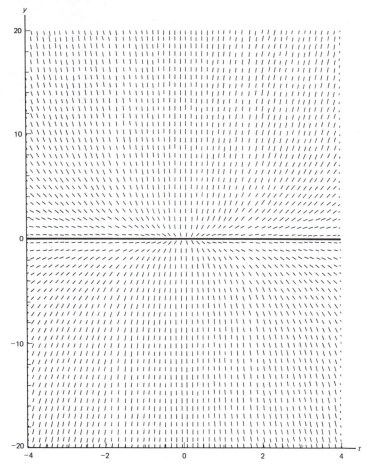

Section 2.2

2. (a) $f = (t^2 + y^2)(2ty)^{-1}$ is undefined at $t_0 = 0$, $y_0 = 1$.   (b) The theorem does not apply since $f = 2yt^{-1}$ is undefined at $t_0 = 0$, $y_0 = 0$. However, $y_1 = t^2$ and $y_2(t)$ are both solutions.

4. $y(t) \equiv 1$ is a solution, as is $y(t) = \begin{cases} -1, & t < -\pi \\ \cos t, & -\pi \le t \le 0 \\ 1, & t > 0. \end{cases}$

6. (a) The Existence and Uniqueness Theorem holds in the region R defined by $|t| < 2$, $|y| < 2$. The closed region S defined by $|t| \le 1$, $|y| \le 1$ is a subset of R and contains the initial point $(0, 0)$. Then use the Extension Principle.   (b) The rate y' is imaginary for $|y| > 2$.

Section 2.3

2. *Hint*: It is shown in Appendix B that the upper bound on the round-off error begins to increase as h decreases when $h = (r/K)^{1/2}$, where r is an upper bound for the round-off error at each step and K is an upper bound on $\frac{1}{2}|y''(t)|$, $0 \le t \le 1$. Since $y_{j+1} = y_j - y_jh$, each step involves just two arithmetic operations. Estimate r according to the number of decimal points carried by the computer or calculator you are using. Since $y''(t) = e^{-t}$, one may set $K = \frac{1}{2}$. Thus one may estimate $(r/K)^{1/2}$ for a given computer and compare the actual value of h where the round-off errors generated by the computer may begin to increase the total error.

4. (a) $y_N(T) = \left(1 + \dfrac{2T}{N}\right)^N \to e^{2T}$ as $N \to \infty$.   (b) $y_N(T) = \left(1 - \dfrac{T}{N}\right)^N \to e^{-T}$ as $N \to \infty$.

Section 2.4

2. Let $y_0 = 1$, $h = 0.2$, $f_n = -y_n^3 + (nh)^2$.
(a) Use (3): $y_5 \approx 0.812466$.   (b) Use (4): $y_5 \approx .820602$.   (c) Use (6) and (7): $y_5 \approx .817324$.

CHAPTER 3

Section 3.1

2. (a) $y = -\frac{1}{2} + Ce^{t^2}$.   (b) $y = 1 + e^{2t} + Ce^t$.   (c) $y = 1 + Ce^{\cos t}$.
(d) $y = e^{-t} + Ce^{-3t/2}$.   (e) $y = \frac{1}{2} + Ce^{-t^2/2}$.   (f) $y = 1 + t^2 e^{-t}/2 + Ce^{-t}$.
(g) $y = e^{-\sin t}\left[C + 7\displaystyle\int^t e^{\sin s}ds\right]$.   (h) $y = \exp\left(-(\cosh t)/7\right)\left[C + \frac{1}{7}\displaystyle\int^t \cosh s \, e^{(\cosh s)/7} \, ds\right]$.

4. $y = Ct + \dfrac{1}{n}\, t^{n+1}$, $n > 0$, $y = Ct + t \ln t$, $n = 0$.   **6.** $2085.   **8.** $z = \exp\left(-\frac{1}{2}\displaystyle\int^t a(s) \, ds\right)$.

Section 3.2

2. $|y(t)| \to \infty$ as $t \to 0$.   **4.** Equation cannot be normalized on any interval containing $\pi/2$.
6. (a) 12 lb.   (b) 85.3 lb.   (c) 118 lb.
8. (a) $C = C_0 + B_0(1 - e^{-K_B t}) + \dfrac{A_0}{K_B - K_A}\, [K_B(1 - e^{-K_A t}) - K_A(1 - e^{-K_B t})]$.   (b) 226 g.

Section 3.3

2. In 48 years.   **4.** (a) 7.5 years.   (b) 25 years.   (c) 30 years.
6. (a) If $H > rK/4$, the roots of the quadratic in P, $r(1 - P/K)P - H$, are complex and the quadratic is negative for all P.
(b)

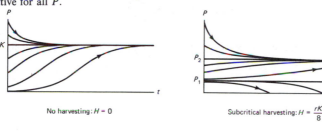

No harvesting: $H = 0$ Subcritical harvesting: $H = \dfrac{rK}{8}$

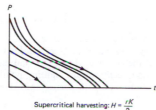

Supercritical harvesting: $H = \dfrac{rK}{2}$

(c) If $0 < H < rK/4$, rate function has two positive roots by the quadratic formula (roots are equal if $H = rK/4$).
8. (a) $y = g + 1/z$. Differentiate to get *ODE* in z; using the fact that y and g satisfy the Riccati equation.   (b) $y = 1 + (1 - t + Ce^{-t})^{-1}$.   (c) $y = e^t + (Ce^{-3t} - e^{-t}/2)^{-1}$.
(d) $y = t + t(C - t^5/5)^{-1}$.   (e) $P = 1200 + 800 (800Ce^{0.005t} - 1)^{-1}$.
(g) $y = (1 - tCe^{-t})(1 - Ce^{-t})^{-1}$.

Section 3.4

2. $C_0 r_1 \leq r_0/10$.   **4.** $y = te^{1/t}$ is a solution which $\to \infty$ as $t \to \infty$. Principle does not apply since $p(t)$ $\equiv t^{-2} \geq p_0 > 0$, all $t \geq 1$, for no positive p_0.
6. (a) $|y|e^{y^2/2} = Ke^{-t}$. As $t \to \infty$, right side $\to 0$. Hence $|y| \to 0$.   (b) Minimum of $1 - y/(1 + y^2)$ is $\frac{1}{2}$. Thus $y' \geq \frac{1}{2}$.   (c) Since $y(t) \geq t/2 + C$, $y \to \infty$ as $t \to \infty$.

CHAPTER 4

Section 4.1

2. (a) $y = (1 - \ln x)^{-1}$, $0 < x < e$.   (b) $y - \ln (y + 1) = x^2/2 + 2 \ln x + e^{1/2} - 2$.   (c) $x = \frac{1}{a} \ln \left| \frac{by}{a - by} \right|$.   (d) $\ln y = 2 - e^{-x}$.   (e) $y = 2 + \ln (1 + t^3)$, $t > -1$.   (f) $y^2 = -x^2 + 5$.
(g) $y = \ln 2 - \ln (1 + e^{-x^2})$
4. (a) $y = (2\pi B)^{-1/2} \exp ((x - A)^2/2B)$.   (b) $y = \frac{1}{C} e^{-x/C}$ (c) $y = x^n e^{-x/C}/C^{1+n}n!$.
6. (a) Loss rate is proportional to numbers on both sides.   (b) Divide y' by x' and integrate.
(c)

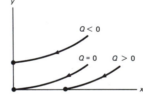

(d) If $Q > 0$, conventional force x wins. If $Q = 0$, neither wins. If $Q < 0$, guerrilla force wins.
(e) Immediate from $Q > 0$.

Section 4.2

2. $4y^2 = -x^4 + Cx^2$.
4. (a) $x = \tan \left(\frac{x + y}{2} \right) + C$.   (b) Let $z = 2x + y$ and show that $dz/dx = \frac{3z + 5}{2z + 3}$.
(c) Let $z = x + 2y$ and find a differential equation for z.
(d) Let $u = e^{-y}$. Then $u' = -e^{-y}y'$ and $-u' + u = xe^x$. Solve this linear first-order equation for u.
6. $r = ae^{-\theta}/\sqrt{2}$ if the origin is at the center of the square and, at $\theta = 0$, one vertex is at $r = a/\sqrt{2}$.
8. Coast Guard boat travels 3 miles directly toward sighting point and then follows path $r = e^{\theta/\sqrt{8}}$, where sighting point is the origin and initial path corresponds to $\theta = 0$.

Section 4.3

2. Use Problem 1 for (a) $-$ (d); (e) and (g) follow from (b) and (d). For (f) let α be a scalar, expand $\|u + \alpha v\|^2$ in α, and set $\alpha = -u \cdot v/\|v\|^2$.
4. (a) Orthogonal if $3w_1 - w_2 + 2w_3 = 0$, in plane if $w_1 - 7w_2 - 5w_3 = 0$.   (b) The unique α and w are $\frac{1}{2}$ and $2\mathbf{i} - \frac{3}{2}\mathbf{j} + \frac{5}{4}\mathbf{k}$.
6. 22 mi north, 7.88 mi east, 1.39 mi above.   **8.** $mx'' = -mg + nRT/x$.

Section 4.4

2. (a) $y = (-3t/2 + 1)^{2/3}$.   (b) $y^2 = 12e^{2t} + 4$.   (c) $y = g\,(1 - t - e^{-t}) + h$.
4. (a) $v_s = gt$, $v_f = mgk_f^{-1}(1 - \exp(-k_f t/m))$, $v_r = (mg/k_r)^{1/2} \tanh((gk_r/m)^{1/2}t)$. Then use Maclaurin expansions in t.
(b) From the Maclaurin expansions show that $v_f = gt - k_f gt^2/2m + $ [higher order in t], while $v_r = gt - k_r g^2 t^3/3m + $ [higher order in t]. Show that for t close to 0, $v_s > v_r > v_f$, but v_r and v_f are close to $v_s = gt$.   **6.** $\tilde{v}_0 = \sqrt{\tilde{G}M/R}$.

Section 4.5

2. All are exact but (d).   (a) $x^2 + y^2 + 4xy = C$.   (b) $x^2y + y^2 - x^3 = C$.
(c) $2x \cos 2y - 2x^3y^2 + \sin 2y = C$.   (d) $x^2 - y^2 = Cx$.   (e) $x^2 - 3xy + y^2 = C$.
(f) $x^2y^2 + 2xy + x = C$.
4. (a) $\alpha = 3$, $2x^3y + x^2y^2 = C$.   (b) $\alpha = 1$, $x^2 + e^{2xy} = C$.
6. (a) $y \cos^2 x + 3 \sin x = C$.   (b) $6xy^4 - y^6 = C$.
10. (a) See 8(a): $e^{y/x} + y^2 = C$.   (b) $a = -2$, $b = 1$, $xy^3 - x^{-1}y^2 = c$.

CHAPTER 5

Section 5.1

2. (a) $y = C_1 \ln t + C_2$.   (b) $y = a + t^2/2$, $b = 0$.   (c) The normalized form of $P(D)y = 2t$ does not have coefficients continuous at the origin.
4. (a) $y = e^{2t} - e^t$.   (b) $y = (e \cos 1)e^{-t} \cos t + (e \sin 1)e^{-t} \sin t + t - 1$.
8. (a) $y = C_1 e^{-t} + C_2 e^{3t}$; C_1, C_2 in **R**.   (b) $y = C_1 e^{-2t} \cos 3t + C_2 e^{-2t} \sin 3t$; C_1, C_2 in **R**.
(c) $y = C_1 e^{\sqrt{2}(1+i)t} + C_2 e^{-\sqrt{2}(1+i)t}$; C_1, C_2 in **C**.   (d) $y = C_1 e^{-t} + C_2 e^{-it}$; C_1, C_2 in **C**.

Section 5.2

2. (a) $y = C_1 e^{2t} + C_2 e^{-t}$; C_1, C_2 in **R**.   (b) $y = C_1 e^{2t} \cos t + C_2 e^{2t} \sin t$; C_1, C_2 in **R**.
(c) $y = C_1 e^{-t} + C_2 t e^{-t}$; C_1, C_2 in **C**.   (d) $y = C_1 e^{(-1+i)t} + C_2 e^{(-1-i)t}$; C_1, C_2 in **C**.
(e) $y = C_1 e^{(1+i\sqrt{7})t/2} + C_2 e^{(1-i\sqrt{7})t/2}$; C_1, C_2 in **C**.   (f) $y = C_1 t^2 + C_2 t$; C_1, C_2 in **R**.
4. (a) $y = Ce^{-t}$; C in **R**.   (b) $y = 0$.
6. (a) $y = e^{-t} \sin t + e^{-t} \cos t$.   (b) $y \equiv 0$.   (c) $y = \frac{3}{5}e^{-3(t+1)} + \frac{2}{5}e^{2(t+1)}$.   (d) $y = -\frac{1}{5}e^{-t} \cos t - \frac{2}{5}e^{-t} \sin t + \frac{1}{5}e^t$.   (e) $y = -\frac{1}{2}t^2 + \frac{3}{2}$.
8. Wronskian and Independence Theorem applies only to equations of the form $(D^2 + a(t)D + b(t))y = 0$, where $a(t)$, $b(t)$ are in $C^0(I)$; in this case $I = \mathbf{R}$, but $a(t)$ and $b(t)$ are not continuous on I.
12. (a) *Hint*: The roots of the characteristic polynomial are $-a/2 \pm (a^2 - 4b)^{1/2}/2$. Consider separately the cases $a^2 > 4b$, $a^2 = 4b$, $a^2 < 4b$.
14. (c) Let z be as in 13(b); show that uz satisfies ODE for v in 14(b).

18. (a). *Hint*: Let y_1, y_2 be independent solutions, α, β, $\alpha < \beta$, consecutive zeros of y_2. Use Problem 17 and show that $W[y_1, y_2](t)$ has a fixed sign, $\alpha \leq t \leq \beta$, and that $y_2'(\alpha)$ and $y_2'(\beta)$ have opposite signs. Show that $y_1(\alpha)$ and $y_1(\beta)$ must have opposite signs, and conclude that y_1 has a zero between α and β. Then interchange the roles of y_1 and y_2 in the argument.

Section 5.3

2. (a) $\operatorname{Re}[e^{2it} + ie^{it}]$.   (b) $\operatorname{Re}[\frac{1}{2}t(1 - e^{2it})]$.   (c) $\operatorname{Re}[t^2ie^{-2it} - \frac{1}{2}(1 + t)(1 + e^{2it})]$.
(d) $\operatorname{Re}\left[\frac{i}{4}\left(3e^{-it} - e^{-3it}\right)\right]$.

4. (a) $y = k_1 \cos t + k_2 \sin t + t \sin t + \cos t \ln |\cos t|$.   (c) $y = k_1 e^{-4t} + k_2 e^{3t} - \frac{1}{10} e^{-2t} - \frac{1}{6} - \frac{1}{6} e^{2t}$.   (e) $y = k_1 \cos 2t + k_2 \sin 2t + (\frac{1}{4}t^2 - \frac{1}{8}) + \frac{3}{64} (8t^2 \sin 2t - 3 \sin 2t + 4t \cos 2t)$.
6. (a) $y = e^{2t} - \frac{1}{2}e^{-2t} + (-\frac{1}{2} + 2t)$.   (b) $y = 4 \cos 3t + \sin 3t + (9t^2 - 2) - 2 \cos 4t$.
(c) $y = -2 \cos t - 4 \sin t + 2e^{2t}$.   (d) $y = 2e^t - e^{-t} - 2e^{-t} \sin t$.
(e) $y = 8e^{2t} - 24e^t + 4t^2 + 12t + 14 + 2e^{-t}$.
8. $y = -\frac{4}{5}e^{2t} + (-\frac{2}{5} - \frac{1}{10} (\cos 1 - 3 \sin 1)) e^{t-1}$.

Section 5.4

2. (a) $q = 10^{-3} + (r_2 e^{r_1 t} - r_1 e^{r_2 t})/9r_1 r_2(r_2 - r_1)$ where r_1, $r_2 = -25 \pm \frac{1}{3}\sqrt{4625}$. $I(t) = q'(t)$.
4. $q(t) = -\frac{9}{26}e^{-t} \cos 3t - \frac{7}{26}e^{-t} \sin 3t + \frac{1}{26} (9 \cos 2t + 6 \sin 2t)$.

Section 5.5

2. (a) ± 2 rad/sec.   (b) ± 4 rad/sec^2.   **4.** $mx'' = \frac{\alpha}{b - x} - kx$.

8. $K(t, s) = e^{(s-t)/10} \sin (t - s)$. $x_p = \int_0^t K(t,s)f(s) \, ds = \sum_{j=0}^{n-1} \int_{2j\pi}^{(2j+1)\pi} K(t, s) \, ds + \int_{2n\pi}^t K(t, s) \, ds$, if $2n\pi \leq t \leq (2n + 1)\pi$, but the sum only if $(2n - 1)\pi \leq t \leq 2n\pi$.

Section 5.6

6. (a) Spring constant $k = 773$ N/m; damping constant $c = 0.307$ N-s/m.
(b) $B(\omega^*) = A/[(k - \omega^{*2})^2 + \omega^{*2}c^2]^{1/2}$ is amplitude of the response at resonant frequency $\omega^* = (k - c^2/2)^{1/2}$, where k and c are as in part (a).

CHAPTER 6

Section 6.1

2. (a) Not a subspace.   (b) Not a subspace.   (c) A subspace.   (d) Not a subspace.
4. *Hint*: Let the reals $a_0, \ldots, a_n, b_1, \ldots, b_n$ be such that $a_0 + \sum_1^n (a_k \cos kt + b_k \sin kt) \equiv 0$.

Multiply by $\sin jt$ or $\cos jt$ and integrate from $-\pi$ to π.
6. (a) $\{y = a(-2t + 1):$ all a in $\mathbf{R}\}$ is the solution set. It is a subspace of dimension one.
(b) $y = t^2 - t$ is the only solution. It is not a subspace.   (c) $y = t^2 - 2at + a$, a an arbitrary constant, are all solutions. The solution set is not a subspace.

Section 6.2

6. (a) 156.   (b) 118.

Section 6.3

2. (a) $y(t) = 12i - 6t - it^2 + C_1 e^{it} + C_2 t e^{it} + C_3 t^2 e^{it}$.
(b) $y(t) = (C_1 + C_2 t + C_3 t^2)e^{(2+i)t} + (C_4 + C_5 t + C_6 t^2)e^{(2-i)t} + (C_7 + C_8 t)e^{it} + (C_9 + C_{10}t)e^{-it}$.
(c) $y(t) = (C_1 + C_2 t)e^{3t} + (C_3 + C_4 t)e^t + (C_5 + C_6 t + C_7 t^2)e^{(-1-i)t} + (C_8 + C_9 t + C_{10}t^2)e^{(-1+i)t}$.

4. (a) $y(t) = \frac{1}{3}e^{-t} + e^{t/2}\left(\dfrac{1}{\sqrt{3}}\sin\left(\dfrac{\sqrt{3}t}{2}\right) - \frac{1}{3}\cos\left(\dfrac{\sqrt{3}t}{2}\right)\right)$.

(b) $y(t) = \frac{1}{8}e^{-t} + \frac{7}{8}e^t \cos 2t + \frac{1}{8}e^t \sin 2t$.   (c) $y(t) = (\frac{17}{25} + \frac{3}{5}t)e^{-2t} + \frac{19}{25}\sin t - \frac{17}{25}\cos t$.

6. (a) $c_1 e^t + c_2 e^{-t} + c_3 e^{-2t} + \frac{1}{6}\displaystyle\int_{t_0}^{t} [2e^{-2(t-s)} - 3e^{-(t-s)} + e^{t-s}]f(s)\,ds$.

(b) $(2 - 3k)e^{-t} + ke^t + (2k - 1)e^{-2t} + \frac{1}{6}\displaystyle\int_0^t [2e^{-2(t-s)} - 3e^{-(t-s)} + e^{t-s}]f(s)\,ds$, for all constants k.

(c) $a(t^{-1} - 1) + b(t - 1) + \displaystyle\int_1^t [\frac{1}{2}t^{-1} + \frac{1}{2}ts^{-2} - s^{-1}]f(s)\,ds$, for all values of the constants a, b.

8. *Hint:* Differentiate the dependency equation $(n - 1)$ times and form the matrix equation $W[y_1, \ldots, y_n]C = 0$, where C is column vector $[c_1\ c_2 \ldots c_n]^T$, and use Singularity Theorem.
10. *Hint:* Use mathematical induction on the number of distinct constants $\lambda_1, \lambda_2, \ldots, \lambda_m$.
12. *Hint:* Use the Abel formula for each Wronskian (see Problem 11).

Section 6.4

2. (a) Range $A = \text{span}\left\{\begin{bmatrix}2\\1\\0\end{bmatrix}, \begin{bmatrix}1\\0\\1\end{bmatrix}\right\}$.   (b) Null space of $A = \text{span}\left\{\begin{bmatrix}5\\-3\\0\\4\end{bmatrix}, \begin{bmatrix}-1\\-1\\2\\0\end{bmatrix}\right\}$.

(c) Rank $= 2$, nullity $= 2$, rank $+$ nullity $=$ dimension of domain $= 4$.
6. If $y = [a\ \ b]^T$, then $Ax = y$ has a solution $x = [\frac{1}{4}(a + b + 5)\ \ \frac{1}{4}(a - 3b - 7)\ \ 1]^T$. So there is a solution of $Ax = y$ for every y in $\mathbf{R}^2$. Also $N(A^T) = (0, 0)$, and hence is trivial, verifying the Fredholm Alternative.
8. (a) *Hint:* The operator M is itself an invertible operator on W.
10. (a) $\lambda = -n^2\pi^2$, $n = 1, 2, 3, \ldots$, and to each λ_n corresponds a solution set spanned by $\{\sin (\sqrt{-\lambda}t)\}$.
(b) $\lambda = (n + \frac{1}{2})^2$, $n = 0, 1, 2, \ldots$, and to each λ_n corresponds a solution set spanned by $\{\cos \sqrt{\lambda}t\}$.
(c) $\lambda = n^2$, $n = 0, 1, 2, \ldots$. The solution set corresponding to $\lambda = 0$ is spanned by the constant function $y_0 \equiv 1$. The solution set for λ_n, $n = 1, 2, \ldots$ is $\{c_1 \cos nt + c_2 \sin nt\colon$ for arbitrary constants $c_1, c_2\}$.

12. (a) $G(t, s) = \dfrac{(t - s)^{n-1}}{(n - 1)!}$.   (b) $G(t, s) = (t^3 - s^3)/(3t^2)$ is the Green's kernel for the operator

$Q(D) = D^2 + \dfrac{2}{t} D - \dfrac{2}{t^2}$.   (c) $G(t, s) = \frac{1}{6}e^{t-s} - \frac{1}{2}e^{-(t-s)} + \frac{1}{3}e^{-2(t-s)}$.

CHAPTER 7

Section 7.1

2. (a) $(1 + e^{-\pi s})(1 + s^2)^{-1}$, $s_0 = 0$.   (b) $e^{-s}(1 + s)/s^2$, $s_0 = 0$.   (c) $(1 - e^{-s} - se^{-s})/s^2$, $s_0 = 0$.

Section 7.2

2. (a) $y = \frac{1}{5}e^{3t} + \frac{4}{5}e^{-2t}$.   (b) $y = \frac{3}{2}\sin t - \frac{1}{2}t\cos t$.   (c) $y = e^t \sin t$.
(d) $y = \frac{1}{9}e^t + \frac{8}{9}e^{-2t} + \frac{8}{3}\cdot te^{-2t}$.

(e) $y = \begin{cases} e^t(1-t), & t < 1, \\ (1 - 2e^{-1})e^t + te^{t-1}, & t \geq 1. \end{cases}$

(f) $y = \begin{cases} \frac{1}{3}e^{-3t} + e^t - \frac{1}{3}, & t < 1 \\ c_1 e^{-3t} + c_2 e^t, t \geq 1, c_1 = \frac{1}{12}(4 - 3e^3), c_2 = 1 - e^{-1}/12. \end{cases}$

4. $q = H(t - a)E_0 C[1 - \cos(t - a)]$.

6. $A = (A - h/k)e^{360k} + he^{330k}/k$.

8. $x(t) = \frac{1}{3} + \sum\limits_{j=1}^{\infty} \left(\frac{4}{3}\right)^j H(t - j), \quad t \geq 0$.

Section 7.3

2. (a) $1 + H(t - 1)$ (b) $\sqrt{3}\, H(t - 2) \sin \frac{1}{\sqrt{3}}(t - 2)$.   (c) $t^{n-1}e^t/(n - 1)!$.

(d) $t + H(t - 1)(1 - t)$.   (e) $t + (2 - t)H(t - 1)$.   (f) $t(e^{-2t} - e^{-3t})$.

4. (a) $y = -\frac{1}{100}e^{-5t} + \frac{1}{4}e^{-t} - \frac{6}{25} + \frac{1}{5}t$.   (b) $y = \frac{1}{4}(e^t - e^{-t} - 2te^t)$.

(c) $y = -\frac{2e^t}{5}\cos t + \frac{1}{5}e^t \sin t + \frac{1}{5}\sin t + \frac{2}{5}\cos t$.   (l) $y = \int_0^t (t - u)e^{-2(t-u)}f(u)\, du$, where f

denotes the triangular wave.

6. Use induction on n to show that $L(t^n e^{rt}) = n!(s - r)^{-1-n}$,

Section 7.4

2. Collision occurs in all cases.

4. (a) Follows directly from the assumptions.   (b) *Hint*: Let $u = x_j(t) - x_{j+1}(t)$. Then the model equation is $v'_{j+1}(t + T) = \lambda u'(t)/u(t)$; integrate each side with respect to time, using the fact that in a traffic jam $v = 0$ and $u = l$.   (d) $v_0 = \lambda \ln(a/l)$, where a is the separation between cars, for case

(b), and $v_0 = \frac{\lambda}{m - 1}[l^{1-m} - a^{1-m}]$, for case (c).

Section 7.5

2. (a) $\frac{1}{18}\int_0^t \sin 3(t - u)u \sin 3u\, du$.

(b) First write transform as $\frac{(s + 2)^2}{[(s + 2)^2 + 3^2]^2} = \left[\frac{s + 2}{(s + 2)^2 + 3^2}\right]^2$. Then use I.2 with $a = -2$, II.6, and

convolution.

(c) Write as $\frac{1}{(s + 2)^2}\frac{s}{(s + 2)(s + 1)}$; use II.4, II.10, and convolution.   (e) $\int_0^t (t - u)e^{-u}\, du$.

(h) $\int_0^t \frac{1}{2}(t - u)^2 H(u - 3)f(u - 3)\, du$.

4. (a) $g = \frac{1}{2}e^{-3t}\sin 2t$.   (b) $g = te^{-t/6}$.   (c) $Lg = \frac{1}{s + 1}\frac{1}{(s - \frac{1}{2})^2 + \frac{3}{4}}$; use II.3, II.5, I.2 and

convolution to find g.

6. Follows from definition of (*).

Section 7.6

4. Use definition of δ.

6. $x(t) = A \sin \omega t + \sin \sqrt{k}t + \frac{2}{\sqrt{k}} H(t - 2) \sin \sqrt{k}(t - 2)$. Note that k is the spring constant.

CHAPTER 8

Section 8.1

2. (a) $(-\infty, +\infty)$.   (b) $(-2, 2)$.   (c) $(-\infty, +\infty)$.

4. (a) $\sum_{n=1}^{\infty} \frac{2n}{(n-1)!} x^n$.   (b) $\sum_{n=0}^{\infty} (n+2)(n+1)a_{n+2}x^n$.   (c) $\sum_{n=2}^{\infty} \frac{(-1)^n x^n}{n(n-1)}$.

6. (a) $a_n = 1 - \frac{2}{n}$.   (b) $a_n = \frac{a_0}{(n+1)!}$.   (c) $a_n = \frac{a_0}{n!}$ for even n. $a_n = \frac{a_1}{n!}$ for odd n.

8. $y = \sum_{0}^{\infty} \frac{(2x)^{2n+1}}{(2n+1)!} + 2 \sum_{0}^{\infty} \frac{(2x)^{2n}}{(2n)!}$ or $y = \sinh 2x + 2 \cosh 2x$.

10. $a_{100} = -1.00041 \times 10^{-6}$, approximately.

Section 8.2

2. (a) No singular points.   (b) $x = 0$.   (c) $x = 0$.   (d) $x = 1$.   (e) $x = \pm 1$.
(f) $x = 0$.

4. $y = a_0 \left[1 + \sum_{1}^{\infty} \frac{(-1)^n x^{3n}}{(3n)(3n-1)(3n-3) \cdots 5 \cdot 3 \cdot 2} \right] +$

$a_1 \left[x + \sum_{1}^{\infty} \frac{(-1)^n x^{3n+1}}{(3n+1)(3n)(3n-2) \cdots 6 \cdot 4 \cdot 3} \right] a_0, a_1$ arbitrary.

6. Let C_n, $n = 0, 1, 2, 3$, denote the class of all polynomial solutions of the Chebyshev equation of order n. Then $C_0 = A$, $C_1 = Ax$, $C_2 = A(2x^2 - 1)$, $C_3 = A(4x^3 - 3x)$, where A is arbitrary constant.

8. (b) $y = \frac{1}{3} x^3 + \frac{1}{63} x^7 + \frac{2}{2079} x^{11} + \cdots$.

10. $\left| y(1) - (1 - \frac{1}{3} + \frac{1}{18} - \frac{1}{162} + \frac{1}{1944}) \right| < 0.0001$.

Section 8.3

2. $y = c_1 \left(\frac{x}{2} \ln \frac{1+x}{1-x} - 1 \right) + c_2 x$.

Section 8.4

2. (a) $x = \pm 1$ are the only singularities, both regular ones.   (b) $x = 1$ is the only singularity, but it is not regular.   (c) $x = 0$, $x = \pm 1$ only singularities, $x = 0$, $x = -1$, regular, $x = 1$ not.
(d) $x = 0$, $x = \pm 1$ only singularities, $x = 0$ not regular, $x = \pm 1$ are regular.
4. (a) $y = k_1 x^2 + k_2 x^3 + (x + \frac{5}{6})$.   (b) $y = k_1 x^3 + k_2 x^{-2} - \frac{1}{2} x^2$.   **6.** (b) $y_1 = x^{-1}$, $y_2 = x^{\sqrt{2}}$,
$y_3 = x^{-\sqrt{2}}$.

Section 8.5

2. $y = a_0 + a_1 x + a_2 x^2 + a_3 x^3 + \cdots$; $a_0 = 0$, $a_1 = -1$, $a_2 = -\frac{3}{4}$, $a_3 = -\frac{1}{6}$.
4. The normalized equation does not have continuous coefficients near the origin.

6. (b) $y = 1 - \alpha x - \frac{\alpha}{4} (1 - \alpha) x^2 + \cdots + \frac{(-\alpha)(1-\alpha) \cdots (n-1-\alpha)}{(n!)^2} x^n + \cdots$

Section 8.6

4. (a) Integration by parts (b) Prove by induction

(c) *Hint*: Let $x = u^2$ in $\int_0^\infty x^{z-1}e^{-x}\,dx$ to show that $\Gamma(\tfrac{1}{2}) = 2\int_0^\infty e^{-u^2}\,du$. Then note that $\Gamma^2(\tfrac{1}{2})/4 =$

$\left(\int_0^\infty e^{-u^2}\,du\right)\left(\int_0^\infty e^{-v^2}\,dv\right) = \int_0^\infty\int_0^\infty e^{-u^2-v^2}\,du\,dv$. Evaluate the double integral by polar coordinates.

6. *Hint*: Use recursion formula (13) to show that J_p and J_{p+1} cannot both vanish at the same point, and show that J_{p+1} must vanish at least once between any two successive zeros of J_p. The same property with roles of J_p and J_{p+1} interchanged is proven by recursion formula (12).

8. *Hint*: Mimic the proof in the problem set of Section 5.2 for $p = 0$ in the case $0 < p < \tfrac{1}{2}$. Treat $p = \tfrac{1}{2}$ separately (it is very easy) and then consider $p > \tfrac{1}{2}$.

Section 8.7

2. $J_0(x) = 1 - \dfrac{x^2}{4} + \dfrac{x^4}{64} + \cdots$, where $x = 6e^{-t}$.

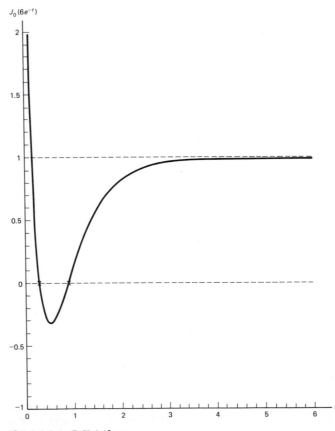

4. (b) $y(x) = e^x[C_1 J_1(x) + C_2 Y_1(x)]$.

6. (c)(i) $y(x) = x^{1/2}[C_1 J_{1/(4\sqrt{2})}(x^2/2) + C_2 J_{-1/(4\sqrt{2})}(x^2/2)]$.

 (ii) $y(x) = x^{1/2}[C_1 J_{1/6}(x^3/3) + C_2 J_{-1/6}(x^3/3)]$.

CHAPTER 9

Section 9.1

2. $x_1'' + 9x_1 = \cos t$. $x_1 = C_1 \cos 3t + C_2 \sin 3t + \frac{1}{8} \cos t$. $x_2 = -3C_1 \sin 3t + 3C_2 \cos 3t - \frac{1}{8} \sin t$.

4.

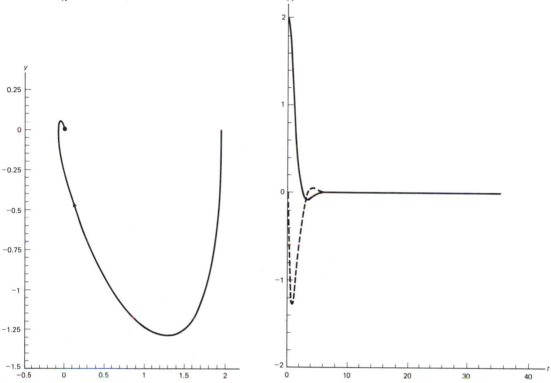

6. (a) $x_1 = C_1 e^{-2t} + 3C_2 e^{2t}$, $x_2 = -C_1 e^{-2t} + C_2 e^{2t}$.   (b) $x_1 = C_1 e^{-t} + C_2 e^t$, $x_2 = C_1 e^{-t} + C_2 e^t$.

8. *Hint:* The equations may be written in terms of the derivative operator $D = d/dt$ as $(D - 3)y_1 + 2(D + 2)y_2 = 0$, $2(D + 1)y_1 + (D - 1)y_2 = 0$. Apply $D - 1$ to the first equation, $-2(D + 2)$ to the second, and add to obtain $(3D^2 + 16D + 5)y_1 = 0$. Solve to find y_1, and substitute into one of the earlier equations and solve for y_2. Substitute y_1 and y_2 into the original equations to verify that the solutions are correct. You will have to set one of the three "constants of integration" equal to zero to get the equations to check. Why should there be only two arbitrary constants in the general solutions for y_1 and y_2? How did the third constant enter the formula?   *Answer:* $y_1 = C_1 e^{-t/3} + C_2 e^{-5t}$, $y_2 = C_1 e^{-t/3} - 4C_2 e^{-5t}/3$.

10. (a) $I' = -V_2/L + V_1/L$, $V_2' = I/C - V_2/RC$.   (b) $V_2'' + V_2'/RC + V_2/LC = V_1/LC$.

(c) $V_2 = -\frac{2}{5}e^{-t} (\cos t + 3 \sin t) + \frac{2}{5} (\cos t + 2 \sin t)$, $I = -\frac{2}{5}e^{-t} (2 \cos t + \sin t) + \frac{1}{5} (4 \cos t + 3 \sin t)$.

Section 9.2

2. $t = \frac{1}{a - b} \ln a \frac{2a - b}{a}$, $b \neq a$, $2a$.

4. (a) $\ln (1 + x) = \sum_{1}^{\infty} \frac{(-1)^{n+1}}{n} x^n \approx x$, for small x.   (b) Set $x = kA/(k_1K)$ in part (a).
(c) Using the result in part (b), $A/K \approx 0.000102$. Uncertainty of 3.4×10^9 A/K years.
6. $5.9 \times 10^{-5} \leqq A/K \leqq 1.17 \times 10^{-4}$.

Section 9.3

2. (a) $(0, 0)$, $(1, 1)$.   (b) $(a, 0)$, $(0, b)$, a, b any real numbers.   (c) No equilibria.
(d) $(n\pi/3, -1)$, $n = 0, \pm 1, \pm 2, \ldots$ (e) (a, a), a any real number.
4. (b) $r = (1 + Ce^{-2t})^{-1/2}$.   (c) *Hint:* Do not solve for r as a function of t, but observe that $r(t)$ is
an increasing function of t if $0 < r < 1$ or if $r > 2$, while it is a decreasing function if $1 < r < 2$.
Note that $\theta = t + c$, that $r = 0$ gives an equilibrium point in the x_1x_2-plane, and that $r = 1$ and $r = 2$ define cycles.

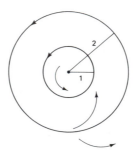

6. See Figure 9.13 with $\omega^2 = 1$.
8. (a) $\theta = A \cos \left(\sqrt{\frac{g}{L}} \, t - \delta \right)$. $\theta^2 + \frac{L}{g} (\theta')^2 = A^2$, which is the equation of an ellipse.   (c) Show
that the graph in the $\theta\theta'$-plane of $L(\theta')^2 = 2g(\cos \theta - \cos \theta_0)$ is symmetric through the origin and the
two axes. Then show that in the first quadrant the graph is an arc that falls from the point $\theta = 0$, $\theta' = [2g(1 - \cos \theta_0)/L]^{1/2}$ to the point $\theta = \theta_0$, $\theta' = 0$, where it has a vertical tangent.   (d) Integrate
over the "quarter" of the orbit in the first quadrant.   (e) Use $\cos \phi \, d\phi = (1 - k^2 \sin^2 \phi)^{1/2} \, d\theta/2k$.

Section 9.4

2–12 (Equilibrium points only).
2. $(0, 0)$.   **3.** $(0, 0)$, $(5, 5)$.   **4.** $(0, 0)$.   **5–8.** $(n\pi, 0)$, $n = 0, \pm 1, \pm 2, \ldots$ **9.** $(0, 0)$
10. $(0, 0)$, $(-2, -1)$.   **11.** $(0, 0)$.   **12.** $r = 0$ and $r = 1$, $\theta = 0$, π.

Section 9.5

2. (a) $(0, -e/c)$, $((ac + be)/ad, a/b)$; the latter is inside the quadrant.   (b) $(0, d/f)$, $(0, 0)$, $(a/b, 0)$, $(1/\Delta)(af - cd, bd - ae)$, where $\Delta = bf - ce$; the fourth point inside the quadrant if $af > cd$, $bf > ce$, and $bd > ae$—alternatively, if every ">" is replaced by "<."
4. Let $K = [(c + H)/d][(a - H)/b + (c + H)/d]^{-1}$ and show that $dK/dH > 0$.
6. (a) Show that $f'(y) = g'(x) = 0$ at $y = a/b$, $x = c/d$. Use L'Hôpital's rule to show that $f(y)$, $g(x) \rightarrow 0$ as $x, y \rightarrow \infty$.   (c) Look at the graphs in part (a).   (d) Use (c).
8. x is the satiable predator, y is the prey whose natural law is logistic; $(0, 0)$, $(0, 1)$, $(\frac{1}{2}, \frac{1}{2})$ are equilibrium points.
10. (a) $(0, 0)$, $(0, 0.9)$, $(\frac{3}{16}, \frac{3}{4})$.   (b) Harvesting helps the prey and harms the predator.
12. Each species in isolation would follow a logistic law. Their interaction is mutually harmful (the minus sign in $-xy$ and $-2xy$); $(0, 0)$, $(0, \frac{1}{2})$, $(1, 0)$ are equilibrium points.

CHAPTER 10

Section 10.1

2. $\det [A - \lambda E] = \det \begin{bmatrix} a_{11} - \lambda & \cdots & a_{1n} \\ & \ddots & \vdots \\ & & \ddots \\ 0 & \cdots & a_{nn} - \lambda \end{bmatrix} = (a_{11} - \lambda) \cdots (a_{nn} - \lambda).$

4. $A^k v = A^{k-1} A v = A^{k-1} \lambda v = \cdots = \lambda^k v$ if $A v = \lambda v$. If $A = \begin{bmatrix} 2 & 0 \\ 0 & -3 \end{bmatrix}$, then A has eigenvalues 2 and -3 while $A^2 = \begin{bmatrix} 4 & 0 \\ 0 & 9 \end{bmatrix}$ has eigenvalues 4, 9 whose square roots are 2, -2, 3, -3; therefore, $\mu^{1/2}$ is *not* necessarily an eigenvalue of A if μ is an eigenvalue of A^2.

6. $p(\lambda) = (-1)^n (\lambda - \lambda_1) \cdots (\lambda - \lambda_n) = \det [A - \lambda E -] = \det \begin{bmatrix} a_{11} - \lambda & \cdots & a_{1n} \\ & \ddots & \vdots \\ a_{n1} & \cdots & a_{nn} - \lambda \end{bmatrix}$

$= (-1)^n \lambda^n + \cdots + \det A.$ Thus $(-1)^n (-\lambda_1) \cdots (-\lambda_n)$
$= (-1)^{2n} \lambda_1 \cdots \lambda_n = \lambda_1 \cdots \lambda_n = \det A.$

8. A is nonsingular if and only if $\det A \neq 0$, if and only if $\lambda_1 \cdots \lambda_n \neq 0$, if and only if no eigenvalue is zero.

10. (a) Any singleton set is independent.   (b) *Hint:* Suppose that $B_i = \{v^i\}$ and that there are scalars $c_1, \ldots, c_k$ for which $c_1 v^1 + \cdots + c_k v^k = 0$. Apply A to each side to obtain $c_1 \lambda_1 v^1 + \cdots + c_k \lambda_k v^k = 0$, multiply the first equation by λ_k and subtract from the second. Apply the induction hypothesis to the resulting equation.   (c) *Hint:* If $B_1 \cup \cdots \cup B_p$ is dependent, there is a union $C_1 \cup \cdots \cup C_p$ of singleton sets $C_i \subset B_i$ which is dependent.

Section 10.2

2. $x = C_1 \begin{bmatrix} 1 \\ -1 \\ 1 \end{bmatrix} e^t + C_2 \begin{bmatrix} 1 \\ 1 \\ 0 \end{bmatrix} e^{-t} + C_3 \begin{bmatrix} 1 \\ 0 \\ 1 \end{bmatrix} e^{-t}.$

4. (a) $A \rightarrow \begin{bmatrix} -1 & -1 \\ 0 & -1 \end{bmatrix}$, $x = C_1 \begin{bmatrix} 1 \\ 2 \end{bmatrix} e^{-t} + C_2 \left(\begin{bmatrix} 2 \\ 4 \end{bmatrix} t e^{-t} + \begin{bmatrix} 0 \\ -2 \end{bmatrix} e^{-t} \right).$

(b) $A \rightarrow \begin{bmatrix} 0 & -2 \\ 0 & 0 \end{bmatrix}$, $x = C_1 \begin{bmatrix} 2 \\ 1 \end{bmatrix} + C_2 \left(\begin{bmatrix} 0 \\ 1 \end{bmatrix} - 2t \begin{bmatrix} 2 \\ 1 \end{bmatrix} \right).$

6. (a) *Hint:* Let $e^{\lambda t} v = w(t) + i z(t)$, where w and z are real vector functions. Since $(e^{\lambda t} v)' = A(e^{\lambda t} v)$, we have that $w' + i z' = A(w + iz) = Aw + iAz$. Then match the real terms on both sides and the complex terms to obtain $w' = Aw$, $z' = Az$. Then show that w and z are independent.

(b) $x = C_1 e^{-2t} \left(\begin{bmatrix} 0 \\ 1 \end{bmatrix} \cos t + \begin{bmatrix} 1 \\ 0 \end{bmatrix} \sin t \right) + C_2 e^{-2t} \left(\begin{bmatrix} 1 \\ 0 \end{bmatrix} \cos t + \begin{bmatrix} 0 \\ -1 \end{bmatrix} \sin t \right).$

(c) $x = C_1 \begin{bmatrix} 1 \\ -1 \\ -1 \end{bmatrix} e^{-2t} + C_2 e^{-t} \left(\begin{bmatrix} -1 \\ 0 \\ 2 \end{bmatrix} \cos \sqrt{2}\, t - \begin{bmatrix} \sqrt{2} \\ \sqrt{2} \\ 0 \end{bmatrix} \sin \sqrt{2}\, t \right)$

$+ C_3 e^{-t} \left(\begin{bmatrix} \sqrt{2} \\ \sqrt{2} \\ 0 \end{bmatrix} \cos \sqrt{2}\, t - \begin{bmatrix} -1 \\ 0 \\ 2 \end{bmatrix} \sin \sqrt{2}\, t \right).$

Section 10.3

2. (a)

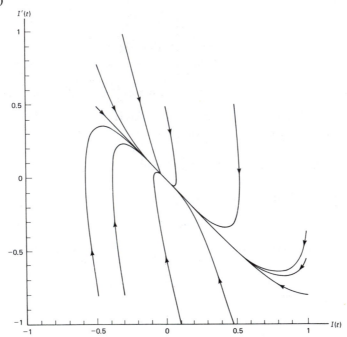

(d)

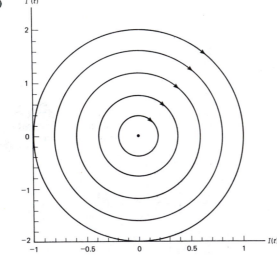

4. $x = c_1 \begin{bmatrix} 0 \\ 0 \\ 1 \end{bmatrix} e^t + c_2 e^{-t} \left(\begin{bmatrix} 0 \\ 1 \\ 0 \end{bmatrix} \cos t + \begin{bmatrix} -1 \\ 0 \\ 0 \end{bmatrix} \sin t \right) + c_3 e^{-t} \left(\begin{bmatrix} -1 \\ 0 \\ 0 \end{bmatrix} \cos t - \begin{bmatrix} 0 \\ 1 \\ 0 \end{bmatrix} \sin t \right) .$

Section 10.4

2. (a) $e^{tA} = \begin{bmatrix} 1 & t & t + \frac{1}{2}t^2 \\ 0 & 1 & t \\ 0 & 0 & 1 \end{bmatrix} \cdot \quad x = \begin{bmatrix} 1 + 5t + \frac{3t^2}{2} \\ 2 + 3t \\ 3 \end{bmatrix} .$

(b) $e^{tA} = \begin{bmatrix} e^{2t} & 0 & 0 \\ 0 & e^{-3t} & 0 \\ 0 & 0 & e^{7t} \end{bmatrix} , \quad x = \begin{bmatrix} e^{2t} \\ 2e^{-3t} \\ 3e^{7t} \end{bmatrix} .$

(c) $e^{tA} = \frac{1}{3} \begin{bmatrix} e^{-t} + 4e^{4t} & -4e^t + 4e^{4t} & 0 \\ e^t - e^{4t} & 4e^t - e^{4t} & 0 \\ 0 & 0 & 3e^t \end{bmatrix}$

4. (a) $x = \begin{bmatrix} \cos 2t & \sin 2t \\ -\sin 2t & \cos 2t \end{bmatrix} \begin{bmatrix} a \\ b + \frac{1}{2} \end{bmatrix} - \begin{bmatrix} 0 \\ \frac{1}{2} \end{bmatrix}$

(b) $e^{tA} = \frac{1}{2} \begin{bmatrix} -e^{-t} + 3e^t & e^{-t} - e^t \\ -3e^{-t} + 3e^t & 3e^{-t} - e^t \end{bmatrix} .$

(c) $e^{tA} = \begin{bmatrix} \cos t + 2\sin t & -5\sin t \\ \sin t & \cos t - 2\sin t \end{bmatrix} .$

(d) $e^{tA} = e^{-t} \begin{bmatrix} \cos 2t & -2\sin 2t \\ \frac{1}{2}\sin 2t & \cos 2t \end{bmatrix} .$

(e) $e^{tA} = \begin{bmatrix} e^{2t} + te^{2t} & -te^{2t} & te^{2t} \\ -t + te^{2t} + e^{2t} & e^t - te^{2t} & te^{2t} \\ -e^t + e^{2t} & e^t - e^{2t} & e^{2t} \end{bmatrix} ,$

$x = e^{tA} \begin{bmatrix} a \\ b \\ c \end{bmatrix} + e^{tA} \int_0^t e^{-sA} \begin{bmatrix} f_1(s) \\ f_2(s) \\ f_3(s) \end{bmatrix} ds .$

6. (a) $e^{t(A+B)} = \begin{bmatrix} e^t & e^t - 1 \\ 0 & 1 \end{bmatrix} , \quad e^{tB}e^{tA} = \begin{bmatrix} e^t & te^t \\ 0 & 1 \end{bmatrix} , \quad e^{tA}e^{tB} = \begin{bmatrix} e^t & t \\ 0 & 1 \end{bmatrix} .$

8. (b) (i) $\Phi(t, t_0) = e^{\cos t_0 - \cos t}, \ T = 2\pi, \ B = 1, \ \Phi(t, t_0)$ has period 2π in t.

(ii) $\Phi(t, t_0) = e^{\int_{t_0}^t a(s)\, ds}$, period is T, $B = e^{\int_{t_0}^{t_0+T} a(s)\, ds}$, $\Phi(t, t_0)$ does not necessarily have period T in t.

(iii) $\Phi(t, t_0) = e^{\cos t_0 - \cos t} \begin{bmatrix} 1 & 0 \\ t - t_0 & 1 \end{bmatrix} . \ T = 2\pi, \ B = \begin{bmatrix} 1 & 0 \\ 2\pi & 1 \end{bmatrix} ,$

$\Phi(t, t_0)$ does not have period 2π in t.

10. Let $\Phi = \Phi(t, t_0)$. $(\Phi c)' - A\Phi c - F = \Phi' c + \Phi c' - A\Phi c - F = A\Phi c + \Phi c' - A\Phi c - F = \Phi c' - F = 0$ if and only if $\Phi c' = F$, which holds if and only if $c' = \Phi(t_0, t)F(t)$, where $\Phi(t, t_0)^{-1} = \Phi(t_0, t)$. This holds if and only if $c(t) = \int_{t_0}^t \Phi(t_0, s)F(s)\, ds$, where $c(t_0)$ must be 0 since we require that $x(t_0) = 0$.

12. (a) $L[x'](s) = L[x](s) - x^0$. Thus we have $sL[x](s) - x^0 = AL[x](s) + L[F](s)$.   (b) $sE - A$ is invertible if $|s| > \max_{i \leq j \leq n} |\lambda_j|$, where $\lambda_1, \ldots, \lambda_n$ are the eigenvalues of A.   (c) From part (a),

$L[x] = (sE - A)^{-1}x^0 + (sE - A)^{-1}L[F]$. Thus applying L^{-1}, $x = L^{-1}(sE - A)^{-1}x^0 + L^{-1}(sE - A)^{-1}L[F]$. Comparing with (10), $L^{-1}(sE - A)^{-1} = \Phi(t, 0) = e^{tA}$ and $L^{-1}[(sE - A)^{-1}]L[F] = e^{tA} * F(t)$.

Section 10.5

2. (a) $\alpha < -\frac{3}{2}$.   (b) $-1 < \alpha < 0$.   (c) $\alpha < -\frac{2}{3}$.
4. *Hint*: Solve the equivalent scalar equation $x_1'' + x_1 = \cos t$.
6. (a) If $x = te^{\lambda t}v$, $x' = e^{\lambda t}v + \lambda te^{\lambda t}v$, while $Ax + e^{\lambda t}v = Ate^{\lambda t}v + e^{\lambda t}v = te^{\lambda t}Av + e^{\lambda t}v = \lambda te^{\lambda t}v + e^{\lambda t}v$. Thus if $\lambda = i\beta$, β real, $x = te^{i\beta t}v$ is an unbounded solution.
(b) Let $\lambda = \alpha + i\beta$, $\alpha > 0$, be an eigenvalue, v a corresponding eigenvector. Then $x = e^{(\alpha + i\beta)t}v$ is an unbounded solution.
8. (a) (i) All roots have negative real parts.
     (ii) $-(v)$ Some roots have nonnegative real parts.
(b) The Routh Array is

1	3	a
2	1	0
$\frac{5}{2}$	a	0
$\frac{\frac{5}{2} - 2a}{\frac{5}{2}}$	0	
a		

Thus we require $a > 0$, $\frac{5}{2} > 2a$, or $0 < a < 1.25$.
10. Let $x^3 = x^1 - x^2$. Then $(x^3)' = (x^1)' - (x^2)' = A(x^1 - x^2) = Ax^3$. Thus $x^3(t) \to 0$ as $t \to \infty$. This contradicts the assumption that x^1 is unbounded and x^2 is bounded. Hence all solutions are unbounded if one is.

14. $x(t)$ is a forced oscillation if and only if $x(0) = x(T) = e^{TA}x(0) + \int_0^T e^{(T-s)A}F(s)\,ds$, that is, if and only if $[E - e^{TA}]x(0) = \int_0^T e^{(T-s)A}F(s)\,ds$. The latter equation is a matrix equation, $Bx(0) = b$, for the unknown $x(0)$. Show that B is invertible since if B were singular then for some nonzero vector w, $Bw = 0$, and hence $y' = Ay$, $y(0) = w$, has a solution of period T.

Section 10.6

2. (a)

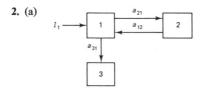

(b) All compartments are connected to the "environment," compartment 3, and wash out as $t \to \infty$.
4. (a) $x_1 = \frac{1}{2}(1 + e^{-2t})$, $x_2 = \frac{1}{2}(1 - e^{-2t})$, $x_3 = 0$. The system is closed.
(b) $x_1 = e^{-t}$, $x_2 = te^{-t}$, $x_3 = \frac{1}{2}t^2 e^{-t}$. The system is open.
6. The column sums are zero if the environment is included.

CHAPTER 11

Section 11.1

2. (a) Eigenvalues are $\pm 4i$, hence neutrally stable.   (b) Solutions are $x_1 = c_1$, while $x_2 = (2t + c_2)^{-1/2} \to 0$ as $t \to \infty$, hence neutrally stable.
4. (a) $K = x_1 x_2^3$, unstable.   (b) $K = 25x_1^2 + x_2^2$, neutrally stable.
(c) $K = x_1^4 + x_2^4$, neutrally stable.

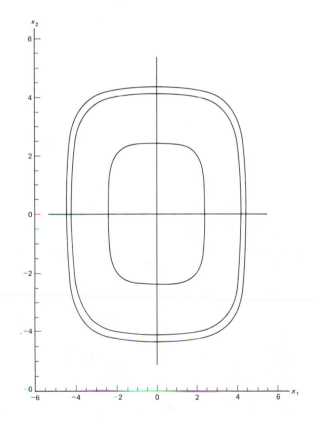

(d) $K = \frac{x_2}{x_1} - \sin x_1$, unstable.

6. (a) $K = \frac{1}{2}x_2^2 + \int_0^{x_1} g(s)\,ds$ is an integral. Hence there are no attractors.   (b) $\frac{1}{2}x_2^2 + 5x_1^2 + \frac{1}{4}x_1^4 =$ constant. The origin is neutrally stable.   (c) $\frac{1}{2}x_2^2 + 5x_1^2 - \frac{1}{4}x_1^4 =$ constant. The origin is neutrally stable. The critical points $(\pm\sqrt{10},\, 0)$ are unstable.

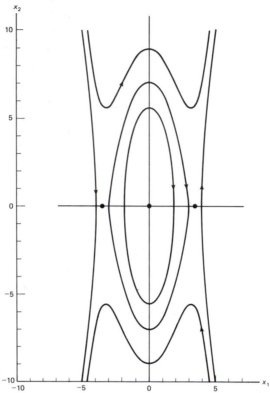

(d) $K = \int_0^{x_2} f(s)\,ds - \int_0^{x_1} g(s)\,ds$ is an integral.
8. (a) Since $K'(x) \equiv 0$, $K(x) = C$ is an integral "surface," which in this case is the "sphere" $x_1^2 + \cdots + x_n^2 = C$ in $\mathbf{R}^n$.

Section 11.2

2. (a) $V = 3x_1^2 + 4x_2^2$, asymptotically stable.   (b) $V = 9x_1^2 + x_2^2$, neutrally stable.   (c) $V = x_1^2 + x_2^2$, asymptotically stable.   (d) $V = -2x_1^2 + 4x_1x_2 + x_2^2$, unstable.   (e) $V = x_1^2 + x_2^2$, asymptotically stable.   (f) $V = x_1^2 + x_2^2$, asymptotically stable.   (g) $V = x_1^2 - x_2^2$, unstable.
4. (a) $V = 2x_1^2 + x_2^4$, asymptotically stable.   (b) $V = x_1^4 + x_2^4$, asymptotically stable.
6. (a) $V' = 2x^TF(x) \leq 0$ [< 0].   (b) Outward ray from 0 along x makes an obtuse angle with the vector $F(x)$ which has initial point at x.

8. (a) Use the Chain Rule to find $\frac{d}{dt} H(x_1, \ldots, x_n; y_1, \ldots, y_n)$.   (d) Impossible; system is Hamiltonian.

Section 11.3

2. (a) Asymptotically stable.   **(b)** Unstable   **(c)** Unstable.   **(d)** Unstable.
4. (a)

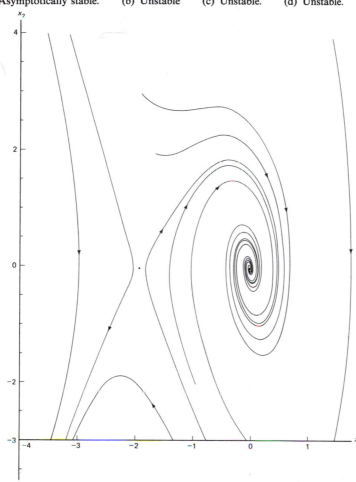

(b)

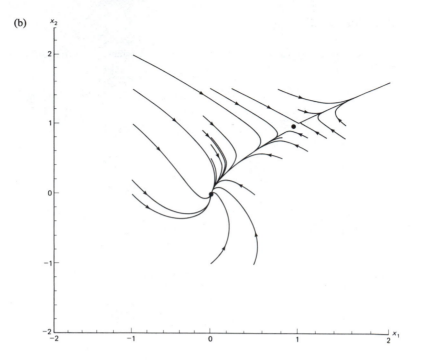

(c)

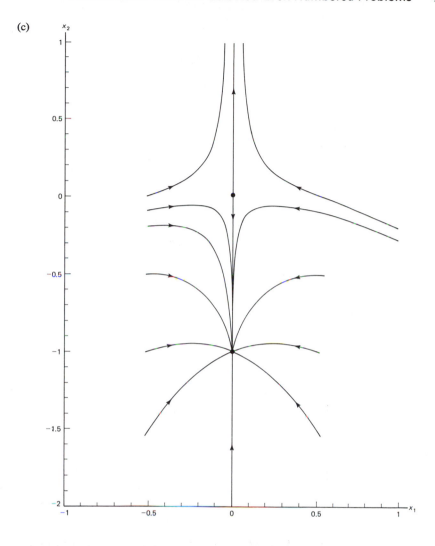

(d)

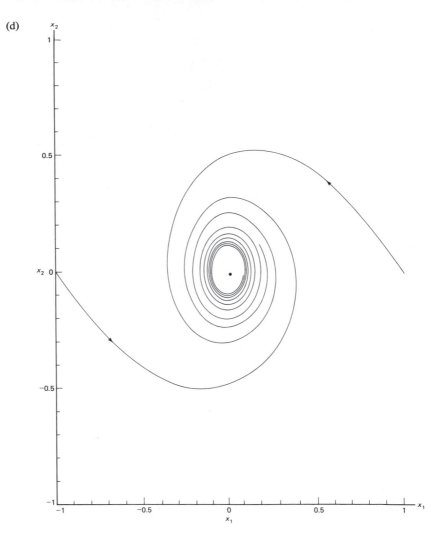

6. Equilibrium points P_n are at $x_2 = 0$, $x_1 = n\pi$. Corresponding eigenvalues of the linear part of the system at P_n are $-\frac{1}{2}a \pm \frac{1}{2}(a^2 + (-1)^n 4b)^{1/2}$. Stable focus or node if n is odd, unstable saddle if n is even.

8. $B^T = \int_0^\infty (e^{A^T t} e^{At})^T \, dt = \int_0^\infty (e^{At})^T (e^{A^T t})^T \, dt = \int_0^\infty e^{A^T t} e^{At} \, dt = B$.

10. (a) $P_1 = -(x_1^2 + x_2^2)x_1$, $P_2 = -(x_1^2 + x_2^2)x_2$.
(b) $P_1 = (x_1^2 + x_2^2)x_1$, $P_2 = (x_1^2 + x_2^2)x_2$.   (c) $P_1 = (x_1^2 + x_2^2)x_2$, $P_2 = -(x_1^2 + x_2^2)x_1$.
(d) Linearized system is not asymptotically stable.

CHAPTER 12

Section 12.1

2. (a) $r' = 0$ exactly at $r = 0$, $r = 1$. $r' > 0$ otherwise.   (b) $r' = 0$ at $r = 0$, $r = 1$; $r' < 0$ if $0 < r < 1$ and $r' > 0$ if $r > 1$—hence $r = 1$ is a repeller.   (c) Cycles at $r = 1/(n\pi)$, $n = 1, 2, \ldots$; r' changes sign as r passes through these values.

4. Show that $x' = y - x(|x| - 2)$, $y' = -2x$ satisfies (a) – (c) of the van der Pol Cycle Theorem, although f is not differentiable.

Section 12.2

2. (a) No cycles since $x' > 0$.   (b) No cycles since $\frac{\partial}{\partial x}(x') + \frac{\partial}{\partial y}(y') > 0$.   (c) $r = (n\pi)^{1/2}$, $n = 1, 2, \ldots$ give cycles.

4. $\oint [g\,dx - f\,dy] = \iint \left(\frac{\partial f}{\partial x} + \frac{\partial g}{\partial y} \right) dx\,dy = \int [gf\,dt - fg\,dt] = 0.$

6. $\frac{\partial}{\partial x}(x') + \frac{\partial}{\partial y}(y') > 0$ inside circle.

8. Use Poincaré–Bendixson Test (see also Example 12.6). Note that $F \cdot \nabla K > 0$ implies F points towards increasing K.

10. (a) Use proof of Bendixson's Negative Criterion, replacing f and g by Kf and Kg in that proof.

Section 12.3

2. In each case show that the eigenvalues of the coefficient matrix of the linear part of the rate functions at $(0, 0)$ have the form $\lambda = \alpha(\epsilon) \pm i\beta(\epsilon)$ where $\alpha(0) = 0$, $\alpha'(0) > 0$, $\beta(0) > 0$. Write the system in polar coordinates. Then check that in a neighborhood of the origin r' is negative if $\epsilon = 0$ (except at $r = 0$).

(a) $\lambda = \epsilon \pm 2i$ meets the above conditions, $r' = -r^3$ at $\epsilon = 0$.   (b) $\lambda = \epsilon \pm 3i$ and $r' = -r^7$ at $\epsilon = 0$.   (c) $\lambda = \frac{1}{2}\epsilon \pm i\sqrt{1 - \epsilon^2/4}$, and $rr' = -x^4 - y^4$.

4. Let $x' = v$, $v' = -x + \epsilon v - v^3$. Argue that since $r' = -r^3 \sin^4 \theta \leqq 0$ for $\epsilon = 0$, the origin is asymptotically stable for $\epsilon = 0$. [*Hint*: Use the Three-Fold Way.]

6. (a)

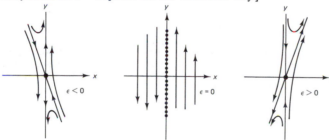

(b)

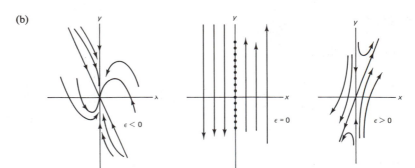

$\epsilon < 0$ $\epsilon = 0$ $\epsilon > 0$

CHAPTER 13

Section 13.1

2. (a) $(c_1 e^{kx} + c_2 e^{-kx})(c_3 e^{ckt} + c_4 e^{-ckt})$.   (b) $(c_1 e^{kx} + c_2 e^{-kx})e^{k^2 Kt}$.   (c) $(c_1 e^{kx} + c_2 e^{-kx})(c_3 e^{iky} + c_4 e^{-iky})$.   (d) $e^{\lambda x^2/2}(c_1 y^{(1+\sqrt{1-4\lambda})/2} + c_2 y^{(1-\sqrt{1-4\lambda})/2})$.   (e) $(c_1 e^{kx} + c_2 e^{-kx})e^{(1+k^2)y}$.
(f) $(c_1 e^{kx} + c_2 e^{-kx})(c_3 y^{\alpha_1} + c_4 y^{\alpha_2})$, where α_1, α_2 are the roots of $\alpha^2 + (2h)\alpha - a^2 k^2 = 0$.
4. (a) Solvable only when G has the form $G(y) = ke^{-y^2/2}$, where k is a constant. In that case all solutions given by $u = f(x)e^{-y^2/2} + F(y)$, where f is any C^2-function such that $f(0) = 0$, $f'(0) = k$.
(b) $u = e^{(x^2-y^2)/2} + y^2 - 1$ is the unique solution.

Section 13.2

2. $f(x + iy) + g(x - iy)$, for arbitrary C^2-functions f and g. For example, put $f(t) \equiv e^{kt}$, for k an arbitrary constant, and $g \equiv 0$, and obtain the real-valued solutions $e^{kx} \cos ky$ and $e^{kx} \sin ky$.

Section 13.3

2. (a) $u(x, t) = \dfrac{\tilde{f}(x + ct) + \tilde{f}(x - ct)}{2} + \dfrac{1}{2c} \displaystyle\int_{x-ct}^{x+ct} \tilde{g}(s)\, ds$, where $\tilde{f}$, $\tilde{g}$ are obtained by extending f and g into $[-L, 0]$ as even functions, and then into **R** as periodic functions with period $2L$.
(b) $u(x, t)$ has same form as in part (a), but $\tilde{f}$, $\tilde{g}$ are obtained from f and g as follows. Extend f, g into $[L, 2L]$ as even functions about $x = L$, and then extend these functions into $[-2L, 0]$ as odd functions, and finally into **R** as periodic functions with period $4L$.

Section 13.5

6. (c) The pair of vectors $v_1 = (1, 1, 3)$, $v_2 = (14, -8, -2)$ is an orthogonal basis for the plane, while $w = (1, 2, -1)$ is orthogonal to the plane. Thus $\|(\langle(1, 1, 1), w\rangle/\|w\|^2)w\| = \langle(1, 1, 1), w\rangle/\|w\| = 2/\sqrt{6}$ is the distance between the point $(1, 1, 1)$ and the plane: $x + 2y - z = 0$.
8. *Hint:* Write $p_n = f_n - f$, $q_m = g_m - g$, and observe that $|\langle f_n, g_m\rangle - \langle f, g\rangle| = |\langle p_n + f, q_m + g\rangle - \langle f, g\rangle|$. Then use the Triangle Inequality and the Cauchy–Schwartz Inequality.
10. (a) $\text{FS}[f] = 2\dfrac{A}{\pi} \displaystyle\sum_{k=1}^{\infty} \dfrac{\sin kx}{k}$.   (b) $\text{FS}[f] = \dfrac{A}{2} + \dfrac{4A}{\pi^2} \displaystyle\sum_{k=0}^{\infty} \dfrac{\cos (2k + 1)x}{(2k + 1)^2}$.
(c) $\text{FS}[f] = \dfrac{\pi}{2} - \dfrac{4}{\pi} \displaystyle\sum_{k=0}^{\infty} \dfrac{\cos (2k + 1)x}{(2k + 1)^2}$.

Section 13.6

2. (a) $\text{FS}[2x - 2] = -2 + 4 \sum_{k=1}^{\infty} \frac{(-1)^{k+1}}{k} \sin kx.$ (b) $\text{FS}[x^2] = \frac{\pi^2}{3} + 4 \sum_{k=1}^{\infty} (-1)^k \frac{\cos kx}{k^2}.$

(c) $\text{FS}[a + bx + cx^2] = a + \frac{c\pi^2}{3} + 2b \sum_{k=1}^{\infty} \frac{(-1)^{k+1}}{k} \sin kx + 4c \sum_{k=1}^{\infty} \frac{(-1)^k}{k^2} \cos kx.$

(d) $\text{FS}[\sin \pi x] = \frac{2 \sin \pi^2}{\pi} \sum_{k=1}^{\infty} \frac{(-1)^k k}{\pi^2 - k^2} \sin kx.$

(e) $\text{FS}[|x| + e^x] = \frac{\pi}{2} - \frac{4}{\pi} \sum_{k=0}^{\infty} \frac{\cos (2k+1)x}{(2k+1)^2} + \frac{2 \sinh \pi}{\pi} \sum_{k=0}^{\infty} (-1)^k \frac{\cos kx - k \sin kx}{1 + k^2}.$

(g) $\text{FS}[f] = \frac{AB}{\pi} + \frac{2A}{\pi} \sum_{k=1}^{\infty} \frac{\sin kB \cos kx}{k}.$ (h) $\text{FS}[f] = \frac{2A}{\pi} \sum_{k=1}^{\infty} (\cos kB - 1) \frac{\sin kx}{k}.$

4. (a) *Hint*: Show first that $\sin \frac{m\pi x}{c}$ is odd about $x = \frac{c}{2}$ when m is even.

6. (a) $\text{FS}[\sin x] = \sin x, -\pi < x < \pi.$ (b) $\text{FS}[\sin x] = \frac{8}{\pi} \sum_{k=1}^{\infty} \frac{(-1)^k k}{1 - 4k^2} \sin 2kx, -\frac{\pi}{2} < x < \frac{\pi}{2}.$

(c) $\text{FS}[\sin x] = \frac{8}{\pi} \sum_{k=1}^{\infty} \frac{(-1)^k k}{4k^2 - 9} \sin \frac{2kx}{3}, -\frac{3}{2}\pi < x < \frac{3}{2}\pi.$

8. $I(t) = \pi C/2 \frac{2}{\pi} \sum_{|k|=1}^{\infty} \frac{e^{i(2k-1)t}}{(2k-1)^2(C^{-1} - L(2k-1)^2 + R(2k-1)i)}$

where $R = 10 \ \Omega, L = 0.5 \ H, C = 10^{-4} \ F.$

10. *Hint*: Break up the interval $[a, a + 2T]$ into two subintervals by finding an integer k such that $a \le (2k + 1)T < a + 2T$. Then translate coordinates to take $[a, (2k + 1)T]$ onto one part of the interval $[-T, T]$. Translate coordinates again to take $[(2k + 1)T, a + 2T]$ onto the other part of $[-T, T]$.

14. The limit is zero because the integral is π times the nth Fourier coefficient of the continuous function $e^{\sin x}(x^5 - 7x + 1)^{52}.$

16. From Problem 10(c) we have $\text{FS}[|x|]$. Evaluating this series at $x = 0$ and using the Pointwise

Convergence Theorem, we see that $0 = \frac{\pi}{2} - \frac{4}{\pi} \sum_{k=0}^{\infty} \frac{1}{(2k + 1)^2}$, from which part (a) follows.

Section 13.7

2. (a) $u = a_0 + b_0 t + \sum_{n=1}^{\infty} \cos \frac{n\pi x}{L} \left[A_n \cos \frac{n\pi a}{L} t + B_n \sin \frac{n\pi a}{L} t \right]$

$a_0 = \frac{1}{L} \int_0^L f(x) \, dx, \ b_0 = \frac{1}{L} \int_0^L g(x) \, dx$

$A_n = \frac{2}{L} \int_0^L f(x) \cos \frac{n\pi x}{L} \, dx, \ B_n = \frac{2}{L} \int_0^L g(x) \cos \frac{n\pi x}{L} \, dx.$

4. $u(x, t) = \sum_{\substack{n=1 \\ n \ne n^*}}^{\infty} \left[\sin \frac{n\pi x}{L} (A_n e^{r_{1n}t} + B_n e^{r_{2n}t}) \right] + \sin \frac{b^2 x}{2a} (A_{n^*} t e^{rt} + B_{n^*} t e^{rt})$

where $r_{1n} = -\frac{b^2 + \sqrt{b^4 - 4\left(\frac{an\pi}{L}\right)^2}}{2}, \quad r_{2n} = -\frac{b^2 - \sqrt{b^4 - 4\left(\frac{an\pi}{L}\right)^2}}{2}.$

$r = -\frac{b^2}{2}, \ n^* = \frac{Lb^2}{2|a|\pi}$ (n^* comes into the expression only if n^* is a positive integer).

$A_n = \frac{r_{2n} M_n - P_n}{r_{2n} - r_{1n}}$, all n; $B_n = \frac{P_n - r_{1n} M_n}{r_{2n} - r_{1n}}$, $B_{n^*} = P_{n^*} - A_{n^*} r,$

where $M_n = \frac{2}{L} \int_0^L f(x) \sin \frac{n\pi x}{L} \, dx$, and $P_n = \frac{2}{L} \int_0^L g(x) \sin \frac{n\pi x}{L} \, dx.$

Section 13.8

2. (a) $u = 10 + 10x - \dfrac{40}{\pi} \sum\limits_{\text{odd } n} \sin (n\pi x) \, e^{-\lambda_n Kt}, \; \lambda_n = n^2\pi^2, \; n = 1, 2, \ldots$.

(b) As $t \to \infty$, $e^{-\lambda_n Kt} \to 0$ and $u \to 10 + 10x$.

4. $u = \sum\limits_{n=1}^{\infty} \left[\dfrac{(-1)^n 2}{Kn^3\pi^3} - \dfrac{6}{2 + Kn^2\pi^2} e^{-2t} \left(\dfrac{1}{n\pi} - \dfrac{(-1)^n}{n\pi} \right) \right] \sin n\pi x$.

6. (a) nth term $\not\to 0$ as $n \to \infty$.   (b) The heat equation is valid only for t increasing from the initial time. If initial data are *not* smooth, then only for $t > 0$ is equation meaningful.

(c) $u = \dfrac{1}{\pi} \sin \dfrac{\pi x}{L} \exp(-K\pi^2 t/L^2)$, defined for $-\infty < t < \infty$. Here the initial data are smooth enough to support the reversal of time.

Section 13.9

2. $u = A_0 + B_0 \ln r + \sum\limits_{n=1}^{\infty} \{ r^n [A_n \cos n\theta + A_n^* \sin n\theta] + r^{-n}[B_n \cos n\theta + B_n^* \sin n\theta] \}$,

where $A_0 + B_0 \ln \rho = \dfrac{1}{2\pi} \displaystyle\int_{-\pi}^{\pi} f(\theta) \, d\theta, \quad A_0 + B_0 \ln R = \dfrac{1}{2\pi} \displaystyle\int_{-\pi}^{\pi} g(\theta) \, d\theta$,

$A_n \rho^n + B_n \rho^{-n} = \dfrac{1}{\pi} \displaystyle\int_{-\pi}^{\pi} f(\theta) \cos n\theta \, d\theta, \quad A_n R^n + B_n R^{-n} = \dfrac{1}{\pi} \displaystyle\int_{-\pi}^{\pi} g(\theta) \cos n\theta \, d\theta$,

$A_n^* \rho^n + B_n^* \rho^{-n} = \dfrac{1}{\pi} \displaystyle\int_{-\pi}^{\pi} f(\theta) \sin n\theta \, d\theta, \quad A_n^* R^n + B_n^* R^{-n} = \dfrac{1}{\pi} \displaystyle\int_{-\pi}^{\pi} g(\theta) \sin n\theta \, d\theta$.

4. (a) $u = -1 + 2P_2(\cos \phi)\rho^2$.   (b) $u = \dfrac{1}{2} + \dfrac{1}{2} \sum\limits_{m=0}^{\infty} \{P_{2m}(0) - P_{2m+2}(0)\} P_{2m+1}(\cos \phi) r^{2m+1}$.

(c) $u = \sum\limits_{\text{even } n} A_n \rho^n P_n (\cos \phi), \quad A_n = \dfrac{n+1}{2n+3} \{P_{2n}(0) - P_{2n+2}(0)\} + \dfrac{n}{2n-1} \{P_{2n-2}(0) - P_{2n}(0)\}$.

$\left[\text{Note: } P_{2n}(0) = \dfrac{(-1)^n (2n)!}{2^{2n}(n!)^2} . \right]$

6. (a) $u = 2 \sum\limits_{n=1}^{\infty} \dfrac{\sinh (k_n(1-z)) \, J_0(k_n r)}{\sinh k_n \; J_1(k_n)}$, where $\{k_n : n = 1, 2, \ldots\}$ are the positive zeros of $J_0(x)$.

8. (a) *Hint*: Suppose that u_1 and u_2 are solutions and set $w = u_1 - u_2$. Then $\nabla^2 w = 0$ in G and $\partial w / \partial n = 0$ on ∂G. Apply the two-dimensional Divergence Theorem to w^2: $\displaystyle\int_{\partial G} \nabla(w^2) \cdot \mathbf{n} \, ds =$

$\displaystyle\int_{\text{cl}G} \text{div} (\nabla w^2) \, dA$. But $\nabla(w^2) = 2w\nabla w$ and $\nabla w \cdot \mathbf{n} = \partial w/\partial n = 0$ on ∂G. On the other hand,

$\displaystyle\int_{\text{cl}G} \text{div} (\nabla w^2) \, dA = \displaystyle\int_{\text{cl}G} 2[w\nabla^2/w + \nabla w \cdot \nabla w] \, dA = 2 \displaystyle\int_{\text{cl}G} \|\nabla w\|^2 \, dA$. Thus $\nabla w \equiv 0$ throughout cl G. Thus $w \equiv$ constant on cl G.

(b) *Hint*: Apply the planar divergence theorem to u: $\displaystyle\int_{\text{cl}G} \text{div} (\nabla u) \, dA = \displaystyle\int_{\text{cl}G} \nabla^2 u = \displaystyle\int_{\partial G} \nabla u \cdot \mathbf{n} \, ds =$

$\displaystyle\int_{\partial G} \partial u / \partial n \, ds = \displaystyle\int_{\partial G} f(s) \, ds$.   (c) Solve the Neumann Problem above if G is the unit disk in the plane and $f(\theta) = \sin \theta$, while $u = 0$ at $\theta = 0, r = 1$.

APPENDIX

Section A.1

2. (a) If $y \neq 0$, then $y = (ax + c)^{1/a}, \; a = 1 - \dfrac{m}{n}$ if $\dfrac{m}{n} > 1$, which cannot pass through $(0, 0)$ for any constant c. Note that $y \equiv 0$ is the only solution.

(b) $y \equiv 0$ is one solution, $y_k = \begin{cases} 0, & t \leq k \\ (at - ak)^{1/a}, & t > k \end{cases}$ is another for any $k > 0$. Show that the right- and left-hand derivatives are equal at $t = k$ (needed to show that y_k is differentiable).

Section A.2

2. $y_1 = 2 + 2t^2$, $y_2 = 2 + 2t^2 + t^4$, $y_3 = 2 + 2t^2 + t^4 + \frac{1}{3}t^6$. $y_n = \sum_{j=0}^{n} \frac{2}{j!} t^{2j}$, which is the $(n + 1)$st partial sum of the Maclaurin series of the exact solution $y = 2e^{t^2}$.

4. $y_n = 1 - t + \cdots + (-1)^{n-1}t^{n-1}$, the nth partial sum of the Maclaurin series of the exact solution e^{-t}.

6. $y_{n+1}(t) = 1 + \int_0^t (1 + y_n^2(s)) \, ds$. If y_n is a polynomial of degree $2^n - 1$, then $\int y_n^2$ is a polynomial of degree $2(2^n - 1) + 1 = 2^{n+1} - 1$. The exact solution is $y = \tan t$, $-\pi/2 < t < \pi/2$. Although $y_n(t)$ is defined for all t, the limit as $n \to \infty$ is not defined for all t.

8. $y(t) = \lim y_n(t)$ is the solution of $y' = -y^3$, $y(0) = 3$.

Section A.3

2. $y' = 2ty^2$, $y > 0$, has solutions $y = (c - t^2)^{-1}$.

Section A.4

2. $|y(t) - \bar{y}(t)| \le |a - \bar{a}|e^{t/10} + 5(e^{t/10} - 1)$, $t \ge 0$.

Index

e^{A+A}

stability

singular

indicial polynomial

III. Transforms of Graphically Defined Functions

$f(t)$	$g(s) = L[f] = \int_0^\infty e^{-st} f(t)\, dt$

Triangular wave function: period a

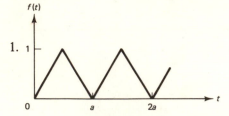

1. $\dfrac{2}{as^2} \tanh\left(\dfrac{as}{4}\right)$

Square wave function: period $2a$

 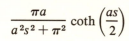

2. $\dfrac{1}{s} \tanh\left(\dfrac{as}{2}\right)$

Rectified sine wave function: period a

$f(t) = \left| \sin \dfrac{\pi t}{a} \right|$

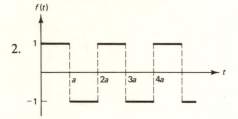

3. $\dfrac{\pi a}{a^2 s^2 + \pi^2} \coth\left(\dfrac{as}{2}\right)$

Half-rectified sine wave function: period $2a$

$f(t) = \begin{cases} \sin \dfrac{\pi t}{a}, & 2na \leqq t \leqq (2n+1)a \\[2mm] 0, & (2n-1)a \leqq t \leqq 2na \end{cases}$

4. $\dfrac{\pi a}{(a^2 s^2 + \pi^2)(1 - e^{-as})}$

Sawtooth wave function: period a

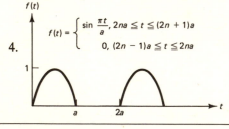

5. $\dfrac{1}{as^2} - \dfrac{e^{-as}}{s(1 - e^{-as})}$